units

magnitude &
direction

P. 326 - vector products
278, 272 - accel
277, 271 - velocity
196 - # of planets
193 - limiting ratio
122 - backlash
115 - sm. # teeth

Mechanisms and Dynamics of Machinery

Mechanisms and Dynamics of Machinery

Third Edition

Hamilton H. Mabie

Professor of Mechanical Engineering,
Virginia Polytechnic Institute
Blacksburg, Virginia

Fred W. Ocvirk

Late Professor of Mechanical Engineering
Cornell University
Ithaca, New York

John Wiley and Sons, Inc.

New York · London · Sydney · Toronto

Library of Congress Cataloging in Publication Data:

Mabie, Hamilton Horth, 1914–
Mechanisms and dynamics of machinery.

Includes bibliographical references and index.
1. Mechanical movements. 2. Machinery, Kinematics of. 3. Machinery, Dynamics of. I. Ocvirk, Fred W., joint author. II. Title.

TJ175.M123 1975 621.8 74-30405
ISBN 0-471-55935-0

Printed in the United States of America

10 9 8 7 6 5 4 3 2

Preface to Third Edition

This edition has been delayed for several years because of the sad and untimely death of my coauthor F. W. Ocvirk in 1967.

The major changes in this edition have been in Chapter 10, "Kinematics of Machinery," and in Chapter 11, "Force Analysis of Machinery." In Chapter 10, the following material has been added: Velocity and acceleration analysis by unit vectors, analytical solution of relative velocity and acceleration equations by unit vectors, extension of graphical differentiation to computer solutions, spatial linkage analysis by complex vectors. Graphical velocity and acceleration analysis has been retained together with analysis by complex vectors.

In Chapter 11, the following material has been added: Force analysis using transverse and radial components treated graphically and with unit vectors, superposition using unit vectors, linkage analysis by method of virtual work, linkage motion analysis from dynamic characteristics. Graphical analysis by superposition has been retained as well as force analysis by complex vectors.

SI units have been introduced throughout the book with the exception of the chapters on gearing. SI standards on gears do not exist at present.

The author is indebted to the following fellow members of the Mechanical Engineering Department at Virginia Polytechnic Institute and State University for their helpful suggestions in the preparation of this edition: N. S. Eiss, J. P. Mahaney, H. P. Marshall, L. D. Mitchell, R. G. Mitchiner, L. A. Padis, and H. H. Robertshaw. Acknowledgment is also given to the reviewers of this text for their thorough evaluation.

HAMILTON H. MABIE

Blacksburg, Virginia
January 1975

Preface to Second Edition

Part 1 Mechanisms

The material of the first edition has been retained with the exception of that on automotive-type cams which was eliminated in order to provide room for new material. The treatment of analytical cam design in Chapter 3 has been expanded to include a disk cam with oscillating roller follower. The graphical design of three-dimensional cams has also been added. In Chapter 4, the latest AGMA standard has been inserted for spur gear teeth. More material on backlash has been introduced together with its method of calculation. In Chapter 5, some of the material has been rearranged and new sketches added to make the material on nonstandard gears more readable.

In Chapter 6, the latest Gleason standard for bevel gears has been inserted. The material on planetary gear trains in Chapter 7 has been revised and expanded to include the tabulation method of solution and the generalized equation method of relative velocities. The method for checking to see whether a planetary gear train can be assembled has been greatly expanded using material from the notes of Professor G. B. DuBois of Cornell University. Chapter 8 on computing mechanisms has been expanded and information on scale determination has been added. In Chapter 9 on synthesis, Raven's method of using complex variables for mechanism synthesis has been introduced. A graphical method for the simple synthesis of a four-bar linkage has also been given.

I am indebted to those who have used the first edition and who have made helpful comments that have assisted in the preparation of this second edition.

HAMILTON H. MABIE

Albuquerque, New Mexico
April 1963

Part 2 Dynamics of Machinery

Of the changes made since the first edition, the principal ones appear in Chapters 10 and 11 on kinematic analysis and force analysis.

Many users of the first edition expressed a preference for separated treatments of velocity and acceleration analyses to allow a longer period of reflection on velocity problems before undertaking the relatively more complicated relationships in acceleration determination. Accordingly, the separation is made to some degree in this edition, and the sections on velocity analysis by instantaneous centers are placed ahead of the sections on acceleration. As a result of the separation, the presentation of ideas and insight on the constraint of motion in mechanisms is changed, and new problems appear in the illustrative examples.

Always a challenge to the teacher and the student is the development of acceleration relationships in which the Coriolis component appears. Readers will find here a development different from that in the first edition. Also, a discussion is included on alternate solutions (and their limitations) to problems normally solved using the Coriolis vector.

For the complex cases of kinematic analysis, the method of "auxiliary points" is now combined with the "three-line construction" method of the first edition in order to eliminate the trial-and-error aspect of the latter method.

The most important new material appearing in this edition is the numerical method of analysis using the complex number representation of vectors. The treatment of velocity and acceleration problems is patterned after Raven's method, which is also used in synthesis problems in Chapters 3 and 9 of Part 1. Kinematic equations for several typical linkages are determined by this method, beginning with the simple case of a particle on a rotor and ending with the four-bar linkage. Examples are included to show the problems which arise in making numerical evaluations from the equations.

As a natural extension of the kinematic analysis, the method of complex numbers is also used in the force analysis of linkages, and an illustrative example of the force analysis of the four-bar mechanism is included.

I, too, am grateful for the many suggestions and comments which the users of the first edition have so kindly offered Dr. Mabie and me. I wish particularly to thank R. K. Malhotra for his thesis work showing the practicability of force analysis by complex numbers.

FRED W. OCVIRK

Ithaca, New York
April 1963

Preface to First Edition

This text has been used for several years at Cornell University in mimeographed form in courses entitled Mechanisms and Dynamics of Machinery, two consecutive courses of three credit hours each. It has been the opinion of the authors that in order to meet the demands of a rapidly expanding technology these courses should include such topics as analytical cam design, nonstandard spur gears, computing mechanisms, synthesis, and vibrations. By using the text in mimeographed form, it has been possible to determine the amount of such advanced work that can be included in an undergraduate course of this nature. Furthermore, it has been our experience that students are more interested if the elementary work is streamlined and more time is given to advanced topics. As an example of this procedure, the graphical method of cam design is briefly presented to illustrate the principles involved in designing representative cams and then the subject of analytical cam design is introduced. We have found that the students are more interested in analytical rather than in graphical cam design even though the analytical method is much more difficult.

Increasing course content without increasing the number of credit hours requires each of the subjects to be treated efficiently. To this end, the authors have attempted to make the text as brief as possible in order to conserve the student's reading time and yet cover thoroughly the fundamentals of mechanisms and dynamics of machinery. It is assumed that the student beginning Part 1 (Mechanisms) will have completed courses in physics and calculus, and that the student beginning Part 2 (Dynamics of Machinery) will also have completed courses in statics and dynamics. For this reason the derivations of fundamental relationships treated in the prerequisites are not repeated.

Part 1: Mechanisms

The theory of spur gears is thoroughly developed including material on involutometry. After the general theory has been presented, the subject of standard spur gearing is introduced. Unlike most other texts on mechanisms this book also contains a chapter on nonstandard spur gears. It is felt that this material is important not only for the understanding of nonstandard spur gears but also because it leads to a better understanding of standard spur gears.

The material on bevel, helical, and worm gearing is presented completely and yet concisely. In reference to bevel gears, only the Gleason system for straight bevel gears is included. Although the proportions of the older bevel gear system are occasionally found in practice, discussion in this text is limited to the Gleason system because it is the one that has been adopted by the American Gear Manufacturers' Association and American Standards Association.

With the increasing demand for computers and control systems, the importance of computing elements is increasing constantly. For this reason a chapter on mechanical computing elements is included. As the chapter heading states, this is only an *introduction* to computing mechanisms. Obviously, the subject is too broad and is growing too rapidly to be covered completely in a text of this nature. However, it is felt that this is such an important field that its principles should be presented in an undergraduate course. The students who have used this text in mimeographed form have been very enthusiastic about the work on computing mechanisms.

The final chapter in Part 1 is entitled Introduction to Synthesis. Here again it is possible only to introduce the subject and to present two of the many methods of synthesis for the four-bar linkage. It is felt that this subject is growing in importance also, and although it is not possible to go into great detail an introduction to synthesis outlines the principles, the difficulties, and some of the applications.

Part 2: Dynamics of Machinery

The scope and level of Part 2 on machinery dynamics are intended to be suitable for a required undergraduate course without the implication that dynamics is a subject to be reserved for advanced undergraduates or for graduate students. The text is written from the point of view that relatively few fundamental principles are needed to recognize and solve dynamical problems associated with basic types of machines and mechanisms. Principles involving Coriolis acceleration, gyroscopic action, vibration and vibration transmissibility, and critical speeds are included as elementary and necessary material. Applications are made to turbomachinery to demonstrate the ramifications of high speed, and to reciprocating machinery and other linkages to demonstrate the simplicity and straightforwardness of analytical methods which may appear complex in a multilink system.

In the chapter on kinematics of machinery, velocities and accelerations are taken up simultaneously rather than in independent subdivisions of the text. Problems involving the Coriolis acceleration and the kinematics of rolling elements are emphasized. Instantaneous centers of zero velocity are taken up after the principles of relative motion are fully developed because it is felt by this method students better understand the definition of instantaneous center with respect to both velocity and acceleration and better understand the relationship among relative instantaneous centers.

The chapter on force analysis of machinery is intended to demonstrate the effect of acceleration on the forces and stresses related to the design of a machine part. Besides resultant inertia forces, determinations are made of the distribution of inertia force for a given machine part in motion. An attempt has been made to point out the dynamic effects on critical mechanical elements such as bearings, pins, gear teeth, and cam surfaces. Flywheel action and gyroscopic action are included in this chapter because of their relationship with the force analysis of engines. Simplicity of force analysis is shown and is based on the use of free-body diagrams, on the method of superposition, and on static force systems which are the equivalent of dynamic systems.

The material in Chapter 12 on the balancing of rotating and reciprocating machines is traditional and represents an introduction to vibrations, where the harmonic motions are rigid body motions rather than motions allowed by flexible members. An attempt has been made to show the effect of the dynamic shaking caused by unbalance on elements such as bearings, rotor shafts, and crankshafts in bending, and engine housings and mountings.

In the chapter on vibrations, basic mechanics is reviewed to show the source of forced vibrations and the behavior of bodies in vibration. Emphasis is placed on the numerical determinations of vibration amplitude, of transmitted force, and of critical speeds of rotors. Single degree of freedom cases are explored fully, and cases of several degrees of freedom are pointed out primarily with respect to rotor shafts in bending and torsion.

Acknowledgments

The authors wish first to express their gratitude to their present colleagues in the Department of Machine Design at Cornell University for their understanding and encouragement. Particular thanks are due A. D. McIntyre, F. H. Raven, T. M. Sedgwick, and S. S. Strong, who have taught from this text in mimeographed form during the last two years and who have given valuable assistance in its revision. Acknowledgement is also given to former members of the staff, J. E. K. Foreman, J. F. Hamilton, R. T. Hinkle, S. G. Holt, W. S. Neef, G. A. Nothmann, F. Saltz, S. S. Thomas, and W. A. Wheeler, for their contribution to the great number of problems included. Appreciation is expressed to Mrs. Evelynn Richards for typing and mimeographing the many revisions from which this text was eventually developed. We would also like to express our gratitude to our reviewers and to Mr. Wells Coleman of Gleason Works for his suggestions on bevel gears.

HAMILTON H. MABIE
FRED W. OCVIRK

Ithaca, New York
December 1965

Contents

Part 1 Mechanisms

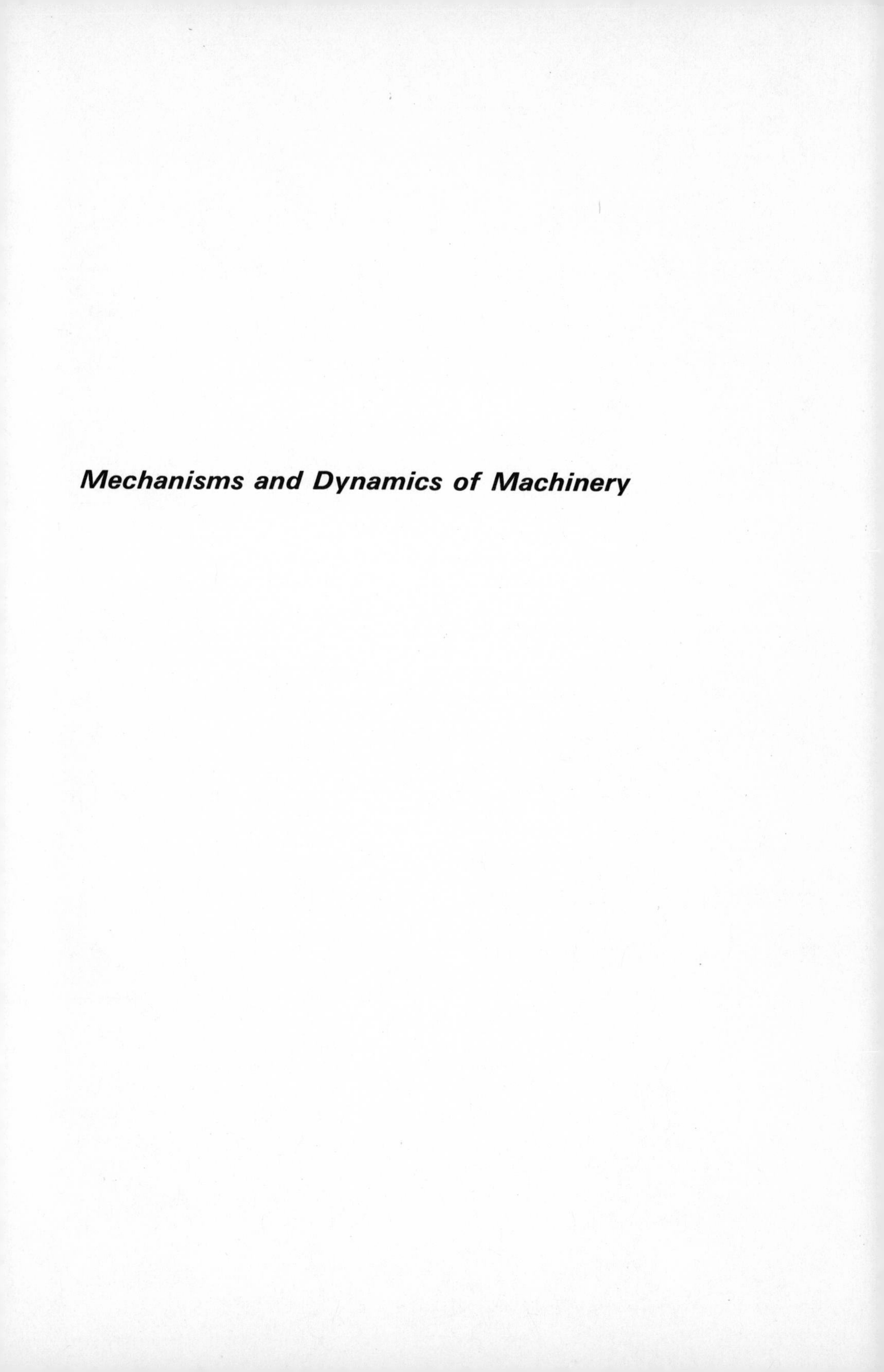

Mechanisms and Dynamics of Machinery

Part 1

MECHANISMS

1

Introduction

1.1 Introduction to the Study of Mechanisms. The study of mechanisms is very important. With the tremendous advances made in the design of instruments, automatic controls, and automated equipment, the study of mechanisms takes on new significance. *Mechanisms* may be defined as that division of machine design which is concerned with the kinematic design of linkages, cams, gears, and gear trains. *Kinematic design* is design on the basis of motion requirements in contrast to design on the basis of strength requirements. An example of each of the mechanisms listed above will be given in order to present a comprehensive picture of the components to be studied.

A sketch of a linkage is shown in Fig. 1.1. This particular arrangement is known as the slider-crank mechanism. Link 1 is the frame and is stationary, link 2 the crank, link 3 the connecting rod, and link 4 the slider. A common application of this linkage is in the internal-combustion engine where link 4 becomes the piston (Fig. 1.2).

Figure 1.3 shows the sketch of a cam and follower. The cam rotates at a constant angular velocity, and the follower moves up and down. On the upward motion the follower is driven by the cam, and on the return motion by the action of gravity or of a spring. Cams are used in many machines, but one of the most common is the automotive engine where two cams are used per cylinder to operate the intake and exhaust valves, also shown in Fig. 1.2. A three-dimensional cam is shown in Fig. 1.4. In this cam, the

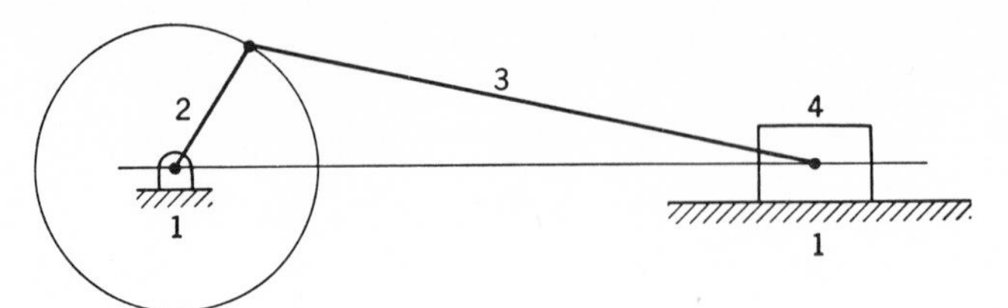

Figure 1.1

motion of the follower depends not only upon the rotation of the cam but also upon the axial motion of the cam.

Gears are used in many applications to transmit motion from one shaft to another with a constant angular velocity ratio. Figure 1.5 shows several commonly used gears.

In some cases, the desired reduction in angular velocity is too great to achieve using only two gears. When this occurs, several gears must be connected together to give what is known as a *gear train*. Figure 1.6 shows a gear train where the speed is stepped down in going from gear 1 to gear 2

Figure 1.2 Chevrolet V-8 engine showing slider-crank mechanism. General Motors Corp.

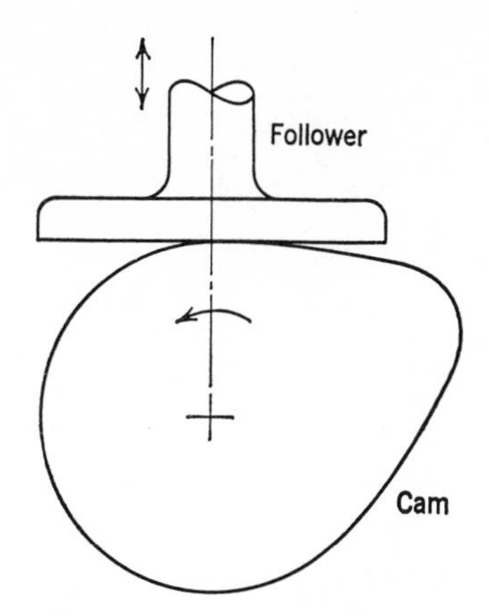

Figure 1.3 Two-dimensional cam.

and again in going from gear 3 to gear 4. Gear 1 is the driver, and gears 2 and 3 are mounted on the same shaft. In many gear trains, it is necessary to be able to shift gears in and out of mesh so as to obtain different combinations of speeds. A good example of this is the automobile transmission where three speeds forward and one in reverse are obtained by shifting two gears.

In devices such as instruments and automatic controls, obtaining the correct motion is all-important. The power transmitted by the elements may be so slight as to be negligible, which allows the components to be proportioned primarily on the basis of motion, strength being of secondary importance.

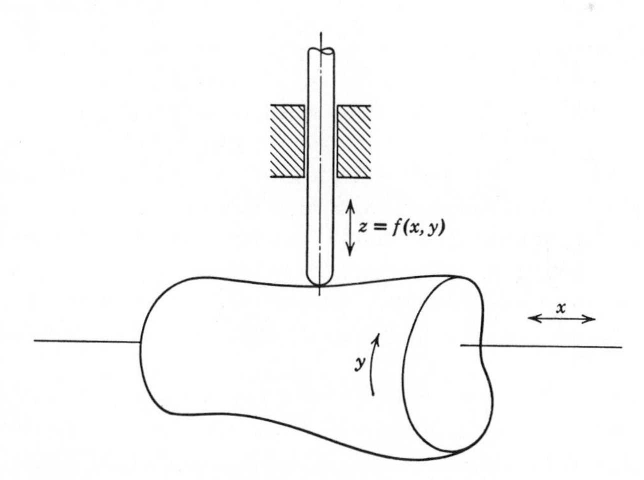

Figure 1.4 Three-dimensional cam.

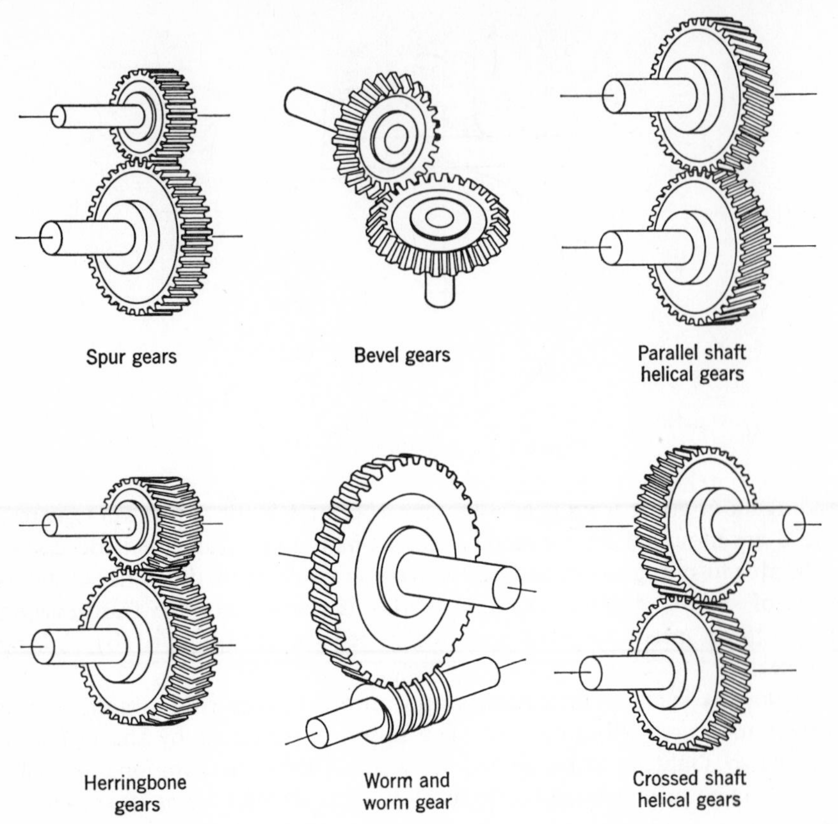

Figure 1.5

There are other machines, however, where the kinematic analysis is only one step in the design. After it has been determined how the various machine components will act to accomplish the desired work, the forces acting upon these parts must be analyzed. From this the physical size of the parts may be determined. A machine tool is a good example; its strength and rigidity are more difficult to attain than the desired motions.

It is important at this time to define the terms used in the study of mechanisms. This is done in the following section.

1.2 Mechanism, Machine. In the study of mechanisms the terms mechanism and machine will be used repeatedly. These are defined as follows:

A *mechanism* is a combination of rigid or resistant bodies so formed and connected that they move upon each other with definite relative motion. An example is the crank, connecting rod, and piston of an internal-combustion engine as shown diagrammatically in Fig. 1.1.

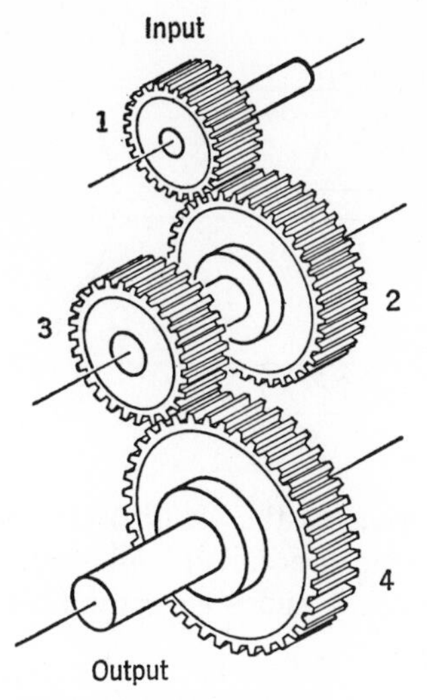

Figure 1.6 Gear train.

A *machine* is a mechanism or collection of mechanisms which transmit force from the source of power to the resistance to be overcome. An example is the complete internal-combustion engine.

1.3 Motion. In dealing with the study of mechanisms, it is necessary to define the various types of motion produced by these mechanisms.

Plane Motion. TRANSLATION. When a rigid body so moves that the position of each straight line of the body is parallel to all of its other positions, the body has motion of translation.

1. Rectilinear translation. All points of the body move in parallel straight line paths. When the body moves back and forth in this manner, it is said to reciprocate. This is illustrated in Fig. 1.7, where the slider 4 reciprocates between the limits B' and B''.

2. Curvilinear translation. The paths of the points are identical curves parallel to a fixed plane. Figure 1.8 shows the mechanism that was used in

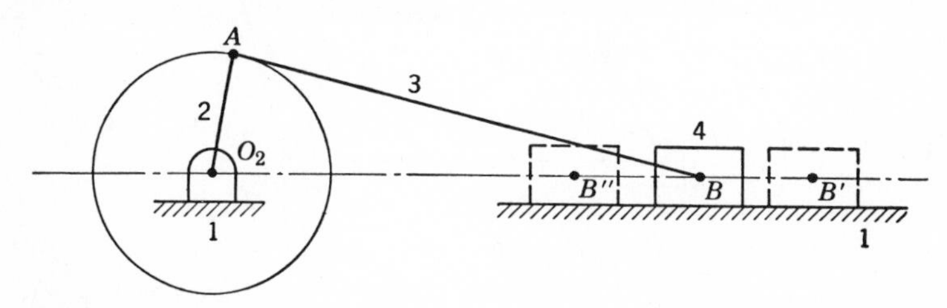

Figure 1.7

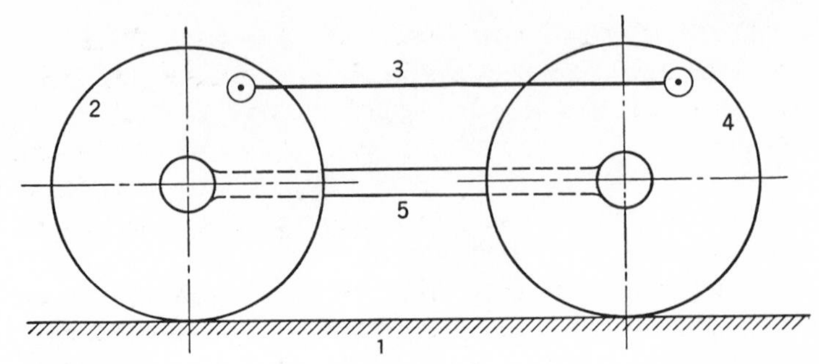

Figure 1.8

connecting the drive wheels of the steam locomotive. In this mechanism, link 3 has curvilinear translation, and all points in the body trace out identical cycloids as wheels 2 and 4 roll along track 1. Link 5 moves with rectilinear translation.

ROTATION. If each point of a rigid body having plane motion remains at a constant distance from a fixed axis which is perpendicular to the plane of motion, the body has motion of rotation. If the body rotates back and forth through a given angle, it is said to oscillate. This is shown in Fig. 1.9 where link 2 rotates and link 4 oscillates between the position B' and B''.

ROTATION AND TRANSLATION. Many bodies have motion which is a combination of rotation and translation. Link 3 in Fig. 1.7, links 2 and 4 in Fig. 1.8, and link 3 in Fig. 1.9 are examples of this type of motion.

Helical Motion. When a rigid body moves so that each point of the body has motion of rotation about a fixed axis and at the same time has translation parallel to the axis, the body has helical motion. An example of helical motion is the motion of a nut as the nut is screwed onto a bolt.

Spherical Motion. When a rigid body moves so that each point of the body has motion about a fixed point while remaining at a constant distance from it, the body has spherical motion.

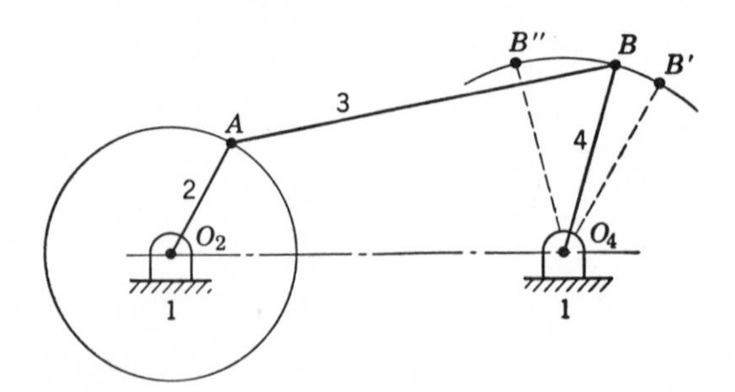

Figure 1.9

1.4 Cycle, Period, and Phase of Motion. When the parts of a mechanism have passed through all the possible positions they can assume after starting from some simultaneous set of relative positions and have returned to their original relative positions, they have completed a *cycle* of motion. The time required for a cycle of motion is the *period*. The simultaneous relative positions of a mechanism at a given instant during a cycle are a *phase*.

1.5 Pairing Elements. The geometrical forms by which two members of a mechanism are joined together so that the relative motion between these two members is consistent are known as *pairing elements*. If the joint by which two members are connected has surface contact such as a pin joint, the connection is known as a *lower pair*. If the connection takes place at a point or along a line such as in a ball bearing or between two gear teeth in contact, it is known as a *higher pair*. A pair that permits only relative rotation is a *revolute* or *turning pair*, and one that allows only sliding is a *sliding pair*. A turning pair can be either a lower or a higher pair depending upon whether a pin and bushing or a ball bearing is used for the connection. A sliding pair will be a lower pair as between a piston and cylinder wall.

1.6 Link, Chain. A *link* is a rigid body having two or more pairing elements by means of which it may be connected to other bodies for purposes of transmitting force or motion. Generally a link is a rigid member with provision at each end for connection to two other links. This may be extended, however, to include three, four, or even more connections. Figures 1.10*a*, *b*, and *c* show these arrangements. Perhaps the extreme case of a multiply connected link is the master rod in a nine-cylinder radial aircraft engine as seen in Fig. 1.10*d*.

A well-known example of a link with three connections is the bell crank which can be arranged as shown in Fig. 1.11*a* or Fig. 1.11*b*. This link is

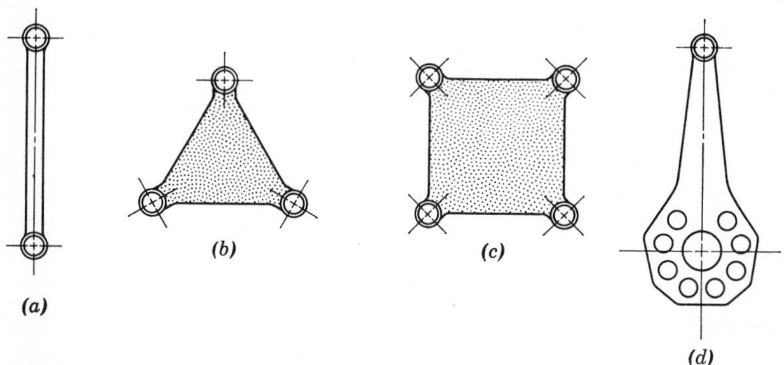

Figure 1.10

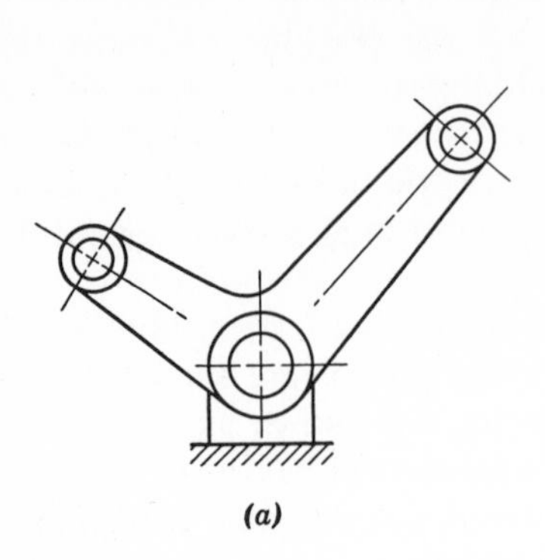

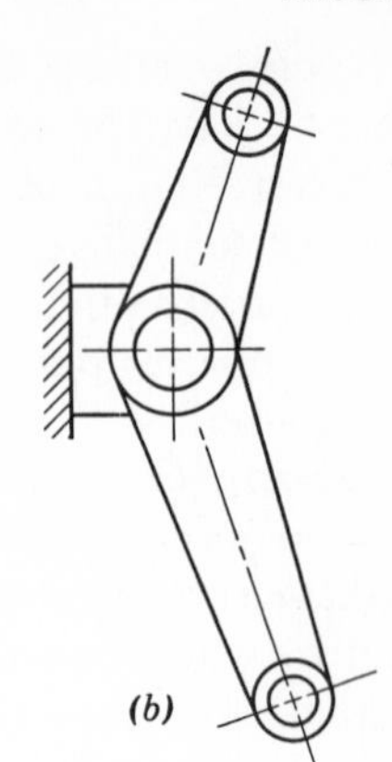

Figure 1.11

generally used for motion reduction and can be proportioned for a given ratio with a minimum of distortion of the required motion.

When a number of links is connected by means of pairs, the resulting system is a kinematic *chain*. If these links are connected in such a way that no motion is possible, a locked chain (structure) results. A constrained chain is obtained when the links are so connected that, no matter how many motion cycles are passed through, the relative motion will always be the same between the links. It is also possible to connect links so that an unconstrained chain results, which means that the motion pattern will vary from time to time depending on the amount of friction present in the joints. If one of the links of a constrained chain is made a fixed link, the result is a mechanism.

1.7 Inversion. If in a mechanism, the link which was originally fixed is allowed to move and another link becomes fixed, the mechanism is said to be inverted. The inversion of a mechanism does not change the motion of its links relative to each other but does change their absolute motions (relative to the ground).

1.8 Transmission of Motion. In the study of mechanisms, it is necessary to investigate the method in which motion may be transmitted from one member to another. Motion may be transmitted in three ways: (*a*) direct contact between two members such as between a cam and follower or between two gears, (*b*) through an intermediate link or connecting rod, and (*c*) by a flexible connector such as a belt or chain.

The angular velocity ratio is determined for the case of two members in direct contact. Figure 1.12 shows cam 2 and follower 3 in contact at point P. The cam has clockwise rotation, and the velocity of point P as a point on body 2 is represented by the vector $\mathbf{PM}_2$. The line NN' is normal to the two surfaces at point P and is known as the *common normal*, the *line of transmission*, or the *line of action*. The common tangent is represented by TT'. The vector $\mathbf{PM}_2$ is broken into two components, $\mathbf{Pn}$ along the common normal and $\mathbf{Pt}_2$ along the common tangent. Because of the fact that the cam and the follower are rigid members and must remain in contact, the normal component of the velocity of P as a point on body 3 must be equal to the normal component of P as a point on body 2. Therefore, knowing the direction of the velocity vector of P as a point on body 3 to be perpendicular to the radius O_3P and its normal component, it is possible to find the velocity $\mathbf{PM}_3$ as shown in the sketch. From this vector, the angular velocity of the follower may be determined from the relation $V = R\omega$, where V equals the linear velocity of a point moving along a path of radius R and ω equals the angular velocity of the radius R.

In direct-contact mechanisms, it is often necessary to determine the velocity of sliding. From the sketch this can be seen to be the vector difference between the tangential components of the velocities of the points of contact. This difference is given by the distance t_2t_3 because the component $\mathbf{Pt}_3$ is opposite in direction to that of $\mathbf{Pt}_2$. If t_2 and t_3 fall on the same side of P, then the distance will subtract. If the contact point P should fall on the line of centers, then the vectors $\mathbf{PM}_2$ and $\mathbf{PM}_3$ will be equal and in the same direction. The tangential components must also be equal and in the same direction so that the velocity of sliding will be zero. The two members will then have pure rolling motion. Thus, it may be said that *the condition for pure rolling is that the point of contact shall lie on the line of centers.*

For the mechanism of Fig. 1.12, the motion between the cam and the follower will be a combination of rolling and sliding. Pure rollings can only take place where the point of contact P falls on the line of centers. However,

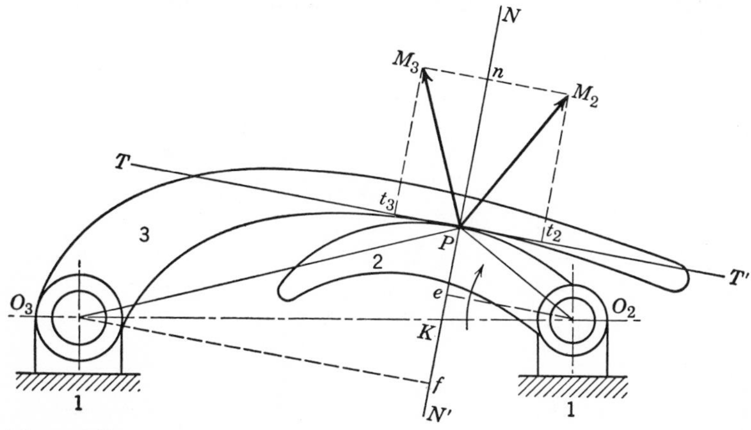

Figure 1.12

contact at this point may not be possible because of the proportions of the mechanism. Pure sliding cannot occur between cam 2 and follower 3. For this to happen, a point on one link, within the limits of its travel, has to come in contact with all the successive points on the active surface of the other link.

It is possible to determine a relation so that the angular velocity ratio of two members in direct contact can be determined without going through the geometrical construction outlined above. From O_2 and O_3 drop perpendiculars upon the common normal striking it at e and f, respectively. The following relations will be seen to hold:

$$\omega^2 = \frac{PM_2}{O_2P} \qquad \text{and} \qquad \omega_3 = \frac{PM_3}{O_3P}$$

$$\frac{\omega_3}{\omega_2} = \frac{PM_3}{O_3P} \times \frac{O_2P}{PM_2}$$

From the fact that triangles PM_2n and O_2Pe are similar,

$$\frac{PM_2}{O_2P} = \frac{Pn}{O_2e}$$

Also, PM_3n and O_3Pf are similar triangles; therefore,

$$\frac{PM_3}{O_3P} = \frac{Pn}{O_3f}$$

Therefore,

$$\frac{\omega_3}{\omega_2} = \frac{Pn}{O_3f} \times \frac{O_2e}{Pn} = \frac{O_2e}{O_3f}$$

With the common normal intersecting the line of centers at K, triangles O_2Ke and O_3Kf are also similar; therefore,

$$\frac{\omega_3}{\omega_2} = \frac{O_2e}{O_3f} = \frac{O_2K}{O_3K} \tag{1.1}$$

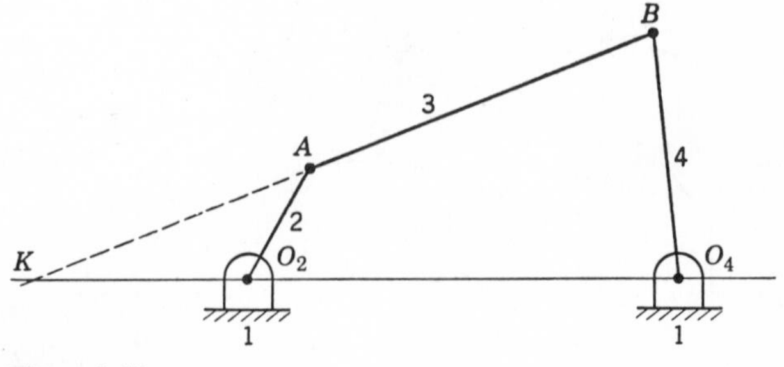

Figure 1.13

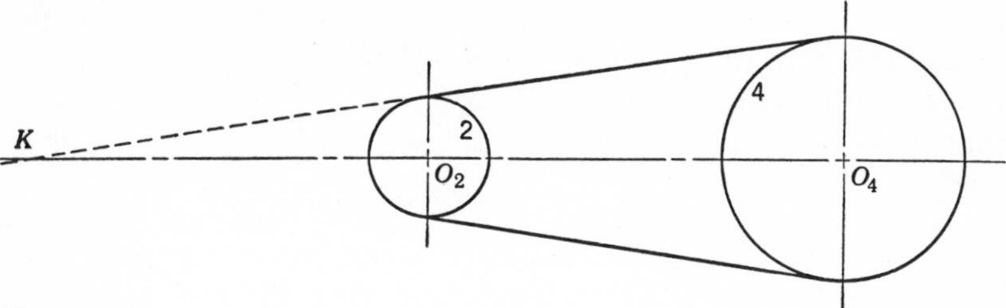

Figure 1.14

Therefore, for a pair of curved surfaces in direct contact, the angular velocities are inversely proportional to the segments into which the line of centers is cut by the common normal. From this it can be seen that *for constant angular velocity ratio the common normal must intersect the line of centers in a fixed point.*

It is also possible to derive the above relations for the transmission of motion through an intermediate link or connecting rod and for the transmission of motion through a flexible connector. Figures 1.13 and 1.14 show these two cases, respectively, where the angular velocity ratio is given by

$$\frac{\omega_4}{\omega_2} = \frac{O_2K}{O_4K} \tag{1.2}$$

In Fig. 1.14, the ratio ω_4/ω_2 is independent of the center distance O_2O_4.

Problems

1.1 (*a*) If $\omega_2 = 20$ rad/min, calculate the angular velocity of link 3 for the two cases shown in Fig. 1.15. (*b*) Calculate the maximum angle and the minimum angle of the follower with the horizontal.

1.2 Lay out the mechanisms for Problem 1.1 to full scale and graphically determine the velocity of sliding between links 2 and 3. Use a velocity scale of 1 in. = 10 in./min.

1.3 If $\omega_2 = 20$ rad/min for the mechanism shown in Fig. 1.15, using graphical construction, determine the angular velocities of link 3 for one revolution of the cam in 60° increments starting from the position where $\omega_3 = 0$. Plot ω_3 versus cam angle θ letting the scale of ω_3 be 1 in. = 2.0 rad/min and the scale of θ be $\frac{1}{2}$ in. = 60°.

1.4 Prove for the linkage shown in Fig. 1.13 that the angular velocities of the driven and driver links are inversely proportional to the segments into which the line of centers is cut by the line of transmission.

1.5 Prove for the belt and pulleys shown in Fig. 1.14 that the angular velocities of the pulleys are inversely proportional to the segments into which the line of centers is cut by the line of transmission.

1.6 In a linkage as shown in Fig. 1.13, the crank 2 is 1.90 cm long and rotates at a constant angular velocity of 15 rad/sec. Link 3 is 3.80 cm long and link 4 is 2.50 cm

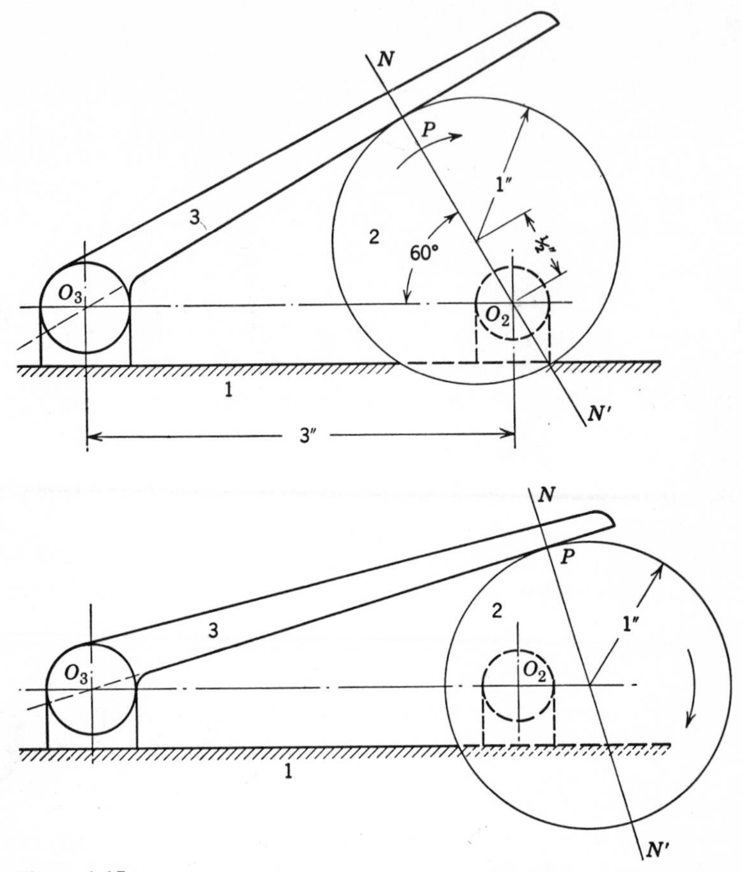

Figure 1.15

long. The distance between centers O_2 and O_4 is 5.10 cm. Graphically determine the angular velocity of link 4 when link 2 is counterclockwise 45° from the horizontal. State whether or not ω_4 is constant.

1.7 A pulley of 10 cm diam drives one of 20 cm diam by means of a belt. If the angular velocity of the drive pulley is 65 rad/sec and the center distance between pulleys 40 cm, graphically determine the speed of the 20 cm pulley. Will its speed be constant?

2

Linkages

2.1 Four-Bar Linkage. One of the simplest and most useful mechanisms is the four-bar linkage. A sketch of this linkage is shown in Fig. 2.1. Link 1 is the frame or ground and is generally stationary. Link 2 is the driver which may rotate completely or may oscillate. In either case, link 4 will oscillate. If link 2 rotates completely, then the mechanism is transforming rotary motion into oscillatory motion. If the crank oscillates, then the mechanism multiplies oscillatory motion.

When link 2 is rotating completely, there is no danger of the linkage locking. However, if link 2 oscillates, care must be taken in proportioning the links to avoid dead points so that the mechanism will not stall in its extreme positions. These dead points will occur when the line of action of the driving force is directed along link 4. This condition is shown in Fig. 2.2.

If the four-bar mechanism is designed so that link 2 can rotate completely but link 4 is made the driver, dead points will occur, and it is necessary to provide a flywheel to pass through these dead points.

In addition to possible dead points in a four-bar linkage, it is necessary to consider the *transmission angle* which is the angle between the connecting or coupler link 3 and the output link 4. This is shown in Fig. 2.3*a* as angle γ.

An equation for the transmission angle can be derived by applying the Law of Cosines to triangles AO_2O_4 and ABO_4 as follows:

$$z^2 = r_1^2 + r_2^2 - 2r_1r_2 \cos \theta_2$$

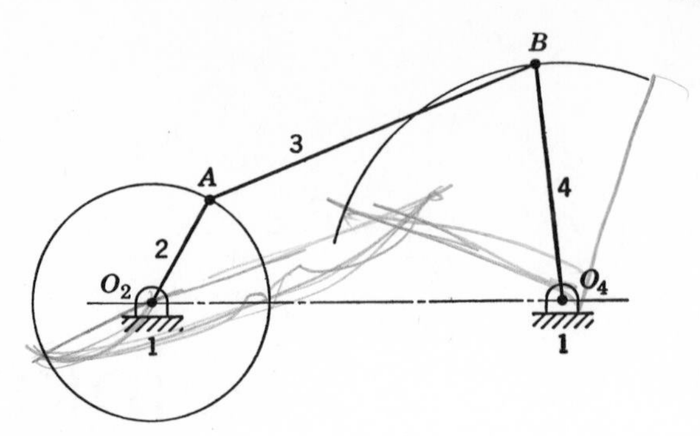

Figure 2.1

also

$$z^2 = r_3^2 + r_4^2 - 2r_3r_4 \cos \gamma$$

Therefore,

$$r_1^2 + r_2^2 - 2r_1r_2 \cos \theta_2 = r_3^2 + r_4^2 - 2r_3r_4 \cos \gamma$$

and

$$\cos \gamma = \frac{r_1^2 + r_2^2 - r_3^2 - r_4^2 - 2r_1r_2 \cos \theta_2}{-2r_3r_4} \tag{2.1}$$

In general, the maximum transmission angle should not be greater than 140° and the minimum not less than 40° if the linkage is to be used to transmit high forces. If the transmission angle becomes less than 40°, the linkage tends to bind because of friction in the joints; also links 3 and 4 tend to align and may lock. It is especially important to check transmission angles when linkages are designed to operate close to dead points. An illustration of the minimum and the maximum transmission angle for a four-bar linkage is shown in Fig. 2.3*b* by γ' and γ'', respectively. In this mechanism, link 2 rotates completely and link 4 oscillates.

The four-bar linkage may take other forms as shown in Fig. 2.4. In Fig. 2.4*a* the mechanism has been crossed and will give the same type of motion as in Fig. 2.1. In Fig. 2.4*b* opposite links are all the same length and, therefore,

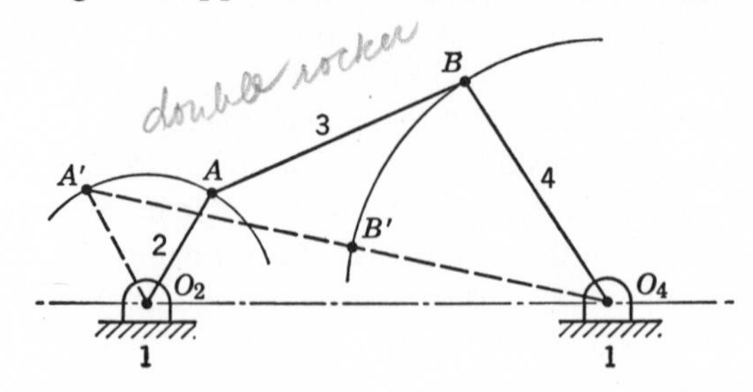

Figure 2.2

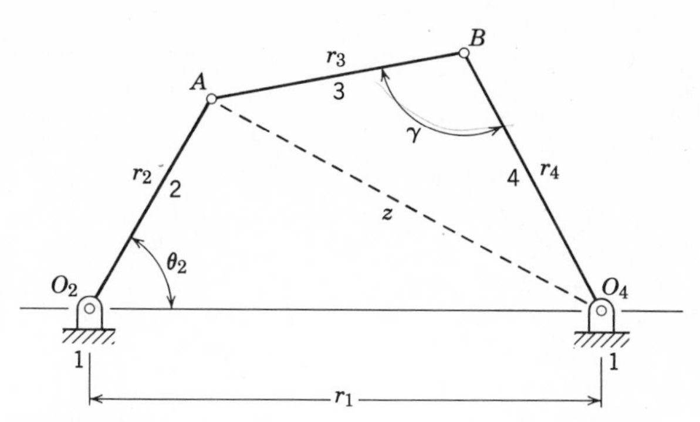

Figure 2.3*a*

always remain parallel; both links 2 and 4 rotate completely. This type of motion was characteristic of a steam locomotive drive. Figure 2.4*c* shows another arrangement whereby both the driver and follower rotate continuously. This form of the four-bar linkage is the basis for the drag-link mechanism which will be discussed under the subject of quick-return mechanisms. For rotation of crank 2 at a constant angular velocity, link 4 will rotate at a nonuniform rate. In order to prevent locking of the mechanism, certain relations must be maintained between the links:

$$O_2A \quad \text{and} \quad O_4B > O_2O_4$$
$$(O_2A - O_2O_4) + AB > O_4B$$
$$(O_4B - O_2O_4) + O_2A > AB$$

The second and third relation can be derived from the triangles $O_4A'B'$

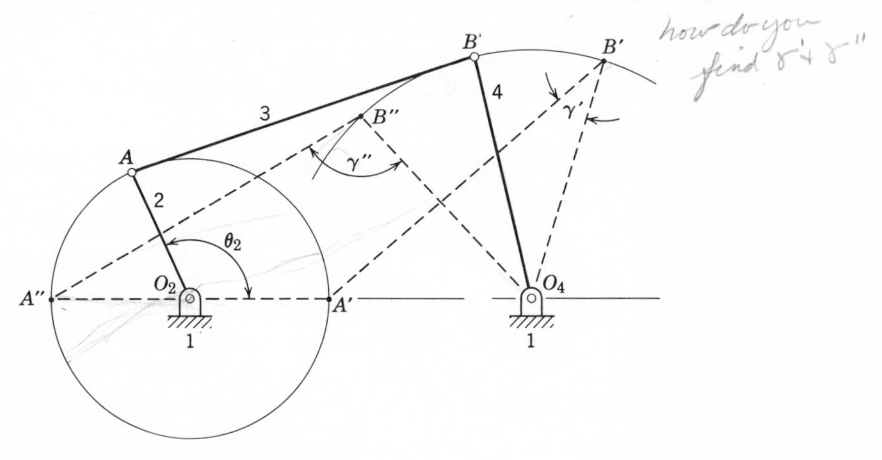

Figure 2.3*b*

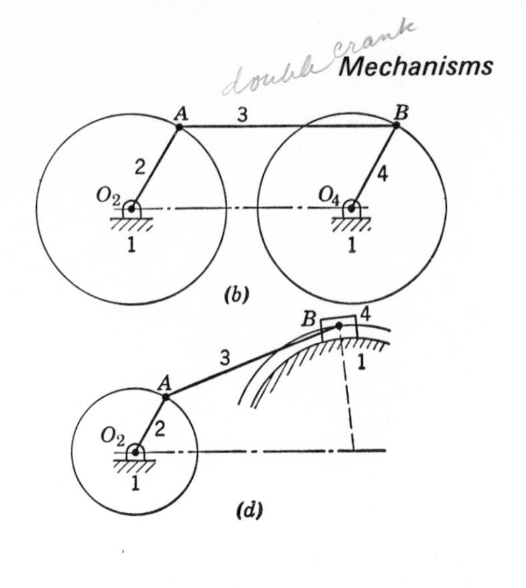

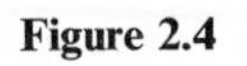
Figure 2.4

and $O_2A''B''$ respectively, and the fact that the sum of two sides of a triangle must be greater than the third side.

Figure 2.4*d* shows an arrangement whereby link 4 of Fig. 2.1 has been replaced by a sliding block. The motion of the two linkages is identical.

The motion of the four-bar linkage is often characterized by the term *crank rocker* to indicate that crank 2 rotates completely and link 4 oscillates as in Fig. 2.4*a*. In a similar manner, the term *double crank* means that both link 2 and link 4 rotate completely as in Figs. 2.4*b* and *c*. The term *double rocker* indicates that both link 2 and link 4 oscillate as shown in Fig. 2.2.

As a means of determining whether a four-bar linkage will operate as a crank rocker, a double crank, or a double rocker, Grashoff's Law can be applied. This law[1] states that if the sum of the lengths of the longest link and the shortest link is *less than* the sum of the lengths of the other two, there will be formed:

1. Two different crank rockers when the shortest link is the crank and either of the adjacent links is the fixed link.
2. A double crank when the shortest link is the fixed link.
3. A double rocker when the link opposite the shortest is the fixed link.

Also, if the sum of the lengths of the longest and the shortest links is *greater than* the sum of the lengths of the other two, only double-rocker mechanisms will result. Also, if the sum of the longest and shortest links is *equal to* the sum of the other two, the four possible mechanisms are similar to those of

[1] R. S. Hartenberg and J. Denavit, *Kinematic Synthesis of Linkages*, McGraw-Hill Book Company, 1964.

1, 2, and 3 above. However, in this last case the center lines of the links can become colinear so that the driven link can change direction of rotation unless some means is provided to avoid it. Such a linkage is shown in Fig. 2.4*b* where the links become colinear along the line of centers O_2O_4. At this position, the direction of rotation of the driven link 4 could change unless inertia carried link 4 through this point.

2.2 Slider-Crank Mechanism. This mechanism is widely used and finds its greatest application in the internal-combustion engine. Figure 2.5*a* shows a sketch in which link 1 is the frame (considered fixed), link 2 the crank, link 3 the connecting rod, and link 4 the slider. With the internal-combustion engine, link 4 is the piston upon which gas pressure is exerted. This force is transmitted through the connecting rod to the crank. It can be seen that there will be two dead points during the cycle, one at each extreme position of piston travel. In order to overcome these, it is necessary to attach a flywheel to the crank so that the dead points can be passed. This mechanism is also used in air compressors where an electric motor drives the crank which in turn drives the piston that compresses the air.

In considering the slider crank, it is often necessary to calculate the displacement of the slider and its corresponding velocity and acceleration. Equations for displacement, velocity, and acceleration are derived using Fig. 2.5*b*.

$$
\begin{aligned}
x &= R + L - R\cos\theta - L\cos\phi \\
&= R(1 - \cos\theta) + L(1 - \cos\phi) \\
&= R(1 - \cos\theta) + L\left[1 - \sqrt{1 - (R/L)^2 \sin^2\theta}\right]
\end{aligned}
\tag{2.2}
$$

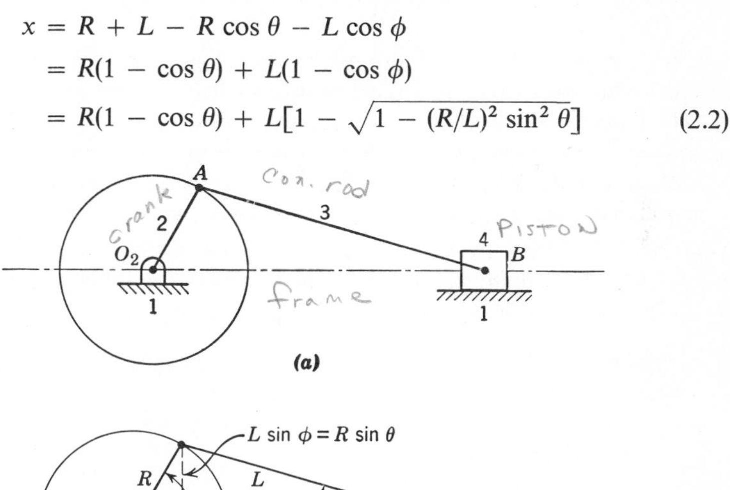

Figure 2.5

In order to simplify the above expression, the radical can be approximated by replacing it with the series

$$(1 \pm B^2)^{1/2} = 1 \pm \frac{1}{2} B^2 - \frac{B^4}{2 \cdot 4} \pm \frac{1 \cdot 3B^6}{2 \cdot 4 \cdot 6} - \frac{1 \cdot 3 \cdot 5B^8}{2 \cdot 4 \cdot 6 \cdot 8} \pm \cdots$$

where $B = (R/L) \sin \theta$.

In general, it is sufficiently accurate to use only the first two terms of the series.

Therefore,

$$\sqrt{1 - \left(\frac{R}{L}\right)^2 \sin^2 \theta} = 1 - \frac{1}{2}\left(\frac{R}{L}\right)^2 \sin^2 \theta \qquad \text{(approximately)}$$

and

$$x = R(1 - \cos \theta) + \frac{R^2}{2L} \sin^2 \theta$$

where $\theta = \omega t$ because ω is constant

$$V = \frac{dx}{dt} = R\omega \left[\sin \theta + \frac{R}{2L} \sin 2\theta \right] \tag{2.3}$$

$$A = \frac{d^2x}{dt^2} = R\omega^2 \left[\cos \theta + \frac{R}{L} \cos 2\theta \right] \tag{2.4}$$

It is possible to fix some link other than 1 on the slider crank and thus obtain three inversions, which are shown in Fig. 2.6. In Fig. 2.6*a* the crank is held fixed and all of the other links are allowed to move. This gives a mechanism that was used in early aircraft engines. They were known as rotary engines because the crank was stationary and the cylinders rotated about the crank. A more modern application of this inversion is in the Whitworth mechanism which will be discussed under quick-return mechanisms. Figure 2.6*b* shows an inversion in which the connecting rod is held fixed. This inversion is used on donkey steam engines and is also the basis

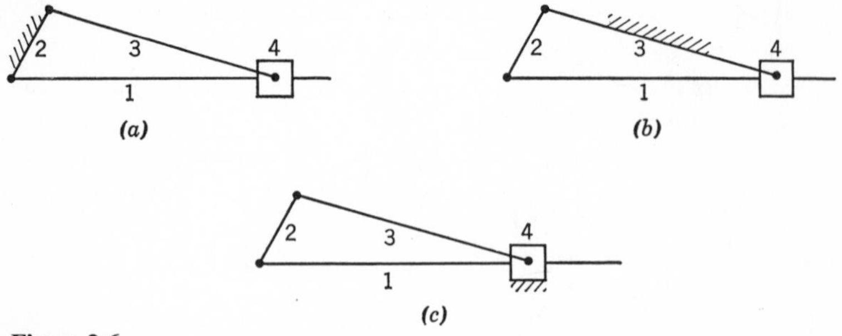

Figure 2.6

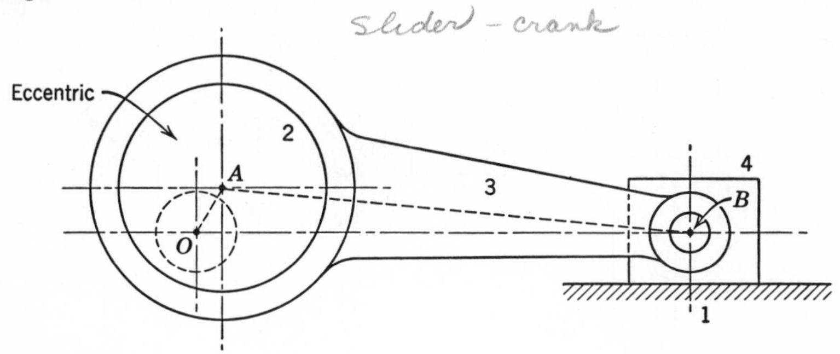

Figure 2.7

for the crank-shaper mechanism to be discussed later. The third inversion where the slider is held fixed, Fig. 2.6*c*, is sometimes used in the hand farm pump.

A variation of the slider-crank mechanism can be affected by increasing the size of the crank pin until it is larger than the shaft to which it is attached and at the same time offsetting the center of the crank pin from that of the shaft. This enlarged crank pin is called an *eccentric* and can be used to replace the crank in the original mechanism. Figure 2.7 shows a sketch where point A is the center of the eccentric and point O the center of the shaft. The motion of this mechanism with the equivalent crank length OA is identical with that of the slider crank. One serious disadvantage of this mechanism, however, is the problem of proper lubrication between the eccentric and the rod. This limits the amount of power that can be transmitted.

2.3 Scotch Yoke. This mechanism is one which will give simple harmonic motion. Its early application was on steam pumps, but it is now used as a mechanism on a test machine to produce vibrations. It is also used as a sine-cosine generator for computing elements. Figure 2.8*a* shows a sketch of this mechanism. Figure 2.8*b* shows the manner in which simple harmonic motion is generated. The radius r rotates at a constant angular velocity ω_r, and the projection of the point P upon the x axis (or y axis) moves with simple harmonic motion. The displacement from where the circle cuts the x axis and increasing to the left is

$$x = r - r\cos\theta_r \qquad \text{where } \theta_r = \omega_r t \tag{2.5}$$

Therefore,

$$x = r(1 - \cos\omega_r t)$$

$$V = \frac{dx}{dt} = r\omega_r \sin\omega_r t = r\omega_r \sin\theta_r \tag{2.6}$$

$$A = \frac{d^2x}{dt^2} = r\omega_r^2 \cos\omega_r t = r\omega_r^2 \cos\theta_r \tag{2.7}$$

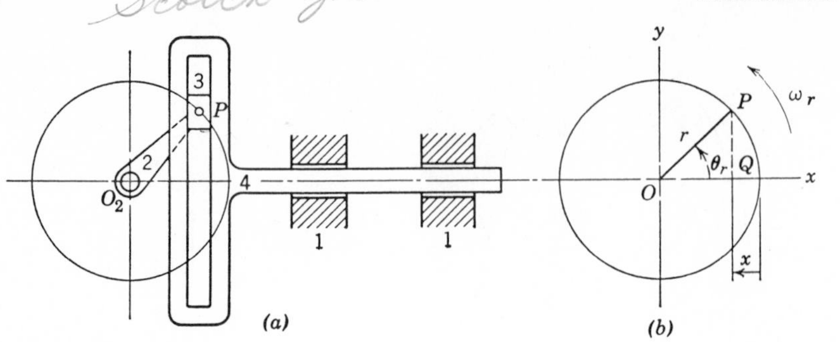

Figure 2.8

Another mechanism which will give simple harmonic motion is a circular cam (eccentric) with a flat-faced radial follower. This is discussed in the following chapter.

2.4 Quick-Return Mechanisms. These mechanisms are used on machine tools to give a slow cutting stroke and a quick return stroke for a constant angular velocity of the driving crank and are combinations of simple linkages such as the four-bar linkage and the slider-crank mechanism. An inversion of the slider crank in combination with the conventional slider crank is also used. In the design of quick-return mechanisms, the ratio of the crank angle for the cutting stroke to that for the return stroke is of prime importance and is known as the *time ratio*. To produce a quick return of the cutting tool, this ratio must obviously be greater than unity and as large as possible. As an example, the crank angle for the cutting stroke for the mechanism shown in Fig. 2.11 is labeled α and that for the return stroke is labeled β. With the assumption that the crank operates at a constant speed, the time ratio is, therefore, α/β which is much greater than unity.

There are several types of quick-return mechanisms which are described as follows:

Drag Link. This is developed from the four-bar linkage and is shown in Fig. 2.9. For a constant angular velocity of link 2, link 4 will rotate at a nonuniform velocity. Ram 6 will move with nearly constant velocity over most of the upward stroke to give a slow upward stroke and a quick downward stroke when driving link 2 rotates clockwise.

Whitworth. This is a variation of the first inversion of the slider crank in which the crank is held fixed. Figure 2.10 shows a sketch of the mechanism, and both links 2 and 4 make complete revolutions.

Crank Shaper. This mechanism is a variation of the second inversion of the slider crank in which the connecting rod is held fixed. Figure 2.11 shows the arrangement in which link 2 rotates completely and link 4 oscillates.

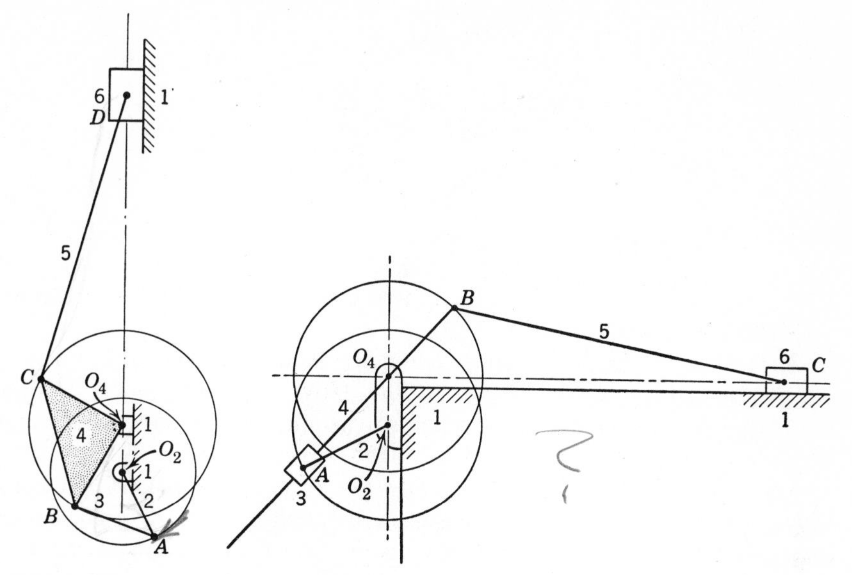

Figure 2.9

Figure 2.10

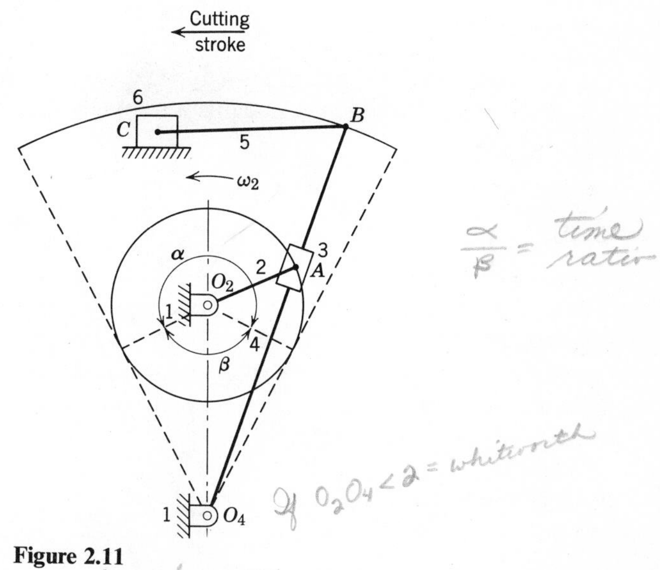

Figure 2.11

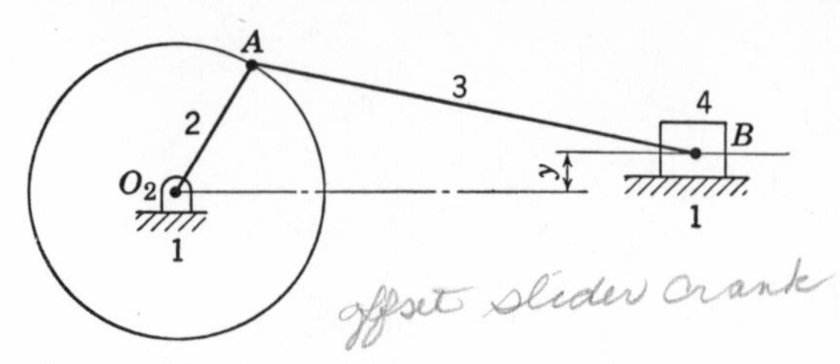

Figure 2.12

If the distance O_2O_4 is shortened until it is less than the crank, the mechanism will revert to the Whitworth.

Offset Slider Crank. The slider crank can be offset as shown in Fig. 2.12, which will give a quick return motion. However, the amount of quick return is very slight, and the mechanism would only be used where space was limited and the mechanism had to be simple.

2.5 Toggle Mechanism. This mechanism has many applications where it is necessary to overcome a large resistance with a small driving force. Figure 2.13 shows the mechanism; links 4 and 5 are of the same length. As the angles α decrease and links 4 and 5 approach being colinear, the force F required to overcome a given resistance P decreases as shown by the following relation:

$$\frac{F}{P} = 2 \tan \alpha \tag{2.8}$$

It can be seen that for a given F as α approaches zero, P approaches infinity. A stone crusher utilizes this mechanism to overcome a large resistance with

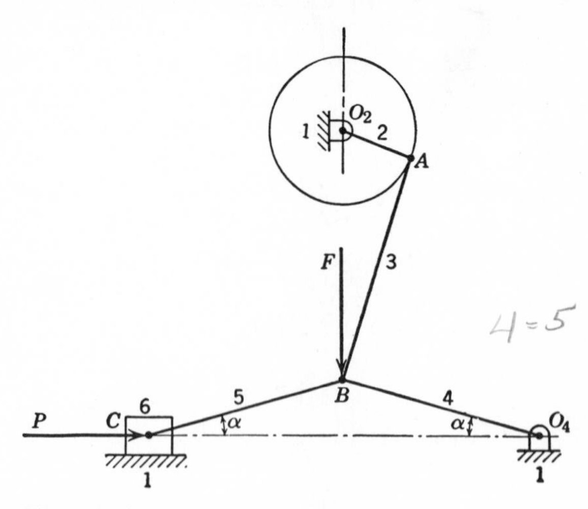

Figure 2.13

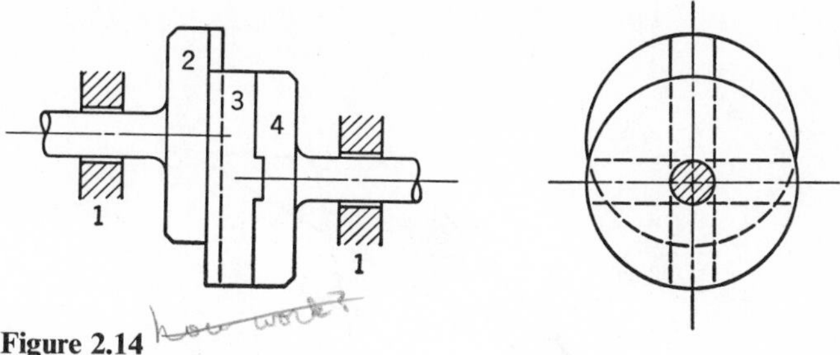

Figure 2.14

a small force. This mechanism can be used statically as well as dynamically, as is seen in numerous toggle clamping devices for holding work pieces.

2.6 Oldham Coupling. This mechanism provides a means for connecting two parallel shafts which are out of line a small amount so that a constant angular velocity ratio can be transmitted from the drive shaft to the driven shaft. A sketch is shown in Fig. 2.14. This mechanism is an inversion of the Scotch yoke.

2.7 Straight-Line Mechanisms. As the name suggests, these mechanisms are designed so that a point on one of the links will move in a straight line. This straight line will be either an approximate or a theoretically correct straight line, depending on the mechanism.

An example of an approximate straight-line mechanism is the Watt, which is shown in Fig. 2.15. Point P is so located that the segments AP and BP are inversely proportional to the lengths O_2A and O_4B. Therefore, if links 2 and 4 are equal in length, point P must be the midpoint of link 3. Point P will trace out a path in the form of a figure 8. Part of this path will very nearly approach a straight line.

The Peaucellier mechanism is one which will generate an exact straight line. Figure 2.16 shows a sketch where links 3 and 4 are equal. Links 5, 6, 7, and 8 are also equal, and link 2 equals the distance O_2O_4. Point P will trace out an exact straight-line path.

Straight-line mechanisms have many applications; notable among these

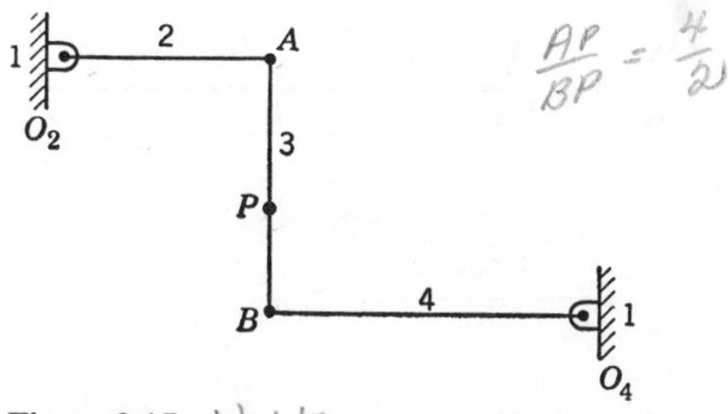

Figure 2.15

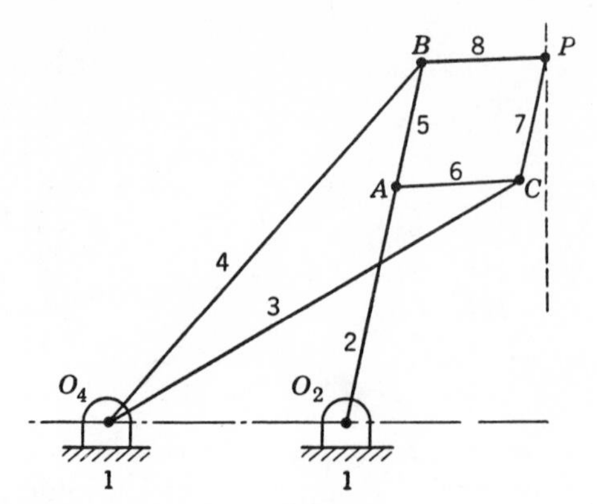

Figure 2.16

are the mechanisms for engine indicators and for electrical switch gear equipment.

2.8 Pantograph. This mechanism is used as a copying device. When one point is made to follow a certain path, another point on the mechanism will trace out an identical path that is enlarged or reduced. Figure 2.17 shows a sketch. Links 2, 3, 4, and 5 form a parallelogram, and point *P* is on an extension of link 4. Point *Q* is on link 5 at the intersection of a line drawn from *O* to *P*. As point *P* traces out a path, point *Q* will trace out a similar path to a reduced scale.

This mechanism finds many applications in copying devices, particularly in engraving or profiling machines. One use of the profiling machine is in making dies or molds. Point *P* serves as a finger and traces out the contour of a templet while a rotating endmill is placed at *Q* to machine the die to a smaller scale.

2.9 Chamber Wheels. This mechanism takes several forms, which fall into two classifications. The first type consists of two lobed wheels operating within a casing. The Roots blower, as shown in Fig. 2.18, is an example of this type. The rotors are cycloids and are driven by a pair of meshing gears of equal size in back of the case. In the modern application,

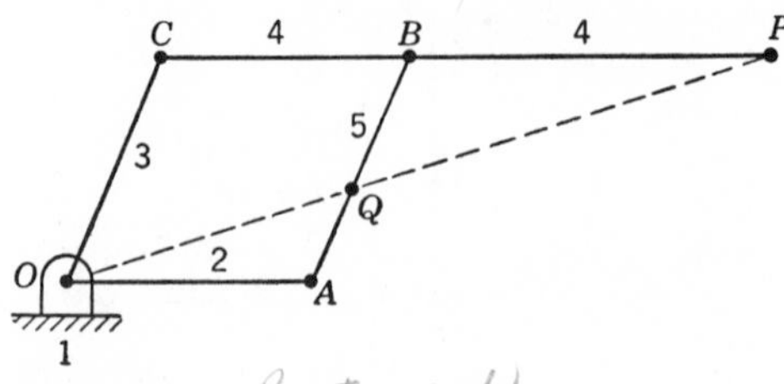

Figure 2.17

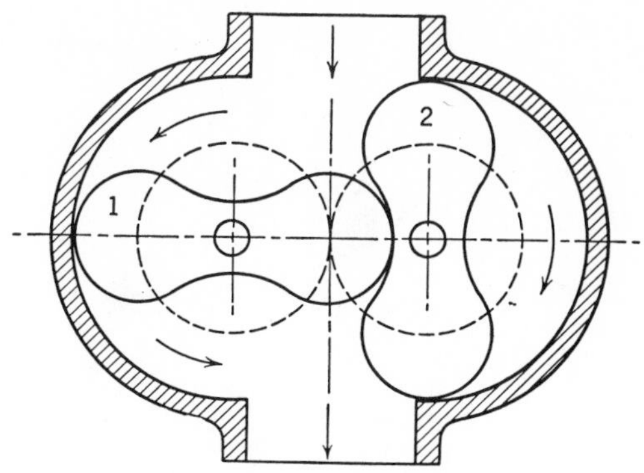

Figure 2.18

the Roots blower has three lobes on each rotor and is used for a low-pressure supercharger on Diesel engines.

The other class of chamber wheels has only one rotor placed eccentrically within the casing and is generally a variation of the slider-crank mechanism. Figure 2.19 shows a sketch of this type. The mechanism shown was originally designed for a steam engine, but its modern application is in the form of a pump.

Another example of the second type of chamber wheel is shown in Fig. 2.20 which illustrates the principle of the *Wankel engine*. In this mechanism the expanding gases act upon the three-lobed rotor, which revolves directly on the eccentric and transmits torque to the output shaft through the eccentric which is integral with the shaft. The phase relation between the rotor and the rotation of the eccentric shaft is maintained by a pair of internal and external gears (not shown) so that the orbital motion of the rotor is properly controlled.

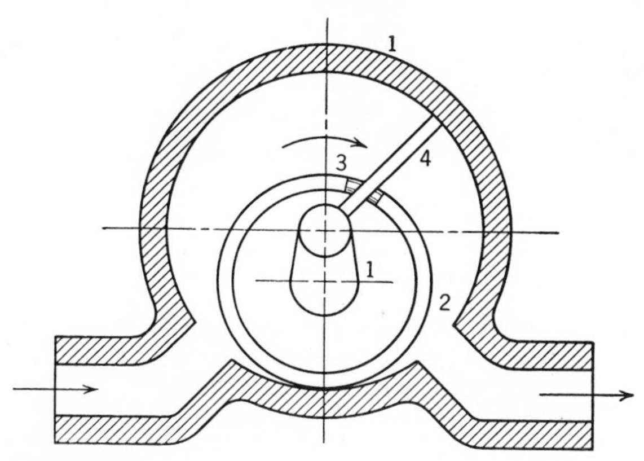

Figure 2.19

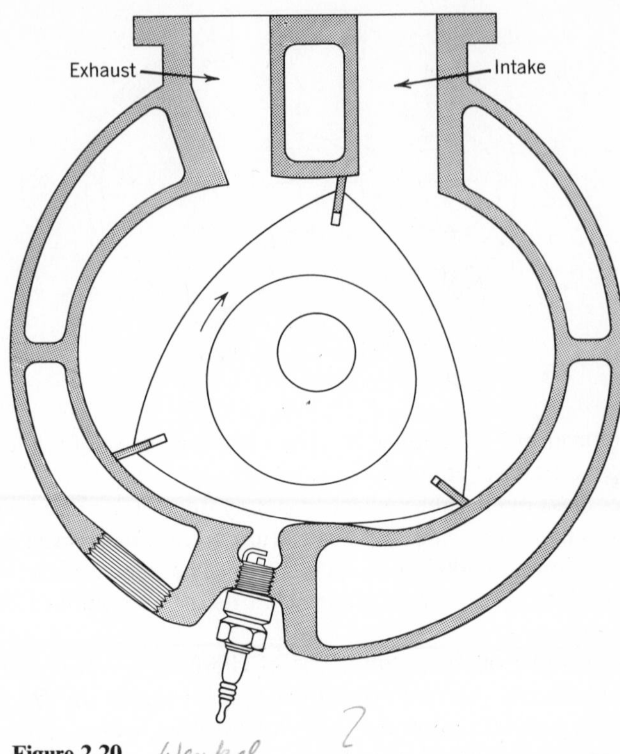

Figure 2.20 Wankel ?

2.10 Hooke's Coupling. This coupling is used to connect two intersecting shafts. It is also known as a universal joint and has its widest use in the automotive field. A sketch of the coupling is shown in Fig. 2.21, and a commercial model is illustrated in Fig. 2.22. In Fig. 2.21, link 2 is the driver and link 4 the follower. Link 3 is a cross piece that connects the two yokes. It can be shown[2] that, although both shafts must complete a revolution in the same length of time, the angular velocity ratio of the two shafts is not constant during the revolution but varies as a function of the angle β between the shafts and of the angle of rotation θ of the driver. The relation is given as

$$\frac{\omega_4}{\omega_2} = \frac{\cos\beta}{1 - \sin^2\beta \sin^2\theta} \tag{2.9}$$

A plot of this equation in polar coordinates for a quarter revolution of the driving shaft is shown in Fig. 2.23, which clearly indicates the effect of a large angle β between the shafts.

[2] J. E. Shigley, *Kinematic Analysis of Mechanisms*, Second Edition, McGraw-Hill Book Company, 1969.

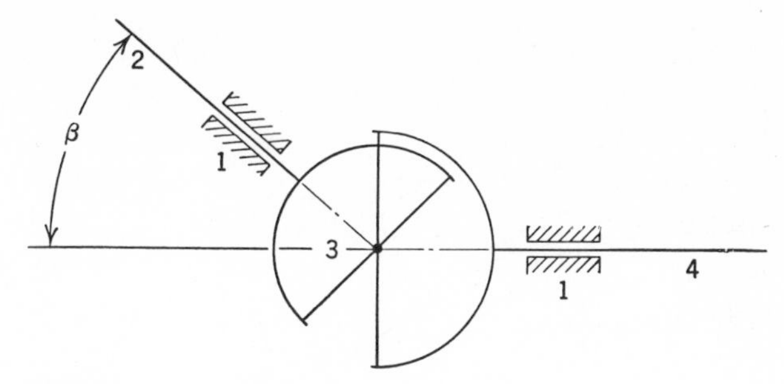

Figure 2.21

It is possible to connect two shafts by two Hooke's couplings and an intermediate shaft such that the uneven velocity ratio of the first coupling will be canceled out by the second. Figure 2.24 shows this application when the two shafts 2 and 4, which are to be connected, do not lie in the same plane. The connection must be made so that the driver and driven shafts 2 and 4 make equal angles β with the intermediate shaft 3. Also the yokes on shaft 3 must be connected in such a way that when one yoke lies in the plane of shafts 2 and 3, the other yoke lies in the plane of shafts 3 and 4. If the two shafts to be connected lie in the same plane, then the yokes on the intermediate shaft will be parallel. An application of two universal joints connecting shafts that lie in the same plane is the Hotchkiss automotive drive, which is used on most of the cars today.

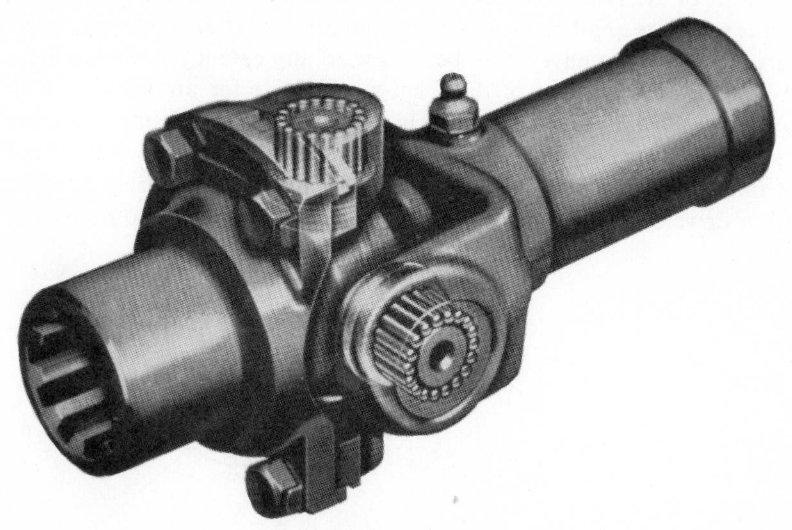

Figure 2.22 Hooke-type universal joint. (Courtesy of Mechanics Universal Joint Division, Borg–Warner Corp.)

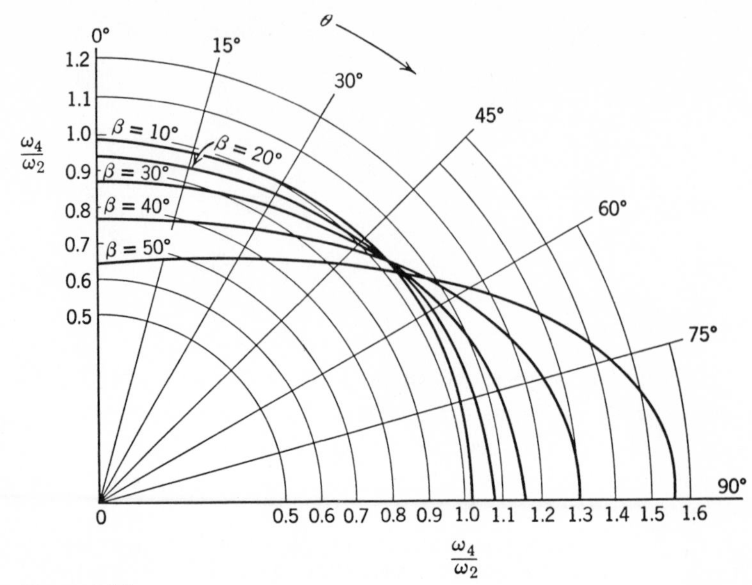

Figure 2.23

2.11 Constant Velocity Universal Joints. Engineers have been considering for many years the development of a single universal joint capable of transmitting a constant velocity ratio. Several joints which were variations of the Hooke principle were proposed, one as early as 1870, with the intermediate shaft reduced to zero length. As far as is known, however, joints of this design have never been used to any extent commercially.

With the development of the front-wheel drive for automotive vehicles, the need for a universal joint which was capable of transmitting constant angular velocity ratio was increased. It was true that two Hooke's couplings and an intermediate shaft could be used, but this was not entirely satisfactory. With a drive such as is required on a front wheel of an automobile, where

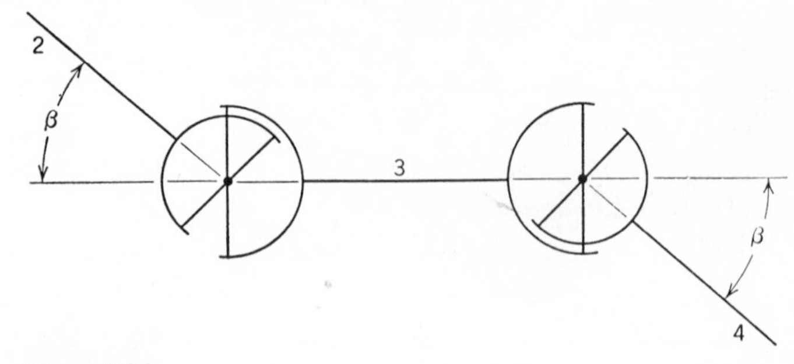

Figure 2.24

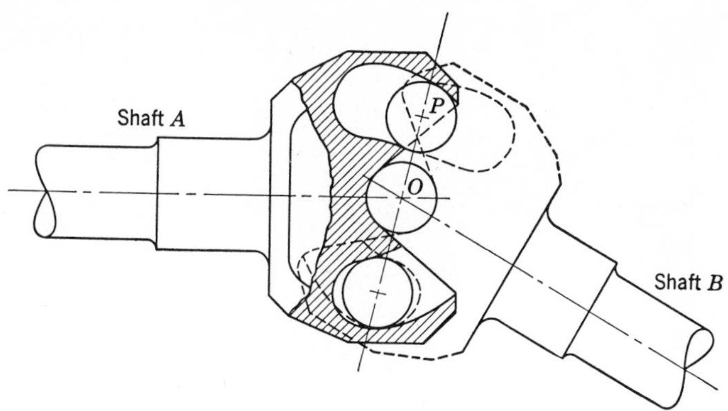

Figure 2.25

the angle β is sometimes quite large, the changing conditions made it almost impossible to obtain constant angular velocity ratio. The need for a constant velocity universal joint was met by the introduction of the Weiss and the Rzeppa joints in this country and by the Tracta joint in France. The Weiss joint was first patented in 1925, the Rzeppa in 1928, and the Tracta in 1933. Operation of these joints is not based on the same principle as the Hooke coupling.

A *Bendix–Weiss joint* is shown in Fig. 2.25. As shown in the figure, grooves that are symmetrical with respect to each other about the center lines of the shafts are formed in the surfaces of the prongs of the yokes, and four steel balls are located between these prongs at a point where the axes of the grooves in one yoke intersect the axes of the grooves in the other yoke. Power is transmitted from the driver to the follower through these balls. A fifth ball with a slot provides for locking of the parts in assembly as well as for taking end thrust. In operation, the balls will automatically shift their positions as the angular displacement of the two shafts is varied, so that the plane containing the centers of the balls will always bisect the angle between the two shafts. It can be proved[3] that constant angular velocity ratio will result from this condition. A photograph of a Bendix–Weiss joint is shown in Fig. 2.28.

A bell-type *Rzeppa joint* is shown in Fig. 2.26. The joint consists of a spherical housing and an inner race with corresponding grooves in each part. Six steel balls inserted in these grooves transmit torque from the driver to the follower. The grooves are made concentric with the intersection O of the shaft axes. The six balls are carried in a cage whose position is controlled by a rod. One end of this rod engages a socket in the end of

[3] H. H. Mabie, "Constant Velocity Joints," *Machine Design*, May 1948.

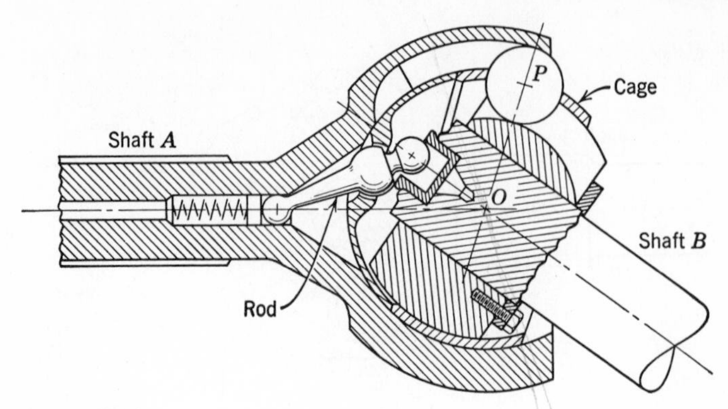

Figure 2.26

shaft B, and the other end slides in a hole in the end of shaft A. A spherical enlargement on the rod engages the cage.

If the shaft B is deflected with respect to the shaft A, it must pivot about O as a center because the unit is concentric about O. Through the motion of this shaft the rod is actuated, in turn moving the cage, and hence the balls, through approximately half of the angle turned through by the shaft B. Although it is possible to prove geometrically that the angle between the shafts is exactly bisected by the plane of the centers of the balls for one and

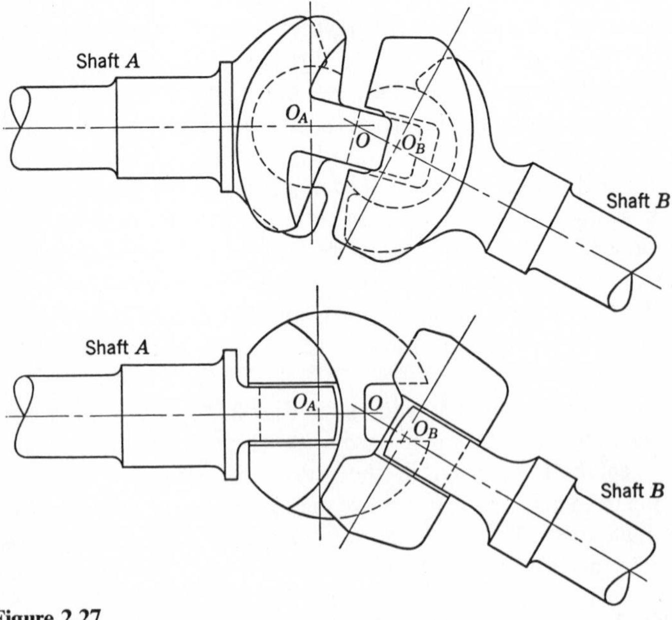

Figure 2.27

only one angle (other than zero) between the shafts, depending upon the proportions of the piloting mechanism, the deviations are so slight for other angles up to about 40° as to be negligible. Therefore, for all practical purposes, the plane of the centers of the balls bisects the angle between the two shafts, and the joint transmits a constant angular velocity ratio. A photograph of a Rzeppa joint is shown in Fig. 2.28.

A *Tracta joint* is shown in Fig. 2.27; it consists of four parts: two shafts with forked ends and two hemispherical parts, one of which has a tongue and the other a groove to receive the tongue. In addition, each of the hemispherical bodies is provided with a groove that permits the connection of a fork. The forks subtend an angle greater than 180° so as to be self-locking when assembled. The tongue and the tongue groove are at right angles to the grooves which admit the forks. By means of the union of the tongue and groove when the joint is assembled, the axes of the hemispherical parts must always remain in the same plane. When the joint is assembled, the forks are free to rotate about the axes of the hemispherical bodies, which lie in the plane of the tongue and groove.

The joint is held in proper alignment for industrial application by two spherical housings not shown. When assembled, these provide a ball-joint type of housing that support the shafts so that their axes will intersect at all times at a point equidistant from the centers of the hemispherical members. With this alignment the Tracta joint will transmit motion with a constant velocity ratio. A photograph of a Tracta joint is shown in Fig. 2.28.

2.12 Intermittent Motion Mechanisms. There are many instances where it is necessary to convert continuous motion into intermittent motion. One of the foremost examples is the indexing of a work table on a machine tool so as to bring a new work piece before the cutters with each index of the table. There are several ways of accomplishing this type of motion.

Geneva Wheel. This mechanism is very useful in producing intermittent motion because the shock of engagement is minimized. Figure 2.29 shows a sketch where plate 1, which rotates continuously, contains a driving pin P that engages in a slot in the driven member 2. In the sketch, 2 is turned one-quarter revolution for each revolution of plate 1. The slot in member 2 must be tangential to the path of the pin upon engagement in order to reduce shock. This means that angle O_1PO_2 will be a right angle. It can also be seen that angle β is one half of the angle turned through by member 2 during the indexing period. For the case shown β is 45°.

It is necessary to provide a locking device so that when member 2 is not being indexed, it will not tend to rotate. One of the simplest ways of accomplishing this is to mount a locking plate upon plate 1 whose convex surface will mate with the concave surface of member 2 except during the

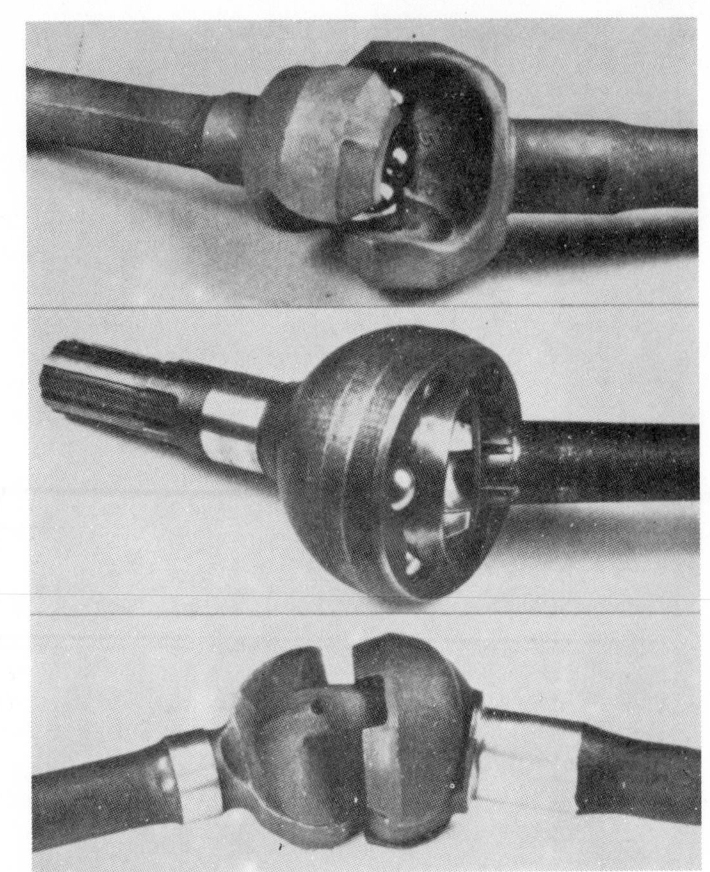

Figure 2.28 Constant velocity universal joints: Bendix–Weiss, Rzeppa, and Tracta.

indexing period. It is necessary to cut the locking plate back to provide clearance for member 2 as it swings through the indexing angle. The clearance arc in the locking plate will be equal to twice the angle α.

If one of the slots in member 2 is closed, then plate 1 can make only a limited number of revolutions before the pin P strikes the closed slot and motion ceases. This modification is known as the Geneva stop and is used in watches and similar devices to prevent overwinding.

Ratchet Mechanism. This mechanism is used to produce intermittent circular motion from an oscillating or reciprocating member. Figure 2.30 shows the details. Wheel 4 is given intermittent circular motion by means of arm 2 and driving pawl 3. A second pawl 5 prevents 4 from turning backward when 2 is rotated clockwise in preparation for another stroke. The line of action PN of the driving pawl and tooth must pass between centers

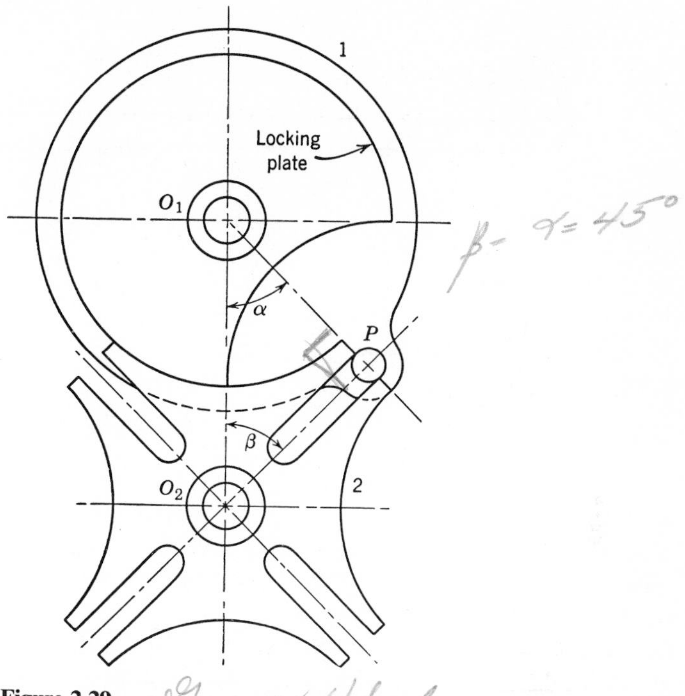

Figure 2.29

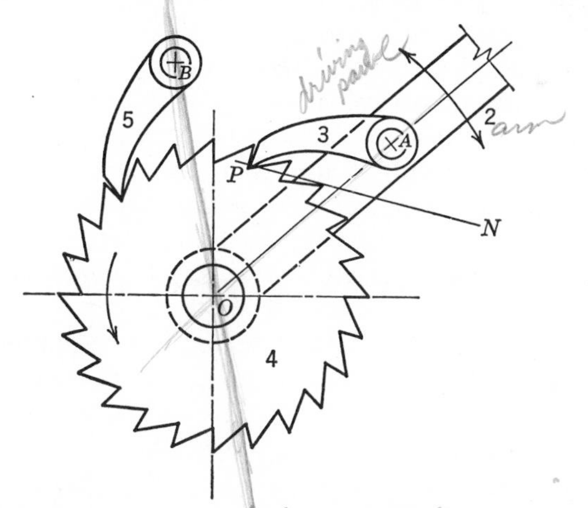

Figure 2.30

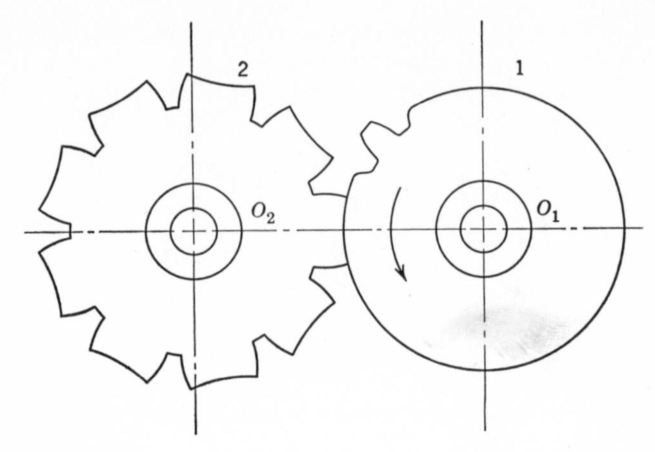

Figure 2.31

O and A as shown in order to have the pawl 3 remain in contact with the tooth. The line of action (not shown) for the locking pawl and tooth must pass between centers O and B. This mechanism has many applications, particularly in counting devices.

Intermittent Gearing. This mechanism finds application where the loads are light and shock is of secondary importance. The driving wheel will carry one tooth and the driven member a number of tooth spaces to produce the required indexing angle. Figure 2.31 shows this arrangement. A locking device must be employed to prevent wheel 2 from rotating when not indexing. One method is shown in the figure; the convex surface of wheel 1 mates with the concave surface between the tooth spaces on member 2.

Escapements. This type of mechanism is one in which a toothed wheel, to which torque is applied, is allowed to rotate in discrete steps by the action of a pendulum. Because of this action, the mechanism can be used as a timing device, and as such finds its widest application in clocks and watches. A second application is its use as a governor to control displacement, torque, or velocity.

There are many types of escapements, but the one that is used in watches and clocks because of its high accuracy is the *balance-wheel escapement* shown in Fig. 2.32.

The balance wheel and hairspring constitute a torsional pendulum with a fixed period (time of oscillation through one cycle). The escape wheel is driven by a mainspring and gear train (not shown) and has intermittent clockwise rotation as governed by the lever. For every complete oscillation of the balance wheel, the lever allows the escape wheel to advance one tooth. The escape wheel, therefore, counts the number of times the balance wheel oscillates and also supplies energy through the lever to the balance wheel to make up for friction and windage losses.

In order to study the motion of this mechanism through one cycle, consider the lever held against the left banking pin by the escape wheel tooth A acting on the left pallet stone. The balance wheel rotates counterclockwise so that its jewel strikes the lever driving it clockwise. The motion of the lever causes the left pallet stone to slip past and unlock the escape wheel tooth A. The wheel now rotates clockwise, with the top of tooth A giving an impulse to the bottom of the left pallet stone as it slides under it. From this impulse the lever now begins to drive the jewel, thereby giving energy to the balance wheel to maintain its motion.

After the escape wheel rotates a short distance, it comes to rest again as tooth B strikes the right pallet stone, which has been lowered due to rotation of the lever. The lever strikes the right banking pin and stops, but the

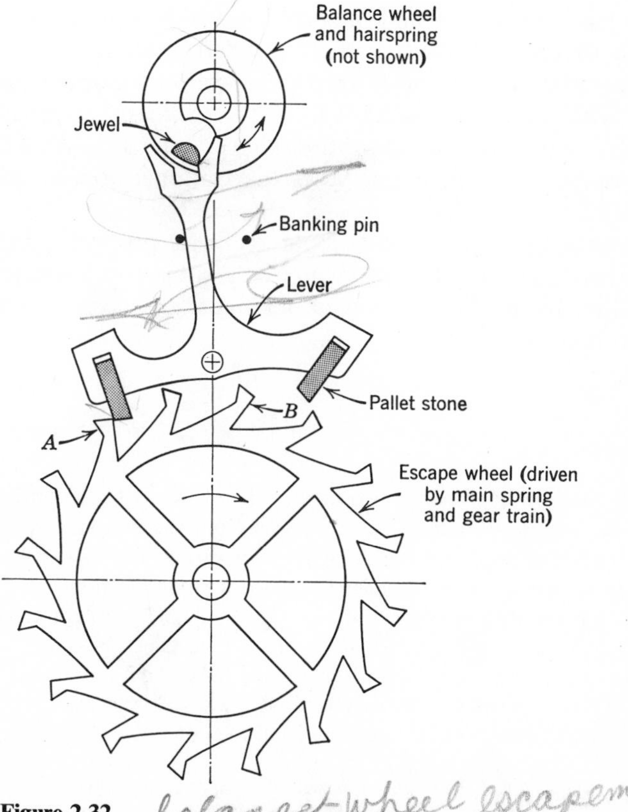

Figure 2.32 balance wheel escapement

balance wheel continues to rotate until its energy is overcome by tension in the hairspring, pivot friction, and air resistance.

The force of escape wheel tooth *B* on the right pallet stone keeps the lever locked against the right banking pin. The balance wheel completes its swing, reverses direction, and returns with clockwise motion. The jewel now strikes the left side of the lever notch and drives the lever counter-clockwise. This action unlocks tooth *B*, which gives an impulse to the lever through the right pallet stone. After a short rotation of the escape wheel, it comes to rest again as tooth *A* strikes the left pallet stone.

The balance wheel escapement is also known as the detached lever escapement because the balance wheel is free and out of contact with the lever during most of its oscillation. Because of this relative freedom of the balance wheel, the escapement has an accuracy of $\pm 1\%$.

For more information on escapements and their applications consult one of the many references on the subject.[4]

2.13 Synthesis. In the linkages studied in this chapter, the proportions have been given and the problem has been to analyze the motion produced by the linkage. It is quite a different matter, however, to start with a required motion and to try to proportion a mechanism to give this motion. This procedure is known as the *synthesis of mechanisms*. Undoubtedly, many problems in synthesis have been solved by trial and error, but it has only been in recent years that rational approaches have been developed.

There have been many methods of synthesis proposed, both graphical and analytical, and their study would be a subject in itself. In Chapter 9, Introduction to Synthesis, several methods are given to illustrate the principles involved.

Problems

2.1 In the four-bar linkage shown in Fig. 2.1, let $O_2O_4 = 2$ in., $O_2A = 2\frac{1}{2}$ in., $AB = 1\frac{1}{2}$ in., and $O_4B = 1\frac{3}{4}$, $2\frac{3}{4}$, and $3\frac{1}{4}$ in. Sketch the mechanism full size for the three sets of dimensions, and determine for each case whether links 2 and 4 rotate or oscillate. In the case of oscillation determine the limiting positions.

2.2 In the four-bar linkage shown in Fig. 2.1, link 2 is to rotate completely and link 4 oscillate through an angle of 75°. Link 4 is to be 11.4 cm long, and when it is at one extreme position the distance O_2B is to be 10.2 cm and at the other extreme position 22.9 cm. Determine the length of links 2 and 3, and draw the mechanism to scale as a check. Determine the maximum and the minimum transmission angle.

[4] A. L. Rawlings, *The Science of Clocks and Watches*, Second Edition, Pitman, 1948.
T. K. Steele, "Clock-Escapement Mechanisms," *Product Engineering*, January 1957, p. 179.

2.3 If for the drag-link mechanism shown in Fig. 2.4*c*, $O_2A = 7.62$ cm, $AB = 10.2$ cm, and $O_4B = 12.7$ cm, what can be the maximum length of O_2O_4 for proper operation of the linkage?

2.4 In the four-bar mechanism shown in Fig. 2.33, the guide is part of the fixed link and its centerline is a circular arc of radius R. Draw the mechanism full size and using graphical construction, determine the magnitude of the angular velocity ω_4 of the slider when the mechanism is in the phase shown and ω_2 is 1 rad/sec. Give the sense of ω_4.

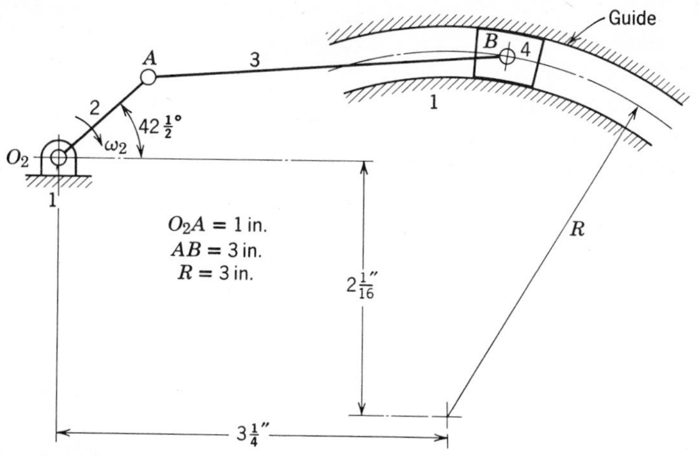

Figure 2.33

2.5 Considering the slider-crank mechanism shown in Fig. 2.5*b*, derive equations for the displacement, velocity, and acceleration of the slider as a function of R, L, θ, ω, and ϕ. Do not make approximations. Let ω be constant.

2.6 The approximate equation for the displacement of the slider in the slider-crank mechanism is $x = R(1 - \cos\theta) + (R^2/2L)\sin^2\theta$, and $\theta = \omega t$ because ω is constant. Derive the equations for the velocity and acceleration of the slider if ω is not constant.

2.7 Write a computer program to calculate the slider displacement, velocity, and acceleration of the slider crank shown in Fig. 2.5. Use both the exact equations and the approximate equations. Let $R = 2$ in., $L = 8$ in., $n_2 = 2400$ r/min. Calculate displacement, velocity, and acceleration at 10° intervals of θ from 0 to 360°.

2.8 A slider-crank mechanism has a crank length R of 5 cm and operates at 250 rad/sec. Calculate the maximum values of velocity and acceleration and determine at what crank angles these maximums occur for connecting rod lengths of 20, 23, and 25 cm. Use approximate equations, and assume ω constant.

2.9 Write a computer program to compare simple harmonic motion of the Scotch yoke (Fig. 2.8) with the motion of the slider crank. Let $n = 1800$ r/min, $R = 2$ in., $L = 8$ in. for the slider crank, and $r = 2$ in. for the Scotch yoke. Vary the angle θ from 0 to 360° (ccw) and calculate displacement, velocity, and acceleration at each value of θ. Use approximate equations for the slider crank, and assume ω constant.

2.10 In the mechanism shown in Fig. 2.34, neglect the connecting rod effect (assume connecting rod infinitely long) and determine an expression for the relative motion of the two sliders. This relation should be a function of time and reduced to a single trigonometric term. ω is constant.

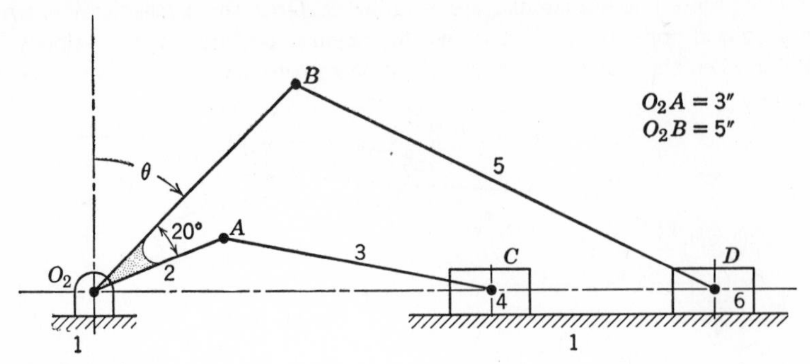

Figure 2.34

2.11 If link 2 in the Scotch yoke shown in Fig. 2.8*a* rotates at 100 r/min, determine the maximum velocity and maximum acceleration of link 4 if its stroke is 10 cm.

2.12 Shown in Fig. 2.35 is a modified Scotch yoke mechanism in which the guide of the yoke is a circular arc of radius r. R is the crank radius. Derive an expression for the displacement x of the yoke (link 4) in terms of θ, R, and r. Indicate the displacement x on the sketch.

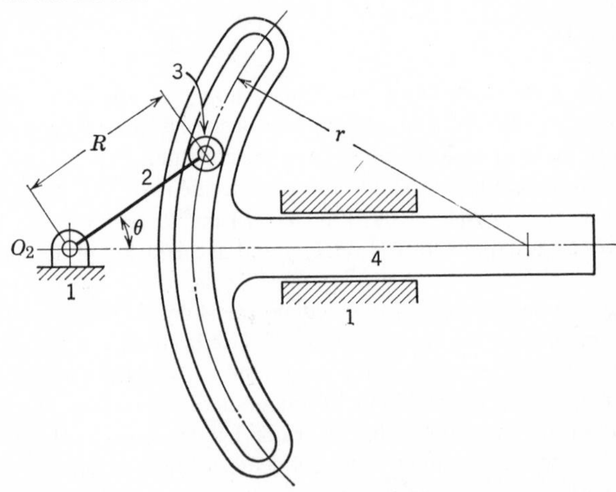

Figure 2.35

2.13 Considering the drag-link quick-return mechanism shown in Fig. 2.9, determine the velocity (fpm) of the slider 6 for a complete revolution of the crank 2 in 45° increments. The crank rotates clockwise at 100 r/min. Use a scale of 4 in. = 12 in., and let $O_2O_4 = 3$ in., $O_2A = 4\frac{1}{2}$ in., $AB = 5\frac{1}{2}$ in., $BC = 8\frac{1}{2}$ in., $O_4B = 6$ in., $O_4C = 6$ in., and $CD = 18\frac{1}{2}$ in. Determine ω_4 graphically using the principle of the transmission of motion, and then calculate the velocity of the slider using the slider-crank equation.

2.14 Using the proportions of the drag-link quick-return mechanism given in Problem 2.13, graphically determine the length of stroke of slider 6, and the time ratio of advance to return. Use a scale of 3 in. = 12 in.

2.15 For the Whitworth quick-return mechanism shown in Fig. 2.10, graphically determine the length of stroke of slider 6 and the time ratio of advance to return. Use a scale of 3 in. = 12 in., and let $O_2O_4 = 2\frac{1}{2}$ in., $O_2A = 5$ in., $O_4B = 5$ in., and $BC = 18$ in.

2.16 For the crank-shaper mechanism shown in Fig. 2.11, graphically determine the length of stroke and the time ratio of advance to return. Use a scale of 3 in. = 12 in., and let $O_2O_4 = 16$ in., $O_2A = 6$ in., $O_4B = 26$ in., $BC = 12$ in., and the distance from O_4 to the path of $C = 25$ in.

2.17 Design a Whitworth quick-return mechanism to have a length of stroke of 12 in. and a time ratio of 11/7. Use a scale of 3 in. = 12 in.

2.18 Design a crank-shaper mechanism to have a length of stroke of 12 in. and a time ratio of 11/7. Use a scale of 3 in. = 12 in.

2.19 For the quick-return mechanism shown in Fig. 2.36, derive an expression for the displacement x of the slider (link 5) as a function only of θ of the driving link (link 2) and the constant distances shown.

2.20 Shown in Fig. 2.37 is a quick-return mechanism in which link 2 is the driver.

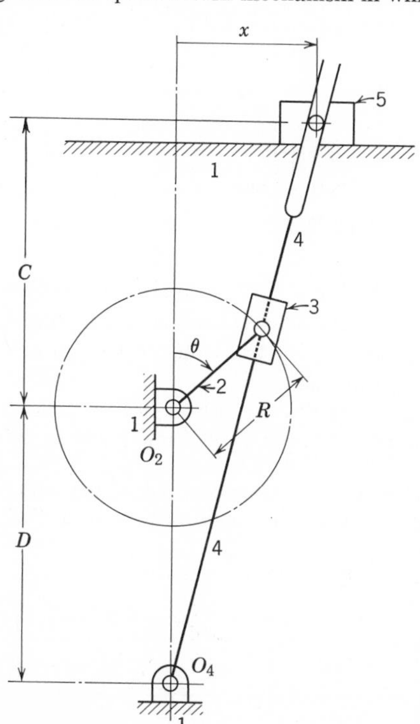

Figure 2.36

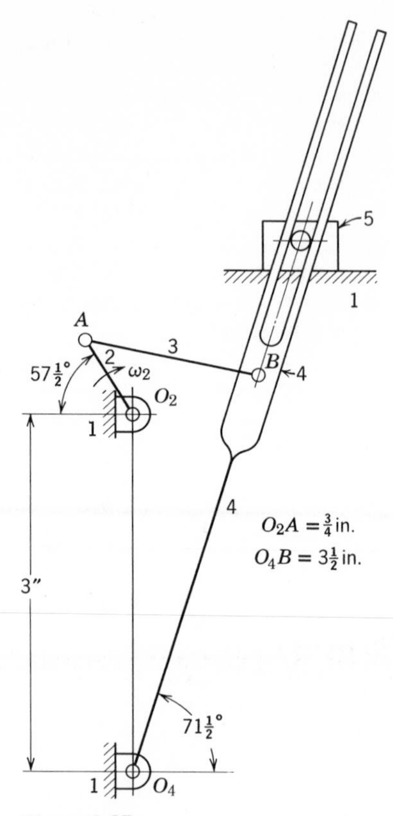

Figure 2.37

Link 5 moves to the right during a working stroke and to the left during a quick-return stroke. Draw the mechanism to full scale and using graphical construction, determine (*a*) the angular velocity ratio ω_4/ω_2 when the mechanism is in the phase shown, and (*b*) the time ratio of the mechanism.

2.21 Derive equations for displacement, velocity, and acceleration for the offset slider-crank mechanism shown in Fig. 2.12. They should be in form similar to Eqs. 2.2, 2.3, and 2.4.

2.22 Calculate the length of the crank and of the connecting rod for an offset slider-crank mechanism to satisfy the conditions shown in Fig. 2.38.

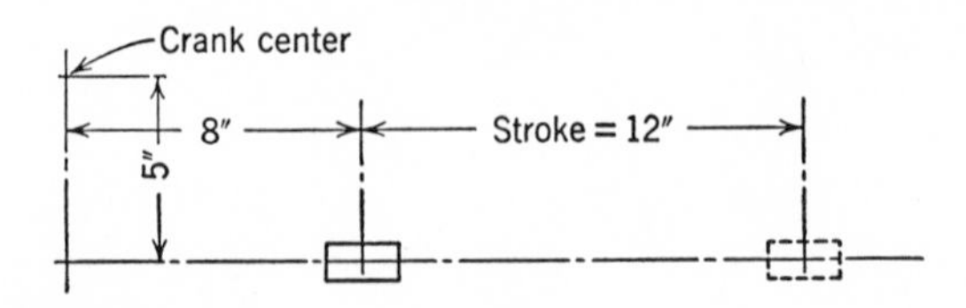

Figure 2.38

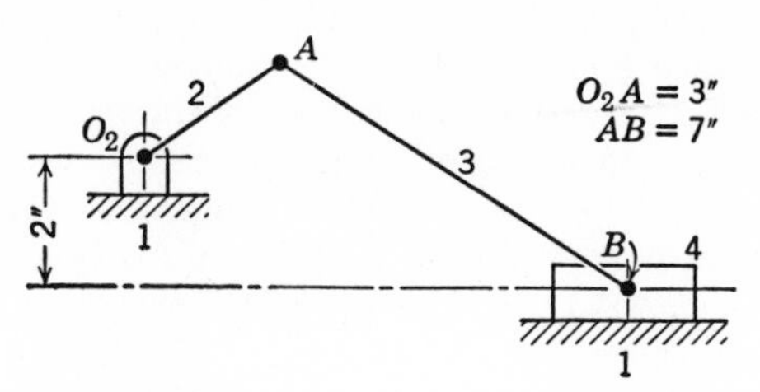

Figure 2.39

2.23 For the offset slider-crank mechanism shown in Fig. 2.39 calculate (*a*) the length of stroke of slider 4, (*b*) the distance O_2B when the slider is in its extreme left position, and (*c*) the time ratio of working stroke to return stroke.

2.24 Referring to Fig. 2.13 and considering only links 4, 5, and 6 of the toggle mechanism shown, write a computer program to illustrate the force development of this mechanism. Let F be a constant value of 10 lb. *Suggestion:* Use Eq. 2.8 and vary α from 10° to near 0°.

2.25 Plot the path of point P in the Watt straight-line mechanism shown in Fig. 2.15 if $O_2A = 2$ in., $O_4B = 3$ in., $AP = 1\frac{1}{2}$ in., $BP = 1$ in., and links 2 and 4 are perpendicular to link 3.

2.26 Referring to Fig. 2.15, graphically determine the proportions of the Watt straight-line mechanism that will give an approximate straight-line motion of point P over a length of 5 in.

2.27 Prove that point P in the Peaucellier straight-line mechanism shown in Fig. 2.16 will trace true straight-line motion.

2.28 Prove that points P and Q in the pantograph shown in Fig. 2.17 will trace similar paths.

2.29 In the pantograph shown in Fig. 2.40 point Q is to trace a 7.60 cm path while P traces a 20.3 cm path. If OP is to have a maximum working distance of 39.4 cm, design a pantograph to give the required motion using a scale of 1 cm = 3 cm. Draw the mechanism in its two extreme positions, and dimension the links.

2.30 A Hooke coupling connects two shafts at an angle of 135° ($\beta = 45°$) as shown

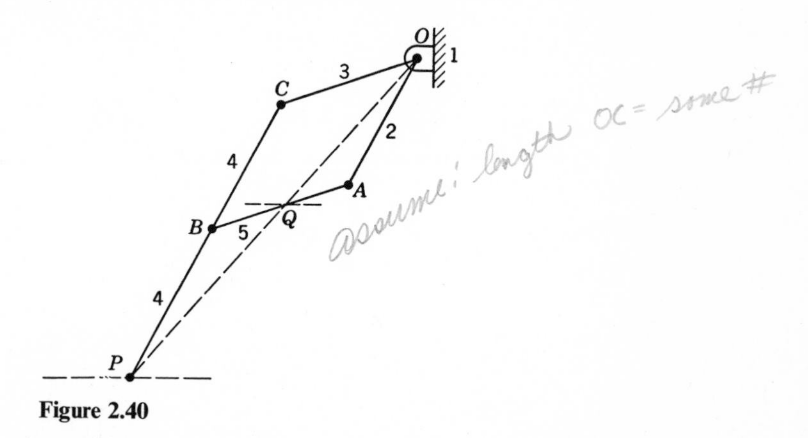

Figure 2.40

in Fig. 2.21. If the angular velocity of the drive shaft is constant at 100 r/min, calculate the maximum and minimum velocity of the driven shaft.

2.31 Derive equations that describe the angular displacement and angular velocity of the driven member of a Geneva mechanism (Fig. 2.29) from the point where the driving pin engages the driven wheel to the point of disengagement. Find $\beta = f(\alpha)$, $d\beta/d\alpha = f(\alpha)$, and use $(d\beta/d\alpha)(d\alpha/d\tau) = d\beta/d\tau$ to determine an equation for the angular velocity of the driven member.

2.32 Using the equations derived in Problem 2.31, write a computer program and calculate the values of β and ω_2 for α varying from 60° to 0° in increments of 10°. Let α at the point of first contact = 60°, $O_1P = 1\frac{3}{4}$ in., $O_1O_2 = 3\frac{1}{2}$ in., $n_1 = 1000$ r/min (constant).

2.33 Lay out a Geneva-wheel mechanism to satisfy the following conditions: The driver is to rotate continuously while the driven member rotates intermittently, making a quarter revolution for every revolution of the driver. The distance between the centers of the driving and driven shafts is to be $3\frac{1}{2}$ in. Let the diameter of the driving pin be $\frac{3}{8}$ in. The diameters of the driving and driven shafts are to be $\frac{5}{8}$ in. and 1 in. with keyway for a $\frac{3}{16} \times \frac{3}{16}$ in. and $\frac{1}{4} \times \frac{1}{4}$ in. key, respectively. Show a hub on each member with the hub on the driver shown in back of the plate. Let the diameters of the hubs be $1\frac{3}{4}$ to two times the diameters of the bores. Dimension the angles α and β.

3

Cams

Cams play a very important part in modern machinery and are extensively used in internal combustion engines, machine tools, mechanical computers, instruments, and many other applications. A cam may be designed in two ways: (*a*) to assume the required motion for the follower and to design the cam to give this motion, or (*b*) to assume the shape of the cam and to determine what characteristics of displacement, velocity, and acceleration this contour will give.

The first method is a good example of synthesis. In fact, designing a cam mechanism from the desired motion is an application of synthesis that can be solved every time. However, after the cam is designed, it may be difficult to manufacture. This difficulty of manufacture is eliminated in the second method by making the cam symmetrical and by using shapes for the cam contours that can be generated. This is the type of cam that is used in automotive applications where the cam must be produced accurately and cheaply.

Only the design of cams with specified motion will be treated. For the automotive type cam where the contour is specified the reader is referred to the reference below.[1] Cams with specified motion can be designed graphically and in certain cases analytically. Graphical procedure will be considered first.

[1] H. A. Rothbart, *Cams*, John Wiley and Sons, 1956.

Graphical Cam Design

3.1 Disk Cam with Radial Follower. Figure 3.1 shows a disk cam with a radial flat-faced follower. As the cam rotates at a constant angular velocity in the direction shown, the follower moves upwards a distance of 1 in. with the displacements shown in half a revolution of the cam. The return motion is to be the same. In order to determine the cam contour graphically, it will be necessary to invert the mechanism and to hold the cam stationary while the follower moves around it. This will not affect the relative motion between the cam and the follower, and the procedure is as follows:

1. Rotate the follower about the center of the cam in a direction opposite to that of the proposed cam rotation.
2. Move the follower radially outward the correct amount for each division of rotation.
3. Draw the cam outline tangent to the polygon that is formed by the various positions of the follower face.

Unfortunately, in the last step, there is no graphical way of determining

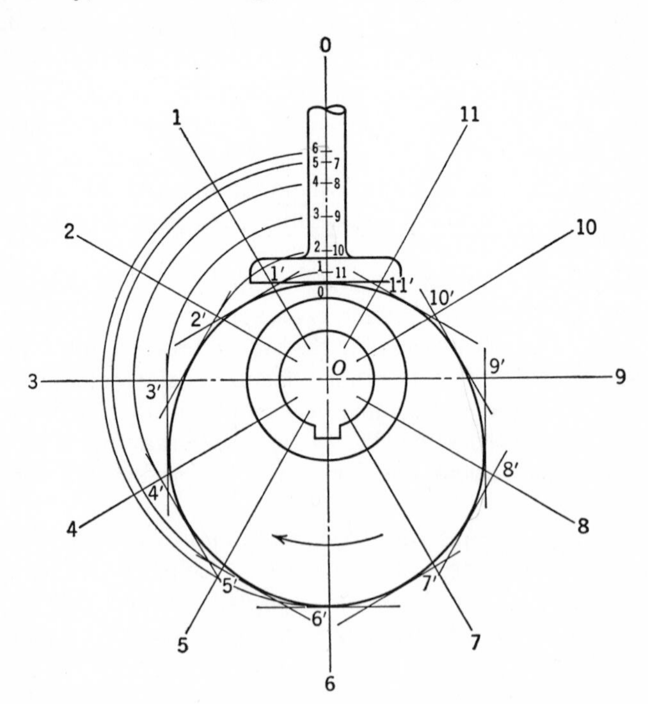

Figure 3.1 Disk cam with radial flat-faced follower.

the contact point between the cam and the follower, and it must be determined by eye with the use of a French curve. The length of the follower face must also be determined by trial. Occasionally a combination of displacement scale and minimum radius of cam is selected that gives a cam profile with a sharp corner or cusp. This cusp can be eliminated by modifying the displacement scale or by increasing the minimum radius of the cam.

Figure 3.2*a* shows the same type of cam with a roller follower. With this type of follower the center of the roller will move with the prescribed motion. The principles of construction are the same as for the flat-faced follower with the exception that the cam is drawn tangent to the various positions of the roller follower. From Fig. 3.2*a* it can also be seen that the line of action from the cam to the follower cannot be along the axis of the follower except when the follower is dwelling (no motion up or down). This produces a side thrust on the follower and may result in deflection and jamming of the follower stem. The angle between the line of action and the center line of the follower is known as the *pressure angle*, and the maximum must be kept as small as possible, especially in light mechanisms. At the present time this maximum has arbitrarily been set at 30°. Although it is possible to measure the maximum pressure angle from the graphical construction of a cam, it is often difficult to determine this maximum analytically. For this reason a nomogram for finding maximum pressure angles is given in a later section on analytical cam design. The pressure angle for any radial flat-faced follower is a constant. For the follower shown in Fig. 3.1 where the follower face is at right angles to the stem, the pressure angle is zero so that the side thrust on the follower is negligible compared to that on a roller follower. Pressure angles may be reduced by increasing the minimum radius of the cam so that the follower travels a greater linear distance on the cam for a given rise. This is analogous to increasing the length of an inclined plane for a given rise in order to decrease the angle of ascent. Also, in a cam with a roller follower, the radius of curvature of the pitch surface must be larger than the radius of the roller; otherwise, the cam profile will become pointed.

In both the flat-faced and the roller follower, the follower stems are sometimes offset instead of being radial as shown in Figs. 3.1 and 3.2*a*. This may be done for structural reasons or, in the case of the roller follower, for the purpose of reducing the pressure angle on the upward stroke. It should be noted, however, that, although the pressure angle on the upward stroke is reduced, the pressure angle on the downward stroke is increased. Figure 3.2*b* shows a cam designed with the follower offset and with the same displacement scale and minimum cam radius as in Fig. 3.2*a*. If the line of motion of an offset flat-faced follower is parallel to a radial line on the cam, the same cam will result as for a radial follower. However, the length of the follower face must be increased to take care of the offset.

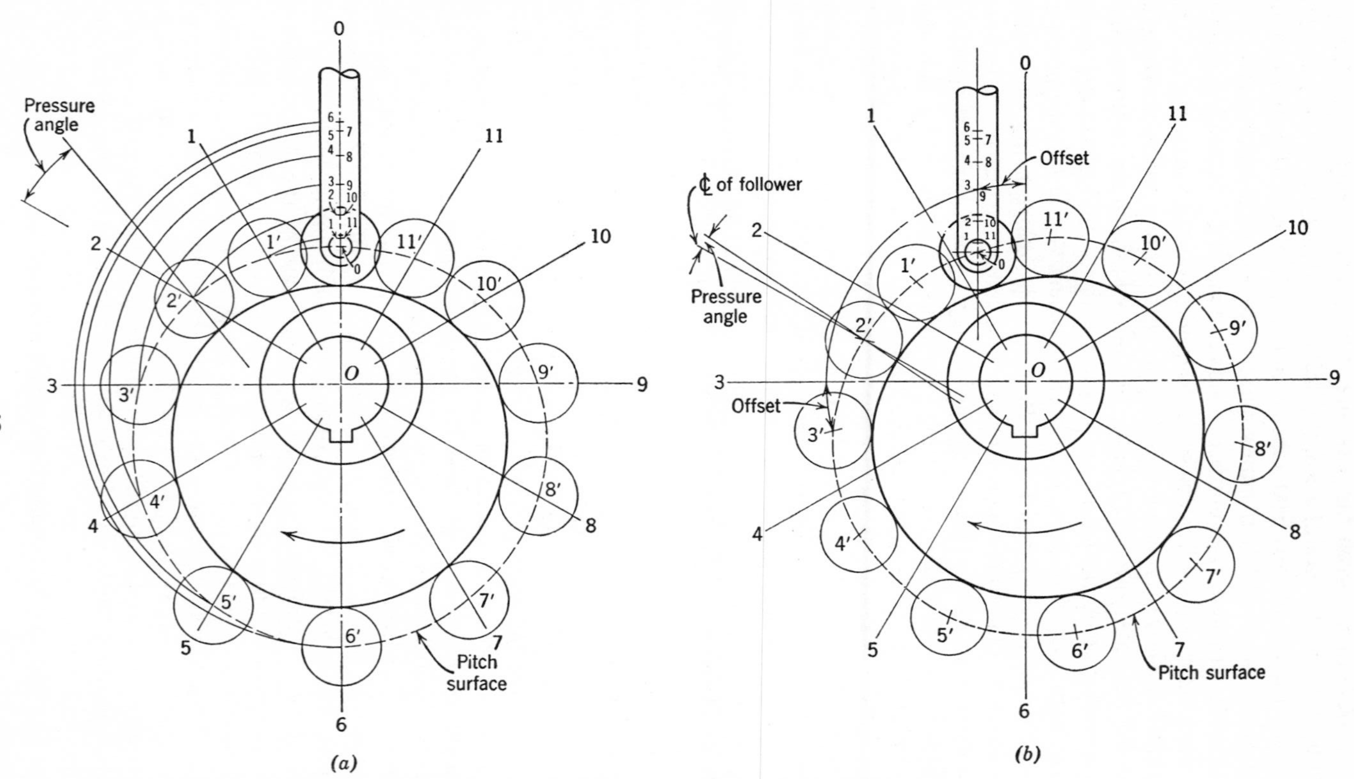

Figure 3.2 (*a*) Disk cam with radial roller follower. (*b*) Disk cam with offset roller follower.

3.2 Disk Cam with Oscillating Follower. Figure 3.3 shows a disk cam with an oscillating flat-faced follower. Using the same principle of construction as in the disk cam with radial follower, the follower is rotated about the cam. At the same time the follower must be rotated about its own center through the required displacement angle for each position. There are several ways of rotating the follower about its own center. The method shown in Fig. 3.3 is to use the intersection of two radii (for example, point 3′) to determine a point on the rotated position of the follower face. The first of these two radii (cam center to position 3 on displacement scale) is swung from the cam center. The second radius (follower center to displacement scale) is swung from the follower center which has been rotated into position 3. The intersection of these two radii gives point 3′. Because an infinite number of lines can be drawn through point 3′, it is necessary to have additional information to locate the correct position of the follower face through point 3′. As shown in the figure, this was supplied by a circle

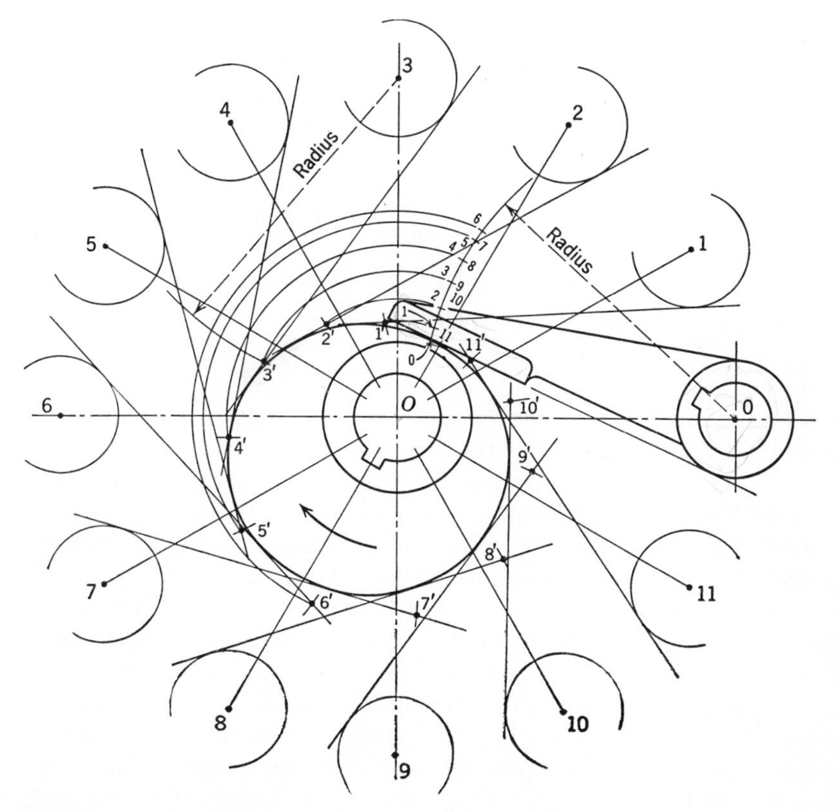

Figure 3.3 Disk cam with oscillating flat-faced follower.

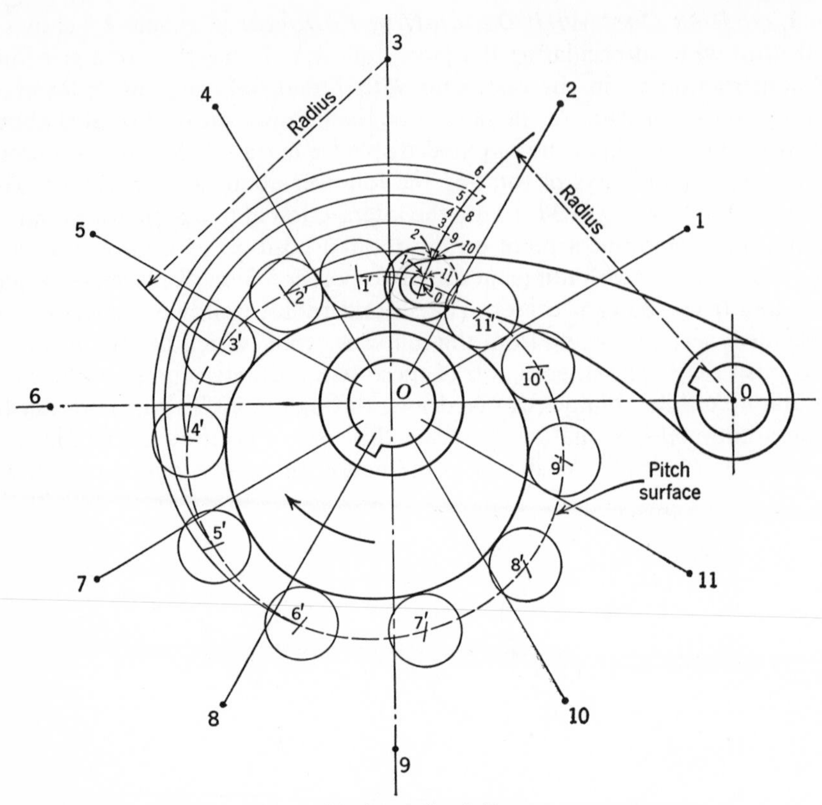

Figure 3.4 Disk cam with oscillating roller follower.

tangent to the follower face which has been extended in the zero position. For the follower design shown, this circle happens to coincide with the outside of the follower hub. The radius of this circle is then swung from each of the rotated positions of the follower center. For position 3 the follower face is drawn through point 3′ tangent to the rotated circle of the follower hub. By repeating this process, the polygon of the various positions of the follower face is obtained from which the cam is drawn.

Figure 3.4 shows a disk cam with oscillating roller follower. The procedure for determining the points labeled with primes (for example, point 3′) is similar to that of Fig. 3.3. In this case, however, the primed points are the centers of the rotated roller follower. After these roller circles are drawn, the cam can now be drawn tangent to them. It should be noted that in an actual design, smaller cam divisions would be used so that the circles would intersect each other to minimize cam contour error. It should also be mentioned that the same procedure can be used in designing a cam with an oscillating roller follower as was used for a cam with an offset translating follower.

Although most of the cams in use are of the types mentioned, there are many others, some of which find wide application. Three of these are discussed in the following sections.

3.3 Positive-Return Cam. With a disk cam and radial follower, it is often necessary to return the follower in a positive manner and not by the action of gravity or of a spring. Figure 3.5 shows a cam of this type where the cam positively controls the motion of the follower not only during the outward motion but also on the return stroke. Of necessity, the return motion must be the same as the outward motion, but in the opposite direction. This cam is also spoken of as a constant-breadth cam.

This type of cam may also be designed using two roller followers instead

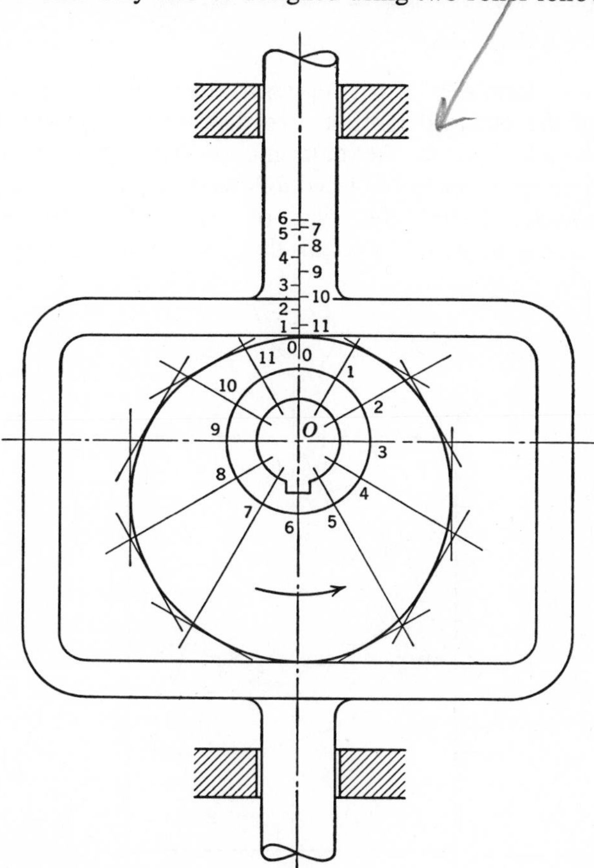

Figure 3.5 Positive-return cam.

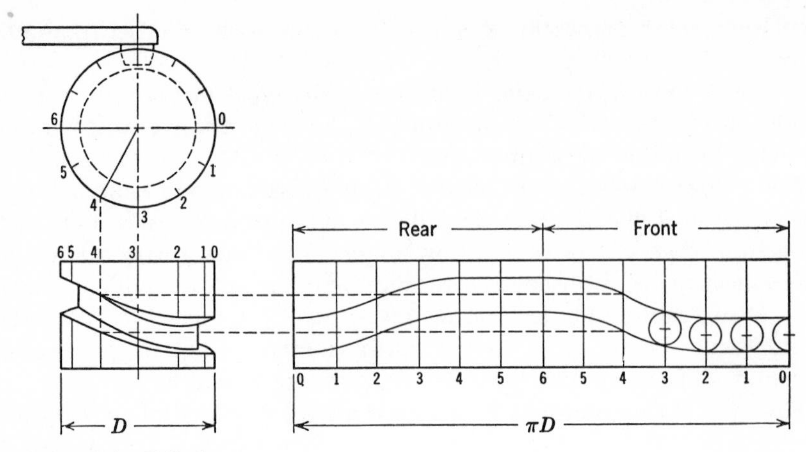

Figure 3.6 Cylinder cam.

of the flat-faced followers. If it is necessary to have the return motion independent of the outward motion, two disks must be used, one for the outward motion and one for the return motion. These double-disk cams can be used with either roller or flat-faced followers.

3.4 Cylinder Cam. This type of cam finds many applications, particularly on machine tools. Perhaps the most common example, however,

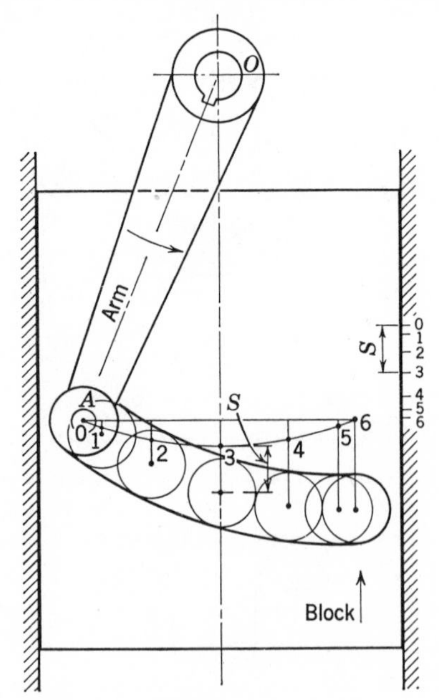

Figure 3.7 Inverse cam.

is on level-winding fishing reels. Figure 3.6 shows a sketch where the cylinder rotates completely about its axis imparting motion to a follower which is guided by a groove in the cylinder.

3.5 Inverse Cam. Occasionally it is advantageous to reverse the role of the cam and the follower and to let the follower drive the cam. This inversion finds application in sewing machines and other mechanisms of similar nature. Figure 3.7 shows a sketch of a plate cam where the arm oscillates, causing reciprocation of the block by action of a roller in the cam groove.

Types of Follower Motion

Before a cam contour can be determined it is necessary to select the motion with which the follower will move in accordance with the requirements of the system. If operation is to be at slow speed, the motion may be any one of several common motions, for example, parabolic (constant acceleration and deceleration), parabolic with constant velocity, simple harmonic, or cycloidal.

Parabolic motion has the lowest theoretical acceleration for a given rise and cam speed for the motions listed, and for this reason has been used for many cam profiles. However, in slow-speed work this has little significance. Parabolic motion may or may not have equal intervals of acceleration and deceleration, depending on requirements. Parabolic motion may also be modified to include an interval of constant velocity between the acceleration and deceleration; this is often spoken of as *modified constant velocity*.

Simple harmonic motion has the advantage that with a radial roller follower the maximum pressure angle will be smaller than with parabolic motion with equal time intervals or with cycloidal motion. This will allow the follower to be less rigidly supported and more overhung in its construction. Less power will also be needed to operate the cam. For these reasons simple harmonic motion is to be preferred over the other types.

After the follower motion has been selected, it is necessary to determine the displacement scale and to mark it off on the follower axis as shown in Fig. 3.1. The scale increments may be calculated, but they are more easily determined graphically by plotting a displacement-time graph.

In plotting the displacement-time graph it is necessary to first determine the inflection point if the motion is parabolic or a modification thereof. For simple harmonic and cycloidal motion, the inflection point is automatically determined by the method of generation of the curve. The inflection point for parabolic motion will be at the midpoint of the displacement scale and of the time scale if the intervals are equal. To find the inflection points

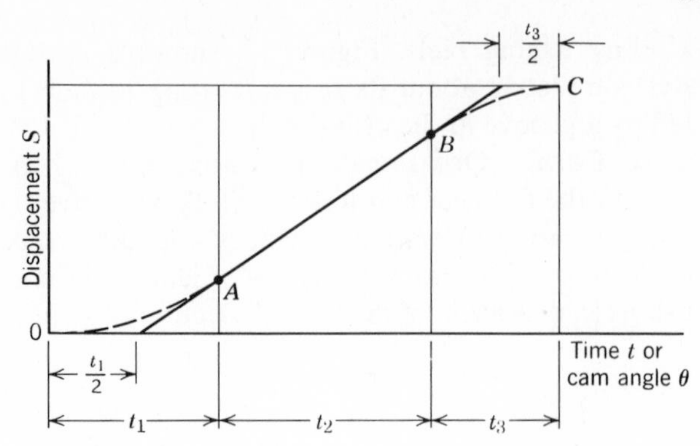

Figure 3.8

where the parabolic motion has been modified is a little more complicated, as shown below.

Consider a point moving with modified constant velocity where it starts from rest with constantly accelerated motion, next has constant velocity, and finally comes to rest with constantly decelerated motion. The inflection points may be found by specifying the time intervals or the displacement intervals corresponding to each type of motion. Figure 3.8 shows a graphical means for finding the inflection points A and B when the time intervals are given. Figure 3.9 shows the construction for displacement intervals. From the relations $S = \frac{1}{2}At^2$, $V = At$, and $S = Vt$, it is possible to prove the validity of the construction shown in Figs. 3.8 and 3.9.

After the inflection points have been determined, as for example in Fig. 3.9, the constantly accelerated portion $0A$ of the displacement curve is constructed as shown in Fig. 3.10 where the displacement L (corresponding to S_1 of Fig. 3.9) is divided into the same number of parts as is the time scale,

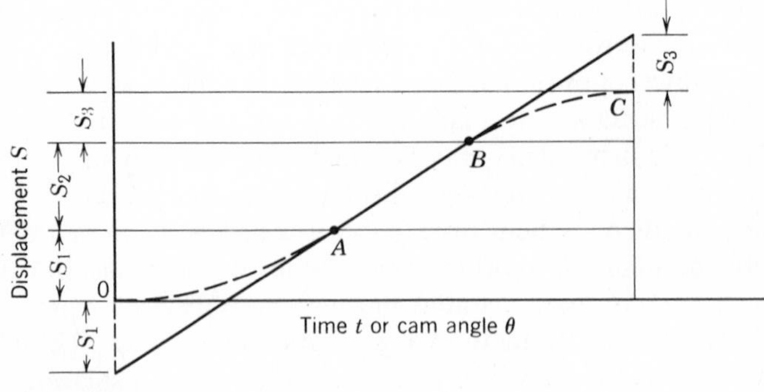

Figure 3.9

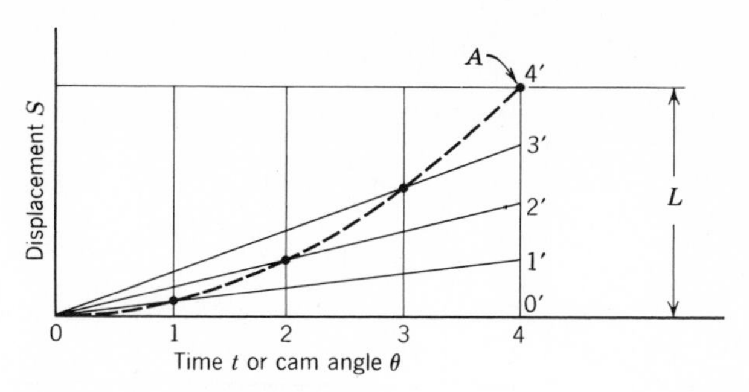

Figure 3.10 Parabolic motion.

in this case four. The deceleration portion BC of the curve in Fig. 3.9 is constructed in a similar manner for its particular displacement S_3 and corresponding time interval.

Figure 3.11 shows simple harmonic motion $[S = r(1 - \cos \omega_r t)]$ for a displacement L in six time intervals. In Fig. 3.11 it should be noted that if the cam rotates through half a revolution while the follower moves through the displacement L, the angular velocity ω_r of the rotational radius r equals the angular velocity ω of the cam, and the equation for follower displacement can be written as $S = r(1 - \cos \omega t) = r(1 - \cos \theta)$. If the cam rotates only a quarter revolution for the displacement L, $\omega_r = 2\omega$ and $S = r(1 - \cos 2\theta)$. Therefore, it can be seen that the relation between ω_r and ω can be expressed as

$$\frac{\omega_r}{\omega} = \frac{180°}{\text{degrees of cam rotation for follower rise } L}$$

A circular cam (eccentric) will impart simple harmonic motion to a radial flat-faced follower because the contact point between the cam and follower is always over the geometric center of the cam.

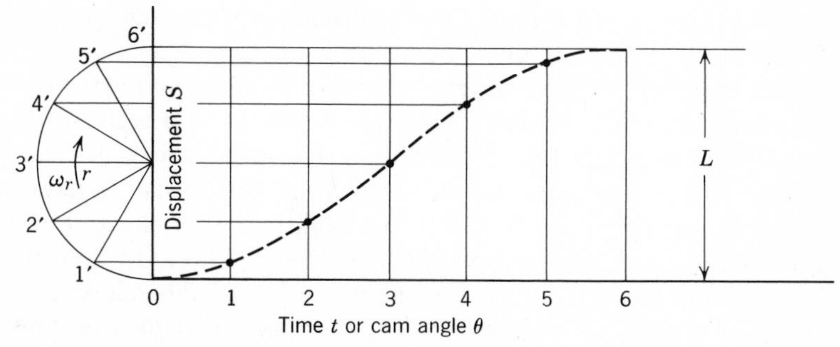

Figure 3.11 Simple harmonic motion.

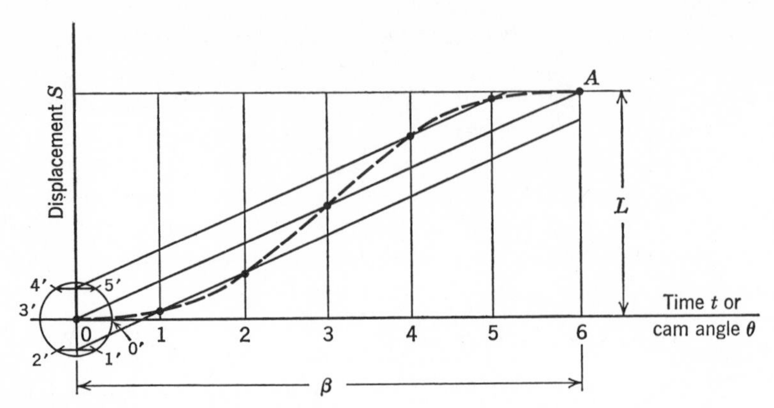

Figure 3.12 Cycloidal motion.

Figure 3.12 shows the construction for cycloidal motion

$$\left[S = L\left(\frac{\theta}{\beta} - \frac{1}{2\pi}\sin 2\pi\frac{\theta}{\beta}\right)\right]$$

for a displacement L in six time intervals. The radius of the construction circle is $L/2\pi$. The circumference of this circle is divided into the same number of parts as is the time scale, in this case six. The six marks on the circumference are projected horizontally onto the vertical diameter of the circle. The marks on the vertical diameter are then projected parallel to $0A$ to the corresponding line on the time axis.

For high-speed cams the selection of the motion of the cam follower must be based not only on the displacement but also on the forces acting on the system as a result of the motion selected. For many years cam design was concerned only with moving a follower through a given distance in a certain length of time. Speeds were low so that accelerating forces were unimportant. With the trend toward higher machine speeds, however, it has become necessary to consider the dynamic characteristics of the system and to select a cam contour that will minimize the dynamic loading.

As an example of the importance of dynamic loading, consider parabolic motion. On the basis of inertia forces this motion would seem to be very desirable because of its low acceleration. However, the fact cannot be overlooked that the acceleration increases from zero to its constant value almost instantaneously, which results in a high rate of application of load. The rate of change of acceleration is determined by the third derivative of the displacement and has been given the name "jerk." Therefore jerk is an indication of the impact characteristic of the loading; it may be said that impact has jerk equal to infinity. Lack of rigidity and backlash in the system also tend to increase the effect of impact loading. In parabolic motion where jerk is infinite, this impact occurs twice during the cycle and has the effect of a sharp blow on the system, which may set up undesirable vibrations as well as causing structural damage.

As a means of avoiding infinite jerk and its deleterious effect on the cam train, a system of cam design has been developed by Kloomok and Muffley[2] that utilizes three analytic functions: (*a*) cycloid (and half-cycloid), (*b*) harmonic (and half-harmonic), and (*c*) eighth-power polynomial. A plot of the displacement, velocity, and acceleration curves of these functions are given in Figs. 3.13, 3.14, and 3.15. The curves have continuous derivatives at all intermediate points. Therefore, acceleration changes gradually, and jerk is finite. Infinite jerk is avoided at the ends by matching accelerations. It should be noted that the velocities will also match because no discontinuities can appear in the displacement time curve. As an example, when a rise follows a dwell, the zero acceleration at the end of the dwell is matched by selecting a curve having zero acceleration at the start of the rise. The acceleration required at the end of the rise is determined by the succeeding condition. If a fall follows immediately, the rise can end in a fairly high value of deceleration because this can be matched precisely by a curve having the same deceleration for the start of the fall.

The selection of profiles to suit particular requirements is made according to the following criteria:

1. The cycloid provides zero acceleration at both ends of the action. Therefore, it can be coupled to a dwell at each end. Because the pressure angle is relatively high and the acceleration returns to zero unnecessarily, two cycloids should not be coupled together.
2. The harmonic provides the lowest peak acceleration and pressure angle of the three curves. Therefore, it is preferred when the acceleration at both start and finish can be matched to the end acceleration of the adjacent profiles. Because acceleration at the midpoint is zero, the half-harmonic can often be used where a constant-velocity rise follows an acceleration. The half-harmonic can also be coupled to a half-cycloid or to a half-polynomial.
3. The eighth-power polynomial has a nonsymmetrical acceleration curve and provides a peak acceleration and pressure angle intermediate between the harmonic and cycloid.

Example 3.1

A roller follower is to move through a total displacement and return with no dwells in the cycle. Because of the operation performed by the mechanism,

[2] M. Kloomok and R. V. Muffley, "Plate Cam Design—with Emphasis on Dynamic Effects," *Prod. Eng.*, February 1955.

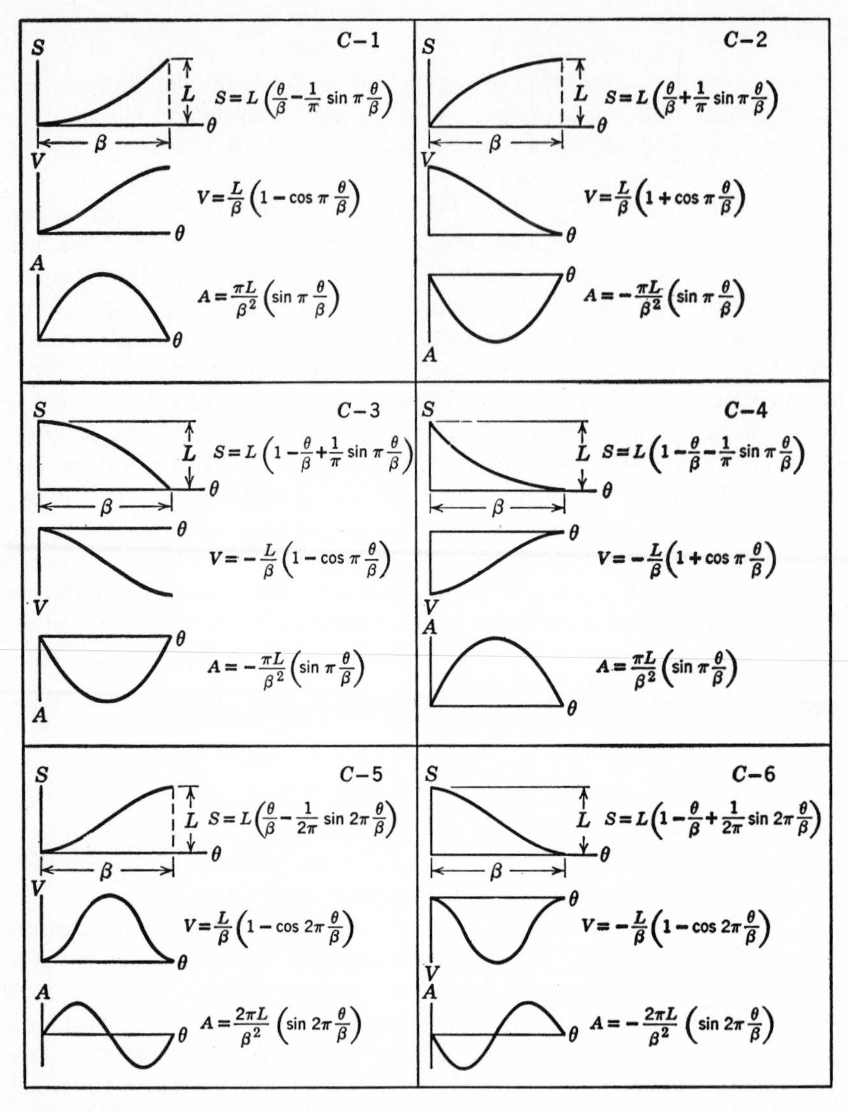

Figure 3.13 Cycloidal motion chracteristics. S = displacement, inches; V = velocity, inches per degree; A = acceleration, inches per degree^2. (From M. Kloomok and R. V. Muffley, "Plate Cam Design—with Emphasis on Cam Effects," *Prod. Eng.*, February 1955.)

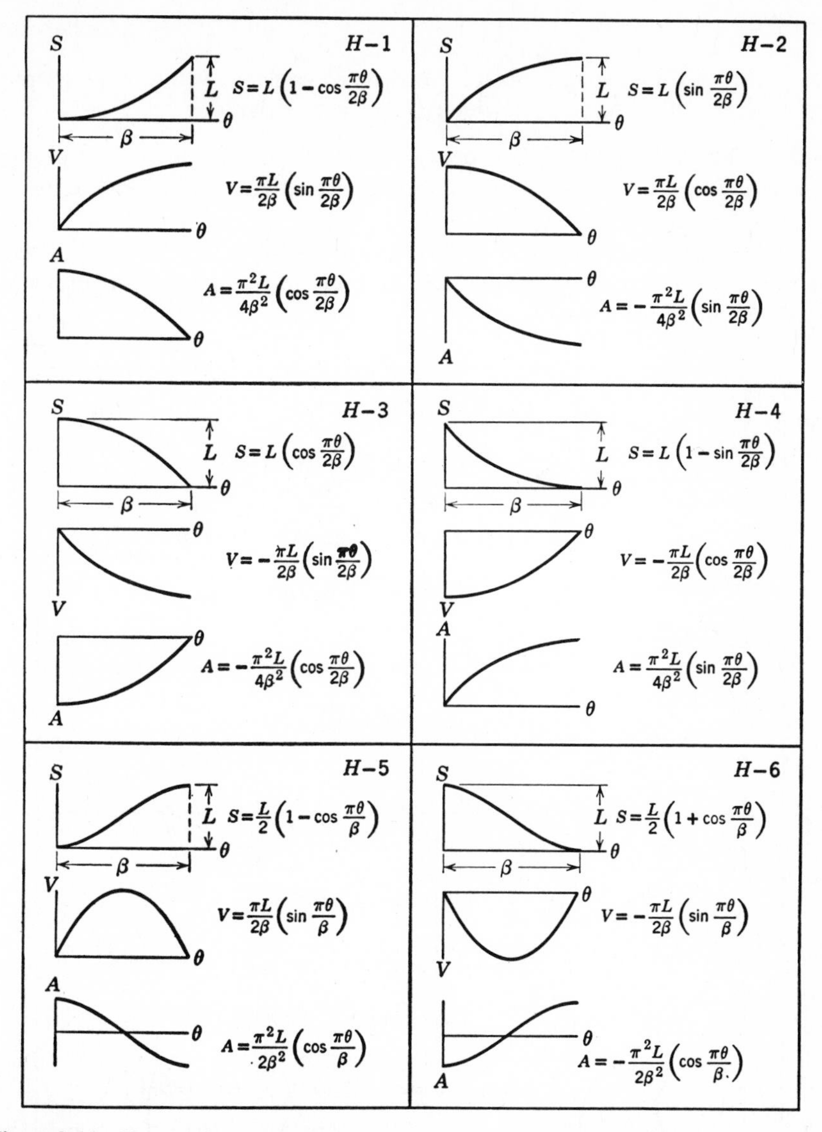

Figure 3.14 Harmonic motion characteristics. S = displacement, inches; V = velocity, inches per degree; A = acceleration, inches per degree^2. (From M. Kloomok and R. V. Muffley, "Plate Cam Design—with Emphasis on Dynamic Effects" *Prod. Eng.*, February 1955.)

a portion of the outward motion must be at constant velocity. Determine the motion curves to be used. Referring to Fig. 3.16*a*:

AB: Use half-cycloid *C*-1 to provide zero acceleration at start of motion *A* and at *B* where connection is made to constant velocity portion of curve.

BC: Constant velocity.

CD: Use half-harmonic *H*-2, which will couple at *C* to the constant velocity section with zero acceleration and provide minimum pressure angle over the rest of the curve.

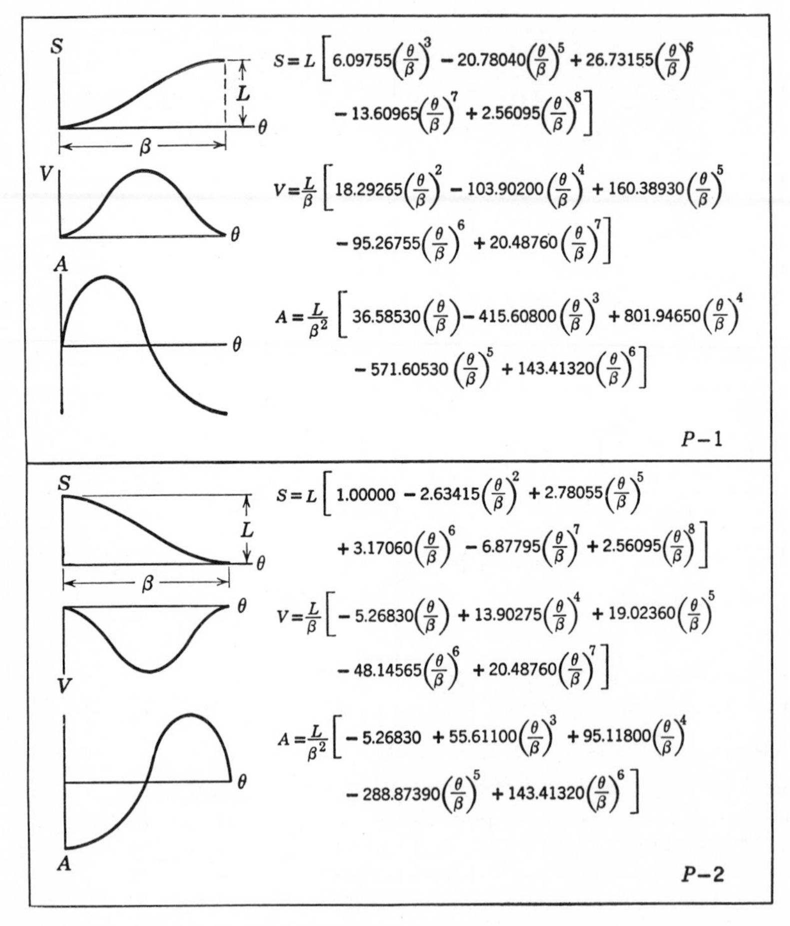

Figure 3.15 Eighth-power polynomial motion characteristics. S = displacement, inches; V = velocity, inches per degree; A = acceleration, inches per degree^2. (From M. Kloomok and R. V. Muffley, "Plate Cam Design—with Emphasis on Dynamic effects," *Prod. Eng.*, February 1955.)

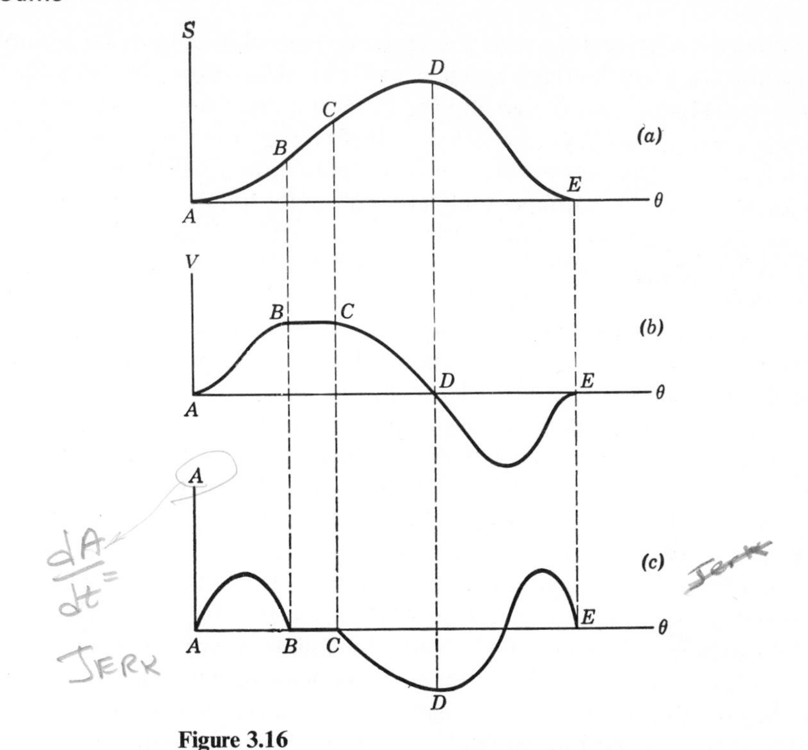

Figure 3.16

DE: Use polynomial *P*-2 to match the deceleration of the harmonic at *D* and to provide a zero acceleration juncture at the end of the cycle at *E*.

The velocities and accelerations are matched, and their curves are shown in Figs. 3.16*b* and *c*. From Fig. 3.16*c*, it can be seen that jerk is not infinite any place in the cycle.

Cam Production

The graphical method of cam design is limited to slow-speed applications. The production of this type of cam depends upon the accuracy of the design layout and upon the method used in following this layout as a template. At one extreme, the layout of the cam is scribed on a steel plate and the cam cut out with a band saw. At the other extreme is production by a milling cutter whose motion is guided by a tracer moving over the cam outline on a copy of the layout drawing. The drawing over which this tracer moves may be made several times actual size in order to improve the accuracy of copying. In either of these cases, the cam profile must be finished by hand.

Graphical design and the resulting copying method of production are not sufficiently accurate for high-speed cams. For this reason, attention has turned to analytical cam design and the method it offers for cam generation. If it is possible to calculate the follower displacements for small increments of cam rotation, the cam profile can be generated on a milling machine or on a jig borer with the cutter assuming the role of the follower. If the follower is to be a roller, the axis of the cutter will be perpendicular to the plane of the cam with the cutter the same size as the roller. If a flat-faced follower is to be used, the axis of the cutter will be parallel to the plane of the cam. In either case, the cutter can be given the correct position corresponding to the cam rotation angle. Naturally, the smaller the increments of cam angle, the better the cam finish will be. Generally, increments of 1° are used, which leave tiny scallops or flats on the cam that must be removed by hand. Automatic tape-operated cam millers have been developed which index the cam a fraction of a degree, with the cutter advancing by tenths of thousandths of an inch. Although the machine operates in discrete steps, the steps are so minute so as to give the appearance of continual operation. It is hoped that the cam surface finish produced by a machine of this kind will be of such quality as to allow elimination of the hand-finishing operation. This type of machine will also produce a cam much more rapidly than a jig borer when both machines are using the same increments of cam angle.

In the preceding discussions, it was assumed that the cam being generated was the cam to be used in the final application. Where several machines of the same model are being produced and many copies of a cam are needed, it is generally more practical to generate what is known as a master cam and to use this master with a cam duplicating machine. The master cam is often made several times actual size.

Analytical Cam Design

With certain types of cams it is possible to design the cam analytically from the specified motion. Practical analytical design methods have been developed for the disk cam with either a radial flat-faced follower, a radial roller follower, an offset roller follower, an oscillating roller follower, or an oscillating flat-faced follower. The methods for the flat-faced follower, radial roller follower, and oscillating roller follower are given in the following sections.

3.6 Disk Cam with Radial Flat-Faced Follower. The treatment of the flat-faced follower allows the actual cam outline to be determined analytically. In the graphical method, the points of contact between the cam and the follower are unknown, and it is difficult to determine their exact

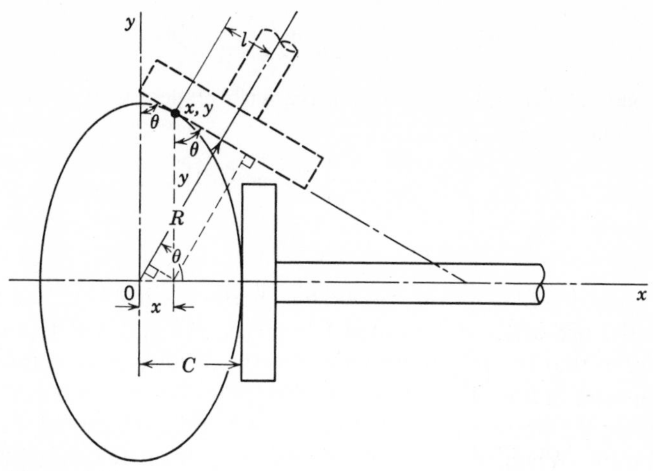

Figure 3.17

location as the cam outline is drawn in. Also the minimum radius of the cam to prevent cusps can only be determined by trial. In the analytical method, which was developed by Carver and Quinn,[3] these disadvantages are overcome, and three valuable characteristics of the cam may be determined: (*a*) parametric equations of the cam contour; (*b*) minimum radius of the cam to avoid cusps; and (*c*) location of the point of contact which gives the length of the follower face. Of these, the first has little practical application, but the other two give information from which the cam can be produced. The development of these characteristics follows.

Figure 3.17 shows a cam with a radial flat-faced follower. The cam rotates with a constant angular velocity. The contact point between the cam and follower is at x,y, which is a distance l from the radial center line of the follower. The displacement of the follower from the origin is given by the following equation:

$$R = C + f(\theta) \tag{3.1}$$

Where the minimum radius of the cam is represented by C, and $f(\theta)$ represents the desired motion of the follower as a function of the angular displacement of the cam.

The equation for the length of contact l can be easily determined from the geometry of Fig. 3.17. From the triangles shown

$$R = y \sin \theta + x \cos \theta \tag{3.2}$$

[3] W. B. Carver and B. E. Quinn, "An Analytical Method of Cam Design," *Mech. Eng.*, August 1945.

and

$$l = y \cos \theta - x \sin \theta \tag{3.3}$$

The right side of Eq. 3.3 is the derivative with respect to θ of the right side of Eq. 3.2. Therefore,

$$l = \frac{dR}{d\theta} = \frac{d}{d\theta}[C + f(\theta)]$$

and

$$l = f'(\theta) \tag{3.4}$$

If the displacement diagram is given by a mathematical equation $S = f(\theta)$, then R and l are easily determined from Eqs. 3.1 and 3.4. From Eq. 3.4 it can be seen that the minimum length of the follower face is independent of the minimum radius of the cam. Also, the point of contact is at its greatest distance from the center line of the follower when the velocity of the follower is a maximum. When the follower moves away from the cam center with positive velocity, l is positive and contact occurs above the axis of the follower in Fig. 3.17. When the follower moves toward the cam center, the velocity is negative and the resulting negative value of l indicates that contact is below the axis of the follower.

To determine the equations for x and y for the cam contour, it is only necessary to solve Eqs. 3.2 and 3.3 simultaneously, which gives

$$x = R \cos \theta - l \sin \theta$$

and

$$y = R \sin \theta + l \cos \theta$$

Substituting the values of R and l from Eqs. 3.1 and 3.4, respectively,

$$x = [C + f(\theta)] \cos \theta - f'(\theta) \sin \theta \tag{3.5}$$

$$y = [C + f(\theta)] \sin \theta + f'(\theta) \cos \theta \tag{3.6}$$

The minimum radius C to avoid a cusp or point on the cam surface can be easily determined analytically. A cusp occurs when both $dx/d\theta$ and $dy/d\theta = 0$. When this happens, a point is formed on the cam as shown at x,y in Fig. 3.18. To demonstrate this, consider that the center line of the follower has rotated through angle θ and that contact between the follower face and the cam occurs at point x,y. When the follower is further rotated through a small angle $d\theta$, the point of contact (x,y) does not change because of the cusp and is still at x,y. Thus it can be seen that $dx/d\theta = dy/d\theta = 0$.

Differentiating Eqs. 3.5 and 3.6,

$$\frac{dx}{d\theta} = -[C + f(\theta) + f''(\theta)] \sin \theta \tag{3.7}$$

$$\frac{dy}{d\theta} = [C + f(\theta) + f''(\theta)] \cos \theta \tag{3.8}$$

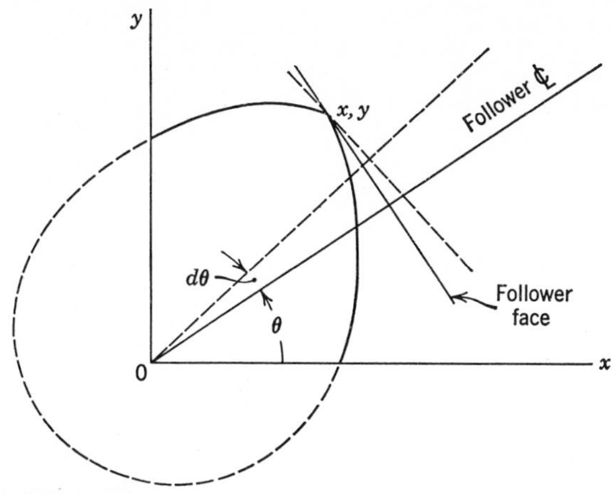

Figure 3.18

Equations 3.7 and 3.8 can become zero simultaneously only when

$$C + f(\theta) + f''(\theta) = 0$$

Therefore, to avoid cusps,

$$C + f(\theta) + f''(\theta) > 0 \tag{3.9}$$

The sum $[f(\theta) + f''(\theta)]$ must be inspected for all values of θ to determine its minimum algebraic value. It is necessary to use the minimum value so that C will be sufficiently large to insure that Eq. 3.9 does not become zero for any value of θ. The sum may be positive or negative. If positive, C will be negative and have no practical significance. In this case, the minimum radius will be determined by the hub of the cam rather than by the function $f(\theta)$.

Points on the cam profile may be determined from Eqs. 3.5 and 3.6, which give the Cartesian coordinates, or by calculating R and l for various values of θ. In general, the second method is easier, but in either case the points have to be connected by use of a French curve to obtain the cam outline. In actual practice, however, it is seldom necessary to draw the cam profile to scale. After the minimum radius C has been determined and the displacements R of the follower calculated, the cam can be generated. For the generating process, the length of the milling cutter must exceed twice the maximum value of l. During cutting, the axis of the milling cutter is parallel to the plane of the cam.

Example 3.2

To illustrate the method of writing the displacement equations consider the following conditions: A flat-faced follower is driven through a total displacement of $1\frac{1}{2}$ in. At the start of the cycle (zero displacement), the follower

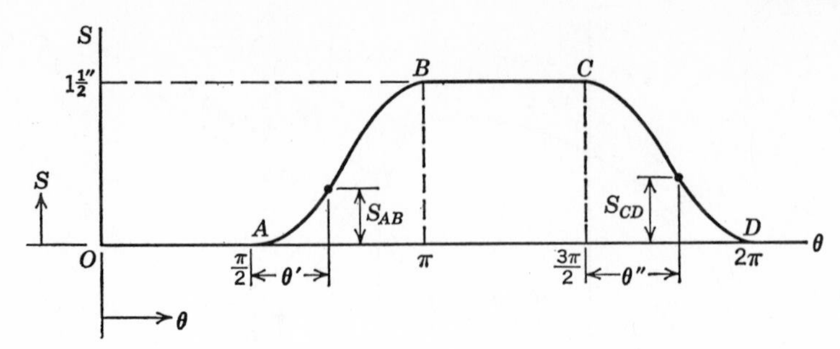

Figure 3.19

dwells for $\pi/2$ rad. It then moves $1\frac{1}{2}$ in. with cycloidal motion (Kloomok and Muffley Curve C-5) in $\pi/2$ rad. The follower dwells for $\pi/2$ rad and returns $1\frac{1}{2}$ in. with cycloidal motion (C-6) in $\pi/2$ rad. A sketch of the displacement diagram is shown in Fig. 3.19.

For the cycloid C-5, the Kloomok and Muffley curves give

$$S = L\left[\frac{\theta}{\beta} - \frac{1}{2\pi}\sin\frac{2\pi\theta}{\beta}\right]$$

It should be mentioned that in writing the relation $S = f(\theta)$, the value of S must always be measured from the abscissa and the value of θ from the ordinate. In the preceding equation, however, θ is measured from point A in Fig. 3.19 and not point O. Therefore, rewrite the equation using θ' as shown in Fig. 3.19.

$$S_{AB} = L\left[\frac{\theta'}{\beta} - \frac{1}{2\pi}\sin\frac{2\pi\theta'}{\beta}\right]$$

It is possible to transfer the origin from point A to point O by substituting the relation

$$\theta' = \theta - \frac{\pi}{2}$$

Therefore,

$$S_{AB} = L\left[\frac{(\theta - \pi/2)}{\beta} - \frac{1}{2\pi}\sin\frac{2\pi(\theta - \pi/2)}{\beta}\right]$$

Substituting $L = 1\frac{1}{2}$ in. and $\beta = \pi/2$ rad,

$$S_{AB} = \frac{3}{\pi}\left(\theta - \frac{\pi}{2}\right) - \frac{3}{4\pi}\sin(4\theta - 2\pi)$$

For the cycloidal curve C-6,

$$S_{CD} = L\left[1 - \frac{\theta''}{\beta} + \frac{1}{2\pi}\sin 2\pi\frac{\theta''}{\beta}\right]$$

where

$$\theta'' = \theta - \frac{3\pi}{2}$$

$$L = 1\tfrac{1}{2} \text{ in.}$$

$$\beta = \frac{\pi}{2}$$

Therefore,

$$S_{CD} = 6 - \frac{3\theta}{\pi} + \frac{3}{4\pi}\sin(4\theta - 6\pi)$$

It should be observed that with the combinations of dwell and cycloidal motion used, velocities and accelerations are matched and jerk is not infinite anywhere in the cycle.

Example 3.3

As an example of how the minimum radius C and the length of the follower face are determined, consider a flat-faced radial follower, which moves out and back 2 in. with simple harmonic motion for half a revolution of the cam. Two motion cycles of the follower occur for one revolution of the cam.

Only one displacement equation is necessary to specify the follower motion throughout the entire cycle.

$$S = r(1 - \cos\theta_r),$$ [4]

where r is the rotational radius and θ_r the angle turned through by the rotational radius for generation of harmonic motion (see Fig. 3.11). For the given data,

$$r = 1 \text{ in.}$$

$$\theta_r = 2\theta$$

Therefore,

$$S = f(\theta) = 1 - \cos 2\theta$$

$$f'(\theta) = 2\sin 2\theta$$

and

$$f''(\theta) = 4\cos 2\theta$$

To find the minimum radius, the sum $C + f(\theta) + f''(\theta)$ must be greater than zero. Substituting for $f(\theta)$ and $f''(\theta)$ and simplifying

$$C + 1 + 3\cos 2\theta > 0$$

[4] The Kloomok and Muffley equation $S = (L/2)(1 - \cos \pi\theta/\beta)$ for the harmonic H-5 could also be used.

The sum of $(1 + 3 \cos 2\theta)$ will be a minimum at $\theta = \pi/2$, which gives

$$C + 1 - 3 > 0$$

or

$$C > 2 \text{ in.}$$

The length of the follower face is determined from

$$l = f'(\theta) = 2 \sin 2\theta$$

$$l_{max} = 2 \text{ in.}$$

Because the motion is symmetrical, the theoretical length of the follower face is 2 in. on each side of the center line. An additional amount must be added to each side of the follower to prevent contact from occurring at the very end of the face.

3.7 Disk Cam with Radial Roller Follower. The analytical determination of the pitch surface of a disk cam with a radial roller follower presents no difficulties. In Fig. 3.20 the displacement of the center of the follower from the center of the cam is given by the following equation:

$$R = R_0 + f(\theta) \tag{3.10}$$

where R_0 is the minimum radius of the pitch surface of the cam and $f(\theta)$ is the radial motion of the follower as a function of cam angle. Once the value of R_0 is known it is an easy matter to determine the polar coordinates of the centers of the roller follower from which the cam may be generated.

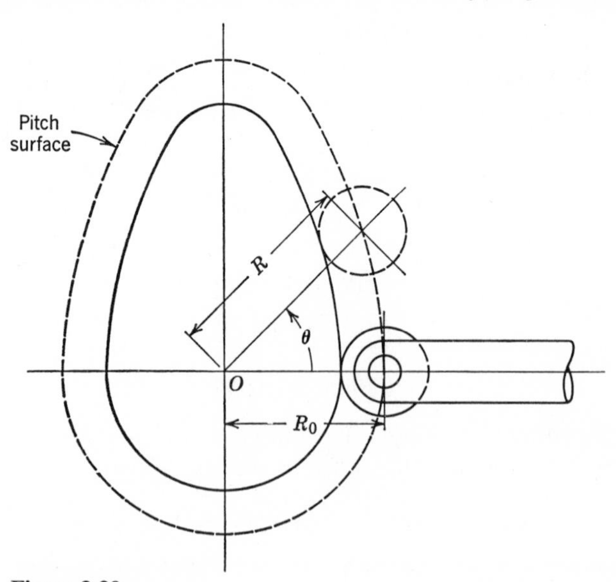

Figure 3.20

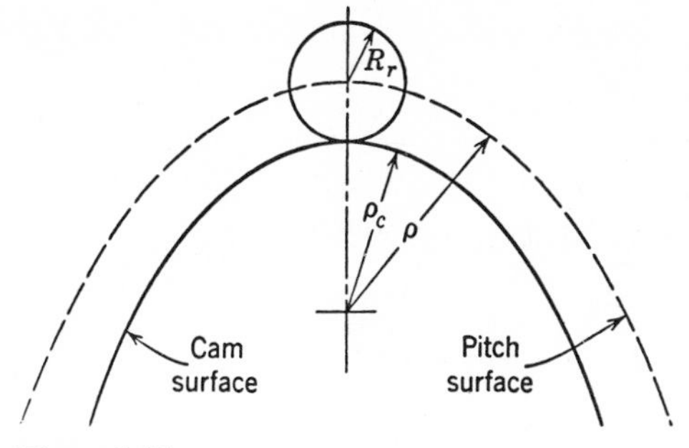

Figure 3.21

A method for checking this type of cam for pointing has been developed by Kloomok and Muffley,[5] which considers the radius of curvature ρ of the pitch surface and the radius of the roller R_r. These values are shown in Fig. 3.21 together with the radius of curvature ρ_c of the cam surface. If in Fig. 3.21 ρ is held constant and R_r is increased, ρ_c will decrease. If this is continued until R_r equals ρ, then ρ_c will be zero and the cam becomes pointed as shown in Fig. 3.22*a*. As R_r is further increased, the cam becomes undercut as shown in Fig. 3.22*b* and the motion of the follower will not be as prescribed. Therefore, in order to prevent a point or an undercut from occurring on the cam profile, R_r must be less than $\rho_{\min}$ where $\rho_{\min}$ is the minimum value of ρ over the particular segment of profile being considered. If there are several types of motion through which the follower passes, each case must be checked separately. Because it is impossible to undercut a concave portion of a cam, only the convex portions need to be investigated.

The radius of curvature at a point on a curve expressed in polar coordinates can be given by

$$\rho = \frac{[R^2 + (dR/d\phi)^2]^{3/2}}{R^2 + 2(dR/d\phi)^2 - R(d^2R/d\phi^2)}$$

where $R = f(\phi)$ and the first two derivatives are continuous. This equation can be used for finding the radius of curvature of the pitch surface of the cam. For this case, $f(\theta) = f(\phi)$. From Eq. 3.10

$$R = R_0 + f(\theta)$$

$$\frac{dR}{d\theta} = f'(\theta)$$

$$\frac{d^2R}{d\theta^2} = f''(\theta)$$

[5] M. Kloomok and R. V. Muffley, "Plate Cam Design—Radius of Curvature," *Prod. Eng.*, September 1955.

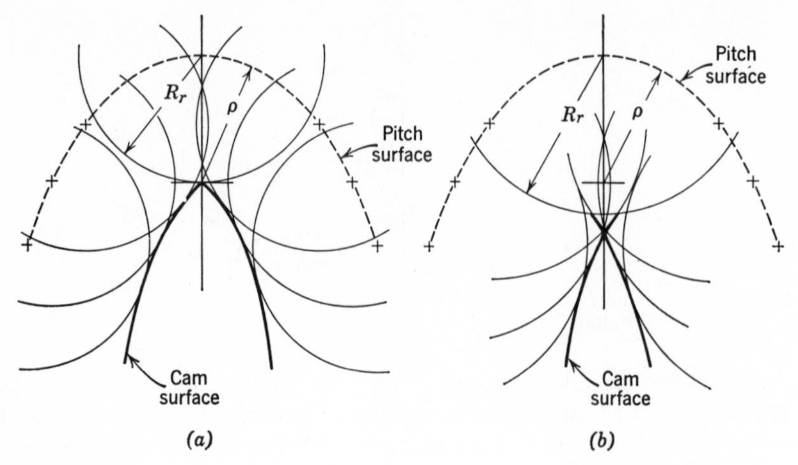

Figure 3.22

Therefore,

$$\rho = \frac{\{R^2 + [f'(\theta)]^2\}^{3/2}}{R^2 + 2[f'(\theta)]^2 - R[f''(\theta)]} \tag{3.11}$$

Equation 3.11 may be evaluated to find the expression for ρ for a particular type of motion. However, in order to prevent points or undercuts on the cam profile, ρ_{min} must be determined. Differentiation of Eq. 3.11 with its various functions to obtain minima gives very complex transcendental equations. For this reason, three sets of curves are given that show the plot of ρ_{min}/R_0 versus β for various values of L/R_0. In these curves, β is the total angular rotation of the cam for a complete event, and L is the lift. Figure 3.23 shows the graph for cycloidal motion, Fig. 3.24 for simple harmonic motion, and Fig. 3.25 for eighth-power polynomial motion. By means of these curves it is possible to determine whether or not ρ_{min} is greater than R_r.

Example 3.4

A radial roller follower is to move through a total displacement of $L = 0.60$ in. with cycloidal motion while the cam rotates $\beta = 30°$. The follower dwells for 45° and then returns with cycloidal motion in 70°. Check the cam for pointing or undercutting if the radius R_r of the roller is 0.25 in. and the minimum radius R_0 of the pitch surface is 1.50 in.

$$\frac{L}{R_0} = \frac{0.60}{1.50} = 0.40$$

The outward motion will govern because of its smaller β. Therefore, from

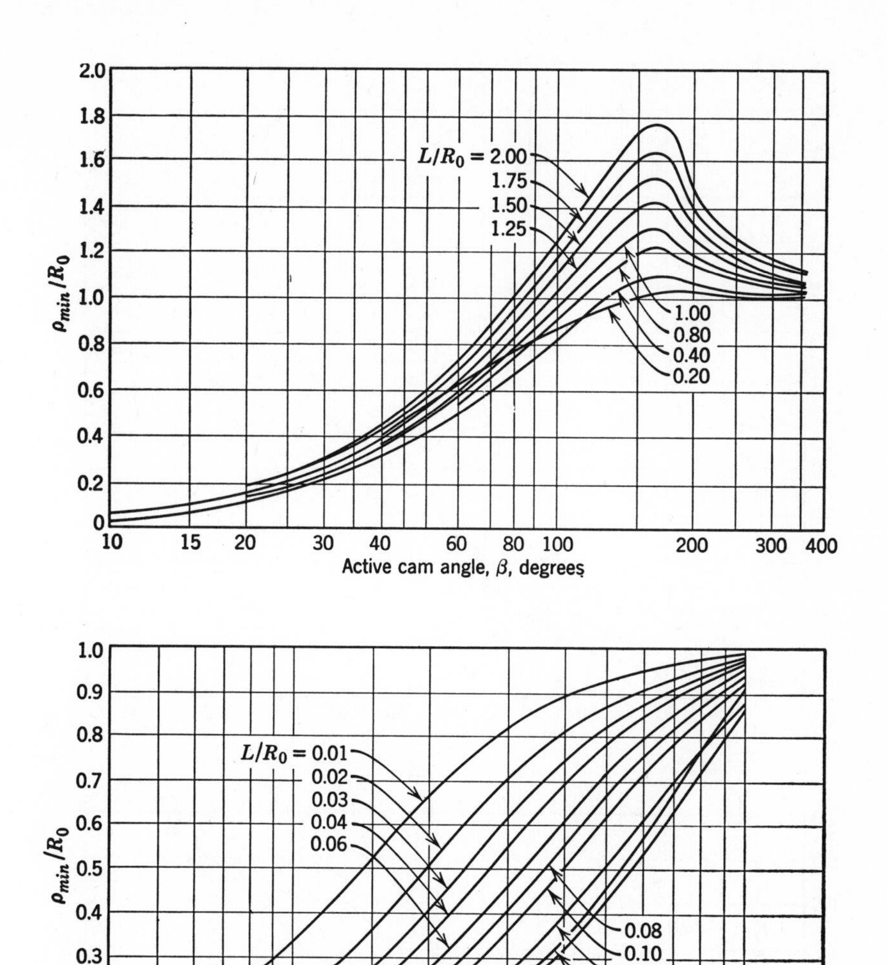

Figure 3.23 Cycloidal motion. (From M. Kloomok and R. V. Muffley, "Plate Cam Design—Radius of Curvature," *Prod. Eng.*, September 1955.)

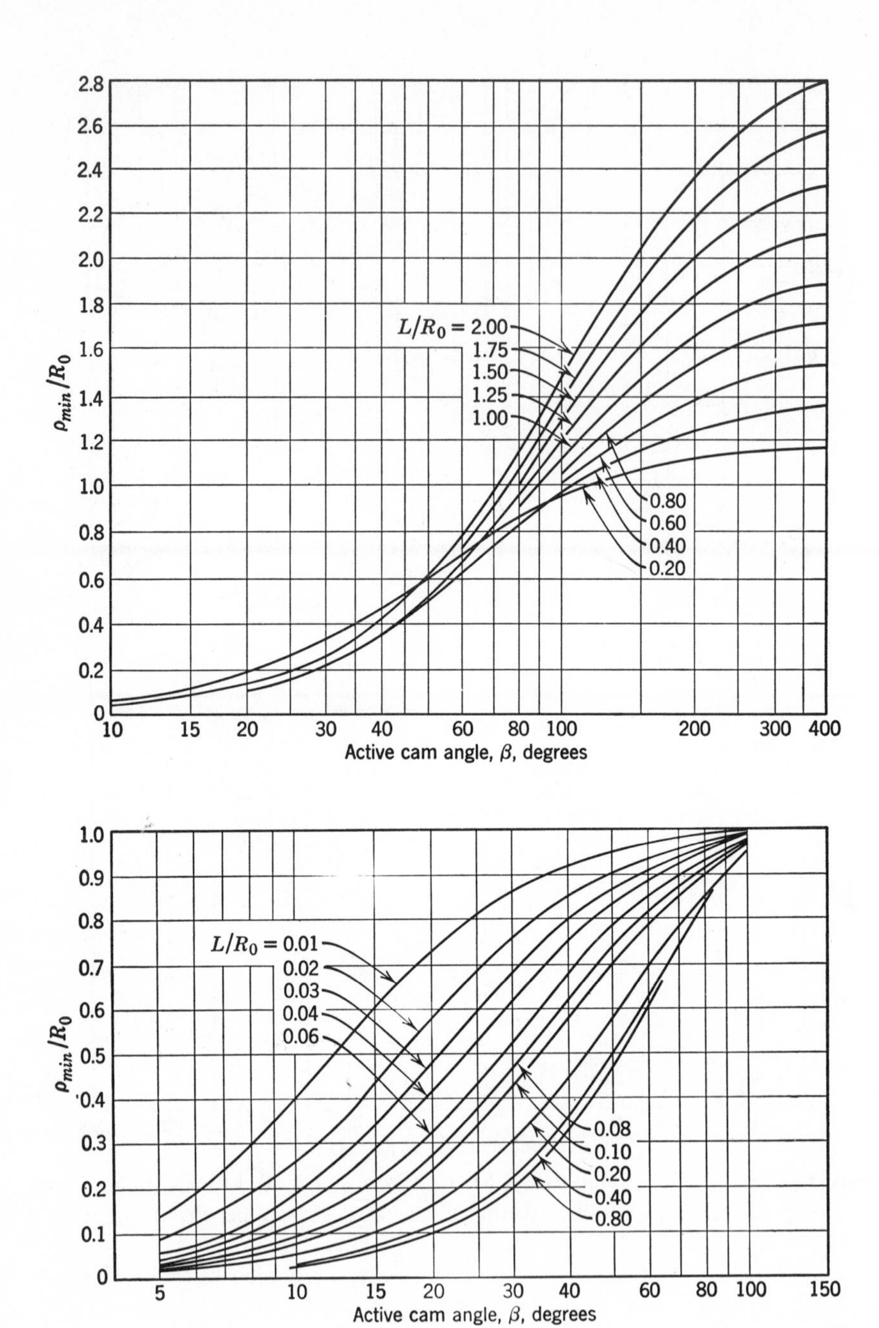

Figure 3.24 Harmonic motion. (From M. Kloomok and R. V. Muffley, "Plate Cam Design—Radius of Curvature," *Prod. Eng.*, September 1955.)

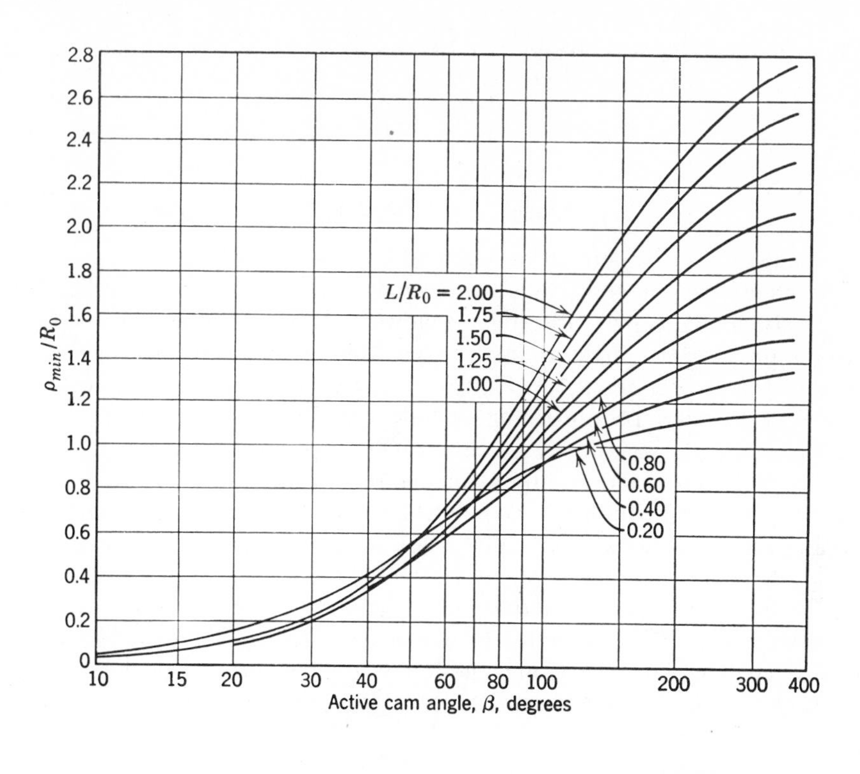

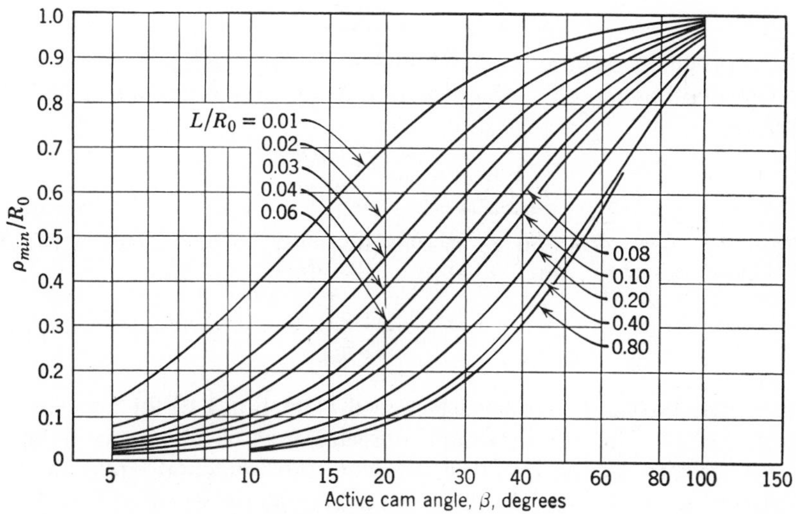

Figure 3.25 Eighth-power polynomial motion. (From M. Kloomok and R. V. Muffley, "Plate Cam Design—Radius of Curvature," *Prod. Eng.*, September 1955.)

Fig. 3.23 for $L/R_0 = 0.40$ and $\beta = 30°$,

$$\frac{\rho_{\min}}{R_0} = 0.22$$

and

$$\rho^{\min} = 0.22 \times 1.50 = 0.33 \text{ in.}$$

The cam will not be pointed or undercut because $\rho_{\min} > R_r$.

As mentioned previously, pressure angle is an important consideration when designing cams with roller followers. It is necessary to keep the maximum pressure angle as small as possible, and to date this maximum has arbitrarily been set at 30°. However, higher values are occasionally used when conditions permit. Although it is possible to make a layout of the cam and measure the maximum pressure angle, analytical methods are to be preferred. Several methods are available, one of which has been developed by Kloomok and Muffley,[6] whereby the pressure angle can be determined analytically for either a radial roller follower or an oscillating roller follower. Only the radial roller follower will be treated here.

For the disk cam and radial roller follower shown in Fig. 3.26, the pressure angle OCA is denoted by α and the center of the cam by O. The cam is assumed stationary, and the follower center rotates clockwise from position C to C' through a small angle $\Delta\theta$. From the sketch,

$$\alpha' = \tan^{-1}\frac{C'E}{CE}$$

As $\Delta\theta$ approaches zero, angles OCE and ACC' approach 90°. At the same time CD approaches CF, which equals $R\,\Delta\theta$, and both approach CE. Therefore,

$$\lim_{\Delta\theta\to 0}\alpha' = \tan^{-1}\frac{1}{R}\frac{dR}{d\theta}$$

Because the sides of α and α' become mutually perpendicular when $\Delta\theta$ approaches zero, α' becomes equal to α. Therefore,

$$\alpha = \tan^{-1}\frac{1}{R}\frac{dR}{d\theta} \tag{3.12}$$

An expression for α may be determined from Eq. 3.12 for any type of motion. To solve for the maximum pressure angle is often very difficult, however, because of the resulting complex transcendental equation. For this reason, Kloomok and Muffley use a nomogram developed by E. C.

[6] M. Kloomok and R. V. Muffley, "Plate Cam Design—Pressure Angle Analysis," *Prod. Eng.*, May 1955.

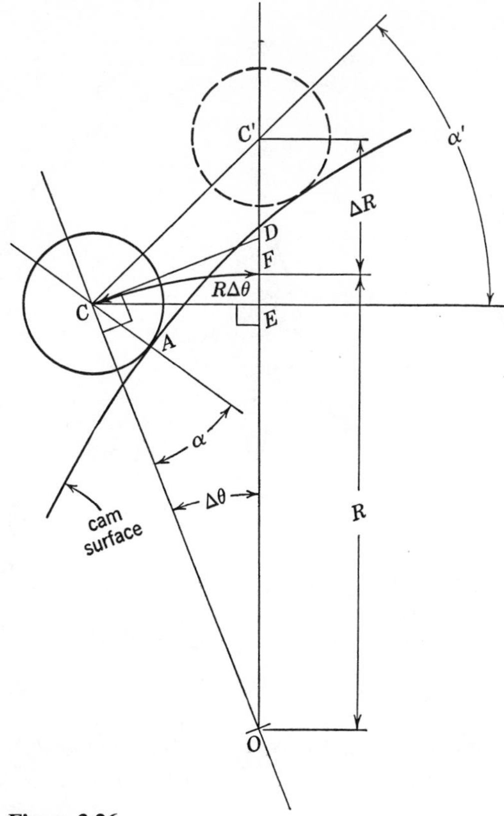

Figure 3.26

Varnum,[7] which is given in Fig. 3.27; β and L/R_0 are parameters as previously defined. From this chart the maximum value of pressure angle may be determined for the three types of motion.

Example 3.5

A radial roller follower is to move through a total displacement of 0.75 in. with cycloidal motion while the cam rotates 45°. The follower dwells for 30° and then returns with cycloidal motion in 60°. Find the value of R_0 to limit α_{max} to 30°. The outward motion will govern because of its smaller β.

For $\beta = 45°$ and $\alpha_{max} = 30°$,

$$\frac{L}{R_0} = 0.26 \qquad \text{(from Fig. 3.27)}$$

[7] E. C. Varnum, "Circular Nomogram Theory and Construction Technique," *Prod. Eng.*, August 1951.

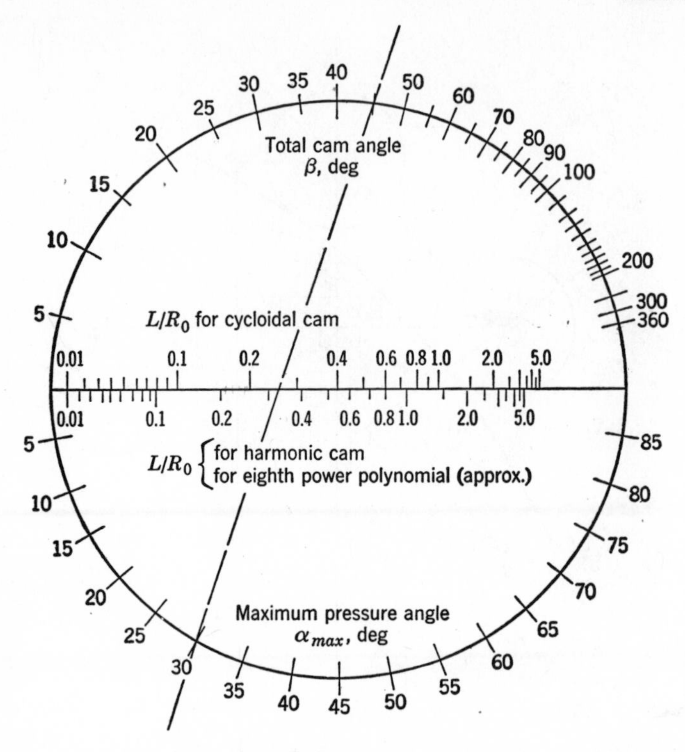

Figure 3.27 Nomogram to determine maximum pressure angle. (Courtesy of E. C. Varnum, Barber–Colman Company.)

Therefore,

$$R_0 = \frac{0.75}{0.26} = 2.88 \text{ in.}$$

If space does not permit such a value of R_0, β can be increased and the cam run faster to maintain the lift time at a constant value.

3.8 Disk Cam with Oscillating Roller Follower. In Fig. 3.28 is seen the start of a layout of a disk cam with an oscillating roller follower. The displacement angle ψ is a function of the cam angle θ. Although the cam rotates through the angle θ for the displacement angle ψ, the radius R rotates through the angle ϕ. By specifying values of R and ϕ, it is possible to generate the cam.[8]

From Fig. 3.28 it can be seen that

$$\phi = \theta - \lambda \tag{3.13}$$

[8] L. S. Linderoth, "Calculating Cam Profiles," *Machine Design*, July 1951.

where

$$\lambda = \beta - \Gamma \tag{3.14}$$

Angle β is a constant for the system, and its equation can be derived from triangle OAO' as

$$\cos \beta = \frac{S^2 + R_0^2 - l^2}{2SR_0} \tag{3.15}$$

where S, R_0, and l are fixed dimensions.

Angle Γ is a function of R; its equation can be derived from triangle OBO' as

$$\cos \Gamma = \frac{S^2 + R^2 - l^2}{2SR} \tag{3.16}$$

An equation for R can also be written from triangle OBO' as follows:

$$R^2 = l^2 + S^2 - 2lS \cos (\psi + \Sigma) \tag{3.17}$$

Angle Σ is a constant determined from triangle OAO' as

$$\cos \Sigma = \frac{l^2 + S^2 - R_0^2}{2lS} \tag{3.18}$$

and angle ψ is the displacement angle for a particular cam angle θ. Therefore, from the preceding equations, the values of R and ϕ can be calculated for given values of cam angle θ and their corresponding angles of displacement ψ.

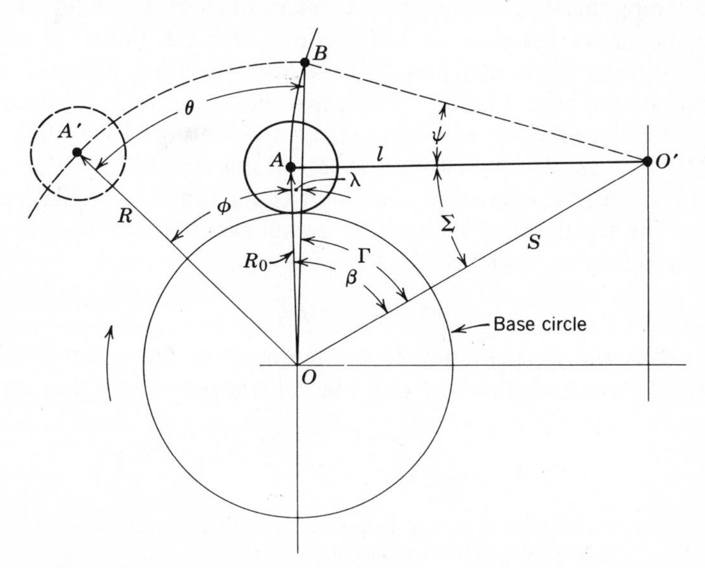

Figure 3.28

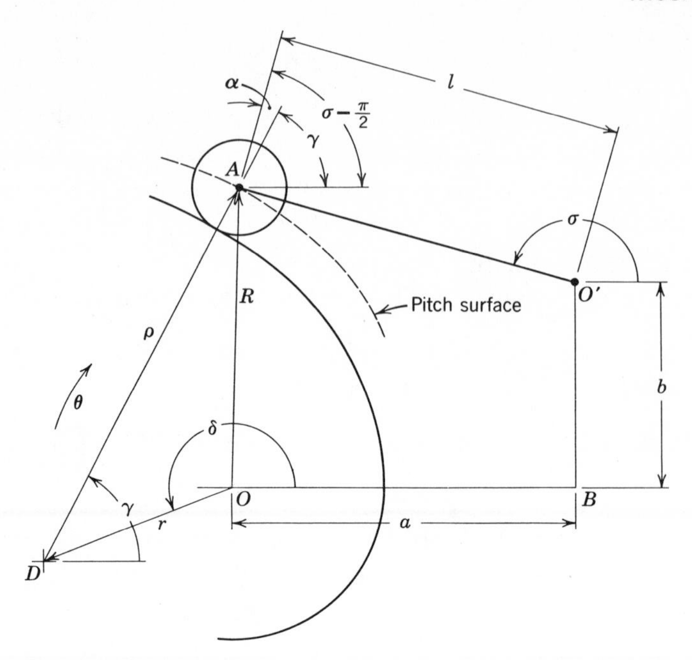

Figure 3.29

In designing this type of cam, it is necessary to check for undercutting and for the maximum pressure angle. Equations for the radius of curvature and the pressure angle can best be developed by using Raven's[9] method of complex variables. Figure 3.29 shows the sketch of a disk cam and oscillating roller follower with the radius of curvature of the pitch surface designated as ρ and the pressure angle as α. Point O is the center of the cam, point D the center of curvature, and point O' the center of oscillation of the follower. The angular displacement of the follower from the horizontal is σ, which is given by the equation

$$\sigma = \sigma_0 + f(\theta) \tag{3.19}$$

where $f(\theta)$ is the desired angular displacement of the follower from the reference angle σ_0 (not shown). From Fig. 3.29 the pressure angle α is given by

$$\alpha = \sigma - \frac{\pi}{2} - \gamma$$

[9] F. H. Raven, "Analytical Design of Disk Cams and Three-Dimensional Cams by Independent Position Equations," *ASME Transactions*, Series E, Vol. 26, No. 1, pp. 18–24, March 1959.

Substituting Eq. 3.19 for σ,

$$\alpha = [\sigma_0 + f(\theta)] - \frac{\pi}{2} - \gamma \tag{3.20}$$

In order to obtain an expression for angle γ, two independent position equations are written for the center of the roller A. One is written by following the path O to D to A, and the other by going from O to B to O' to A. The equation for the first path (O–D–A) is given by

$$\begin{aligned}\mathbf{R} &= re^{i\delta} + \rho e^{i\gamma} \\ &= r(\cos\delta + i\sin\delta) + \rho(\cos\gamma + i\sin\gamma)\end{aligned} \tag{3.21}$$

The equation for the second path (O–B–O'–A) is given by

$$\begin{aligned}\mathbf{R} &= a + bi + le^{i\sigma} \\ &= a + bi + l(\cos\sigma + i\sin\sigma)\end{aligned} \tag{3.22}$$

By separating real and imaginary parts of Eqs. 3.21 and 3.22, it follows that

$$r\cos\delta + \rho\cos\gamma = a + l\cos\sigma \tag{3.23}$$

$$r\sin\delta + \rho\sin\gamma = b + l\sin\sigma \tag{3.24}$$

Differentiating Eqs. 3.23 and 3.24 with respect to θ,

$$-r\sin\delta\frac{d\delta}{d\theta} - \rho\sin\gamma\frac{d\gamma}{d\theta} = -l\sin\sigma\frac{d\sigma}{d\theta}$$

$$r\cos\delta\frac{d\delta}{d\theta} + \rho\cos\gamma\frac{d\gamma}{d\theta} = l\cos\sigma\frac{d\sigma}{d\theta}$$

For an infinitesimal rotation of the cam, ρ may be considered to remain constant. Thus point D, the center of curvature of the cam at the point of contact, and $\mathbf{r}$ may be regarded as fixed to the cam for an incremental rotation $d\theta$. Therefore, the magnitude of $d\delta$ is equal to $d\theta$, and since δ decreases as θ increases, it follows that $d\delta/d\theta = -1$. Also, $d\sigma/d\theta = f'(\theta)$. Therefore,

$$r\sin\delta - \rho\sin\gamma\frac{d\gamma}{d\theta} = -lf'(\theta)\sin\sigma \tag{3.25}$$

$$-r\cos\delta + \rho\cos\gamma\frac{d\gamma}{d\theta} = lf'(\theta)\cos\sigma \tag{3.26}$$

Eliminating $d\gamma/d\theta$ from Eqs. 3.25 and 3.26,

$$\tan\gamma = \frac{r\sin\delta + lf'(\theta)\sin\sigma}{r\cos\delta + lf'(\theta)\cos\sigma}$$

The terms $r \cos \delta$ and $r \sin \delta$ can be evaluated from Eqs. 3.23 and 3.24 to give

$$\tan \gamma = \frac{b + l \sin \sigma[1 + f'(\theta)]}{a + l \cos \sigma[1 + f'(\theta)]} \tag{3.27}$$

which, when substituted in Eq. 3.20, will give the pressure angle α. To find α_{max}, it will be necessary to work out design charts similar to those given by Kloomok and Muffley in footnote 6.

In order to find the radius of curvature ρ, it is necessary to first differentiate Eq. 3.27 with respect to θ. Substituting $d\gamma/d\theta$ from Eq. 3.26 and with the aid of Eqs. 3.19, 3.23, and 3.27, the following equation for ρ is obtained:

$$\rho = \frac{[C^2 + D^2]^{3/2}}{(C^2 + D^2)[1 + f'(\theta)] - (aC + bC)f'(\theta) + (a \sin \sigma - b \cos \sigma)lf''(\theta)} \tag{3.28}$$

where

$$C = a + l \cos \sigma[1 + f'(\theta)]$$
$$D = b + l \sin \sigma[1 + f'(\theta)]$$

To avoid undercutting, ρ must be greater than the radius of the roller. Therefore it must be possible to determine ρ_{min} for each portion of the cam profile. In order to do this, it will be necessary to work out design charts similar to those given by Kloomok and Muffley in footnote 5.

3.9 Three-Dimensional Cams. This type of cam is difficult to design and to manufacture but finds wide application in gun-fire control mechanisms. A sketch of a three-dimensional cam is shown in Fig. 3.30,

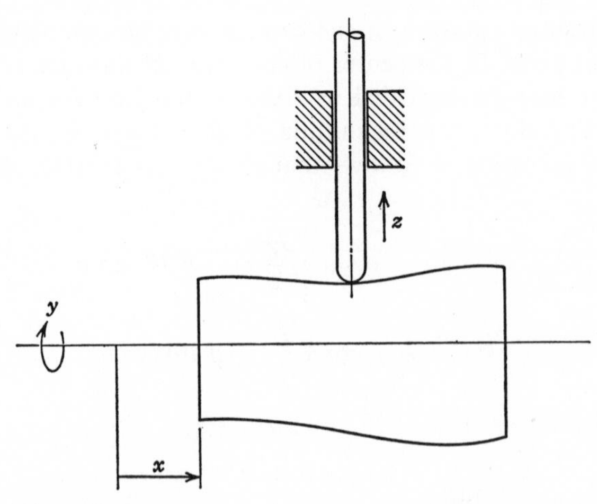

Figure 3.30

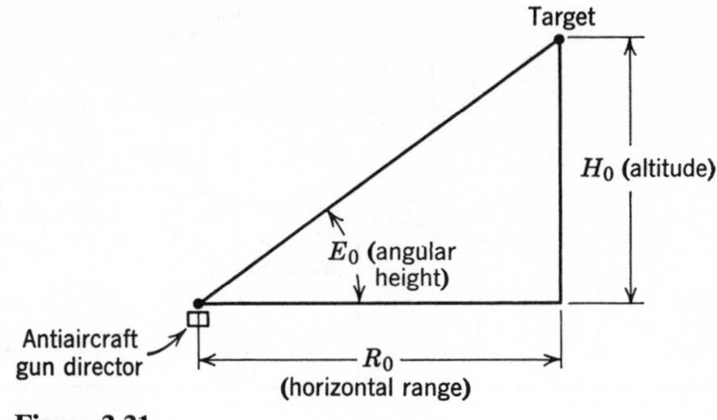

Figure 3.31

where the displacement z of the follower is a function of both the rotation y and the translation x of the cam.

A problem that can be solved very simply by the use of a three-dimensional cam is that of computing the horizontal range of an aerial target given the angular height and the altitude of the target.[10] The horizontal range is computed from the triangle shown in Fig. 3.31 from the relation $R_0 = H_0 \cot E_0$. This relation is solved in the gun director by a three-dimensional cam that rotates in proportion to the altitude H_0 of the target and translates in proportion to the horizontal range R_0. The displacement of the follower represents angular height E_0. After the altitude of the target has been set into the cam by rotation, translation of the cam will produce follower lift indicating angular height. When this angular height matches that obtained from elevation tracking, the correct horizontal range will be indicated.

To illustrate the method of designing a three-dimensional cam, consider a target moving toward a gun director at an altitude of 8000 ft. Figure 3.32

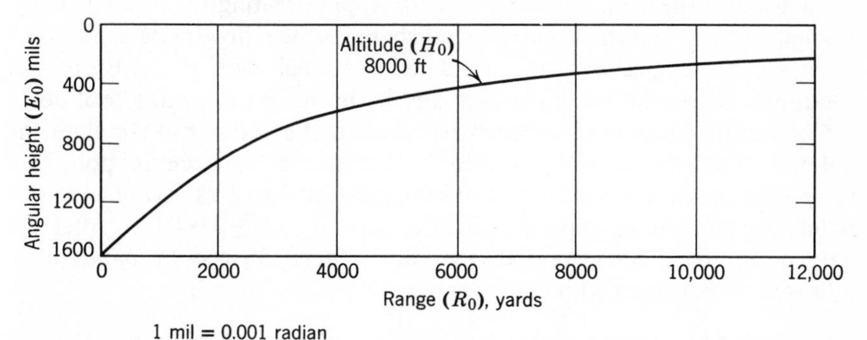

Figure 3.32

[10] T. J. Hayes, *Elements of Ordnance*, John Wiley and Sons, 1938.

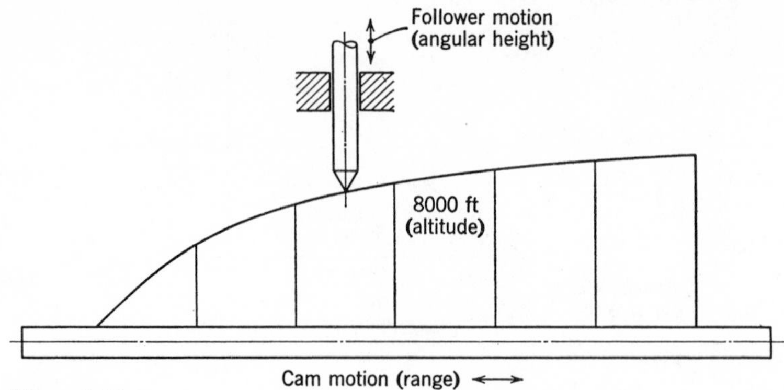

Figure 3.33

shows a plot of angular height versus horizontal range for this altitude. If this curve is used for the contour of a two-dimensional cam, horizontal translation of the cam will move the follower so that it indicates the correct angular height of the target. Such a cam is shown in Fig. 3.33.

In order to determine angular heights at altitudes other than 8000 ft, it is necessary to plot additional curves and to produce their corresponding plate cams. If these plates are then arranged around the cam shaft, the cam of Fig. 3.34*a* results. Rotation of this cam will place the correct altitude plate under the follower. As the number of plates is gradually increased to include more altitudes, the cam finally becomes solid as seen in Fig. 3.34*b*.[11]

For determination of the pressure angle and radius of curvature of a three-dimensional cam, the reader is referred to the work of F. H. Raven.[12]

The production of a three-dimensional cam is very difficult because of the accuracy and hand finishing required. After the displacements of the follower have been specified for the desired increments of rotation and translation of the cam, a casting is made approximating the desired shape. Using a cutting tool the same size and shape as the follower, the cam blank is set up in a cam miller, and a cut made at each data point. By proper rotation and translation of the cam and by bringing the cutting tool down the correct displacement for each data point, the cutter will simulate the follower in its relation with the cam. In this manner, an accurate point will be spotted on the cam contour. According to Rothbart,[13] as many as 15,000 points are sometimes required with the accuracy of ± 0.0004 in. After the data points have been spotted, the cam is next finished by hand filing, followed by polishing with emery paper.

[11] "Principles of Operation, Prediction Mechanism, and Ballistic Cam of Sperry M-7 Anti-aircraft Director," Sperry Gyroscope Company, December 1946.

[12] *Ibid.*

[13] H. A. Rothbart, *Cams*, John Wiley and Sons, 1956.

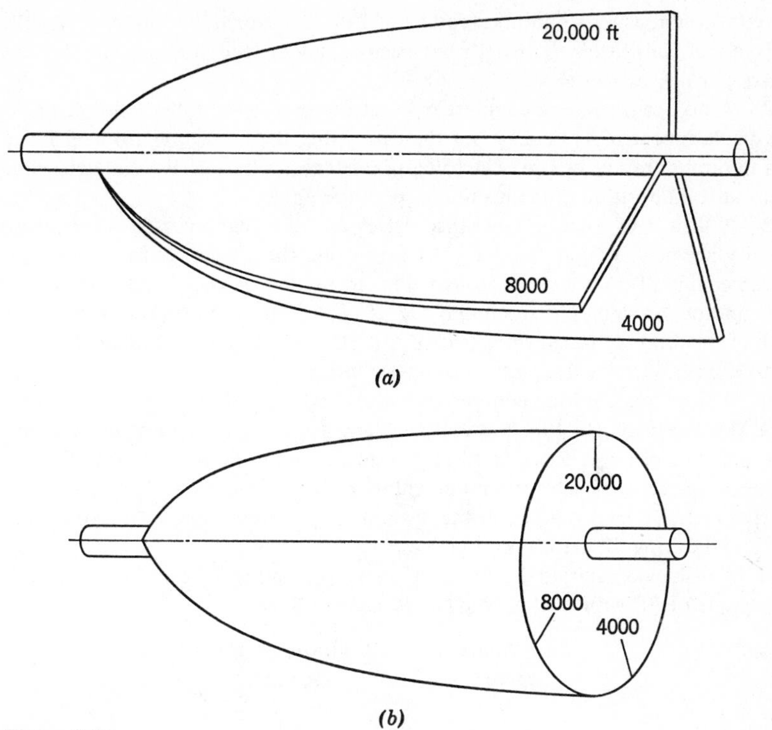

Figure 3.34

Problems

3.1 A disk cam rotating clockwise drives a radial flat-faced follower through a total displacement of $1\frac{1}{2}$ in. with the following lift figures:

Cam Angle, degrees	Lift, in.
0	0.00
30	0.10
60	0.37
90	0.75
120	1.13
150	1.40
180	1.50
210	1.40
240	1.13
270	0.75
300	0.37
330	0.10
360	0.00

Lay out the cam using a minimum radius of 1 in. Determine the length of the follower face (symmetrical). After the length has been found by trial, add $\frac{1}{8}$ in. to each end to positively insure proper contact.

3.2 A disk cam rotating counterclockwise drives a radial roller follower through a total displacement of $1\frac{1}{2}$ in. Lay out the cam using the lift figures from Problem 3.1 and a minimum radius of 1 in. The roller diameter is to be $\frac{7}{8}$ in. By trial determine the magnitude and position of the maximum pressure angle.

3.3 A disk cam rotating clockwise drives an offset flat-faced follower through a total displacement of $1\frac{1}{2}$ in. Lay out the cam using the lift figures from Problem 3.1. The center line of the follower is offset $\frac{1}{2}$ in. to the left of and parallel to the vertical center line of the cam. The minimum radius of the cam is to be 1 in. Determine the length of the follower face (symmetrical). After the length has been found by trial, add $\frac{1}{8}$ in. to each end to positively insure proper contact.

3.4 A disk cam rotating counterclockwise drives an offset roller follower through a total displacement of $1\frac{1}{2}$ in. Lay out the cam using the lift figures in Problem 3.1. The center line of the follower is offset $\frac{1}{2}$ in. to the right of and parallel to the vertical center line of the cam. The minimum radius of the cam is to be 1 in. and the roller diameter $\frac{7}{8}$ in. By trial determine the maximum pressure angle during the outward motion and during the return motion.

3.5 A disk cam rotating clockwise drives an oscillating flat-faced follower through a total angle of 20° with the following displacement figures:

Cam Angle, degrees	Follower Angle, degrees
0	0.0
30	1.5
60	5.5
90	10.0
120	14.5
150	18.5
180	20.0
210	18.5
240	14.5
270	10.0
300	5.5
330	1.5
360	0.0

Lay out the cam using a minimum radius of 3.0 cm. The center of the hub of the follower is to be 8.0 cm to the right of the center and on the horizontal center line of the cam similar to Fig. 3.3. The distance from the center of the follower hub to the arc of the displacement scale is to be 7.0 cm. Determine the length of the follower face. After the length has been found by trial, add 0.3 cm to each end to positively insure proper contact. Assuming a bore of 1.6 cm, a hub of 2.5 cm, and a 0.5 cm square key, draw in the rest of the follower to reasonable proportions.

3.6 A disk cam rotating counterclockwise drives an oscillating roller follower through a total angle of 20°. Lay out the cam using the displacement figures from Problem 3.5 and a minimum radius of 1 in. The center of the hub of the follower is to be 3 in. to the right of the center and on the horizontal center line of the cam similar to Fig. 3.4. The diameter of the roller is to be $\frac{3}{4}$ in., and the distance from the center of the hub to the center of the roller $2\frac{7}{8}$ in. Using a bore of $\frac{5}{8}$ in., a hub of 1 in., and a $\frac{3}{16} \times \frac{3}{16}$ in. key, draw in the rest of the follower to reasonable proportions.

3.7 A positive return cam rotating clockwise drives a flat-faced yoke as shown in Fig. 3.5. The lift figures for the outward motion are as follows:

Cam Angle, degrees	Lift, cm
0	0.00
30	0.127
60	0.432
90	0.965
120	1.70
150	2.34
180	2.54

Lay out the cam using a minimum radius of 2.5 cm. Using reasonable proportions, complete the sketch of the follower.

3.8 A positive-return cam rotating counterclockwise drives a yoke with roller followers. Lay out the cam using the lift figures from Problem 3.7 for the outward motion and a minimum radius of 2.5 cm. The diameters of the rollers are to be 1.9 cm. Using reasonable proportions, complete the yoke that carries the roller followers.

3.9 An oscillating roller follower moving through a total angle of 60° drives an inverse cam as shown in Fig. 3.7 with the following displacement figures:

Follower Angle, degrees	Cam Displacement, in.
0.0	0.00
4.5	0.06
16.0	0.24
30.0	0.50
44.0	0.76
55.5	0.94
60.0	1.00

Lay out the groove in the cam block if the cam is to move upward and to the right at an angle of 45° as the follower moves counterclockwise. The follower moves symmetrically about the vertical center line. The distance from the center of the roller follower to the center of oscillation is 3 in. and the diameter of the roller $\frac{5}{8}$ in. The cam block is 3 × 4 in.

3.10 Prove that the method for finding inflection points when the time intervals are known as shown in Fig. 3.8 is correct.

3.11 Prove that the method for finding inflection points when the displacement intervals are known as shown in Fig. 3.9 is correct.

3.12 Prove that the method of construction for parabolic motion shown in Fig. 3.10 is correct.

3.13 Draw the displacement-time graph for a follower that rises through a total displacement of $1\frac{1}{2}$ in. with constant acceleration for a three-sixteenth revolution, constant velocity for a quarter revolution, and constant deceleration for a quarter revolution of the cam. After dwelling for a sixteenth revolution, the follower returns with simple harmonic motion in a quarter revolution of the cam. Use an abscissa 4 in. long.

3.14 Draw the displacement-time graph for a follower that rises 1.9 cm with simple harmonic motion in a quarter revolution, dwells for an eighth revolution, and then rises 1.9 cm with simple harmonic motion in a quarter revolution of the cam. The follower dwells for a sixteenth revolution and then returns 3.8 cm with parabolic motion in a quarter revolution, followed by a dwell for a sixteenth revolution of the cam. Use an abscissa 16 cm long.

3.15 Draw the displacement-time graph for a follower that rises $1\frac{1}{2}$ in. in a half revolution of the cam such that the first $\frac{3}{8}$ in. is constant acceleration, constant velocity for the next $\frac{3}{4}$ in., and constant deceleration for the remaining $\frac{3}{8}$ in. The return motion is simple harmonic in a half revolution of the cam. Use an abscissa 6 in. long.

3.16 Draw the displacement-time graph for a follower that moves through a total displacement of 3.2 cm with constant acceleration for 90° and constant deceleration for 45° of cam rotation. The follower returns 1.6 cm with simple harmonic motion in 90°, dwells for 45°, and returns the remaining 1.6 cm with simple harmonic motion in 90° of cam rotation. Use an abscissa 16 cm long.

3.17 The radial flat-faced follower shown in Fig. 3.35 is made to reciprocate by the action of a circular disk cam rotating about the axis at O_2. (a) Derive expressions for the follower displacement R and for the distance l to the point of contact in terms of the

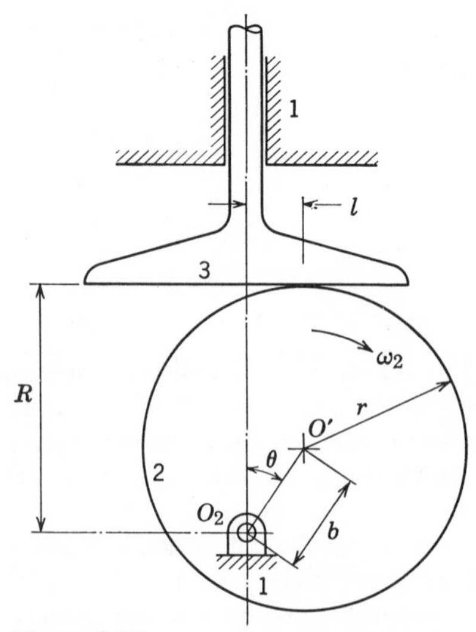

Figure 3.35

cam angle θ, the radius r of the cam, and the offset distance b. (b) Make a sketch of the displacement R versus cam angle θ for one cycle of the cam. Label the lift L of the follower as the distance from the follower's lowest position to its highest position. Give the magnitude of L. (c) Name the type of follower motion produced by the cam.

3.18 A radial follower is actuated by a cam rotating at 1 rad/sec. The follower starts from rest and moves through a distance of 5.0 cm with simple harmonic motion while the cam turns 120°. The follower dwells for the next 120° and then returns with simple harmonic motion during the remaining 120° of cam rotation. Using an abscissa of 15 cm and increments of cam angle of 30°, plot the displacement, velocity, acceleration, and jerk curves on the same axis.

3.19 From the relation for simple harmonic motion derive the equation for displacement S for motion classification H-5 of Fig. 3.14.

3.20 Derive equations which will allow the use of the Kloomok and Muffley equations to be used to determine follower velocities and accelerations when the angular velocity of the cam is not constant.

3.21 A follower is to have cyclical motion according to the displacement curve shown in Fig. 3.36. The displacement and velocity requirements are as follows:

Point A	Point B	Point C
$S = L$	$S = 0$	$S = L$
$V = 0$	$V = 0$	$V = 0$

Recommend the curves to be used for the displacement graph and the relation between β_1 and β_2 to match accelerations at point B and at points A and C.

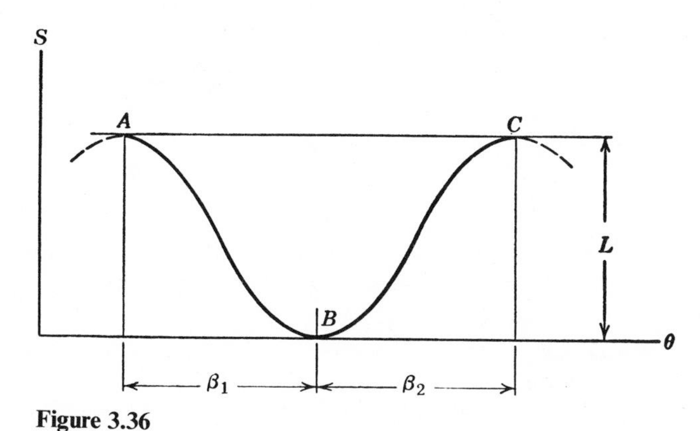

Figure 3.36

3.22 A follower dwells and then passes through the motion cycle shown in Fig. 3.37 and dwells again. The motion requirements are as follows:

Point A	Point B	Point C
$S = 0$	$S = L$	$S = 0$
$V = 0$	$V = 0$	$V = 0$
$A = 0$	$A = A_1$	$A = 0$

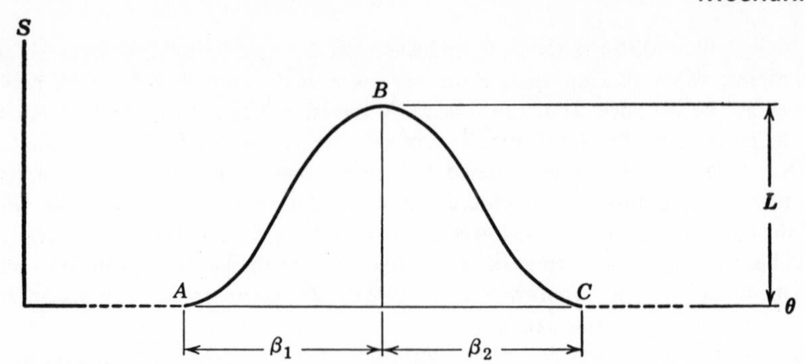

Figure 3.37

Recommend the curves to be used for the displacement graph and the relation between β_1 and β_2 to match accelerations at point B.

3.23 A follower dwells, rises with acceleration, rises with constant velocity, rises with deceleration, and then dwells again as shown in Fig. 3.38. The motion requirements are as follows:

Point A	Point B	Point C	Point D
$S = 0$	$S = L_1$	$S = L_1 + L_2$	$S = L_1 + L_2 + L_3$
$V = 0$	$V = V_1$	$V = V_1$	$V = 0$
$A = 0$	$A = 0$	$A = 0$	$A = 0$

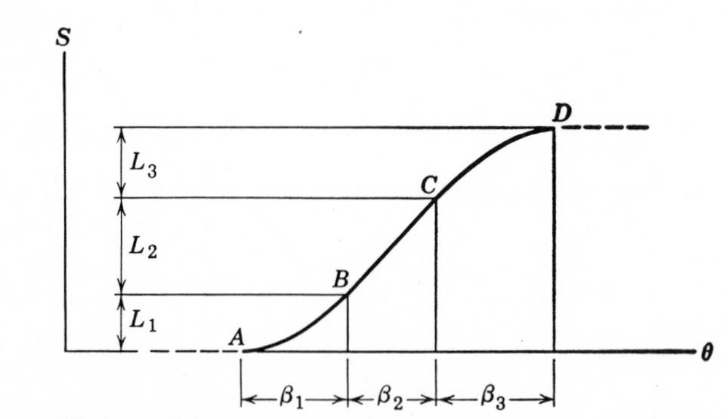

Figure 3.38

Recommend the curves to be used for the displacement graph and the relation between β_1, β_2, and β_3 to match velocities at points B and C.

3.24 For the motion given in Fig. 3.16*a* of Example 3.1 let β_1 denote the angle for curve AB, β_2 the angle for BC, β_3 the angle for CD, and β_4 the angle for DE. Also let L_1 be the rise AB, L_2 the rise BC, L_3 the rise CD, and L_4 the fall DE. Determine the relation between β_3 and β_4 to match accelerations at point D.

3.25 Determine (*a*) the relation between the cam angles β_1, β_2 and the lifts L_1, L_2 to match a cycloidal curve C-1 with a constant velocity curve, and (*b*) the relation to match a constant velocity curve with curve C-4.

3.26 Set up the equations for relating the lifts L_1, L_2 and cam angles β_1, β_2 for transferring from: (*a*) Cycloidal to harmonic motion; (*b*) Cycloidal to constant velocity; (*c*) Harmonic to cycloidal; (*d*) Harmonic to constant velocity.
Transfer motions only when acceleration equals zero.

3.27 Determine (*a*) the relation between cam angles β_1, β_2 and lifts L_1, L_2 to match a cycloidal curve C-1 with a harmonic curve H-2, and (*b*) the relation to match curve H-3 with curve C-4.

3.28 Determine (*a*) the relation between cam angles β_1, β_2 and lifts L_1, L_2 to match a harmonic curve H-1 with a cycloidal curve C-2, and (*b*) the relation to match curve C-3 with curve H-4.

3.29 Determine (*a*) the relation between cam angles β_1, β_2 and lifts L_1, L_2 to match a harmonic curve H-1 with a constant velocity curve, and (*b*) the relation to match a constant velocity curve with curve H-4.

3.30 A follower is to have a period of constant velocity motion during its outward travel and again on its return. Is it possible to use harmonic curves with these constant velocity curves and not have jerk become infinite? If so, recommend the curves to be used and sketch the displacement graph showing the motions.

3.31 Determine (*a*) the relation between cam angles β_1, β_2 and lifts L_1, L_2 to match a harmonic curve H-5 with an eighth-power polynomical curve P-2, and (*b*) the relation to match a harmonic curve H-2 to an eighth-power polynomial curve P-2.

3.32 Select a combination of cycloidal motion, harmonic, and eighth-power polynomial that will not produce infinite jerk.

3.33 Determine (*a*) the relation between cam angles β_1, β_2 and lifts L_1, L_2 to match an eighth-power polynomial curve P-1 with a harmonic curve H-6, and (*b*) the relation to match an eighth-power polynomial curve P-1 with a harmonic curve H-3.

3.34 Select a combination of harmonic motion and eighth-power polynomial that will not produce infinite jerk.

3.35 A follower moves with harmonic motion (H-1) a distance of 2.5 cm in $\pi/4$ rad of cam rotation. The follower then moves 2.5 cm more with cycloidal motion (C-2) to complete its displacement. The follower dwells and returns 2.5 cm with cycloidal motion (C-3) and then moves the remaining 2.5 cm with harmonic motion (H-4) in $\pi/4$ rad. (*a*) Find the intervals of cam rotation for the cycloidal motions and the dwell by matching velocities and accelerations. (*b*) Determine the equation for S as a function of θ for each type of motion. These equations should be written so that the displacement measured from the zero position can be calculated for any cam angle by use of the proper equation.

3.36 In the displacement diagram of Fig. 3.39 (see p. 90) it is desired to achieve the full lift of 1.5 in. of a radial flat-faced follower by matching the cycloidal curve C-1 with the harmonic curve H-2. (*a*) Using the data given in the diagram, determine the angle β_2 for the harmonic event so that both velocity and acceleration of the follower will be matched at B where the two events are joined. (*b*) Determine the maximum theoretical length of follower face needed for the two events shown.

3.37 A disk cam drives a radial flat-faced follower with simple harmonic motion. The follower moves out and back in one revolution of the cam. If the total displacement is 5.0 cm and the minimum radius 2.5 cm, determine the parametric (x and y) equations

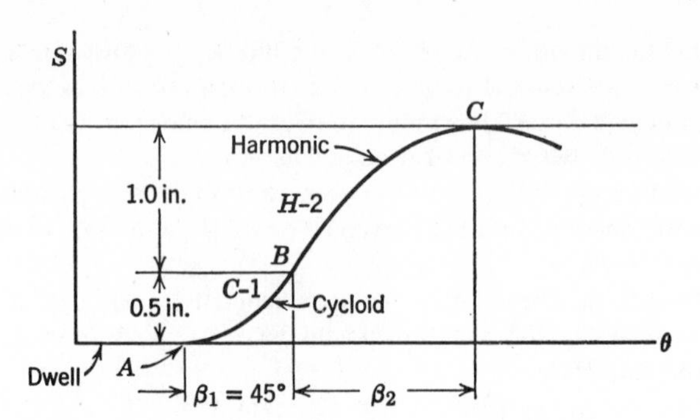

Figure 3.39

of the cam contour. Eliminate the parameter to obtain the equation of the curve, which is the cam contour. Determine the theoretical length of the follower face.

3.38 A flat-faced radial follower is driven through a total displacement of 1.6 in. The follower moves upwards 0.40 in. with constant acceleration for 60°, 0.80 in. with constant velocity for 60°, and the remaining 0.40 in. with constant deceleration for another 60° of cam rotation. The follower dwells for 45° and returns with simple harmonic motion as the cam completes its revolution.

For each type of motion write an equation expressing displacement S as a function of cam angle θ. These equations should be written so that the displacement, measured from the zero position, can be calculated for any cam angle by the use of the proper equation. Calculate the minimum radius C and the maximum length of contact l_{max} for each type of motion. Specify the minimum radius of the cam and the length of the follower face.

3.39 A flat-faced radial follower is driven through a total displacement of 3.8 cm. The follower moves upwards 2.5 cm with constant acceleration for 120° and the remaining 1.3 cm with constant deceleration for 60° of cam rotation. The follower returns with simple harmonic motion in 90° and dwells for the remainder of the revolution of the cam. Complete the solution as outlined in Problem 3.38.

3.40 In the sketch shown in Fig. 3.40 the disk cam is used to position the radial flat-faced follower in a computing mechanism. The cam profile is to be designed to give a follower displacement S for a counterclockwise cam rotation θ according to the function $S = k\theta^2$ starting from dwell. For 60° cam rotation from the starting position, the lift of the follower is 1.0 cm. By analytical methods determine the distances R and l when the cam has been turned 45° from the starting position. Also calculate whether cusps in the cam profile would occur in the total cam rotation of 60°.

3.41 A radial roller follower is driven through a total displacement of 2.5 cm with simple harmonic motion in a half revolution of the cam. The return motion is the same in a half revolution of the cam. Using a minimum radius R_0 of the pitch surface of 3.8 cm and a roller diameter of 1.9 cm, calculate a set of lift figures for the center of the follower for 15° increments of cam angle and lay out the cam full size. Calculate pressure angles to determine contact points.

3.42 A radial roller follower moves through a total displacement of 5.0 cm with cycloidal motion in 180° of cam rotation. The follower dwells for the next 90° and then returns 5.0 cm with cycloidal motion in 90° of cam rotation. Using a minimum radius R_0 of the pitch surface of 2.5 cm, calculate with a computer the displacement, velocity, acceleration, and pressure angle for the follower for each 10° rotation of the cam.

3.43 A radial roller follower is to move through a total displacement of $L =$ 0.75 in. with harmonic motion while the cam rotates $\beta = 30°$. Check the cam for pointing if the radius R_r of the roller is 0.25 in. and the minimum radius R_0 of the pitch surface is 1.875 in.

3.44 A radial roller follower is to move through a total displacement of $L =$ 0.65 cm with harmonic motion while the cam rotates $\beta = 45°$. The radius R_r of the roller is 0.65 cm. Determine the limiting R_0 that will give a pointed cam profile during this event.

3.45 A radial roller follower moves through a total displacement of $L = 0.75$ in. with cycloidal motion while the cam rotates $\beta = 30°$. Determine the radius of curvature ρ of the pitch surface when $\theta = 15°$. The radius R_r of the roller is 0.25 in., and R_0 is 1.875 in.

3.46 A radial roller follower is to move through a total displacement of $L =$ 1.9 cm with harmonic motion while the cam rotates $\beta = 30°$. Find the value of R_0 to limit α_{max} to 30°.

3.47 By use of Eq. 3.12 and the appropriate expressions for R and $dR/d\theta$, develop the equation for α for cycloidal motion. Using the data from Example 3.5, calculate the pressure angle α when $\theta = 22.5°$.

3.48 A radial roller follower is to move through a total displacement of $L =$ 1.6 cm with cycloidal motion while the cam rotates $\beta = 30°$. Assuming $R_0 = 38$ cm, determine α_{max}. If α_{max} is too large and if space requirements dictate that R_0 cannot be increased, make other recommendations to limit α_{max} to 30°.

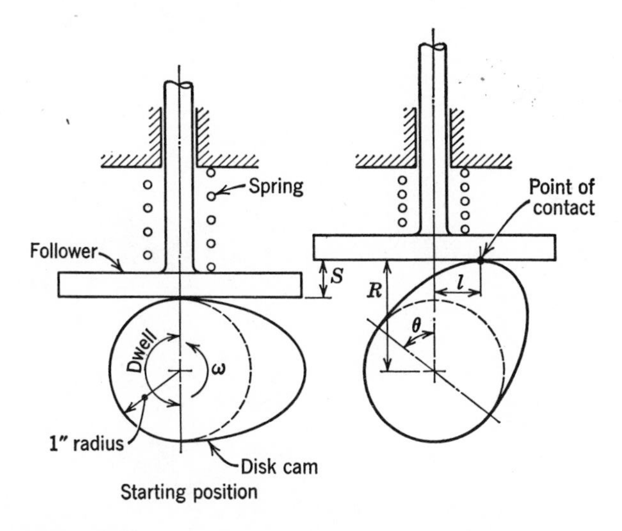

Figure 3.40

3.49 Using the displacement figures of Problem 3.5, calculate the values of R and ϕ for a disk cam with oscillating roller follower. The cam is to rotate counterclockwise and has a minimum radius of 1 in. The diameter of the roller is to be $\frac{3}{4}$ in. and the distance from the center of the hub of the follower to the center of the roller is $2\frac{7}{8}$ in. The center of the hub is 3 in. to the right of the center of the cam. Let the zero position of the follower fall on the vertical centerline of the cam. Lay out the cam full size from the calculated values of R and ϕ, and check it graphically.

3.50 In the preceding problem (3.49) $\psi = 0.174\ (1 - \cos\theta)$ rad approximately. Using this relation, calculate the pressure angle at position 3.

3.51 Using the relation for ψ as a function of θ as given in Problem 3.50 and data from Problem 3.49, calculate the pressure angle for position 0 and check graphically.

3.52 Using the relation for ψ as a function of θ from Problem 3.50 and data from Problem 3.49, calculate the radius of curvature for position 2.

3.53 A three-dimensional cam is to be designed to solve the equation $Q = 0.05ah^{1/2}$, where Q is flow (ft^3/sec), a is the orifice area ($in.^2$), and h is the pressure head (ft). The area a is to be set into the cam by rotation and the head h by translation of the cam. The follower displacement or lift will determine Q. (*a*) Calculate a set of lift figures Q where a varys from 1.0 to 1.5 by tenths and h = 1, 4, 9, 16, 25, 36, 49. (*b*) Using a scale of 1 in. = 0.20 ft^3/sec for Q and 1 in. = 8 ft for h, construct a vertical axial section through the cam. Let the position of a = 1.0 be at the top of the cam. (*c*) Construct a transverse section through the cam at h = 25 ft and at h = 49 ft.

4

Spur Gears

4.1 Introduction to Involute Spur Gears. In considering two curved surfaces in direct contact, it has been shown that the angular velocity ratio is inversely proportional to the segments into which the line of centers is cut by the line of action or common normal to the two surfaces in contact. If the line of action always intersects the line of centers at a fixed point, then the angular velocity ratio remains constant. This is the condition that is desired when two gear teeth mesh together: The angular velocity ratio must be constant. It is possible to assume the form of the tooth on one gear and by applying the above principle (the common normal intersects the line of centers at a fixed point), to determine the outline of the mating tooth. Such teeth would be considered *conjugate teeth*, and the possibilities are limited only by one's ability to form the teeth. Of the many shapes possible, only the cycloid and the involute have been standardized. The cycloid was used first but has been replaced by the involute for all applications except watches and clocks. The involute has several advantages, the most important of which are its ease of manufacture and the fact that the center distance between two involute gears may vary without changing the velocity ratio. The involute system of gearing is discussed in detail in the following paragraphs. A pair of involute spur gears is shown in Fig. 4.1.

Consider two pulleys connected by a crossed wire as shown in Fig. 4.2. It is evident that the two pulleys will turn in opposite directions and that the angular velocity ratio will be constant provided the wire does not slip

Figure 4.1 Spur gears. (Courtesy of Philadelphia Gear Works.)

and will depend upon the inverse ratio of diameters. It is also seen that the angular velocity ratio will not change when the center distance is changed. For convenience, assume that one side of the wire is removed and a piece of cardboard is attached to wheel 1 (Fig. 4.3*a*). Place a pencil at a point Q on the wire and turn wheel 2 counterclockwise. Relative to the ground, point Q will trace a straight line, whereas relative to wheel 1, Q will trace an involute on the cardboard. The same involute could be generated by cutting the wire at Q and unwrapping the wire from the wheel 1, keeping the wire taut. If a cardboard is now attached to wheel 2 (Fig. 4.3*b*) and the process is repeated, an involute is generated on the cardboard of wheel 2. If the cardboards are now cut along the involute, one side of a tooth is formed on both wheels 1 and 2. The involute on wheel 1 can now be used to drive the involute on wheel 2. The angular velocity ratio will be constant because the line of action, which by the method of constructing the involute is normal to the involutes at the point of contact Q, cuts the line of centers at a fixed point. As in the case of the pulleys with the crossed wire, the angular velocity ratio is inversely proportional to the diameters of the wheels. If the center distance is changed, involute 1 will still drive involute 2, but a different

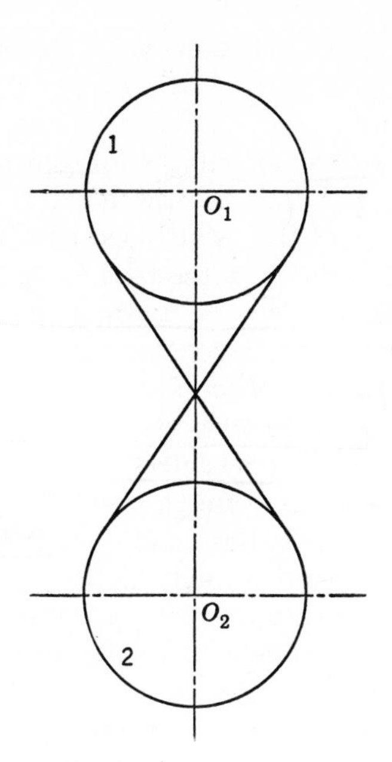

Figure 4.2

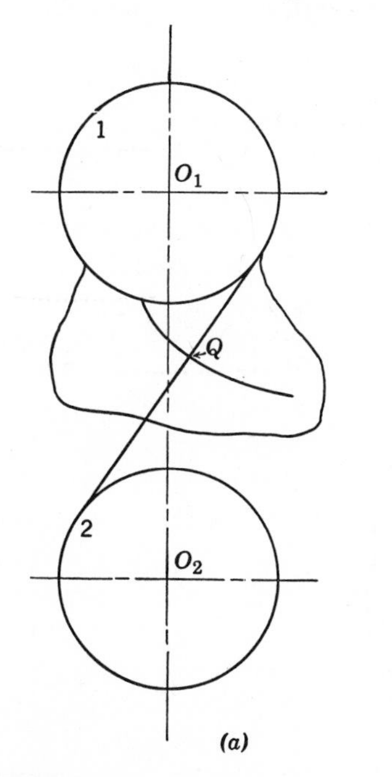

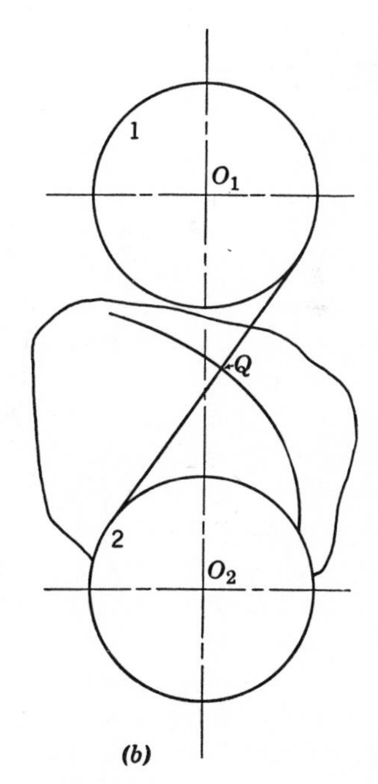

Figure 4.3

portion of the two involutes will now be in contact from that originally. As long as the diameters of the wheels are not changed, the velocity ratio will be the same as before.

The circles that were used as a basis for generating the involutes are known as *base circles*, and they are the heart of the involute system of gearing. In Fig. 4.4 the angle that is included by a line perpendicular to the line of action through the center of the base circle and a line from O_1 to Q (or O_2 to Q) is known as the *involute pressure angle* and is a dimension of the point on the involute at which contact is taking place. If in Fig. 4.4 the point of intersection of the line of action and the line of centers is labeled P, the angular velocity ratio will be inversely proportional to the segments into which this point divides the line of centers.

It is possible to draw circles through point P using first O_1 as a center and then O_2. Figure 4.5 shows this condition. Point P is called the pitch point, and the circles which pass through this point are known as *pitch circles*. It can be proved that, as involute 1 drives involute 2, the two pitch circles will move together with pure rolling action. Because the segments into which point P divides the line of centers are now the radii of the pitch circles, the angular velocity ratio is inversely proportional to the radii of the two pitch circles. If the diameter of pitch circle 1 is D_1 and that of circle 2 is D_2, then $\omega_1/\omega_2 = D_2/D_1$. It will be shown in a later section that the number of teeth

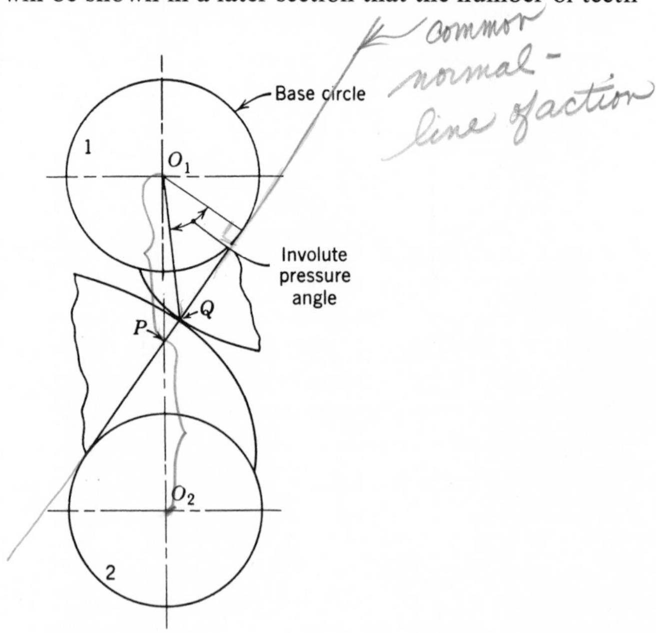

Figure 4.4

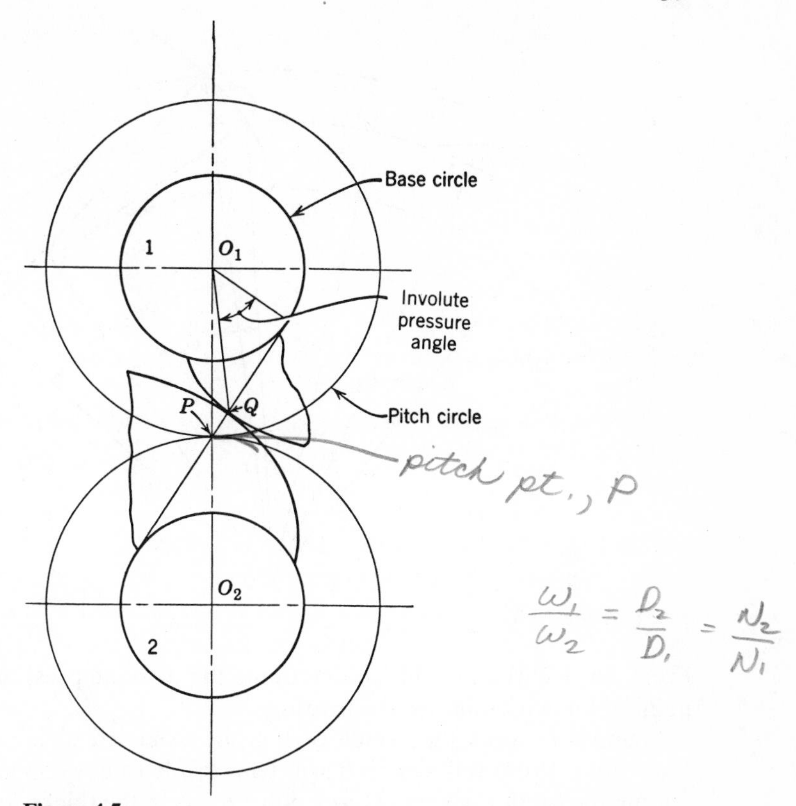

Figure 4.5

on a gear is directly proportional to the pitch diameter. Therefore, $\omega_1/\omega_2 = D_2/D_1 = N_2/N_1$.

4.2 Involutometry. In considering the involute for a tooth form, it is necessary to be able to calculate certain properties of the involute.

Figure 4.6 shows an involute which has been generated from a base circle of radius R_b. The involute contains two points A and B with corresponding radii R_A and R_B and involute pressure angles ϕ_A and ϕ_B. It is an easy matter to work out a relationship for the above factors because the base circle radius remains constant no matter which point is under consideration. Therefore,

$$R_b = R_A \cos \phi_A \tag{4.1}$$

or

$$R_b = R_B \cos \phi_B$$

and

$$\cos \phi_B = \frac{R_A}{R_B} \cos \phi_A \tag{4.2}$$

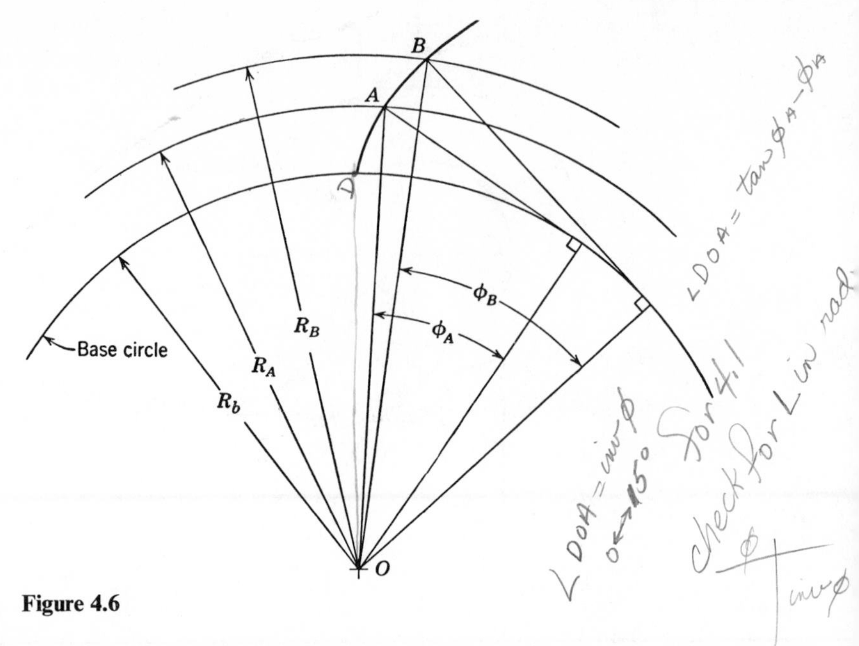

Figure 4.6

From Eq. 4.2 it is possible to determine the involute pressure angle at any point of known radius on the involute.

Figure 4.7 shows the sketch of Fig. 4.6 extended to include the whole gear tooth. From this sketch it will be possible to develop an equation for finding the tooth thickness at any point *B*, given the thickness at point *A*.

From the principle of the generation of an involute, arc *DG* is equal to length *BG*. Therefore,

$$\angle DOG = \frac{\text{arc } DG}{OG} = \frac{BG}{OG}$$

$$\tan \phi_B = \frac{BG}{OG}$$

Thus,

$$\angle DOG = \tan \phi_B$$

Also

$$\angle DOB = \angle DOG - \phi_B$$

$$= \tan \phi_B - \phi_B$$

It can also be shown that

$$\angle DOA = \tan \phi_A - \phi_A$$

The expression $(\tan \phi - \phi)$ is called an *involute function* and is sometimes written inv ϕ. It is easy to calculate the involute function when the angle is known; ϕ is expressed in radians. However, it is very difficult to convert from inv ϕ to ϕ, and for this reason extensive tables of involute functions have been published. See Appendix 1.

Referring again to Fig. 4.7,

$$\angle DOE = \angle DOB + \frac{\frac{1}{2}t_B}{R_B}$$

$$= \text{inv}\ \phi_B + \frac{t_B}{2R_B}$$

Also

$$\angle DOE = \angle DOA + \frac{\frac{1}{2}t_A}{R_A}$$

$$= \text{inv}\ \phi_A + \frac{t_A}{2R_A}$$

From the above relations,

$$t_B = 2R_B\left[\frac{t_A}{2R_A} + \text{inv}\ \phi_A - \text{inv}\ \phi_B\right] \tag{4.3}$$

By means of Eq. 4.3 it is possible to calculate the tooth thickness at any point on the involute, given the thickness at some other point. An interesting application of this equation is to determine the radius at which the tooth becomes pointed.

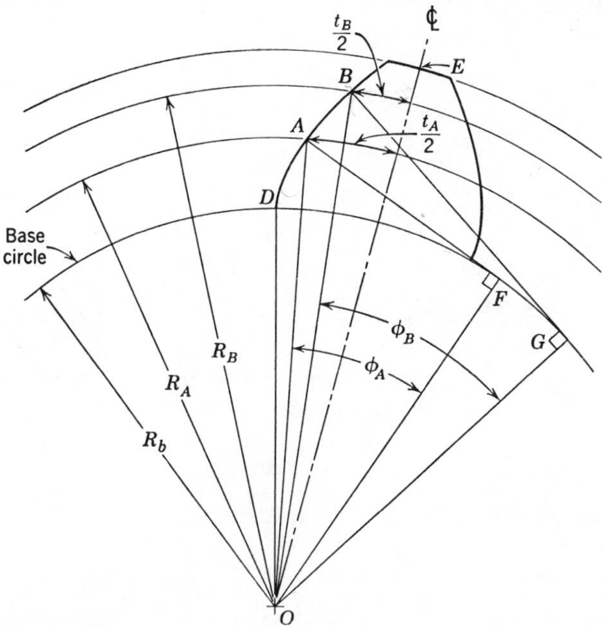

Figure 4.7

4.3 Spur Gear Details. In order to continue the study of involute gearing, it is necessary to define the basic elements of a gear as shown in Figs. 4.8*a* and *b*. It should also be mentioned that the smaller of two gears in mesh is called the *pinion*; the pinion is generally the driver. If the pitch radius R of a gear becomes infinite, a rack results as seen in Figs. 4.8*c* and 4.9. The side of the rack tooth is a straight line, which is the form taken by an involute when it is generated on a base circle of infinite radius. From Fig. 4.8*a*, the *base pitch* p_b is the distance from a point on one tooth to the corresponding point on the next tooth measured on the base circle. The *circular pitch* p is defined in the same way except that it is measured on the pitch circle. The *addendum* a and *dedendum* b are radial distances measured as shown. The portion of the flank below the base circle is approximately a radial line. The tooth curve is the line of intersection of the tooth surface and the pitch surface.

Although it is impossible to show on the sketches of Fig. 4.8, backlash is an important consideration in gearing. *Backlash* is the amount by which the width of a tooth space exceeds the thickness of the engaging tooth on the pitch circles. Theoretically, backlash should be zero, but practically some allowance must be made for thermal expansion and tooth error. Unless otherwise stated, zero backlash is to be assumed in this text. In a later section, the method for calculating backlash for a change in center distance will be given.

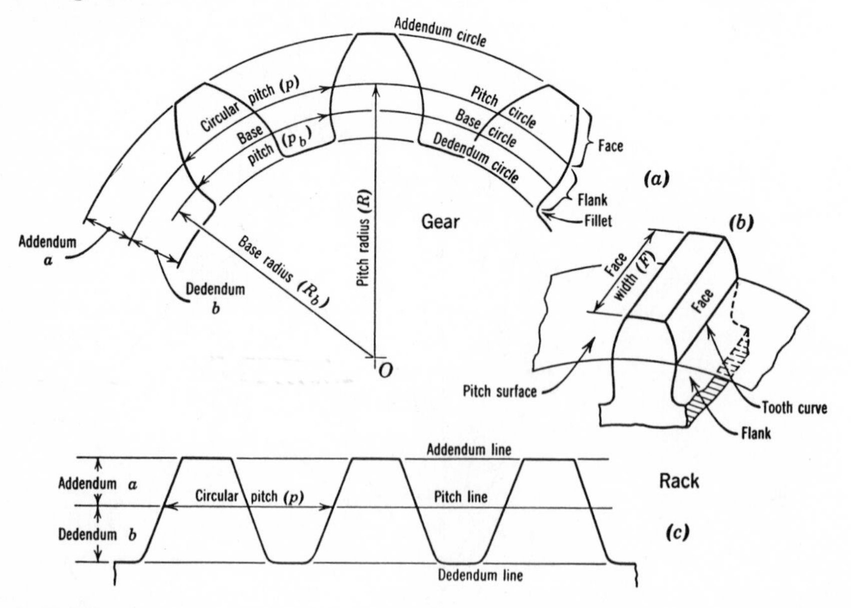

Figure 4.8

Figure 4.9 Spur pinion and rack. (Models courtesy of Illinois Gear & Machine Company.)

4.4 Characteristics of Involute Action. From the discussion of the generation of the involute, it was seen that the common normal to the two involute surfaces is tangent to the two base circles. This common normal is also referred to as the *line of action*. The beginning of contact occurs where the line of action intersects the addendum circle of the driven gear, and the end of contact where the line of action intersects the addendum circle of the driver. That this occurs is evident from Fig. 4.10, which shows a pair of teeth just coming into contact and the same pair as they later go out of contact (shown dotted). Point A is the beginning of contact and point B the end of contact. The path of the point of contact is therefore along the straight line APB. Point C is where the tooth profile (gear 1) at the beginning of contact cuts the pitch circle. Point C' is where the profile at the end of contact cuts the pitch circle. Points D and D' are similar points on gear 2. The arcs CC' and DD' are called *arcs of action* and must be equal for pure rolling action of the pitch circles to take place as mentioned earlier. The angles of motion are generally broken into two parts as shown in Fig. 4.10, where α is the *angle of approach* and β the *angle of recess*. The angle of approach does not in general equal the angle of recess. For continuous driving to take place, the arc of action must be equal to or greater than the circular pitch. If this is true, then a new pair of teeth will come into action before the preceding pair goes out of action.

The ratio of the arc of action to the circular pitch is known as the *contact ratio*. The contact ratio for involute gears is also equal to the ratio of the *length of action* (that is, the distance from the beginning to the end of contact measured on the line of action) to the base pitch and is generally calculated in this way as will be shown later. Considered physically the contact ratio is the average number of teeth in contact. If, for example, the ratio is 1.60, it does not mean that there are 1.60 teeth in contact. It means that there are alternately one pair and two pairs of teeth in contact, and

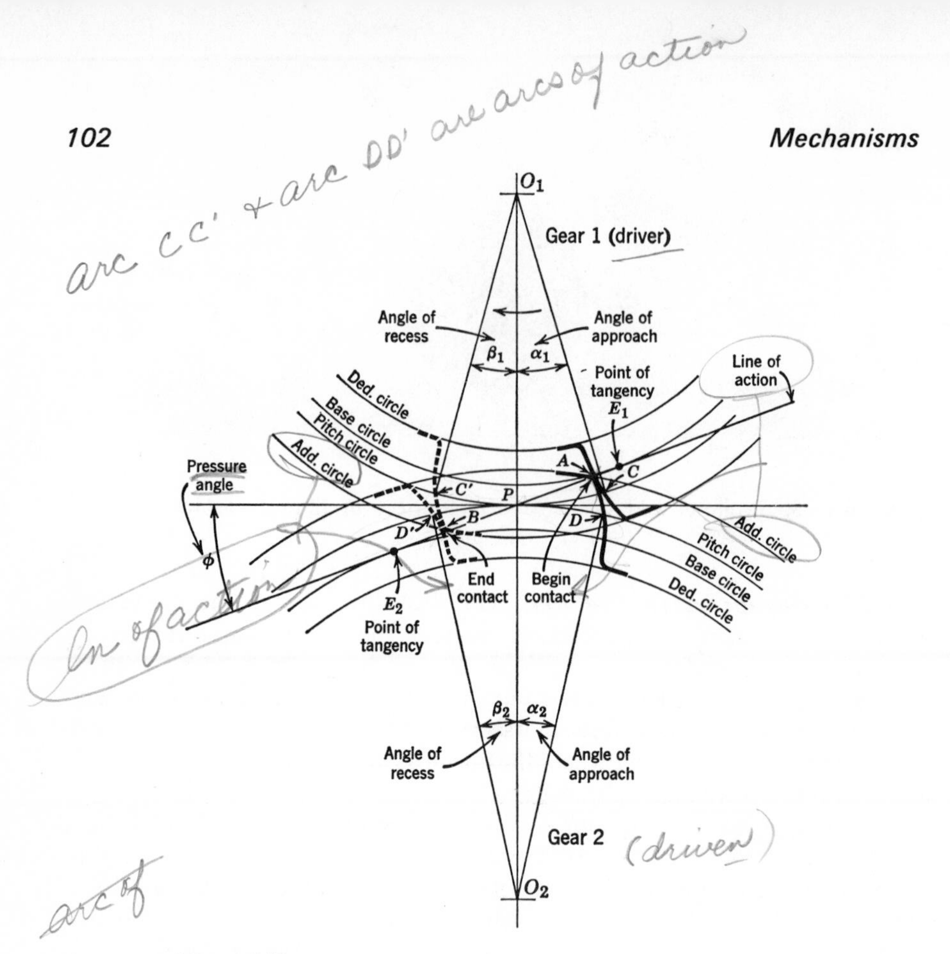

Figure 4.10

on a time basis the number averages 1.60. The theoretical minimum value of the contact ratio is 1.00. This value, of course, must be increased for actual operating conditions. Although it is difficult to quote specific values because of the many conditions involved, 1.40 has been used as a practical minimum with 1.20 for extreme cases. It should be noted, however, that the lower the contact ratio, the higher the degree of accuracy needed in machining the profiles to secure quiet running.

Figure 4.10 also shows an angle ϕ, which is formed by the line of action and a line perpendicular to the line of centers at the pitch point P. This angle is known as the *pressure angle of the two gears in mesh* and must be differentiated from the involute pressure angle of a point on an involute. When the two gears are in contact at the pitch point, the pressure angle of the gears in mesh and the involute pressure angles of the two involutes in contact at the pitch point will be equal. These angles can be seen in Fig. 4.11.

An equation for the length of action Z can be derived from Fig. 4.11,

where

A = beginning of contact
B = end of contact
E_1 and E_2 = points of tangency of line of action and base circles
R_O = outside radius
R_b = base radius
ϕ = pressure angle
C = center distance

From the figure,

$$Z = AB = E_1B + E_2A - E_1E_2$$

Therefore,

$$Z = \sqrt{(R_{O_1})^2 - (R_{b_1})^2} + \sqrt{(R_{O_2})^2 - (R_{b_2})^2} - C \sin \phi \qquad (4.4)$$

The base pitch p_b is given by

$$p_b = \frac{2\pi R_b}{N} \qquad (4.5)$$

where

R_b = base radius
N = number of teeth

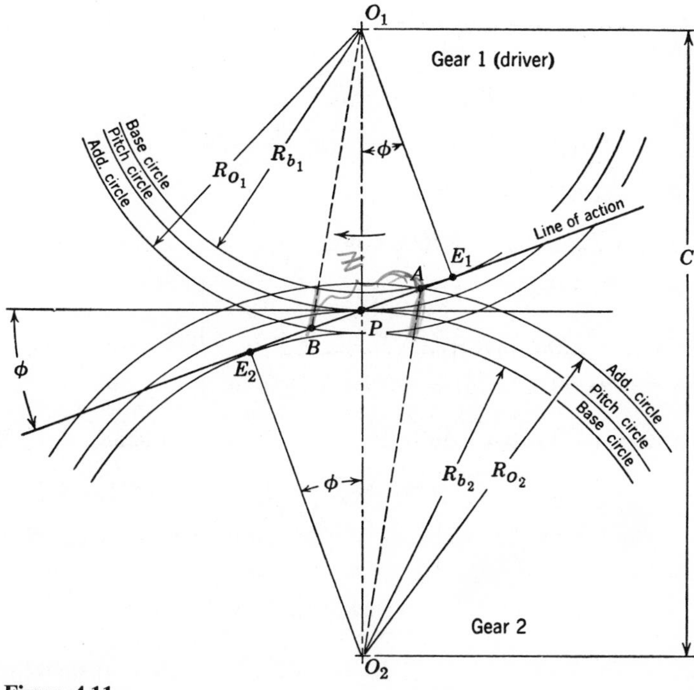

Figure 4.11

The contact ratio m_p is then

$$m_p = \frac{Z}{p_b} \tag{4.6}$$

If it seems odd to calculate contact ratio by dividing a straight-line measurement by a circular measurement, consider the sketches in Fig. 4.12. In Fig. 4.12*a* is shown two adjacent teeth on one gear of a mating pair. The base pitch p_b is dimensioned on the base circle in accordance with its definition. A straight-line distance on the line of action is also designated p_b. From the manner in which two adjacent involutes would be generated, it can be seen that the two distances marked p_b must be equal. Therefore, the base pitch can also be considered as the normal distance between the corresponding sides of adjacent teeth. Figure 4.12*b* illustrates how the base pitch is measured on a rack.

Example 4.1

A pinion of 24 teeth drives a gear of 60 teeth at a pressure angle of 20°. The pitch radius of the pinion is 1.5000 in., and the outside radius is 1.6250 in. The pitch radius of the gear is 3.7500 in., and the outside radius 3.8750 in. Using the sketches in Figs. 4.10 and 4.11, calculate the length of action, contact ratio, and angles of approach and recess for the pinion and gear.

Solution. From Fig. 4.11,

$$Z = AB = E_1B + E_2A - E_1E_2$$

$$Z = \sqrt{(R_{O_1})^2 - (R_{b_1})^2} + \sqrt{(R_{O_2})^2 - (R_{b_2})^2} - C \sin\phi$$

$$R_{O_1} = 1.6250 \text{ in.}$$

$$R_{b_1} = R_1 \cos\phi = 1.5000 \cos 20° = 1.4095 \text{ in.}$$

$$R_{O_2} = 3.8750 \text{ in.}$$

$$R_{b_2} = R_2 \cos\phi = 3.75 \cos 20° = 3.5238 \text{ in.}$$

$$C \sin\phi = (1.50 + 3.75) \sin 20° = 1.7956 \text{ in.}$$

$$Z = \sqrt{1.6250^2 - 1.4095^2} + \sqrt{3.8750^2 - 3.5238^2} - 1.7956$$

$$= \sqrt{2.6406 - 1.9867} + \sqrt{15.0156 - 12.4172} - 1.7956$$

$$= 0.8099 + 1.6115 - 1.7956 = 0.6258 \text{ in.}$$

Therefore,

$$Z = AB = 0.6258 \text{ in.}$$

$$m_p = \frac{Z}{p_b} \quad \text{and} \quad p_b = \frac{2\pi R_{b_1}}{N_1} = \frac{2\pi \times 1.4095}{24} = 0.3689 \text{ in.}$$

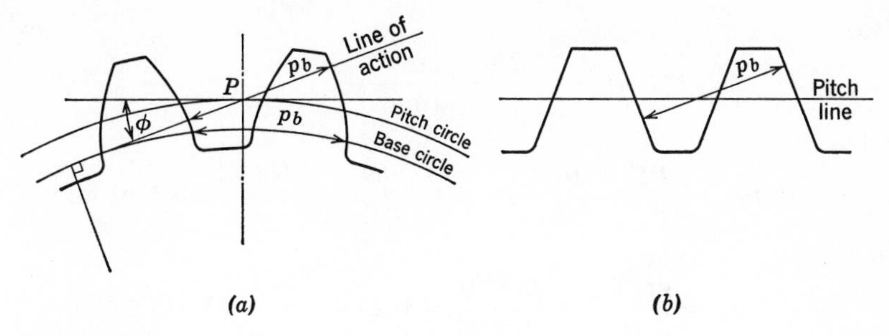

Figure 4.12

Therefore,

$$m_p = \frac{0.6258}{0.3689} = 1.6964$$

From the above calculations,

$$E_1B = \sqrt{(R_{O_1})^2 - (R_{b_1})^2} = 0.8099 \text{ in.}$$
$$E_1A = E_1B - AB = 0.8099 - 0.6258 = 0.1841 \text{ in.}$$
$$E_1P = R_1 \sin \phi = 1.5000 \sin 20° = 0.5130 \text{ in.}$$
$$AP = E_1P - E_1A = 0.5130 - 0.1841 = 0.3289 \text{ in.}$$
$$PB = AB - AP = 0.6258 - 0.3289 = 0.2969 \text{ in.}$$

The contact ratio m is also equal to the arc of action CC' divided by the circular pitch p.

$$m_p = \frac{\text{arc } CC'}{p} \quad \text{and} \quad p = \frac{2\pi R_1}{N_1} = \frac{2\pi \times 1.5000}{24} = 0.3927 \text{ in.}$$

Therefore,

$$\text{arc } CC' = p \times m_p = 0.3987 \times 1.6964 = 0.6662 \text{ in.}$$

From Fig. 4.10 it is known that arc DD' must equal arc CC' so that arc DP = arc CP and arc PD' = arc PC'. The arc of approach CP of gear 1 can be found from the following ratio:

$$\frac{AP}{AB} = \frac{\text{arc } CP}{\text{arc } CC'}$$

Therefore,

$$\text{arc } CP = \frac{AP \times \text{arc } CC'}{AB} = \frac{0.3289 \times 0.6662}{0.6258} = 0.3501 \text{ in.}$$

Also,

$$\frac{PB}{AB} = \frac{\text{arc } PC'}{\text{arc } CC'}$$

so that

$$\text{arc } PC' = \frac{PB \times \text{arc } CC'}{AB} = \frac{0.2969 \times 0.6662}{0.6258} = 0.3161 \text{ in.}$$

Therefore,

$$\alpha_1 = \frac{\text{arc } CP}{R_1} = \frac{0.3501}{1.5000} = 0.2334 \text{ rad} = 13.373°$$

$$\alpha_2 = \frac{\text{arc } DP}{R_2} = \frac{0.3501}{3.7500} = 0.0934 \text{ rad} = 5.349°$$

$$\beta_1 = \frac{\text{arc } PC'}{R_1} = \frac{0.3161}{1.5000} = 0.2107 \text{ rad} = 12.074°$$

$$\beta_2 = \frac{\text{arc } PD'}{R_2} = \frac{0.3161}{3.7500} = 0.0843 \text{ rad} = 4.829°$$

As a check,

$$\alpha_1 + \beta_1 = \frac{\text{arc } CC'}{R_1} = \frac{0.6662}{1.5000} = 0.4441 \text{ rad} = 25.447°$$

$$\alpha_2 + \beta_2 = \frac{\text{arc } DD'}{R_2} = \frac{0.6662}{3.7500} = 0.1777 \text{ rad} = 10.179°$$

Therefore,

$$\alpha_1 + \beta_1 = 13.373° + 12.074° = 25.447°$$
$$\alpha_2 + \beta_2 = 5.349° + 4.829° = 10.178°$$

It is also possible to calculate the angles of approach and recess using involutometry. The equation for the angle of approach α_2 of gear 2 is derived as follows using the sketch of Fig. 4.13.

$$\alpha_2 = \theta + \phi_D - \phi$$

where

$$\begin{aligned}\theta &= (\phi_A + \text{inv } \phi_A) - (\phi_D + \text{inv } \phi_D)\\ &= (\phi_A + \tan \phi_A - \phi_A) - (\phi_D + \tan \phi_D - \phi_D)\\ &= \tan \phi_A - \tan \phi_D\end{aligned}$$

Making the substitution for θ

$$\alpha_2 = \tan \phi_A - \tan \phi_D + \phi_D - \phi$$

From the fact that point D is a point on the involute at the pitch circle,

$$\phi_D = \phi$$

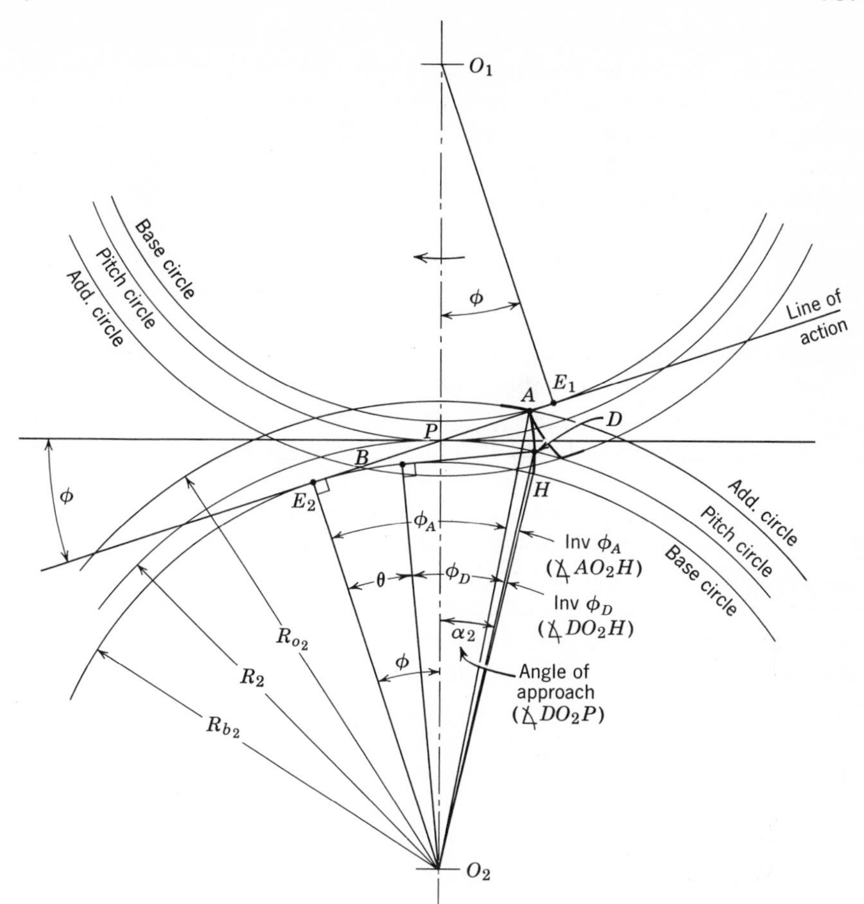

Figure 4.13

Therefore,

$$\alpha_2 = \tan \phi_2 - \tan \phi$$

Equations for α_1, β_1, and β_2 can be developed in a similar manner using the appropriate sketches.

4.5 Interference in Involute Gears. It was mentioned previously that an involute starts at the base circle and is generated outward. It is therefore impossible to have an involute inside the base circle. The line of action is tangent to the two base circles of a pair of gears in mesh, and these two points represent the extreme limits of the length of action. These two points are spoken of as *interference points*. If the teeth are of such proportion that the beginning of contact occurs before the interference point is met, then the involute portion of the driven gear will mate with a noninvolute portion of the driving gear, and *involute interference* is said to occur. This condition is shown in Fig. 4.14. E_1 and E_2 show the interference points that should limit the length of action. A shows the beginning of contact and B the end of contact. It is seen that the beginning of contact occurs before the interference point E_1 is met; therefore, interference is present. The tip of the driven tooth will gouge out or undercut the flank of the driving tooth as shown by the dotted line. There are several ways of eliminating interference, one of which is to limit the addendum of the driven gear so that it passes through the interference point E_1, thus giving a new beginning of contact. If this is done in this case, interference will be eliminated.

Involute interference is undesirable for several reasons. Interference and the resulting undercutting not only weaken the pinion tooth but may also remove a small portion of the involute adjacent to the base circle, which may cause a serious reduction in the length of action.

The conditions for interference of a rack and a pinion will now be discussed. In Fig. 4.15 is shown a pinion and rack in mesh. The point of tangency of the line of action and the base circle of the pinion is labeled as the interference point E, the same as in the case for the pinion and gear. The interference point therefore fixes the maximum addendum for the rack for the pressure angle shown. With the rack addendum as shown in Fig. 4.15, contact begins

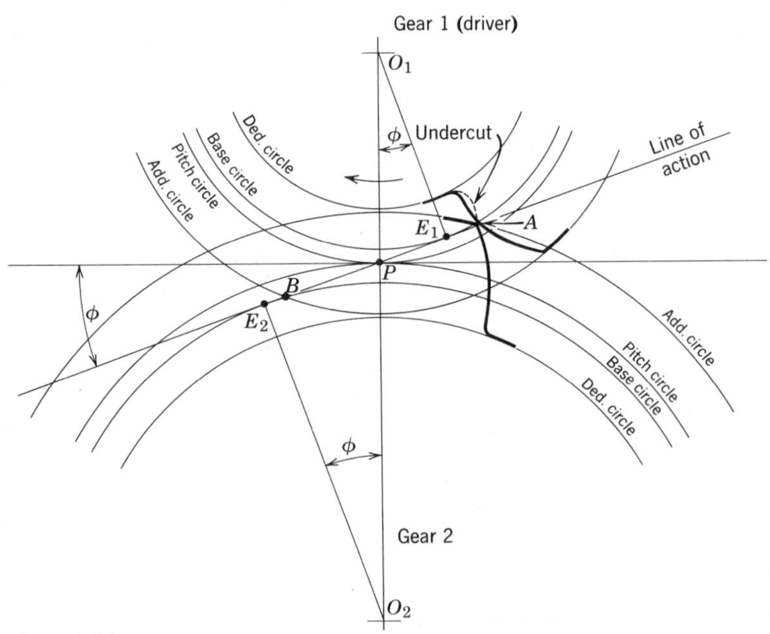

Figure 4.14

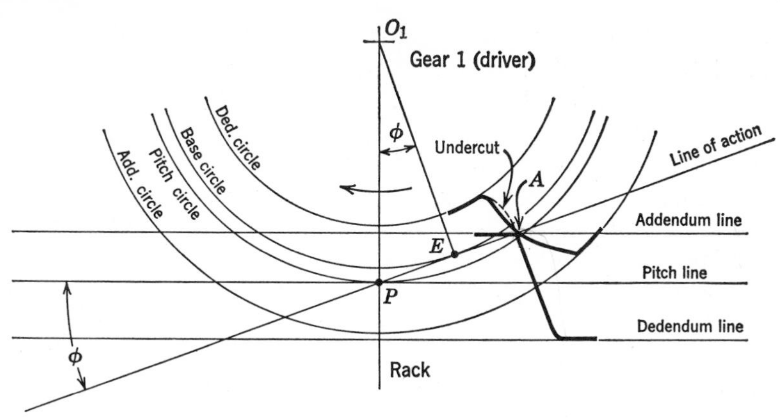

Figure 4.15

at A and undercutting will occur as shown by the dotted line. If the addendum of the rack extends only to the line that passes through the interference point E, then the interference point becomes the beginning of contact and interference is eliminated.

It can also be seen from Fig. 4.15 that, if a gear of finite radius with the same addendum as the rack (rack addendum now passing through interference point) were to mesh with the pinion, the beginning of contact would occur on the line of action some place between the pitch point P and the interference point E. Therefore, there would be no possibility of interference between the pinion and gear. It may be concluded then that, if the number of teeth on the pinion is such that it will mesh with a rack without interference, it will mesh without interference with any other gear having the same or a larger number of teeth.

Although involute interference and its resulting undercutting should be avoided, a small amount might be tolerated if it does not reduce the contact ratio for a pair of mating gears below a suitable value. However, the problem of determining the length of action when undercutting has occurred is difficult, and it cannot be calculated from Eq. 4.4. One method has been developed by Spotts to which the reader is referred.[1] It can be seen from Fig. 4.11 and Eq. 4.4 that if the value of either radical is greater than $C \sin \phi$, interference will exist.

4.6 Interchangeable Gears. So far no attempt has been made to bring up the question of interchangeable gears; the foregoing discussions applied to spur gears in general. Tied in with the question of interchangeability is the manner in which the gears are to be cut. There are several ways

[1] M. F. Spotts, "How to Predict Effects of Undercutting Hobbed Spur Gear Teeth," *Machine Design*, April 19, 1956.

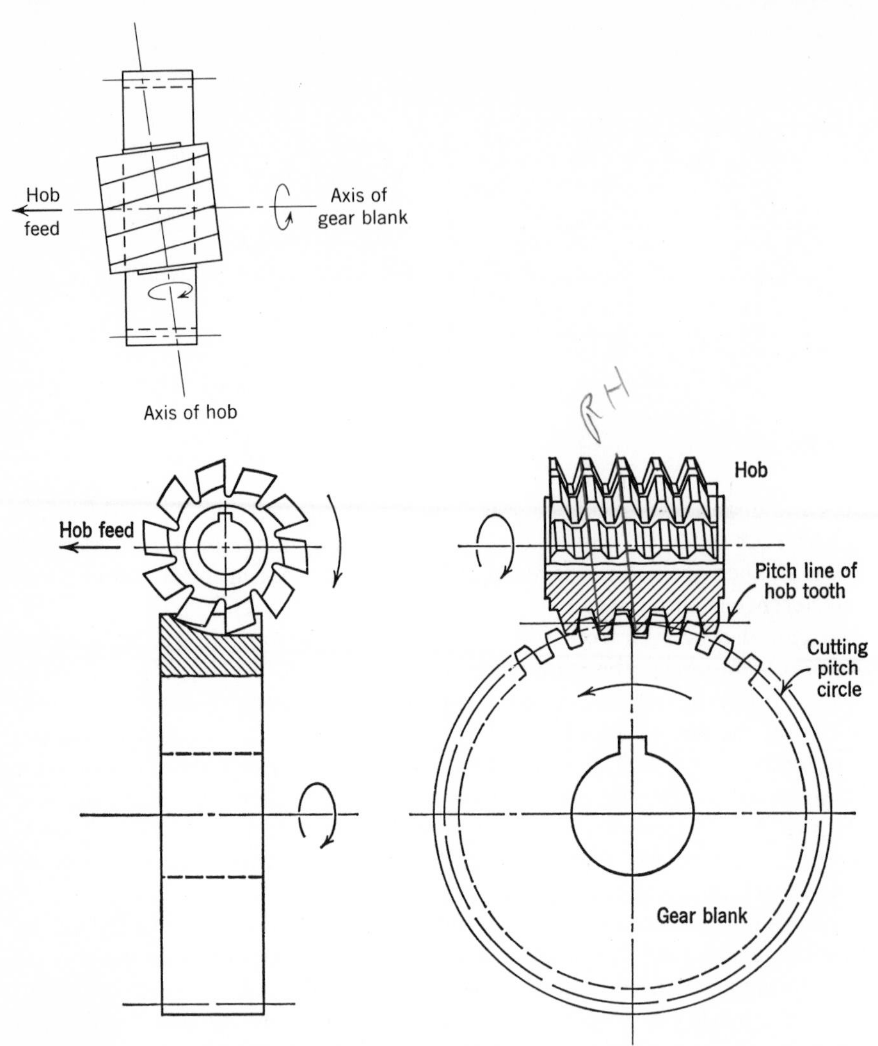

Figure 4.16 Generating a spur gear with a hob

of generating spur gears, the two most common are the method of hobbing and the Fellows method of shaping. The principle of hobbing and that of the Fellows method are illustrated in Fig. 4.16 and Fig. 4.17, respectively. As these methods of cutting were developed, a means of classifying the cutters and the gears which were cut by them was sought. The classification adopted was that of specifying the ratio of number of teeth to the pitch diameter. This ratio was given the name *diametral pitch.* The diametral pitch can be expressed mathematically as follows:

$$P = \frac{N}{D} \tag{4.7}$$

where

N = number of teeth
D = pitch diameter

For the purpose of specifying cutting tools, the values of diametral pitch were taken as whole numbers with certain exceptions. The following are commonly used diametral pitches:

$1, 1\frac{1}{4}, 1\frac{1}{2}, 1\frac{3}{4}, 2, 2\frac{1}{4}, 2\frac{1}{2}, 2\frac{3}{4}, 3, 3\frac{1}{2}, 4, 5, 6, 7,$
$8, 9, 10, 12, 14, 16, 18, 20, 22, 24, 26, 28, 30,$
$48, 64, 72, 80, 96, 120$

Finer pitches may be specified by even increments up to 200. Pitches commonly used for precision gears for instruments are 48, 64, 72, 80, 96, and 120. For tool economy, gears are generally cut using one of the common

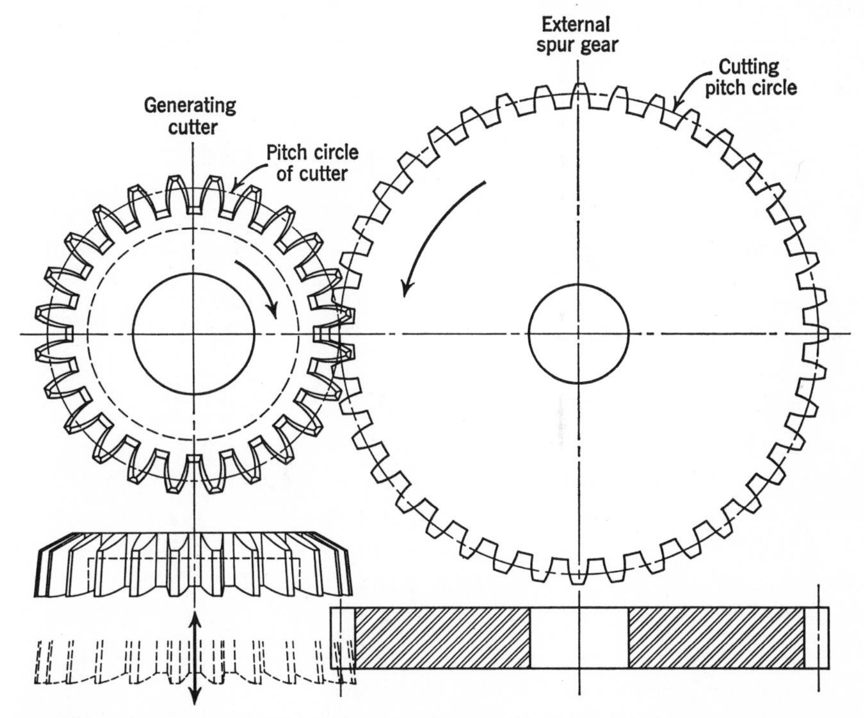

Figure 4.17 Fellows method of gear generating. (Courtesy of Fellows Gear Shaper Company.)

pitches listed above. It is possible to cut gears so that the diametral pitch is not one of the above numbers. This might require a special cutter, but generally it can be done with one of the above tools at a special setting. This will be discussed in Chapter 5.

When cutters were standardized, a pressure angle of $14\frac{1}{2}°$ was adopted. This was a carryover from the process of casting gears which used $14\frac{1}{2}°$ because sin $14\frac{1}{2}°$ approximates $\frac{1}{4}$, which was convenient for pattern layout. Later a pressure angle of 20° was also adopted. Both the $14\frac{1}{2}$ and the 20° have been used for many years, but the tendency in recent years has been toward using 20° in preference to $14\frac{1}{2}°$. It will be shown in a later section that it is possible to have a pinion with fewer teeth without undercutting when using 20° instead of $14\frac{1}{2}°$. As a result of the trend toward higher pressure angles, the AGMA (American Gear Manufacturers Association) has adopted

Table 4.1 TOOTH PROPORTIONS—INVOLUTE SPUR GEARS

	Coarse Pitch (1–19.99 P) AGMA 201.02 August 1968 20° or 25° Full Depth	Fine Pitch (20–200 P) AGMA 207.06 November, 1974 20° Full Depth
Addendum (a)	$\frac{1.000}{P}$	$\frac{1.000}{P}$
Dedendum (b)	$\frac{1.250}{P}$	$\frac{1.200}{P} + 0.002$ (min)
Clearance (c) (dedendum − addendum)	$\frac{0.250}{P}$	$\frac{0.200}{P} + 0.002$ (min)*
Working depth (h_k) (twice addendum)	$\frac{2.000}{P}$	$\frac{2.000}{P}$
Whole depth (h_t) (addendum + dedendum)	$\frac{2.250}{P}$	$\frac{2.200}{P} + 0.002$ (min)
Fillet radius of basic rack (r_f)	$\frac{0.300}{P}$	Not given
Tooth thickness (t)	$\frac{1.5708}{P}$	$\frac{1.5708}{P}$

* For shaved or ground teeth, $c = 0.350/P + 0.002$ (min).

Although the latest AGMA standards are given in Table 4.1, many gears and cutters are available which conform to the older (and now obsolete) ASA Standard B.6-1932. For this reason, the major tooth proportions of these systems are given in Table 4.2.

Table 4.2

	$14\frac{1}{2}°$ Full Depth	20° Full Depth	20° Stub
Addendum (a)	$\frac{1.000}{P}$	$\frac{1.000}{P}$	$\frac{0.800}{P}$
Dedendum (b)	$\frac{1.157}{P}$	$\frac{1.157}{P}$	$\frac{1.000}{P}$
Clearance (c)	$\frac{0.157}{P}$	$\frac{0.157}{P}$	$\frac{0.200}{P}$
Fillet radius (r_f)	$\frac{0.209}{P}$	$\frac{0.239}{P}$	$\frac{0.304}{P}$
Tooth thickness (t)	$\frac{1.5708}{P}$	$\frac{1.5708}{P}$	$\frac{1.5708}{P}$

20° and 25° for coarse-pitch gearing (1–19.99 P) and 20° for fine-pitch (20–200 P).

If gears are cut with standard cutters, it is possible to cut them so that the gears will be interchangeable. For this to be possible certain conditions must be met:

1. The diametral pitches must be the same.
2. The pressure angles must be equal.
3. The gears must have the same addendums and the same dedendums.
4. Tooth thickness must equal one-half the circular pitch.

The circular pitch has been defined as the distance measured along the pitch circle from a point on one tooth to the corresponding point on the next tooth. This may be written mathematically as follows:

$$p = \frac{\pi D}{N} \quad \text{also} \quad pP = \pi \tag{4.8}$$

The term *standard gear* is often used and means that the ratio of the number of teeth to pitch diameter is one of the standard values of diametral pitch, and that the tooth thickness must be equal to the tooth space, which equals one-half of the circular pitch. Standard gears are interchangeable gears. *Interchangeable gears* may be defined as those having the same pressure angle, same pitch, and compatible addendum, dedendum, tooth thickness, and tooth space. The spur gears that are offered for sale in gear manufacturers' catalogs are standard gears. However, a great many nonstandard gears are used, notably in automobiles and in aircraft. The proportions of standard equal addendum involute spur gears are given in Table 4.1 (see Fig. 4.8).

4.7 Minimum Number of Teeth to Avoid Interference. The question of interference was considered previously for the meshing of a pinion and gear and for a pinion and rack. From the discussion of Fig. 4.15, it was found that if there were no interference between a pinion and rack, there would be no interference between the pinion and a gear the same size as the pinion or larger. Naturally this is assuming the same tooth proportions for the two cases. When considering a standard gear where the tooth proportions are those given in the tables, it is possible to calculate the minimum number of teeth in a pinion that will mesh with a rack without involute interference. To solve for this limiting case, the addendum line of the rack is passed through the interference point of the pinion.

In Fig. 4.18, the essential features of a pinion and rack for this case are shown. The pitch point is notated by P and the interference point by E. Therefore,

$$\sin\phi = \frac{PE}{R}$$

also

$$\sin\phi = \frac{a}{PE} = \frac{k/P}{PE}$$

where k is a constant that, when divided by the diametral pitch, gives the addendum ($a = k/P$). For the full-depth system $k = 1.00$, and for the stub system $k = 0.80$. Multiplying the two equations for $\sin\phi$ together gives

$$\sin^2\phi = \frac{k}{RP}$$

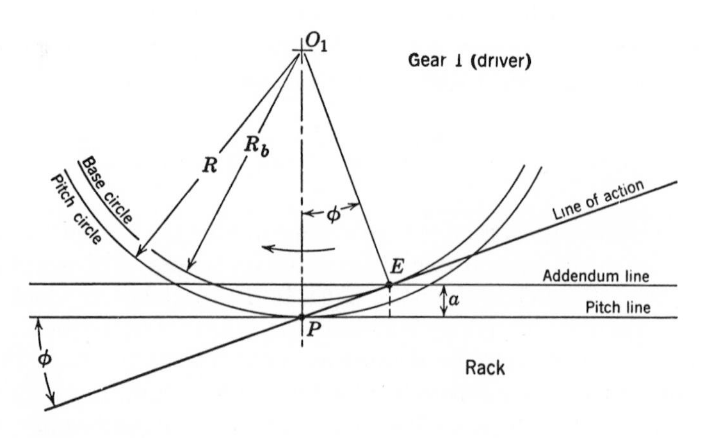

Figure 4.18

But

$$P = \frac{N}{2R} \qquad \text{where } N = \text{number of teeth}$$

Therefore,

$$\sin^2 \phi = \frac{2k}{N}$$

and

$$N = \frac{2k}{\sin^2 \phi} \tag{4.9}$$

From this equation the smallest number of teeth for a pinion that will mesh with a rack without interference can be calculated for any given standard tooth system. These are shown in Table 4.3 for the common systems.

Table 4.3

	$14\frac{1}{2}°$ Full Depth	20° Full Depth	20° Stub	25° Full Depth
N	32	18	14	12

Because these values were calculated for a pinion meshing with a rack, they can also be used as minimums for a pinion meshing with a gear without danger of interference. Because the tooth action of a hob cutting a spur gear is similar to that of a pinion meshing with a rack, the numbers of teeth tabulated above are also the minimums that can be cut by a hob without undercutting.

If the gears are to be produced by some means other than hobbing, for example, Fellows shaping method, the minimum number of teeth that two gears of equal size may have and mesh together without involute interference can be determined from Fig. 4.19. Here the addendum circle of each gear passes through the interference point of the other gear.

$$R_O = R + a \qquad \text{where} \qquad R = \frac{N}{2P} \qquad \text{and} \qquad a = \frac{k}{P}$$

Therefore,

$$R_O = \frac{N}{2P} + \frac{k}{P} = \frac{N + 2k}{2P}$$

$$R_b = R \cos \phi = \frac{N \cos \phi}{2P}$$

$$Z = E_1 E_2 = \sqrt{R_O^2 - R_b^2}$$

$$= \frac{1}{2P} \sqrt{(N + 2k)^2 - (N \cos \phi)^2} \tag{4.10}$$

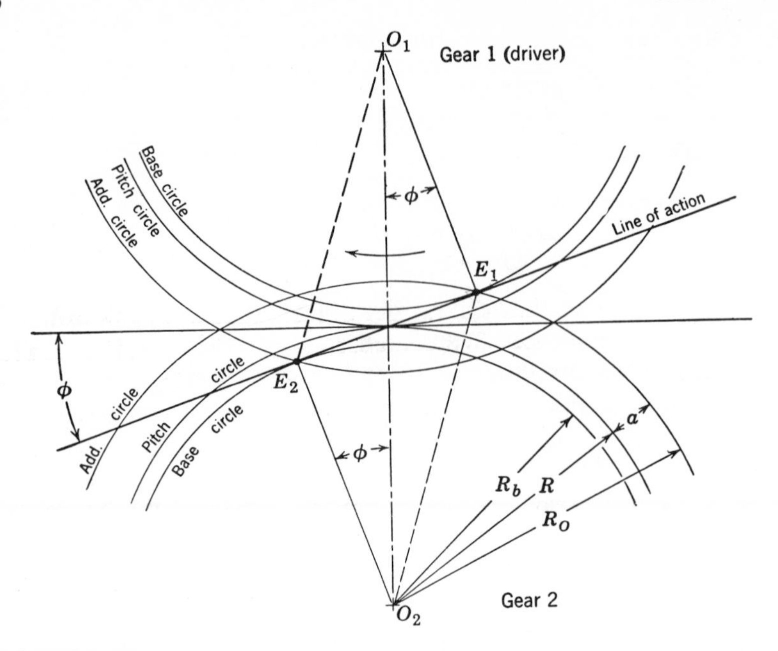

Figure 4.19

Also

$$Z = 2R \sin \phi$$

$$= \frac{2N}{2P} \sin \phi \tag{4.11}$$

Equating Eqs. 4.10 and 4.11,

$$\frac{2N}{2P} \sin \phi = \frac{1}{2P} \sqrt{(N + 2k)^2 - (N \cos \phi)^2}$$

Therefore,

$$3 \sin^2 \phi N^2 - 4kN - 4k^2 = 0 \tag{4.12}$$

From this equation, the smallest number of teeth for two equal gears that will mesh together without involute interference may be determined for any standard tooth system. These values are shown in Table 4.4 for the common systems. The contact ratios (m_p) are also given.

When gears of equal size with the numbers of teeth given in Table 4.4 are cut with a Fellows cutter, they will mesh together without involute interference. If the number of teeth in one gear is held at one of the values given, it is interesting to determine the maximum number the second gear can have without causing interference. It is obvious by comparing the values tabulated

Table 4.4

	$14\frac{1}{2}°$ Full Depth	20° Full Depth	20° Stub	25° Full Depth
N	23 ($m_p = 1.84$)	13 ($m_p = 1.44$)	10 ($m_p = 1.15$)	9 ($m_p = 1.26$)

in Table 4.4 with the minimum number of teeth that will mesh with a rack without interference (Table 4.3) that the second gear cannot approach a rack.

Relations can be developed for this problem from Fig. 4.20, where the addendum circle of gear 2 passes through the interference point of gear 1.

$$R_{O_2} = \sqrt{R_{b_2}^2 + C^2 \sin^2 \phi}$$

Substituting,

$$R_{O_2} = R_2 + a = \frac{N_2}{2P} + \frac{k}{P} = \frac{N_2 + 2k}{2P}$$

$$R_{b_2} = R_2 \cos \phi = \frac{N_2}{2P} \cos \phi$$

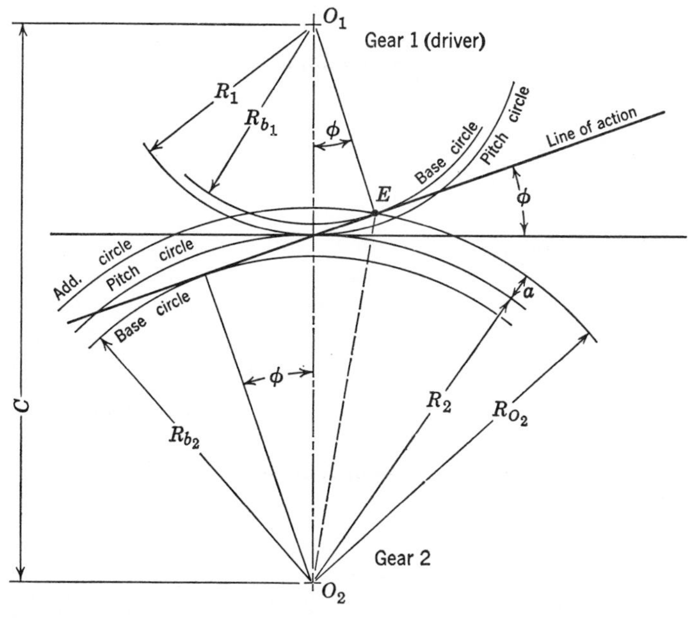

Figure 4.20

Table 4.5

	$14\frac{1}{2}°$ Full Depth	20° Full Depth	20° Stub	25° Full Depth
N_1	23	13	10	9
N_2	26	16	11	13

and

$$C = R_1 + R_2 = \frac{N_1 + N_2}{2P}$$

$$\frac{N_2 + 2k}{2P} = \sqrt{\left(\frac{N_2}{2P}\right)^2 \cos^2 \phi + \left(\frac{N_1 + N_2}{2P}\right)^2 \sin^2 \phi}$$

$$(N_2 + 2k)^2 = N_2^2 \cos^2 \phi + (N_1 + N_2)^2 \sin^2 \phi$$

Expanding and using the relation

$$\sin^2 \phi + \cos^2 \phi = 1,$$

$$N_2 = \frac{4k^2 - N_1^2 \sin^2 \phi}{2N_1 \sin^2 \phi - 4k} \tag{4.13}$$

From this equation can be determined the largest gear (N_2) that can be meshed with a given gear (N_1) without interference. These values are shown in Table 4.5, using as N_1 the values found previously for equal gears.

If Eq. 4.13 is rewritten as

$$2N_1 \sin^2 \phi - 4k = \frac{4k^2 - N_2^2 \sin^2 \phi}{N_2} \tag{4.13a}$$

and N_2 approaches a rack to become infinite, the right-hand side of the equation approaches zero and Eq. 4.9 is obtained to give the number of teeth N_1 that will mesh with a rack without interference. It is also interesting to note that if a value of N_1 greater than those given in Table 4.3 for meshing with a rack without interference is substituted in Eq. 4.13, a negative and impossible value of N_2 is obtained.

4.8 Determination of Backlash. In Fig. 4.21*a* is shown the outline of two standard gears meshing at the standard center distance

$$C = \frac{N_1 + N_2}{2P}$$

with zero backlash. The pitch circles at which these two gears operate are the pitch circles at which they were cut, and their radii are given by $R = N/2P$.

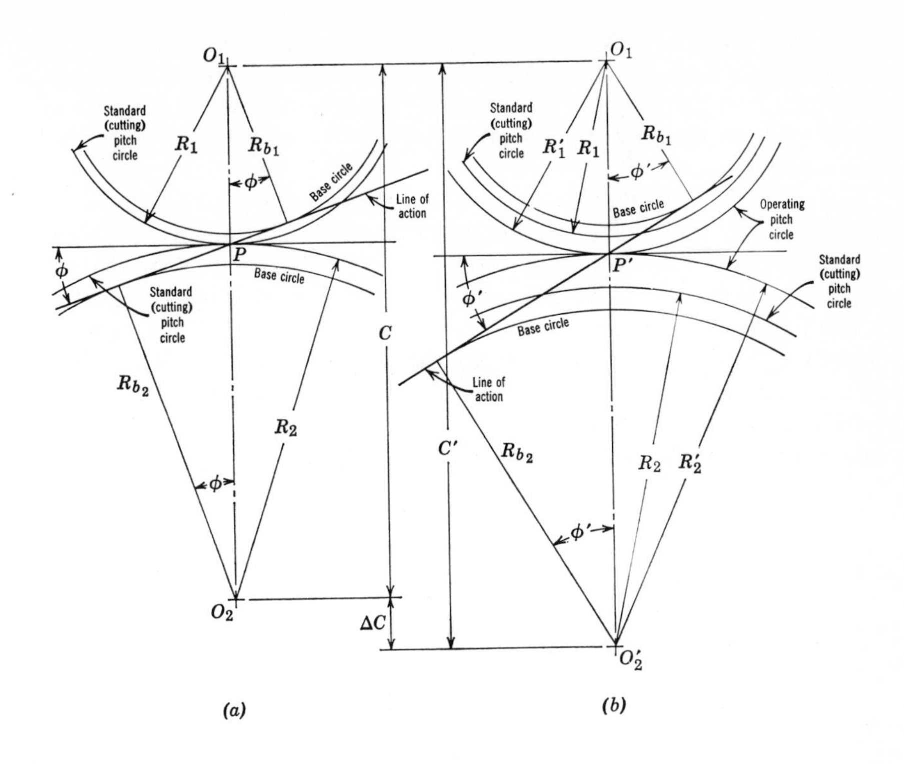

O_1
Standard (cutting) pitch circle
R_1
R_{b_1}
ϕ
Base circle
Line of action
P
Base circle
Standard (cutting) pitch circle
R_{b_2}
R_2
C
O_2
ΔC
(a)
O_1
Standard (cutting) pitch circle
R_1' R_1
R_{b_1}
ϕ'
Base circle
Operating pitch circle
Standard (cutting) pitch circle
P'
Base circle
Line of action
C'
R_{b_2}
R_2 R_2'
O_2'
(b)

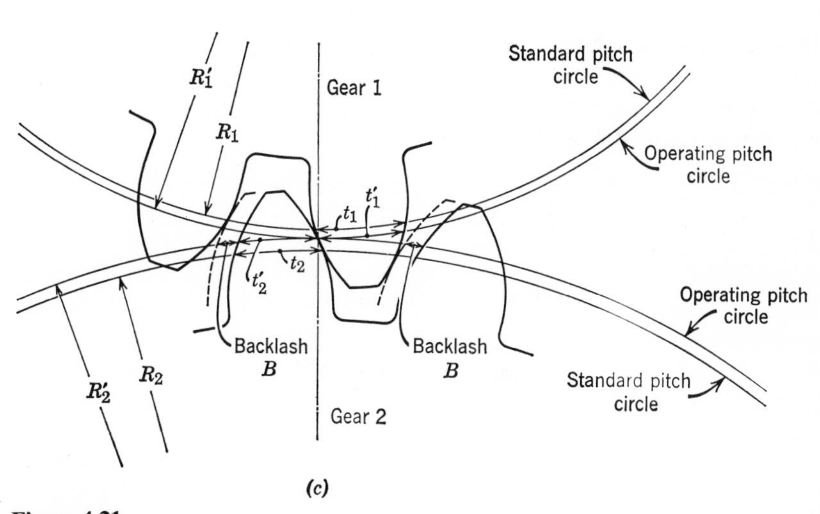

R_1'
R_1
Gear 1
Standard pitch circle
Operating pitch circle
t_1'
t_1
t_2'
t_2
Backlash B
Backlash B
Operating pitch circle
Standard pitch circle
R_2'
R_2
Gear 2
(c)

Figure 4.21

The cutting pitch circles are also known as *standard pitch circles.* The pressure angle ϕ at which the gears will operate is the pressure angle at which they were cut, that is, $14\frac{1}{2}°$, 20°, or 25°. In other words, the cutting and operating pitch circles are identical as well as the cutting and operating pressure angles.

Figure 4.21*b* shows the condition where the two gears have been pulled apart a distance ΔC to give a new center distance C'. The line of action now crosses the line of centers at a new pitch point P'. It can be observed that the standard or cutting pitch circles (radii R_1 and R_2) are no longer tangent to each other. Also the pitch point P' divides the center distance C' into segments which are inversely proportional to the angular velocity ratio. These segments become the radii R_1' and R_2' of new pitch circles that are tangent to each other at point P'. These circles are known as *operating pitch circles*, and equations for their radii can be determined from

$$\frac{\omega_1}{\omega_2} = \frac{N_2}{N_1} = \frac{R_2'}{R_1'}$$

and

$$R_1' + R_2' = C'$$

to give

$$R_1' = \left(\frac{N_1}{N_1 + N_2}\right)C'$$

and

$$R_2' = \left(\frac{N_2}{N_1 + N_2}\right)C'$$

In addition to the change in pitch circles the pressure angle also increases. Angle ϕ' is known as the *operating pressure angle* and is larger than the cutting pressure angle ϕ. An equation for the operating pressure angle ϕ' can easily be derived from Fig. 4.21*b* as follows:

$$C' = \frac{R_{b_1} + R_{b_2}}{\cos\phi'} = (R_1 + R_2)\frac{\cos\phi}{\cos\phi'} = C\frac{\cos\phi}{\cos\phi'}$$

or

$$\cos\phi' = \frac{C}{C'}\cos\phi \qquad (4.14)$$

Also

$$\begin{aligned}\Delta C &= C' - C \\ &= C\frac{\cos\phi}{\cos\phi'} - C \\ &= C\left(\frac{\cos\phi}{\cos\phi'} - 1\right)\end{aligned} \qquad (4.15)$$

When the gears are operated under the condition of Fig. 4.21*b*, backlash will be present as shown in Fig. 4.21*c*. The angular velocity ratio will not be affected as long as the gears remain in mesh. If the direction of rotation is reversed, however, lost motion will be encountered. An equation for backlash may be derived from the fact that the sum of the tooth thicknesses plus backlash must be equal to the circular pitch all measured on the operating pitch circle. From Fig. 4.21*c* the following equation may be written:

$$t_1' + t_2' + B = \frac{2\pi R_1'}{N_1} = \frac{2\pi R_2'}{N_2} \tag{4.16}$$

where

t' = tooth thickness on operating pitch circle
B = backlash
R' = radius of operating pitch circle
N = number of teeth

From Eq. 4.3 developed in the section on involutometry,

$$t_1' = 2R_1'\left[\frac{t_1}{2R_1} + \text{inv}\,\phi - \text{inv}\,\phi'\right]$$

$$= \frac{R_1'}{R_1}t_1 - 2R_1'(\text{inv}\,\phi' - \text{inv}\,\phi) \tag{4.17}$$

$$t_2' = 2R_2'\left[\frac{t_2}{2R_2} + \text{inv}\,\phi - \text{inv}\,\phi'\right]$$

$$= \frac{R_2'}{R_2}t_2 - 2R_2'(\text{inv}\,\phi' - \text{inv}\,\phi) \tag{4.18}$$

where

t = tooth thickness on standard or cutting pitch circle ($t = p/2 = \pi/2P$)
R = radius of standard or cutting pitch circle ($R = N/2P$)
ϕ = cutting pressure angle ($14\frac{1}{2}°$, 20°, 25°)
ϕ' = operating pressure angle

Also

$$\frac{R_1}{R_1'} = \frac{R_2}{R_2'} = \frac{C}{C'} \tag{4.19}$$

and

$$C' = R_1' + R_2' \tag{4.20}$$

Substituting Eqs. 4.17, 4.18, 4.19, and 4.20 into Eq. 4.16 and remembering that

$$\frac{2\pi R}{N} = p = \frac{\pi}{P},$$

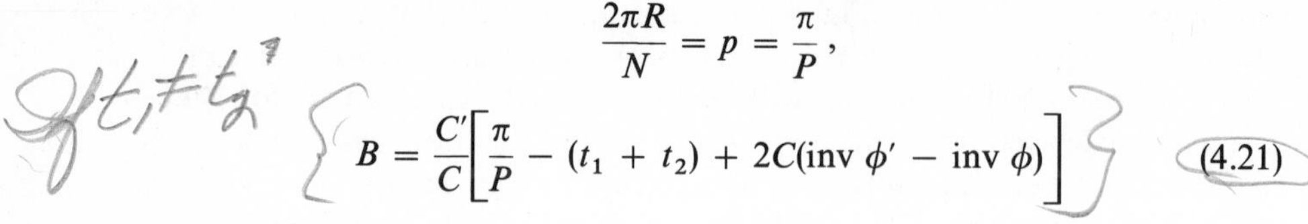

$$B = \frac{C'}{C}\left[\frac{\pi}{P} - (t_1 + t_2) + 2C(\text{inv}\,\phi' - \text{inv}\,\phi)\right] \tag{4.21}$$

For standard gears,

$$t_1 = t_2 = \frac{p}{2} = \frac{\pi}{2P}$$

and Eq. 4.21 simplifies to

$$B = 2C'(\text{inv}\,\phi' - \text{inv}\,\phi) \tag{4.22}$$

Equation 4.21 should be used if the gears are not standard, that is, if $t_1 \neq t_2$. Nonstandard gears will be presented in Chapter 5. Recommended values for backlash can be found in the AGMA *Gear Handbook*, Volume 1, 390.03 (January 1972).

4.9 Internal (Annular) Gears. In many applications an internal involute gear is meshed with a pinion instead of using two external gears in order to derive certain advantages. Perhaps the most important advantage is that of compactness of the drive. Also for the same tooth proportions internal gears will have greater length of contact, greater tooth strength, and lower relative sliding between meshing teeth than external gears.

In an internal gear, the tooth profiles are concave instead of convex as in an

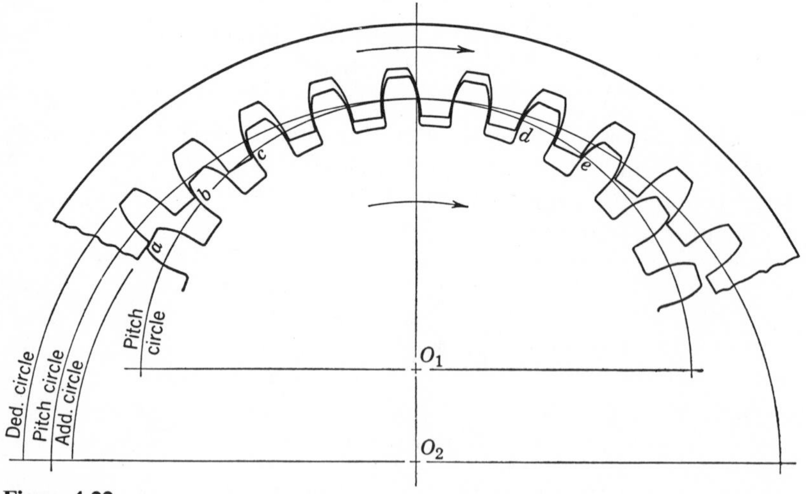

Figure 4.22

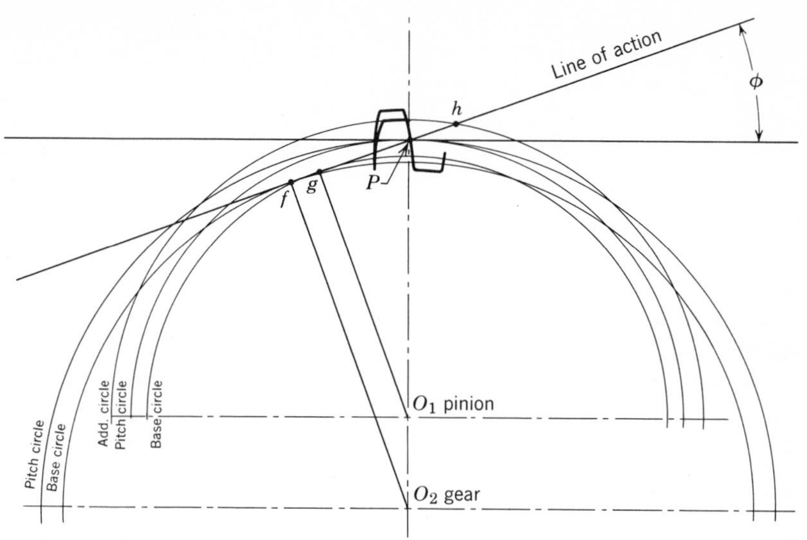

Figure 4.23

external gear. Because of this shape, a type of interference can occur that is not possible in an external gear or a rack. This interference is known as *fouling* and occurs between inactive profiles as the teeth go in and out of mesh. Fouling will occur when there is not sufficient difference between the numbers of teeth on the internal gear and on the pinion. Figure 4.22 shows a pinion meshing with an internal gear. They are so nearly the same size that fouling occurs at points *a*, *b*, *c*, *d*, and *e*. When an internal gear is cut, a Fellows cutter is used with two less teeth than the gear being cut. This automatically relieves the tips of the internal gear teeth to prevent fouling at points *a*, *b*, *c*, *d*, and *e*. Involute interference can also occur between active profiles the same as in external gears. This will be discussed in the following paragraph.

Figure 4.23 shows two teeth in contact from Fig. 4.22 with the line of action tangent to the base circle of the gear at point *f* and tangent to the base circle of the pinion at point *g*. An involute profile for the gear can begin at point *f*, but the involute for the pinion cannot begin until point *g*. Point *g* is, therefore, the first possible point of contact without involute interference and determines the maximum addendum of the gear. Point *h*, the intersection of the addendum circle of the pinion and the line of action, is the end of contact, and the length of action is *gPh*. It should be mentioned that the relation $P = N/D$ holds for an internal gear as well as for an external gear.

4.10 Cycloidal Gears. Even though the cycloidal gear has been largely replaced by the involute, the cycloidal profile possesses certain advantages worthy of note. These are discussed briefly below. For a detailed

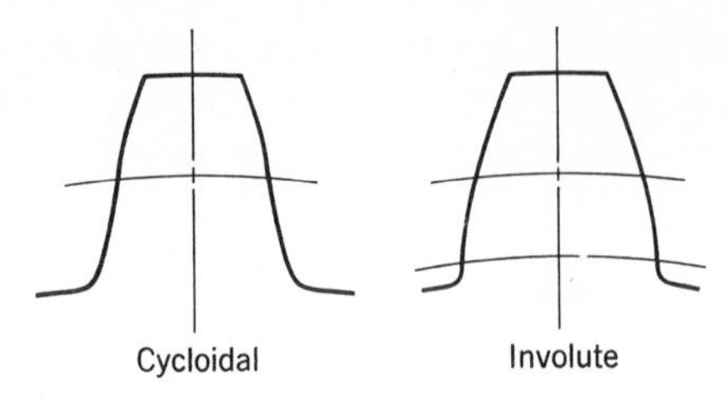

Figure 4.24

treatment of cycloidal gears the reader is referred to one[2] of the many excellent references on the subject.

Cycloidal gears do not have interference, and a cycloidal tooth is in general stronger than an involute tooth because it has spreading flanks in contrast to the radial flanks of an involute tooth. Also cycloidal teeth have less sliding and, therefore, less wear. Figure 4.24 shows a cycloidal gear tooth and for comparison an involute tooth. However, an important disadvantage of cycloidal gearing is the fact that for a pair of cycloidal gears there is only one theoretically correct center distance for which they will transmit motion at a constant angular velocity ratio. Another disadvantage is that, although it is possible to hob a cycloidal gear, the hob is not as easily made as an involute hob because cycloidal rack teeth are not straight sided as are involute rack teeth. For this reason involute gears can be produced more accurately and more cheaply than cycloidal gears.

Involute gears have completely replaced cycloidal gears for power transmission. However, cycloidal gears are extensively used in watches, clocks, and certain instruments in cases where the question of interference and strength is a prime consideration. In watches and clocks the gear train from the power source to the escapement steps up or increases its angular velocity ratio with the gear driving the pinion. In a watch this step-up may be as high as 5000:1.[3] The gears will, therefore, be so small that in order to avoid using excessively small teeth it is necessary to use pinions (the driven gears in this case) having as few as 6 or 7 teeth. The tooth profile of these pinions must also be capable of action over 60° of rotation. For this purpose cycloidal gears are used in preference to involute gears. The problem of center distance and angular velocity ratio is not important in this case because the whole train as governed by the escapement comes to rest and then starts again several times per second. The operation of the train thus involves such large changes of momentum that the effect of tooth form on change of momentum is negligible. The effect of tooth form on consistency of velocity ratio is thus unimportant in itself.

[2] C. D. Albert and F. S. Rogers, *Kinematics of Machinery*, John Wiley and Sons, 1931.

[3] W. O. Davis, *Gears for Small Mechanisms*, N.A.G. Press, 1953.

For winding, hand-setting, and minute-to-hour reduction trains, the pinion drives the gear, and either cycloidal or involute gearing may be used. However, American watches generally use involute gears for this train.

Problems*

4.1 An involute is generated on a base circle having a radius R_b of 4 in. As the involute is generated, the angle which corresponds to inv ϕ varies from 0 to 15°. For increments of 3° for this angle calculate the corresponding pressure angle ϕ and radius R for points on the involute. Plot this series of points in polar coordinates and connect with a smooth curve to give the involute.

4.2 Write a computer program for Problem 4.1 letting R_b = 3, 4, and 5 in. Determine the corresponding values of pressure angle ϕ and radius R for each value of R_b.

4.3 The thickness of an involute gear tooth is 0.314 in. at a radius of 3.5 in. and a pressure angle of $14\frac{1}{2}°$. Calculate the tooth thickness and radius at a point on the involute which has a pressure angle of 25°.

4.4 If the involutes which form the outline of a gear tooth are extended they will intersect and the tooth become pointed. Determine the radius at which this occurs for a tooth which has a thickness of 0.262 in. at a radius of 4 in. and a pressure angle of 20°.

4.5 The thickness of an involute gear tooth is 0.196 in. at a radius of 2.0 in. and a pressure angle of 20°. Calculate the tooth thickness on the base circle.

4.6 The pitch radii of two spur gears in mesh are 2.00 and 2.50 in., and the outside radii are 2.25 and 2.75 in., respectively. The pressure angle is 20°. Make a full-size layout of these gears as shown in Fig. 4.10, and label the beginning and end of contact. The pinion is the driver and rotates clockwise. Determine and label the angles of approach and recess for both gears. Draw the involutes necessary for finding α and β by the approximate method given in the Appendix.

4.7 A pinion of 2.00 in. pitch radius rotates clockwise and drives a rack. The pressure angle is 20°, and the addendum of the pinion and of the rack is 0.20 in. Make a full-size layout of these gears, and label the beginning and end of contact. Determine and label the angle of approach and recess for the pinion. Draw the involutes necessary for finding α and β by the approximate method given in the Appendix.

4.8 Two equal spur gears of 48 teeth mesh together with pitch radii of 4.00 in. and addendums of 0.167 in. If the pressure angle is $14\frac{1}{2}°$, calculate the length of action Z and the contact ratio m_p.

4.9 Contact ratio is defined either as the arc of action divided by the circular pitch or as the ratio of the length of action to the base pitch. Prove that

$$\frac{\text{Arc of action}}{\text{Circular pitch}} = \frac{\text{Length of action}}{\text{Base pitch}}$$

4.10 Derive an equation for the length of action Z for a pinion driving a rack in

* Because AGMA Standards have not been developed for gears with SI units, there are no problems in Chapters 4, 5, 6, and 7 employing SI units.

terms of the pitch radius R, the base radius R_b, the addendum a, and the pressure angle ϕ.

4.11 A pinion with a pitch radius of 1.50 in. drives a rack. The pressure angle is $14\frac{1}{2}°$. Calculate the maximum addendum possible for the rack without having involute interference on the pinion.

4.12 A 12-pitch, 20° full-depth pinion of 24 teeth drives a 40-tooth gear. Calculate the pitch radii, base radii, addendum, dedendum, and tooth thickness on the pitch circle.

4.13 An 8-pitch, 25° full-depth pinion of 18 teeth drives a 45-tooth gear. Calculate the pitch radii, base radii, addendum, dedendum, and the tooth thickness on the pitch circle.

4.14 A 120-pitch, 20° full-depth pinion of 42 teeth drives a gear of 90 teeth. Calculate the contact ratio.

4.15 If the radii of a pinion and gear are increased so that each becomes a rack, the length of action theoretically becomes a maximum. Determine the equation for the length of action under these conditions and calculate the maximum contact ratio for $14\frac{1}{2}°$, 20°, and 25° full-depth systems.

4.16 A 4-pitch, 20° stub pinion of 20 teeth drives a rack. Calculate the pitch radius, base radius, working depth, whole depth, and the tooth thickness of the rack on the pitch line.

4.17 A 20° full-depth rack has an addendum of 0.25 in. Calculate the base pitch, and show it as a dimension on a full-size sketch of a portion of the rack.

4.18 Determine the number of teeth in a $14\frac{1}{2}°$ full-depth involute spur gear so that the base circle diameter will equal the dedendum circle diameter.

4.19 Determine the following for a pair of standard spur gears in mesh: (*a*) An equation for the center distance C as a function of the numbers of teeth and diametral pitch. (*b*) The various combinations of 20° full-depth gears that can be used to operate at a center distance of 5.00 in. with an angular velocity ratio of 3:1. The diametral pitch is not to exceed 12 and the gears are not to be undercut. The gears are to be hobbed.

4.20 A 6-pitch, 25° full-depth pinion with 30 teeth drives a rack. Calculate the length of action and the contact ratio.

4.21 A 2-pitch, 20° full-depth pinion with 24 teeth drives a rack. If the pinion rotates counterclockwise at 360 rpm, determine graphically the velocity of sliding between the pinion tooth and the rack tooth at the beginning of contact, pitch point, and at the end of contact. Use a scale of 1 in. = 10 fps.

4.22 Two shafts whose axes are 8.5 in. apart are to be coupled together by standard spur gears with an angular velocity ratio of 1.5:1. Using a diametral pitch of 6, select two pairs of gears to best fit the above requirements. What change in the given data would have to be allowed if each set were to be used?

4.23 An 8-pitch, $14\frac{1}{2}°$ full-depth hob is used to cut a spur gear. The hob is right handed with a lead angle of 2° 40′, a length of 3.00 in., and an outside diameter of 3.00 in. Make a full-size sketch of the hob cutting a 48-tooth spur gear. The gear blank is $1\frac{1}{2}$ in. wide. Show the pitch cylinder of the hob on top of the gear blank with the pitch helix of the hob in correct relation to the pitch element of the gear tooth. Show three tooth elements on the gear and $1\frac{1}{2}$ turns of the thread on the hob; position these elements by means of the normal circular pitch. Label the axis of the hob and gear blank, the lead angle of hob, and the direction of rotation of the hob and gear blank.

4.24 For a pressure angle of 22.5° in the full-depth system, calculate the minimum number of teeth in a pinion to mesh with a rack without involute interference. Also calculate the number of teeth in a pinion to mesh with a gear of equal size without involute interference.

4.25 An 8-pitch, 20° pinion with 24 teeth drives a 56-tooth gear. Determine the outside radii so that the addendum circle of each gear passes through the interference point of the other. Calculate the value of k for each gear.

4.26 Two equal 5-pitch, 20° gears mesh together such that the addendum circle of each gear passes through the interference point of the other. If the contact ratio is 1.622 calculate the number of teeth and the outside radius on each gear.

4.27 Two equal 20° involute gears are in mesh at the standard center distance. The addendum circle of each gear passes through the interference point of the other. Derive an equation for k as a function of N, where N is the number of teeth and k is a constant which when divided by the diametral pitch gives the addendum.

4.28 In the sketch of a standard gear shown in Fig. 4.25, the teeth are 20° full depth. If the pitch diameter is 4.80 in. and the diametral pitch is 5, calculate the radius of the pin which contacts the profile at the pitch point. Compute the diameter m measured over two opposite pins.

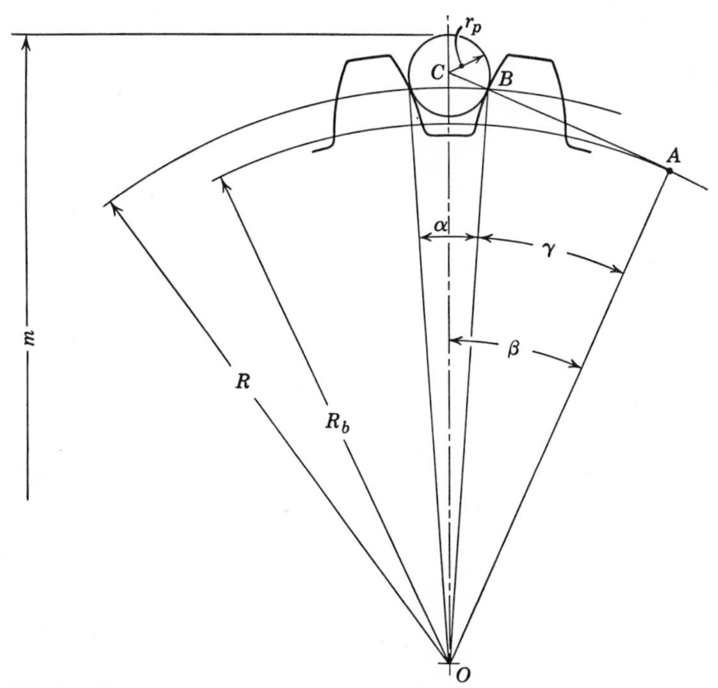

Figure 4.25

4.29 A 10-pitch, $14\frac{1}{2}°$ full-depth pinion with 40 teeth meshes with a rack with no backlash. If the rack is pulled out 0.07 in. calculate the backlash produced.

4.30 A 12-pitch, 20° full-depth pinion of 18 teeth drives a gear of 54 teeth. If the center distance at which the gears operate is 3.05 in., calculate the operating pressure angle.

4.31 A 10-pitch, $14\frac{1}{2}°$ full-depth pinion with 36 teeth drives a gear with 60 teeth. If the center distance is increased by 0.025 in., calculate (*a*) the radii of the operating pitch circles, (*b*) the operating pressure angle, and (*c*) the backlash produced.

4.32 A 4-pitch, 20° stub pinion of 24 teeth drives a gear of 40 teeth. Calculate (*a*) the maximum theoretical distance that these gears can be drawn apart and still mesh together with continuous driving, and (*b*) the backlash on the new pitch circles when the gears are drawn apart the amount calculated in part *a*.

4.33 A pinion with 24 teeth has a tooth thickness of 0.255 in. at a cutting pitch radius of 1.50 in. and a pressure angle of 20°. A gear having 40 teeth has a tooth thickness of 0.230 in. at a cutting pitch radius of 2.50 in. and a pressure angle of 20°. Calculate the pressure angle and center distance if these gears are meshed together without backlash.

4.34 A 10-pitch, 25° pinion of 15 teeth drives a gear of 45 teeth. Using a computer calculate the backlash produced when the center distance is increased from 3.000 to 3.030 in. by increments of 0.001 in.

4.35 A 96-pitch pinion of 34 teeth drives a gear with 60 teeth. If the center distance is increased by 0.005 in., compare the backlash produced with pressure angles of $14\frac{1}{2}$, 20, and 25°.

5

Nonstandard Spur Gears

5.1 Theory of Nonstandard Spur Gears. The most serious defect of the involute system of gearing is the possibility of interference between the tip of the gear tooth and the flank of the pinion tooth when the number of teeth in the pinion is reduced below the minimum for that system of gearing.

When interference does occur, the interfering metal is removed from the flank of the pinion tooth by the cutter when the teeth are generated. This removal of metal by the cutter is known as *undercutting* and would normally occur unless steps are taken to prevent it. If the cutter did not remove this metal, the two gears would not rotate when meshed together because the gear causing the interference would jam on the flank of the pinion. Actually, however, the gears will be able to rotate freely together because the flank of the pinion has been undercut. This undercutting, however, not only weakens the pinion tooth but may also remove a small portion of the involute adjacent to the base circle, which may cause a serious reduction in the length of action.

The attempt to eliminate interference and its consequent undercutting has led to the development of several nonstandard systems of gearing, some of which require special cutters. However, two of these systems have been successful and find wide application because standard cutters can be used to generate the teeth. In the first method when the pinion is being cut, the cutter is withdrawn a certain amount from the blank so that the addendum of the basic rack passes through the interference point of the pinion. This will eliminate the undercutting, but the width of the tooth will be increased with a

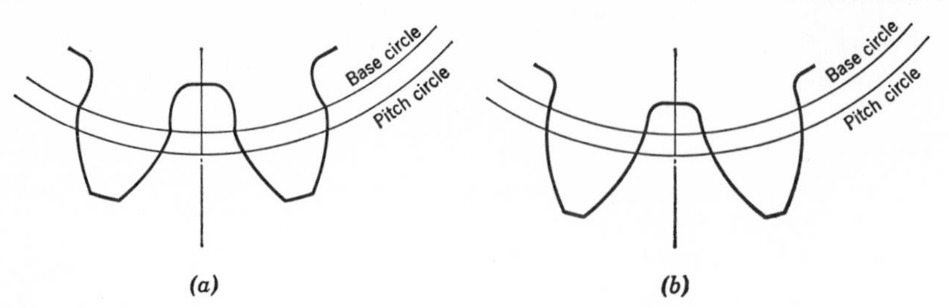

Figure 5.1

corresponding decrease in the tooth space. This is illustrated in Fig. 5.1, where (*a*) shows undercut teeth, and (*b*) the teeth resulting when the cutter has been withdrawn. When this pinion (Fig. 5.1*b*) is now mated with its gear, it will be found that the center distance has been increased because of the decreased tooth space. It can no longer be calculated directly from the diametral pitch and numbers of teeth and is therefore not considered standard. The pressure angle at which the gears operate also increases. This method of eliminating interference is known as the *extended center distance system*.

The withdrawal of the cutter need not be limited to the pinion blank alone but may be applied to both the pinion and gear if conditions warrant.

A variation of the extended center distance system is the practice of advancing the cutter into the gear blank the same amount that it will be withdrawn from the pinion blank. This will result in an increased addendum for the pinion and a decreased addendum for the gear; the increase in pinion addendum will equal the decrease in gear addendum. The dedendums will also change on both the pinion and gear so that the working depth will be the same as if the gears were standard. The center distance remains standard as well as the pressure angle. This system is known as the *long and short addendum system*.

Because of the change in the tooth proportions, the thickness of the gear tooth on the pitch circle is reduced and that of the pinion increased. Because of the fact that pinion teeth are weaker than gear teeth when both are made of the same material, the long and short addendum system will tend to equalize the tooth strengths. The long and short addendum system can be applied only when the interference occurs on one gear of a pair in mesh. Also, this system cannot be applied when the gears are equal or nearly equal in size because, although it would eliminate the interference on one gear, it would accentuate it on the other gear.

These two methods were developed primarily as a means of eliminating interference. However, they are also used extensively for improving contact ratio, for changing tooth shape to increase the strength of a tooth even though interference may not be present, and for fitting gears to nonstandard center distances.

The two systems can be applied to spur, helical, and bevel gearing. In fact, the standard tooth system for bevel gears is a long and short addendum system.

Formulas for the application of these two systems to spur gears cut with a hob will now be developed.

5.2 Extended Center Distance System. Figure 5.2*a* in solid line shows a rack cutting a pinion where the pinion has fewer teeth than the minimum allowed to prevent interference. The rack and pinion are meshing at the standard center distance, with the standard pitch line of the rack tangent to the standard or cutting pitch circle of the pinion. The addendum line of the rack falls above the interference point E of the pinion so that the flanks of the pinion teeth are undercut as shown. In order for the rack tooth to cut the necessary clearance at the root of the pinion tooth, its height would have to be increased. To simplify the sketch, this additional height is shown (dotted) on only one tooth. The same layout may be used to illustrate the action of a hob cutting the pinion because kinematically a rack tooth and a hob tooth are identical.

To avoid undercutting, the rack is withdrawn a distance e so that the addendum line of the rack passes through the interference point E. This condition is shown dotted in Fig. 5.2*a*, and results in the rack cutting a pinion

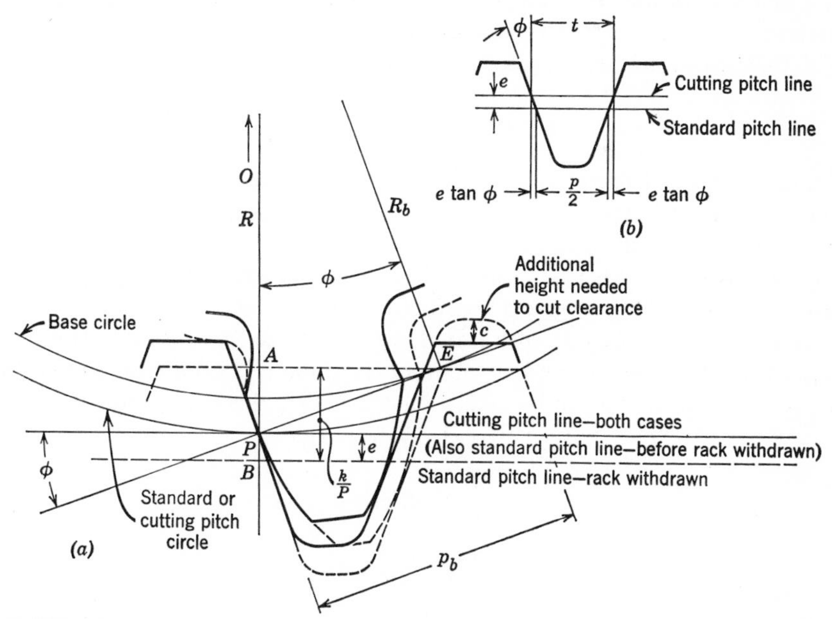

Figure 5.2

with a wider tooth than before. As the rack is withdrawn, the outside radius of the pinion must also be increased (by machining a larger pinion blank) to allow the clearance between the tip of the pinion tooth and the root of the rack tooth to remain the same. The same clearance is used regardless of whether a gear is standard or nonstandard. To show the change in the pinion tooth more clearly, the rack in Fig. 5.2*a* was withdrawn downward to the right to keep the left side of the tooth profile the same in both cases. When two gears, one or both of which have been generated with the cutter offset, are meshed together, the center distance will be greater than the standard center distance. In addition, the pressure angle at which these gears operate will be greater than the cutting pressure angle.

As mentioned previously, when a standard pinion is generated by the rack, the standard pitch line of the rack is tangent to the cutting pitch circle of the pinion. In this case, the standard pitch line is also the cutting pitch line. When the rack is withdrawn a distance e, however, the standard pitch line is no longer tangent to the cutting pitch circle; therefore, it cannot serve as the cutting pitch line. A new line on the rack will therefore act as a cutting pitch line. Figure 5.2*b* shows more clearly the two pitch lines on the rack when it is cutting a nonstandard tooth. From Fig. 5.2*a*, it can be seen that the cutting pitch circle on the pinion remains the same regardless of whether the pinion being cut is standard or nonstandard.

The width of the enlarged pinion tooth on its cutting pitch circle can be determined from the tooth space of the rack on its cutting pitch line. From Fig. 5.2*b*, this thickness can be expressed by the following equation:

$$t = 2e \tan \phi + \frac{p}{2} \tag{5.1}$$

Equation 5.1 can therefore be used to calculate the tooth thickness on the standard or cutting pitch circle of a gear generated by a standard cutter offset an amount e; e will be negative if the hob is advanced into the gear blank. This equation can also be used to determine the amount a cutter must be fed into a gear blank to give a specified amount of backlash.

In Fig. 5.2 the rack was withdrawn just enough so that the addendum line passed through the interference point of the pinion. It is possible to develop an equation so that e can be determined to satisfy this condition.

$$\begin{aligned} e &= AB + OA - OP \\ &= \frac{k}{P} + R_b \cos \phi - R \\ R_b &= R \cos \phi \\ R &= \frac{N}{2P} \end{aligned}$$

Therefore,

$$e = \frac{k}{P} - R(1 - \cos^2 \phi)$$

$$= \frac{1}{P}\left(k - \frac{N}{2}\sin^2 \phi\right) \tag{5.2}$$

There are two equations that were developed in the section on involutometry (Chapter 4) that find particular application in the study of nonstandard gearing

$$\cos \phi_B = \frac{R_A}{R_B} \cos \phi_A \tag{5.3}$$

$$t_B = 2R_B\left[\frac{t_A}{2R_A} + \text{inv}\,\phi_A - \text{inv}\,\phi_B\right] \tag{5.4}$$

By means of these equations it is possible to determine the pressure angle and tooth thickness at any radius R_B if the pressure angle and tooth thickness are known at some other radius R_A. For nonstandard gears, the reference thickness that corresponds to the thickness t_A in Eq. 5.4 is the tooth thickness on the cutting pitch circle, which can easily be calculated for any cutter offset by Eq. 5.1. The reference pressure angle that corresponds to ϕ_A is the pressure angle of the cutter. The radius at this pressure angle is the radius of the cutting pitch circle.

When two gears, gear 1 and gear 2, which have been cut with a hob offset e_1 and e_2 respectively, are meshed together, they will operate on new pitch circles of radii R'_1 and R'_2 and at a new pressure angle ϕ'. The thickness of the teeth on the operating pitch circles can be expressed as t'_1 and t'_2 and can easily be calculated from Eq. 5.4. These dimensions are shown in Fig. 5.3 together with the thickness of the teeth t_1 and t_2 on the cutting pitch circles of radii R_1 and R_2.

An equation will now be developed for determining the pressure angle ϕ' at which these two gears will operate.

$$\frac{\omega_2}{\omega_1} = \frac{N_1}{N_2} = \frac{R'_1}{R'_2} \tag{5.5}$$

and

$$t'_1 + t'_2 = \frac{2\pi R'_1}{N_1} = \frac{2\pi R'_2}{N_2} \tag{5.6}$$

Substituting Eq. 5.4 for t'_1 and t'_2,

$$2R'_1\left[\frac{t_1}{2R_1} + (\text{inv}\,\phi - \text{inv}\,\phi')\right] + 2R'_2\left[\frac{t_2}{2R_2} + (\text{inv}\,\phi - \text{inv}\,\phi')\right] = \frac{2\pi R'_1}{N_1}$$

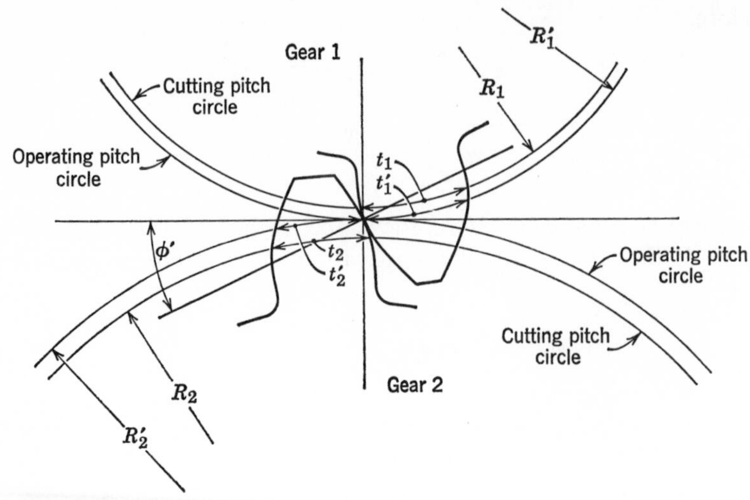

Figure 5.3

Dividing by $2R_1'$,

$$\left[\frac{t_1}{2R_1} + (\text{inv } \phi - \text{inv } \phi')\right] + \frac{R_2'}{R_2'}\left[\frac{t_2}{2R_2} + (\text{inv } \phi - \text{inv } \phi')\right] = \frac{\pi}{N_1}$$

$$\frac{t_1}{2R_1} + \frac{R_2'}{R_1'}\frac{t_2}{2R_2} = \frac{\pi}{N_1} + \left(1 + \frac{R_2'}{R_1'}\right)(\text{inv } \phi' - \text{inv } \phi)$$

Substituting Eq. 5.5 and $2R = N/P$,

$$\frac{t_1 P}{N_1} + \frac{N_2}{N_1}\frac{t_2 P}{N_2} = \frac{\pi}{N_1} + \frac{N_1 + N_2}{N_1}(\text{inv } \phi' - \text{inv } \phi)$$

Multiplying by N_1/P,

$$t_1 + t_2 = \frac{\pi}{P} + \frac{N_1 + N_2}{P}(\text{inv } \phi' - \text{inv } \phi)$$

Substituting Eq. 5.1 for t_1 and t_2,

$$2e_1 \tan \phi + \frac{p}{2} + 2e_2 \tan \phi + \frac{p}{2} = \frac{\pi}{P} + \frac{N_1 + N_2}{P}(\text{inv } \phi' - \text{inv } \phi)$$

$$2 \tan \phi(e_1 + e_2) + p = \frac{\pi}{P} + \frac{N_1 + N_2}{P}(\text{inv } \phi' - \text{inv } \phi)$$

Substituting $p = \pi/P$ and solving for inv ϕ',

$$\text{inv } \phi' = \text{inv } \phi + \frac{2P(e_1 + e_2)\tan \phi}{N_1 + N_2} \tag{5.7}$$

or

$$e_1 + e_2 = \frac{(N_1 + N_2)(\text{inv } \phi' - \text{inv } \phi)}{2P \tan \phi} \tag{5.7a}$$

Using Eq. 5.7, it is possible to determine the pressure angle ϕ' at which the two gears will operate after having been cut with a hob offset e_1 and e_2, respectively. To calculate the increase in the center distance (over the standard center distance C) due to the increased pressure angle, Eq. 4.15 can be used and is repeated here:

$$\Delta C = C\left[\frac{\cos \phi}{\cos \phi'} - 1\right] \tag{5.8}$$

Very often it is necessary to design gears to fit a predetermined center distance. In this case, the pressure angle is fixed by the conditions of the problem, and it is necessary to determine the hob offsets e_1 and e_2. The sum $(e_1 + e_2)$ can be determined from Eq. 5.7*a*. *However, it should be noted that the sum of* e_1 *and* e_2 *does not equal the increase in the center distance over the standard center distance.* Unfortunately, there is no way of rationally determining e_1 and e_2 independently of each other. Because of this, the values are usually selected by assuming one of them or by using some empirical relation such as letting e_1 and e_2 vary inversely (or directly if $e_1 + e_2$ is negative) with the numbers of teeth in the gears in an attempt to strengthen the pinion teeth. However, this method of selecting e_1 and e_2 generally does not yield pinion and gear teeth of anywhere near equal strength. In an attempt to correct this situation, a method was developed by Walsh and Mabie[1] for determining the hob offset e_1 from the value of $e_1 + e_2$ for a pair of spur gears designed to operate at a nonstandard center distance. By using a digital computer, it was possible to adjust e_1 and e_2 for various velocity ratios and changes in center distance so that the stress in the pinion teeth was approximately equal to that in the gear teeth.

Because of the complexity of the problem, the results had to be given in the form of design charts. These show curves of $e_1/(e_1 + e_2)$ versus $N_2/(N_1 + N_2)$ for various changes in center distance. These design charts were developed for a cutting pressure angle ϕ of 20°, full-depth teeth ($k = 1$), and coarse pitch. Although the charts were plotted for data based on a diametral pitch of one, they can be used for any diametral pitch up to 19.99 (end of coarse pitch). The charts were also plotted for $N_1 = 18$ and N_2 from 18 to 130 teeth. Where N_1 takes on other values, a very slight error (less than 4%) is

[1] E. J. Walsh and H. H. Mabie, "A Simplified Method for Determining Hob Offset Values in the Design of Nonstandard Spur Gears," *Proceedings*, Second OSU Applied Mechanism Conference, Stillwater, Oklahoma, October 1971.

introduced. A sample chart is shown in Fig. 5.4 for changes in center distance $\Delta C = 1.175$ to 1.275 in. for $P = 1$.

Example 5.1

A pinion and gear of 20 and 30 teeth respectively are to be cut by a 5-pitch, 20° full-depth hob to operate on a center distance of 5.25 in. with no backlash. Determine the value of e_1 and e_2 to give teeth of the proper thickness so that the strengths of the pinion teeth and gear teeth will be approximately equal. The standard center distance is given by:

$$C = \frac{N_1 + N_2}{2P} = \frac{20 + 30}{2 \times 5}$$

$$= 5.00 \text{ in.}$$

Operating pressure angle:

$$\cos \phi' = \frac{C}{C'} \cos \phi = \frac{5.0}{5.25} \cos 20°$$

$$\phi' = 26.50°$$

Change in center distance:

$$\nabla C = C' - C = 5.25 - 5.00$$

$$= +0.25 \text{ in.}$$

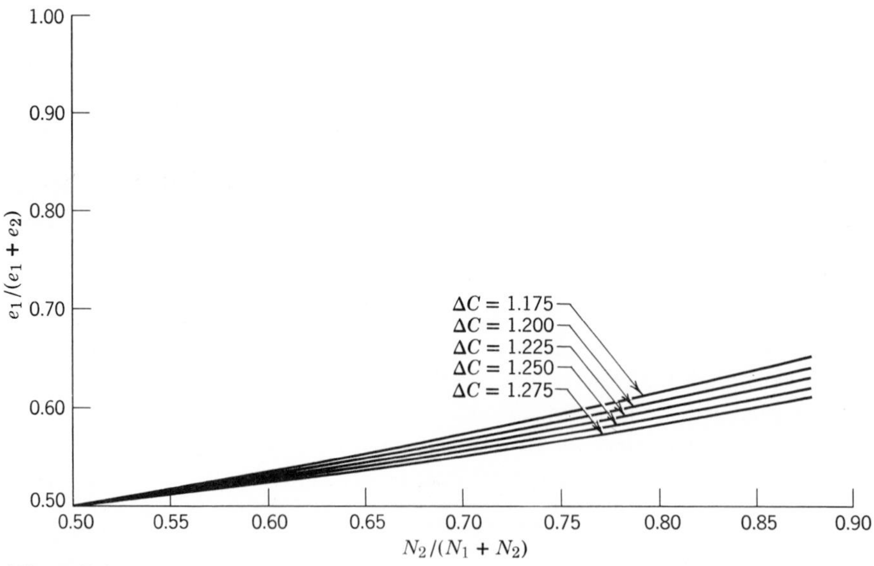

Figure 5.4

The value of ∇C must be multiplied by the diametral pitch because the charts are based on $P = 1$.

$$\Delta C = \nabla C \times P = 0.25 \times 5$$
$$= 1.25$$

Also

$$\frac{N_2}{N_1 + N_2} = \frac{30}{20 + 30} = 0.60$$

Therefore from Fig. 5.4,

$$\frac{e_1}{e_1 + e_2} = 0.543$$

Calculating the value of $e_1 + e_2$ from Eq. 5.7*a*,

$$e_1 + e_2 = \frac{(N_1 + N_2)(\text{inv }\phi' - \text{inv }\phi)}{2P \tan \phi}$$
$$= \frac{(20 + 30)(\text{inv } 26.5^\circ - \text{inv } 20^\circ)}{2 \times 5 \tan 20^\circ}$$
$$= 0.29073 \text{ in.}$$

Combining these results,

$$e_1 = 0.543(e_1 + e_2)$$
$$= 0.543(0.29073)$$
$$= 0.15787 \text{ in.}$$

and

$$e_2 = 0.13286 \text{ in.}$$

Although it is not practical to go through all of the calculations necessary to find the stresses in the pinion and gear teeth, it is interesting to note that

$$S_1 = \frac{9.959F_n}{F}$$

and

$$S_2 = \frac{9.991F_n}{F}$$

where

$$F_n = \text{normal load at tooth tip}$$
$$F = \text{tooth face width}$$

In addition to the charts for positive changes in center distance, as

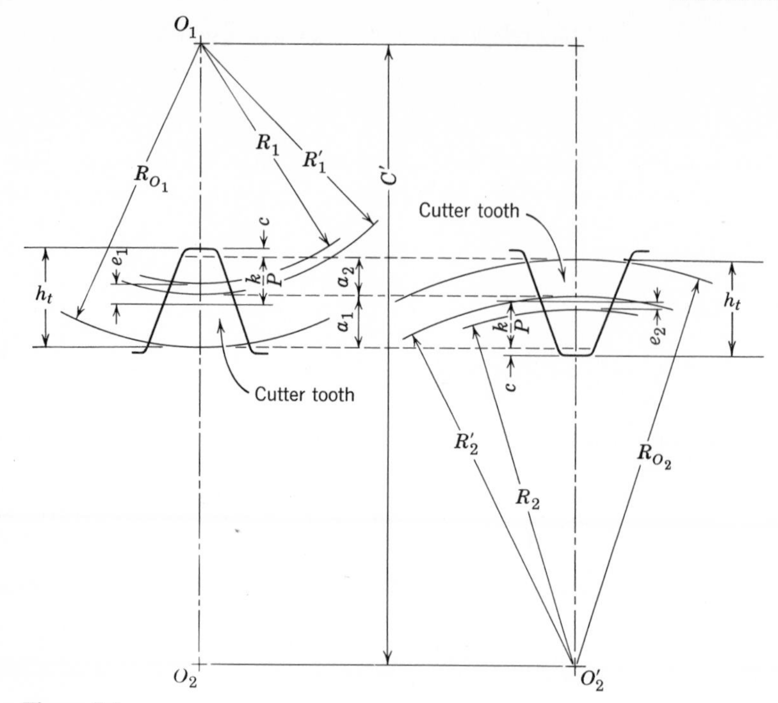

Figure 5.5

illustrated in Fig. 5.4, a series of charts is also given in the paper for negative changes in center distance.

Another approach to the solution of the problem of determining e_1 and e_2 has been developed by Siegel and Mabie.[2] By this method e_1 and e_2 are selected for a particular application to give tooth proportions which yield a maximum ratio of recess action to approach action, and, at the same time, give a contact ratio m_p of 1.20 or higher. This system is based on the fact that a pair of gears mesh more smoothly coming out of contact than they do going into contact. Therefore, it is advantageous to have as high a ratio of recess to approach action as possible especially for gears for instrument application.

It is not possible to calculate the addendum and dedendum of an extended center distance gear unless information is available regarding the gear with which it is to mesh. Figure 5.5 shows two gears that are to mesh at a given

[2] R. E. Siegel and H. H. Mabie, "Determination of Hob Offset Values for Nonstandard Gears Based on Maximum Ratio of Recess to Approach Action," *Proceedings*, Third OSU Applied Mechanism Conference, Stillwater, Oklahoma, November 1973.

center distance C'. The gears are to be cut with a hob which is offset e_1 on the pinion and e_2 on the gear. It is necessary to calculate the outside diameter of each gear and the depth of cut. The center line of gear 2 has been moved to the right so that a cutter tooth may be shown engaged with each blank. Knowing the center distance, the cutting pitch radii, the hob offsets, and the tooth form and diametral pitch of the hob, it is possible to write equations for the outside radii as follows:

$$R_{O_1} = C' - R_2 - e_2 + \frac{k}{P}$$

$$R_{O_2} = C' - R_1 - e_1 + \frac{k}{P} \qquad (5.9)$$

It should be noted from the sketch that the addendums of the two gears are not equal to each other nor is either one equal to the k/P of the hob.

An equation for the depth of cut can also be easily developed from the sketch.

$$h_t = R_{O_1} + R_{O_2} - C' + c \qquad (5.10)$$

where c is obtained from Tables 4.1 or 4.2.

5.3 Long and Short Addendum System. If the cutter is advanced into the gear blank the same amount that it is withdrawn from the pinion, then $e_2 = -e_1$ and from Eq. 5.7, $\phi' = \phi$. Therefore the pressure angle at which the gears will operate is the same as the pressure angle at which they were cut. Because there is no change in the pressure angle, $R_1' = R_1$ and $R_2' = R_2$, and the gears will operate at the standard center distance.

The addendum of the pinion is increased to $k/P + e$ and the addendum of the gear reduced to $k/P - e$. The tooth thicknesses on the cutting pitch circles can be readily calculated from Eq. 5.1, keeping in mind that the gear tooth thickness decreases the same amount that the pinion tooth increases. As has been mentioned previously, there are conditions under which the long and short addendum system will not work properly. In order for the long and short addendum system to be successful, Professor M. F. Spotts of Northwestern University has worked out that for $14\frac{1}{2}°$ gears, the sum of the teeth in the gears must be at least 64, and for 20° gears the sum of the teeth should be at least 34. For 25° gears the sum of the teeth should not be less than 24.

The proportions of gears cut by a pinion cutter for either of these two systems will not be the same as when cut by a hob. The preceding formulas apply only to gears cut by a hob or by a rack-type cutter. However, formulas for gears cut by pinion cutters can be developed using the principles above.

Example 5.2

Two spur gears of 12 teeth and 15 teeth respectively are to be cut by a 20° full-depth 6-pitch hob. Determine the center distance at which to operate the gears to avoid undercutting.

$$e_1 = \frac{1}{P}\left(k - \frac{N}{2}\sin^2\phi\right)$$

$$= \tfrac{1}{6}(1.00 - \tfrac{12}{2}\sin^2 20)$$

$$= 0.04968 \text{ in.}$$

$$e_2 = \tfrac{1}{6}(1.00 - \tfrac{15}{2}\sin^2 20)$$

$$= 0.02045 \text{ in.}$$

$$\text{inv}\,\phi' = \text{inv}\,\phi + \frac{2P(e_1 + e_2)\tan\phi}{N_1 + N_2}$$

$$= 0.01490 + \frac{2 \times 6(0.04968 + 0.02045)\tan 20^\circ}{12 + 15}$$

$$= 0.01490 + 0.01134$$

$$= 0.02624$$

From the table of involute functions,

$$\phi' = 23.97^\circ$$

$$R_1' = \frac{R_1 \cos\phi}{\cos\phi'} = \frac{1 \times 0.9397}{0.9135} = 1.0286 \text{ in.}$$

$$R_2' = \frac{R_2 \cos\phi}{\cos\phi'} = \frac{1.25 \times 0.9397}{0.9135} = 1.2858 \text{ in.}$$

and

$$C' = R_1' + R_2' = 2.3144 \text{ in.}$$

Example 5.3

Two $14\frac{1}{2}^\circ$ 8-pitch, full-depth spur gears of 32 and 48 teeth are operating together on the standard center distance of 5 in. In order to affect a change in the speed ratio, it is desired to replace the 32-tooth gear with one of 31 teeth. The tooth thickness on the cutting pitch circle of the 48-tooth gear and the 5-in. center distance are to remain unchanged.

Determine the value of e_1 that will give teeth of the proper thickness to mesh with the 48-tooth gear.

$$R_1 = \frac{N}{2P} = \frac{31}{2 \times 8} = 1.9375 \text{ in.}$$

$$R_2 = 3 \text{ in.}$$

$$R_1' = \frac{N_1}{N_1 + N_2} C'$$

$$R_1' = \tfrac{31}{79} \times 5 = 1.9621 \text{ in.}$$

$$R_2' = \frac{N_2}{N_1 + N_2} C'$$

$$R_2' = \tfrac{48}{79} \times 5 = 3.0379 \text{ in.}$$

$$R_1' = \frac{R_1 \cos \phi}{\cos \phi'}$$

$$\cos \phi' = \frac{R_1 \cos \phi}{R_1'} = \frac{1.9375 \times \cos 14\tfrac{1}{2}^\circ}{1.9621}$$

$$\phi' = 17.06^\circ$$

$$e_1 + e_2 = \frac{(N_1 + N_2)(\text{inv } \phi' - \text{inv } \phi)}{2P \tan \phi}$$

$$e_1 + e_2 = \frac{(31 + 48)(0.009120 - 0.0055448)}{2 \times 8 \times 0.25862}$$

$$e_1 + e_2 = \frac{79 \times 0.003575}{16 \times 0.25862} = \frac{0.282425}{4.13792}$$

$$e_2 = 0$$

$$e_1 = 0.06825 \text{ in.}$$

5.4 Recess-Action Gears. Another interesting type of nonstandard gears is recess-action gears, so called because most or all of the action between teeth takes place during the recess portion of contact. The long and short addendum system is a form of recess-action gears. It is known that the recess portion of contact of a pair of gears is much smoother than the approach portion. It was on this basis that recess-action gears were developed, and it has been found that these gears wear longer and operate

Table 5.1 TOOTH PROPORTIONS, RECESS-ACTION GEARS
(Pressure angle $\phi = 20°$)

	Semi Recess-Action		Full Recess-Action	
	Driver	Follower	Driver	Follower
Addendum (a)	$\frac{1.500}{P}$	$\frac{0.500}{P}$	$\frac{2.000}{P}$	0
Dedendum (b)	$\frac{0.796}{P}$	$\frac{1.796}{P}$	$\frac{0.296}{P}$	$\frac{2.296}{P}$
Pitch diameter (D)	$\frac{N}{P}$	$\frac{N}{P}$	$\frac{N}{P}$	$\frac{N}{P}$
Outside radius (R_o)	$\frac{N+3}{2P}$	$\frac{N+1}{2P}$	$\frac{N+4}{2P}$	$\frac{N}{2P}$
Tooth thickness (t)	$\frac{1.9348}{P}$	$\frac{1.2068}{P}$	$\frac{2.2987}{P}$	$\frac{0.8429}{P}$

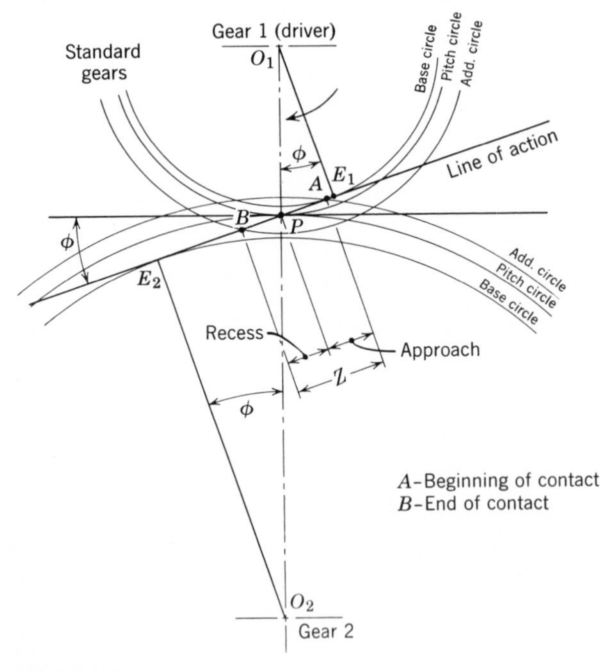

Figure 5.6a

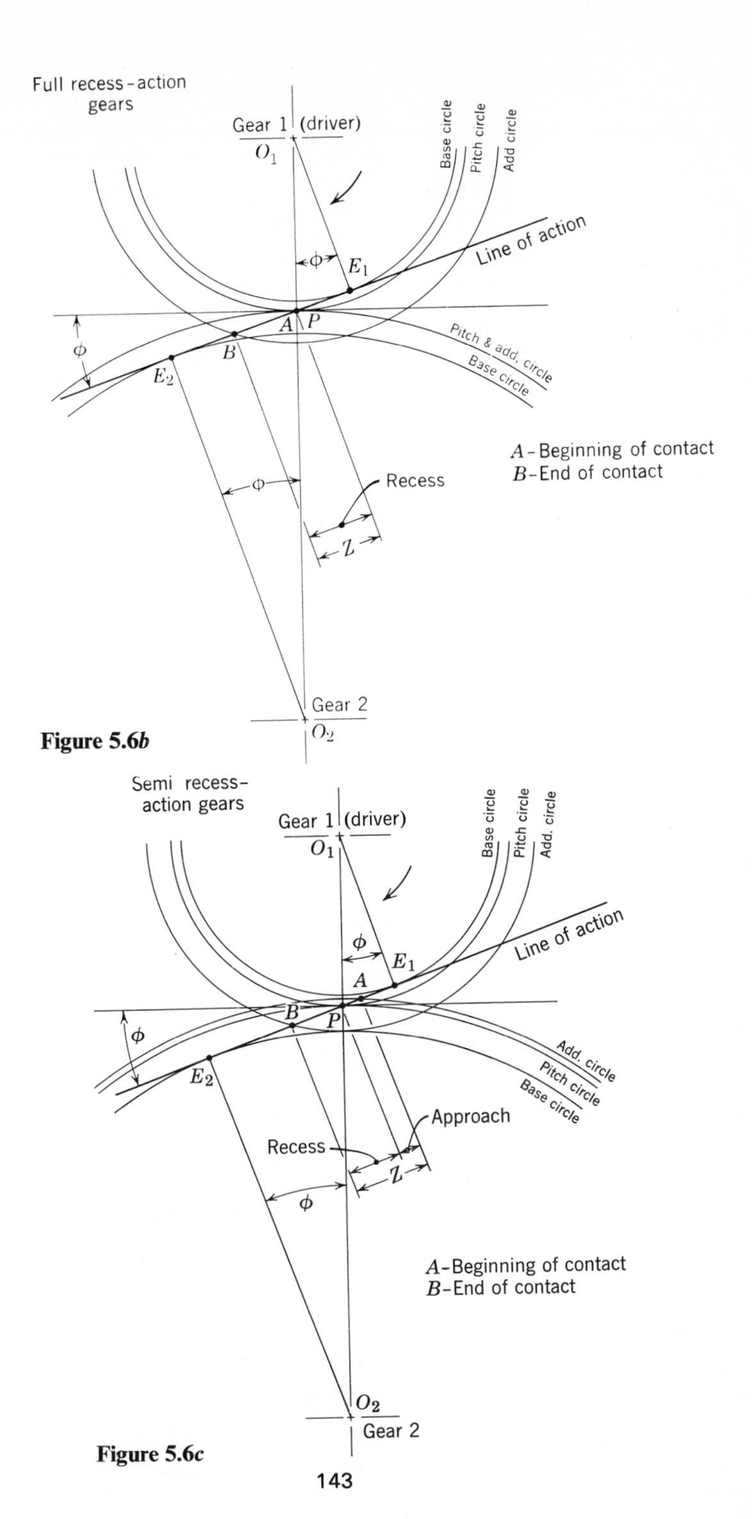

Figure 5.6b

Figure 5.6c

with less friction, vibration, and noise than gears with teeth of standard proportions.[3]

Recess-action gears can be machined using standard hobs and cutters, and their tooth form is the same as standard gears and they mesh at the same center distance. Therefore, a pair of recess action gears can be used to replace a pair of standard spur gears with no change in center distance.

The strength of recess-action gears is approximately the same as for standard gears. However, a recess-action gear must be designed to operate either as a driver or as a follower; it cannot be designed to operate as both. A recess-action pinion, however, can drive a follower in either direction, that is, it can change direction of rotation during an operating cycle. Also, the gears can be used for a speed-increasing drive as well as for a speed-reducing drive but the power flow must always be in the same direction. If the power flow changes direction during operation, binding in the tooth contact area will occur with resulting high friction and wear. Because of these limitations, recess-action gears cannot be used as idlers operating at standard center distances.

There are two types of recess-action gears: (*a*) full recess-action where all of the contact is recess and (*b*) semi recess-action. In order for a pair of recess-action gears to have adequate contact ratio, little or no undercut, and teeth not be pointed, full recess-action gears cannot have less than 20 teeth in the driver nor less than 27 teeth in the follower. For semi recess-action gears, however, the minimum number of teeth in the driver is reduced to 10 and that in the follower to 20. Full recess-action gears are to be preferred because all of the action is in the recess portion. However, many times the large number of teeth required for full recess-action limits their use, and semi recess-action gears must be used instead.

Table 5.1 shows the proportions for the two systems of recess-action gears. To give a comparison between these gears and standard gears, Fig. 5.6 has been drawn which shows the addendum, pitch, and base circles and length of action of (*a*) standard gears, (*b*) full recess-action gears, and (*c*) semi recess-action gears in mesh. In Fig. 5.6*b* for the full recess-action system the pitch circle of the follower (gear 2) becomes the addendum circles because the addendum is zero. Therefore, the approach portion of tooth contact is zero, and all of the length of action is in the recess portion. Figure 5.6*c* for semi recess-action shows the recess portion considerably larger than the approach portion for this system.

References

Albert, C. D., and F. S. Rogers, *Kinematics of Machinery*, John Wiley and Sons, 1931.

Buckingham, E., *Spur Gears*, McGraw-Hill Book Company, 1928.

[3] E. K. Buckingham, "Recess Action Gears," *Prod. Eng.*, pp. 82–89, June 8, 1964.

Spotts, M. F., *Design of Machine Elements*, first edition, Prentice-Hall, 1948.
Steeds, W., *Involute Gears*, Longmans, Green & Company, 1948.

Problems

5.1 A 12-tooth pinion is to be cut with a 2-pitch, 20° full-depth hob. Make a layout of the theoretical rack and pinion tooth at the standard setting as shown in Fig. 5.2*a*. Draw the pinion involute by the approximate method but do not draw the flanks of the pinion tooth. Show the effect on the pinion tooth of withdrawing the basic rack until its addendum line just passes through the interference point. This layout should be shown dotted and superimposed upon the first sketch with the side of the rack tooth passing through the pitch point. Label the base circle, cutting pitch circle, hob offset, pressure angle, and pitch lines (cutting and standard) of the rack.

5.2 A 24-tooth pinion is to be cut with a 10-pitch, $14\frac{1}{2}°$ full-depth hob. Calculate the minimum distance the hob will have to be withdrawn to avoid undercutting. Calculate the radius of the cutting pitch circle and the tooth thickness on the cutting pitch circle.

5.3 A 26-tooth gear is to be cut with a 7-pitch, 20° full-depth hob. Calculate the maximum distance the hob can be advanced into the gear blank without causing undercutting. Calculate the radius of the cutting pitch circle and the tooth thickness on the cutting pitch circle.

5.4 A 20-tooth gear is cut by a 4-pitch, $14\frac{1}{2}°$ full-depth hob that has been withdrawn 0.10 in. Determine if this hob offset is enough to eliminate undercutting. If so, calculate the tooth thickness on the cutting pitch circle and on the base circle.

5.5 A 35-tooth gear is to be cut with a 4-pitch, $14\frac{1}{2}°$ full-depth hob. Calculate the change in cutter setting from its standard position to give a tooth thickness of 0.400 in. on a circle for which the pressure angle is 20°.

5.6 A 20-tooth pinion is to be cut with a 6-pitch, 20° full-depth hob. What would be the change in cutter setting from its standard position to give a tooth thickness of 0.274 in. on a circle for which the pressure angle is $14\frac{1}{2}°$?

5.7 A 20-tooth pinion is to be cut with a 6-pitch, 20° full-depth hob. Calculate the minimum tooth width that can be produced on a circle for which the pressure angle is $14\frac{1}{2}°$. The tooth is not to be undercut.

5.8 A pinion of 11 teeth and a gear of 14 teeth were cut with an 8-pitch, 20° full-depth hob. To avoid undercutting the hob was withdrawn 0.0446 in. on the pinion and 0.0227 in. on the gear. Calculate the pressure angle and the center distance at which these gears will operate when meshed together. Determine the difference between the center distance calculated above and the standard center distance, and compare with $e_1 + e_2$.

5.9 Prove that

$$(e_1 + e_2) > \Delta C \qquad \text{for} \qquad \phi' > \phi$$

and that

$$(e_1 + e_2) < \Delta C \qquad \text{for} \qquad \phi' < \phi$$

5.10 A pinion of 15 teeth and a gear of 21 teeth are to be cut with a 6-pitch, $14\frac{1}{2}°$

full-depth hob to operate on a center distance of 3.20 in. Determine whether these gears can be cut without undercutting to operate at this center distance.

5.11 Using the data from Example 5.2, calculate the outside radii of the gear blanks, the depth of cut, and the contact ratio.

5.12 A pinion and gear of 13 teeth and 24 teeth respectively are to be cut by a 4-pitch, 20° full-depth hob to operate at a center distance of 4.83 in. Calculate the pressure angle at which the gears will operate and the value of e_1 and e_2. Let e_1 and e_2 vary inversely as the number of teeth. Check e_1 to see if it is large enough to prevent undercutting. Determine the outside radii of the gear blanks, the depth of cut, and the contact ratio.

5.13 Using the data from Example 5.3, check to see if the value of e_1 is large enough to avoid undercutting. Calculate the outside radii of the gear blanks, the depth of cut, and the contact ratio.

5.14 A 12-tooth pinion has a tooth thickness of 0.2608 in. on its cutting pitch circle. A 32-tooth gear that meshes with the pinion has a tooth thickness of 0.1880 in. on its cutting pitch circle. If both gears have been cut by a 7-pitch, 20° full-depth hob, calculate the hob offset e used in cutting each gear and the pressure angle at which the gears operate.

5.15 A nonstandard 35-tooth pinion has a tooth thickness of 0.188 in. at a radius of 2.50 in. and a pressure angle of 20°. The pinion meshes with a rack at the 2.50 in. radius with zero backlash. If the rack is 7-pitch, 20° full-depth, calculate the distance from the center of the pinion to the standard pitch line of the rack.

5.16 An 11-tooth pinion is to drive a 23-tooth gear at a center distance of 2.00 in. If the gears are to be cut by a 9-pitch, 20° full-depth hob, calculate the value of e_1 and e_2 so that the beginning of contact during cutting of the pinion occurs at the interference point of the pinion.

5.17 A 10-pitch, 20° full-depth pinion with 20 teeth drives a gear with 30 teeth at a center distance of 2.50 in. It is necessary to replace these gears with a pair that will give a velocity ratio of $1\frac{1}{3}$:1 and yet maintain the same center distance. Using the same diametral pitch hob as the original gears, select a pair of gears for the job which vary as little as possible from standard gears. Determine the hob offsets, the outside radii, and the depth of cut.

5.18 It is necessary to connect two shafts whose center distance is 3.90 in. with a pair of spur gears having a velocity ratio of 1.25:1. Using a 10-pitch, $14\frac{1}{2}$° full-depth hob, recommend a pair of gears for the job whose angular velocity ratio will approach 1.25:1 as closely as possible and not be undercut. Calculate the hob offsets, the outside diameters, depth of cut, and the contact ratio.

5.19 A pinion and gear of 27 and 39 teeth respectively are to be cut with a 6-pitch, $14\frac{1}{2}$° full-depth hob to give long and short addendum teeth. The hob is offset 0.03 in. Determine for each gear the pitch diameter, the outside diameter, the depth of cut, and the tooth thickness on the pitch circle.

5.20 A pair of long and short addendum gears of 18 and 28 teeth are cut with a 4-pitch, 20° full-depth hob that has been offset 0.06 in. Compare the contact ratio of these gears with the contact ratio of a pair of standard gears of the same pitch and numbers of teeth.

5.21 A 20-pitch, 20° full-depth pinion with 30 teeth is to mesh with a 40-tooth gear at the standard center distance. If 0.004 in. backlash is required, calculate the amount the hob must be fed into the pinion and into the gear to give this backlash. Assume both gears to be thinned the same amount.

5.22 An 8-pitch, 25° pinion with 20 teeth is to mesh with a 40-tooth gear at a center distance of 3.80 in. If the hob is pulled out 0.0352 in. when cutting the pinion and 0.0165 in. when cutting the gear, calculate the backlash produced.

5.23 A pair of long and short addendum gears of 18 and 30 teeth respectively cut with a 6-pitch, 25° hob are designed to give zero backlash when the hob is offset 0.05 in. Calculate the value of e_1 and e_2 if these gears are modified to give a backlash of 0.005 in., assuming that both gears are thinned the same amount.

5.24 A 12-pitch, 20° pinion of 18 teeth drives a gear of 42 teeth. If these gears are semi recess-action gears, calculate the ratio of the recess action to the approach action.

5.25 A pair of semi recess-action gears mesh together without blacklash. The pinion has 20 teeth and the gear 48 teeth. If the gears are cut with a 10-pitch, 20° full-depth hob, calculate the contact ratio.

5.26 A pair of recess-action gears are to be designed to mesh together without backlash. The pinion is to have 20 teeth and the gear 44 teeth and the gears are to be cut with an 8-pitch, 20° full-depth hob. Calculate whether a contact ratio of 1.40 can be attained using semi or full recess-action gears or both.

5.27 A 10-pitch, 20° pinion of 24 teeth drives a gear of 40 teeth. The gears have semi recess-action and the length of action $Z = 0.468$ in. Calculate the ratio of recess action to approach action.

6

Bevel, Helical, and Worm Gearing

Bevel Gearing

6.1 Theory of Bevel Gears. Bevel gears (Fig. 6.1) are used to connect shafts whose axes intersect. The *shaft angle* is defined as the angle between the center lines which contains the engaging teeth. Although the shaft angle is usually 90°, there are many bevel gear applications that require shaft angles of greater or less than this amount.

The pitch surface of a bevel gear is a cone. When two bevel gears mesh together, their cones contact along a common element and have a common apex where the shaft center lines intersect. The cones roll together without slipping and have spherical motion. Each point in a bevel gear remains at a constant distance from the common apex.

Figure 6.2 shows an axial section of a pair of bevel gears in mesh with the shafts at right angles. Because the pitch cones roll together without slipping, the angular velocity ratio is inversely proportional to the diameters of the bases of the cones. These cone diameters become the pitch diameters of the gears. The angular velocity ratio can then be expressed as $\omega_1/\omega_2 = D_2/D_1 = N_2/N_1$ as in the case of spur gears. The relation $P = N/D$ also holds as in spur gears.

In making a sketch of a pair of spur gears in mesh, it was a simple matter, knowing the pitch diameters, to draw the pitch circles in their correct position. In the case of bevel gears, however, pitch angles as well as pitch diameters

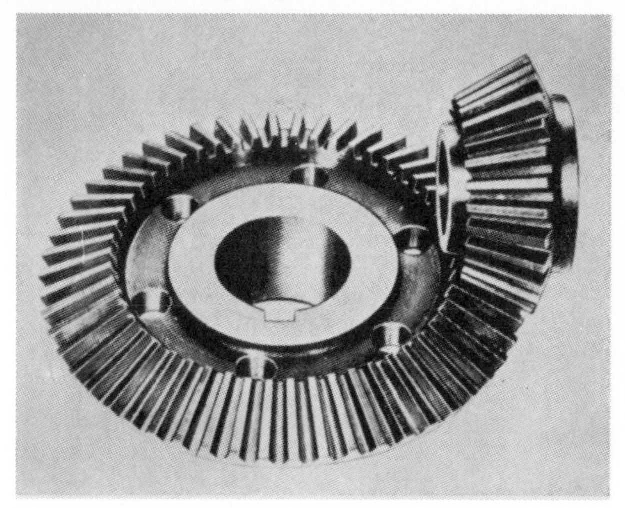

Figure 6.1 Straight bevel gears. (Courtesy of Gleason Works.)

have to be considered. Equations for the pitch angles are derived below; A_o is the length of the pitch cone element.

$$\sin \Gamma_1 = \frac{D_1}{2A_o} = \sin (\Sigma - \Gamma_2)$$

$$\sin \Gamma_1 = \sin \Sigma \cos \Gamma_2 - \cos \Sigma \sin \Gamma_2$$

$$\frac{\sin \Gamma_1}{\sin \Sigma \sin \Gamma_2} = \frac{\cos \Gamma_2}{\sin \Gamma_2} - \frac{\cos \Sigma}{\sin \Sigma}$$

$$\frac{1}{\sin \Sigma}\left[\frac{\sin \Gamma_1}{\sin \Gamma_2} + \cos \Sigma\right] = \frac{1}{\tan \Gamma_2}$$

Also

$$\frac{\sin \Gamma_1}{\sin \Gamma_2} = \frac{D_1}{D_2}$$

Therefore,

$$\tan \Gamma_2 = \frac{\sin \Sigma}{\cos \Sigma + D_1/D_2} = \frac{\sin \Sigma}{\cos \Sigma + N_1/N_2} \tag{6.1}$$

Similarly,

$$\tan \Gamma_1 = \frac{\sin \Sigma}{\cos \Sigma + N_2/N_1} \tag{6.2}$$

Although Eqs. 6.1 and 6.2 were derived for gears with shafts at right angles, these equations also apply to bevel gears with any shaft angle.

In making a layout of a pair of bevel gears in mesh, the position of the

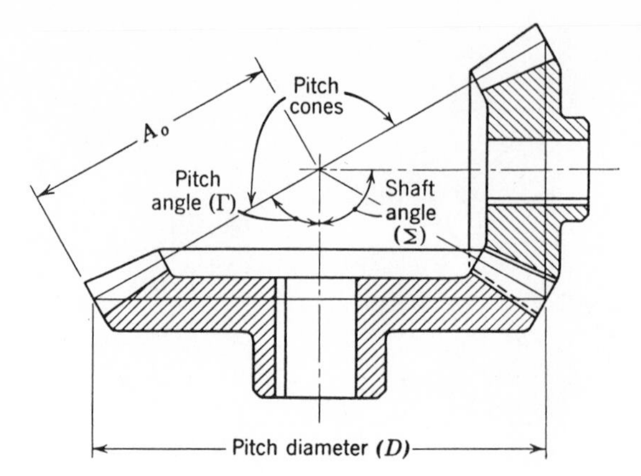

Figure 6.2

common pitch cone element can be determined graphically if the angular velocity ratio and the shaft angle are known.

As has been mentioned, the pitch cones of a pair of bevel gears have spherical motion. Therefore, in order to have the large ends of the bevel gear teeth match perfectly when in mesh, they should lie in the surface of a sphere whose center is the apex of the pitch cones and whose radius is the common pitch cone element. It is not customary, however, to make the back of a bevel gear spherical, so it is made conical as shown in Fig. 6.3. This cone is known as the *back cone* and is tangent to the theoretical sphere at the pitch diameter. The elements of the back cone are therefore perpendicular to those of the pitch cone. For all practical purposes, the surface of the back cone and the surface of the sphere are identical in the region of the ends of the bevel gear teeth. The distances from the apex of the pitch cones to the outer ends of the teeth at any point except the pitch point are not equal, so the end surfaces of meshing teeth will not be quite flush. However, this variation is slight and does not affect tooth action.

All the proportions of the tooth of a bevel gear are figured at the large end of the tooth. This will be amplified in a later section. When it is necessary to show the outline of the large end of the tooth, use is made of the fact that the profile of the bevel gear tooth closely corresponds to that of a spur gear tooth having a pitch radius equal to the back cone element and a diametral pitch equal to that of the bevel gear. This spur gear is called the *equivalent spur gear*, and this section through the bevel gear is known as the *transverse section.*

In addition to the general type of bevel gears shown in Fig. 6.2, there are the following three special types:

1. *Miter Gears*. The gears are of equal size and the shaft angle is 90°.

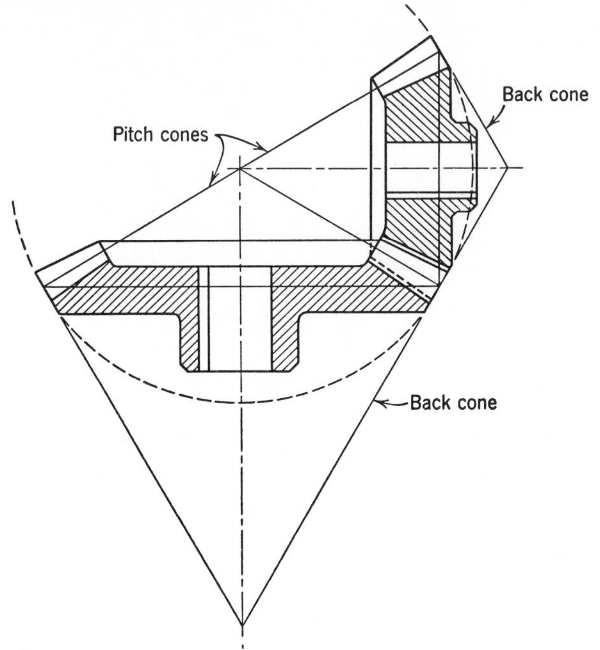

Figure 6.3

2. *Angular Bevel Gears*. The shaft angle is greater or less than 90°. A sketch is shown in Fig. 6.4.

3. *Crown Gear*. The pitch angle equals 90°, and the pitch surface becomes a plane. A sketch is shown in Fig. 6.5.

Up to the present time, the discussion has dealt primarily with the general theory and types of bevel gears. We are now ready to consider the form of the bevel gear tooth.

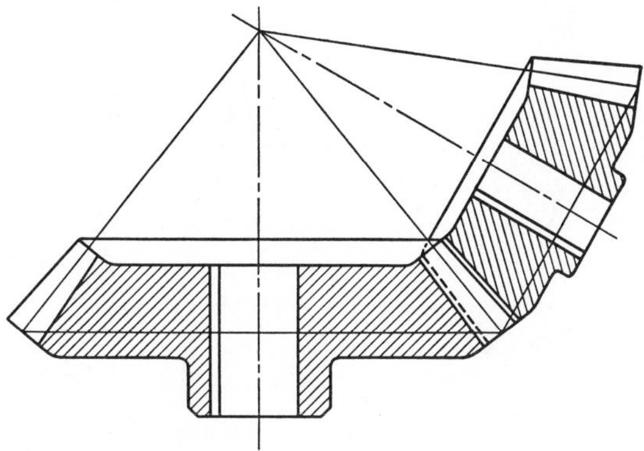

Figure 6.4

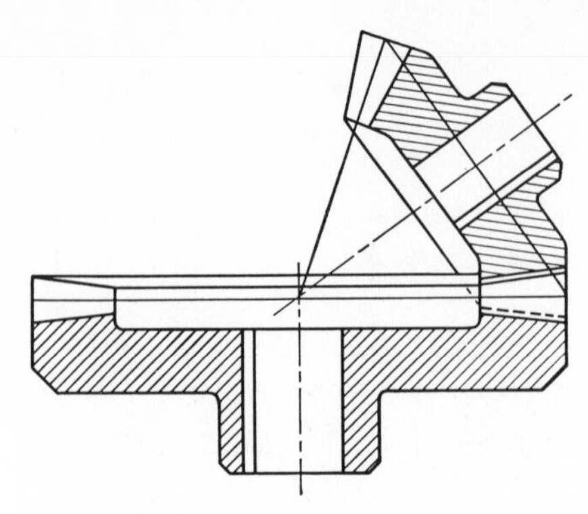

Figure 6.5

As was seen from the study of Chapter 4, the involute profile of a spur gear was easily generated from a base circle and took the form of a cylindrical involute when the thickness of the gear was considered. The involute form is not used for bevel gears, however, because the base surface would be a cone. This means that, when a plane is rolled on this base cone, a line in the plane generates a spherical involute. A spherical involute is impractical to manufacture.

The bevel gear system that has been developed is one in which the teeth are generated conjugate to a crown gear having teeth with flat sides. The crown gear therefore bears the same relation to bevel gears as a rack does to spur gears. Figure 6.6 shows the sketch of a theoretical crown gear. The sides of the teeth lie in planes which pass through the center of the sphere. When the crown gear is meshed with a conjugate gear, the complete path of

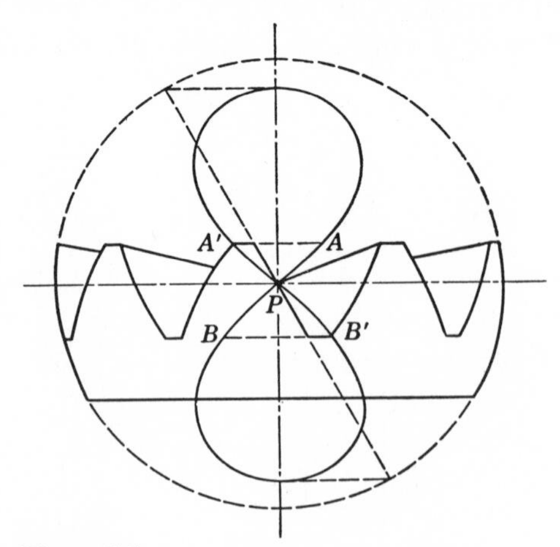

Figure 6.6

contact in the surface of the sphere is in the form of a figure 8. Because of this, the teeth in the crown gear and in the conjugate gear are called *octoid teeth*. Only a portion of the path is used, and for teeth of the height shown the path of contact is either APB or $A'PB'$.

6.2 Bevel Gear Details. In order to consider the details of a bevel gear, an axial section of a pair of Gleason straight-tooth bevel gears is shown in Fig. 6.7*a*. The Gleason system has been adopted as the standard for bevel gears. As seen in the sketch the dedendum elements are drawn toward the apex of the pitch cones. The addendum elements, however, are drawn parallel to the dedendum elements of the mating member, thus giving a constant clearance and eliminating possible fillet interference at the small ends of the teeth. Elimination of this possible interference allows larger edge radii to be used on the generating tools, which will increase tooth strength through increased fillets. The large ends of the teeth are proportioned according to the long and short addendum system discussed in Chapter 5 so that the addendum on the pinion will be greater than that on the gear. Long addendums are used on the pinion primarily to avoid undercut, to balance tooth wear, and to increase tooth strength. The Gleason standard for the proportions of straight bevel gear teeth is given in a later section. Figure 6.7*b* is the transverse section A–A showing the tooth profiles.

The addendum and dedendum are measured perpendicularly to the pitch cone element at the outside of the gear; therefore the dedendum angle is given by

$$\tan \delta = \frac{b}{A_o} \tag{6.3}$$

Because the addendum element is not drawn toward the apex of the pitch cones, the addendum angle α must be determined indirectly. It can be shown that the addendum angle of the pinion will equal the dedendum angle of the gear. Likewise, the addendum angle of the gear will equal the dedendum angle of the pinion. The face angle and the root angle are therefore

$$\Gamma_O = \Gamma + \alpha \tag{6.4}$$

$$\Gamma_R = \Gamma - \delta \tag{6.5}$$

Because the back angle is equal to the pitch angle, the outside diameter of a bevel gear is

$$D_O = D + 2a \cos \Gamma \tag{6.6}$$

The face width of a bevel gear is not determined by the kinematics of tooth action but by requirements of manufacture and load capacity. If the tooth is made too great a proportion of the cone distance A_o, manufacturing

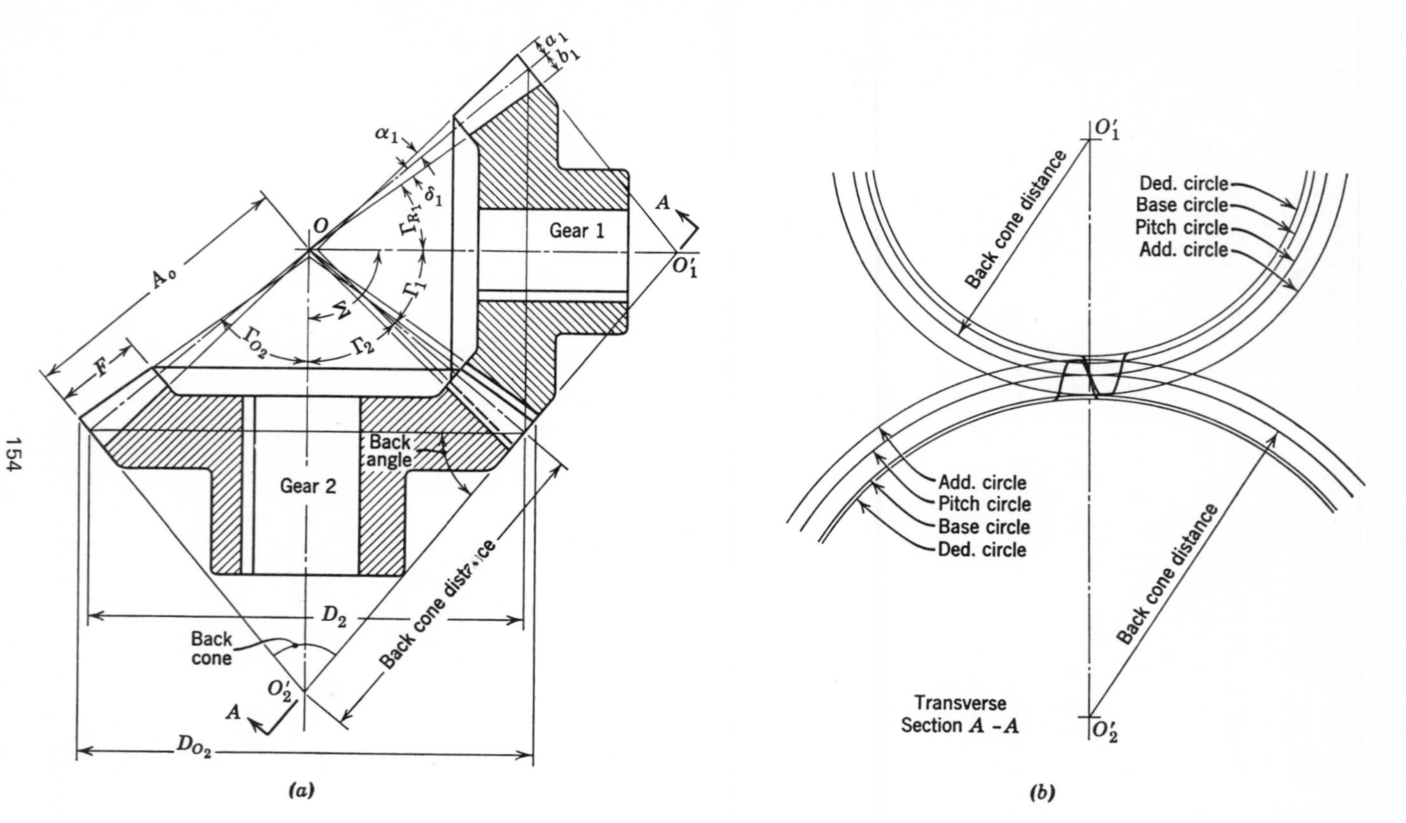

Figure 6.7 D = pitch diameter. D_O = outside diameter. A_O = pitch cone distance. F = face width, a = addendum. b = dedendum Σ = shaft angle. Γ = pitch angle. α = addendum angle. δ = dedendum angle. Γ_O = face angle. Γ_R = root angle.

difficulties are encountered so that the face width is limited as follows:

$$F < \frac{A_o}{3} \quad \text{or} \quad \frac{10}{P} \quad \text{(whichever is smaller)} \tag{6.7}$$

Although integral diametral pitches are frequently used on bevel gears, there is not the same necessity for restricting designs on this account since tooling for bevel gears is not limited to standard pitches as in the case of spur gearing.

6.3 Gleason Straight Bevel Gear Tooth Proportions. (For straight bevel gears with axes at right angles and 13 or more pinion teeth.)

1. Numbers of teeth:

   16 or more teeth in the pinion
   15 teeth in pinion and 17 or more teeth in gear
   14 teeth in pinion and 20 or more teeth in gear
   13 teeth in pinion and 30 or more teeth in gear

2. Pressure angle, $\phi = 20°$.
3. Working depth, $h_k = 2.000/P$.
4. Whole depth, $h_t = 2.188/P + 0.002$.
5. Addendum:

$$\text{Gear:} \quad a_G = \frac{0.540}{P} + \frac{0.460}{P(N_2/N_1)^2}$$

$$\text{Pinion:} \quad a_P = \frac{2.00}{P} - a_G$$

6. Dedendum:

$$\text{Gear:} \quad b_G = \frac{2.188}{P} - a_G$$

$$\text{Pinion:} \quad b_P = \frac{2.188}{P} - a_P$$

7. Circular thickness (tooth thickness on pitch circle):

$$\text{Gear:} \quad t_G = \frac{p}{2} - (a_P - a_G)\tan\phi \ \text{(approximately)}^1$$

$$\text{Pinion:} \quad t_P = p - t_G$$

where p is the circular pitch.

[1] To obtain the exact value, a set of curves is necessary which is not suitable for inclusion here. See Gleason, *Design Manual*.

6.4 Angular Straight Bevel Gears. The proportions of angular straight bevel gears can be determined from the same relations as given for bevel gears at right angles with the following exceptions:

1. The limiting numbers of teeth cannot be taken from item one in Section 6.3. Each application must be examined separately for undercutting with the aid of a chart in Gleason's *Design Manual*. This chart shows a plot of maximum pinion dedendum angle for no undercut versus pitch angle. Curves are given for several pressure angles.
2. The pressure angle is determined in conjunction with the preceding item.
3. In determining the gear addendum from item five in Section 6.3, it is necessary to use an equivalent 90° bevel gear ratio for the ratio N_2/N_1.

$$\text{Equivalent } 90^\circ \text{ ratio} = \sqrt{\frac{N_2 \cos \Gamma_1}{N_1 \cos \Gamma_2}}$$

For a crown gear ($\Gamma = 90°$) this ratio equals infinity.

For angular bevel gears where the shaft angle is greater than 90° and the pitch angle of the gear is also greater than 90°, an internal bevel gear results. In this case, the calculations should be referred to the Gleason Works to determine whether the gears may be cut.

6.5 Zerol Bevel Gears. In addition to straight bevel gears, there are two other types of bevel gears, one of which is the Zerol bevel. Zerol bevel gears have curved teeth with zero spiral angle at the middle of the face width as shown in Fig. 6.8 and have the same thrust and tooth action as straight bevel gears. They may therefore be used in the same mountings. The advantage of the Zerol gear over the straight bevel is that it can have its tooth surfaces ground. Also the Zerol gear has localized tooth contact, that is, contact over only the central portion of the tooth instead of along the entire tooth, whereas the straight bevel may or may not, depending upon the bevel gear generator used. Modern straight bevel gear generators produce a tooth with localized bearing by curving the teeth along their length ever so

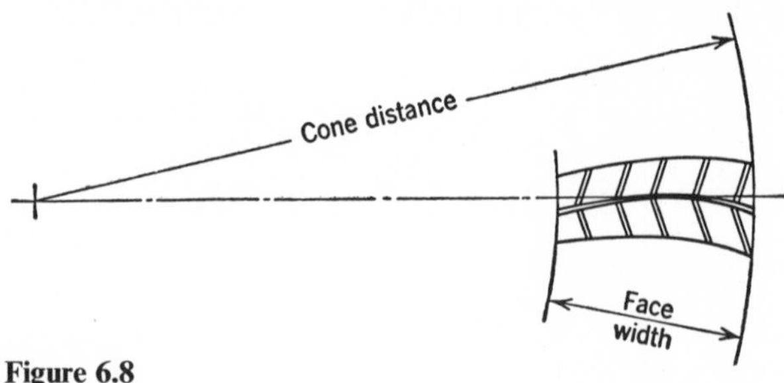

Figure 6.8

(a)
(b)

Figure 6.9 (*a*) Coniflex bevel gears showing localized contact. (*b*) Zerol bevel gears showing localized contact. (Courtesy of Gleason Works.)

slightly. Mating teeth are therefore slightly convex so that contact takes place near the middle of the tooth. A straight bevel gear with this feature is known as a Coniflex gear. The localized contact allows a slight amount of adjustment during assembly and some displacement due to deflection under operating loads without concentrating the load on the ends of the teeth. Photographs of Coniflex and Zerol bevel gears showing localized contact appear in Fig. 6.9.

6.6 Spiral Bevel Gears. The second type is the spiral bevel gear, which has obliquely curved teeth. Figure 6.10*a* shows a section of a pair of teeth in contact, and Fig. 6.10*b* shows the tooth spiral of one gear. The teeth are given a spiral angle such that the face advance (Fig. 6.10*b*) is greater than the circular pitch, which results in continuous pitch line contact in the plane of the axes of the gears. This makes it possible to obtain smooth operation with a smaller number of teeth in the pinion than with straight or Zerol bevel gears which do not have continuous pitch line contact. Also, in spiral bevel gears, contact between the teeth begins at one end of the tooth and progresses obliquely across the face of the tooth. This is in contrast to the tooth action of straight or Zerol bevel gears, where contact takes place all at once across the entire width of face. Therefore, for these reasons spiral bevel gears have smoother action than either straight or Zerol bevel gears and are especially suitable for high-speed work. As shown in Fig. 6.10*a*, spiral bevel gears have localized tooth contact, which is easily controlled by varying the radii of curvature of mating teeth. Spiral bevel gears can also have their tooth surfaces ground. Figure 6.11 shows a pair of spiral bevel gears in mesh.

6.7 Hypoid Gears. At one time, spiral bevel gears were used exclusively for automotive rear axle drive gears (ring gear and pinion). In 1925, Gleason introduced the hypoid gear, which has replaced the spiral bevel for this application. Hypoid gears are similar in appearance to spiral bevel gears with the exception that the axis of the pinion is offset from that of the gear so that the axes no longer intersect. See Fig. 6.12. To take care of this offset and still maintain line contact, the pitch surface of a hypoid gear approaches a hyperboloid of revolution rather than a cone as in bevel gears. In automotive applications the offset is advantageous because it allows the drive shaft to be lowered, resulting in a lower slung body. In addition, hypoid pinions are stronger than spiral bevel pinions. The reason for this is that hypoid gears can be designed so that the spiral angle of the pinion is larger than that of the gear. This results in a larger pinion diameter, and hence stronger, than the corresponding spiral bevel pinion. Another difference is that hypoid gears have sliding action along the teeth, whereas spiral bevel gears do not. Hypoid gears operate more quietly and can be used for higher reduction ratios than spiral bevel gears. Hypoid gears can also be ground.

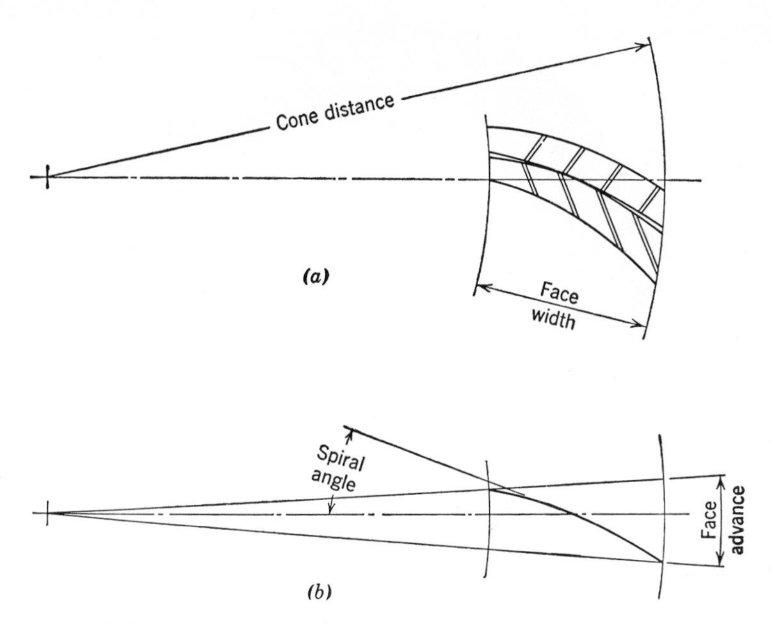

Figure 6.10

Figure 6.11 Spiral bevel gears. (Courtesy of Gleason Works.)

Figure 6.12 Hypoid gears. (Courtesy of Gleason Works.)

The tooth form for Zerol bevel, spiral bevel, and hypoid gears is the long and short addendum system except when both gears have the same number of teeth. Standards similar to that given for straight bevel gears have been developed for these systems and may be found in the Gleason *Design Manual* for bevel and hypoid gears.

Helical Gearing

6.8 Theory of Helical Gears. If a plane is rolled on a base cylinder, a line in the plane parallel to the axis of the cylinder will generate the surface of an involute spur gear tooth. If the generating line is inclined to the axis, however, the surface of a helical gear tooth will be generated. These two conditions are shown in Figs. 6.13*a* and *b*, respectively.

Helical gears are used to connect parallel shafts and nonparallel, nonintersecting shafts. The former are known as *parallel helical gears* and the latter as *crossed helical gears*. See Figs. 6.14*a* and *b*.

In determining the tooth proportions of a helical gear for either crossed or parallel shafts, it is necessary to consider the manner in which the teeth are to be cut. If the gear is to be hobbed, all dimensions are figured in a plane which is normal to the tooth pitch element, and the diametral pitch and the pressure

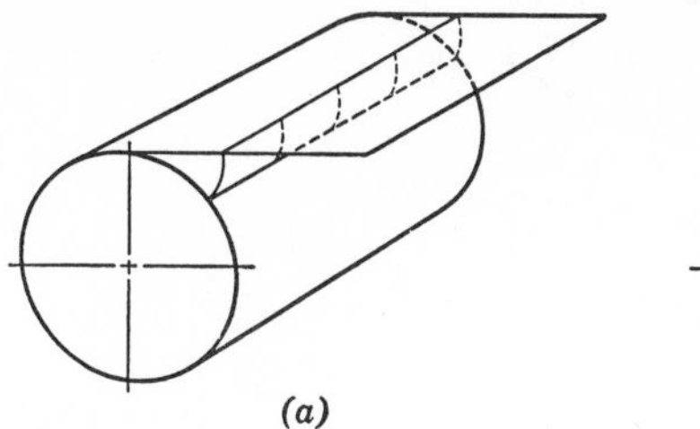

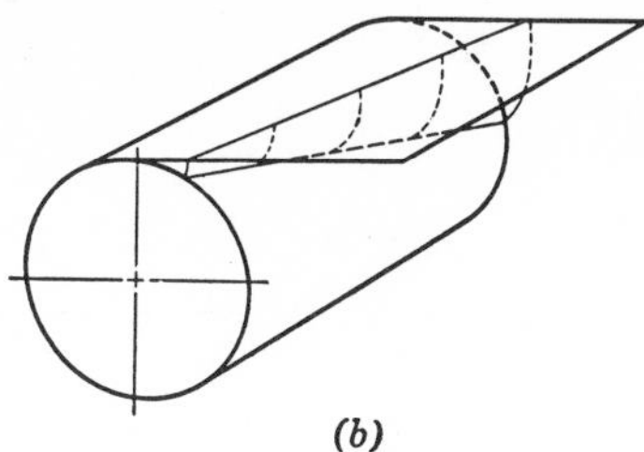

Figure 6.13

angle are standard values in that plane. Because the cutting action of a hob occurs in the normal plane, it is possible to use the same hob to cut both helical gears and spur gears of a given pitch; in a spur gear the normal plane and the plane of rotation are identical. Figure 6.15 shows a sketch of a helical gear with the circular pitch measured in the normal plane and in the plane of rotation. From the sketch

$$p_n = p \cos \psi = \frac{\pi \cos \psi}{P} \tag{6.8}$$

where P = diametral pitch in plane of rotation (also known as *transverse diametral pitch*).

Figure 6.14 Helical gears (*a*) for parallel shafts and (*b*) for crossed shafts. (Courtesy of D. O. James Gear Manufacturing Company.)

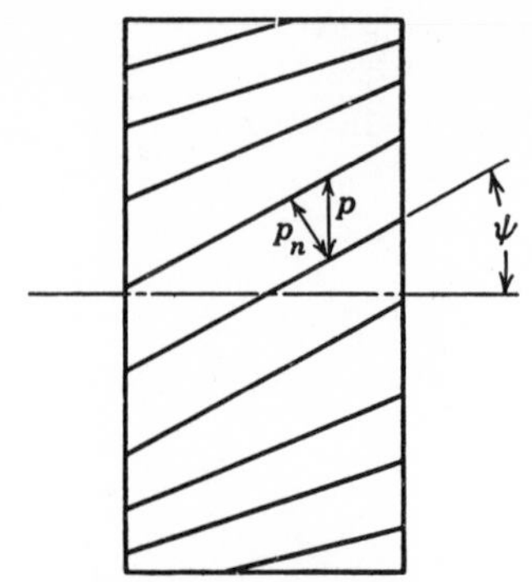

Figure 6.15 p_n = normal circular pitch. p = circular pitch in plane of rotation. ψ = helix angle.

When a helical gear is cut by a hob, the normal circular pitch p_n of Fig. 6.15 becomes equal to the circular pitch of the hob. From this and from the fact that $p = \pi/P$ the following relation can be written:

$$p_n = \frac{\pi}{P_n}$$

where P_n = normal diametral pitch and is equal to the diametral pitch of the hob. Substituting for p_n in Eq. 6.8,

$$P = P_n \cos \psi \tag{6.9}$$

Also by substituting $P = N/D$ in Eq. 6.9,

$$D = \frac{N}{P_n \cos \psi} \tag{6.10}$$

While it is not the intent to go into detail concerning the forces acting on a helical gear, it is necessary to consider them in determining the relation between pressure angle in the plane of rotation ϕ and the normal pressure angle ϕ_n and the helix angle ψ. From Fig. 6.16 showing these forces,

$$\tan \phi = \frac{F_s}{F_t} \text{(plane } OABH\text{)}$$

$$\tan \phi_n = \frac{F_s}{OD} \text{(plane } ODC\text{)}$$

$$OD = \frac{F_t}{\cos \psi} \text{(plane } OADG\text{)}$$

Substituting,

$$\tan \phi_n = \frac{F_s \cos \psi}{F_t}$$

and

$$\tan \phi = \frac{\tan \phi_n}{\cos \psi} \tag{6.11}$$

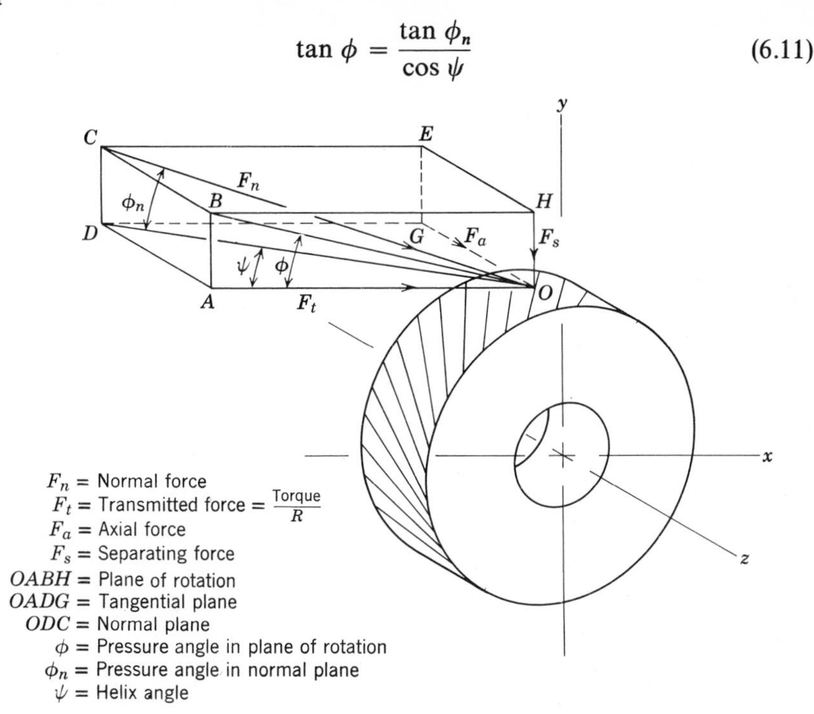

Figure 6.16

It is also interesting to consider the effect of the helix angle on the number of teeth that can be cut by a hob on a helical gear without undercutting. Referring to Fig. 4.17 (spur gears), one can develop the equation for the minimum number of teeth for helical gears cut by a hob in the following manner:

$$\sin \phi = \frac{PE}{R}$$

also

$$\sin \phi = \frac{a}{PE} = \frac{k/P_n}{PE}$$

Multiplying the two equations together,

$$\sin^2 \phi = \frac{k}{RP_n}$$

where

$$R = \frac{N}{2P}$$

Table 6.1

	N		
ψ, degrees	ϕ_n, degrees		
	$14\frac{1}{2}$	20	25
0 (spur gears)	32	18	12
5	32	17	12
10	31	17	11
15	29	16	11
20	27	15	10
23	26	14	10
25	25	14	9
30	22	12	8
35	19	10	7
40	15	9	6
45	12	7	5

(Courtesy of AGMA). Extracted from USA Standard System—Tooth Proportions for Fine-Pitch Spur and Helical Gears (USAS B6.7-1967), with the permission of the publisher, The American Gear Manufacturers Association, 1330 Massachusetts Avenue N.W., Washington, D.C. 20005.

and

$$P_n = \frac{P}{\cos \psi}$$

Therefore,

$$\sin^2 \phi = \frac{2k \cos \psi}{N}$$

and

$$N = \frac{2k \cos \psi}{\sin^2 \phi} \tag{6.12}$$

A table has been compiled by AGMA (207.05, June 1971) to give the minimum number of teeth that can be hobbed on a helical gear without undercutting. These are given in Table 6.1 as a function of helix angle ψ and normal pressure angle ϕ_n for full-depth teeth.

If it is necessary to use a pinion smaller than those given in Table 6.1, the pinion may be cut without undercutting by withdrawing the hob in a manner similar to that shown for spur gears in Chapter 5. An equation which is equivalent to Eq. 5.2 for spur gears can be derived for helical gears as

$$e = \frac{1}{P_n}\left[k - \frac{N \sin^2 \phi}{2 \cos \psi}\right] \tag{6.13}$$

The value of e given by Eq. 6.13 is the amount the hob will have to be withdrawn in order to have the addendum line of the rack or hob just pass through the interference point of the pinion being cut.

Although most hobs are designed to have the diametral pitch a standard value in the normal plane, there are hobs produced that have the diametral pitch, a standard value in the plane of rotation. These hobs are known as *transversal hobs*, and the pitch in the plane of rotation is known as the *transverse diametral pitch*.

If the gear is to be cut by the Fellows gear shaper method, the dimensions are considered in the plane of rotation, and the diametral pitch and the pressure angle are standard values in that plane. When a helical gear is cut by a Fellows cutter, the circular pitch p of Fig. 6.15 becomes equal to the circular pitch of the cutter so that the following relations apply:

$$p = \frac{\pi D}{N} = \frac{\pi}{P} \tag{6.14}$$

and

$$P = \frac{N}{D} \tag{6.15}$$

In the Fellows method, the same cutter cannot be used to cut both helical and spur gears.

The features discussed apply to helical gears with parallel shafts and with crossed shafts. The two types will now be considered separately.

6.9 Parallel Helical Gears. For parallel helical gears to mesh properly the following conditions must be satisfied:

1. Equal helix angles.
2. Equal pitches.
3. Opposite hand, that is, one gear with a left-hand helix and the other with a right-hand helix.

The velocity ratio is

$$\frac{\omega_1}{\omega_2} = \frac{D_2}{D_1} = \frac{N_2}{N_1} \tag{6.16}$$

The spur gear equation for center distance

$$C = \frac{(N_1 + N_2)}{2P}$$

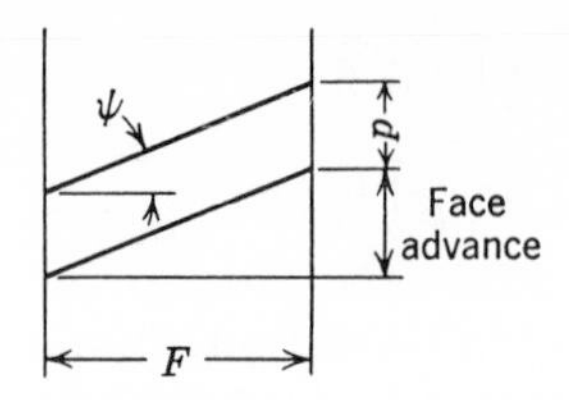

Figure 6.17

can also be used for parallel helical gears provided P is the diametral pitch in the plane of rotation.

In a parallel helical gear, the face width is made large enough so that for a given helix angle ψ the face advance is greater than the circular pitch, as illustrated in Fig. 6.17. It will give continuous contact in the axial plane as the gears rotate. This ratio (face advance to circular pitch) may be considered as a contact ratio. From Fig. 6.17, it can be seen that to have the face advance just equal the circular pitch, the face width would have to equal $p/\tan\psi$. To provide a margin of safety, the AGMA recommends that this limiting face width be increased by at least 15%, which results in the following equation

$$F > \frac{1.15p}{\tan\psi} \tag{6.17}$$

In addition to the contact ratio resulting from the twist of the teeth, parallel helical gears will also have a contact ratio in the plane of rotation the same as spur gears. The total contact ratio will therefore be the sum of these two values, and is greater than that for spur gears.

Helical gears connecting parallel shafts have line contact similar to spur gears. However, in spur gears the contact line is parallel to the axis, whereas in helical gears it runs diagonally across the face of the tooth. Parallel helical gears have smoother action and hence less noise and vibration than spur gears, and are therefore to be preferred for high-speed work. The reason for the smoother action is that the teeth come into contact gradually beginning at one end of the tooth and progressing across the tooth surface, whereas in spur gears contact takes place simultaneously over the entire face width. The disadvantage of parallel helical gears is in the end thrust produced by the tooth helix. If this end thrust is so large that it cannot be conveniently carried by the bearings, it may be counterbalanced by using two helical gears of opposite hand or by using a herringbone gear which is in effect a double helical gear cut on one blank. Figure 6.18 shows a photograph of a herringbone gear.

Example 6.1

As an example of parallel helical gears, consider that, in order to reduce the noise in a gear drive, two 16-pitch, 20° full-depth spur gears of 30 and 80 teeth are to be replaced by helical gears. The center distance and the angular velocity ratio must remain the same. Determine the helix angle, the outside diameters, and the face width of the new gears. Assume the helical gears to be cut by a 16-pitch, 20° full-depth hob. From the spur gear data,

$$C = \frac{N_1 + N_2}{2P} = \frac{30 + 80}{2 \times 16} = 3.4375 \text{ in.}$$

$$\frac{\omega_1}{\omega_2} = \frac{N_2}{N_1} = \frac{80}{30} = \frac{40}{15}$$

Figure 6.18 Herringbone gears. (Courtesy of D. O. James Gear Manufacturing Company.)

For the helical gears,

$$P_n = 16 \qquad C = \frac{N_1 + N_2}{2P} \quad \text{or} \quad P = \frac{N_1 + N_2}{2C}$$

$$P < 16$$

$$N_2 = \frac{40}{15} N_1 \qquad P = \frac{(\frac{55}{15})N_1}{2(\frac{55}{16})} = \frac{8}{15} N_1$$

$$C = 3.4375 \text{ in.} \quad (3\tfrac{7}{16} \text{ in.})$$

By trial find numbers of teeth and P.

N_1	N_2	P	Remarks
30	80	16	Original spur gears
29	77.33	15.47	N_2 not whole number
28	74.67	14.93	N_2 not whole number
27	72	14.40	Satisfactory to use

Therefore, let

$$N_1 = 27$$

$$N_2 = 72$$

$$\cos \psi = \frac{P}{P_n} = \frac{14.40}{16} = 0.9000$$

$$\psi = 25.84°$$

There are other combinations of numbers of teeth and helix angle that will satisfy the conditions, but the one listed should be selected because it will give the smallest helix angle.

The outside diameters of the two gears are

$$D_{O_1} = D_1 + 2a = \frac{N_1}{P} + 2\left(\frac{k}{P_n}\right) = \frac{27}{14.4} + 2\left(\frac{1}{16}\right) = 2.000 \text{ in.}$$

$$D_{O_2} = D_2 + 2a = \frac{N_2}{P} + 2\left(\frac{k}{P_n}\right) = \frac{72}{14.4} + 2\left(\frac{1}{16}\right) = 5.125 \text{ in.}$$

Note that the addendum was calculated using the diametral pitch of the hob (P_n).

The face width is

$$F > \frac{1.15p}{\tan \psi}$$

$$p = \frac{\pi}{P} = \frac{\pi}{14.4} = 0.2185 \text{ in.}$$

Therefore,

$$F > \frac{(1.15)(0.2185)}{\tan 25.84°} > 0.5189 \text{ in.}$$

Use

$$F = \frac{9}{16} \text{ in.}$$

6.10 Crossed Helical Gears. For crossed helical gears to mesh properly there is only one requirement, that is, they must have common normal pitches. Their pitches in the plane of rotation are not necessarily and not usually equal. Their helix angles may or may not be equal, and the gears may be of the same or of opposite hand. The velocity ratio is

$$\frac{\omega_1}{\omega_2} = \frac{N_2}{N_1} = \frac{D_2 \cos \psi_2}{D_1 \cos \psi_1} \tag{6.18}$$

If Σ is the angle between two shafts connected by crossed helical gears, and ψ_1 and ψ_2 the helix angles of the gears, then

$$\Sigma = \psi_1 \pm \psi_2 \tag{6.19}$$

The plus and minus signs apply, respectively, when the gears have the same or the opposite hand. Equation 6.19 is illustrated in Fig. 6.19 showing pairs of crossed helical gears in and out of mesh.

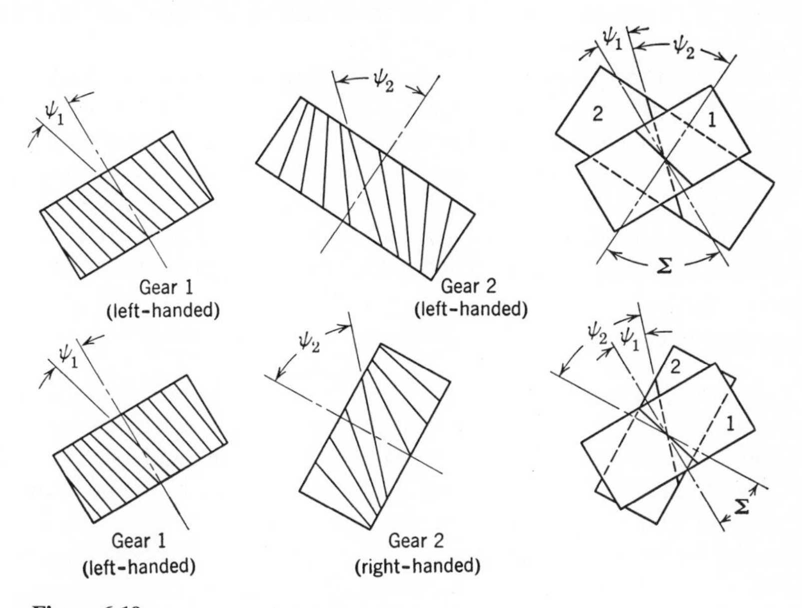

Figure 6.19

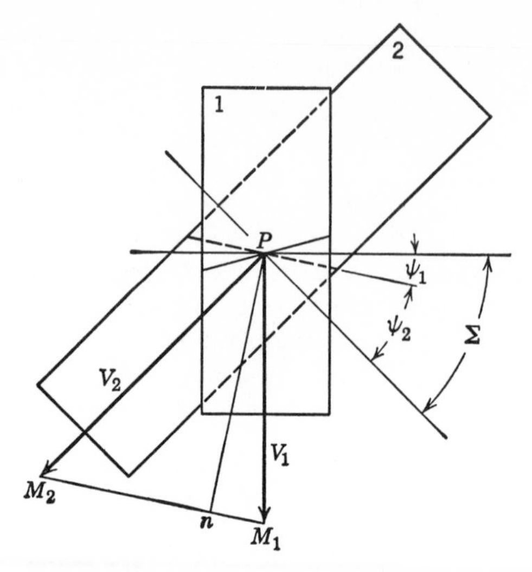

Figure 6.20

The action of crossed helical gears is quite different from that of parallel helical gears. Crossed helical gears have point contact. In addition, sliding action takes place along the tooth, which is not present in parallel helical gears. For these reasons, crossed helical gears are used to transmit only small amounts of power. An application of these gears is on the distributor drive on an automotive engine.

Using the principle of the velocity of sliding developed in Chapter 1, it is possible to determine the tooth helices across the faces of two crossed helical gears provided the peripheral velocity of the pitch point of each gear is known. Figure 6.20 shows this construction, where V_1 and V_2 are known, and it is required to find the tooth helices and helix angles for these velocities and given shaft angle. The two helices in contact at point P are parallel to the line M_1M_2. This contact occurs on the bottom of gear 1 and on the top of gear 2.

Example 6.2

To illustrate crossed helical gears, consider a pair of gears connecting two shafts at an angle of 60° with a velocity ratio of 1.5:1. The pinion has a normal diametral pitch of 6, a pitch diameter of 7.75 in., and a helix angle of 35°. Determine the helix angle and the pitch diameter of the gear and the numbers of teeth on both the pinion and the gear.

To find the helix angle of the gear, assume both gears have the same hand. Then

$$\Sigma = \psi_1 + \psi_2 \qquad \text{where} \qquad \Sigma = 60° \qquad \text{and} \qquad \psi_1 = 35°$$

Therefore,

$$\psi_2 = 25°$$

The pitch diameter of the gear can be determined as follows:

$$P_n = \frac{N_1}{D_1 \cos \psi_1} = \frac{N_2}{D_2 \cos \psi_2}$$

$$\frac{\omega_1}{\omega_2} = \frac{N_2}{N_1} = \frac{D_2 \cos \psi_2}{D_1 \cos \psi_1}$$

$$D_2 = \frac{D_1 \cos \psi_1}{\cos \psi_2} \times \frac{\omega_1}{\omega_2} = \frac{(7.75)(0.8192)(1.5)}{(0.9063)}$$

$$D_2 = 10.5 \text{ in.}$$

The numbers of teeth on the pinion and on the gear are

$$N_1 = P_n D_1 \cos \psi_1 = (6)(7.75)(\cos 35°)$$

$$N_1 = 38$$

$$N_2 = N_1 \frac{\omega_1}{\omega_2} = (38)(1.5)$$

$$N_2 = 57$$

6.11 Worm Gearing. If a tooth on a helical gear makes a complete revolution on the pitch cylinder, the resulting gear is known as a *worm*. The mating gear for a worm is designated as a *worm gear* or worm wheel; however, the worm gear is not a helical gear. A worm and worm gear are used to connect nonparallel, nonintersecting shafts usually at right angles. See Fig. 6.21. The gear reduction is generally quite large. The relation between a spur or helical gear and its hob during cutting is similar to the relation between a worm and worm gear. Worms that are true involute helical gears may be used to drive spur or helical gears, but point contact obviously results, which is unsatisfactory from the standpoint of wear. It is possible, however, to secure line contact by mating the worm with a worm gear that has been cut with a hob having the same diameter and same form of tooth as the worm. If this is done, the worm and worm gear will be conjugate, but the worm will not have involute teeth. Figure 6.22*a* shows a sketch of a worm where λ is the lead angle, ψ the helix angle, p_x the axial pitch, and D the pitch diameter. The axial pitch of the worm is the distance between corresponding points of adjacent threads measured parallel to the axis.

In considering the characteristics of a worm, the lead is of primary importance and may be defined as the axial distance that a point on the helix of the worm will move in one revolution of the worm. The relation between the

(a)

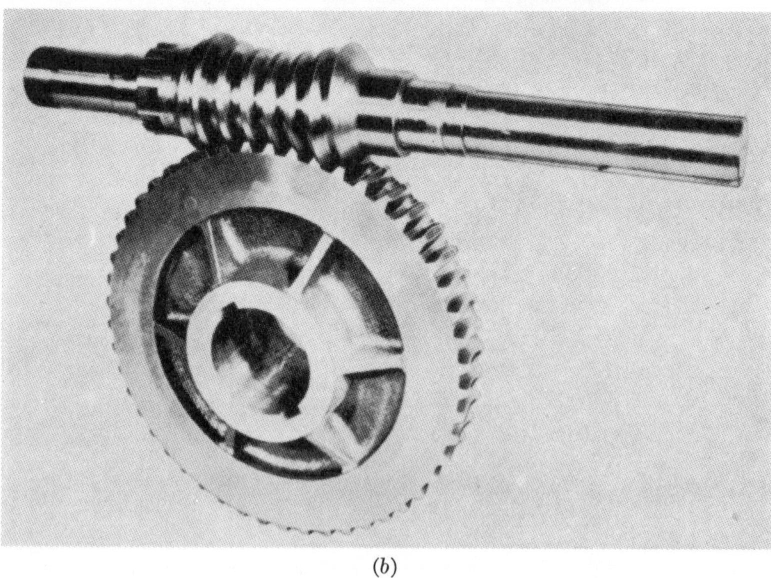

(b)

Figure 6.21 (a) Worm and worm gear. (Courtesy of Foote Brothers Gear & Manufacturing Corp.) (b) Hourglass worm and worm gear. (Courtesy of Cone Drive Gears, Division of Michigan Tool Company.)

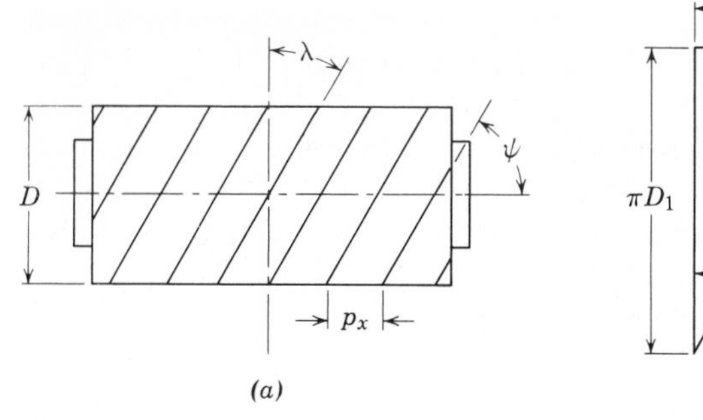

Figure 6.22

lead and axial pitch is

$$l = p_x N_1 \tag{6.20}$$

where N_1 is the number of threads (or teeth) wrapped on the pitch cylinder of the worm. A worm may be obtained with one to ten threads.

If a complete revolution of a thread on a worm is unwrapped, a triangle results as shown in Fig. 6.22*b*. From the figure it can be seen that

$$\tan \lambda = \frac{l}{\pi D_1} \tag{6.21}$$

where D_1 is the diameter of worm.

The diameter of a worm gear can be calculated from

$$D_2 = \frac{pN_2}{\pi} \tag{6.22}$$

where N_2 is the number of teeth in worm gear. The velocity ratio is

$$\frac{\omega_1}{\omega_2} = \frac{N_2}{N_1} = \frac{D_2 \cos \psi_2}{D_1 \cos \psi_1} \tag{6.23}$$

and

$$\frac{\omega_1}{\omega_2} = \frac{\pi D_2}{l} \tag{6.24}$$

for shafts at right angles.

For a worm and worm gear with shafts *at right angles* to mesh properly, the following conditions must be satisfied:

1. Lead angle of worm = helix angle of worm gear.
2. Axial pitch of worm = circular pitch of worm gear.

A worm and worm gear drive may or may not be reversible depending on the application. When used as a drive for a hoist, it is necessary that the unit be self-locking and driven only by the worm. However, if a worm drive is used for an automotive drive, it is necessary that the drive be reversible and that the worm gear be able to drive the worm. If the lead angle of the worm is greater than the friction angle of the surfaces in contact, the drive will be reversible. The coefficient of friction μ and the friction angle ϕ are related by the equation $\mu = \tan\phi$. A worm and worm gear are considered self-locking when the lead angle of the worm is less than 5°.

Example 6.3

As an example of worm gearing, consider a triple-threaded worm driving a worm gear of 60 teeth; the shaft angle is 90° as shown in Fig. 6.23. The circular pitch of the worm gear is $1\frac{1}{4}$ in., and the pitch diameter of the worm is 3.80 in. Determine the lead angle of the worm, the helix angle of the worm gear, and the distance between shaft centers.

The lead angle of the worm can be found from

$$l = p_x N_1 \qquad \text{and} \qquad p_x = p$$

$$l = 1.25 \times 3 = 3.75 \text{ in.}$$

$$\tan\lambda = \frac{l}{\pi D_1} = \frac{3.75}{\pi \times 3.80} = 0.314$$

Therefore,

$$\lambda = 17.4^\circ$$

The helix angle of the worm gear = the lead angle of the worm. Therefore,

$$\psi_2 = 17.4^\circ$$

The center distance is found by

$$D_2 = \frac{pN_2}{\pi} = \frac{(1.25)(60)}{\pi} = 23.9 \text{ in.}$$

$$C = \frac{D_1 + D_2}{2} = \frac{3.8 + 23.9}{2} = 13.85 \text{ in.}$$

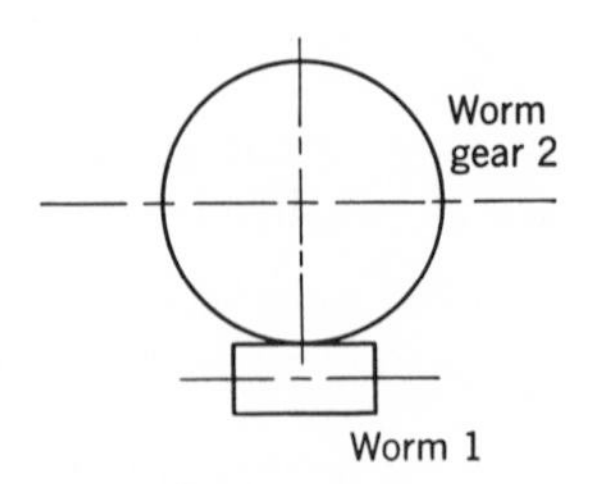

Figure 6.23

Problems

6.1 A pair of bevel gears have a velocity ratio of ω_1/ω_2, and the shaft center lines intersect at any angle Σ. If distances x and y are laid off from the intersection point along the shaft axes in the ratio ω_1/ω_2, prove that the diagonal of a parallelogram with sides x and y will be the common pitch cone element of the bevel gears.

6.2 A Gleason crown bevel gear of 24 teeth and a diametral pitch of 5 is driven by a 16-tooth pinion. Calculate the pitch diameter and pitch angle of the pinion, the addendum and dedendum, the face width, and the pitch diameter of the gear. Make a full-size axial sketch of the pinion and gear in mesh using reasonable proportions for the hubs and webs as shown in Fig. 6.7*a*.

6.3 A Gleason crown bevel gear of 48 teeth and a diametral pitch of 12 is driven by a 24-tooth pinion. (*a*) Calculate the pitch angle of the pinion and the shaft angle. (*b*) Make a sketch (to scale) of the pitch cones of the two gears in mesh. Show the back cones of each gear and label the pitch cones and the back cones.

6.4 A pair of Gleason miter gears have 20 teeth and a diametral pitch of 4. Calculate the pitch diameter, the addendum and dedendum, the face width, the pitch cone distance, the face angle, the root angle, and the outside diameter. Make a full-size axial sketch of the gears in mesh using reasonable proportions for the hub and web as shown in Fig. 6.7*a*. Dimension the drawing with the values calculated.

6.5 A Gleason 6-pitch, straight bevel pinion of 21 teeth drives a gear of 27 teeth. The shaft angle is 90°. Calculate the pitch angles, the addendums and dedendums, and the face width of each gear. Make a full-size axial sketch of the gears in mesh using reasonable proportions for the hubs and webs as shown in Fig. 6.7*a*.

6.6 A Gleason 4-pitch, straight bevel pinion of 14 teeth drives a gear of 20 teeth. The shaft angle is 90°. Calculate the addendum and dedendum, circular tooth thickness for each gear, and the pitch and base radii of the equivalent spur gears. Make a full-size sketch of the equivalent gears showing two teeth in contact as in Fig. 6.7*b*.

6.7 A Gleason 5-pitch, straight bevel pinion of 16 teeth drives a gear of 24 teeth. The shaft angle is 45°. After making the necessary calculations, lay out a full-size axial sketch of the pinion and gear in mesh using reasonable proportions for the hubs and webs as shown in Fig. 6.7*a*.

6.8 A pair of Gleason bevel gears mesh with a shaft angle of 75°. The diametral pitch is 10, and the numbers of teeth in the pinion and gear are 30 and 40, respectively. (*a*) Calculate the pitch angles and the addendums and dedendums of the pinion and gear. (*b*) Make a full-size sketch of the pitch cones and the back cones of the two gears in mesh. Label the pitch cones, back cones, and pitch angles of both gears. (*c*) Mark off (double size) the addendum and dedendum of the pinion on the sketch and clearly label them.

6.9 Prove with the aid of a suitable sketch that in a Gleason straight tooth bevel gear the addendum angle of the pinion equals the dedendum angle of the gear and that $\Gamma_o = \Gamma + \alpha$.

6.10 A 14-tooth helical gear is to be cut by a 10-pitch, 20° full-depth hob. Calculate the following: (*a*) The minimum helix angle which this gear must have in order to be cut at the standard setting without undercutting. (*b*) The amount the hob will have to be withdrawn to avoid undercutting if the helix angle of the gear is made 20°.

6.11 A 12-tooth helical pinion is to be cut with an 8-pitch, 20° full-depth hob. If the helix angle is to be 20°, calculate the amount the hob will have to be withdrawn to avoid undercutting.

6.12 Two equal spur gears of 48 teeth, 1 in. face width, and 6 diametral pitch mesh together in the drive of a fatigue tester. Calculate the helix angle of a pair of helical gears to replace the spur gears if the face width, center distance, and velocity ratio are to remain the same. Use the following cutters: (*a*) Fellows of 6 diametral pitch, (*b*) Hob of 6 normal diametral pitch.

6.13 Two standard spur gears were cut with a 10-pitch, 20° full-depth hob to give a velocity ratio of 3.5:1 and center distance of 6.75 in. Helical gears are to be cut with the same hob to replace the spur gears keeping the center distance and angular velocity ratio the same. Determine the helix angle, numbers of teeth, and face width of the new gears keeping the helix angle to a minimum.

6.14 Two standard spur gears are to be replaced by helical gears. The spur gears were cut by an 8-pitch, 20° full-depth hob, the velocity ratio is 1.75:1 and the center distance 5.5 in. The helical gears are to be cut with the same hob and maintain the same center distance. The helix angle is to be between 15° and 20° and the velocity ratio between 1.70 and 1.75. Find the numbers of teeth, helix angle, and velocity ratio.

6.15 In a proposed gear drive, two standard spur gears (16 diametral pitch and 20° full-depth) with 36 and 100 teeth are meshed at the standard center distance. It is decided to replace these spur gears with helical gears having a helix angle of 22° and the same numbers of teeth. Determine the change in center distance required if the helical gears are cut (*a*) with a 16-pitch, 20° full-depth hob, (*b*) with a 16-pitch, 20° Fellows cutter.

6.16 A pair of helical gears for parallel shafts are to be cut with an 8-pitch, 25° full-depth hob. The helix angle is to be 20° and the center distance between 6.00 and 6.25 in. The angular velocity ratio is to approach as closely as possible 2:1. Calculate the circular pitch and the diametral pitch in the plane of rotation. Determine the numbers of teeth, pitch diameters, and center distance to satisfy the above conditions.

6.17 A 10-pitch, 20-tooth spur pinion drives two gears, one of 36 teeth and the other of 48 teeth. It is desired to replace all three gears with helical gears and to change the velocity ratio between the 20-tooth gear shaft and the 48-tooth gear shaft to 2:1. The velocity ratio and the center distance between the 20-tooth gear shaft and the 36-tooth gear shaft is to remain the same. Using an 8-pitch, 20° stub hob and keeping the helix angle as low as possible, determine the number of teeth, helix angle and hand, face width, and outside diameter for each gear. Calculate the change in center distance between the shafts that originally mounted the 20- and 48-tooth gears.

6.18 A 12-pitch, 24-tooth spur pinion drives two gears, one of 36 teeth and the other of 60 teeth. It is necessary to replace all three gears with helical gears keeping the same velocity ratios and center distances. Using a 16-pitch, 20° stub hob and keeping the helix angle as low as possible, determine the number of teeth, helix angle and hand, face width, and outside diameter for each gear.

6.19 Two parallel shafts are to be connected by a pair of helical gears (gears 1 and 2). The angular velocity ratio is to be 1.25:1 and the center distance 4.5 in. In addition gear 2 is to drive a helical gear 3 whose shaft is at right angles to shaft 2. The angular velocity ratio between gears 2 and 3 is to be 2:1. Using a 9-pitch, 20° full-depth hob, determine the number of teeth, helix angle, and pitch diameter of each gear and find center distance C_{23}.

6.20 Two parallel shafts are to be connected by a pair of helical gears (gears 1 and 2). The angular velocity ratio is to be 1.75:1 and the center distance 2.75 in. In addition gear 2 is to drive a third helical gear (gear 3) with an angular velocity ratio of 2:1. Three hobs are available for cutting the gears: Hob A (7 pitch, 20° full-depth), Hob B (9 pitch, 20° full-depth), and Hob C (12 pitch, 20° full-depth). (*a*) Choose the hob which will result in the smallest helix angle, Ψ. (*b*) Which hob will permit the shortest center distance, C_{23}, between shafts 2 and 3 while maintaining a helix angle *less than* 35°?

6.21 The formula for the center distance between two spur or parallel helical gears is given by $C = (N_1 + N_2)/2P$, where C is dependent upon the number of gear teeth N_1 and N_2 and the diametral pitch P. Show that C_{23} is independent of P for three gears (spur, parallel, helical) in mesh whose center distance, C_{12}, and angular velocity ratios, ω_1/ω_2 and ω_2/ω_3, are known.

6.22 Two 18-pitch, 20° full-depth spur gears of 36 and 90 teeth are to be replaced by helical gears. The center distance and the velocity ratio are to remain the same. If the width of the gears cannot exceed $\frac{1}{2}$ in. because of space limitations, determine a pair of helical gears for this job keeping the helix angle as small as possible. Use an 18-pitch, 20° full-depth hob, and determine the numbers of teeth, helix angle, face width, and outside diameters.

6.23 Two 18-pitch, 20° full-depth spur gears of 32 and 64 teeth are to be replaced by helical gears. The center distance and velocity ratio are to remain the same. If the width of the gears cannot exceed $\frac{7}{16}$ in. because of space limitations, determine which of the following hobs should be used keeping the helix angle as small as possible: Hob A (18-pitch, 20° full-depth) or Hob B (20-pitch, 20° full-depth). In addition, determine the numbers of teeth, helix angle, face width, and outside diameters.

6.24 Two parallel shafts are to be connected by a pair of helical gears (gears 1 and 2). The angular velocity ratio is to be $1\frac{1}{3}$:1 and the center distance 3.50 in. Considering that hobs are available from 6 to 12 pitch (inclusive), tabulate the numbers of teeth, helix angle, and face width for the various combinations (of N_1 and N_2) that will satisfy the given conditions. What is the best selection for this drive? Why? Let 15 be the lowest number of teeth for the smaller gear at $P_n = 6$.

6.25 Two shafts crossing at right angles are to be connected by helical gears. The angular velocity ratio is to be $1\frac{1}{2}$:1 and the center distance 5.00 in. Assuming the gears to have equal helix angles, calculate the diametral pitch of a cutter to generate 20 teeth on the pinion if the cutter is (*a*) a hob and (*b*) a Fellows cutter.

6.26 The following helical gears, cut with a 12-pitch, 20° full-depth hob, are meshed without backlash.

Gear 1—36 teeth, right-hand, 30° helix angle
Gear 2—72 teeth, left-hand, 40° helix angle

Determine the shaft angle, the angular velocity ratio, and the center distance.

6.27 Two shafts crossed at right angles are connected by helical gears (gears 1 and 2), cut with a 12-pitch, 20° full-depth hob. Both gears are right-hand and the angular velocity ratio is 1.5:1. $D_2 = 5.196$ in. and $\Psi_1 = 60°$. A design modification requires a reduction of the outside diameter (o.d.) of gear 1 by 0.25 in. to provide clearance for a new component. Assuming that the same hob must be used for cutting any new gears, show that the o.d. of gear 1 can be reduced without changing the velocity ratio, the shaft angle, and the numbers of gear teeth N_1 and N_2. The o.d. of gear 2 and the center distance may be altered if necessary. In the analysis, calculate and compare the following data for both the original and the new gears: C_{12}, D_1, D_2, N_1, N_2, Ψ_1, Ψ_2.

6.28 A 21-tooth helical gear of 6 normal diametral pitch is to drive a spur gear. The angular velocity ratio is to be 2:1 and the angle between the shafts 45°. Determine the pitch diameters for the two gears and the helix angle for the helical gear. Make a full-size sketch of the two gears (pitch cylinders) in contact similar to Fig. 6.20 with the pinion on top; the width of the gears is to be 1 in. Show the tooth elements in contact and also a tooth element on top of the pinion. Label and dimension the helix angle and the shaft angle.

6.29 Two crossed shafts are to be connected by helical gears. The angular velocity ratio is to be $1\frac{1}{2}$:1 and the center distance 8.50 in. If one gear is available from a previous job with 30 teeth, 30° helix angle and 5 normal diametral pitch, calculate the shaft angle that must be used. Let both gears be of the same hand and let the 30-tooth gear be the pinion.

6.30 Two crossed shafts are connected by helical gears. The velocity ratio is 1.8:1 and the shaft angle 45°. If $D_1 = 2.31$ in. and $D_2 = 3.73$ in., calculate the helix angles if both gears have the same hand.

6.31 Two shafts crossed at right angles are to be connected by helical gears. The angular velocity ratio is to be $1\frac{1}{2}$:1 and the center distance 5.00 in. Select a pair of gears for this application to be cut by the Fellows method.

6.32 Two crossed shafts are connected by helical gears. The velocity ratio is 3:1, the shaft angle 60°, and the center distance 10.00 in. If the pinion has 35 teeth and a normal diametral pitch of 8, calculate the helix angles and pitch diameters if the gears are of the same hand.

6.33 A helical pinion of 2.00 in. pitch diameter drives a helical gear of 3.25 in. as shown in Fig. 6.20, $\Sigma = 30°$. Let the velocity of the pitch point of gear 1 be represented by a vector 2 in. long and that of gear 2 by a vector 3 in. long. Using a face width of 1 in. for the gears, graphically determine the tooth element on the top of each gear, the helix angle and the hand of each gear, and the velocity of sliding.

6.34 An 8-pitch, $14\frac{1}{2}°$ full-depth hob is used to cut a helical gear. The hob is right-handed with a lead angle of 2° 40′, a length of 3.00 in., and an outside diameter of 3.00 in. Make a full-size sketch of the hob cutting a 47-tooth right-hand helical gear with a helix angle of 20°. The gear blank is $1\frac{1}{2}$ in. wide. Show the pitch cylinder of the hob on top of the gear blank with the pitch helix of the hob in correct relation to the pitch element of the gear tooth. Show three tooth elements on the gear and $1\frac{1}{2}$ turns of the thread on the hob; position these elements by means of the normal circular pitch. Label the axis of the hob and gear blank, the lead angle of hob, the helix angle of the gear, and the direction of rotation of the hob and gear blank.

6.35 Repeat Problem 6.34 with a left-hand helical gear.

6.36 A double-threaded worm having a lead of 2.00 in. drives a worm gear with a velocity ratio of 20:1; the angle between the shafts is 90°. If the center distance is 9.00 in., determine the pitch diameter of the worm and worm gear.

6.37 A worm and worm gear with shafts at 90° and a center distance of 7.00 in. are to have a velocity ratio of 18:1. If the axial pitch of the worm is to be $\frac{1}{2}$ in., determine the maximum number of teeth in the worm and worm gear that can be used for the drive and their corresponding pitch diameters.

6.38 A worm and worm gear connect shafts at 90°. Derive equations for the diameters of the worm and worm gear in terms of the center distance C, velocity ratio (ω_1/ω_2), and lead angle λ.

6.39 A worm and worm gear with shafts at 90° and a center distance of 6.00 in. are to have a velocity ratio of 20:1. If the axial pitch of the worm is to be $\frac{1}{2}$ in., determine the smallest diameter worm that can be used for the drive.

6.40 A four-threaded worm drives a 60-tooth worm gear with shafts at 90°. If the center distance is 8.00 in. and the lead angle of the worm 20°, calculate the axial pitch of the worm and the pitch diameters of the two gears.

6.41 A four-threaded worm drives a 48-tooth worm gear having a pitch diameter of 7.64 in. and a helix angle of 20°. If the shafts are at right angles, calculate the lead and the pitch diameter of the worm.

6.42 A six-threaded worm drives a worm gear with an angular velocity ratio of 8:1 and a shaft angle of 80°. The axial pitch of the worm is $\frac{1}{2}$ in. and the lead angle 20°. Calculate the pitch diameters of the worm and worm gear and the circular pitch of the gear.

6.43 A five-threaded worm drives a 33-tooth worm gear with a shaft angle of 90°. The center distance is 2.75 in. and the lead angle 20°. Calculate the pitch diameters, the lead, and the axial pitch of the worm.

6.44 A worm and worm gear with shafts at 90° and a center distance of 3.10 in. are to have a velocity ratio of 7:1. Using a lead angle of 20°, determine the pitch diameters and numbers of teeth for the gears. Make the axial pitch a simple fraction.

6.45 A worm and worm gear with shafts at 90° and a center distance of 3.00 in. are to have a velocity ratio of 30:1. Determine a pair of gears for the job, and specify the numbers of teeth, pitch diameters, and lead angle. Make the axial pitch a simple fraction.

7

Gear Trains

7.1 Introduction to Gear Trains. Often it is necessary to combine several gears and to obtain by so doing what is known as a *gear train*. Given the input angular velocity to a gear train, it is important to be able to determine easily the angular velocity of the output gear and its direction of rotation. The ratio of the input angular velocity to the output angular velocity is known as the *angular velocity ratio* and is expressed as ω_{in}/ω_{out}.

Figure 7.1 shows a pinion driving an external spur gear and a pinion driving an internal spur gear. In both cases, the angular velocity ratio is inversely proportional to the number of teeth as indicated. The external gears rotate in opposite directions, and the internal gear rotates in the same direction as its pinion. This is indicated by a minus sign on the velocity ratio in the first case and by a plus sign in the second case. Up to the present time it has been unnecessary to assign an algebraic sign to the angular velocity ratio of a pair of gears. However, when gears are combined to give a gear train, it is important to consider the sign because it indicates direction of rotation. This is especially true in the analysis of planetary gear trains.

Occasionally it is necessary to change the direction of rotation of a gear without changing its angular velocity. This can be done by placing an *idler gear* between the driven and the driver gear. When an idler gear is used, the direction of rotation is changed but the velocity ratio remains the same.

It can be shown that the *angular velocity ratio* of a gear train where all

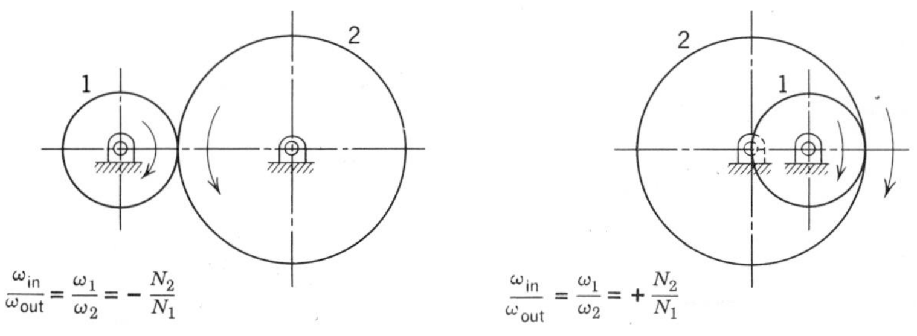

Figure 7.1

gears have fixed axes of rotation is the product of the numbers of teeth of all the driven gears divided by the product of the numbers of teeth of all the driving gears. This relation is given in equation form by

$$\frac{\omega_{\text{in}}}{\omega_{\text{out}}} = \frac{\omega_{\text{driver}}}{\omega_{\text{driven}}} = \frac{\text{Product of teeth of driven gears}}{\text{Product of teeth of driving gears}} \tag{7.1}$$

To illustrate the use of Eq. 7.1, consider the gear train of Fig. 7.2 where gear 2 and gear 3 are mounted on the same shaft. The angular velocity ratio is given by

$$\frac{\omega_{\text{in}}}{\omega_{\text{out}}} = \frac{\omega_1}{\omega_4} = +\frac{N_2 \times N_4}{N_1 \times N_3}$$

The plus sign is determined by observation. That the preceding equation is correct can be easily shown.

$$\frac{\omega_1}{\omega_2} = -\frac{N_2}{N_1} \quad \text{and} \quad \frac{\omega_3}{\omega_4} = -\frac{N_4}{N_3}$$

$$\frac{\omega_1}{\omega_2} \times \frac{\omega_3}{\omega_4} = +\frac{N_2}{N_1} \times \frac{N_4}{N_3}$$

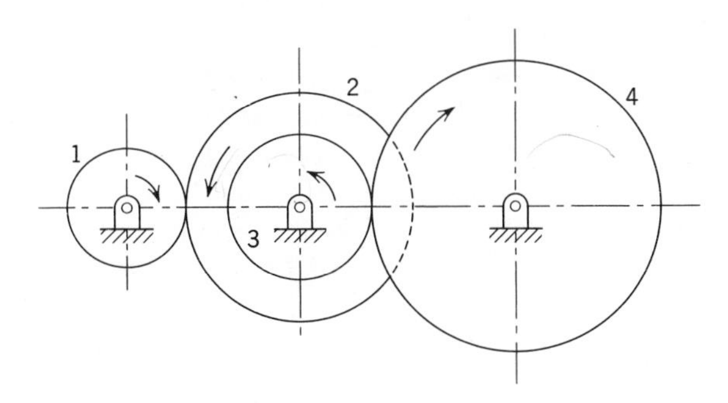

Figure 7.2

But

$$\omega_2 = \omega_3$$

therefore,

$$\frac{\omega_1}{\omega_4} = +\frac{N_2}{N_1} \times \frac{N_4}{N_3}$$

When two gears are fixed to the same shaft as gears 2 and 3 in Fig. 7.2, the gears form *a compound gear*.

Although the angular velocity ratio is used for calculations involving just one pair of gears, it is more convenient when working with a gear train to use the reciprocal of the angular velocity ratio. The reason for this is that the angular velocity of the driver will be known from the speed of the motor,

Figure 7.3 Triple reduction speed reducer. (Courtesy of Jones Machinery, Division of Hewitt–Robins, Inc.)

and it is necessary only to multiply the speed of the driver by a factor to find the speed of the last gear in the train. This reciprocal is known as the *train value* and is given in equation form by

$$\frac{\omega_{\text{driven}}}{\omega_{\text{driver}}} = \frac{\text{Product of teeth of driving gears}}{\text{Product of teeth of driven gears}} \tag{7.2}$$

In general, gear velocities step down so that this value will be less than 1.00. A typical gear train is illustrated in the triple-reduction speed reducer shown in Fig. 7.3.

7.2 Planetary Gear Trains. In order to obtain a desired gear ratio, it is often advantageous to design a gear train so that one of the gears will have planetary motion. With this motion a gear will be so driven that it not only rotates about its own center but at the same time its center rotates about another center. Figures 7.4*a* and *b* show two planetary gear trains, where gear 1 is often referred to as the *sun* gear and gear 2 as the *planet* gear. In Fig. 7.4*a*, arm 3 drives gear 2 about gear 1, which is a fixed external gear. As can be seen, gear 2 rotates about its center *B* while this center rotates about center *A*. As gear 2 rolls on the outside of gear 1, a point on its surface will generate an epicycloid. Figure 7.4*b* shows the case where gear 1 is an internal gear. In this case, a hypocycloid will be generated by a point on the surface of gear 2. Because of the curves generated, a planetary gear train is often referred to as an *epicyclic* or cyclic gear train.

It is more difficult to determine the angular velocity ratio of a planetary gear train than that of an ordinary train because of the double rotation of the planet. The angular velocity ratio may be obtained by the instantaneous center method, the formula method, or by the tabulation method. The instantaneous center method will be reserved for Chapter 10, with the other two methods presented here. The formula method will be treated first.

In Fig. 7.4 let it be required to determine ω_{21} given ω_{31}. It should be noted

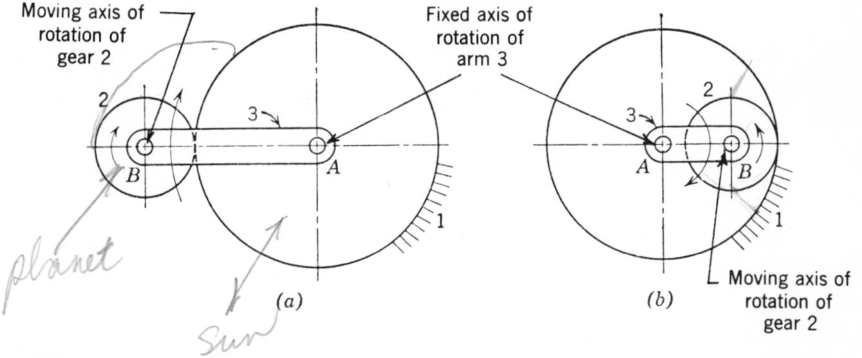

Figure 7.4

that ω_{21} is defined as the angular velocity of gear 2 relative to gear 1 and ω_{31} as the angular velocity of arm 3 relative to gear 1. Because gear 1 is fixed, this is the same as the angular velocity of gear 2 and of arm 3 relative to the ground. In the solution of the problem, ω_{23}/ω_{13} plays an important part.

Consider the gear train in Fig. 7.4*a* to be changed so that arm 3 is stationary instead of gear 1. Arm 3 then becomes the ground, and an ordinary gear train results. The ratio ω_{23}/ω_{13} can therefore be evaluated as $-N_1/N_2$. If the mechanism is now returned to its original condition, that is, arm 3 moving and gear 1 fixed, the ratio ω_{23}/ω_{13} will still be $-N_1/N_2$. The reason for this is that when a mechanism is inverted, the relative motion between links does not change. A solution for ω_{21} in terms of the known quantities, ω_{31} and ω_{23}/ω_{13}, can now be affected by writing an equation for ω_{21} and dividing by ω_{31} as follows:

$$\omega_{21} = \omega_{31} + \omega_{23}$$

$$\frac{\omega_{21}}{\omega_{31}} = 1 + \frac{\omega_{23}}{\omega_{31}} = 1 - \frac{\omega_{23}}{\omega_{13}}$$

Therefore,

$$\omega_{21} = \omega_{31}\left(1 - \frac{\omega_{23}}{\omega_{13}}\right) \tag{7.3}$$

For Fig. 7.4*a*,

$$\frac{\omega_{23}}{\omega_{13}} = -\frac{N_1}{N_2}$$

and

$$\omega_{21} = \omega_{31}\left(1 + \frac{N_1}{N_2}\right) \tag{7.3a}$$

For Fig. 7.4*b*,

$$\frac{\omega_{23}}{\omega_{13}} = +\frac{N_1}{N_2}$$

and

$$\omega_{21} = \omega_{31}\left(1 - \frac{N_1}{N_2}\right) \tag{7.3b}$$

From comparison of Eqs. 7.3*a* and *b*, it is apparent why it is important that the correct algebraic sign of ω_{23}/ω_{13} be substituted into Eq. 7.3.

Consider next the case where all of the gears rotate as well as the arm. This is illustrated in Fig. 7.5, where ω_{31} and ω_{41} are known, and it is required to find ω_{21}. In solving this problem, ω_{24}/ω_{34} is the key ratio because it is the velocity ratio of the two gears relative to the arm and can be easily evaluated. Equations can be written for ω_{24} and ω_{34} and combined so that the ratio

ω_{24}/ω_{34} appears. This is illustrated in the following:

$$\omega_{24} = \omega_{21} - \omega_{41}$$
$$\omega_{34} = \omega_{31} - \omega_{41}$$

Dividing the first equation by the second,

$$\frac{\omega_{24}}{\omega_{34}} = \frac{\omega_{21} - \omega_{41}}{\omega_{31} - \omega_{41}}$$

$$\frac{\omega_{24}}{\omega_{34}}(\omega_{31} - \omega_{41}) = \omega_{21} - \omega_{41}$$

$$\omega_{21} = \left(\frac{\omega_{24}}{\omega_{34}}\right)\omega_{31} + \omega_{41}\left(1 - \frac{\omega_{24}}{\omega_{34}}\right)$$

But

$$\frac{\omega_{24}}{\omega_{34}} = -\frac{N_3}{N_2}$$

therefore,

$$\omega_{21} = \left(-\frac{N_3}{N_2}\right)\omega_{31} + \omega_{41}\left(1 + \frac{N_3}{N_2}\right) \tag{7.4}$$

In deriving Eqs. 7.3 and 7.4, it was seen that in each case the angular velocity ratio of the gears relative to the arm was first obtained, and then the equations of relative velocity written and combined to contain this ratio. Although this method is basic, it means that a new equation must be developed for each planetary system encountered. In order to avoid this repetition, it is possible to derive a general equation which can be applied to any planetary gear train.

Consider Fig. 7.5 again and the equations

$$\omega_{24} = \omega_{21} - \omega_{41}$$
$$\omega_{34} = \omega_{31} - \omega_{41}$$

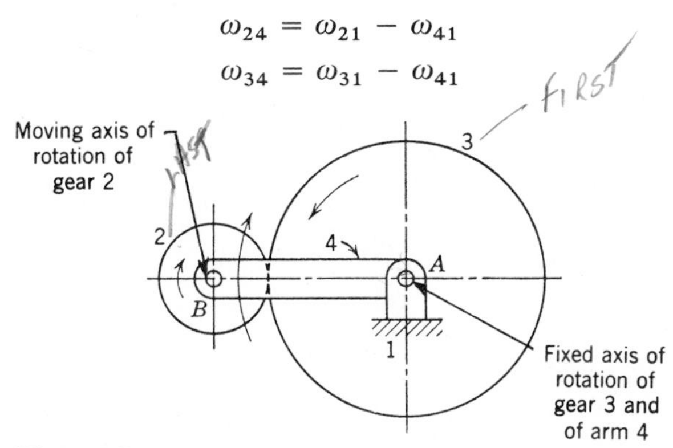

Figure 7.5

and

$$\frac{\omega_{24}}{\omega_{34}} = \frac{\omega_{21} - \omega_{41}}{\omega_{31} - \omega_{41}}$$

If in Fig. 7.5 gear 3 is considered the first gear and gear 2 the last gear, the preceding equation may be written as

$$\frac{\omega_{LA}}{\omega_{FA}} = \frac{\omega_L - \omega_A}{\omega_F - \omega_A} \tag{7.5}$$

where

$\frac{\omega_{LA}}{\omega_{FA}}$ = velocity ratio of last gear to first gear both relative to arm

ω_L = angular velocity of last gear in train relative to fixed link

ω_A = angular velocity of arm relative to fixed link

ω_F = angular velocity of first gear in train relative to fixed link

In using Eq. 7.5, it must be emphasized that the *first gear* and the *last gear* must be gears that mesh with the gear or gears that have planetary motion. Also, the first gear and the last gear must be on parallel shafts because angular velocities cannot be treated algebraically unless the vectors representing these velocities are parallel.

Equation 7.5 will now be used to write the equation for the gear train in Fig. 7.4*a*. Let gear 1 be considered the first gear and gear 2 the last gear:

$$\frac{\omega_{LA}}{\omega_{FA}} = \frac{\omega_L - \omega_A}{\omega_F - \omega_A}$$

$$\frac{\omega_{LA}}{\omega_{FA}} = \frac{\omega_{23}}{\omega_{13}} = -\frac{N_1}{N_2}$$

$$\omega_L = \omega_{21}$$

$$\omega_A = \omega_{31}$$

$$\omega_F = \omega_1 = 0$$

Substituting these values gives

$$-\frac{N_1}{N_2} = \frac{\omega_{21} - \omega_{31}}{0 - \omega_{31}}$$

$$\omega_{21} - \omega_{31} = \left(\frac{N_1}{N_2}\right)\omega_{31}$$

and

$$\omega_{21} = \omega_{31}\left(1 + \frac{N_1}{N_2}\right)$$

which agrees with Eq. 7.3*a*. The application of Eq. 7.5 to a more complicated train is given in the following example.

Example 7.1

If arm 6 and gear 5 in Fig. 7.6 are driven clockwise (viewed from the right end) at 150 and 50 rad/min, respectively, determine ω_{21} in magnitude and direction. Use Eq. 7.5 and let gear 5 be the first gear and gear 2 the last gear.

$$\frac{\omega_{LA}}{\omega_{FA}} = \frac{\omega_L - \omega_A}{\omega_F - \omega_A}$$

$$\frac{\omega_{26}}{\omega_{56}} = \frac{\omega_{21} - \omega_{61}}{\omega_{51} - \omega_{61}}$$

$$\frac{\omega_{26}}{\omega_{56}} = \frac{N_5 \times N_3}{N_4 \times N_2} = \frac{20 \times 30}{28 \times 18} = \frac{25}{21}$$

Therefore,

$$\frac{25}{21} = \frac{\omega_{21} - 150}{50 - 150}$$

$$\omega_{21} = \frac{25}{21}(-100) + 150$$

$$= +30.9 \text{ rad/min}$$

Because the sign of ω_{21} is the same as that of ω_{51} and ω_{61}, ω_{21} is in the same direction, i.e., clockwise viewed from the right end.

Occasionally, it is necessary to analyze a planetary gear train that cannot be solved by a single application of Eq. 7.5 as was done in Example 7.1. For instance, if a fixed internal gear 7 is added to the train of Fig. 7.6 to mesh with

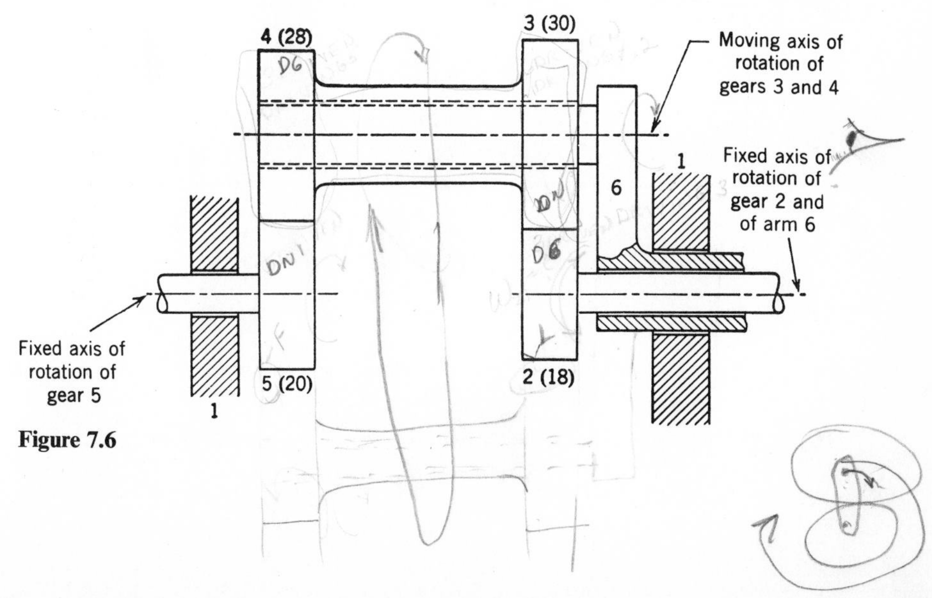

Figure 7.6

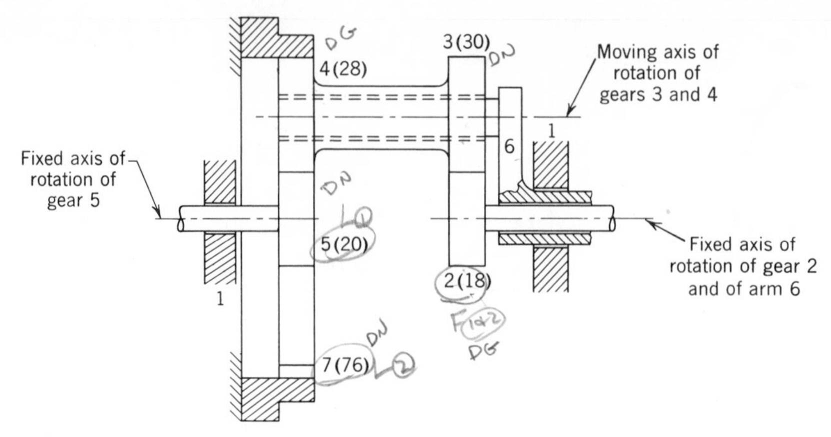

Figure 7.7

gear 4 as shown in Fig. 7.7 and it is required to calculate ω_{51} given ω_{21}, it will be necessary to use Eq. 7.5 twice to solve the problem. The first application of Eq. 7.5 would consider gears 2, 3, 4, 5, and the arm 6 and the second application, gears 2, 3, 4, 7, and the arm 6. This will be illustrated in the following example.

Example 7.2

If ω_{21} rotates clockwise (viewed from the right end) at 60 rad/min, determine ω_{51} and its direction of rotation.

Considering first, gears 2, 3, 4, 5, and arm 6, let gear 2 be the first gear and gear 5 the last gear.

$$\frac{\omega_{LA}}{\omega_{FA}} = \frac{\omega_L - \omega_A}{\omega_F - \omega_A}$$

$$\frac{\omega_{56}}{\omega_{26}} = \frac{\omega_{51} - \omega_{61}}{\omega_{21} - \omega_{61}}$$

$$\frac{\omega_{56}}{\omega_{26}} = \frac{N_2 \times N_4}{N_3 \times N_5} = \frac{18 \times 28}{30 \times 20} = \frac{21}{25}$$

Therefore,

$$\frac{21}{25} = \frac{\omega_{51} - \omega_{61}}{\omega_{21} - \omega_{61}} = \frac{\omega_{51} - \omega_{61}}{60 - \omega_{61}} \tag{a}$$

However, Eq. (*a*) cannot be solved because it contains two unknowns,

ω_{51} and ω_{61}. It is necessary, therefore, to consider gears 2, 3, 4, 7, and arm 6. Let gear 2 be the first gear and gear 7 the last gear.

$$\frac{\omega_{76}}{\omega_{26}} = \frac{\omega_{71} - \omega_{61}}{\omega_{21} - \omega_{61}}$$

$$\frac{\omega_{76}}{\omega_{26}} = -\frac{N_2 \times N_4}{N_3 \times N_7} = -\frac{18 \times 28}{30 \times 76} = -\frac{21}{95}$$

$$-\frac{21}{95} = \frac{\omega_{71} - \omega_{61}}{\omega_{21} - \omega_{61}} = \frac{0 - \omega_{61}}{60 - \omega_{61}} \qquad (b)$$

Solving Eq. (*b*) for ω_{61}

$$-\frac{21}{95}(60 - \omega_{61}) = 0 - \omega_{61}$$

$$\omega_{61} = \frac{21 \times 15}{29} = +10.86 \text{ rad/min}$$

From Eq. (*a*),

$$\frac{21}{25}(60 - \omega_{61}) = \omega_{51} - \omega_{61}$$

$$\frac{21}{25}(60 - 10.86) = \omega_{51} - 10.86$$

Therefore, $\omega_{51} = +52.14$ rad/min, direction of rotation same as that of ω_{21}.

Example 7.3

Consider that in the differential shown in Fig. 7.8, the angular velocity of shaft *A* is 350 rad/min in the direction shown and that of shaft *B* is 2000 rad/min. Determine the angular velocity of shaft *C*.

Use Eq. 7.5 and remember that the first gear and the last gear selected for the equation must be gears that mesh with the gears that have planetary motion. Therefore, let gear 4 be the first gear and gear 7 be the last gear.

$$\frac{\omega_{LA}}{\omega_{FA}} = \frac{\omega_L - \omega_A}{\omega_F - \omega_A}$$

$$\frac{\omega_{78}}{\omega_{48}} = \frac{\omega_{71} - \omega_{81}}{\omega_{41} - \omega_{81}}$$

$$\frac{\omega_{78}}{\omega_{48}} = -\frac{N_4 \times N_6}{N_5 \times N_7} = -\frac{30 \times 24}{64 \times 18} = -\frac{5}{8}$$

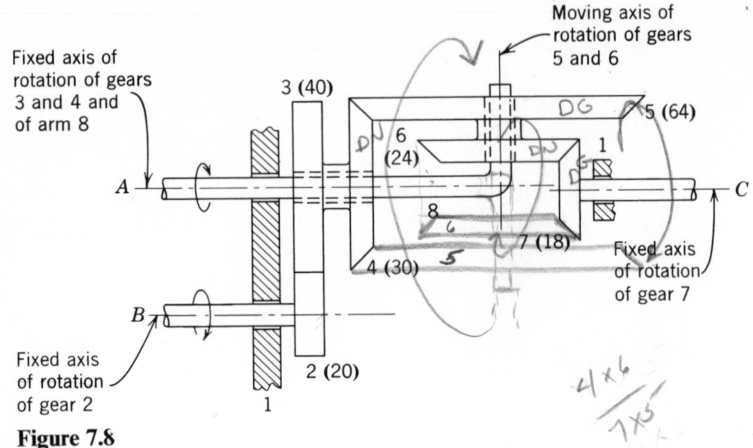

Figure 7.8

Also

$$\omega_{41} = \omega_{31} = \omega_B \times \frac{N_2}{N_3} = 2000 \times \frac{20}{40}$$

$$= 1000 \text{ rad/min, same direction as } \omega_A$$

and

$$\omega_{81} = \omega_A = 350 \text{ rad/min}$$

Making the substitutions,

$$-\frac{5}{8} = \frac{\omega_{71} - 350}{1000 - 350}$$

$$\omega_{71} = -\frac{5}{8}(650) + 350$$

$$= -406.3 + 350$$

$$= -56.3 \text{ rad/min, opposite direction to } \omega_A$$

The tabulation method is another convenient way of solving planetary gear problems. To illustrate its use, consider the gear train of Fig. 7.4*a* and the following procedure:

1. Disconnect gear 1 from the ground and lock it to arm 3 together with gear 2. There can now be no relative motion among members 1, 2, and 3.
2. Rotate arm 3 (and gears 1 and 2) one positive revolution about center *A*.
3. Unlock gears 1 and 2 from arm 3. *Holding arm 3 fixed*, rotate gear 1 one negative revolution. Gear 2, therefore, rotates $+N_1/N_2$ revolutions.

The results of steps 2 and 3 are entered in Table 7.1 together with the total number of revolutions made by each member of the train relative to the ground. It can be seen from the "Total" line of Table 7.1 that with gear 1

Table 7.1

	Gear 1	Gear 2	Arm 3
Motion with arm relative to frame (item 2)	+1	+1	+1
Motion relative to arm (item 3)	−1	$+\frac{N_1}{N_2}$	0
Total motion relative to frame	0	$1+\frac{N_1}{N_2}$	+1

stationary, gear 2 turns $(1 + N_1/N_2)$ revolutions for one revolution of arm 3. This agrees with Eq. 7.3*a*.

Two examples will be given to illustrate the use of the tabular method.

Example 7.4

Consider that arm 4 of Fig. 7.9 rotates counterclockwise at 50 rad/min. Determine ω_{21} in magnitude and direction. See Table 7.2.

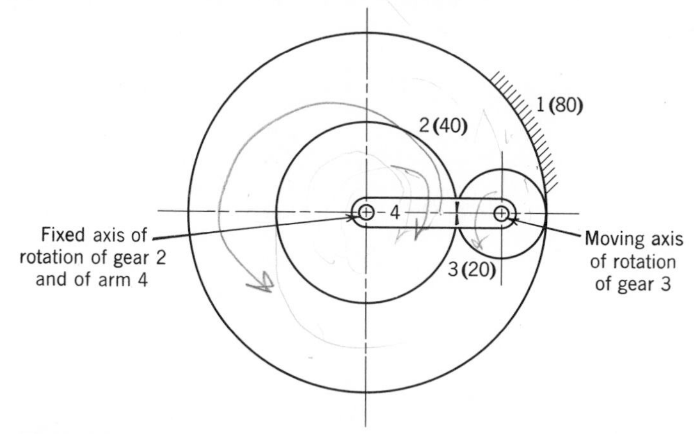

Figure 7.9

Table 7.2

	Gear 1	Gear 2	Gear 3	Arm 4
Motion with arm relative to frame	+1	+1	+1	+1
Motion relative to arm	−1	$+\frac{N_1}{N_2}$	$-\frac{N_1}{N_3}$	0
Total motion relative to frame	0	$1+\frac{N_1}{N_2}$	$1-\frac{N_1}{N_3}$	+1

$$\frac{\omega_{21}}{\omega_{41}} = \frac{1 + N_1/N_2}{1}$$

$$\omega_{21} = \omega_{41}\left(1 + \frac{N_1}{N_2}\right) = 50\left(1 + \frac{80}{40}\right)$$

$$= 150 \text{ rad/min (ccw)}$$

One distinct advantage of the tabular method is the fact that more than one ratio can be obtained from a solution. In Example 7.4, if it had been necessary to determine the value of ω_{31}, this could have easily been accomplished from the data in the table.

Example 7.5

Example 7.1 and Fig. 7.6 will now be worked by the tabular method. Because all of the gears in this train rotate, it is easier to work with the actual speeds of gear 5 and arm 6 in the table rather than with one revolution as in Example 7.4. Because arm 6 rotates 150 rad/min, this must be the number of turns to which the entire train is subjected when locked together for line 1 of Table 7.3 (because of the 0 for arm 6 in line 2). With +150 for gear 5 in line 1, −100 must be inserted for gear 5 in line 2 in order to give the correct total of +50. With arm 6 stationary in line 2, and gear 5 now rotating a known amount, the rotation of gears 2, 3, and 4 can easily be determined for line 2.

Table 7.3

	Gear 2	Gear 3	Gear 4	Gear 5	Arm 6
Motion with arm relative to frame	+150	+150	+150	+150	+150
Motion relative to arm	$-100\dfrac{N_5 \times N_3}{N_4 \times N_2}$	$+100\dfrac{N_5}{N_4}$	$+100\dfrac{N_5}{N_4}$	−100	0
Total motion relative to frame	$150 - 100\left(\dfrac{N_5 \times N_3}{N_4 \times N_2}\right)$			+50	+150

$$\omega_{21} = 150 - 100\left(\frac{N_5 \times N_3}{N_4 \times N_2}\right) = 150 - 100\left(\frac{20 \times 30}{28 \times 18}\right)$$

$$= 150 - 100 \times \frac{25}{21}$$

$$= +30.9 \text{ rad/min (cw)}$$

Example 7.3 can also be easily worked using the tabulation method.

7.3 Applications of Planetary Gear Trains. Planetary gear trains find many applications in machine tools, hoists, aircraft propeller reduction drives, automobile differentials, automatic transmissions, aircraft servo-drives, and many others. Figure 7.10 shows a diagrammatic sketch of a planetary train used as a reduction between the engine and the propeller in an aircraft power plant. Figure 7.11 shows a photograph of an actual unit. The earlier aircraft engine reduction drives used bevel gears in the planetary train. These were discarded, however, in favor of spur gears because the spur gear planetary drive could transmit more power in a given space.

In Fig. 7.10, the engine drives the internal gear 3. Gear 2 meshes with the fixed gear 1 and with the gear 3 so that it has planetary motion. Arm 4, or planet carrier, which is connected to gear 2, drives the propeller at a slower speed than the engine. The equation for the ratio of engine speed ω_{31} to propeller speed ω_{41} can easily be determined from Eq. 7.5 as follows:

$$\frac{\omega_{31}}{\omega_{41}} = 1 - \frac{\omega_{34}}{\omega_{14}}$$

where

$$\frac{\omega_{34}}{\omega_{14}} = -\frac{N_1}{N_3}$$

Therefore,

$$\frac{\omega_{31}}{\omega_{41}} = 1 + \frac{N_1}{N_3}$$

It is interesting to note that it would be impossible to obtain a velocity ratio as high as 2:1 because this would mean that gear 1 would have to have the same number of teeth as gear 3, which is impossible. In determining the limiting ratio for a given drive, it should be noted that all of the gears will have the same diametral pitch.

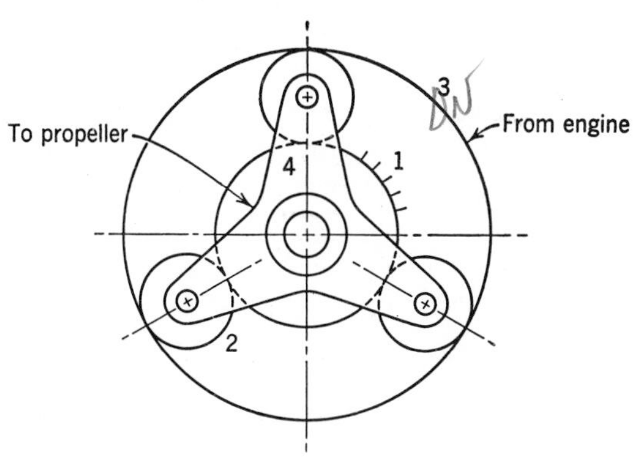

Figure 7.10

Figure 7.11 Planetary reduction unit for aircraft propeller drive. (Courtesy of Foote Brothers Gear & Machinery Corp.)

A planetary gear train used as a differential in an automobile is shown in Fig. 7.12. Figure 7.13 shows a cutaway view of a differential and housing. This mechanism makes it possible for an automobile to turn a corner without the rear wheels slipping. In Fig. 7.12 gear 2 is driven by the engine via the clutch, transmission, and drive shaft. Gear 2 drives gear 3, which is fastened to the carrier 7. If the car is moving straight ahead, gears 4, 5, and 6 turn as a unit with the carrier and there is no relative motion between them. Gears 5 and 6 turn the axles. When the car makes a turn, however, gears 5 and 6 no longer rotate at the same speed and gear 4 has to turn about its own axis, as well as being driven by the carrier, to allow this. It is interesting to note that, if one of the wheels is held stationary and the second is free to rotate, the second wheel will turn at a speed twice that of the carrier. This characteristic is a disadvantage when the car is stuck in snow or mud.

There are many planetary gear train designs and a wide range of possible ratios. The applications mentioned are only two of a wide variety. In many instances it will be found that it is possible to obtain a greater reduction ratio

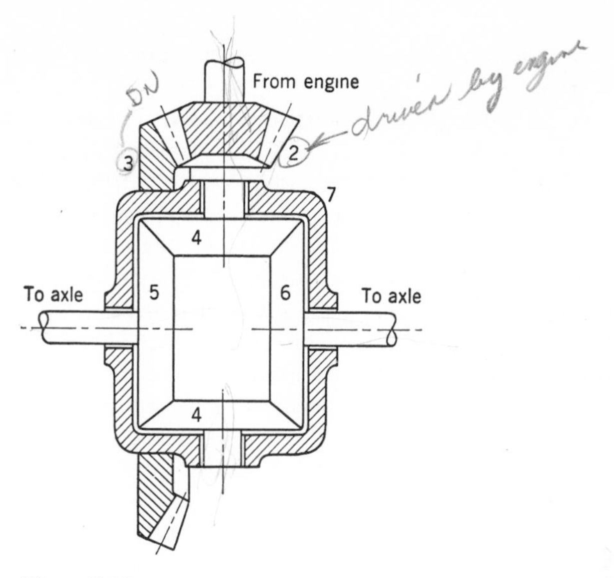

Figure 7.12

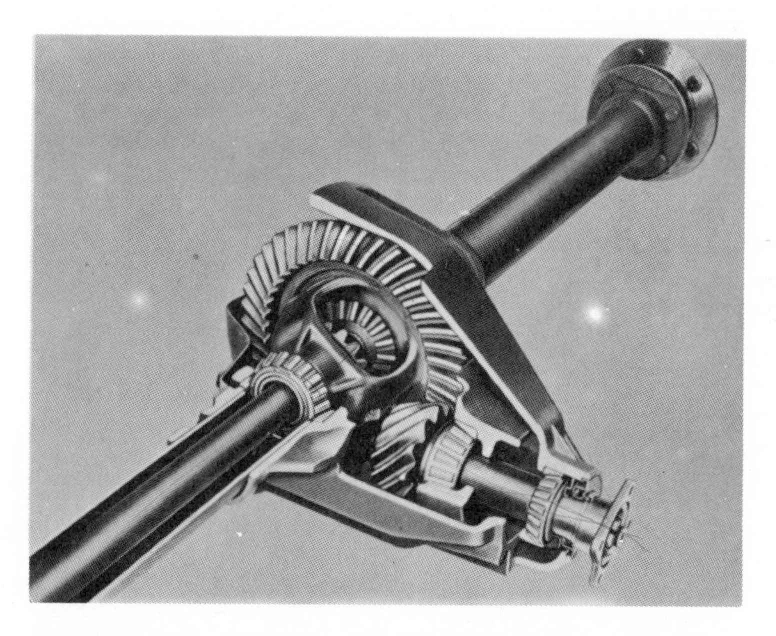

Figure 7.13 Automotive differential. (Courtesy of Gleason Works.)

with a smaller drive using a planetary train than when using an ordinary gear train.

7.4 Assembly of Planetary Gear Trains. When a planetary gear train is designed, the question of assembling the train with equally spaced planets must be considered. With the train illustrated in Fig. 7.14, it is possible that for a given number of teeth in gears 1, 2, and 3 it might not be possible to have three equally spaced planet gears.

In order to determine the number of planets that can be used for a given number of teeth in gears 1, 2, and 3, it is necessary to determine the angle AOB in Fig. 7.15*a* resulting from gear 3 having been rotated a whole number of tooth spaces with gear 1 stationary. The case must also be investigated where gear 3 is stationary, and gear 1 has been rotated a whole number of spaces. This gives angle AOB' as shown in Fig. 7.15*b*. The following method was developed by Professor G. B. DuBois of Cornell University.

Let the numbers of teeth in gears 1, 2, and 3 be N_1, N_2, and N_3. If θ_{31} equals the angular motion of gear 3 after it has been rotated one whole tooth space with respect to gear 1, then

$$\theta_{31} = \frac{1}{N_3} \text{ revolutions}$$

The angular motion of arm 4 with respect to gear 1 when gear 3 has been rotated one tooth space is given by

$$\theta_{41} = \theta_{31} \times \frac{\omega_{41}}{\omega_{31}} \text{ revolutions}$$

From the velocity analysis of the planetary train of Fig. 7.10, which is identical to the one under consideration,

$$\frac{\omega_{41}}{\omega_{31}} = \frac{N_3}{N_3 + N_1}$$

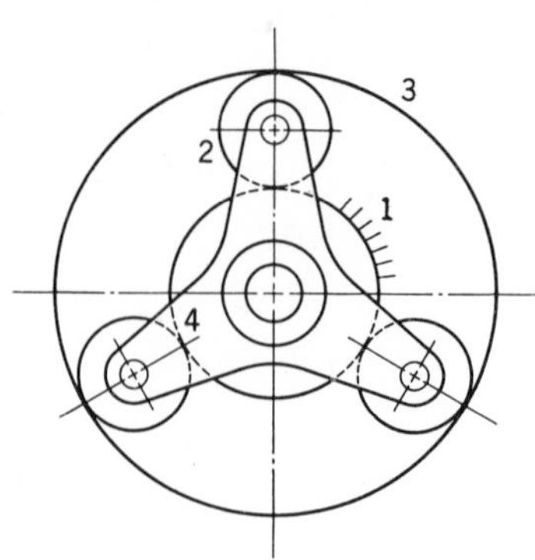

Figure 7.14

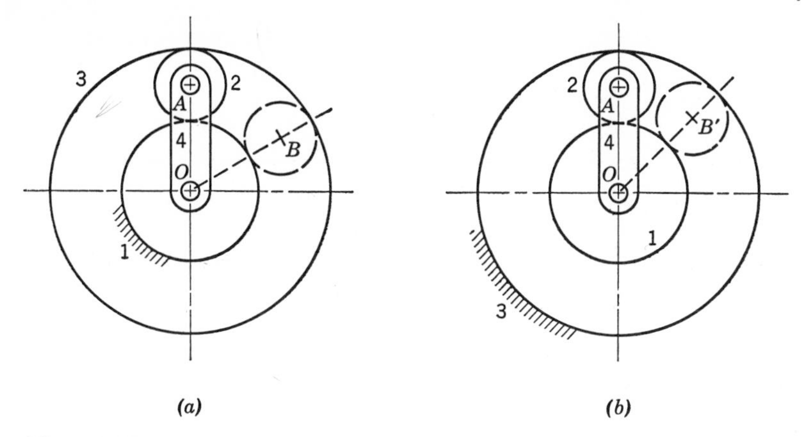

Figure 7.15

Therefore,

$$\theta_{41} = \frac{1}{N_3} \times \frac{N_3}{N_3 + N_1} = \frac{1}{N_3 + N_1} \text{ revolutions}$$

Angle AOB is the angle turned through by arm 4 when gear 3 moves relative to gear 1. If gear 3 moves one tooth space, angle AOB equals θ_{41}. This is the smallest possible angle between planet gears if the planets are allowed to overlap. If gear 3 rotates a whole number of tooth spaces c, then

$$\angle AOB = c(\theta_{41}) = \frac{c}{N_3 + N_1} \text{ revolutions} \tag{7.6}$$

and represents the angle between planets with possible overlapping.

Consider next the case of Fig. 7.15*b* where gear 1 has been rotated one tooth space with gear 3 stationary and it is required to find angle AOB'. If θ_{13} equals the angular motion of gear 1 after it has been rotated one tooth space and θ_{43} equals the resulting motion of arm 4 (both relative to gear 3), then

$$\theta_{13} = \frac{1}{N_1}$$

$$\theta_{43} = \theta_{13} \times \frac{\omega_{43}}{\omega_{13}}$$

But it can be easily derived that

$$\frac{\omega_{43}}{\omega_{13}} = \frac{N_1}{N_1 + N_3}$$

Therefore,

$$\theta_{43} = \frac{1}{N_1} \times \frac{N_1}{N_1 + N_3} = \frac{1}{N_1 + N_3}$$

and

$$\angle AOB' = c(\theta_{43}) = \frac{c}{N_1 + N_3} \tag{7.7}$$

Comparing Eqs. 7.6 and 7.7, it can be seen that arm 4 rotates through the same angle regardless of whether gear 3 or gear 1 rotates one or more tooth spaces.

If angle AOB is the fraction of a revolution between planets, then its reciprocal will be the number of planets. Taking the reciprocal of Eq. 7.6, it is possible to obtain an expression for the number of equally spaced planets around gear 1. If n represents the number of planets, then

$$n = \frac{N_3 + N_1}{c} \tag{7.8}$$

These planets may or may not overlap each other, depending on the value of c.

It is now necessary to determine the maximum number of planets n_{max} that can be used without overlapping. In Fig. 7.16 the outside radii R_{O_2} of two planet gears are shown almost touching at point C. From the figure,

$$n_{max} = \frac{360}{\angle AOB} = \frac{180}{\angle AOC}$$

$$\angle AOC = \sin^{-1} \frac{AC}{OA}$$

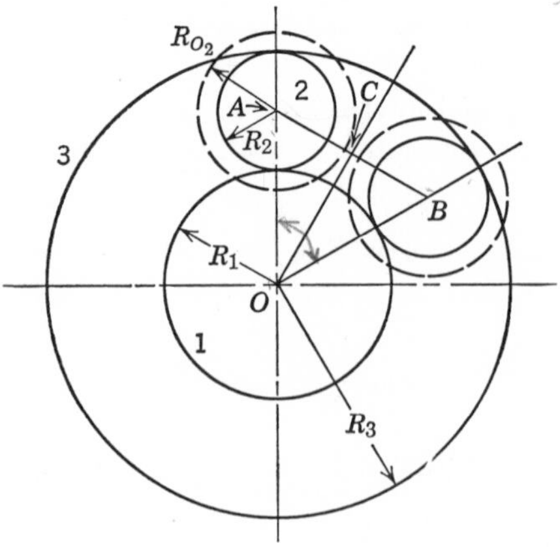

Figure 7.16

where

$$AC > R_{O_2}$$

and

$$OA = R_1 + R_2$$

$$R_{O_2} = R_2 + a = \frac{N_2}{2P} + \frac{k}{P} (k = 1 \text{ for standard full-depth teeth})$$

or

$$R_{O_2} = \frac{N_2 + 2}{2P}$$

and

$$R_1 + R_2 = \frac{N_1 + N_2}{2P}$$

Therefore, for standard full-depth teeth,

$$n_{max} < \frac{180}{\sin^{-1} (N_2 + 2)/(N_1 + N_2)} \tag{7.9}$$

Also from the geometry of Fig. 7.16

$$R_3 = R_1 + 2R_2$$

Because $R = N/2P$ for a standard gear and because the diametral pitches of gears 1, 2, and 3 are equal,

$$N_3 = N_1 + 2N_2$$

For nonstandard gears Eq. 7.9 can be used to give an approximate value of n_{max}. In this case, the fractional value of N_2 resulting from use of the standard equation

$$N_2 = \frac{N_3 - N_1}{2}$$

would be substituted in Eq. 7.9. As a final check, a layout drawing should be made.

Example 7.6

In a planetary gear train similar to Fig. 7.14, gear 1 has 50 teeth and gear 3 has 90 teeth. Determine the number of equally spaced planets that can be used without overlapping. The gears are standard.

$$N_2 = \frac{N_3 - N_1}{2} = \frac{90 - 50}{2} = 20$$

$$n_{max} = \frac{180}{\sin^{-1} (N_2 + 2)/(N_1 + N_2)} = \frac{180}{\sin^{-1} (20 + 2)/(50 + 20)} = 9.8 \text{ planets}$$

Therefore, the number of planets in the gear train cannot exceed 9.

$$n = \frac{N_3 + N_1}{c} = \frac{90 + 50}{c} = \frac{140}{c}$$

The value of c must be a whole number of tooth spaces between planets that, when divided into 140, will give a whole number n. For this case, c can equal 140, 70, 35, 28, or 20. Therefore,

$$n = 1, 2, 4, 5, \text{ or } 7 \text{ equally spaced planets}$$

Problems

7.1 In Fig. 7.17 gear 1 rotates in the direction shown at 240 rpm. Determine the speed of pinion 9 (rpm) and the speed (fpm) and direction of rack 10.

7.2 A hoist is operated by a motor driving a 4-threaded worm that engages a 100-tooth worm gear. The worm gear is keyed to a shaft which also contains a 20-tooth spur pinion. The pinion meshes with a 140-tooth spur gear mounted on the end of the hoisting drum. Make a sketch of the unit, and calculate the speed of hoisting (fpm) if the motor operates at 600 rpm and the drum diameter is 12 in.

7.3 Two slitting rolls A and B for cutting sheet metal are driven by means of the gear train shown in Fig. 7.18. The rolls must operate in the direction shown at a peripheral speed of 45 ips. (*a*) Determine the angular velocity ratio ω_2/ω_3 in order to drive the rolls at the required speed. Gear 1 runs at 1800 rpm. (*b*) Determine the direction of rotation of gear 1 and the hand of worm 6 to give the required rotation of the rolls.

7.4 In the sketch of the press shown in Fig. 7.19, 5 and 6 are single-threaded screws of the opposite hand with 6 threading into 5 as indicated. Gear 4 is fastened to screw 5.

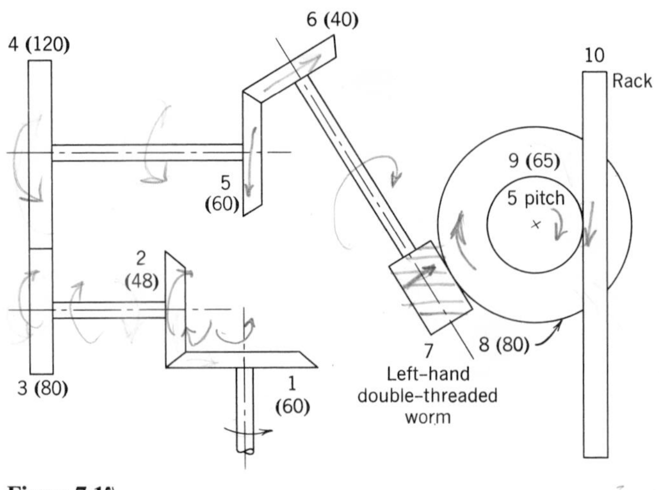

Figure 7.17

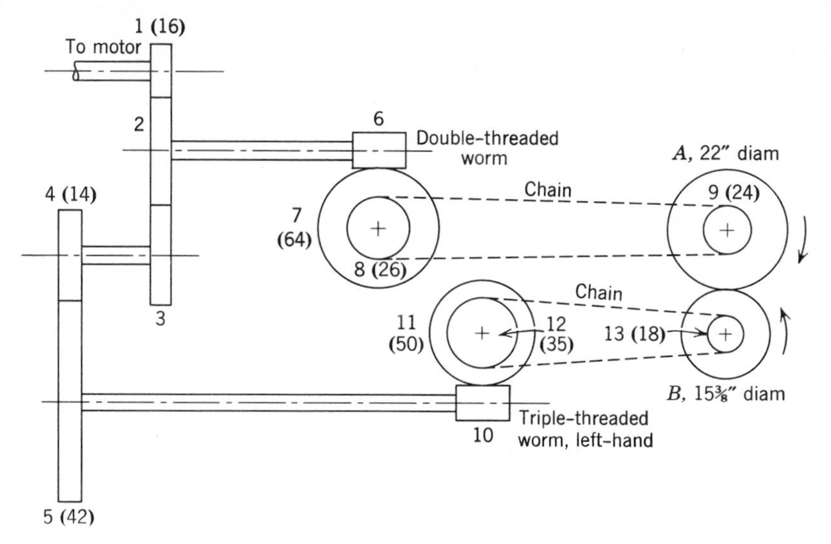

Figure 7.18

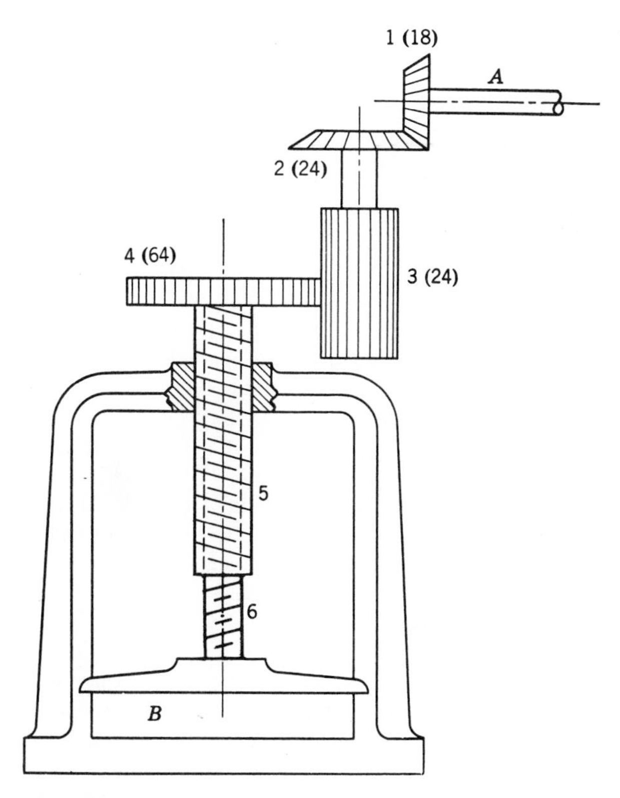

Figure 7.19

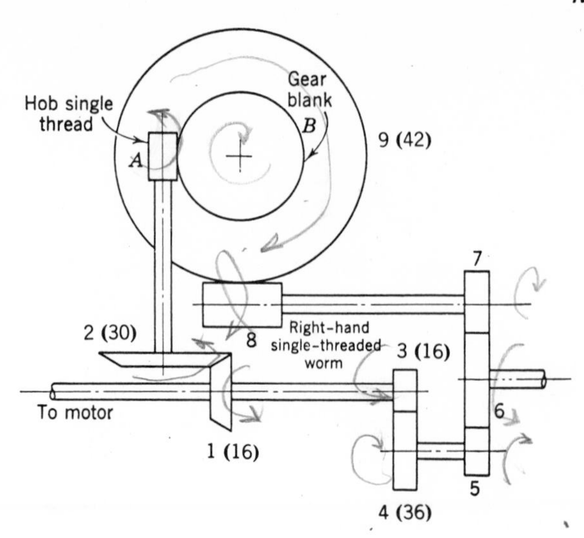

Figure 7.20

Plate B is prevented from turning by a slot in it which engages the frame. If the pitch of 5 is $\frac{1}{4}$ in. and that of 6 is $\frac{1}{8}$ in., determine the direction and the number of turns of shaft A required to lower plate B $\frac{3}{4}$ in.

7.5 The gear train in Fig. 7.20 shows the essential features of the work spindle drive for a gear hobbing machine. Gear blank B and the worm gear 9 are mounted on the same shaft and must rotate together. (*a*) If the gear blank B is to be driven clockwise, determine the hand of the hob A. (*b*) Determine the angular velocity ratio ω_7/ω_5 to cut 72 teeth on the gear blank B.

7.6 A gear train contains shaft A to which is keyed gears 1 and 2, an intermediate shaft B with a sliding compound gear 3, 4, 5 and shaft C to which is keyed gears 6 and 7. The gears are numbered from left to right and are all spur gears with shaft center distances of 12 in. and a diametral pitch of 5. The compound gear can be shifted to the left to give a velocity ratio of 5:1 through gears 1, 4, 3, 6, or to the right to give a velocity ratio 25:9 through gears 2, 4, 5, 7. Draw a sketch of the unit and calculate the number of teeth in each gear if $N_5 = N_2$.

7.7 In the gear train in Fig. 7.21 screws 5 and 6 are single-threaded of opposite hand as shown. Screw 5 has 8 threads per inch and screw 6, 9 threads per inch. Screw 6 threads into screw 5 and screw 5 threads into the frame. Determine the change in x and y in magnitude and direction for one revolution of the handwheel in the direction shown. Gears 1 and 2 are compounded on the handwheel shaft.

7.8 Figure 7.22 shows part of a gear train for a vertical milling machine. Power input is through the pulley and power output through gear 12. Compound gears 1 and 2, 3 and 4, and 10 and 11 can slide as shown to give various combinations of gearing. Determine all of the train values possible between the pulley and gear 12.

7.9 Figure 7.23 shows part of a gear train for a vertical milling machine. Compound gears 1 and 2 can slide so that either gear 1 meshes with gear 5 or gear 2 meshes with

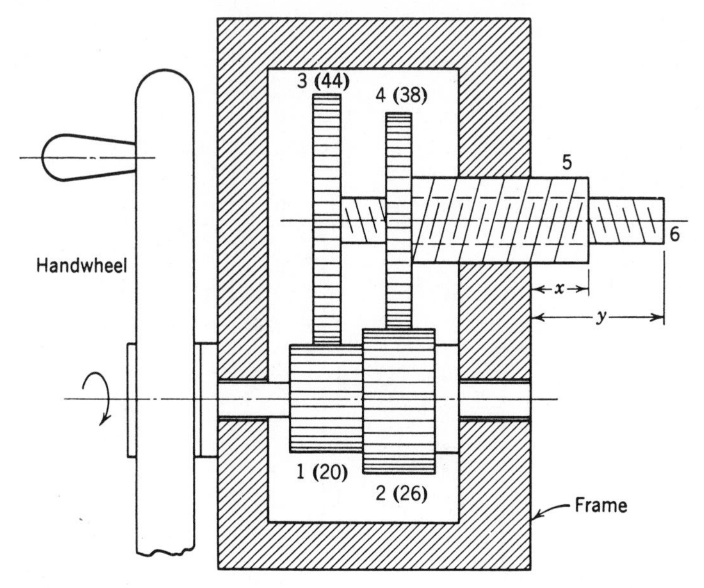

Figure 7.21

gear 3. In the same manner, gear 13 meshes with gear 15 or gear 14 meshes with gear 16. (*a*) With gear 2 meshing with gear 3, determine the two possible spindle speeds for a motor speed of 1800 rpm. Will the spindle rotate in the same or opposite direction to the motor? (*b*) With gear 13 meshing with gear 15 and a spindle speed of 130 rpm, determine the number of teeth for gears 1 and 5 if gears 1, 2, 3, and 5 are standard and have the same diametral pitch.

7.10 A conventional automotive transmission is shown diagrammatically in Fig. 7.24. The transmission of power is as follows: Low gear: Gear 3 shifted to mesh with

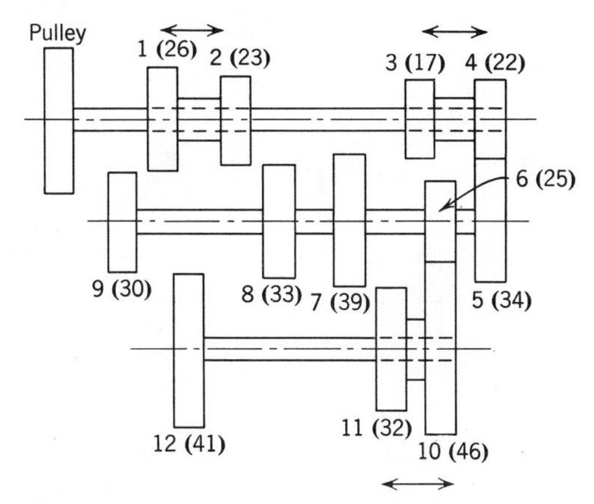

Figure 7.22

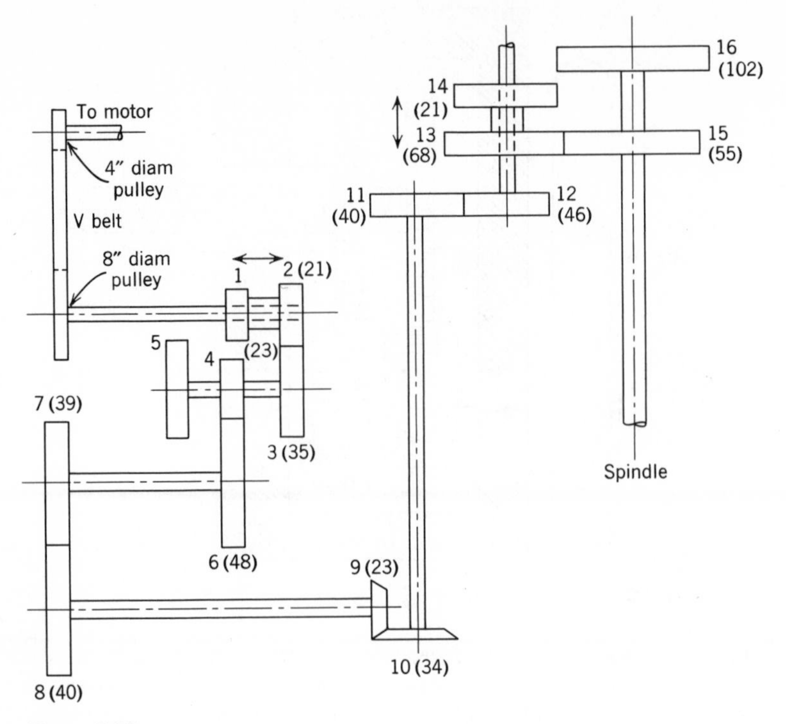

Figure 7.23

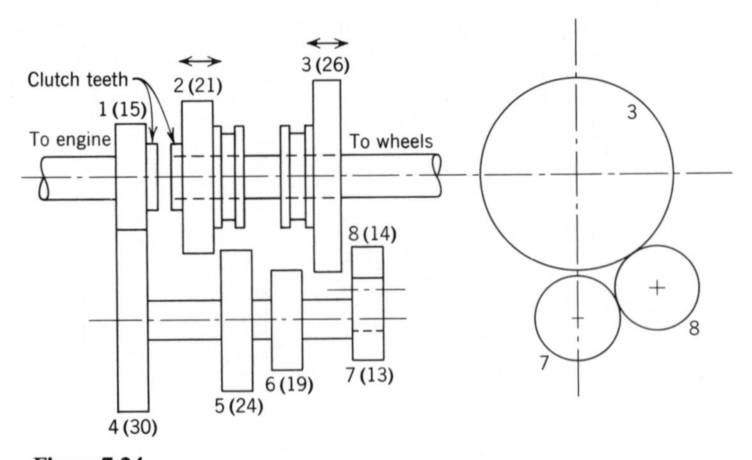

Figure 7.24

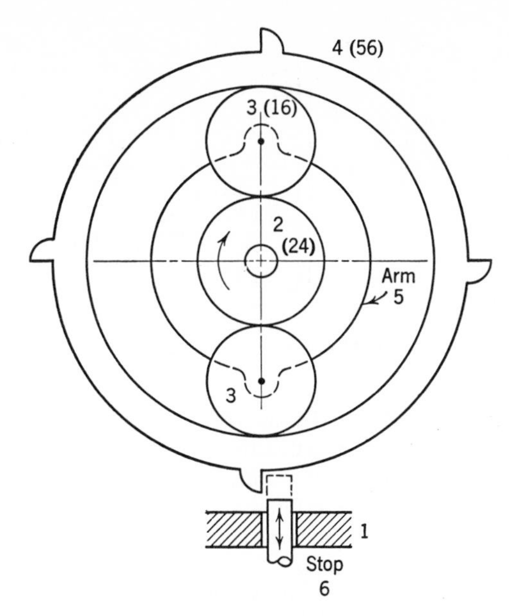

Figure 7.25 Design from *Prod. Eng.*, January 1954.

gear 6. Transmission of power through gears 1, 4, 6, 3. Second gear: Gear 2 shifted to mesh with gear 5. Transmission through gears 1, 4, 5, 2. High gear: Gear 2 shifted so that clutch teeth on end of gear 2 mesh with clutch teeth on end of gear 1. Direct drive results. Reverse gear: Gear 3 shifted to mesh with gear 8. Transmission through gears 1, 4, 7, 8, 3. A car equipped with this transmission has a differential ratio of 2.9:1 and a tire outside diameter of 26 in. Determine the engine speed of the car under the following conditions: (*a*) Low gear and car traveling 20 mph. (*b*) High gear and car traveling 60 mph. (*c*) Reverse gear and car traveling 4 mph.

7.11 In the planetary clutch shown in Fig. 7.25, the stop 6 may be engaged or disengaged. When engaged, a planetary gear train results, and, when disengaged, an ordinary gear train results because arm 5 will remain stationary. If gear 2 rotates in the direction shown at 300 rpm, determine (*a*) the speed of the ring gear 4 when the stop 6 is disengaged as shown, and (*b*) the speed of arm 5 when the stop 6 is engaged with the ring gear 4.

7.12 Considering a bevel gear differential as used in automotive drives, prove that, when one of the rear wheels on a car is jacked up, it will turn twice as fast as the differential carrier.

7.13 If a truck is rounding a right-hand curve at 15 mph, determine the speed in rpm of the differential carrier. The radius of curvature of the curve is 100 ft to the center of the truck, and the truck tread is 6 ft. The outside diameter of the tires is 36 in.

7.14 For the bevel gear planetary drive shown in Fig. 7.26, determine the ratio ω_4/ω_3 when gear 1 is stationary.

7.15 In the ball bearing shown in Fig. 7.27, the inner race 1 is stationary and the

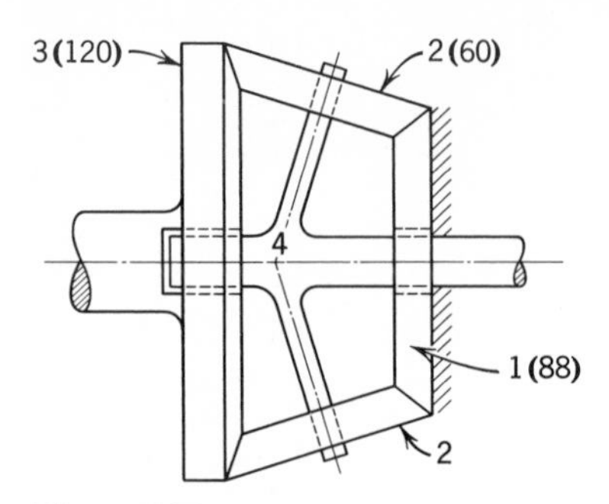

Figure 7.26

outer race 2 rotates with a tubular shaft at 1600 rpm. Assuming pure rolling between the balls and races, determine the speed of the ball retainer 4.

7.16 A mechanism known as Ferguson's paradox is shown in Fig. 7.28. For one revolution of the arm in the direction shown, find the number of revolutions of gears 3, 4, and 5 and their directions of rotation. The gears are nonstandard.

7.17 Shaft A rotates in the direction shown in Fig. 7.29 at 640 rpm. If shaft B is to rotate at 8 rpm in the direction shown, calculate the angular velocity ratio ω_2/ω_4. What would the ratio ω_2/ω_4 have to be in order for shaft B to rotate at 8 rpm in the opposite direction?

7.18 In the mechanism in Fig. 7.30, gear 2 rotates at 60 rpm in the direction shown. Determine the speed and direction of rotation of gear 12.

7.19 A mechanism known as Humpage's gear is shown in Fig. 7.31. Find the angular velocity ratio ω_A/ω_B.

7.20 In the planetary gear train shown in Fig. 7.32 determine the angular velocity ratio ω_2/ω_7. Compare this ratio with that obtained if the arm 4 is connected directly to the output shaft, and gears 5, 6, and 7 are omitted.

7.21 In the gear train for Problem 7.20, gear 2 rotates at 600 rpm in the direction

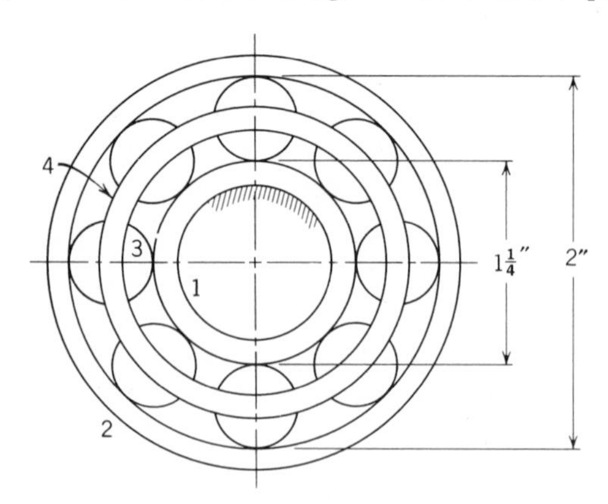

Figure 7.27

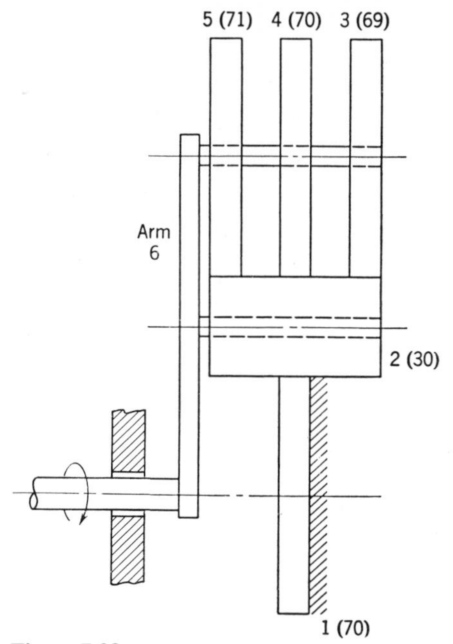

Figure 7.28

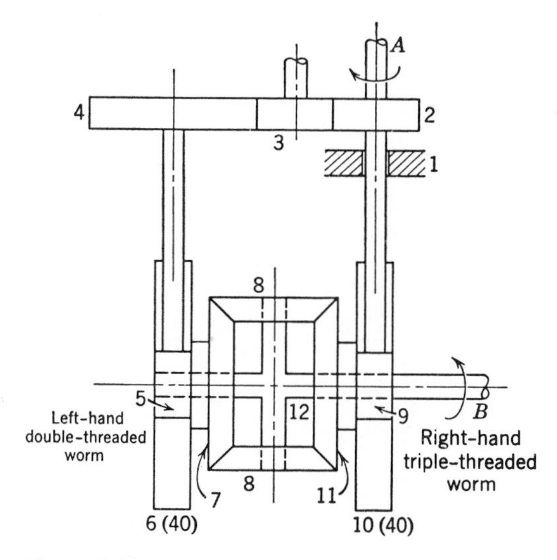

Figure 7.29

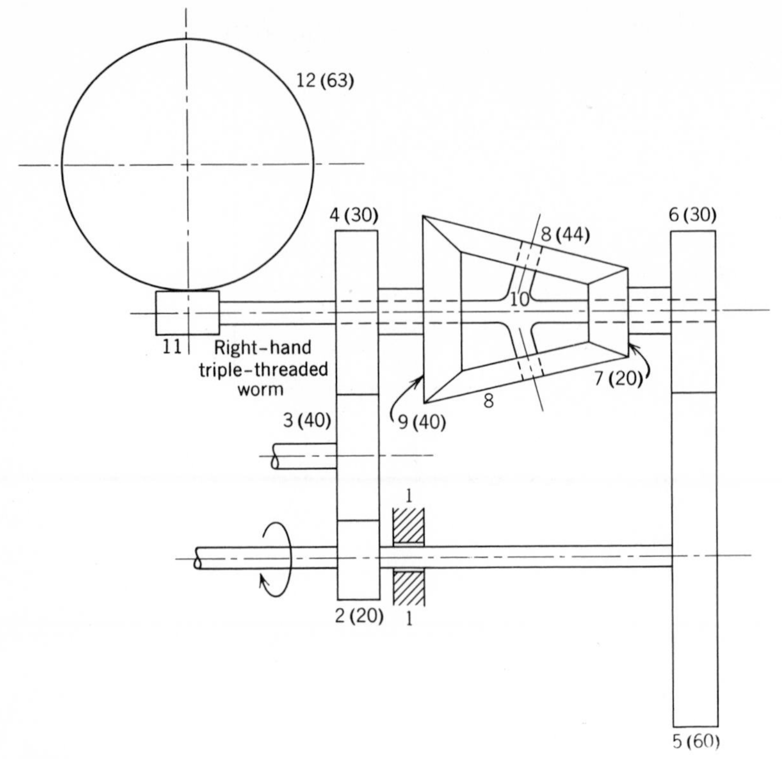

Figure 7.30

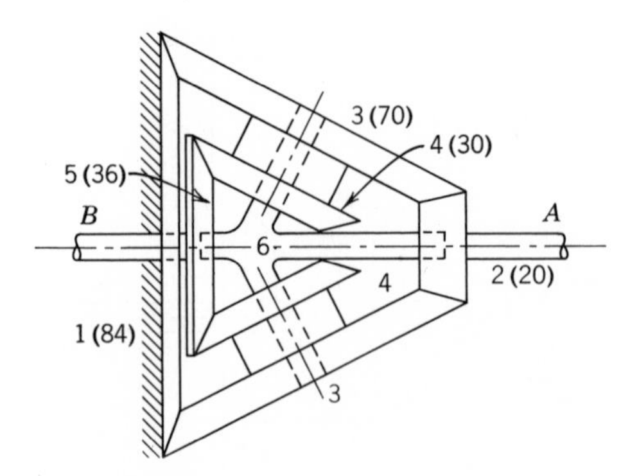

Figure 7.31

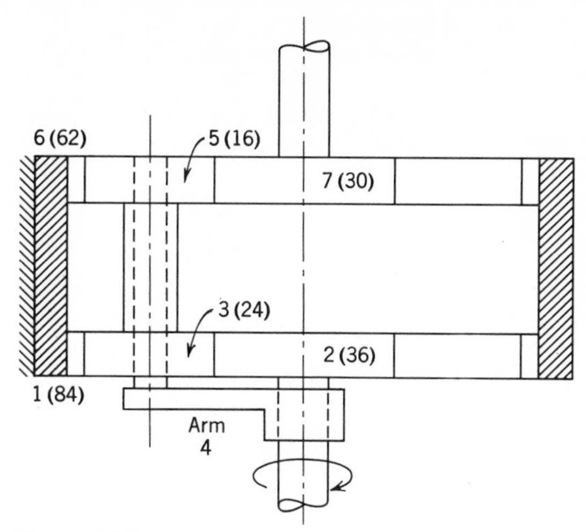

Figure 7.32

shown and gear 1 (and gear 6) rotates at 300 rpm in the opposite direction. Calculate the speed and direction of rotation of gear 7.

7.22 A planetary gear train for a two-speed aircraft supercharger drive is shown in Fig. 7.33. Gear 2 is driven by a 63-tooth gear (not shown) which operates at 2400 rpm. At high speed, gear 2 connects to the supercharger shaft through additional gearing. At low speed, gear 7 is held stationary and shaft B is connected to the supercharger shaft with the same gear ratio as was used between gear 2 and the supercharger shaft. If the supercharger operates at 24,000 rpm at high speed, calculate the low-speed value.

7.23 Figure 7.34 shows the planetary gear and power shaft assembly for an aircraft servo. If shaft A connects to the motor, determine the angular velocity ratio ω_A/ω_B.

7.24 Figure 7.35 shows a planetary gear train for a large reduction. (a) If shaft A connects to the motor, determine the angular velocity ratio ω_A/ω_B. (b) Will gears 2, 3,

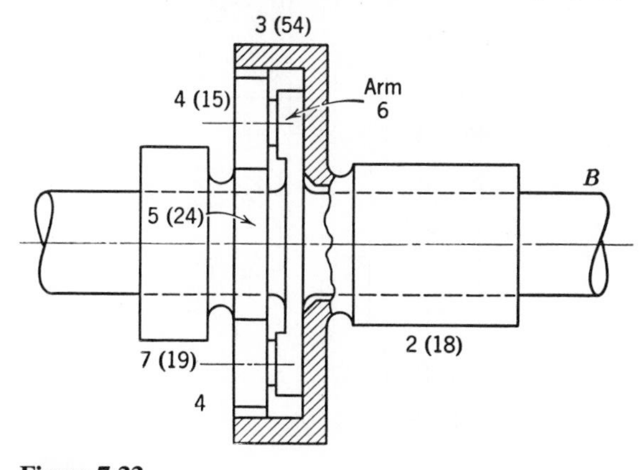

Figure 7.33

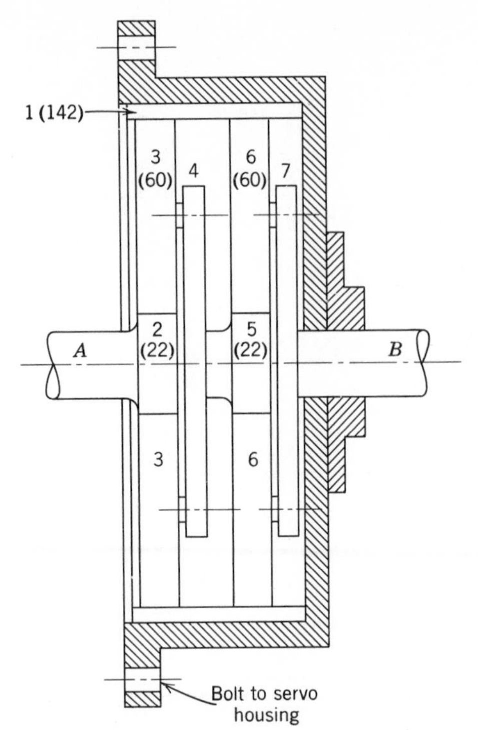

Figure 7.34

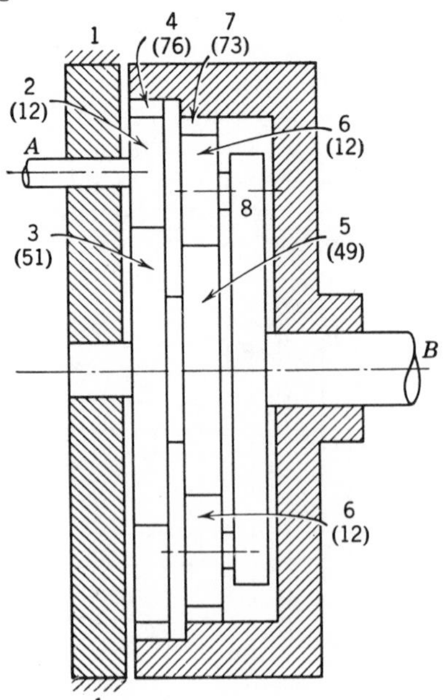

Figure 7.35 Problem by M. F. Spotts, *Prod. Eng.*, December 1954.

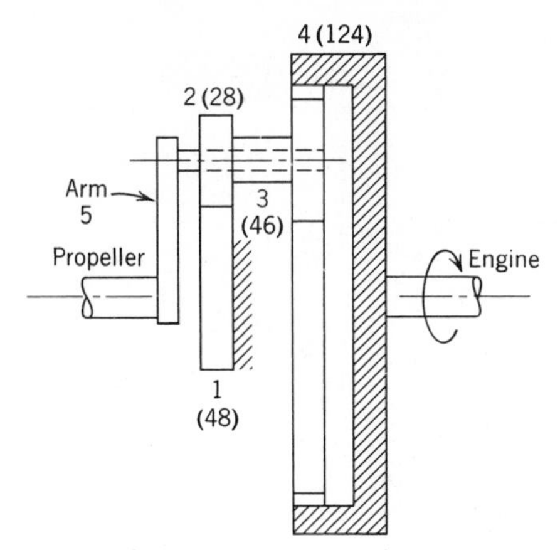

Figure 7.36

and 4 and gears 5, 6, and 7 be standard or nonstandard? Why? (*c*) If the number of teeth in gear 3 is changed from 51 teeth to 52 teeth, calculate the angular velocity ratio ω_A/ω_B.

7.25 An aircraft propeller reduction drive is shown diagrammatically in Fig. 7.36. Determine the propeller speed in magnitude and direction if the engine turns at 2450 rpm in the direction indicated.

7.26 In the planetary reduction unit shown in Fig. 7.37, gear 2 turns at 300 rpm in the direction indicated. Determine the speed and direction of rotation of gear 5.

7.27 In the gear train for Problem 7.26, gear 2 turns at 300 rpm in the direction shown and gear 1 rotates at 50 rpm in the opposite direction. Calculate the speed and direction of rotation of gear 5.

7.28 In the planetary gear train shown in Fig. 7.38, gear 2 turns at 600 rpm in the direction indicated. Determine the speed and direction of rotation of arm 6 if gear 5 rotates at 350 rpm in the same direction as gear 2.

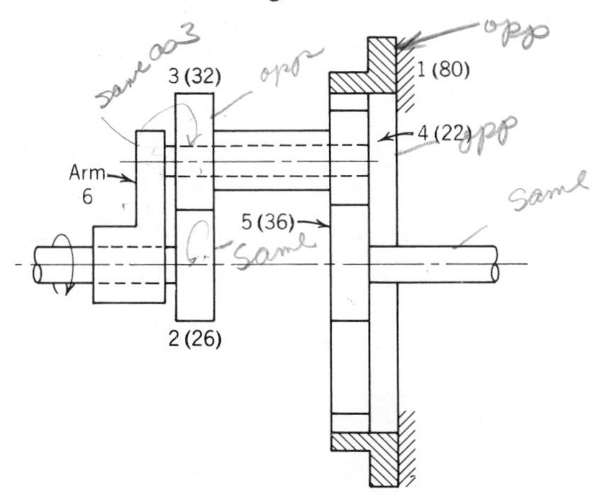

Figure 7.37

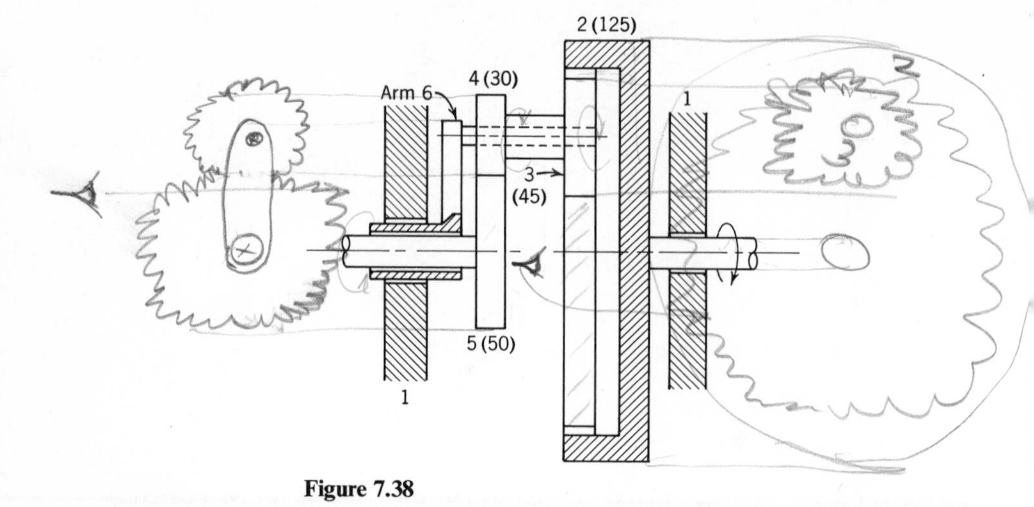

Figure 7.38

7.29 If in the gear train for Problem 7.28, gear 2 rotates at 1000 rpm in the direction shown and gear 5 is held stationary, arm 6 will rotate at 600 rpm in the same direction as gear 2. Determine the speed and direction of rotation which must be given to gear 5 to make arm 6 stand still if gear 2 continues to rotate at 1000 rpm.

7.30 For the gear train of Fig. 7.39, shaft *A* rotates at 300 rpm and shaft *B* at 600 rpm in the directions shown. Determine the speed and direction of rotation of shaft *C*.

7.31 In Fig. 7.40, shaft *A* turns at 100 rpm in the direction shown. Calculate the speed of shaft *B* and give its direction of rotation.

7.32 In the planetary gear train shown in Fig. 7.41, shaft *A* rotates at 450 rpm and shaft *B* at 600 rpm in the directions shown. Calculate the speed of shaft *C* and give its direction of rotation.

7.33 Shaft *A* rotates in Fig. 7.42 at 350 rpm and shaft *B* at 400 rpm in the directions shown. Determine the speed and direction of rotation of shaft *C*.

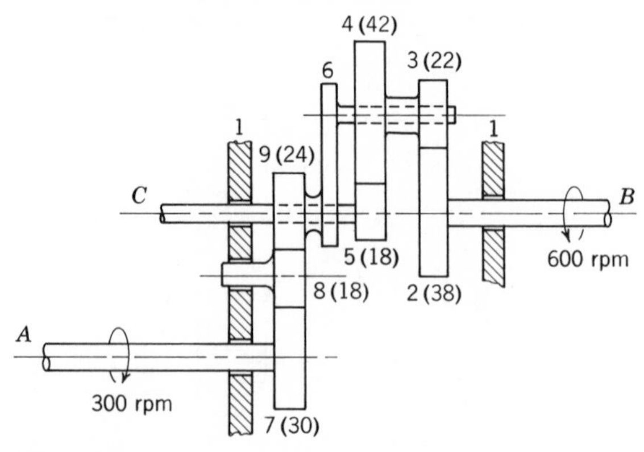

Figure 7.39

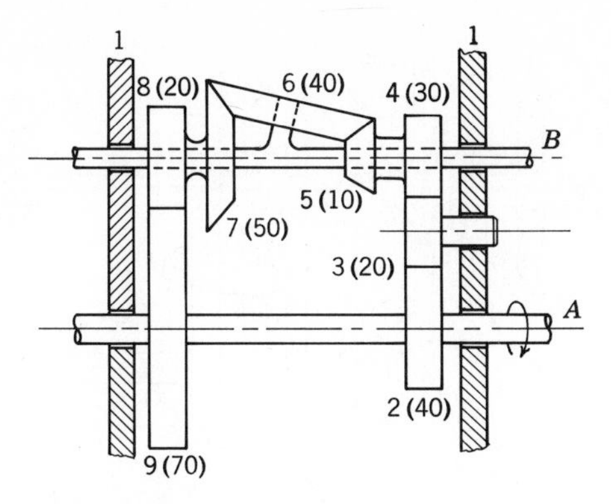

Figure 7.40

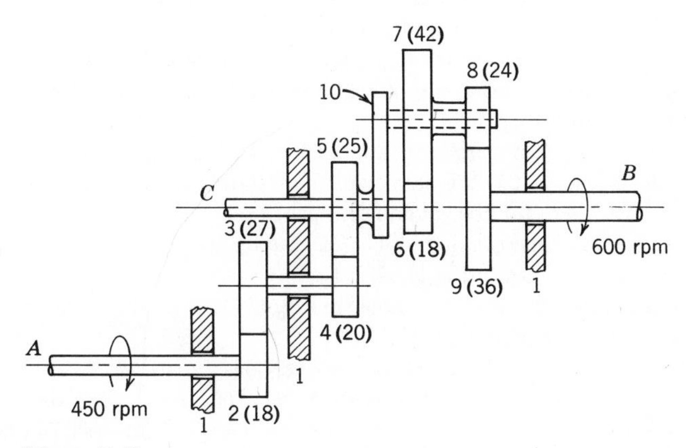

Figure 7.41

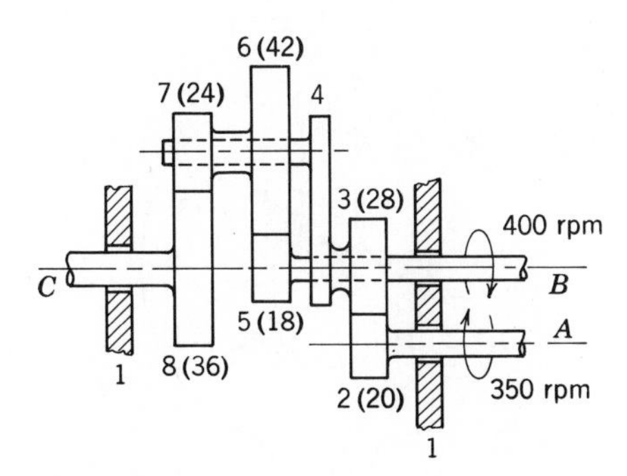

Figure 7.42

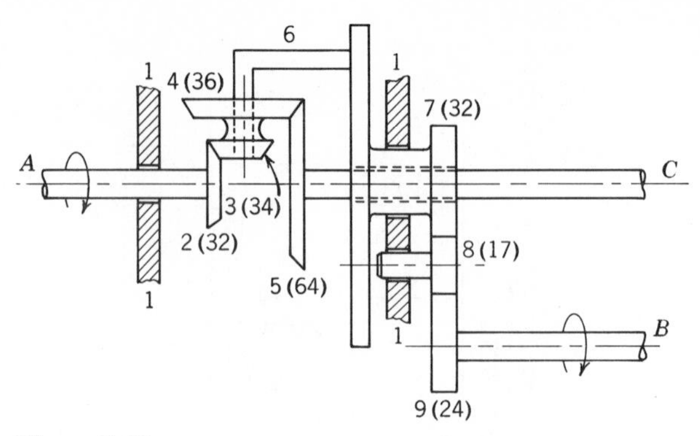

Figure 7.43

7.34 In the bevel gear planetary train shown in Fig. 7.43, shaft A rotates in the direction shown at 1250 rpm and shaft B in the direction shown at 600 rpm. Determine the speed of shaft C in magnitude and direction.

7.35 For the planetary gear train of Fig. 7.33, calculate the maximum number of planets possible without overlapping and the numbers of equally spaced planets that can be used in the train.

7.36 In a planetary train similar to that of Fig. 7.14, gear 1 has 41 teeth, gear 2 has 18 teeth, and gear 3 has 78 teeth. Gears 1 and 3 are standard and gear 2 is nonstandard. Determine the maximum number of equally spaced planets that can be used.

7.37 Calculate the maximum number of equally spaced compound planets that can be used in the gear train of Fig. 7.32.

7.38 For the planetary gear train shown in Fig. 7.37, calculate the maximum number of compound planets that can be used.

7.39 In the planetary gear train shown in Fig. 7.44, the carrier (link 4) is the driving member and the sun gear (link 3) is the driven member. The internal gear is held stationary. The sun gear is to rotate 2.5 times the speed of the carrier. The pitch diameter of the internal gear is to be approximately 11.0 in. (*a*) Design the gear train by determining the numbers of teeth for the internal gear, the sun gear, and the planets using 10 diametral pitch 20° full-depth standard spur gear teeth. Hold the 11.0 in. pitch diameter as closely as possible. (*b*) Determine whether or not three equally spaced planets can be used.

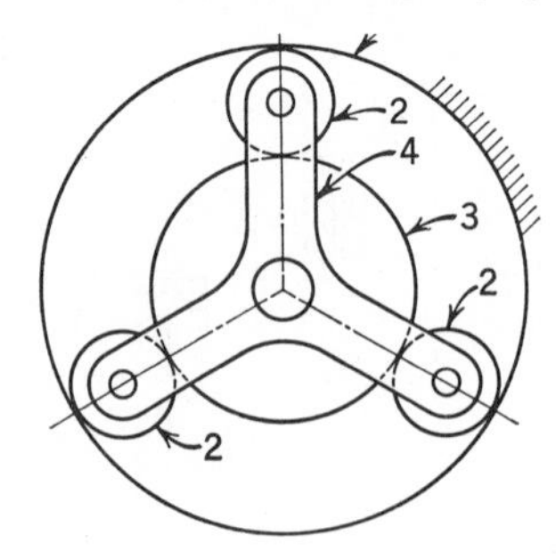

Figure 7.44

8

Computing Mechanisms

The great advances in automatic controls and the trend toward automation have been made possible by the continual development of computing mechanisms and machines. These computers may be divided into two types: digital computers and analog computers.

8.1 Digital Computers. These computers deal with quantities in numerical form and calculate in discrete steps the result of a series of mathematical operations on the input quantities. Generally, digital computers perform mathematical operations by the combination of additions: multiplication is done by repetitive addition, integration by summation, and converging series are substituted for trigonometric functions. Digital computing machines are also made for a wide variety of business applications, that range from small hand calculators to large accounting machines. Digital computers have also been developed for the automatic operation of machine tools, assembling machines, and process control equipment. This has resulted in the large scale industrial applications such as automatic factories.

8.2 Analog Computers. These mechanisms deal with magnitudes instead of with purely numerical values and are essentially continuous in a mathematical sense. They can be applied to the instantaneous or to the continuous solution of specific problems. The inputs and the outputs of this device are represented by physical quantities, and the laws governing the operation of the computer are analogous to the laws governing the process which is to be controlled. Analog computers may be designed to perform the standard operations of algebra and calculus, that is, addition and subtraction, multiplication and division, integration and differentiation, resolution of vectors, and, most important, the generation of mathematical or tabulated functions of one or more independent inputs. Analog computers may be mechanical, electrical, pneumatic, or hydraulic, either separately or in combination. Although this type of computer has wide application to the solution of specific problems, it cannot have as high a degree of accuracy built into it as the digital computer.

Analog computers can be used to build large computing machines for the solution of complicated equations. They are also used as components in the control equipment of guided missiles, navigational devices, telemetering units, fire control equipment, bombsights, and many other systems.

From the viewpoint of the study of mechanisms the mechanical analog computer is of primary interest. The following discusses the mechanisms which are capable of performing simple mathematical operations.

8.3 Addition and Subtraction. There are several devices for adding and subtracting, one of which is shown in Fig. 8.1. Bars 2 and 3 move horizontally over the rollers of bar 4 with pure rolling motion. Owing to this action, bar 4 will also move and its motion is given by

$$S_4 = \frac{S_2 + S_3}{2} \tag{8.1}$$

To add two quantities, bar 2 is set at one of the numbers and bar 3 at the other number. The sum is given by bar 4. Because the motion of bar 4 is one-half that of $S_2 + S_3$, it is said that the output S_4 has a scale factor of $\frac{1}{2}$. To correct for this, the scale of bar 4 must be twice the scale of bar 2 and bar 3. The scales of bars 2 and 3 must be equal and therefore have the same scale factor.

If larger forces have to be transmitted, pinions and racks are used to replace the rollers and bars. To adapt the mechanism for subtraction, negative portions are added to the scales.

Another mechanism which can be used as an adder is shown in Fig. 8.2*a*. This device is known as a linkage differential, and, although it is of simpler form than the mechanism of Fig. 8.1, it is an approximate adder, whereas

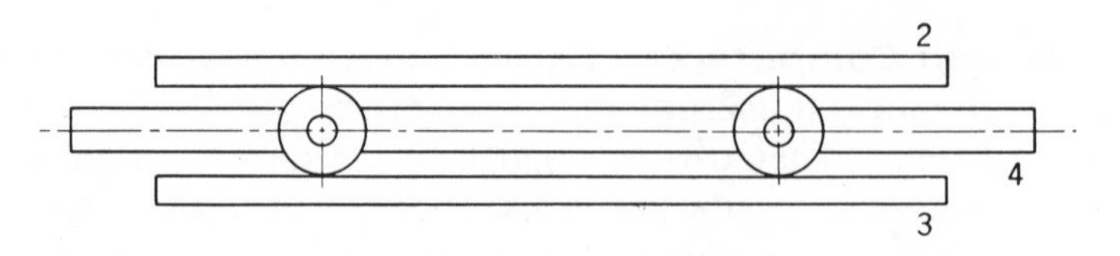

Figure 8.1

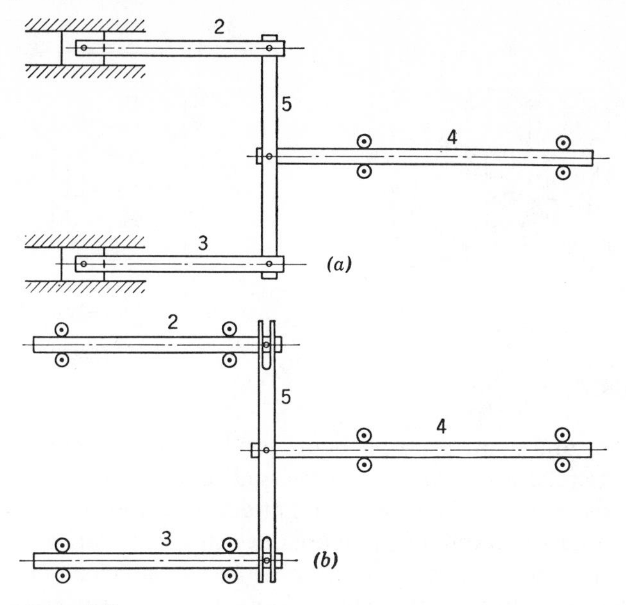

Figure 8.2

the other is exact. The motion of bar 4 is given by the relation

$$S_4 \approx \frac{S_2 + S_3}{2} \tag{8.1a}$$

It can be seen that the angularities of links 2 and 3 will produce an error in the computation. If bar 5 is slotted at its ends and links 2 and 3 are guided as shown in Fig. 8.2*b*, the error will be eliminated.

If it is necessary to add rotations instead of linear quantities, a bevel gear or a spur gear differential may be used as shown in Figs. 8.3*a* and *b*, respectively. From the study of the bevel gear differential as used in automotive drives, it is known that the speed of the arm 5 in Fig. 8.3*a* is the average of the speeds of gears 2 and 4. Therefore,

$$\omega_5 = \frac{\omega_2 + \omega_4}{2} \qquad \text{and} \qquad \theta_5 = \frac{\theta_2 + \theta_4}{2}$$

Gear 6 is fixed to arm 5 so that

$$\theta_6 = \frac{\theta_2 + \theta_4}{2} \tag{8.2}$$

Equation 8.2 also holds for the spur gear differential of Fig. 8.3*b*. The scale factor of $\frac{1}{2}$ for output θ_6 can easily be increased to 1 by meshing gear 6 with another gear to give a step-up of 2:1.

Bevel gear differentials are available commercially in several stock sizes

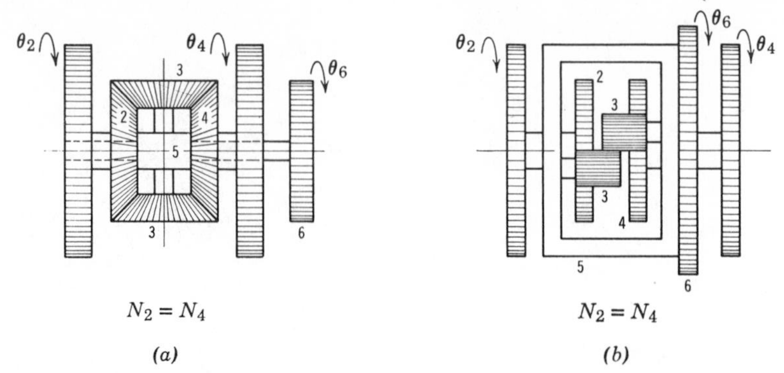

Figure 8.3

for computer and control use; a photograph of one is shown in Fig. 8.4. In the commercial units, the arm is referred to as the *spider* and the bevel pinions as *spider gears*. The spur gears attached to the bevel gears 2 and 4 are known as *end gears* and can be obtained in a wide variety of sizes to fit standard units. Spur gear differentials are also available commercially.

The bevel and spur gear differentials have two sources of error. These are backlash and angular inaccuracy of transmission. Backlash is the angular motion possible between the end gears when one of them and the arm or spider are held fixed. For example, in Fig. 8.3*a* if gear 4 and spider 5 are held fixed, gear 2 can be rotated slightly by the application of a small amount of torque. This lost motion or backlash is the sum of the gear tooth backlash, bearing looseness, and strains in the structural parts of the unit. At the

Figure 8.4 Bevel gear differential for computer. (Courtesy of Librascope, Inc.)

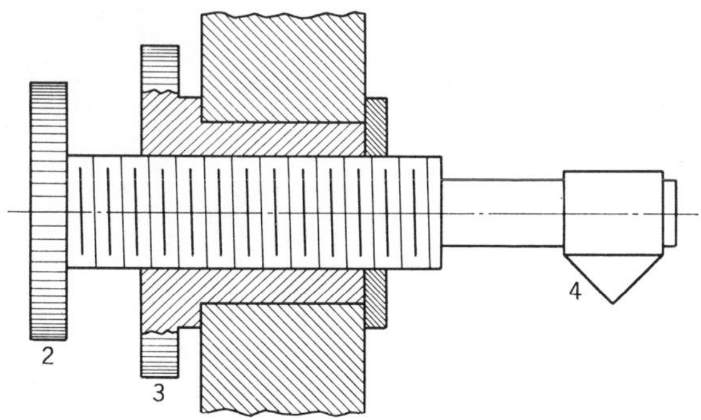

Figure 8.5

present time it is possible to purchase gear differentials having backlash ranging from 5 to 10 min of an angle for torques of 2 to 6 oz-in. at diameters of approximately 1 in.

Inaccuracy of transmission is the nonuniformity of the angular velocity ratio through the unit. Theoretically, the angular velocity ratio between two gears should be constant, but because of pitch circle runout (the amount the pitch circle is eccentric) and tooth-to-tooth errors, there is a slight variation. This inaccuracy results in a position error.

Another means of adding is with a screw differential, which finds application when the inputs are rotary and it is desired to have a linear output. It is also useful when large forces are involved. A sketch of a simple screw differential is shown in Fig. 8.5 where the pointer is constrained so that it will move axially with the screw but will not rotate. The inputs in units of rotation are fed to gears 2 and 3, and the addition of the two quantities is given on a linear scale by pointer 4. In the case of the simple screw shown, the motion of the pointer is the sum of the motions produced by gear 2 rotating while gear 3 is stationary and by gear 2 moving axially but not rotating while gear 3 rotates. The equation for the output S_4 is given by

$$S_4 = \frac{p_x N}{360}(\theta_2 + \theta_3) \tag{8.3}$$

where

θ_2 and θ_3 = angular input degrees
p_x = axial pitch of screw
N = number of threads on screw

8.4 Multiplication and Division. Multiplication and division may be performed in several ways. Figure 8.6 shows a slide multiplier which is based on the theory of similar triangles. In the sketch, bar 2 is moved a distance x_1 from O_4 (link 6 stationary). This will move link 7 a certain distance upward. Link 6 is now moved to the right to position x_2 (link 4 stationary),

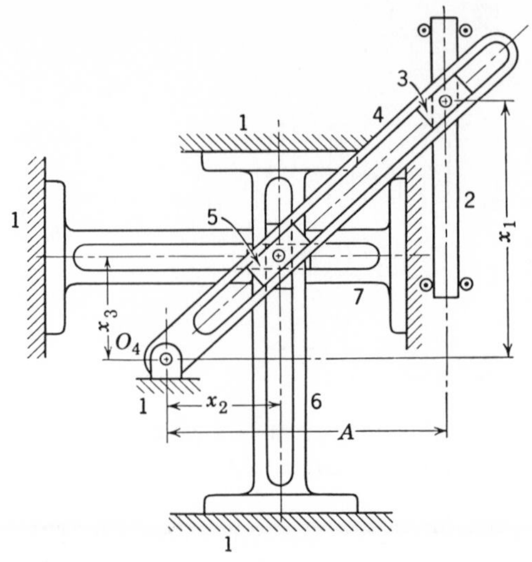

Figure 8.6

which will also move link 7 upward. Therefore, owing to the setting x_1 and x_2, link 7 will be at a distance x_3 from O_4. From similar triangles

$$\frac{x_1}{x_3} = \frac{A}{x_2}$$

Therefore,

$$x_3 = \frac{x_1 x_2}{A} \tag{8.4}$$

In order to have sufficient range this mechanism must be relatively large, which makes it difficult to limit the deflections of the members unless the unit is made quite massive. This together with the difficulty encountered with wear and inaccuracy in the sliders limit the practical applications of this unit. More accurate ways of multiplication will be presented in later sections.

8.5 Integration. A mechanism for integration is shown in Fig. 8.7. Disk 2 rotates driving the balls which are positioned by the ball carriage 3. The balls in turn drive roller 4. Pure rolling action is maintained between the disk and the balls and between the roller and the balls. The input variables are the rate of rotation of disk 2 and the axial displacement r of the balls. The output of roller 4 is the result. The action of the mechanism therefore gives the relation

$$R d\theta_4 = r d\theta_2$$

because the linear distance traveled by the top ball on disk 2 must be equal to that traveled by the bottom ball on roller 4. Integrating the preceding

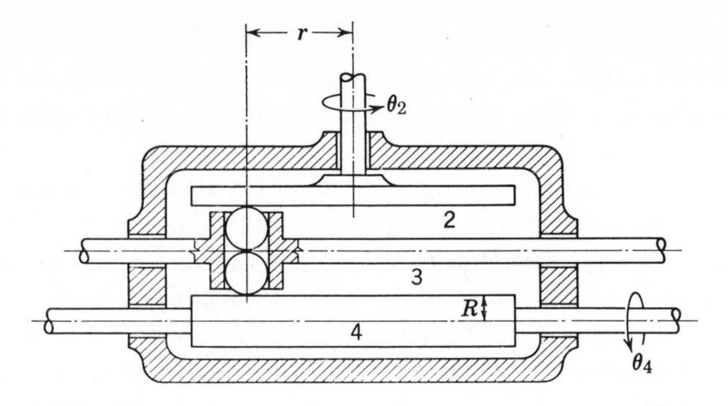

Figure 8.7

equation gives

$$\theta_4 = \frac{1}{R}\int r d\theta_2 \tag{8.5}$$

where r is a function of θ_2. The value $1/R$ is the *integrator constant* and is very important in the design of an integrator system. The unit can also be used as a multiplier by taking r as a constant during each operation. The unit will then generate $\theta_4 = (r/R)\theta_2$.

Equation 8.5 can also be expressed in terms of x, y, and z. Let the rotation θ_2 be represented by x, the ball carriage position r by y, which equals $f(x)$, and the output θ_4 by z. Substituting these quantities into Eq. 8.5 gives

$$z = \frac{1}{R}\int y\, dx \tag{8.5a}$$

These quantities are shown schematically in Fig. 8.8.

In the integrator, input x and output z are shaft rotations, whereas input y is a linear distance from the ball carriage to the center of the disk. To provide

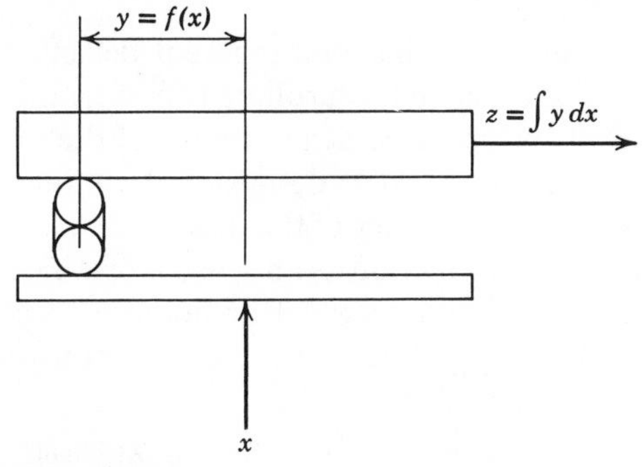

Figure 8.8

the axial motion necessary for y, a lead screw is often used. By so doing, the rotation of the screw, which is proportional to the carriage position, can be used to represent y. Therefore, the input and the output will all be shaft rotations.

In an integrator the scales are more difficult to determine than in the other components discussed so far. In order to determine the scale factor of the output, let S_1 be the scale factor on the disk variable x, S_2 that on the lead screw variable y, and S_3 that on the output variable z. In determining the scale factor of the output, account must be taken of the fact that rotation of the lead screw represents a linear distance y. If the lead screw has n threads per inch and is single threaded, the radial position r of the ball carriage is

$$r = \frac{S_2 y}{n}$$

Substituting scale factors into Eq. 8.5*a* together with the preceding relation gives

$$S_3 z = \frac{1}{R} \int \frac{S_2 y}{n} (S_1\, dx) = \frac{S_1 S_2}{Rn} \int y\, dx \qquad (8.5b)$$

Comparison of Eq. 8.5*b* with the equation to be mechanized, $z = \int y\, dx$, shows that when S_1 and S_2 have given values, S_3 must be given by the following relation:

$$S_3 = \frac{S_1 S_2}{Rn}$$

From this the scale of the output can be determined.

The integrator is a very accurate mechanism, and its size is designated by the diameter of its disk. Commercial models are available which range in size from $1\frac{1}{2}$ to 5 in. The usual accuracy of these units is in the order of 0.5% for input torques up to 1 oz-in. and a disk diameter of $1\frac{1}{2}$ in. However, one manufacturer lists a unit with an accuracy of 0.01% for an input torque of 2 oz-in. and a disk diameter of $1\frac{1}{2}$ in.

The ball-and-disk integrator has two inherent defects: (*a*) output torque is limited because the unit depends on rolling friction, and (*b*) the balls cause slippage (and wear) when operated at the center of the disk. This wear can be reduced by incorporating in the design of the integrator a means for rotating the balls when at the center of the disk.

In addition to its primary function as an integrator, this mechanism can be used to perform other computations which do not directly call for integration. Several examples follow in which the integrator constant $1/R$ has been omitted to simplify the operation.[1]

[1] G. W. Michalec, "Design Guide Analog Computing Mechanisms," *Machine Design*, March 1959.

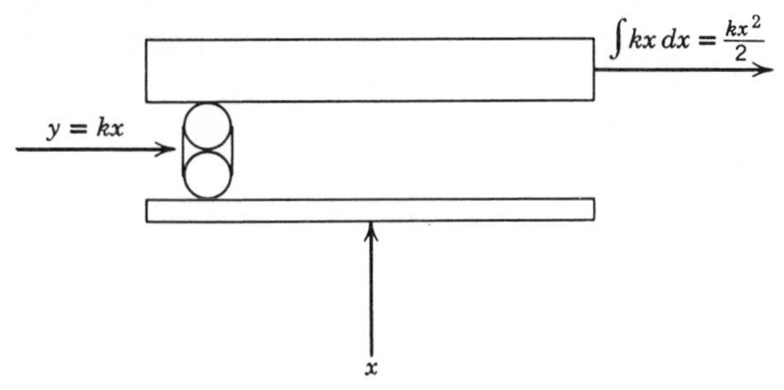

Figure 8.9

The integrator can be used to give an output proportional to the input squared. To do this, the position y of the ball carriage must vary linearly with x as illustrated in Fig. 8.9.

The product of two independent variables can be obtained by the use of two integrators and a differential connected as shown in Fig. 8.10.

Sine and cosine functions can be generated simultaneously by the use of two integrators as seen in Fig. 8.11. The negative sign that would result from $\int \sin\theta\, d\theta$ is eliminated by operating the ball carriages out of phase so that the roller of the second integrator rotates in a direction opposite to that of the roller of the first integrator.

In addition to integration, the ball-and-disk integrator can be used for approximate differentiation.[2] Although differentiation is the opposite of integration, it is not possible to reverse the mechanism to perform this operation. The integrator cannot be made to give a theoretically correct derivative, but by the addition of a differential and two step-up gear ratios it will give results sufficiently accurate for many purposes. The layout for this computation is shown in Fig. 8.12 where u is the output and approximates dy/dx. The differential is connected for subtraction so that the following

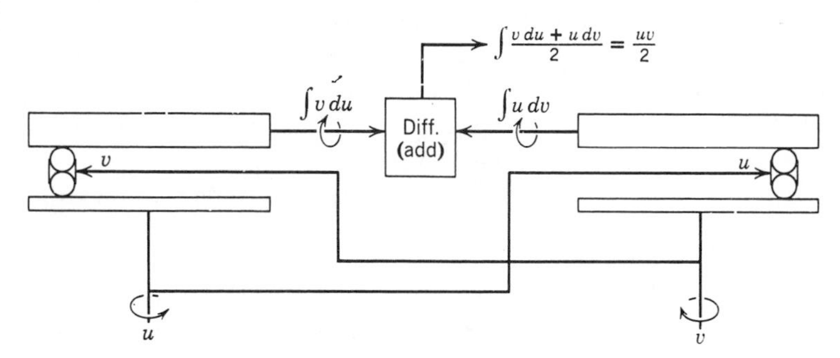

Figure 8.10

[2] *Ibid.*

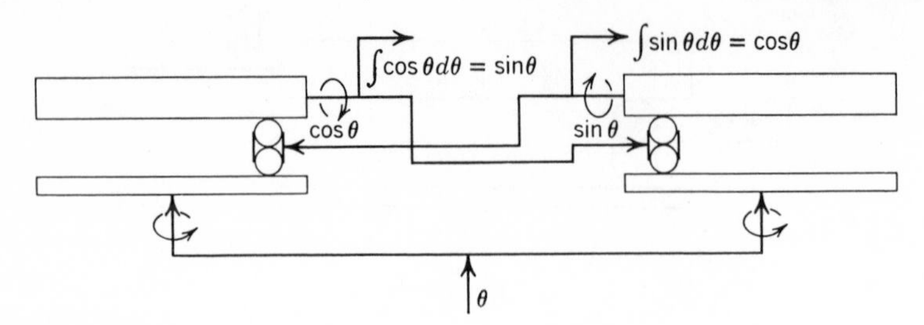

Figure 8.11

equation results for the output:

$$u = \frac{ky - k\int u\,dx}{2}$$

or

$$ky = 2u + k\int u\,dx$$

and

$$y = \frac{2u}{k} + \int u\,dx$$

Differentiating with respect to x,

$$\frac{dy}{dx} = \frac{2}{k}\left(\frac{du}{dx}\right) + u$$

It can be seen that the computer answer u is not the exact derivative of y with respect to x, but is in error by a factor $(2/k)(du/dx)$. To give the correct answer, du/dx would have to be zero. This is an impossible condition because the ball carriage would have to be motionless. However, the error term can be reduced by using as large a value of k as possible without causing too high a response of the ball carriage motion.

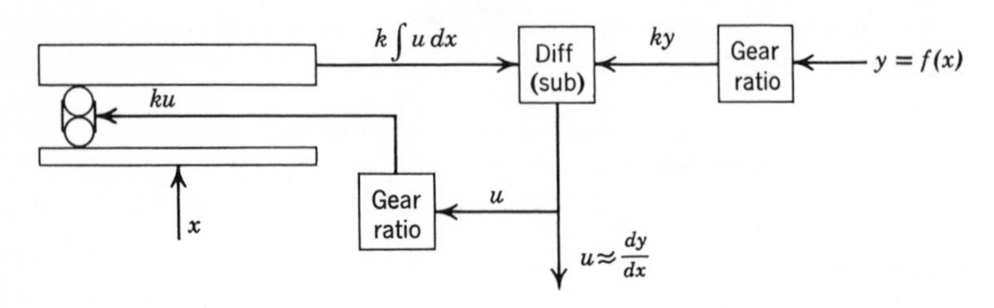

Figure 8.12

8.6 Trigonometric Functions. The sine and cosine functions being smooth and continuous are easily generated by several well-known mechanisms. The Scotch yoke is one of the most common. A sketch of this mechanism is shown in Fig. 2.7. A slider-crank can be used to generate an approximate sine-cosine function. Figure 8.13 shows a sketch of a planetary gear train designed to convert rotary motion into linear sine or cosine movements.

In Fig. 8.13 gear 1 is an internal gear which is fixed. Gear 2 is one-half the pitch diameter of gear 1, and is driven by arm 3. When gear 2 rolls on the inside of gear 1, a point B on the circumference of the pitch circle of gear 2 will trace a hypocycloid. Because the two gears are in the ratio of 2 to 1, the hypocycloid will be a straight line, and point B will travel on a diameter of gear 1 with simple harmonic motion. Therefore link 4 will generate a sine or cosine function.

Figure 8.13*a* shows the generation of a sine function and Fig. 8.13*b* a cosine function. It should be noted that in the first case the zero position is vertical, whereas in the second case the zero position is horizontal.

Sine-cosine generators of the form shown in Fig. 8.13 are available commercially with an accuracy of 0.2%. The outside diameter of these units is about 2 in., and they weigh 2 oz. A photograph of a commercial unit is shown in Fig. 8.14.

A mechanism for generating tangents and secants is shown in Fig. 8.15 where gear 2 is driven through the input angle θ. Link 3 sliding in a groove in gear 2 is pinned to link 4, which is constrained to move vertically. As gear 2 rotates, link 4 moves vertically, and because A is constant, y is a measure of $\tan\theta$ and S a measure of $\sec\theta$. At 45°, y equals A, which will determine the scale of y. At 0° S equals A and at 60° S equals $2A$, which determines the scale of S. Because of the discontinuity of the two functions, the mechanism is useful only for a limited range of input.

If a rotational output is required, it will then be necessary to attach a rack

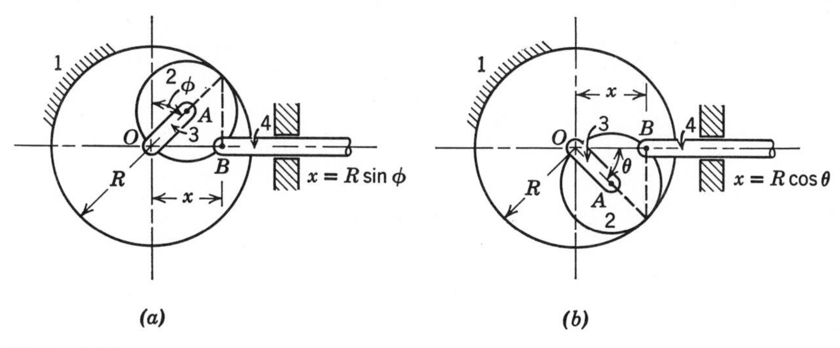

Figure 8.13

Figure 8.14 Sine–cosine generator. (Courtesy of Librascope, Inc.)

to block 4 for the tangent or to block 3 for the secant. A pinion meshing with the rack would give the output.

Trigonometric functions can also be generated by cams which will be discussed later.

A *component resolver* is a mechanism for resolving a vector r into x and y components. In effect, it is a simultaneous sine-cosine generator giving the

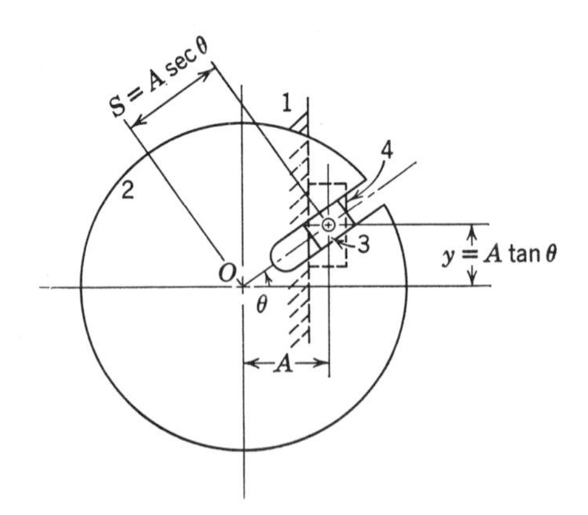

Figure 8.15

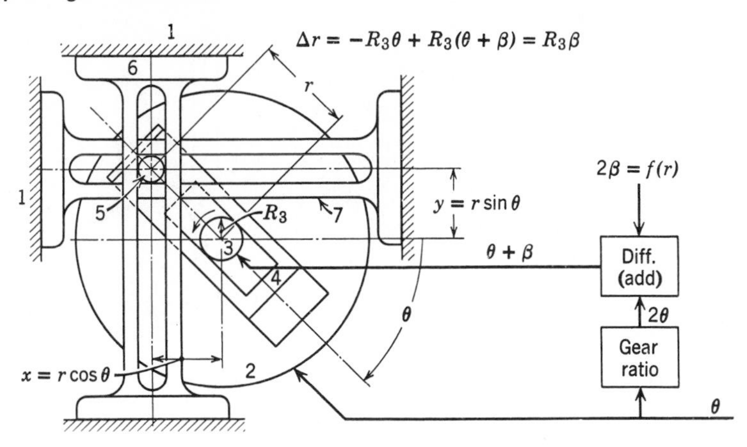

Figure 8.16

functions

$$x = r \cos \theta$$
$$y = r \sin \theta \tag{8.6}$$

where r and θ are independent variables. Figure 8.16 shows a sketch of a component resolver which is seen to be similar to a double Scotch yoke. The only difference is in the fact that on the Scotch yoke r is a fixed distance, whereas on the resolver it is adjustable by means of pinion 3 and rack 4. When input θ is fed into the mechanism, it is also necessary to feed a compensating input into pinion 3 so that the motion θ does not affect r. If pinion 3 were to remain stationary, rotation of gear 2 (and therefore rack 4) would cause axial motion of the rack and a change in r. To provide this compensation, a differential and gear ratio are connected into the circuit as shown.

It should be mentioned that if circular outputs are required for x and y, a rack and output pinion will have to be attached to both links 6 and 7.

In this mechanism rack 4 and pinion 3 may be replaced by attaching pin 5 to a nut that can be adjusted by a screw and bevel gear. With either type of design the units are expensive to build, and because of the high friction of the sliding parts, they must be relatively large in size.

Theoretically, this mechanism can be operated in reverse to compute r and θ for given values of x and y. However, it is generally not practical to do so because of the dead points encountered. Therefore, when the unit is used in this manner as a resultant solver, it is necessary to add a means of passing through these dead points.

8.7 Inversion. In Fig. 8.17 is shown a mechanism for finding reciprocals. Link 2 is pivoted to the ground at O_2 and is slotted at each end to receive links 3 and 5. Link 3 is pivoted to link 4, and link 5 is pivoted to link 6. As link

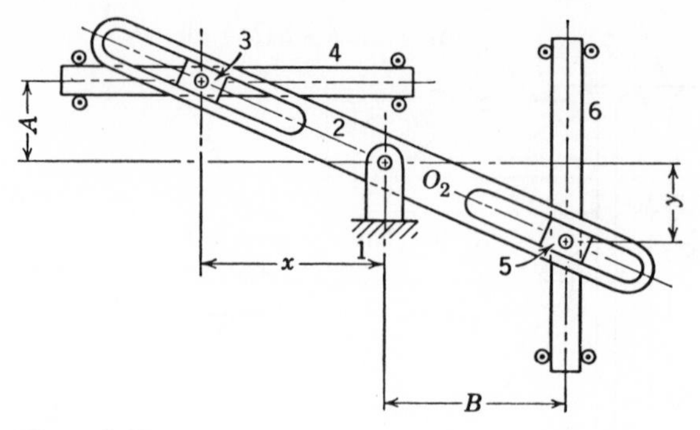

Figure 8.17

2 is rotated, link 4 moves horizontally at a constant distance A from O_2 while link 6 moves vertically at a constant distance B from O_2. From similar triangles

$$\frac{x}{A} = \frac{B}{y}$$

or

$$x = \frac{AB}{y} \tag{8.7}$$

This mechanism can also be used to determine csc θ and cot θ when it is used in conjunction with the mechanisms of Fig. 8.13 and Fig. 8.15.

8.8 Squares, Square Roots, and Square Roots of Products. A mechanism for performing these operations is shown in Fig. 8.18. Link 2 is pivoted to link 7 and is slotted at each end to receive links 3 and 5. Angle ABC is a right angle. Link 3 is pivoted to link 4, and link 5 is pivoted to link 6. Links 4 and 6 are constrained to move horizontally while link 7 moves

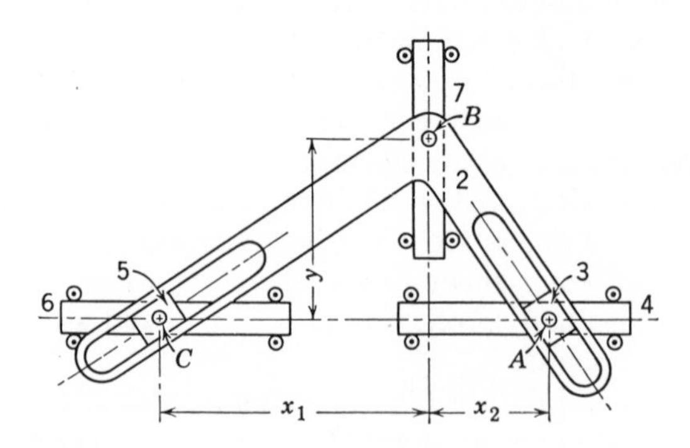

Figure 8.18

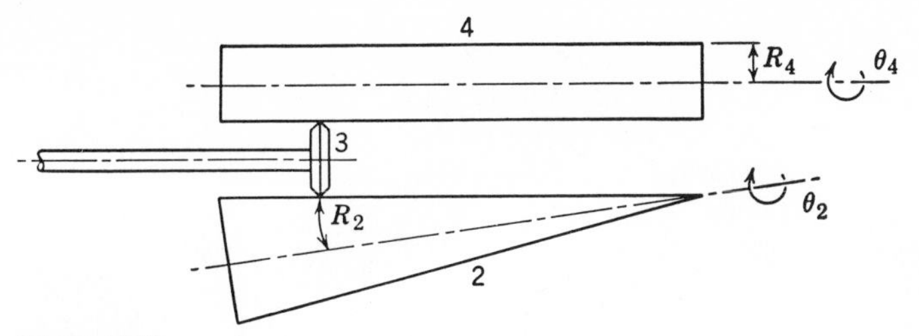

Figure 8.19

vertically. From similar triangles,

$$\frac{y}{x_1} = \frac{x_2}{y}$$

or

$$y^2 = x_1 x_2$$

and

$$y = \sqrt{x_1 x_2} \tag{8.8}$$

In the mechanism x_1, x_2, and y are variables. If x_1 and x_2 are fed into the mechanism, y will read the square root of their product. If x_1 and y or x_2 and y are fed into the mechanism, squares of y divided by x_1 or x_2 will be obtained.

If point A is fixed so that link 4 does not move and x_2 becomes a constant distance D, a square and square root relation results. Therefore,

$$y^2 = Dx_1$$

or

$$y = \sqrt{Dx_1} \tag{8.9}$$

If y is fed into the mechanism, x_1 will read the square of y. However, if x_1 is fed into the mechanism, the square root of x_1 will be given by the y dimension.

A cone and a cylinder may be connected as shown in Fig. 8.19 to give another type of squaring mechanism. The number of revolutions of cylinder 4 is proportional to the square of the number of revolutions of cone 2. Idler 3 is moved axially a distance proportional to the number of revolutions of the cone. If R_2 is the radius of the cone at the point of contact with the idler and R_4 the radius of the cylinder, a small rotation $d\theta_2$ of the cone will give a rotation of the cylinder:

$$d\theta_4 = \frac{R_2}{R_4}\, d\theta_2$$

By an interconnection with the cone, the idler position is determined by the relation

$$R_2 = k\theta_2$$

where k is a constant determined by the cone vertex angle and by the amount

Figure 8.20 Cone and cylinder squaring mechanism. (Courtesy of Librascope, Inc.)

of advance of the idler per unit of θ_2. Therefore

$$d\theta_4 = \frac{k\theta_2}{R_4}\, d\theta_2$$

and

$$\theta_4 = \frac{k\theta_2^2}{2R_4} \tag{8.10}$$

This mechanism may be operated in reverse to obtain the square root of a variable.

A variation of the mechanism of Fig. 8.19 is to eliminate the idler and to use two flexible wires to connect the cone and cylinder. Two wires are necessary to allow rotation in either direction; while one wire winds up, the other wire unwinds. The wires are guided by grooves cut in the cone and cylinder. A photograph of one of these units is shown in Fig. 8.20.

8.9 Computing Cams and Gears. Cams can be designed to generate a wide variety of functions. For this reason, it was considered desirable to discuss them under this heading rather than under the various mathematical operations.

Cams of the general types presented in Chapter 3 are often used for computation. The simplest of these is a disk cam of circular contour (eccentric) imparting simple harmonic motion to a radial flat-faced follower.

There is another type of cam whose application is primarily in computer design. It is known as a *contour cam* and is illustrated in Fig. 8.21. With this type of cam the members roll upon each other without sliding. This facilitates their design for two reasons: (*a*) the contact point P will always lie on the line of centers, and (*b*) both surfaces will roll on each other through the same distance. Making use of these factors, equations for the distance from the cam centers to the contact point can easily be derived.

In Fig. 8.21, R_2 and R_3 are the instantaneous distances from the cam centers to the point of contact, and C is the fixed distance between centers. If cam 2 rotates through a small angle $d\theta_2$ and cam 3 through $d\theta_3$, the point of contact on cam 2 will move through $R_2\, d\theta_2$, and that on cam 3 through $R_3\, d\theta_3$. For pure rolling,

$$R_2\, d\theta_2 = R_3\, d\theta_3$$

Also

$$R_2 + R_3 = C$$

Therefore,

$$R_2 = \frac{C}{1 + (d\theta_2/d\theta_3)} \tag{8.11}$$

and

$$R_3 = \frac{C}{1 + (d\theta_3/d\theta_2)} \tag{8.11}$$

These cams can be used to generate several types of functions, three of which are illustrated on the following page.

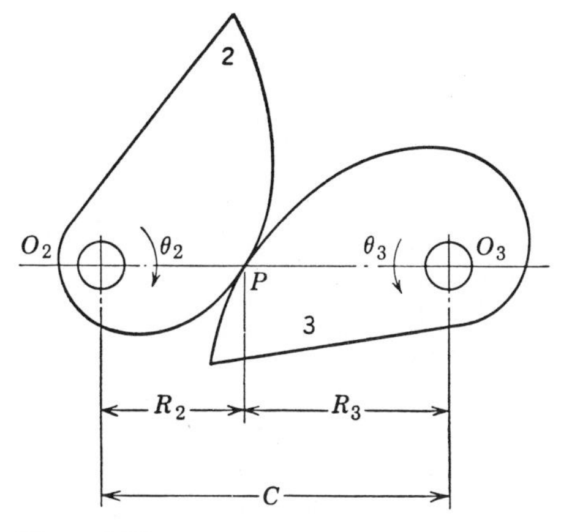

Figure 8.21

1. *Square Function.* To generate the square function,

$$\theta_3 = k\theta_2^2$$

$$\frac{d\theta_3}{d\theta_2} = 2k\theta_2$$

and

$$\frac{d\theta_2}{d\theta_3} = \frac{1}{2k\theta_2}$$

Therefore,

$$R_2 = \frac{2kC\theta_2}{1 + 2k\theta_2}$$

and

$$R_3 = \frac{C}{1 + 2k\theta_2}$$

From the equations for R_2 and R_3, the cam contours may be determined that will generate the given square function. If the cams are operated in reverse, square roots are obtained.

2. *Logarithmic Function.* To generate the logarithm,

$$\theta_3 = \log_{10} \theta_2$$

$$\theta_3 = \frac{1}{2.303} \log_e \theta_2$$

$$\frac{d\theta_3}{d\theta_2} = \frac{1}{2.303\theta_2}$$

and

$$\frac{d\theta_2}{d\theta_3} = 2.303\theta_2$$

Therefore,

$$R_2 = \frac{C}{1 + 2.303\theta_2}$$

and

$$R_3 = \frac{2.303C\theta_2}{1 + 2.303\theta_2}$$

From these equations the cam contours may be determined which will generate the given logarithm. Operation in reverse will give antilogs.

3. *Trigonometric Function.* To illustrate the generation of a trigonometric function, consider

$$\theta_3 = \tan \theta_2$$

$$\frac{d\theta_3}{d\theta_2} = \sec^2 \theta_2$$

and

$$\frac{d\theta_2}{d\theta_3} = \frac{1}{\sec^2 \theta_2} = \cos^2 \theta_2$$

Therefore,

$$R_2 = \frac{C}{1 + \cos^2 \theta_2}$$

and

$$R_3 = \frac{C \cos^2 \theta_2}{1 + \cos^2 \theta_2}$$

If large torques are to be transmitted, the cams can be replaced by gears having pitch surfaces identical to the cam contours. This substitution is possible because of the pure rolling action of the cams. Such gears are known as contour gears or noncircular gears. A photograph of a pair of noncircular gears is shown in Fig. 8.22.

Referring to the equations for R_2 and R_3 developed for the three functions, it is evident that in (1) $R_2 = 0$ when $\theta_2 = 0$ and in (2) $R_3 = 0$ when $\theta_2 = 0$. In (3) $R_3 = 0$ when $\theta_2 = 90°$. When one of the radii goes to zero, an impractical design results. With the functions illustrated, the fact that the scale of θ_2 cannot start at zero in the first two cases nor extend to 90° in the third case will probably not limit the generation of these functions. There are cases, however, where such limitations would prove a disadvantage and a means must be found for eliminating this problem when necessary. Another problem that sometimes arises when designing contour cams is that with certain functions the value of $d\theta_3/d\theta_2$ may become equal to -1, which makes the radii R_2 and R_3 infinite. Either of these problems must be avoided if they occur in the working range of the function. This can be accomplished by

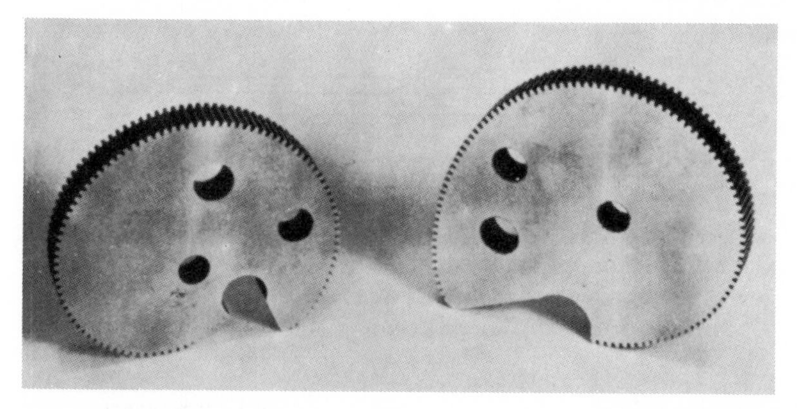

Fig. 8.22 Noncircular gears for computer. (Courtesy of Fellows Gear Shaper Company.)

offsetting the function by a constant which can later be subtracted by a differential. As an example, consider the function

$$\theta_3 = \sin^2 \theta_2$$

and

$$\frac{d\theta_3}{d\theta_2} = 2 \sin \theta_2 \cos \theta_2$$

Therefore,

$$R_2 = \frac{C(2 \sin \theta_2 \cos \theta_2)}{2 \sin \theta_2 \cos \theta_2 + 1}$$

and

$$R_3 = \frac{C}{1 + 2 \sin \theta_2 \cos \theta_2}$$

When θ_2 equals zero, $R_2 = 0$; when θ_2 equals 135°, $d\theta_3/d\theta_2 = -1$. To avoid these conditions, the function may be offset by a constant $k\theta_2$ such that

$$\theta_3' = \sin^2 \theta_2 + k\theta_2$$

and

$$\frac{d\theta_3'}{d\theta_2} = 2 \sin \theta_2 \cos \theta_2 + k$$

After generation of the new function, $k\theta_2$ would be subtracted to give the original function $\theta_3 = \sin^2 \theta_2$. The schematic diagram for this computation is shown in Fig. 8.23.

Another type of cam that is often used for computation is a disk cam with a groove cut into the face. Although the groove could take one of several forms, it is usually a spiral. The output can be in the form of a translating follower or, if teeth are cut in the groove, in the form of a rotating follower. This

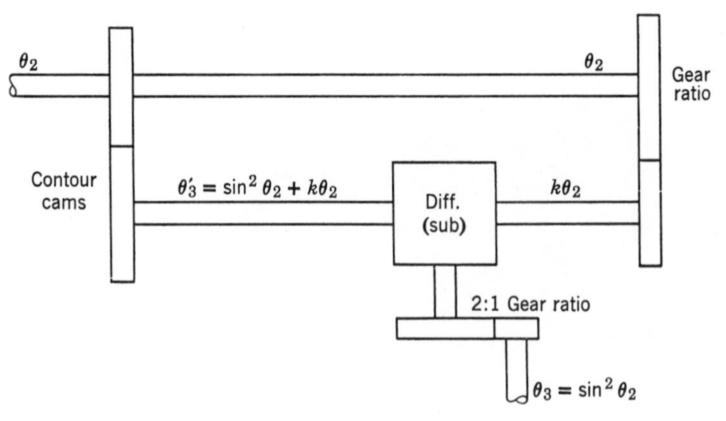

Figure 8.23

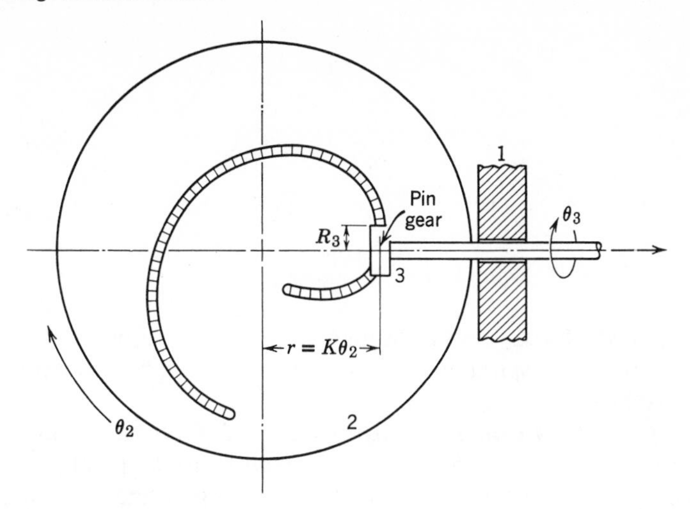

Figure 8.24

second type is known as a *gear cam* or as a *spiral face gear*. If an Archimedes spiral is used for the pitch line of the face gear, a squaring device is obtained. Figure 8.24 shows such a mechanism where θ_2 is the input (radians) and θ_3 the output (radians). R_3 is the pitch radius of the pin gear, and K a constant. Because pure rolling exists between the spiral pitch line and the pitch circle of the pin gear,

$$R_3\, d\theta_3 = r\, d\theta_2$$

From the properties of the Archimedes spiral,

$$r = K\theta_2$$

Therefore,

$$R_3\, d\theta_3 = K\theta_2\, d\theta_2$$

and

$$\theta_3 = \frac{K}{R_3}\int \theta_2\, d\theta_2$$

Because θ_2 will vary from zero to some arbitrary end point depending on the size of the disk,

$$\theta_3 = \frac{K\theta_2^2}{2R_3} \tag{8.13}$$

It should be noted that this equation is similar to Eq. 8.10 for the cone and cylinder squaring mechanism.

The main advantage of this type of cam or gear is that more than one revolution can be imparted to the input, thus providing high accuracy.

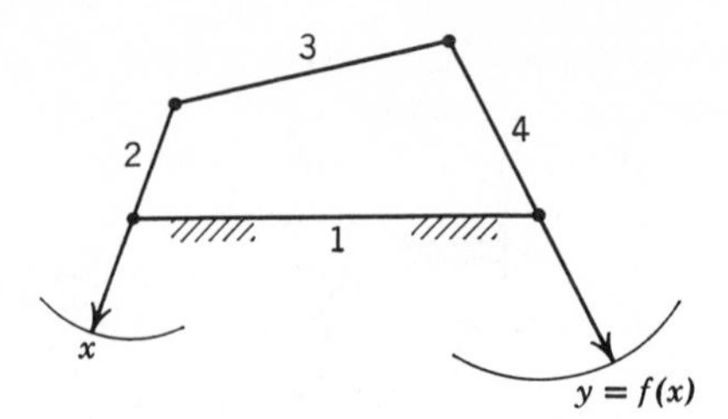

Figure 8.25

According to Rothbart,[3] spirals as high as eight revolutions have been used, and for Archimedes' spiral face gears, accuracies have been obtained of one part in 30,000.

8.10 Linkage Function Generator. A four-bar linkage is sometimes used to generate a function. The computation might be addition, multiplication, raising to a power, taking a logarithm, or generating a trigonometric function. Figure 8.25 shows such a mechanism with pointers attached to illustrate input x with link 2 and output $y = f(x)$ with link 4. In actual operation, however, the input and the output would probably be in the form of shaft rotations.

Function generators of this type are difficult to design, but the cost of production is much less than most of the computing mechanisms discussed previously. Linkages can be made with high precision and have high reliability because of the fact that the pivots are the only place where a malfunction can occur. Unfortunately, there is no way at the present time of determining easily the proportions of a linkage to generate a theoretically correct function at every point on the scale. A method for three-point accuracy will be given in Chapter 9. Methods for four points and five points are given in the paper[4] from which the three-point method was taken.

8.11 Accuracy. In computing elements, there are two main sources of error: (*a*) kinematic or theoretical errors which result from an approximation in the generation of a function, and (*b*) fabrication errors which result from manufacturing tolerances and from clearances in machines parts necessary for successful operation. Of all the elements described, the only ones to have kinematic errors are the adder of Fig. 8.2*a* and the linkage function generator of Fig. 8.25. The integrator may have an error characteristic of friction devices, which is known as *slip error*. All the units will have fabrication errors that must be kept as low as possible commensurate with cost.

[3] H. A. Rothbart, *Cams*, John Wiley and Sons, 1956.

[4] F. Freudenstein, "Approximate Synthesis of Four-Bar Linkages," *Trans. ASME*, Vol. 77, p. 853, 1955.

8.12 Block Diagrams. In working out a design for an analog computer, it is necessary to obtain the relation between the input and output quantities in mathematical form. Once this is accomplished, the designer then proceeds to mechanize the equation using standard computing elements. In determining the manner in which a certain equation is to be solved, the designer makes what is known as a block diagram. The block diagram is made up of a set of symbols, generally squares or circles, in which each symbol represents a separate computation, for example, addition, multiplication, integration. These squares are connected by lines that represent the flow of variables from one operation to the next.

Often it is possible to write an equation in more than one form. This should be done, and the block diagram made for each form of the equation. Then the block diagram should be selected which seems to give the best solution to the problem. As an example, consider the equation

$$y = ax^2 + bx$$

This may also be written

$$y = (ax + b)x$$

or

$$y = \left(a + \frac{b}{x}\right)x^2$$

The block diagrams of these equations are given in Figs. 8.26*a*, *b*, and *c*

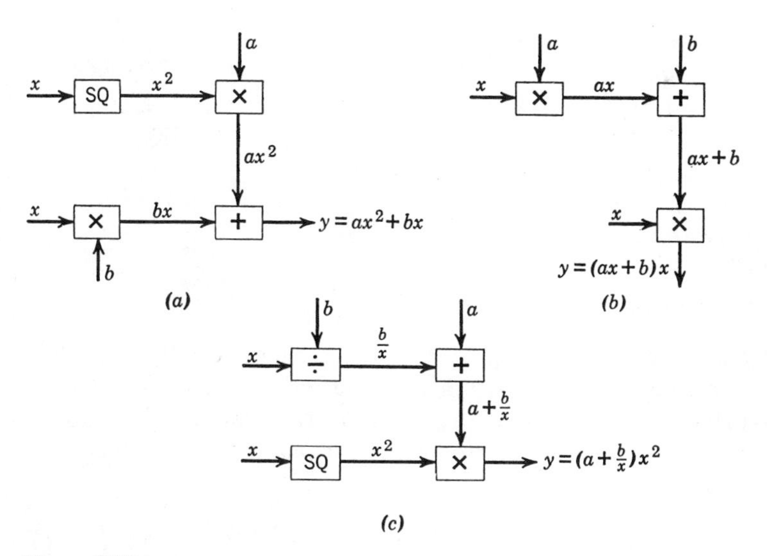

Figure 8.26

where:

$\times$ = a multiplier
$\div$ = a divider
$+$ = an adder (If a differential is used, the output will be divided by 2, which was omitted for simplicity.)
SQ = a squaring mechanism

Considering the three possibilities, it can be seen that the first and third diagram each require four elements, whereas the second requires only three. Also if the value of x should be close to zero, the divider in the layout of Fig. 8.26c would be a source of difficulty. Therefore, the second arrangement seems to be the best choice.

After the block diagram has been selected, it is necessary to determine the individual components for the computer. There may be several alternate devices for performing the same elementary computation, and a choice may be made on the basis of accuracy desired, size and weight limitations, cost, range of variables, calibration requirements, and other factors.

Example 8.1

There are several ways of generating the equation $z = xy$. Figure 8.27 shows a block diagram for the equation using logs (cams), an adder (differential), a gear ratio, and an antilog (cam). It should be mentioned that occasionally the block entitled "gear ratio" is omitted from a block diagram for simplicity, as was done in Fig. 8.26.

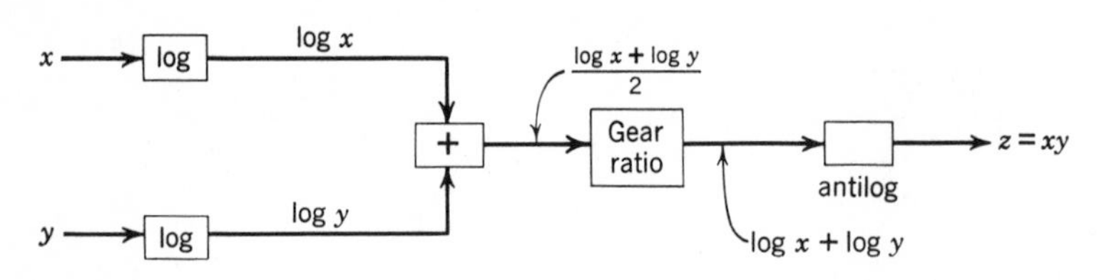

Figure 8.27

Example 8.2

If it is desired to generate $z = xy$ without the use of either logs or a multiplier, the equation can be rewritten using the principle of quarter squares, which states that one fourth the square of the sum of two numbers minus one fourth the square of their difference is equal to their product. Therefore, as shown in Fig. 8.28,

$$z = xy = \frac{(x + y)^2}{4} - \frac{(x - y)^2}{4}$$

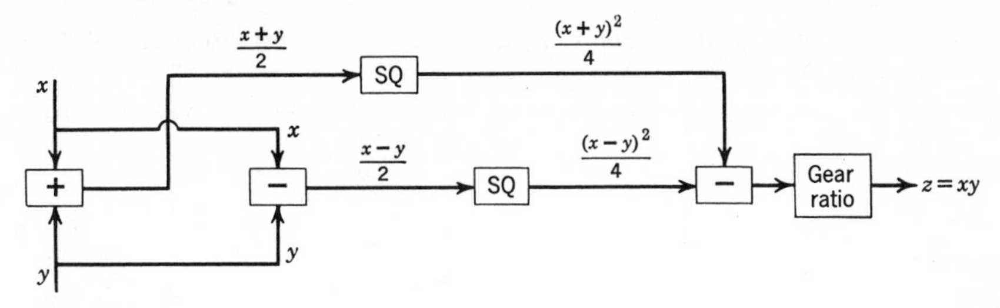

Figure 8.28

In order to show how the various units are connected in a computer, a schematic diagram is given in Fig. 8.29 for the arrangement of the computing elements for Example 8.1.

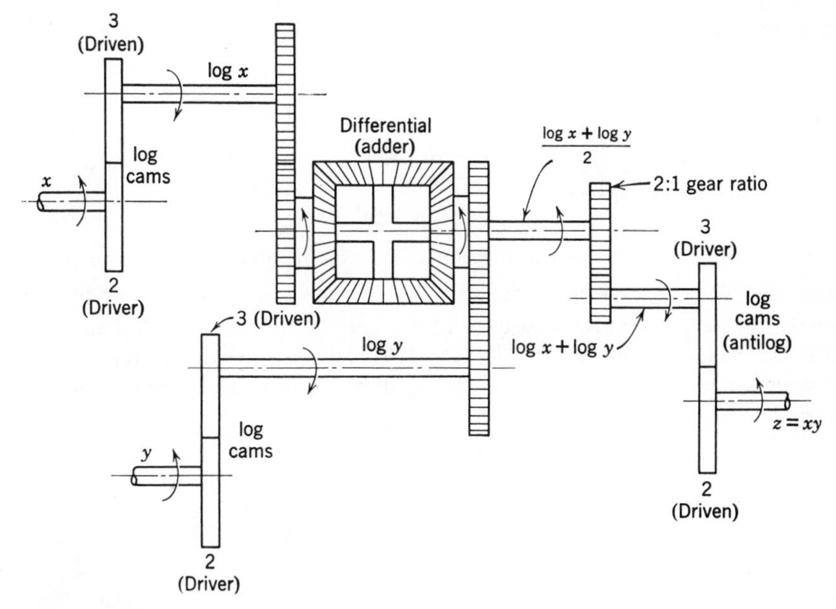

Figure 8.29

References

Billings, J. H., *Applied Kinematics*, third edition, D. Van Nostrand Company, 1953.

Lockenvitz, A. E., J. B. Oliphent, W. C. Wilde, and J. M. Young, "Geared to Compute," *Automation*, August 1955.

Michalec, G. W., "Analog Computing Mechanisms," *Machine Design*, March 19, 1959.

Rothbart, H. A., *Cams*, John Wiley and Sons, 1956.

Soroka, W. W., *Analog Methods in Computation and Simulation*, McGraw-Hill Book Company, 1954.
Svoboda, A., *Computing Mechanisms and Linkages*, McGraw-Hill Book Company, 1948.

Problems

8.1 Develop an expression for determining the amount of error in a computation made with the linkage differential shown in Fig. 8.2*a*.

8.2 For the spur gear differential shown in Fig. 8.3*b*, prove that $\theta_6 = (\theta_2 + \theta_4)/2$.

8.3 In a screw differential for adding as shown in Fig. 8.5, the scales of gears 2 and 3 are 45° = 1 unit. Calculate the scale of S_4 if the axial pitch of the screw is 6 mm and it is double threaded.

8.4 The rate of flow of a liquid through a certain orifice is given by

$$Q = 0.045a\sqrt{h}$$

where

$$Q = \text{flow (ft}^3\text{/sec)}$$
$$a = \text{orifice area (in.}^2\text{)}$$
$$h = \text{pressure head (ft)}$$

A mechanism similar to that of Fig. 8.6 is used for the multiplication after the $\sqrt{h}$ has been obtained. A is taken as 4 in., and x_1 is the $\sqrt{h}$, x_2 is a, and x_3 is Q. If the scale of $\sqrt{h}$ is 1 in. = 3.77 ft$^{1/2}$ and the scale of a is 1 in. = 2.23 in.^{2}, determine the scale of Q.

8.5 In the integrator shown in Fig. 8.7, if the ball carrier is given an input $r = f(t)$ and dr/dt is a constant, show that the number of revolutions θ_4 recorded by the roller is given by

$$\theta_4 = \frac{\omega_2}{R_4}\int_o^t (At + B)\, dt$$

where

$$A = \frac{r_t - r_o}{t} \qquad \text{and} \qquad B = r_o$$

8.6 Make a sketch showing how to connect two integrators so that the output of the second integrator is the reciprocal of the input to the first integrator.

8.7 Using two integrators, connect them so that the output of the second integrator is $z = \int xy\, dx$.

8.8 An integrator is used as a squaring device as shown in Fig. 8.9. If the disk rotates 30° for one unit of x and the scale of y is the same, calculate the scale of the output shaft z. The lead screw is single threaded and has an axial pitch of 6 mm and $R = 0.32$ cm.

8.9 Design a mechanism which will generate a sine and a cosine function simultaneously.

8.10 Prove that if point B in Fig. 8.13*b* moves along the horizontal axis of gear 1 as arm 3 rotates, it will move with *SHM*.

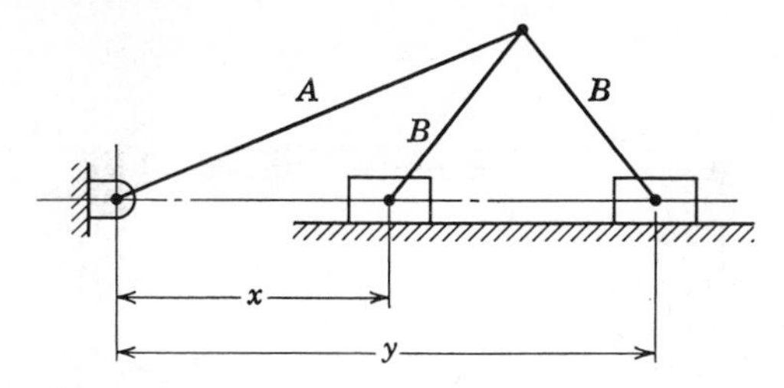

Figure 8.30

8.11 Draw a set of scales for a tangent and secant generator as shown in Fig. 8.15. Letting A be 5 cm, lay out the scale for the tangent and the secant from 0 to 60° in increments of 10°. Show as a dimension the distance A on each scale.

8.12 A modification of the Peaucellier mechanism is shown in Fig. 8.30. Prove that the displacements x and y of the sliders are related by the following equation:

$$xy = k$$

Find the value of the constant k in terms of the link lengths A and B.

8.13 A cone and cylinder squaring mechanism is to be designed as shown in Fig. 8.19. If the radius R_4 of the cylinder is 0.65 cm and the cone vertex angle 15°, determine the amount of advance of the idler per unit of θ_2 to generate the function $\theta_4 = \theta_2^2$.

8.14 Determine the equations for the radii R_2 and R_3 for a pair of contour cams to generate the function $\theta_3 = \sin\theta_2 + k\theta_2$. Repeat for $\theta_3 = ke^{\theta_2}$.

8.15 Determine the equations for the radii R_2 and R_3 for a pair of contour cams to generate the function $\theta_3 = k^{\theta_2}$ where $k > 1$. Repeat for $\theta_3 = \log_e(\theta_2 + k)$.

8.16 The equations for the radii of a pair of contour cams are the following:

$$R_2 = \frac{C\cos\theta_2 + Ck}{1 + \cos\theta_2 + k}$$

$$R_3 = \frac{C}{1 + \cos\theta_2 + k}$$

Determine the function which gives the relationship of the angular displacements θ_2 and θ_3 of the cam axes.

8.17 Determine the equations for the radii R_2 and R_3 for a pair of contour cams to generate the function $\theta_3 = \log_e \cos\theta_2$. Is it possible for either R_2 or R_3 to become zero or infinite? If so, offset the function by a constant to prevent this.

8.18 Determine the equations for the radii R_2 and R_3 for a pair of contour cams to generate the function $\theta_3 = \sin(\theta_2/2)$. Is it possible for either R_2 or R_3 to become zero or infinite? If so, offset the function by a constant to prevent this.

8.19 A pair of contour cams are to be designed to generate the function $\theta_3 = \theta_2^2$, where θ_2 has the range 1 to 10 units. Calculate the radii R_2 and R_3 for a center distance of 3.00 in. with θ_2 varying from 0° to 100° in 10° increments. Make a double-size layout similar to Fig. 8.21 of the cams in contact at position 2. Let the hub diameters be 1 in.

8.20 A pair of contour cams are to be designed to generate the function $\theta_3 = \log_{10}\theta_2$.

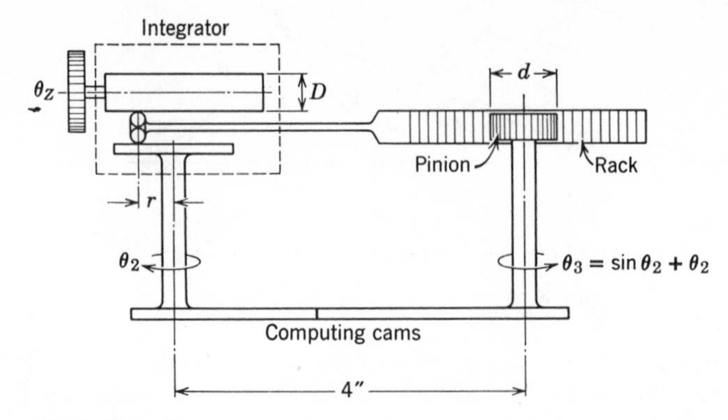

Figure 8.31

The center distance is to be 7.62 cm and θ_2 is to vary from 0° to 360° in 20° increments. Calculate the radii R_2 and R_3, letting the first value of θ_2 be 60° in order to obtain a radian value greater than unity. Make a double-size layout of the cams in contact at the first position.

8.21 A pair of contour cams are to be designed to generate the function $\theta_3 = \tan\theta_2$. The center distance is to be 3.00 in. Calculate the radii R_2 and R_3 with θ_2 varying from 0° to 80° in 10° increments. Make a double-size layout similar to Fig. 8.21 of the cams in contact at position 2.

8.22 Shown in Fig. 8.31 is a computing mechanism consisting of a pair of rolling computing cams, a rack and pinion, and an integrator. The cams have been designed to generate the function $\theta_3 = \sin\theta_2 + \theta_2$. The pitch diameter of the pinion is d, and the diameter of the output roller of the integrator is D. (a) Determine the rotation θ_3 in degrees for a rotation of $\theta_2 = 30°$ from the starting position of the cams. (b) Determine in inches the distance from the point of contact of the cams to the θ_2 axis of rotation for $\theta_2 = 30°$. (c) Derive an expression for the output rotation θ_z of the integrator roller in terms of d, D, and θ_2. Indicate the value of r for the start of the computation when $\theta_2 = 0$.

8.23 A spiral face gear is to be designed to generate $\theta_3 = \theta_2^2$. If the radius of the pin gear equals 0.65 cm, lay out the Archimedes' spiral for the face gear between $\theta = 40°$ and $\theta = 360°$ in 20° increments. Let θ be 0° at the bottom of the vertical center line with θ measured counterclockwise.

8.24 Considering the relation $z = (x + y)^2 - (2x - y)^2$, write this equation in as many forms as possible and draw the block diagram for each form of the equation.

8.25 Make block diagrams for the solution of the following equations:

$$z = \int (ax + b)^2\, dx$$

$$z = \sin\left[\tfrac{1}{2}(x + y)\right] \cos\left[\tfrac{1}{2}(x - y)\right]$$

$$z = \int x^2\sqrt{a^2 - x^2}\, dx$$

8.26 Make block diagrams for the solution of the following equations:

$$E = kT_c(T - T_r) - \frac{k}{2}(T^2 - T_r^2)$$

$$E = \pi(T_r - T) + \int(\sigma_A - \sigma_B)\,dT$$

8.27 Make block diagrams for the solution of the following equations:

$$e = k\left[\frac{R_s + R_c}{R_t + R_l + R_s + R_c}\right]$$

$$Q = KA[T_A^4 - T_B^4]$$

8.28 Make block diagrams for the solution of the following equations:

$$h = \left[\frac{T_m}{T_l}\left(1 + \frac{A_2}{A_1}\right) - 1\right]d - h_o$$

$$q = KA\sqrt{\frac{2g}{\rho}(p_1 - p_2)}$$

8.29 Make a schematic layout of the computing elements necessary to perform the computation in Example 8.2. Use bevel gear differentials for adding or subtracting and contour cams for squaring.

8.30 (*a*) Draw the block diagram for the solution of the following equation:

$$z = \int[(x + y)^2]\,dx$$

(*b*) Make a schematic layout of the elements necessary to perform the computation. Use bevel gear differentials for adding or subtracting and contour cams for squaring. Label all directions of rotation.

9

Introduction to Synthesis

In the study of mechanisms so far, the proportions of a linkage have been given and the problem has been to analyze the motion produced by the linkage. It is quite a different matter, however, to start with a required motion and to try to proportion a mechanism to give this motion. This procedure is known as the *synthesis of mechanisms*. As has been mentioned earlier, designing a cam from the required displacement diagram is the only problem in synthesis that can be solved every time. In the application of synthesis to the design of a mechanism, the problem divides itself into three parts (*a*) the type of mechanism to be used, (*b*) the number of links and connections needed to produce the required motion, and (*c*) the proportions or lengths of the links necessary. These divisions are often referred to as *type*, *number*, and *dimensional* synthesis.

Although designers have been interested in synthesis for many years, perhaps the greatest impetus to this study has come from the development of computing mechanisms. In computer design, it is often necessary to generate arbitrary functions by mechanical means. In some cases, a known mechanism already exists for generating the function, but many times the designer is not so fortunate and must resort to synthesis to solve his problem.

In the application of synthesis, one factor that must be continually kept in mind is that of the accuracy required of the mechanism. Sometimes it is possible to design a linkage that will theoretically generate a given function. Often, however, the designer must be satisfied with an approximation to the

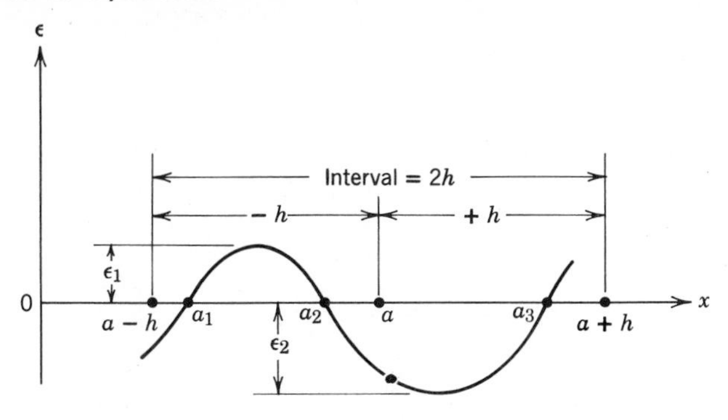

Figure 9.1

given function, and the difference between the function that is desired and the function that is actually produced is known as *structural error*. In addition, there are errors due to manufacture. The error resulting from tolerances in the lengths of the links and bearing clearances is referred to as *mechanical error*. Methods of calculating mechanical error are given by Hartenberg and Denavit[1] and by Garrett and Hall.[2]

In the early development of synthesis, graphical methods played a predominant role. This may have stemmed from the fact that some of the early methods were undoubtedly trial and error which later developed into more rational means. With the continued development of synthesis, several analytical methods have been introduced. Three of these methods will be presented to illustrate the principles involved, the difficulties encountered, and the application of the methods. One graphical method will also be presented.

9.1 Spacing of Accuracy Points. In designing a mechanism to generate a particular function, it is practically impossible to accurately produce the function at more than a few points. These points are known as *accuracy points* and must be so located as to minimize the error generated between these points. As previously mentioned, the error produced is structural error that can be expressed as follows:

$$\varepsilon = f(x) - g(x)$$

where

$$f(x) = \text{desired function}$$
$$g(x) = \text{function actually produced}$$

In Fig. 9.1 is shown a plot of the variation in structural error as a function

[1] R. S. Hartenberg and J. Denavit, *Kinematic Synthesis of Linkages*, McGraw-Hill Book Company, 1964.

[2] R. E. Garrett and A. S. Hall, "Effect of Tolerance and Clearance in Linkage Design," *Trans. ASME*, Vol. 91, No. 1, February 1969.

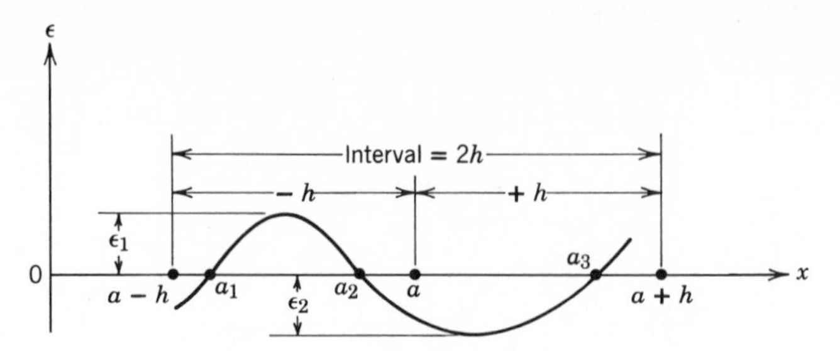

Figure 9.2

is generated over an interval $2h$ with the center of the interval at $x = a$. The error is zero at points a_1, a_2, and a_3, which are the accuracy points mentioned above. From this figure, it can be seen that the maximum error ε_1 produced by the mechanism in going from point a_1 to point a_2 is considerably smaller than the maximum error ε_2 produced in going from a_2 to a_3. By using a theory developed by Chebyshev,[3] it is possible to locate the points a_1, a_2, and a_3 of Fig. 9.1 in such a manner that $\varepsilon_1 = \varepsilon_2$. Figure 9.2 shows this arrangement, and Fig. 9.3 illustrates the method of locating the three accuracy points with Chebyshev spacing. A semicircle is drawn on the x axis with a radius h and center at point a. Half of a regular polygon is then inscribed in the semicircle so that two of its sides are perpendicular to the x axis. Lines drawn perpendicularly to the x axis from the vertices of the half polygon determine the accuracy points a_1, a_2, and a_3. Figure 9.4 shows the construction for four accuracy points. It can be seen that for three accuracy points the polygon is a hexagon and for four accuracy points an octogon. In other words, the number of sides of the polygon is twice the number of accuracy points desired.

9.2 Design of Four-Bar Linkage for Instantaneous Values of Angular Velocity and Acceleration. A method has been developed by Rosenauer[4] by which a four-bar linkage can be designed to give each link a prescribed instantaneous value of angular velocity and of angular acceleration. A description of this method follows.

A four-bar linkage is shown in Fig. 9.5 in which the links are represented by vectors which form a polygon with origin O. It can be seen that link OA forms the angle θ_2 with the horizontal, AB the angle θ_3, CB the angle θ_4, and CO the angle θ_1. Each of these angles is measured in the same direction. If link OA has a length a, AB a length b, CB a length c, and CO a length d,

[3] R. S. Hartenberg and J. Denavit, *Kinematic Synthesis of Linkages*, McGraw-Hill Book Company, 1964.

[4] N. Rosenauer, "Complex Variable Method for Synthesis of Four-Bar Linkages," *Aust. J. Appl. Sci.*, Vol. 5, No. 4, pp. 305–308, 1954.

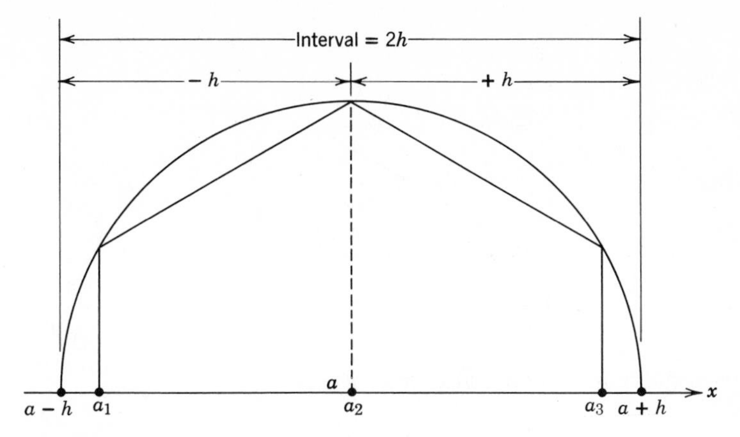

Figure 9.3

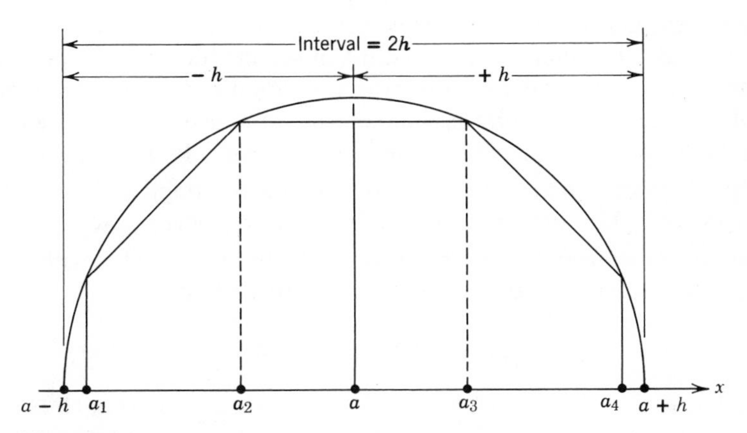

Figure 9.4

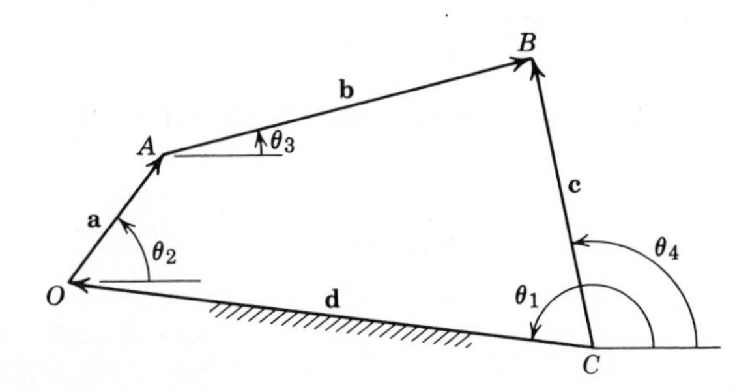

Figure 9.5

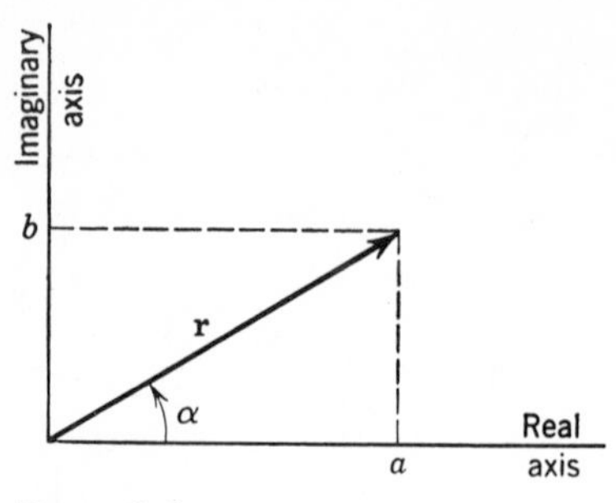

Figure 9.6

the closed polygon can be written vectorially as

$$\mathbf{a} + \mathbf{b} - \mathbf{c} + \mathbf{d} = 0 \tag{9.1}$$

It should be noticed that the vector **c**, which represents link BC, is subtracted in the polygon because the vector is in the direction CB.

As a means of facilitating the handling of vectors numerically, the vectors may be represented by complex numbers. A complex number can be represented graphically by a point in a plane where the real numbers are plotted horizontally and the imaginary numbers vertically. In Fig. 9.6 the point $a + bi$ is shown, where $i = \sqrt{-1}$. By joining the point $a + bi$ with the origin, the complex number can be made to represent a vector whose length r can be calculated from $\sqrt{a^2 + b^2}$. If the angle the vector makes with the real axis is α, the equation of the vector can be expressed as

$$\mathbf{r} = r(\cos \alpha + i \sin \alpha)$$

This relation can be derived from $a + bi$ because $a = r \cos \alpha$, and $b = r \sin \alpha$. From the Maclaurin series development of e^x and sin α and cos α,

$$e^{i\alpha} = \cos \alpha + i \sin \alpha$$

Therefore,

$$\mathbf{r} = re^{i\alpha}$$

The above relation can be applied to the vectors which represent the four-bar linkage.

Therefore,

$$\begin{aligned} \mathbf{a} &= ae^{i\theta_2} \\ \mathbf{b} &= be^{i\theta_3} \\ \mathbf{c} &= ce^{i\theta_4} \\ \mathbf{d} &= de^{i\theta_1} \end{aligned} \tag{9.2}$$

Substituting Eqs. 9.2 into Eq. 9.1,

$$ae^{i\theta_2} + be^{i\theta_3} - ce^{i\theta_4} + de^{i\theta_1} = 0 \tag{9.3}$$

If this equation is differentiated with respect to time and the angular velocities are denoted by

$$\frac{d\theta_2}{dt} = \omega_2, \quad \frac{d\theta_3}{dt} = \omega_3, \quad \frac{d\theta_4}{dt} = \omega_4, \quad \text{and} \quad \frac{d\theta_1}{dt} = 0$$

$$i\omega_2 ae^{i\theta_2} + i\omega_3 be^{i\theta_3} - i\omega_4 ce^{i\theta_4} + 0 \cdot de^{i\theta_1} = 0 \tag{9.4}$$

Differentiating again with respect to time and denoting the angular accelerations by $d\omega_2/dt = \alpha_2$, $d\omega_3/dt = \alpha_3$, $d\omega_4/dt = \alpha_4$, and $d\omega_1/dt = 0$, the following relation is obtained:

$$(i\alpha_2 - \omega_2^2)ae^{i\theta_2} + (i\alpha_3 - \omega_3^2)be^{i\theta_3} - (i\alpha_4 - \omega_4^2)ce^{i\theta_4} + 0 \cdot de^{i\theta_1} = 0 \tag{9.5}$$

If Eqs. 9.3, 9.4, and 9.5 are now changed back into vectorial form, there results

$$\begin{aligned} \mathbf{a} + \mathbf{b} - \mathbf{c} + \mathbf{d} &= 0 \\ \omega_2\mathbf{a} + \omega_3\mathbf{b} - \omega_4\mathbf{c} + 0 \cdot \mathbf{d} &= 0 \\ (i\alpha_2 - \omega_2^2)\mathbf{a} + (i\alpha_3 - \omega_3^2)\mathbf{b} - (i\alpha_4 - \omega_4^2)\mathbf{c} + 0 \cdot \mathbf{d} &= 0 \end{aligned} \tag{9.6}$$

The solution of Eqs. 9.6 can best be carried out by the use of determinants as follows:

$$\mathbf{a} = \frac{\begin{vmatrix} -\mathbf{d} & 1 & -1 \\ 0 & \omega_3 & -\omega_4 \\ 0 & (i\alpha_3 - \omega_3^2) & -(i\alpha_4 - \omega_4^2) \end{vmatrix}}{D}$$

$$= \frac{-\mathbf{d}}{D}[-\omega_3(i\alpha_4 - \omega_4^2) + \omega_4(i\alpha_3 - \omega_3^2)]$$

$$= \frac{\mathbf{d}}{D}[(\omega_3\alpha_4 - \omega_4\alpha_3)i + \omega_4\omega_3(\omega_3 - \omega_4)]$$

In a similar manner,

$$\mathbf{b} = \frac{\begin{vmatrix} 1 & -\mathbf{d} & -1 \\ \omega_2 & 0 & -\omega_4 \\ (i\alpha_2 - \omega_2^2) & 0 & -(i\alpha_4 - \omega_4^2) \end{vmatrix}}{D}$$

$$= \frac{\mathbf{d}}{D}[(\omega_4\alpha_2 - \omega_2\alpha_4)i + \omega_2\omega_4(\omega_4 - \omega_2)]$$

and

$$\mathbf{c} = \frac{\begin{vmatrix} 1 & 1 & -\mathbf{d} \\ \omega_2 & \omega_3 & 0 \\ (i\alpha_2 - \omega_2^2) & (i\alpha_3 - \omega_3^2) & 0 \end{vmatrix}}{D}$$

$$= \frac{\mathbf{d}}{D}[(\omega_3\alpha_2 - \omega_2\alpha_3)i + \omega_2\omega_3(\omega_3 - \omega_2)]$$

Because each vector is multiplied by the same factor $\mathbf{d}/D$, let this factor be -1. This is permissible because **a**, **b**, and **c** are all relative to **d**. A negative 1 is used in order to position the linkage in the positive field. Making this substitution and rearranging so that the imaginary terms come last, the following equations result:

$$\begin{aligned} \mathbf{a} &= \omega_4\omega_3(\omega_4 - \omega_3) + (\omega_4\alpha_3 - \omega_3\alpha_4)i \\ \mathbf{b} &= \omega_2\omega_4(\omega_2 - \omega_4) + (\omega_2\alpha_4 - \omega_4\alpha_2)i \\ \mathbf{c} &= \omega_2\omega_3(\omega_2 - \omega_3) + (\omega_2\alpha_3 - \omega_3\alpha_2)i \end{aligned} \tag{9.7}$$

and

$$\mathbf{d} = \mathbf{c} - \mathbf{a} - \mathbf{b} \qquad \text{from Eq. 9.1}$$

If the real and imaginary parts are denoted by

$$\begin{aligned} a_1 &= \omega_4\omega_3(\omega_4 - \omega_3) & a_2 &= \omega_4\alpha_3 - \omega_3\alpha_4 \\ b_1 &= \omega_2\omega_4(\omega_2 - \omega_4) & b_2 &= \omega_2\alpha_4 - \omega_4\alpha_2 \\ c_1 &= \omega_2\omega_3(\omega_2 - \omega_3) & c_2 &= \omega_2\alpha_3 - \omega_3\alpha_2 \\ d_1 &= c_1 - a_1 - b_1 & d_2 &= c_2 - a_2 - b_2 \end{aligned} \tag{9.8}$$

then the vectors are

$$\begin{aligned} \mathbf{a} &= a_1 + a_2 i \\ \mathbf{b} &= b_1 + b_2 i \\ \mathbf{c} &= c_1 + c_2 i \\ \mathbf{d} &= d_1 + d_2 i \end{aligned} \tag{9.9}$$

The lengths of the links are proportional to

$$\begin{aligned} a &= \sqrt{a_1^2 + a_2^2} \\ b &= \sqrt{b_1^2 + b_2^2} \\ c &= \sqrt{c_1^2 + c_2^2} \\ d &= \sqrt{d_1^2 + d_2^2} \end{aligned} \tag{9.10}$$

Example 9.1

A four-bar linkage is to be designed for the following instantaneous values:

$$\omega_2 = 6 \text{ rad/sec} \qquad \alpha_2 = 0 \qquad \text{for crank } OA$$
$$\omega_3 = 1 \text{ rad/sec} \qquad \alpha_3 = 10 \text{ rad/sec}^2 \qquad \text{for link } AB$$
$$\omega_4 = 3 \text{ rad/sec} \qquad \alpha_4 = 5 \text{ rad/sec}^2 \qquad \text{for link } CB$$

Substituting in Eqs. 9.8,

$$a_1 = 3 \cdot 1(3 - 1) = 6 \qquad a_2 = 3 \cdot 10 - 1 \cdot 5 = 25$$
$$b_1 = 6 \cdot 3(6 - 3) = 54 \qquad b_2 = 6 \cdot 5 - 3 \cdot 0 = 30$$
$$c_1 = 6 \cdot 1(6 - 1) = 30 \qquad c_2 = 6 \cdot 10 - 1 \cdot 0 = 60$$
$$d_1 = 30 - 6 - 54 = -30 \qquad d_2 = 60 - 25 - 30 = 5$$

The links in complex form are

$$\mathbf{a} = 6 + 25i$$
$$\mathbf{b} = 54 + 30i$$
$$\mathbf{c} = 30 + 60i$$
$$\mathbf{d} = -30 + 5i$$

The lengths of the links are proportional to

$$a = \sqrt{6^2 + 25^2} = 25.71$$
$$b = \sqrt{54^2 + 30^2} = 61.78$$
$$c = \sqrt{30^2 + 60^2} = 67.08$$
$$d = \sqrt{30^2 + 5^2} = 30.41$$

Figure 9.7 shows the required linkage $OABC$. The linkage shown in Fig. 9.7 indicates the proper relative lengths of the links to each other and their corresponding angular positions to give the required instantaneous angular velocities and accelerations. The actual lengths of the links could be changed as long as their proportions to each other remain the same; however, the angular positions could not be changed.

9.3 Design of Four-Bar Linkage as a Function Generator. It is often necessary to design a linkage to generate a given function, for example, $y = \log x$. Figure 9.8 shows a four-bar linkage arranged to generate the function $y = f(x)$ over a limited range. As link OA moves between the limits ϕ_1 and ϕ_n with the input x, link BC gives the value of $y = f(x)$ between the limits ψ_1 and ψ_n. It can be seen that in the linkage there are three independent side ratios that define the proportions of the linkage. Also to be

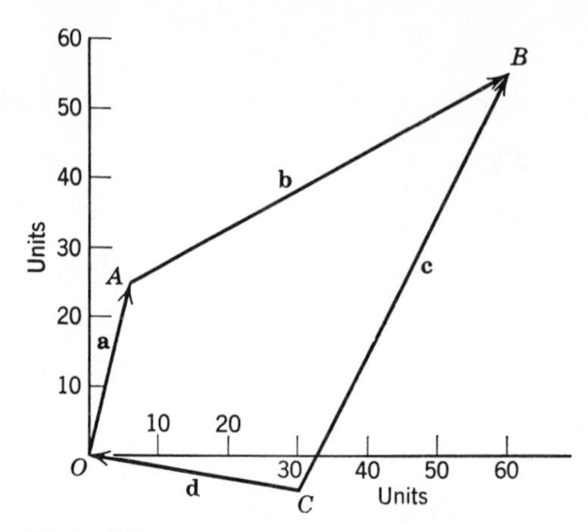

Figure 9.7

considered is the range (and scale factors) of ϕ and ψ and the initial angles ϕ_1 and ψ_1. In all, there are seven variables that must be considered in designing the linkage to generate $y = f(x)$. The magnitude of the task of synthesizing this function is immediately apparent.

A method has been developed by Freudenstein[5] by which a four-bar linkage can be designed to generate a function which is accurate at a finite number of points called *precision points* but which is approximate between these points. In other words, the ideal function and the function actually generated will agree only at the precision points. Between these points the actual function will differ from the ideal by an amount depending upon the distance between the points and upon the nature of the ideal function. Referring again to Fig. 9.8, the function would therefore only be exact at ψ_1 and ψ_n and at a specific number of points in between.

In developing Freudenstein's method, the first step is to determine the relation between ϕ and ψ using the minimum number of side ratios. This relation can be derived considering Fig. 9.9 where a line parallel to link OA has been drawn from point B and a line parallel to link AB has been drawn from point O to give the parallelogram $OABD$. The links form a closed loop, and the sum of the x components of lengths a, b, c must equal length d. In equation form,

$$a \cos(\pi - \phi) + b \cos \alpha + c \cos \psi = d \tag{9.11}$$

Applying the law of cosines to triangle DOC,

$$e^2 = b^2 + d^2 - 2bd \cos \alpha \tag{9.12}$$

Also from triangle DBC,

$$e^2 = a^2 + c^2 - 2ac \cos(\phi - \psi) \tag{9.13}$$

[5] F. Freudenstein, "Approximate Synthesis of Four-Bar Linkages," *Trans. ASME*, Vol. 77, p. 853, 1955.

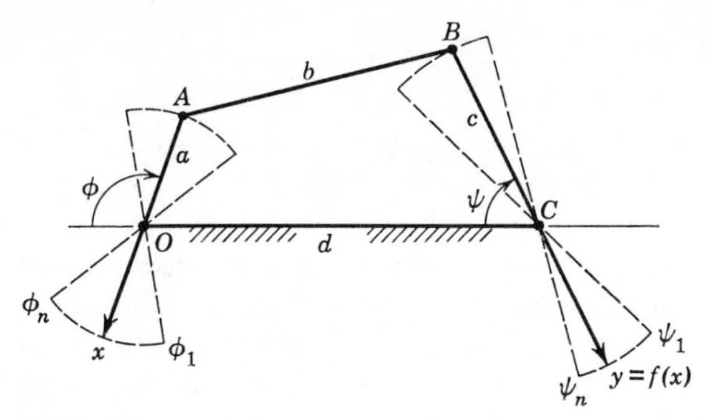

Figure 9.8

Solving Eqs. 9.12 and 9.13 for $b \cos \alpha$ gives

$$b \cos \alpha = \frac{b^2 + d^2 - a^2 - c^2 + 2ac \cos (\phi - \psi)}{2d} \tag{9.14}$$

Substituting Eq. 9.14 into Eq. 9.11 and letting $\cos (\pi - \phi) = - \cos \phi$,

$$a^2 - b^2 + c^2 + d^2 + 2ad \cos \phi - 2cd \cos \psi = 2ac \cos (\phi - \psi) \tag{9.15}$$

Dividing by $2ac$,

$$\frac{a^2 - b^2 + c^2 + d^2}{2ac} + \frac{d}{c} \cos \phi - \frac{d}{a} \cos \psi = \cos (\phi - \psi) \tag{9.16}$$

By letting

$$R_1 = \frac{d}{c}$$

$$R_2 = \frac{d}{a} \tag{9.17}$$

$$R_3 = \frac{a^2 - b^2 + c^2 + d^2}{2ac}$$

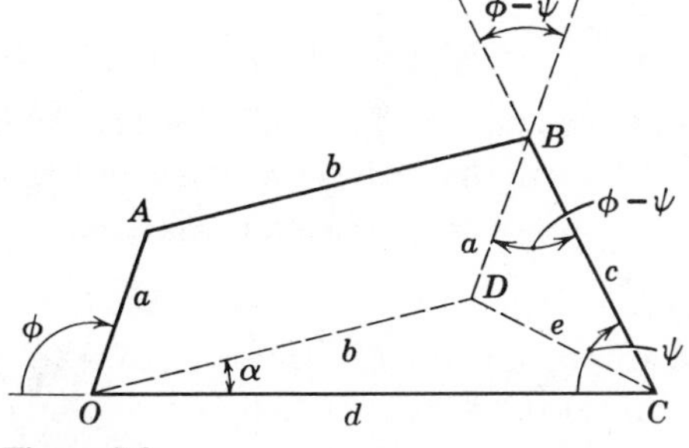

Figure 9.9

Eq. 9.16 becomes

$$R_1 \cos\phi - R_2 \cos\psi + R_3 = \cos(\phi - \psi) \tag{9.18}$$

where R_1, R_2, and R_3 are three independent side ratios. Equation 9.18 gives the simplest relation possible between ϕ and ψ.

Using Eq. 9.18, the method will now be extended to cover the design of a linkage to generate a function which is exact at three points. For greater accuracy four- and five-point approximations have been developed. However, these systems are much more complicated and will not be included here.

The pairs of angles (ϕ, ψ) which correspond to the precision points are substituted into Eq. 9.18, which gives three simultaneous equations. The side ratios can then be determined from the solution of these equations. If the linkage is to pass through (ϕ_1, ψ_1), (ϕ_2, ψ_2), and (ϕ_3, ψ_3), then

$$\begin{aligned} R_1 \cos\phi_1 - R_2 \cos\psi_1 + R_3 &= \cos(\phi_1 - \psi_1) \\ R_1 \cos\phi_2 - R_2 \cos\psi_2 + R_3 &= \cos(\phi_2 - \psi_2) \\ R_1 \cos\phi_3 - R_2 \cos\psi_3 + R_3 &= \cos(\phi_3 - \psi_3) \end{aligned} \tag{9.19}$$

In solving the simultaneous equations 9.19, let

$$\begin{aligned} \cos\phi_1 - \cos\phi_2 &= w_1 \\ \cos\phi_1 - \cos\phi_3 &= w_2 \\ \cos\psi_1 - \cos\psi_2 &= w_3 \\ \cos\psi_1 - \cos\psi_3 &= w_4 \\ \cos(\phi_1 - \psi_1) - \cos(\phi_2 - \psi_2) &= w_5 \\ \cos(\phi_1 - \psi_1) - \cos(\phi_3 - \psi_3) &= w_6 \end{aligned}$$

then

$$\begin{aligned} R_1 &= \frac{w_3 w_6 - w_4 w_5}{w_2 w_3 - w_1 w_4} \\ R_2 &= \frac{w_1 w_6 - w_2 w_5}{w_2 w_3 - w_1 w_4} \\ R_3 &= \cos(\phi_i - \psi_i) + R_2 \cos\psi_i - R_1 \cos\phi_i \qquad \text{where } i = 1, 2, \text{ or } 3 \end{aligned} \tag{9.20}$$

From these side ratios the lengths of the links can be determined from Eqs. 9.17. In determining the lengths of links a and c, a negative sign must be interpreted in a vector sense when drawing the linkage.

Example 9.2

Let it be required to proportion a four-bar linkage to generate $y = x^{1.5}$, where x varies between 1.0 and 4.0. Use Chebyshev spacing, and let $\phi_s = 30°$, $\Delta\phi = 90°$, $\psi_s = 90°$, and $\Delta\psi = 90°$. Assume $d = 1.000$ in.

$$\phi_s = 30^\circ \qquad \psi_s = 90^\circ \qquad x_s = 1.0 \qquad y_s = 1.0$$

$$\Delta\phi = 90^\circ \qquad \Delta\psi = 90^\circ \qquad x_f = 4.0 \qquad y_f = 8.0$$

$$x_1 = 2.5 - 1.5 \cos 30^\circ = 1.201 \qquad y_1 = 1.317$$

$$x_2 = 2.50 \qquad y_2 = 3.96$$

$$x_3 = 2.5 + 1.5 \cos 30^\circ = 3.799 \qquad y_3 = 7.40$$

$$\phi_1 = \phi_s + \frac{x_1 - x_s}{x_f - x_s}\Delta\phi = 30 + \frac{1.201 - 1.0}{4.0 - 1.0} \times 90 = 36.03^\circ$$

$$\phi_2 = \phi_1 + \frac{x_2 - x_1}{x_f - x_s}\Delta\phi = 36.03 + \frac{(2.50 - 1.20)}{3} \times 90 = 75.03^\circ$$

$$\phi_3 = \phi_1 + \frac{x_3 - x_1}{x_f - x_s}\Delta\phi = 36.03 + \frac{(3.799 - 1.20)}{3} \times 90 = 114.0^\circ$$

$$\psi_1 = \psi_s + \frac{y_1 - y_s}{y_f - y_s}\Delta\psi = 90 + \frac{1.317 - 1.0}{8.0 - 1.0} \times 90 = 94.08^\circ$$

$$\psi_2 = \psi_1 + \frac{y_2 - y_1}{y_f - y_s}\Delta\psi = 94.08 + \frac{(3.96 - 1.32)}{7} \times 90 = 128.02^\circ$$

$$\psi_3 = \psi_1 + \frac{y_3 - y_1}{y_f - y_s}\Delta\psi = 94.08 + \frac{(7.40 - 1.32)}{7} \times 90 = 172.25^\circ$$

$$w_1 = \cos\phi_1 - \cos\phi_2 = 0.8087 - 0.2583 = 0.5504$$

$$w_2 = \cos\phi_1 - \cos\phi_3 = 0.8087 + 0.4067 = 1.2154$$

$$w_3 = \cos\psi_1 - \cos\psi_2 = -0.0713 + 0.6159 = 0.5446$$

$$w_4 = \cos\psi_1 - \cos\psi_3 = -0.0713 + 0.9909 = 0.9196$$

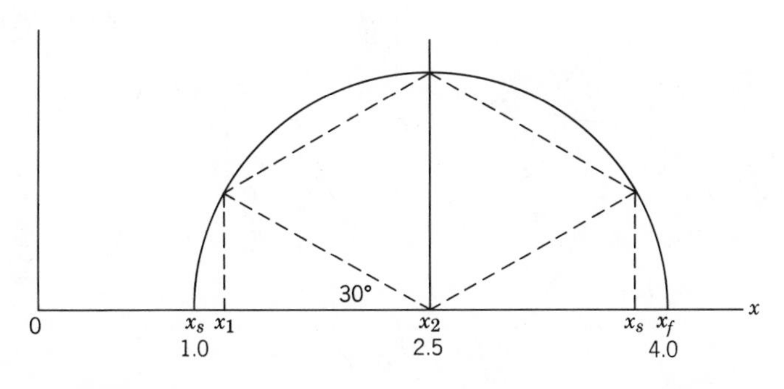

Figure 9.10

$$w_5 = \cos(\phi_1 - \psi_1) - \cos(\phi_2 - \psi_2) = 0.5292 - 0.6019 = -0.0727$$

$$w_6 = \cos(\phi_1 - \psi_1) - \cos(\phi_3 - \psi_3) = 0.5292 - 0.5262 = 0.003$$

$$R_1 = \frac{w_3 w_6 - w_4 w_5}{w_2 w_3 - w_1 w_4} = \frac{(0.545)(0.003) - (0.920)(-0.073)}{(1.215)(0.545) - (0.550)(0.920)}$$

$$R_1 = 0.440$$

$$R_2 = \frac{w_1 w_6 - w_2 w_5}{w_2 w_3 - w_1 w_4} = \frac{(0.550)(0.003) - (1.215)(-0.073)}{(1.215)(0.545) - (0.550)(0.920)}$$

$$R_2 = 0.578$$

$$R_3 = \cos(\phi_1 - \psi_1) + R_2 \cos\psi_1 - R_1 \cos\phi_1$$
$$= 0.5292 + (0.578)(-0.0713) - (0.440)(0.8087)$$
$$= 0.132$$

From Eqs. 9.17 with $d = 1.000$ in.

$$a = \frac{d}{R_2} = \frac{1.000}{0.578} = 1.730 \text{ in.}$$

$$c = \frac{d}{R_1} = \frac{1.000}{0.440} = 2.273 \text{ in.}$$

$$b = [a^2 + c^2 + d^2 - 2acR_3]^{1/2}$$
$$= [1.730^2 + 2.273^2 + 1.00^2 - 2(1.730)(2.273)(0.132)]^{1/2}$$
$$= 2.850 \text{ in.}$$

A sketch of the linkage $OABC$ is shown in Fig. 9.11.

Another method of synthesis using displacement equations has been

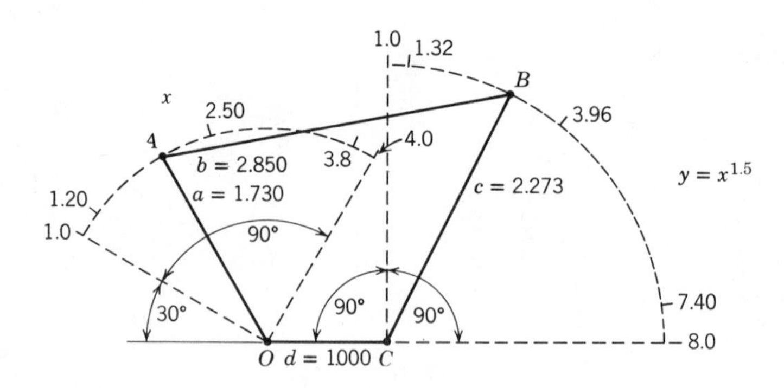

Figure 9.11

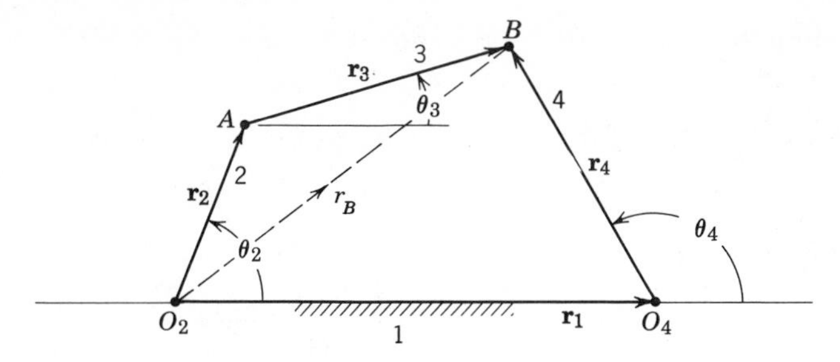

Figure 9.12

developed based on the work of Raven.[6] Consider the four-bar linkage as shown in Fig. 9.12, and let it be required to have θ_4 vary as a function of θ_2. A vector equation in terms of complex numbers can be written for the linkage as follows:

$$\mathbf{r}_B = \mathbf{r}_2 + \mathbf{r}_3 = \mathbf{r}_1 + \mathbf{r}_4$$
$$= r_2 e^{i\theta_2} + r_3 e^{i\theta_3} = r_1 + r_4 e^{i\theta_4} \tag{9.21}$$

The lengths of the links can be made nondimensional by letting

$$R_2 = \frac{r_2}{r_1}, \qquad R_3 = \frac{r_3}{r_1}, \qquad \text{and} \qquad R_4 = \frac{r_4}{r_1}$$

Equation 9.21 may therefore be written

$$R_2 e^{i\theta_2} + R_3 e^{i\theta_3} = 1 + R_4 e^{i\theta_4} \tag{9.22}$$

Writing Eq. 9.22 in terms of real and imaginary parts,

$$R_2 (\cos\theta_2 + i \sin\theta_2) + R_3 (\cos\theta_3 + i \sin\theta_3) = 1 + R_4 (\cos\theta_4 + i \sin\theta_4)$$

Separating the real and imaginary parts and solving for $R_3 \cos\theta_3$ and $R_3 \sin\theta_3$

$$\begin{aligned} R_3 \cos\theta_3 &= 1 + R_4 \cos\theta_4 - R_2 \cos\theta_2 \qquad &\text{(real)} \\ R_3 \sin\theta_3 &= R_4 \sin\theta_4 - R_2 \sin\theta_2 \qquad &\text{(imaginary)} \end{aligned} \tag{9.23}$$

The unknown angle θ_3 can be eliminated from Eqs. 9.23 by squaring the real and imaginary parts and adding

$$1 + R_2^2 - R_3^2 + R_4^2 = 2R_2 \cos\theta_2 - 2R_4 \cos\theta_4 + 2R_2 R_4 \cos(\theta_4 - \theta_2) \tag{9.24}$$

[6] F. H. Raven, "Position, Velocity, and Acceleration Analysis and Kinematic Synthesis of Plane and Space Mechanisms by a Generalized Procedure Called the Method of Independent Position Equations," L.C. Card No. 58–58, University Microfilms, Ann Arbor, Michigan, 1958.

By expanding the term cos $(\theta_4 - \theta_2)$ and rearranging, Eq. 9.24 can be written as

$$1 + R_2^2 - R_3^2 + R_4^2 = 2R_2 \cos\theta_2 + 2R_4(R_2\cos\theta_2 - 1)\cos\theta_4 + 2R_2R_4 \sin\theta_2 \sin\theta_4 \tag{9.25}$$

Solving Eq. 9.25 for θ_4,

$$\sin(\theta_4 + \beta) = \frac{1 + R_2(R_2 - 2\cos\theta_2) - R_3^2 + R_4^2}{2R_4\sqrt{1 + R_2(R_2 - 2\cos\theta_2)}}$$

where

$$\beta = \tan^{-1}\frac{(R_2\cos\theta_2 - 1)}{R_2\sin\theta_2} \tag{9.26}$$

From the complexity of Eq. 9.26 it is obvious that some means other than direct substitution must be employed to proportion the linkage to generate θ_4

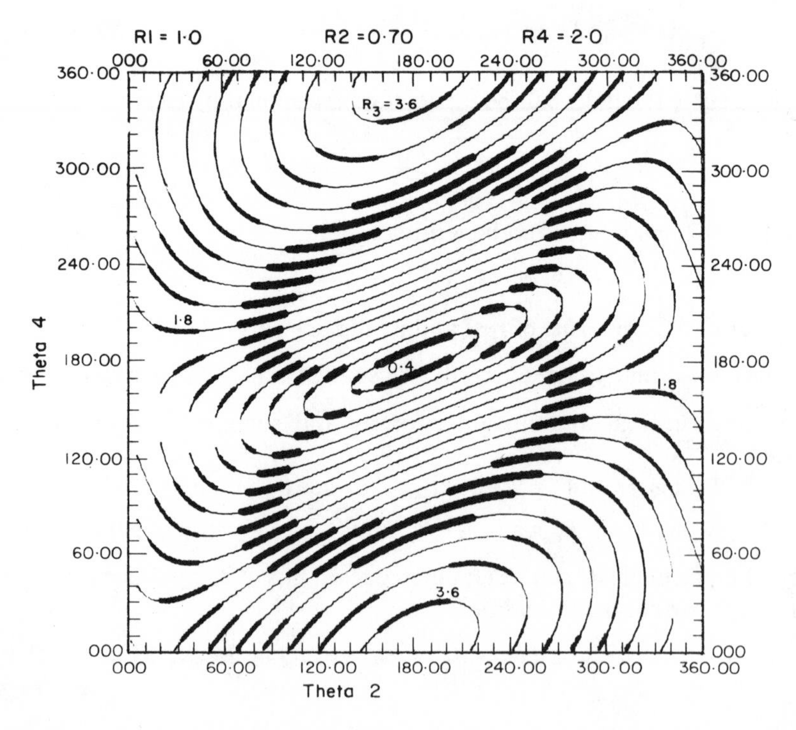

Figure 9.13 (Reprinted with permission from R. S. Brown and H. H. Mabie, "Application of Curve Matching to Designing Four-Bar Mechanisms," *Journal of Mechanisms*, Volume 5, Number 4, 1971 (Winter 1970), p. 566, Pergamon Press Ltd).

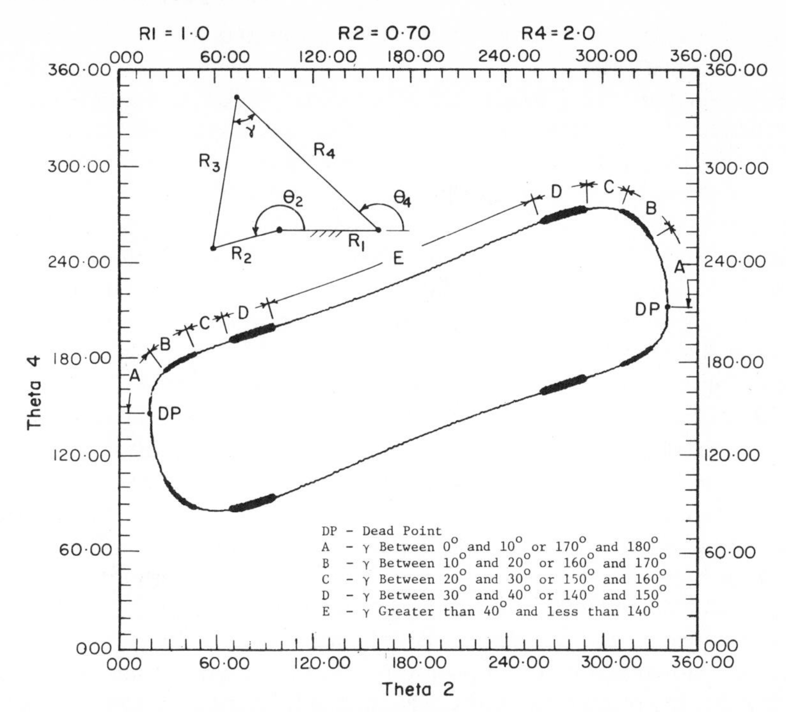

Figure 9.14 (Reprinted with permission from R. S. Brown and H. H. Mabie, "Application of Curve Matching to Designing Four-Bar Mechanisms," *Journal of Mechanisms*, Volume 5, Number 4, 1971, (Winter 1970), p. 567, Pergamon Press Ltd).

as a desired function of θ_2. A method[7] that has been successful is to plot a series of curves of constant R_3 for θ_4 versus θ_2 with R_2 and R_4 given values. Such curves are known as *displacement curves*. To select a linkage to generate a given function, the desired relation of θ_4 versus θ_2 is first plotted on transparent paper, and this curve is then superimposed on the displacement curves. The displacement curve which best fits the desired curve gives the approximate proportions of the linkage. Figure 9.13 shows an example of displacement curves plotted by computer with $R_1 = 1.0$, $R_2 = 0.7$, and $R_4 = 2.0$.

The variation in the width of the lines in Fig. 9.13 indicate values of transmission angles according to the legend given in Fig. 9.14 where only one displacement curve ($R_3 = 1.6$) is shown from Fig. 9.13.

To have a workable system, it is of course necessary to have plots of displacement curves for many combinations of R_2, R_3, and R_4. This system of

[7] R. S. Brown and H. H. Mabie, "Application of Curve Matching to Designing Four-Bar Mechanisms," *Journal of Mechanisms*, Vol. 5, pp. 563–575, 1970.

synthesis is known as *curve matching*, and examples of this method are given in the reference cited.

9.4 Graphical Design of Four-Bar Linkage as a Function Generator. There are many graphical methods of synthesis that have been developed. One method is presented here and others are given in an excellent work by Professor A. S. Hall of Purdue University.[8]

The method[9] to be discussed is one by which the proportions of a four-bar linkage can be found to give a required input-to-output motion at three positions. Figure 9.15 shows the layout where link 2 of known length passes through positions A_1, A_2, and A_3 and drives link 4 (or a pointer attached to it) through the angular positions B_1, B_2, and B_3. The distance O_2O_4 is also known, and it is required to find the lengths of links 3 and 4.

The easiest way to handle the problem is to invert the mechanism so that link 4 is fixed instead of link 1. As the mechanism passes through its cycle, it is evident that point O_2 will trace a circle about point O_4 and that point A will trace a circle about point B. Locating the center of the latter circle determines the position of point B and therefore the lengths of links 3 and 4.

Figure 9.16 shows the graphical construction for determining point B. Link 4 is considered fixed, and link 1 rotates counterclockwise about point O_4 through angles α' and β' which are equal but opposite in direction to α and β. Point O_2 moves through two positions O_2' and O_2'' while point A moves to A_2' and A_3' (the rotated positions of A_2 and A_3). Point A_2' is the intersection of the arc of radius O_2A swung about point O_2' and the arc of radius O_4A_2 swung about O_4. Point A_3' can similarly be determined using the arc of radius O_2A about point O_2'' and the arc of radius O_4A_3 about O_4. With points A_1, A_2', and A_3' available, the perpendicular bisectors of A_1A_2' and $A_2'A_3'$ can be drawn. Their intersection gives point B.

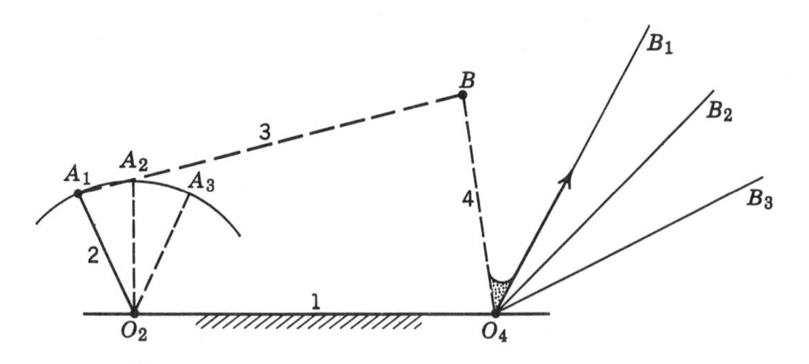

Figure 9.15

[8] A. S. Hall, *Kinematics and Linkage Design*, Prentice-Hall, 1961.

[9] I. E. Kass, "Graphic Linkage Design," *Machine Design*, December 10, 1959.

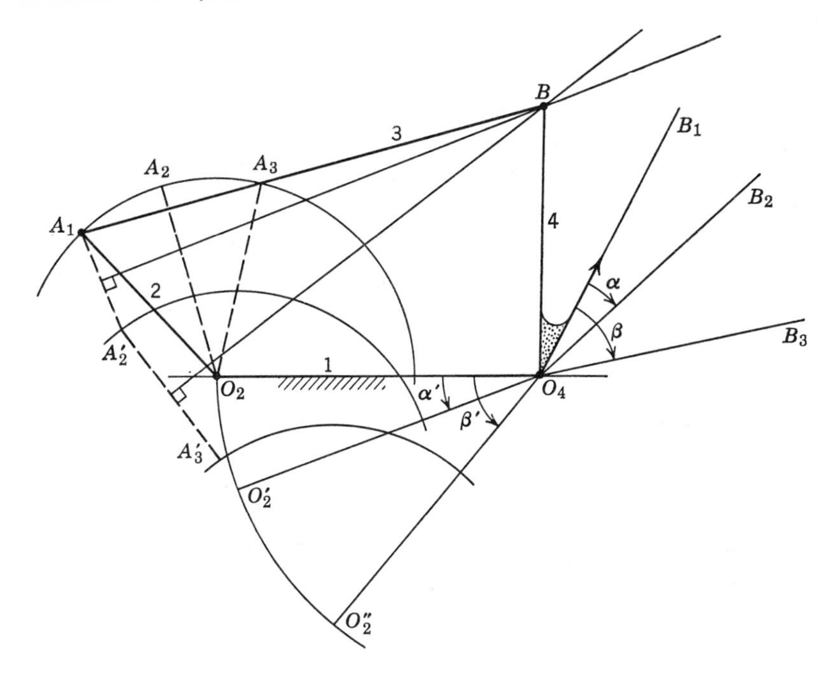

Figure 9.16

It should be mentioned that although a geometric solution is possible, there is no way of telling before a layout is made whether the solution will give a practical mechanism. It must be examined for dead points, reversals, and mechanical advantage. If the solution is impractical, the length or position of link 2 or the length of link 1 must be changed and another trial made.

This method can also be applied to a three-dimensional linkage. The reader is referred to the original article for a description of that method.

In addition to the methods presented, there are many more, both analytical and graphical, which are of interest to the designer. A partial list of these references follows.

References

Hrones, J. A., and G. L. Nelson, *Analysis of the Four-Bar Linkage*, Technology Press, M.I.T., and John Wiley and Sons, 1951.

Pike, E. W., T. R. Silverberg, and P. T. Nickson, "Linkage Layout," *Machine Design*, Vol. 23, pp. 105–110, 194, November 1951.

Rosenauer, N., and A. H. Willis, *Kinematics of Mechanisms*, Associated General Publications, Sidney, Australia, 1953.

Shaffer, B. W., and J. Cochin, "Synthesis of the Quadric Chain When the Position of Two Members Is Prescribed," ASME Paper 53-A-144.

Svoboda, A., *Computing Mechanisms and Linkages*, McGraw-Hill Book Company, 1948.

Problems

9.1 Using Rosenauer's method, design a four-bar linkage to give the following instantaneous values:

$$\omega_2 = 6 \text{ rad/sec} \qquad \alpha_2 = 0 \text{ rad/sec}^2$$
$$\omega_3 = 1 \text{ rad/sec} \qquad \alpha_3 = 8 \text{ rad/sec}^2$$
$$\omega_4 = 4 \text{ rad/sec} \qquad \alpha_4 = 4 \text{ rad/sec}^2$$

Make a sketch of the mechanism to a scale of 1 in. = 20 units.

9.2 Using Rosenauer's method, design a four-bar linkage to give the following instantaneous values:

$$\omega_2 = 6 \text{ rad/sec} \qquad \alpha_2 = 3 \text{ rad/sec}^2$$
$$\omega_3 = 1 \text{ rad/sec} \qquad \alpha_3 = 8 \text{ rad/sec}^2$$
$$\omega_4 = 3 \text{ rad/sec} \qquad \alpha_4 = 5 \text{ rad/sec}^2$$

Make a sketch of the mechanism to a scale of 1 cm = 10 units.

9.3 Using Rosenauer's method, design a four-bar linkage to give the following instantaneous values:

$$\omega_2 = -3 \text{ rad/sec} \qquad \alpha_2 = 0 \text{ rad/sec}^2$$
$$\omega_3 = 1 \text{ rad/sec} \qquad \alpha_3 = 10 \text{ rad/sec}^2$$
$$\omega_4 = 3 \text{ rad/sec} \qquad \alpha_4 = 5 \text{ rad/sec}^2$$

Make a sketch of the mechanism to a scale of 1 in. = 20 units.

9.4 For the method of synthesis developed by Rosenauer a particular set of kinematic conditions gives the vector equations for three of the four links as follows:

$$\mathbf{a} = -30 + 30i$$
$$\mathbf{b} = 0 - 40i$$
$$\mathbf{d} = -20 + 30i$$

(*a*) Sketch the complete linkage to a scale of 1 cm = 10 units. (*b*) Write the vector equation for **c** in terms of real and imaginary numbers.

9.5 A four-bar linkage has been designed so that the vectors representing the links can be expressed by the following equations:

$$\mathbf{a} = 12 + 28i$$
$$\mathbf{b} = 48 + 24i$$
$$\mathbf{c} = 30 + 48i$$
$$\mathbf{d} = -30 - 4i$$

If $\omega_2 = 6$ rad/sec and $\alpha_2 = 0$, calculate ω_4 and α_4 using Rosenauer's method.

9.6 A four-bar linkage has been designed to give the following instantaneous values:

$$\omega_2 = 6 \text{ rad/sec} \qquad \alpha_2 = 0 \text{ rad/sec}^2$$
$$\omega_3 = 1 \text{ rad/sec} \qquad \alpha_3 = 10 \text{ rad/sec}^2$$
$$\omega_4 = 3 \text{ rad/sec} \qquad \alpha_4 = 5 \text{ rad/sec}^2$$

The links in complex form are

$$\mathbf{a} = 6 + 25i$$
$$\mathbf{b} = 54 + 30i$$
$$\mathbf{c} = 30 + 60i$$
$$\mathbf{d} = -30 + 5i$$

(*a*) If link a is changed so that its equation becomes $\mathbf{a} = 6 + 20i$, determine the vector equation of link b assuming that the lengths and positions of links c and d do not change. (*b*) Calculate ω_4 assuming that ω_2 does not change.

9.7 A four-bar linkage has been designed so that the vector representing the links can be expressed by the following equations:

$$\mathbf{a} = 6 + 19i$$
$$\mathbf{b} = 54 + 21i$$
$$\mathbf{c} = 30 + 45i$$
$$\mathbf{d} = -30 + 5i$$

If $\omega_2 = 6$ rad/sec and $\alpha_2 = 3$ rad/sec^2, calculate ω_3, ω_4, α_3, and α_4 using Rosenauer's method.

9.8 Using Freudenstein's method, determine the proportions of a four-bar linkage to generate $y = \tan x$ when x varies between 0° and 45°. Use Chebyshev spacing. Let $\phi_s = 45°$, $\Delta\phi = 90°$, $\psi_s = 90°$, and $\Delta\psi = 90°$. Make a sketch of the linkage letting the ground link d be 1.00 in.

9.9 Using Freudenstein's method, determine the proportions of a four-bar linkage to generate $y = \log_{10} x$, when x varies between 1 and 10. Use Chebyshev spacing. Let $\phi_s = 45°$, $\Delta\phi = 60°$, $\psi_s = 135°$, and $\Delta\psi = 90°$. Make a sketch of the linkage letting the ground link d be 5.00 cm and check for dead points.

9.10 Using the methods of complex variables, derive Eq. 9.15 of Freudenstein's method.

9.11 The crank-shaper mechanism shown in Fig. 9.17 can be used as a function generator to give θ_4 as a function of θ_2. Using complex variables, prove that the relation between θ_4 and θ_2 is given by $\cos\theta_4 + R_2 \sin(\theta_2 - \theta_4) = 0$, where $R_2 = r_2/O_2O_4$.

9.12 Using the relation given in Problem 9.11 for the crank-shaper mechanism of Fig. 9.17, plot θ_4 versus θ_2 for constant values of R_2 of $\frac{1}{2}$, 1, and 2. Let θ_2 and θ_4 both vary from -90 to 270°.

9.13 In a four-bar linkage, the length of link 2 is $1\frac{1}{2}$ in., and it is to rotate clockwise from its initial position (position 1) of 30° above the horizontal to 60° (position 2) to 90° (position 3). As link 2 rotates from position 1 to position 2, link 4 rotates 13°. As link 2 goes from position 2 to position 3, link 4 rotates 20°. If the length of link 1 (O_2O_4) is

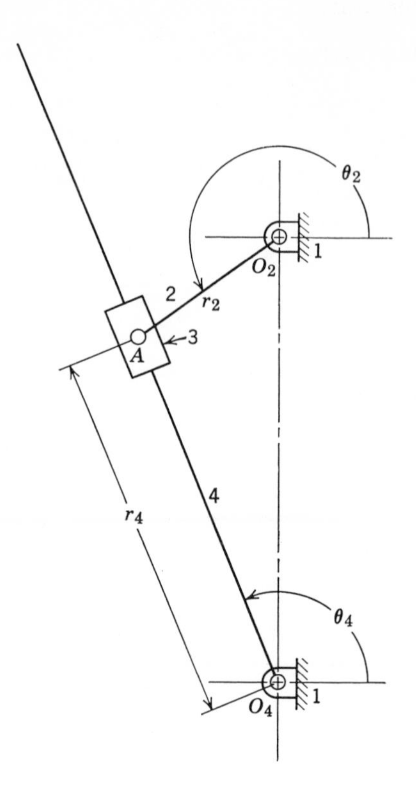

Figure 9.17

2 in., determine graphically the lengths of links 3 and 4. Check the operation of the linkage by drawing it in positions 2 and 3.

9.14 In a four-bar linkage the length of link 2 is 5.0 cm and it is to rotate clockwise from its initial position (position 1) of 60° above the horizontal to 90° (position 2) and to 120° (position 3). As link 2 rotates from position 1 to position 2, link 4 rotates 10°. As link 2 goes from position 2 to position 3, link 4 rotates 15°. If the length of link 1 (O_2O_4) is 7.6 cm, determine graphically the length of links 3 and 4. Check the operation of the linkage by drawing it in positions 2 and 3.

Part 2

DYNAMICS OF MACHINERY

10

Kinematics of Machinery

10.1 Introduction. Because motion is inherent in machinery, kinematic quantities such as velocity and acceleration are of engineering importance in the analysis and design of machine components. Kinematic values in machines have reached extraordinary magnitudes. Rotative speeds, once considered high at 10,000 r/min, are approaching 100,000 r/min. Large rotors of jet engines operate at 10,000 to 15,000 r/min, and small turbine wheels rotate at 30,000 to 60,000 r/min.

Size and rotative speed in rotors are related such that the smaller the size the greater the allowable rotative speed. A more basic quantity in rotors is peripheral speed, which depends on rotative speed and size ($V = \omega R$). Peripheral speeds in turbomachinery are reaching 50,000 to 100,000 ft/min. Peripheral speeds in electric armatures (10,000 ft/min) and automotive crankshafts (3000 ft/min) are lower than in aeronautical rotors. Although the rotor, or crank, speeds of linkage mechanisms are low, the trend is toward higher speeds because of the demand for higher rates of productivity from the machines used in printing, paper making, thread spinning, automatic computing, packaging, bottling, automatic machining, and in numerous other applications.

The centripetal acceleration at a rotor periphery depends on the square of the rotative speed and size ($A^n = \omega^2 R$). In turbines, such accelerations are approaching values of one to three million, ft/sec^2, or about 30,000g to

100,000g, values that may be compared with the acceleration of 10g withstandable by airplane pilots or the 1000g of automotive pistons.

Acceleration is related to force (MA), by Newton's principle, and in turn related to stress and deformation, which may or may not be critical in a machine part, depending on the materials used. The speed of a machine is limited ultimately by the properties of the materials of which it consists and the conditions which influence these properties. High temperature arising from the compression of gases and the combustion of fuels, together with that arising from friction, is a condition in high-speed power machines that influences the strength of the materials. The degree to which the temperature rises also depends on the provisions made for the transmission of heat by coolants such as air, oil, water, or Freon.

The successful design of a machine depends on the exploitation of knowledge in the fields of dynamics, stress analysis, thermodynamics, heat transmission, and properties of materials. However, it is the purpose of this chapter to deal solely with kinematic relationships in machines. In subsequent chapters, acceleration and force are discussed in connection with the determination of forces acting on individual links of a mechanism and in connection with machine balance and vibration.

For bodies rotating about a fixed axis, such as rotors, kinematic values are quickly determined from well-known elementary formulas ($V = \omega R$, $A^n = \omega^2 R$, $A^t = \alpha R$). However, mechanisms such as the slider crank and its inversions are combinations of links consisting not only of a rotor but of oscillating and reciprocating members as well. Because of the relative velocities and relative accelerations among the several members, together with the many geometric relative positions possible, the kinematic analysis of a linkage is relatively complex compared to that of a rotor. The principles and methods illustrated in this chapter are primarily those for the analysis of linkages consisting of combinations of rotors, bars, sliders, cams, gears, and rolling elements.

In the following discussions, the individual links of a mechanism are assumed to be rigid bodies in which the distance between two given particles of a moving link remains fixed. Links which undergo large deformations during motion, such as springs, fall in another category and are analyzed as vibrating members.

Most elementary mechanisms are in plane motion or may be analyzed as such. Mechanisms in which all of the particles move in parallel planes are said to be in *plane motion*. An illustration is a four-bar linkage (Fig. 10.1) consisting of two rockers and a connecting rod. This arrangement is often referred to as a *double-rocker mechanism*.

The motion of a link is expressed in terms of the linear displacements, linear velocities, and linear accelerations of the individual particles which

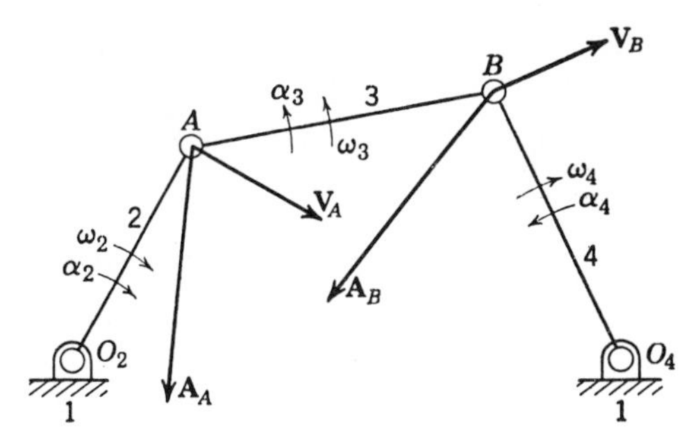

Figure 10.1

constitute the link. However, the motion of a link may also be expressed in terms of angular displacements, angular velocities, and angular accelerations of lines moving with the rigid link.

In Fig. 10.1, the linear velocity $\mathbf{V}_A$ and the linear acceleration $\mathbf{A}_A$ of particle A are shown by the fixed vectors at A. Because of the connecting pin at A, particle A_2 on link 2 and particle A_3 on link 3 have the same motion, and the vectors shown at A represent the motions of both particles. The angular motions of links 2 and 3 are different as given by the angular velocities ω_2, ω_3 and the angular accelerations α_2, α_3. Usually the angular motion of a driving link is known, or assumed, such as ω_2 and α_2 of Fig. 10.1, and the motions of the connecting and driven links are to be determined.

10.2 Linear Motion of a Particle. In useful mechanisms, the particles of the links are constrained to move on given paths, many of which, such as circles and straight lines, are obvious. In Fig. 10.1, the particles of links 2 and 4 are constrained to move on circular paths. The particles of link 3, however, are in motion along generally curvilinear paths less simple than circles or straight lines.

A particle in motion on a curvilinear path is said to be in *curvilinear translation*. The basic kinematic relationships for a particle translating in a plane are well known from the study of mechanics. These are reviewed in the following paragraphs with reference to Fig. 10.2.

The linear velocity $\mathbf{V}_P$ of a particle P is the instantaneous rate of change of the position of the particle, or displacement, with respect to time. Referring to Fig. 10.2*a*, in a small interval of time Δt, the particle is displaced $\Delta\mathbf{S}$ along the curved path from position P to position P'. At the same time, the radius of curvature of the path of the particle changes from $\mathbf{R}$ to $\mathbf{R} + \Delta\mathbf{R}$ and undergoes an angular displacement $\Delta\boldsymbol{\theta}_r$. Therefore, the displacement $\Delta\mathbf{S}$ is made up of two components: one due to the angular displacement $\Delta\boldsymbol{\theta}_r$

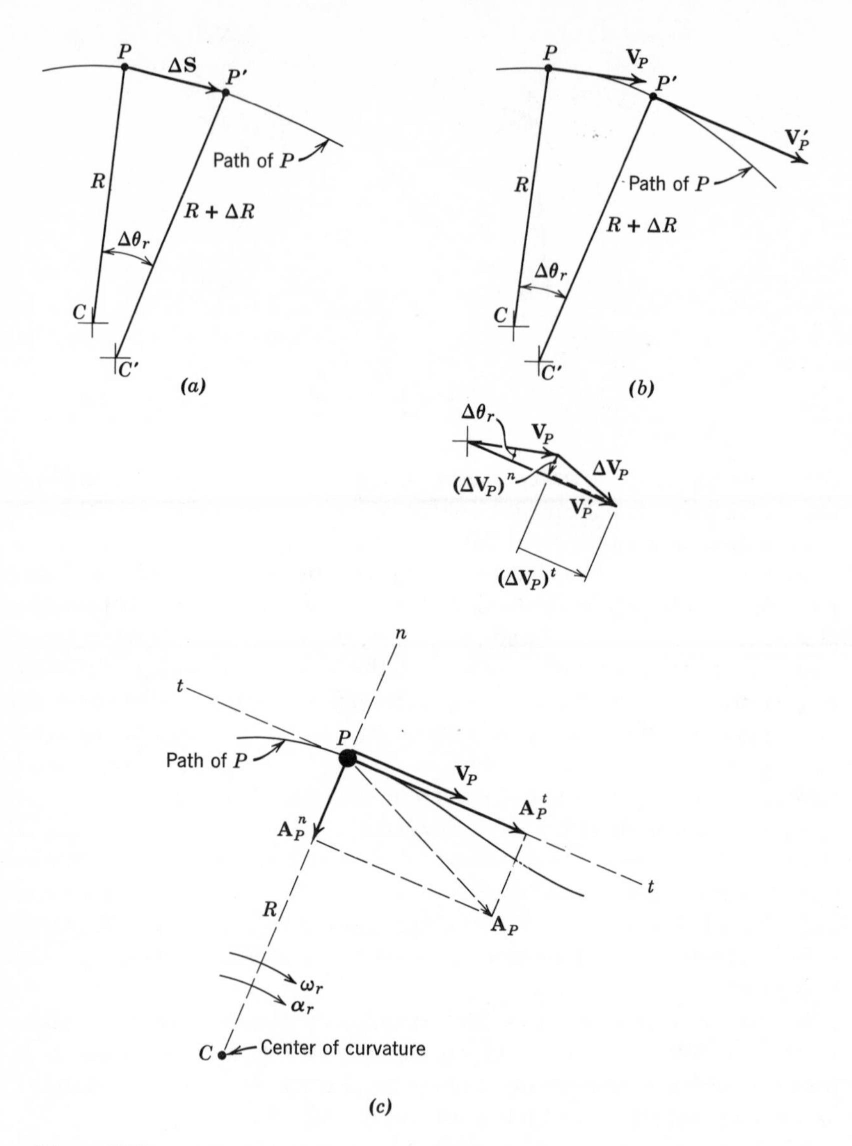

P
ΔS
P′
Path of P
R
R + ΔR
Δθr
C
C′
(a)
P
VP
P′
V′P
Path of P
R
R + ΔR
Δθr
C
C′
(b)
Δθr
VP
ΔVP
(ΔVP)n
V′P
(ΔVP)t
n
t
P
Path of P
VP
AtP
AnP
t
R
AP
ωr
αr
C
Center of curvature
(c)

Figure 10.2

of radius **R** and the other due to the change in length $\Delta\mathbf{R}$. From this displacement, an equation for the velocity $\mathbf{V}_P$ can be determined as follows:

$$\mathbf{V}_P = \lim_{\Delta t \to 0}\left[\frac{\Delta \mathbf{S}}{\Delta t}\right] = \lim_{\Delta t \to 0}\left[\frac{\Delta\boldsymbol{\theta}_r \times \mathbf{R}}{\Delta t} + \frac{\Delta \mathbf{R}}{\Delta t}\right]$$

$$\mathbf{V}_P = \boldsymbol{\omega}_r \times \mathbf{R} + \frac{d\mathbf{R}}{dt} \tag{10.1}$$

where $\boldsymbol{\omega}_r = d\boldsymbol{\theta}_r/dt$ is the instantaneous angular velocity of the radius of curvature and $d\mathbf{R}/dt$ is the rate of change of the radius of curvature with time.

In many applications in kinematics the radius R is constant so that Eq. 10.1 becomes

$$\mathbf{V}_P = \boldsymbol{\omega}_r \times \mathbf{R}$$

and

$$|\mathbf{V}_P| = R\omega_r \tag{10.1a}$$

with the direction of the velocity vector $\mathbf{V}_P$ tangent to the path at point P and with a sense the same as that of the displacement of the particle P.

The linear acceleration $\mathbf{A}_P$ of a particle P is the instantaneous rate of change of its velocity with respect to time. If the path is curvilinear, the change in the velocity vector of the particle in a small time interval Δt may be a change in direction as well as a change in magnitude. Referring to Fig. 10.2*b*, the velocity $\mathbf{V}_P$ of the particle is tangent to the path at P at the time t, and Δt later its velocity is $\mathbf{V}'_P$ and tangent to the path at P'. Thus, the velocity vector has changed in magnitude and has also changed direction by the angular displacement $\Delta\boldsymbol{\theta}_r$. As shown in the vector polygon of Fig. 10.2*b*, the vector change of velocity is $\Delta\mathbf{V}_P$, which may be represented by the component perpendicular vectors $\Delta\mathbf{V}_P^n$ and $\Delta\mathbf{V}_P^t$, the former becoming normal to the path, and the latter becoming tangent to the path, as Δt and $\Delta\boldsymbol{\theta}_r$ approach zero. The rate of change of the normal component of velocity with respect to time is the normal acceleration $\mathbf{A}_P^n$ of the particle as follows:

$$\mathbf{A}_P^n = \lim_{\Delta t \to 0}\left[\frac{\Delta\mathbf{V}_P^n}{\Delta t}\right] = \lim_{\Delta t \to 0}\left[\frac{2\mathbf{V}_P \sin \Delta\theta_r/2}{\Delta t}\right] = \frac{d\boldsymbol{\theta}_r}{dt} \times \mathbf{V}_P$$

Since

$$\mathbf{V}_P = \boldsymbol{\omega}_r \times \mathbf{R} + \frac{d\mathbf{R}}{dt}$$

$$\mathbf{A}_P^n = \frac{d\boldsymbol{\theta}_r}{dt} \times \left[\boldsymbol{\omega}_r \times \mathbf{R} + \frac{d\mathbf{R}}{dt}\right]$$

Therefore,

$$\mathbf{A}_P^n = \boldsymbol{\omega}_r \times (\boldsymbol{\omega}_r \times \mathbf{R}) + \boldsymbol{\omega}_r \times \frac{d\mathbf{R}}{dt} \tag{10.2}$$

If the radius R is constant, Eq. 10.2 becomes:

$$\mathbf{A}_P^n = \boldsymbol{\omega}_r \times (\boldsymbol{\omega}_r \times \mathbf{R})$$

and

$$|\mathbf{A}_P^n| = R\omega_r^2 = V_P\omega_r = \frac{V_P^2}{R} \tag{10.2a}$$

The rate of change of the tangential component of velocity with respect to time is the tangential component of acceleration $\mathbf{A}_P^t$ and is seen to depend on the change in magnitude of velocity as follows:

$$\mathbf{A}_P^t = \lim_{\Delta t \to 0}\left[\frac{\Delta \mathbf{V}_P^t}{\Delta t}\right] = \lim_{\Delta t \to 0}\left[\frac{\mathbf{V}_P' - \mathbf{V}_P}{\Delta t}\right] = \frac{d}{dt}(\mathbf{V}_P)$$

$$= \frac{d}{dt}\left(\boldsymbol{\omega}_r \times \mathbf{R} + \frac{d\mathbf{R}}{dt}\right)$$

$$= \frac{d\boldsymbol{\omega}_r}{dt} \times \mathbf{R} + \boldsymbol{\omega}_r \times \frac{d\mathbf{R}}{dt} + \frac{d^2\mathbf{R}}{dt^2}$$

$$\mathbf{A}_P^t = \boldsymbol{\alpha}_r \times \mathbf{R} + \boldsymbol{\omega}_r \times \frac{d\mathbf{R}}{dt} + \frac{d^2\mathbf{R}}{dt^2} \tag{10.3}$$

where $\boldsymbol{\alpha}_r = d\boldsymbol{\omega}_r/dt$ is the instantaneous angular acceleration of the radius of curvature.

When the radius of curvature R is constant, Eq. 10.3 becomes

$$\mathbf{A}_P^t = \boldsymbol{\alpha}_r \times \mathbf{R}$$

and

$$|\mathbf{A}_P^t| = R\alpha_r \tag{10.3a}$$

The resultant acceleration $\mathbf{A}_P$ can be expressed as follows:

$$\mathbf{A}_P = \mathbf{A}_P^n + \mathbf{A}_P^t$$

$$= \boldsymbol{\omega}_r \times (\boldsymbol{\omega}_r \times \mathbf{R}) + 2\boldsymbol{\omega}_r \times \frac{d\mathbf{R}}{dt}$$

$$+ \boldsymbol{\alpha}_r \times \mathbf{R} + \frac{d^2\mathbf{R}}{dt^2} \tag{10.4}$$

In Fig. 10.2*c* are shown the velocity vector $\mathbf{V}_P$ and the component vectors of acceleration $\mathbf{A}_P^n$ and $\mathbf{A}_P^t$ to show their instantaneous directional orientation with respect to the tangent and normal of the path. It is important to note that the direction of $\mathbf{A}_P^n$ is normal to the path and that its sense is toward the center of curvature C of the path. The direction of $\mathbf{A}_P^t$ is tangent to the path and its sense is for increasing velocity. The resultant acceleration $\mathbf{A}_P$ is the vector sum of $\mathbf{A}_P^n$ and $\mathbf{A}_P^t$ as shown.

Equations 10.1*a*, 10.2*a*, and 10.3*a* are used to calculate only the magnitudes of the vectors describing the linear motion of a particle, and they appear repeatedly in the derivation of kinematic relationships of particles for special cases in mechanisms.

10.3 Angular Motion. Angular velocity and angular acceleration are the first and second derivatives respectively of the angular displacement θ of a line with respect to time t. In machine analysis, the angular motion of a link is expressed by the angular motion of any line visualized fixed to the link. In Fig. 10.3, line AB is in angular motion because of its angular displacement with respect to time. Lines BC and AC undergo the same angular displacements with respect to time as line AB because triangle ABC is fixed in position with link 3 as a rigid body. Since all lines of link 3 have the same angular motion, the angular velocity and angular acceleration of these lines are ω_3 and α_3 of the link, with the subscript denoting the link number.

Angular motion of a link may be the same or different from the angular motions of the radii of curvature of the paths of the individual particles of the link. In Fig. 10.3, since all particles of link 2 are moving on circular paths having a common center of curvature at the fixed center O_2, it is obvious that ω_r and α_r of the radii of curvature of the paths of all particles are equal to the respective angular velocity and angular acceleration ω_2 and α_2 of the link. In the case of the connecting link 3 in Fig. 10.3, which is not rotating about a fixed center, ω_r and α_r of the radius of curvature of the path of any given particle are not the same as ω_3 and α_3 of link 3.

It is an important concept in mechanics that a particle, which has the infinitely small size of a point, may have only *linear* motion (linear velocity and linear acceleration). *Angular* motion is that of a line, and since a particle is a point, not a line, it is not considered to be in angular motion. This concept must be fully understood to understand the *relative* motion among particles. For example, the velocity of the particle on link 2 at O_2 in Fig. 10.3 relative to the velocity of any particle on the fixed link 1 is zero. *Linear* velocity is implied, and it is incorrect to hold that, by virtue of the angular motion of link 2, the particle O_2 has the angular velocity of the link.

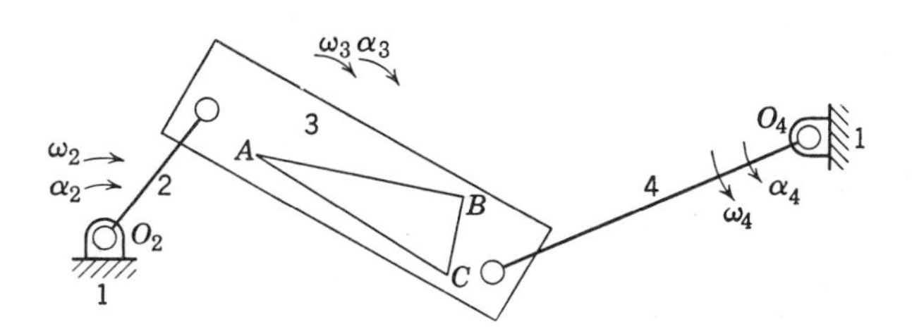

Figure 10.3

10.4 Relative Motion. As will be shown in a later section, the relative motion between particles is very important in the kinematic analysis of mechanisms. In Fig. 10.4*a*, P and Q are particles moving relative to a fixed reference plane at the respective velocities of $\mathbf{V}_P$ and $\mathbf{V}_Q$, and it is necessary to determine the relative velocity $\mathbf{V}_{PQ}$ between the two particles. In determining $\mathbf{V}_{PQ}$ use will be made of the fact that the addition of equal velocities to each particle does not change the relative velocity of the two particles. Therefore, if P and Q are each given a velocity equal and opposite to $\mathbf{V}_Q$, the particle Q becomes stationary in the fixed plane, and P acquires an additional velocity component $-\mathbf{V}_Q$ relative to the fixed plane. The new absolute velocity of $P(\mathbf{V}_P - \mathbf{V}_Q)$, therefore, becomes the relative velocity $\mathbf{V}_{PQ}$ because Q is now fixed relative to the reference plane. This is shown by the vector diagram of Fig. 10.4*b* from which the equation for $\mathbf{V}_{PQ}$ becomes

$$\mathbf{V}_{PQ} = \mathbf{V}_P - \mathbf{V}_Q \tag{10.5}$$

In a similar manner $\mathbf{V}_{QP}$ can be obtained by the addition of $-\mathbf{V}_P$ to each particle. This is shown in Fig. 10.4*c*, and $\mathbf{V}_{QP}$ is given by the equation

$$\mathbf{V}_{QP} = \mathbf{V}_Q - \mathbf{V}_P$$

The vector equation for the acceleration of particle P relative to particle Q is similar in form to Eq. 10.5.

$$\mathbf{A}_{PQ} = \mathbf{A}_P - \mathbf{A}_Q \tag{10.6}$$

The angular motion of a line may be given relative to another line in motion. In Fig. 10.5 the angular velocities ω_2 and ω_3 of the lines on links 2 and 3, respectively, are taken relative to line a–a on the fixed link. If $-\omega_3$ is added to links 2 and 3, link 3 becomes stationary and the new absolute

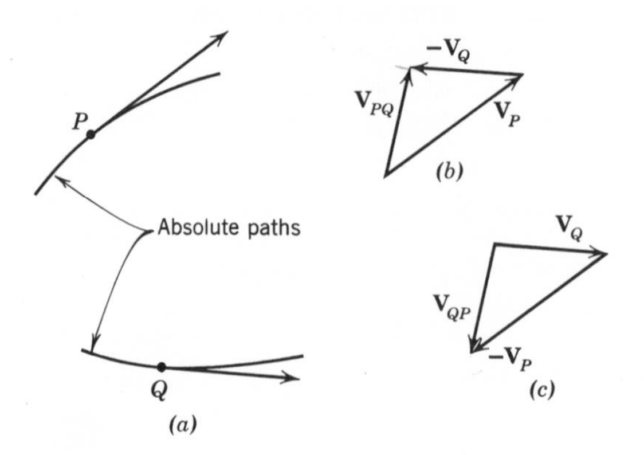

Figure 10.4

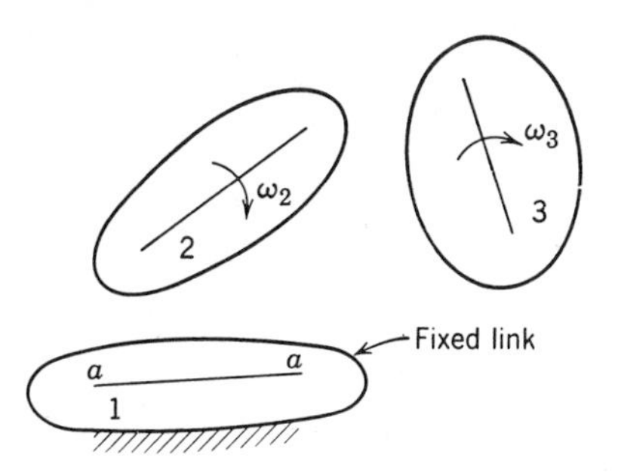

Figure 10.5

velocity of link 2 ($\omega_2 - \omega_3$), therefore, becomes the relative angular velocity ω_{23} because link 3 is now fixed.

Therefore,

$$\omega_{23} = \omega_2 - \omega_3 \tag{10.7}$$

In a similar manner

$$\alpha_{23} = \alpha_2 - \alpha_3 \tag{10.8}$$

10.5 Methods of Velocity and Acceleration Analysis. Of the many methods of determining velocities and accelerations in mechanisms, three find wide usage. These, which will be presented in the following sections, are (*a*) analysis using vector mathematics to express the velocity and acceleration of a point with respect to a moving and a fixed coordinate system; (*b*) analysis using equations of relative motion which are solved graphically by velocity and acceleration polygons; and (*c*) analysis using vector equations written in complex form. In addition, velocities by instant centers will be considered as well as graphical or computer differentiation of displacement–time and velocity–time curves to yield velocities and accelerations, respectively.

Of the methods of velocity and acceleration analysis listed above, the use of either of the first two maintains the physical concept of the problem. However, the third method, using vectors in complex form, tends to become too mechanical in its operation so that the physical aspects of the problem are soon lost. It should also be mentioned that the first and the third methods lend themselves to computer solutions which is a decided advantage if a mechanism is to be analyzed for a complete cycle.

10.6 Velocity and Acceleration Analysis by Vector Mathematics. In Fig. 10.6 the motion of point P is known with respect to the xyz coordinate system which in turn is moving relative to the fixed coordinate system XYZ. The position of point P relative to the XYZ system

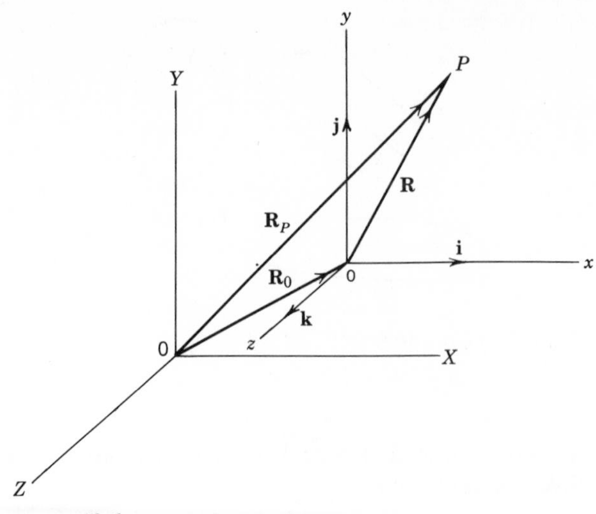

Figure 10.6

can be expressed as

$$\mathbf{R}_P = \mathbf{R}_0 + \mathbf{R} \tag{10.9}$$

If unit vectors **i**, **j**, and **k** are fixed to the x, y, and z axes, respectively,

$$\mathbf{R} = x\mathbf{i} + y\mathbf{j} + z\mathbf{k} \tag{10.10}$$

The velocity of point P relative to the XYZ system may be obtained by differentiating Eq. 10.9 with respect to time to give

$$\mathbf{V}_P = \dot{\mathbf{R}}_P = \dot{\mathbf{R}}_0 + \dot{\mathbf{R}} \tag{10.11}$$

Differentiating Eq. 10.10 with respect to time gives

$$\dot{\mathbf{R}} = (\dot{x}\mathbf{i} + \dot{y}\mathbf{j} + \dot{z}\mathbf{k}) + (x\dot{\mathbf{i}} + y\dot{\mathbf{j}} + z\dot{\mathbf{k}}) \tag{10.12}$$

The term $(\dot{x}\mathbf{i} + \dot{y}\mathbf{j} + \dot{z}\mathbf{k})$ is the velocity of point P relative to the moving coordinate system xyz. For convenience let

$$(\dot{x}\mathbf{i} + \dot{y}\mathbf{j} + \dot{z}\mathbf{k}) = \mathbf{V} \tag{10.13}$$

Consider next the terms in the second parenthesis of Eq. 10.12. From the fact that the velocity of the tip of a vector **r**, which passes through a fixed base point and rotates about the base point with an angular velocity ω, can be shown to be $\mathbf{V} = \boldsymbol{\omega} \times \mathbf{r}$, the velocities of the tips of the unit vectors $\dot{\mathbf{i}}, \dot{\mathbf{j}}, \dot{\mathbf{k}}$ can be expressed as

$$\dot{\mathbf{i}} = \boldsymbol{\omega} \times \mathbf{i}$$

$$\dot{\mathbf{j}} = \boldsymbol{\omega} \times \mathbf{j}$$

$$\dot{\mathbf{k}} = \boldsymbol{\omega} \times \mathbf{k}$$

where ω is the angular velocity of the moving coordinate system xyz relative to the fixed system XYZ. Making the above substitutions,

$$x\dot{\mathbf{i}} + y\dot{\mathbf{j}} + z\dot{\mathbf{k}} = \mathbf{x}(\boldsymbol{\omega} \times \mathbf{i}) + \mathbf{y}(\boldsymbol{\omega} \times \mathbf{j}) + \mathbf{z}(\boldsymbol{\omega} \times \mathbf{k}) = \boldsymbol{\omega} \times (x\mathbf{i} + y\mathbf{j} + z\mathbf{k})$$

and using the relation expressed in Eq. 10.10

$$x\dot{\mathbf{i}} + y\dot{\mathbf{j}} + z\dot{\mathbf{k}} = \boldsymbol{\omega} \times \mathbf{R} \tag{10.14}$$

Equation 10.12 then becomes

$$\dot{\mathbf{R}} = \mathbf{V} + \boldsymbol{\omega} \times \mathbf{R} \tag{10.15}$$

Equation 10.11 can now be rewritten as follows by letting $\mathbf{V}_0 = \dot{\mathbf{R}}_0$ and substituting for $\dot{\mathbf{R}}$ from Eq. 10.15.

$$\mathbf{V}_P = \mathbf{V}_0 + \mathbf{V} + \boldsymbol{\omega} \times \mathbf{R} \tag{10.16}$$

where

$\mathbf{V}_0$ = velocity of origin of xyz system relative to XYZ system
$\mathbf{V}$ = velocity of point P relative to xyz system
$\boldsymbol{\omega}$ = angular velocity of xyz system relative to XYZ system
$\mathbf{R}$ = distance from origin of xyz system to point P

The acceleration of point P relative to the XYZ system may now be obtained by differentiating Eq. 10.16.

$$\mathbf{A}_P = \dot{\mathbf{V}}_P = \dot{\mathbf{V}}_0 + \dot{\mathbf{V}} + \dot{\boldsymbol{\omega}} \times \mathbf{R} + \boldsymbol{\omega} \times \dot{\mathbf{R}} \tag{10.17}$$

To evaluate $\dot{\mathbf{V}}$ it is necessary to differentiate Eq. 10.13

$$\dot{\mathbf{V}} = (\ddot{x}\mathbf{i} + \ddot{y}\mathbf{j} + \ddot{z}\mathbf{k}) + (\dot{x}\dot{\mathbf{i}} + \dot{y}\dot{\mathbf{j}} + \dot{z}\dot{\mathbf{k}}) \tag{10.18}$$

The term $(\ddot{x}\mathbf{i} + \ddot{y}\mathbf{j} + \ddot{z}\mathbf{k})$ is the acceleration of point P relative to the moving coordinate system xyz. Let

$$(\ddot{x}\mathbf{i} + \ddot{y}\mathbf{j} + \ddot{z}\mathbf{k}) = \mathbf{A} \tag{10.19}$$

Considering the terms in the second parenthesis of Eq. 10.18,

$$\dot{x}\dot{\mathbf{i}} + \dot{y}\dot{\mathbf{j}} + \dot{z}\dot{\mathbf{k}} = \dot{\mathbf{x}}(\boldsymbol{\omega} \times \mathbf{i}) + \dot{\mathbf{y}}(\boldsymbol{\omega} \times \mathbf{j}) + \dot{\mathbf{z}}(\boldsymbol{\omega} \times \mathbf{k}) = \boldsymbol{\omega} \times (\dot{x}\mathbf{i} + \dot{y}\mathbf{j} + \dot{z}\mathbf{k})$$

But from Eq. 10.13

$$(\dot{x}\mathbf{i} + \dot{y}\mathbf{j} + \dot{z}\mathbf{k}) = \mathbf{V}$$

Therefore,

$$\dot{x}\dot{\mathbf{i}} + \dot{y}\dot{\mathbf{j}} + \dot{z}\dot{\mathbf{k}} = \boldsymbol{\omega} \times \mathbf{V} \tag{10.20}$$

Equation 10.18 then becomes

$$\dot{\mathbf{V}} = \mathbf{A} + \boldsymbol{\omega} \times \mathbf{V} \tag{10.21}$$

Also from Eq. 10.15

$$\boldsymbol{\omega} \times \dot{\mathbf{R}} = \boldsymbol{\omega} \times \mathbf{V} + \boldsymbol{\omega} \times (\boldsymbol{\omega} \times \mathbf{R}) \tag{10.22}$$

Substituting $\dot{\mathbf{V}}$ from Eq. 10.21 and $\boldsymbol{\omega} \times \dot{\mathbf{R}}$ from Eq. 10.22 into Eq. 10.17 and letting $\mathbf{A}_0 = \dot{\mathbf{V}}_0$, the equation for the acceleration of point P relative to the XYZ system becomes

$$\mathbf{A}_P = \mathbf{A}_0 + \mathbf{A} + 2\boldsymbol{\omega} \times \mathbf{V} + \dot{\boldsymbol{\omega}} \times \mathbf{R} + \boldsymbol{\omega} \times (\boldsymbol{\omega} \times \mathbf{R}) \qquad (10.23)$$

where the term $2\boldsymbol{\omega} \times \mathbf{V}$ is the Coriolis' component of acceleration and

$\mathbf{A}_0$ = acceleration of origin of xyz system relative to the XYZ system
$\mathbf{A}$ = acceleration of point P relative to xyz system[1]
$\boldsymbol{\omega}$ = angular velocity of xyz system relative to XYZ system
$\mathbf{V}$ = velocity of point P relative to xyz system
$\mathbf{R}$ = distance from origin of xyz system to point P

Example 10.1

As an example consider the mechanism shown in Fig. 10.7. Link 2 rotates in the direction shown at a constant angular velocity. The velocity and acceleration of point A are therefore known, and it is necessary to find the velocity and acceleration of point B. Select coordinate axes as shown with point O_2 as the origin of the XY system and point A as the origin of the xy system.

The equation for the velocity of point B can be written from Eq. 10.16 as follows:

$$\mathbf{V}_B = \mathbf{V}_0 + \mathbf{V} + \boldsymbol{\omega} \times \mathbf{R}$$

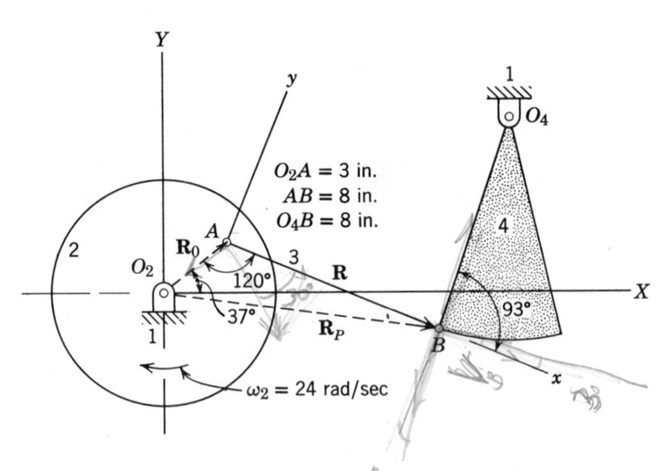

Figure 10.7

[1] It should be noted that in order to specify the normal and tangential components of $\mathbf{A}$, the path of point P relative to the xyz system must be known.

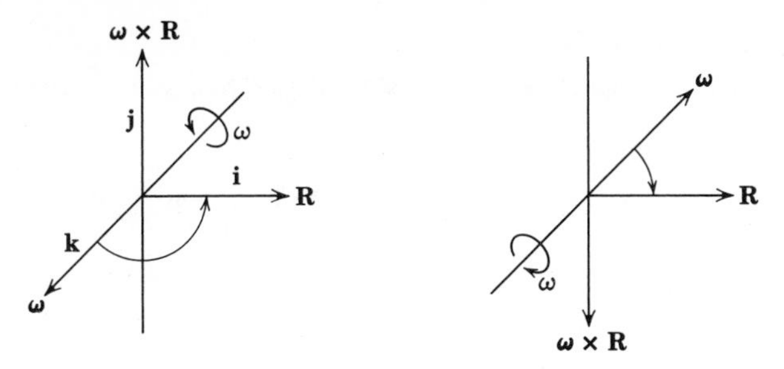

Figure 10.8

where

$\mathbf{V}_B$ = direction perpendicular to O_4B, magnitude unknown
$|\mathbf{V}_0| = |\mathbf{V}_A| = |O_2A)\omega_2 = \frac{3}{12} \times 24 = 6.0$ ft/sec, direction perpendicular to O_2A
$\mathbf{V} = 0$ because B is a fixed point in the xy system
$\boldsymbol{\omega} \times \mathbf{R}$ = direction perpendicular to AB ($\omega = \omega_3$, $R = AB$), magnitude unknown

The direction of $\boldsymbol{\omega} \times \mathbf{R}$ can be determined from the fact that the vector representing $\boldsymbol{\omega}$ will be perpendicular to the xy plane. When $\boldsymbol{\omega}$ is crossed into $\mathbf{R}$, the product $\boldsymbol{\omega} \times \mathbf{R}$ will be in the xy plane and perpendicular to $\mathbf{R}$ by the right-hand rule, This can be shown in Fig. 10.8 where the direction of $\boldsymbol{\omega} \times \mathbf{R}$ is the same regardless of whether ω is clockwise or counterclockwise.

The equation for $\mathbf{V}_B$ can be solved graphically by a polygon or analytically by unit vectors. Solution by the latter method follows where all components have been taken relative to the xy axes.

I.

$$\mathbf{V}_B = \mathbf{V}_0 + \mathbf{V} + \boldsymbol{\omega} \times \mathbf{R}$$
$$\mathbf{V}_B = V_B(\cos 3°\mathbf{i} + \sin 3°\mathbf{j}) = 0.9986V_B\mathbf{i} + 0.0523V_B\mathbf{j}$$
$$\mathbf{V}_0 = \mathbf{V}_A = V_A(\cos 30°\mathbf{i} - \sin 30°\mathbf{j}) = 6(0.8660\mathbf{i} - 0.5000\mathbf{j}) = 5.2\mathbf{i} - 3.0\mathbf{j}$$
$$\mathbf{V} = 0$$
$$\boldsymbol{\omega} \times \mathbf{R} = (\omega \times R)\mathbf{j}$$

Substituting the above relations in the equation for $\mathbf{V}_B$

$$0.9986V_B\mathbf{i} + 0.0523V_B\mathbf{j} = 5.2\mathbf{i} - 3.0\mathbf{j} + (\omega \times R)\mathbf{j}$$

Summing **i** *components*,

$$0.9986V_B\mathbf{i} = 5.2\mathbf{i}$$
$$V_B = 5.21 \text{ ft/sec}$$

Therefore,

$$\mathbf{V}_B = (0.9986)(5.21)\mathbf{i} + (0.0523)(5.21)\mathbf{j} = 5.2\mathbf{i} + 0.271\mathbf{j}$$

Summing **j** *components,*

$$\begin{aligned} 0.0523 V_B\mathbf{j} &= -3.0\mathbf{j} + (\omega \times R)\mathbf{j} \\ (0.0523)(5.21)\mathbf{j} &= -3.0\mathbf{j} + (\omega \times R)\mathbf{j} \\ (\omega \times R) &= 3.271 \text{ ft/sec} \end{aligned}$$

Therefore,

$$\boldsymbol{\omega} \times \mathbf{R} = 3.271\mathbf{j} \text{ ft/sec}$$

and

$$\omega = \omega_3 = \frac{3.271}{R} = \frac{3.271}{\frac{8}{12}} = 4.91 \text{ rad/sec (ccw)}$$

$$\omega_4 = \frac{V_B}{O_4B} = \frac{5.21}{\frac{8}{12}} = 7.82 \text{ rad/sec (ccw)}$$

The equation for the acceleration of point B can be written from Eq. 10.23 as follows:

$$\mathbf{A}_B = \mathbf{A}_0 + \mathbf{A} + 2\boldsymbol{\omega} \times \mathbf{V} + \dot{\boldsymbol{\omega}} \times \mathbf{R} + \boldsymbol{\omega} \times (\boldsymbol{\omega} \times \mathbf{R})$$

where

$$\begin{aligned} |\mathbf{A}_P^n| &= \frac{V_B^2}{O_4B} = \frac{5.21^2}{\frac{8}{12}} = 40.4 \text{ ft/sec}^2\text{, direction from } B \text{ toward } O_4 \\ \mathbf{A}_B^t &= \text{direction perpendicular to } O_4B\text{, magnitude unknown} \\ |\mathbf{A}_0| &= |\mathbf{A}_A| = |\mathbf{A}_A^n| = (O_2A)\omega_2^2 = \tfrac{3}{12} \times 24^2 = 144 \text{ ft/sec}^2\text{, direction from } A \text{ toward } O_2 (\mathbf{A}_A^t = 0) \\ \mathbf{A} &= 0 \text{ because } B \text{ is a fixed point in the } xy \text{ system} \\ 2\boldsymbol{\omega} \times \mathbf{V} &= 0 \text{ because } V = 0 \\ \dot{\boldsymbol{\omega}} \times \mathbf{R} &= \text{direction perpendicular to } AB\text{, magnitude unknown} \\ \boldsymbol{\omega} \times (\boldsymbol{\omega} \times \mathbf{R}) &= -\omega^2\mathbf{R} \text{ direction from } B \text{ toward } A \\ \boldsymbol{\omega} &= 4.91\mathbf{k} \text{ rad/sec from velocity solution} \\ \omega^2 R &= (4.91)^2 \times \tfrac{8}{12} = 16.1 \text{ ft/sec}^2 \end{aligned}$$

The direction of $\dot{\boldsymbol{\omega}} \times \mathbf{R}$ can be determined from the fact that the direction of the vector representing $\dot{\omega}$ will be perpendicular to the xy plane. When $\dot{\boldsymbol{\omega}}$ is crossed into $\mathbf{R}$, the product $\dot{\boldsymbol{\omega}} \times \mathbf{R}$ will be in the xy plane and perpendicular to $\mathbf{R}$. The direction of $\boldsymbol{\omega} \times (\boldsymbol{\omega} \times \mathbf{R})$ can be determined from Fig. 10.9 where ω is counterclockwise as determined from the velocity solution.

The equation for A_B is solved by unit vectors in the following manner:

II.

$$\begin{aligned} \mathbf{A}_B &= \mathbf{A}_0 + \mathbf{A} + 2\boldsymbol{\omega} \times \mathbf{V} + \dot{\boldsymbol{\omega}} \times \mathbf{R} + \boldsymbol{\omega} \times (\boldsymbol{\omega} \times \mathbf{R}) \\ \mathbf{A}_B^n &= A_B^n(-\sin 3°\mathbf{i} + \cos 3°\mathbf{j}) = 40.4(-0.0523\mathbf{i} + 0.9986\mathbf{j}) = -2.1\mathbf{i} + 40.3\mathbf{j} \end{aligned}$$

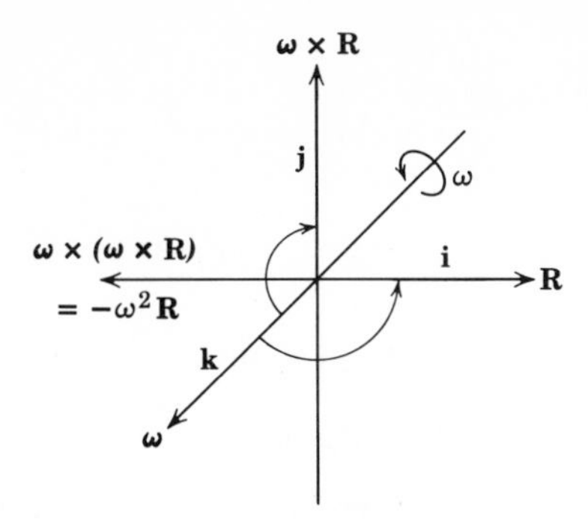

Figure 10.9

$$
\begin{aligned}
\mathbf{A}_B^t &= A_B^t(\cos 3°\mathbf{i} + \sin 3°\mathbf{j}) = 0.9986A_B^t\mathbf{i} + 0.0523A_B^t\mathbf{j} \\
\mathbf{A}_0 &= A_A = A_A^n(-\cos 60°\mathbf{i} - \sin 60°\mathbf{j}) \\
&= 144(-0.500\mathbf{i} - 0.8660\mathbf{j}) = -72\mathbf{i} - 124.8\mathbf{j} \\
\mathbf{A} &= 0 \\
2\boldsymbol{\omega} \times \mathbf{V} &= 0 \\
\dot{\boldsymbol{\omega}} \times \mathbf{R} &= (\dot{\omega} \times R)\mathbf{j} \\
\boldsymbol{\omega} \times (\boldsymbol{\omega} \times \mathbf{R}) &= -16.1\mathbf{i} \text{ ft/sec}^2
\end{aligned}
$$

Substituting the above relations into the equation for $\mathbf{A}_B$,

$$
\begin{aligned}
-2.1\mathbf{i} + 40.3\mathbf{j} + 0.9986A_B^t\mathbf{i} + 0.0523A_B^t\mathbf{j} = -72\mathbf{i} - 124.8\mathbf{j} \\
+ (\dot{\omega} \times R)\mathbf{j} - 16.1\mathbf{i}
\end{aligned}
$$

Summing **i** *components,*

$$
\begin{aligned}
-2.1\mathbf{i} + 0.9986A_B^t\mathbf{i} &= -72\mathbf{i} - 16.1\mathbf{i} \\
0.9986A_B^t\mathbf{i} &= -86.0\mathbf{i} \\
A_B^t &= -86.1 \text{ ft/sec}^2
\end{aligned}
$$

Therefore,

$$
\mathbf{A}_B^t = (0.9986)(-86.1)\mathbf{i} + (0.0523)(-86.1)\mathbf{j} = -86.0\mathbf{i} - 4.5\mathbf{j}
$$

Summing **j** *components,*

$$
\begin{aligned}
40.3\mathbf{j} + 0.0523A_B^t\mathbf{j} &= -124.8\mathbf{j} + (\dot{\omega} \times R)\mathbf{j} \\
40.3\mathbf{j} - 4.6\mathbf{j} &= -124.8\mathbf{j} + (\dot{\omega} \times R)\mathbf{j} \\
(\dot{\omega} \times R) &= 160.5 \text{ ft/sec}^2
\end{aligned}
$$

Therefore,

$$
\dot{\boldsymbol{\omega}} \times \mathbf{R} = 160.5\mathbf{j} \text{ ft/sec}^2
$$

and

$$\dot{\omega} = \alpha_3 = \frac{160.5}{R} = \frac{160.5}{\frac{8}{12}} = 241 \text{ rad/sec}^2 \quad \text{(ccw)}$$

$$\alpha_4 = \frac{A_B^t}{O_4B} = \frac{86.1}{\frac{8}{12}} = 129 \text{ rad/sec}^2 \quad \text{(cw)}$$

$$\mathbf{A}_B = \mathbf{A}_B^n + \mathbf{A}_B^t = -2.1\mathbf{i} + 40.3\mathbf{j} - 86.0\mathbf{i} - 4.5\mathbf{j} = -88.1\mathbf{i} + 35.8\mathbf{j}$$

$$|\mathbf{A}_B| = \sqrt{88.1^2 + 35.8^2} = 95.1 \text{ ft/sec}^2$$

In order to obtain a better understanding of the vectors involved in the velocity and acceleration analysis of the linkage of Example 10.1, a graphical solution of the vector equations will be given. The linkage of Fig. 10.7 is, therefore, redrawn in Fig. 10.10*a*, and polygons shown which give the magnitudes and directions of the vectors which were previously determined analytically. Figure 10.10*b* shows the graphical representation of the velocity equation

$$\mathbf{V}_B = \mathbf{V}_0 + \mathbf{V} + \boldsymbol{\omega} \times \mathbf{R}$$

where

$$\mathbf{V}_0 = \mathbf{V}_A$$

$$\mathbf{V}_A = 0$$

and therefore

$$\mathbf{V}_B = \mathbf{V}_A + \boldsymbol{\omega} \times \mathbf{R}$$

The addition of the vectors $\mathbf{V}_A$ and $\boldsymbol{\omega} \times \mathbf{R}$ to give $\mathbf{V}_B$ can easily be seen in the polygon of Fig. 10.10*b*.

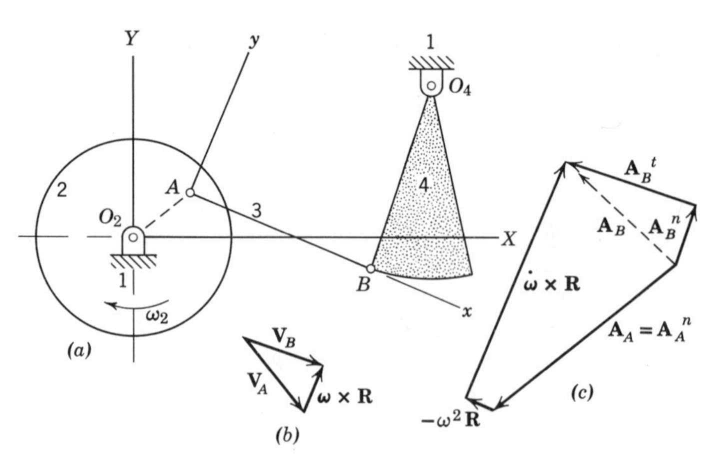

Figure 10.10

Figure 10.10*c* shows the graphical representation of the acceleration equation

$$\mathbf{A}_B = \mathbf{A}_0 + \mathbf{A} + 2\boldsymbol{\omega} \times \mathbf{V} + \dot{\boldsymbol{\omega}} \times \mathbf{R} + \boldsymbol{\omega} \times (\boldsymbol{\omega} \times \mathbf{R})$$

where

$$\mathbf{A}_B = \mathbf{A}_B^n + \mathbf{A}_B^t$$

$$\mathbf{A}_0 = \mathbf{A}_A = \mathbf{A}_A^n(\mathbf{A}_A^t = 0)$$

$$\mathbf{A} = 0$$

$$2\boldsymbol{\omega} \times \mathbf{V} = 0$$

$$\boldsymbol{\omega} \times (\boldsymbol{\omega} \times \mathbf{R}) = -\omega^2\mathbf{R}$$

which results in

$$\mathbf{A}_B^n + \mathbf{A}_B^t = \mathbf{A}_A + \dot{\boldsymbol{\omega}} \times \mathbf{R} - \omega^2\mathbf{R}$$

The addition of these vectors can easily be seen in the polygon of Fig. 10.10*c*.

In comparing the analytical solution with the graphical solution, it is obvious that the graphical is much quicker but less accurate. If the analysis of only one position is required, one would undoubtedly choose the graphical solution. If, however, the analysis of several positions or of a complete cycle is necessary, the analytical solution would be preferred possibly with the aid of a computer.

Example 10.2

As a second example consider the mechanism shown in Fig. 10.11 where the angular velocity of link 2 is constant, and it is required to find the angular velocity and angular acceleration of link 3. The coordinate system *xy* is fixed in link 3 as shown with its origin at point A_3. The system *XY* has its origin at point O_2.

The equation for the velocity of point A_3 cannot be evaluated directly from Eq. 10.16 because, by placing the origin of the *xy* system at point A_3, $\mathbf{V}_0$ equals $\mathbf{V}_{A_3}$, and an identity results. It is necessary, therefore, to write Eq. 10.16 for $\mathbf{V}_{A_2}$ as follows:

$$\mathbf{V}_{A_2} = \mathbf{V}_0 + \mathbf{V} + \boldsymbol{\omega} \times \mathbf{R}$$

where

$|\mathbf{V}_{A_2}| = (O_2A_2)\omega_2 = 2 \times 10 = 20$ in./sec, direction perpendicular to O_2A
$\mathbf{V}_0 = \mathbf{V}_{A_3} =$ direction perpendicular to O_3A_3, magnitude unknown
$\mathbf{V} =$ direction parallel to O_3A_3, magnitude unknown
$\boldsymbol{\omega} \times \mathbf{R} = 0$ because $R = 0$

The equation for $\mathbf{V}_{A_2}$ is solved by unit vectors with all components taken relative to the *xy* axes. ω_3 is calculated from $|\mathbf{V}_{A_3}|$.

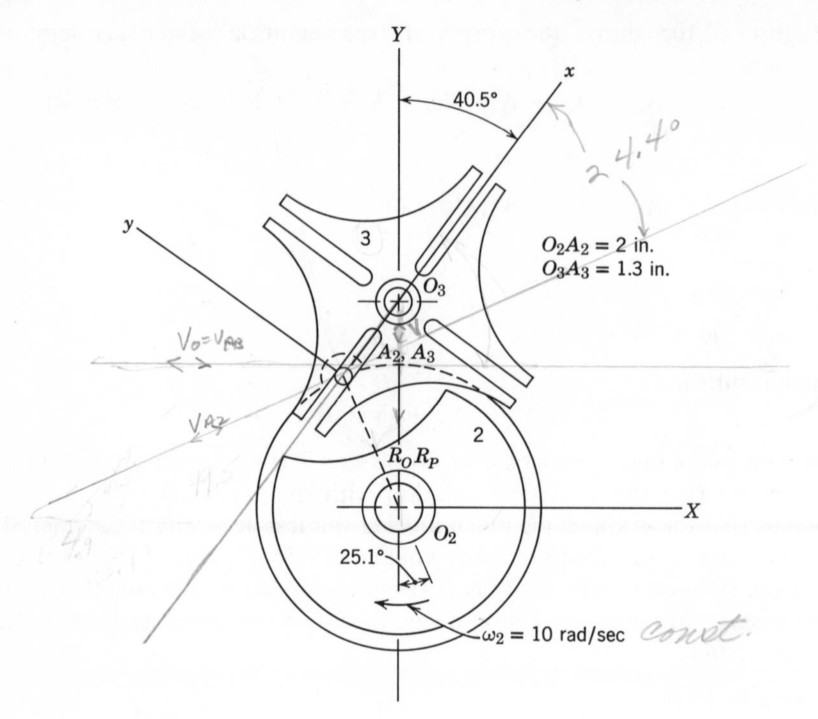

Figure 10.11

I.
$$\mathbf{V}_{A_2} = \mathbf{V}_0 + \mathbf{V} + \boldsymbol{\omega} \times \mathbf{R}$$
$$\mathbf{V}_{A_2} = V_{A_2}(\cos 24.4\mathbf{i} - \sin 24.4\mathbf{j}) = 20(0.9107\mathbf{i} - 0.4131\mathbf{j})$$
$$= 18.214\mathbf{i} - 8.262\mathbf{j}$$
$$\mathbf{V}_0 = \mathbf{V}_{A_3} = -V_{A_3}\mathbf{j}$$
$$\mathbf{V} = V\mathbf{i}$$
$$\boldsymbol{\omega} \times \mathbf{R} = 0$$

Substituting the above relations in the equation for $\mathbf{V}_{A_2}$

$$18.214\mathbf{i} - 8.262\mathbf{j} = -V_{A_3}\mathbf{j} + V\mathbf{i}$$

Summing **i** *components,*

$$18.214\mathbf{i} = V\mathbf{i}$$
$$V = 18.214 \text{ in./sec}$$

Therefore,

$$\mathbf{V} = 18.214\mathbf{i} \text{ in./sec}$$

Summing **j** *components,*

$$-8.262\mathbf{j} = -V_{A_3}\mathbf{j}$$

$$V_{A_3} = 8.262 \text{ in./sec}$$

Therefore,

$$\mathbf{V}_{A_3} = -8.262\mathbf{j} \text{ in./sec}$$

and

$$\omega_3 = \frac{V_{A_3}}{O_3A_3} = \frac{8.262}{1.3} = 6.35 \text{ rad/sec} \quad \text{(ccw)}$$

The equation for the acceleration $\mathbf{A}_{A_2}$ can be written from Eq. 10.23 as follows:

$$\mathbf{A}_{A_2} = \mathbf{A}_0 + \mathbf{A} + 2\boldsymbol{\omega} \times \mathbf{V} + \dot{\boldsymbol{\omega}} \times \mathbf{R} + \boldsymbol{\omega} \times (\boldsymbol{\omega} \times \mathbf{R})$$

where

$|\mathbf{A}^n_{A_2}| = (O_2A_2)\omega_2^2 = 2 \times 10^2 = 200$ in./sec^2, direction from A_2 toward O_2

$\mathbf{A}^t_{A_2} = 0$

$\mathbf{A}_0 = \mathbf{A}_{A_3}$

$|\mathbf{A}^n_{A_3}| = \dfrac{V_{A_3}^2}{O_3A_3} = \dfrac{8.262^2}{1.3} = 52.5$ in./sec, direction from A_3 toward O_3

$\mathbf{A}^t_{A_3} =$ direction perpendicular to O_3A_3, magnitude unknown

$\mathbf{A}^n = 0$ because radius of curvature is infinite (The path of point A_2 relative to the xy system is a straight line along the centerline of the slot.)

$\mathbf{A}^t =$ direction parallel to O_3A_3, magnitude unknown

$|2\boldsymbol{\omega} \times \mathbf{V}| = 2 \times 6.35 \times 18.214 = 232$ in./sec^2, direction along plus y axis (see Fig. 10.12)

$\dot{\boldsymbol{\omega}} \times \mathbf{R} = 0$ because $R = 0$

$\boldsymbol{\omega} \times (\boldsymbol{\omega} \times \mathbf{R}) = 0$ because $R = 0$

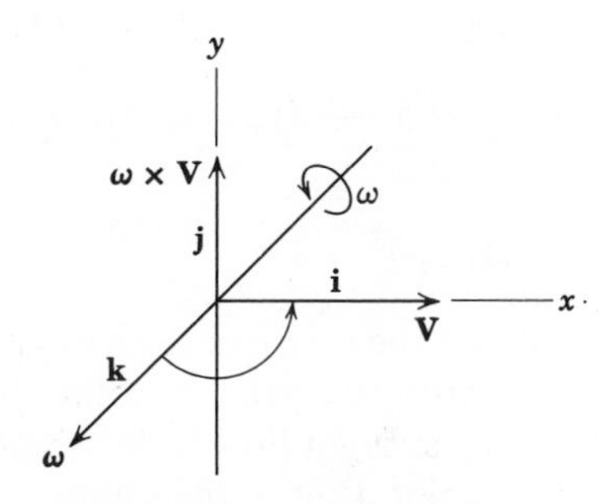

Figure 10.12

The equation for $\mathbf{A}_{A_2}$ is solved by unit vectors and α_3 calculated from $|\mathbf{A}^t_{A_3}|$.

$$
\begin{aligned}
\text{II.} \qquad \mathbf{A}_{A_2} &= \mathbf{A}_0 + \mathbf{A} + 2\boldsymbol{\omega} \times \mathbf{V} + \dot{\boldsymbol{\omega}} \times \mathbf{R} + \boldsymbol{\omega} \times (\boldsymbol{\omega} \times \mathbf{R}) \\
\mathbf{A}^n_{A_2} &= A^n_{A_2}(-\sin 24.4\mathbf{i} - \cos 24.4\mathbf{j}) = 200(-0.4131\mathbf{i} - 0.9107\mathbf{j}) \\
&= -82.62\mathbf{i} - 182.14\mathbf{j} \\
\mathbf{A}^t_{A_2} &= 0 \\
\mathbf{A}_0 &= \mathbf{A}_{A_3} \\
\mathbf{A}^n_{A_3} &= 52.5\mathbf{i} \text{ in./sec}^2 \\
\mathbf{A}^t_{A_3} &= A^t_{A_3}\mathbf{j} \text{ (assume as positive)} \\
\mathbf{A}^n &= 0 \\
\mathbf{A}^t &= A^t\mathbf{i} \\
2\boldsymbol{\omega} \times \mathbf{V} &= 232\mathbf{j} \text{ in./sec}^2 \\
\dot{\boldsymbol{\omega}} \times \mathbf{R} &= 0 \\
\boldsymbol{\omega} \times (\boldsymbol{\omega} \times \mathbf{R}) &= 0
\end{aligned}
$$

Substituting the above relations in the equation for $\mathbf{A}_{A_2}$

$$-82.62\mathbf{i} - 182.14\mathbf{j} = 52.5\mathbf{i} + A^t_{A_3}\mathbf{j} + A^t\mathbf{i} + 232\mathbf{j}$$

Summing **i** *components,*

$$-82.62\mathbf{i} = 52.5\mathbf{i} + A^t\mathbf{i}$$

$$A^t = -135.1 \text{ in./sec}^2$$

Therefore,

$$\mathbf{A}^t = -135.1\mathbf{i} \text{ in./sec}^2$$

Summing **j** *components,*

$$-182.14\mathbf{j} = A^t_{A_3}\mathbf{j} + 232\mathbf{j}$$

$$A^t_{A_3} = -413 \text{ in./sec}^2$$

Therefore,

$$\mathbf{A}^t_{A_3} = -413\mathbf{j} \text{ in./sec}^2$$

$$\mathbf{A}_{A_3} = \mathbf{A}^n_{A_3} + \mathbf{A}^t_{A_3} = 52.5\mathbf{i} - 413\mathbf{j}$$

$$|\mathbf{A}_{A_3}| = \sqrt{52.5^2 + 413^2} = 416 \text{ in./sec}^2$$

$$\alpha_3 = \frac{A^t_{A_3}}{O_3A_3} = \frac{413}{1.3} = 318 \text{ rad/sec}^2 \quad \text{(ccw)}$$

It should be mentioned that the origin of the xy system was taken at point A_3 with point A_2 as P because the path of point A_2 relative to point A_3 (and hence the xy system) is a straight line. If the origin of the xy system had been taken at point A_2 with point A_3 as P, the solution would have been more difficult because the path of A_3 relative to A_2 is not readily known.

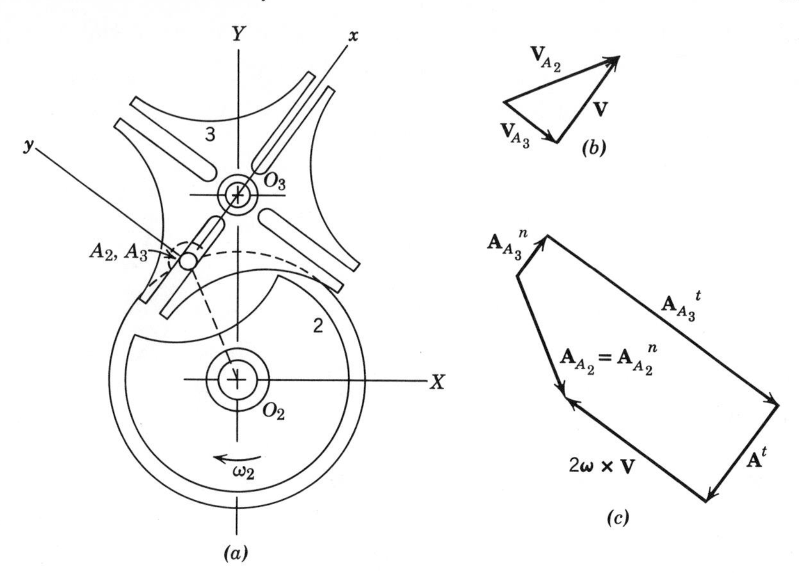

Figure 10.13

In order to present the graphical solution of Example 10.2, the linkage of Fig. 10.11 is redrawn in Fig. 10.13*a*. Figure 10.13*b* shows the graphical representation of the velocity equation

$$\mathbf{V}_{A_2} = \mathbf{V}_0 + \mathbf{V} + \boldsymbol{\omega} \times \mathbf{R}$$

where

$$\mathbf{V}_0 = \mathbf{V}_{A_3}$$

$$\boldsymbol{\omega} \times \mathbf{R} = 0$$

and therefore

$$\mathbf{V}_{A_2} = \mathbf{V}_{A_3} + \mathbf{V}$$

The addition of the vectors $\mathbf{V}_{A_3}$ and $\mathbf{V}$ to give $\mathbf{V}_{A_2}$ can easily be seen in the polygon of Fig. 10.13*b*. The value of $\boldsymbol{\omega}_3$ is calculated in the same manner as in the analytical solution, namely

$$\omega_3 = \frac{V_{A_3}}{O_3A_3} = 6.35 \text{ rad/sec} \quad \text{(ccw)}$$

Figure 10.13*c* shows the graphical representation of the acceleration equation

$$\mathbf{A}_{A_2} = \mathbf{A}_0 + \mathbf{A} + 2\boldsymbol{\omega} \times \mathbf{V} + \dot{\boldsymbol{\omega}} \times \mathbf{R} + \boldsymbol{\omega} \times (\boldsymbol{\omega} \times \mathbf{R})$$

where

$$\mathbf{A}_{A_2} = \mathbf{A}^n_{A_2}(\mathbf{A}^t_{A_2} = 0)$$
$$\mathbf{A}_0 = \mathbf{A}_{A_3} = \mathbf{A}^n_{A_3} + \mathbf{A}^t_{A_3}$$
$$\mathbf{A} = \mathbf{A}^t(\mathbf{A}^n = 0)$$
$$\dot{\boldsymbol{\omega}} \times \mathbf{R} = 0$$
$$\boldsymbol{\omega} \times (\boldsymbol{\omega} \times \mathbf{R}) = 0$$

and, therefore,

$$\mathbf{A}^n_{A_2} = \mathbf{A}^n_{A_3} + \mathbf{A}^t_{A_3} + \mathbf{A}^t + 2\boldsymbol{\omega} \times \mathbf{V}$$

The addition of the four vectors to give $\mathbf{A}_{A_2}$ can easily be seen in the polygon of Fig. 10.13*c*. The value of $\boldsymbol{\alpha}_3$ is calculated in the same manner as in the analytical solution, namely

$$\alpha_3 = \frac{A^t_{A_3}}{O_3A_3} = 318 \text{ rad/sec}^2 \quad \text{(ccw)}$$

10.7 Graphical Determination of Velocity in Mechanisms. By graphical methods the determination of the linear velocities of all particles of a mechanism may be quickly determined with relatively little calculation as illustrated in several examples that follow. However, a fundamental insight on the relative motion of the particles in the mechanism is needed.

In Fig. 10.14 are shown three types of linkages, in which the driving link (link 2) is the same, but in which the motion transmitted to the driven link depends on a different type of constraint. In Fig. 10.14*a* motion constraint is achieved through pin connections, in 10.14*b* by sliding in a guide, and in 10.14*c* by rolling contact. The absolute velocity of any particle on link 2 is quickly determined if the driving angular velocity ω_2 is known. The magnitude of $\mathbf{V}_A$, for example, may be calculated from Eq. 10.1*a*.

$$|\mathbf{V}_A| = R\omega_r$$
$$= (O_2A)\omega_2$$

The direction of $\mathbf{V}_A$ is known to be tangent to the circular path of A, and the sense of $\mathbf{V}_A$ is known from the sense of ω_2. However, to determine the linear velocity of any particle on the driven links or followers, a knowledge of the relative motion of pairs of particles is required.

10.8 Relative Velocity of Particles in Mechanisms. Referring to Eq. 10.5 and Fig. 10.4, the relative velocity $\mathbf{V}_{PQ}$ of one particle relative to another may be determined from the vector difference of the absolute velocities $\mathbf{V}_P$ and $\mathbf{V}_Q$ provided the absolute velocities are known. However, in a linkage analysis, only one of the absolute velocities is usually known and the other is to be determined. The unknown absolute velocity $\mathbf{V}_P$,

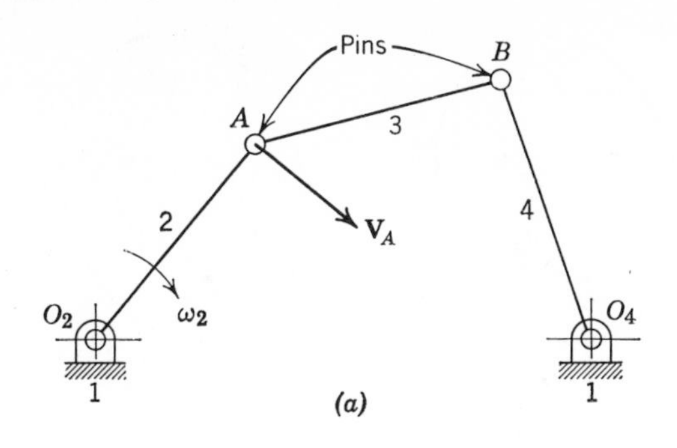

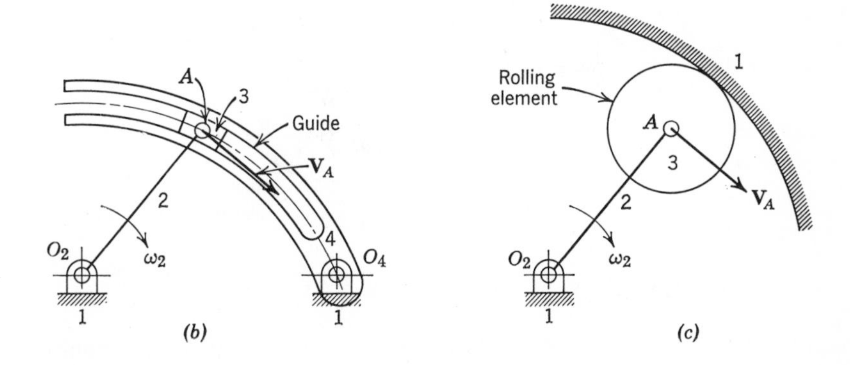

Figure 10.14

for example, may be determined from Eq. 10.5 in the following form:

$$\mathbf{V}_P = \mathbf{V}_Q + \mathbf{V}_{PQ} \tag{10.24}$$

Although $\mathbf{V}_Q$ may be known, it is necessary that the relative velocity $\mathbf{V}_{PQ}$ also be known. In linkages the motions of particles P and Q are not independent as in Fig. 10.4 but are constrained relative to each other so that their relative motion is controlled. In the following section, the basic types of motion constraint are discussed to show the determination of the magnitude, direction, and sense of $\mathbf{V}_{PQ}$.

10.9 Relative Velocity of Particles in a Common Link. Considering the rigid body (link 3) in Fig. 10.15*a*, any particle such as Q may be at the absolute velocity $\mathbf{V}_Q$ and the link at an absolute angular velocity ω_3. If observations are made relative to Q, then Q is at rest as shown in Fig. 10.15*b*. However, since particle Q has no angular motion, the angular velocity ω_3 of the link relative to Q is unchanged. Therefore, as in Fig. 10.15*b*,

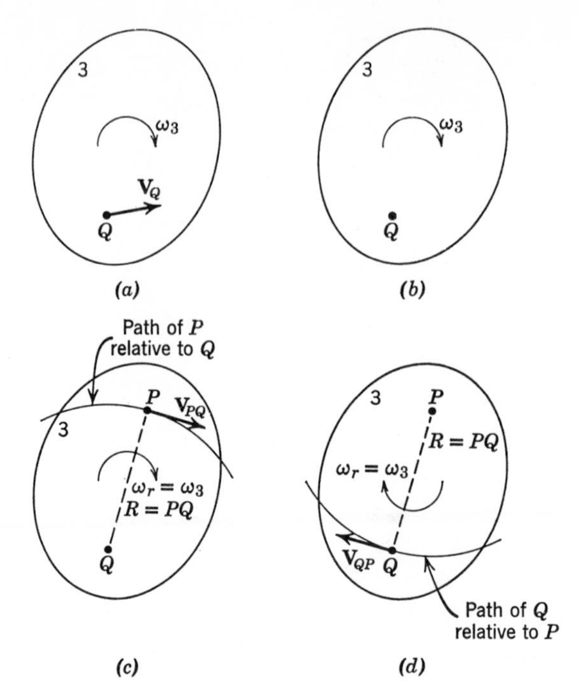

Figure 10.15

relative to Q, the link rotates at the absolute angular velocity ω_3 about Q as if Q were a fixed center.

Relative to Q, any other particle in the link such as P is constrained to move on a circular path as shown in Fig. 10.15c because the link is a rigid body and the distance PQ is fixed. The relative velocity $\mathbf{V}_{PQ}$ of P relative to Q is tangent to the relative path as shown. Since the radius of curvature R of the relative path is equal to PQ and the angular velocity of the radius of curvature ω_r is equal to ω_3, the magnitude of $\mathbf{V}_{PQ}$ may be determined from Eq. 10.1a as follows:

$$|\mathbf{V}_{PQ}| = (PQ)\omega_3 \tag{10.25}$$

Because the relative path is circular, dR/dt is zero.

It is to be observed from Eq. 10.25 that if the link has no absolute angular velocity, the relative velocity $\mathbf{V}_{PQ}$ of any two particles of the link is zero. The link is then in pure translation, and the absolute velocities of all particles of the link are identical.

In Fig. 10.15c, the direction of $\mathbf{V}_{PQ}$ is tangent to the relative circular path and is shown as a fixed vector at P. The sense of $\mathbf{V}_{PQ}$ is determined by making its turning sense about Q the same as the sense of ω_3. In Fig. 10.15d is shown the vector $\mathbf{V}_{QP}$ denoting the velocity of Q relative to P. It may be seen that relative to P the angular velocity ω_3 of link 3 is the same in magnitude and sense as relative to Q. Therefore, the magnitudes of $\mathbf{V}_{QP}$ and $\mathbf{V}_{PQ}$ are the same. Their directions are also the same since both are normal to the line PQ. However, the sense of $\mathbf{V}_{QP}$ is opposite to that of $\mathbf{V}_{PQ}$.

As illustrated in the following example, Eqs. 10.24 and 10.25 and the knowledge of the direction and sense of the relative velocity of two particles in a given link are necessary in the kinematic analysis of mechanisms.

Example 10.3

Link 2 of the four-bar linkage of Fig. 10.16*a* is the driving link having a uniform angular velocity ω_2 of 30 rad/sec. For the phase shown, draw the velocity polygon and determine the velocity $\mathbf{V}_B$ of point B, the angular velocities ω_3 and ω_4, and the relative angular velocities ω_{32} and ω_{43}. Also determine the velocity images of all links to show how the linear velocity of any point in the linkage may be determined. Velocity equations can be written as follows:

I. $\mathbf{V}_B = \mathbf{V}_A + \mathbf{V}_{BA}$

where

V_B = direction perpendicular to O_4B, magnitude unknown
$V_A = (O_2A)\omega_2 = (10.2)30 = 306$ cm/sec, direction perpendicular to O_2A
V_{BA} = direction perpendicular to BA, magnitude unknown

Measured on polygon $V_B = 152$ cm/sec and $V_{BA} = 318$ cm/sec

$$\omega_3 = \frac{V_{BA}}{BA} = \frac{318}{20.3} = 15.7 \text{ rad/sec} \quad \text{(ccw)}$$

$$\omega_4 = \frac{V_B}{O_4B} = \frac{152}{7.62} = 19.9 \text{ rad/sec} \quad \text{(ccw)}$$

$$\omega_{32} = \omega_3 - \omega_2 = 15.7 - (-30) = 45.7 \text{ rad/sec} \quad \text{(ccw)}$$

$$\omega_{43} = \omega_4 - \omega_3 = 19.9 - 15.7 = 4.2 \text{ rad/sec} \quad \text{(ccw)}$$

II. $\mathbf{V}_C = \mathbf{V}_A + \mathbf{V}_{CA}$
III. $\mathbf{V}_C = \mathbf{V}_B + \mathbf{V}_{CB}$

where

V_C = direction unknown, magnitude unknown
V_{CA} = direction perpendicular to CA, magnitude unknown
V_{CB} = direction perpendicular to CB, magnitude unknown

Measured on polygon $V_C = 305$ cm/sec, $V_{CA} = 160$ cm/sec, $V_{CP} = 239$ cm/sec

Solution. Equation I expresses $\mathbf{V}_B$ in terms of $\mathbf{V}_A$ and $\mathbf{V}_{BA}$. As indicated, the components $\mathbf{V}_B$ and $\mathbf{V}_{BA}$ are known only in direction while $\mathbf{V}_A$ is known in magnitude, sense, and direction. In constructing the velocity polygon Fig. 10.16*b* starting with the right side of Eq. I, the vector $\mathbf{V}_A$ is drawn from pole O_v, and its tip is labeled "*A*". Next, add the direction of $\mathbf{V}_{BA}$ starting at point *A*. As can be seen, it is impossible to complete the solution using only these two components. Therefore, consider the left side of the equation and draw the direction of $\mathbf{V}_B$ from O_v. The intersection of the direction of $\mathbf{V}_B$ and the direction of $\mathbf{V}_{BA}$ completes the polygon. Arrowheads are now added to the vectors $\mathbf{V}_B$ and $\mathbf{V}_{BA}$ so that the addition of the vectors of the polygon agrees with the addition of the terms of Eq. I. The tip of the vector $\mathbf{V}_B$ is is labeled "*B*".

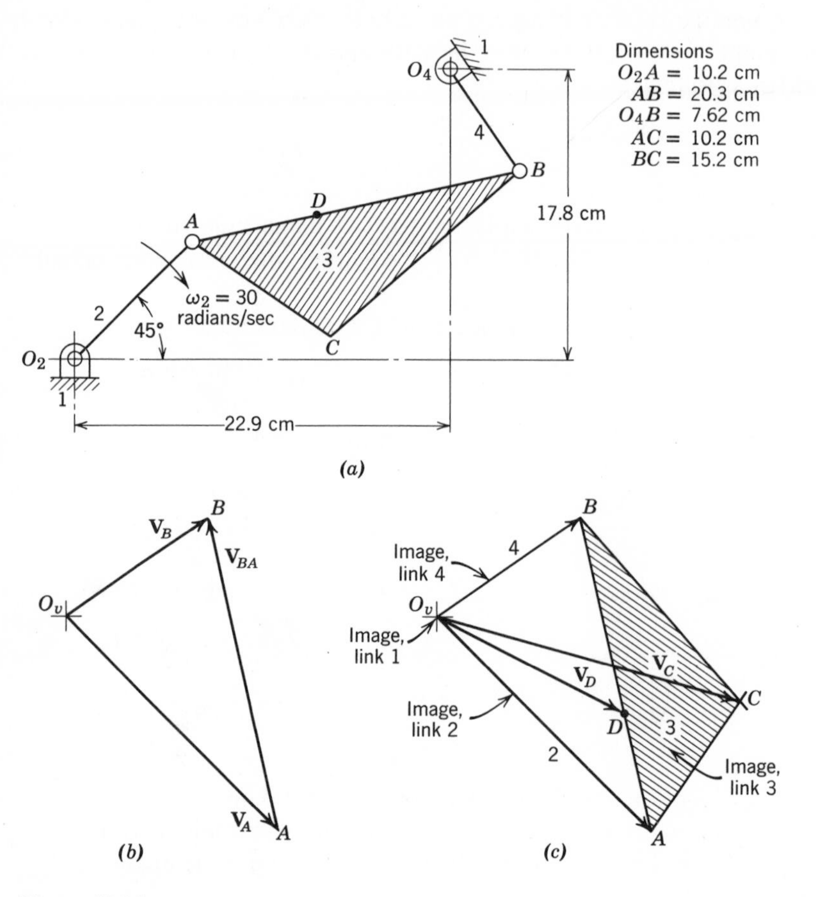

Figure 10.16

The magnitudes and senses of ω_3 and ω_4 can now be determined from V_{BA} and V_B, respectively, as shown. The values of ω_{32} and ω_{43} can also be determined as indicated.

To determine $\mathbf{V}_C$ it is necessary to use Eqs. II and III, which give the relations between $\mathbf{V}_C$ and $\mathbf{V}_A$ and $\mathbf{V}_B$. The directions of $\mathbf{V}_{CA}$ and $\mathbf{V}_{CB}$ are known as indicated. The velocity vectors $\mathbf{V}_A$ and $\mathbf{V}_B$ are redrawn in Fig. 10.16*c* to give a clearer diagram. Use Eq. II and draw the direction of the vector $\mathbf{V}_{CA}$ from point A in Fig. 10.16*c*. Next, consider Eq. III and draw the direction of the vector $\mathbf{V}_{CB}$ from point B. The intersection of the direction of $\mathbf{V}_{CA}$ and the direction of $\mathbf{V}_{CB}$ completes the polygon. This intersection is point C, which gives $\mathbf{V}_C$. The vector addition in the polygon is checked to see that it agrees with that of Eqs. II and III.

The shaded triangle ABC of Fig. 10.16*c* is known as the velocity image of link 3, and as such has the same shape as link 3. The velocity of any point D as shown on link 3 can be determined by locating its corresponding position on the velocity image of link 3. The vector from O_v to D is $\mathbf{V}_D$ (208 cm/sec) as shown in Fig. 10.16*c*. The velocity image of link 1 is at the pole O_v because link 1 is fixed and has zero velocity. The images of links 2 and 4 are lines O_vA and O_vB, respectively, which correspond to O_2A and O_4B, respectively, in the configuration diagram.

In the analysis above, the angular velocity ω_3 was determined from the relation

$$\omega_3 = \frac{V_{BA}}{BA}$$

It should also be mentioned that after the velocity image of link 3 has been completed, ω_3 can also be found from

$$\omega_3 = \frac{V_{CA}}{CA} = \frac{V_{CB}}{CB} = \frac{V_{DA}}{DA}$$

In other words, all relative velocities of points on a link are proportional to the distances between these points.

10.10 Relative Velocity of Coincident Particles on Separate Links. In many mechanisms such as in Fig. 10.14*b*, constraint of relative motion is achieved by guiding a particle P on one link along a prescribed path relative to another link by a guiding surface. Such constraint is to be found in cams and the inversions of the slider-crank, where a surface on one link controls the motion of a particle on another link by relative sliding or rolling.

In Fig. 10.17, particle P_3 on link 3 is in motion along a curvilinear path traced on link 2 because of the guiding slot in link 2. The path of P_3 relative to link 2 is shown with tangent t–t and normal n–n constructed at P_3. Consider

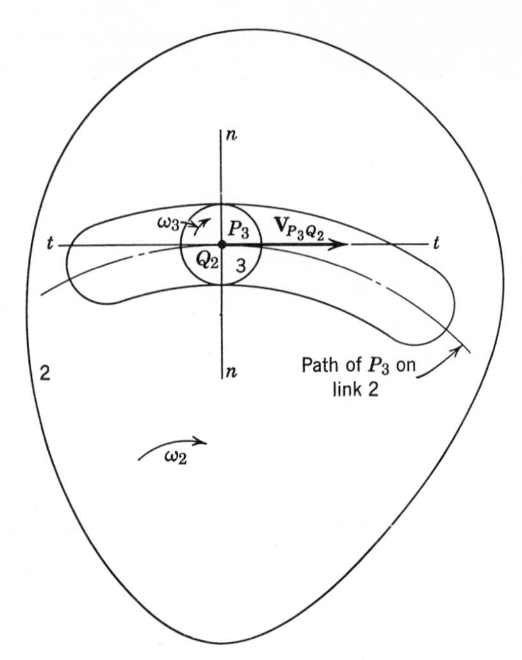

Figure 10.17

a particle Q_2 on link 2 which is coincident in position with particle P_3 on link 3. It may be seen that regardless of the absolute angular velocities ω_2 and ω_3 of links 2 and 3, the guide constrains the motion of P_3 so that it cannot displace relative to Q_2 in the normal direction n–n, and therefore there cannot be a relative velocity of the two particles in this direction. However, the guide permits freedom for particle P_3 to displace relative to Q_2 in the tangential direction t–t, and therefore the relative velocity $\mathbf{V}_{P_3Q_2}$ can be only in the tangential direction of the guide.

In mechanisms where guiding constraint is utilized, the knowledge that the relative velocity of coincident particles can be only in the tangential direction of the guide is sufficient to solve velocity problems as illustrated in the following example.

Example 10.4

The disk cam of Fig. 10.18*a* drives an oscillating roller follower and a radial point follower simultaneously. The cam rotates counterclockwise at a constant angular velocity ω_2 of 10 rad/sec. Springs (not shown) are used to maintain contact of the followers with the cam. For the phase shown, determine the velocity $\mathbf{V}_{A_4}$ of point A_4 on the oscillating follower and the velocity $\mathbf{V}_{B_5}$ of point B_5 on the point follower. Velocity equations can be written as follows:

I. $\mathbf{V}_{A_4} = \mathbf{V}_{A_2} + \mathbf{V}_{A_4A_2}$

where

V_{A_4} = direction perpendicular to O_4A_4, magnitude unknown
$V_{A_2} = (O_2A_2)\omega_2 = (2.5)10 = 25$ in./sec, direction perpendicular to O_2A_2
$V_{A_4A_2}$ = direction parallel to straight side of cam, magnitude unknown

Measured on polygon of Fig. 10.18*b* $V_{A_4} = 12.3$ in./sec and $V_{A_4A_2} = 26.3$ in./sec

II. $\mathbf{V}_{B_5} = \mathbf{V}_{B_2} + \mathbf{V}_{B_5B_2}$

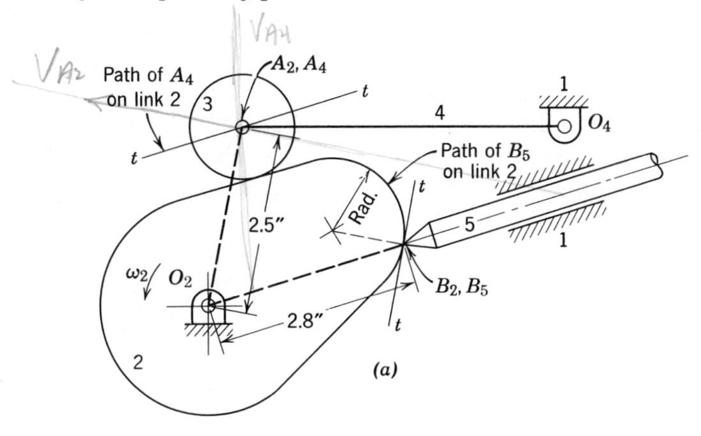

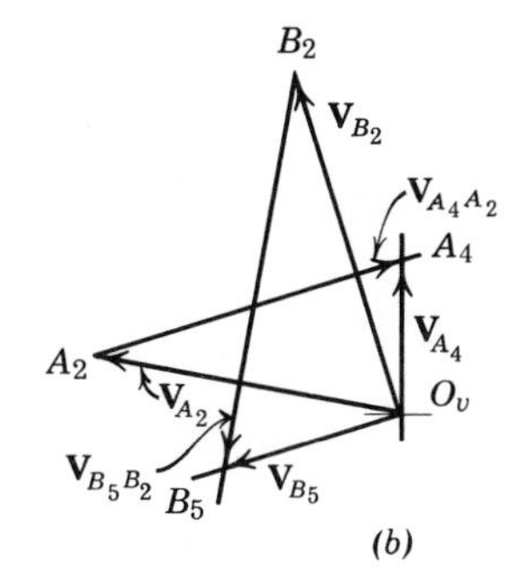

Figure 10.18

where

V_{B_5} = direction along centerline of follower 5, magnitude unknown

V_{B_2} = $(O_2B_2)\omega_2 = (2.8)10 = 28$ in./sec, direction perpendicular to O_2B_2

$V_{B_5B_2}$ = direction along tangent to cam contour at point B_2, magnitude unknown

Measured on polygon of Fig. 10.18*b*, $V_{B_5} = 14.7$ in./sec and $V_{B_5B_2} = 31.6$ in./sec.

Solution. Considering first the oscillating follower link 4, it can be seen that the straight side of the cam 2 is a guiding surface which constrains point A_4 on link 4 to follow a straight-line path relative to link 2. Point A_2 on link 2 and point A_4 on link 4 are coincident, and Eq. I shows the relation of their velocities. As indicated, the components, $\mathbf{V}_{A_4}$ and $\mathbf{V}_{A_4A_2}$ are known in direction while $\mathbf{V}_{A_2}$ is known in magnitude, sense, and direction.

The construction of the velocity polygon of Fig. 10.18*b* is started with the right side of Eq. I, and the vector $\mathbf{V}_{A_2}$ is drawn from pole O_v with its tip labeled "A_2". Next, add the direction of $\mathbf{V}_{A_4A_2}$ starting at point A_2. Because it is impossible to complete the solution using only these two components, consider the left side of the equation and draw the direction of $\mathbf{V}_{A_4}$ from O_v. The intersection of the direction of $\mathbf{V}_{A_4}$ and the direction of $\mathbf{V}_{A_4A_2}$ completes the polygon. Arrowheads are now added to the vectors $\mathbf{V}_{A_4}$ and $\mathbf{V}_{A_4A_2}$ so that the addition of the vectors of the polygon agrees with the addition of the terms of Eq. I. The tip of vector $\mathbf{V}_{A_4}$ is labeled "A_4".

The velocity polygon for the determination of the velocity $\mathbf{V}_{B_5}$ of the point follower can be drawn in a similar manner from Eq. II. Points B_2 and B_5 are coincident, and, as indicated, the components $\mathbf{V}_{B_5}$ and $\mathbf{V}_{B_5B_2}$ are known

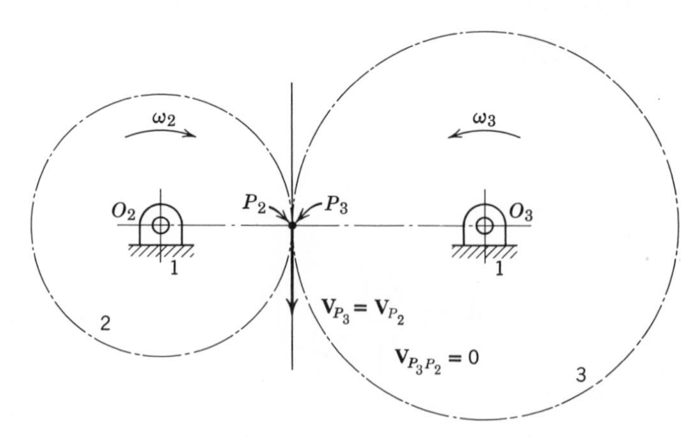

Figure 10.19

in direction while $\mathbf{V}_{B_2}$ is known in magnitude, sense, and direction. Figure 10.18*b* shows the polygon drawn from the same pole point O_v as was the polygon for the determination of $\mathbf{V}_{A_4}$.

10.11 Relative Velocity of Coincident Particles at the Point of Contact of Rolling Elements. A third type of constraint in mechanisms is that which occurs because one link is constrained to roll on another link without slipping at the point of contact. In Fig. 10.19 are shown the rolling pitch circles of a pair of gears in mesh with the coincident particles at the point of contact, P_3 on link 3 and P_2 on link 2. Because the circles have pure rolling contact, these particles have identical velocities so that $\mathbf{V}_{P_3} = \mathbf{V}_{P_2}$, and the relative velocity between the two particles will be zero. Example 10.5 which follows illustrates the use of this principle.

Example 10.5

In Fig. 10.20*a* is shown a mechanism consisting of three bars, two gears, and a rack. The velocity $\mathbf{V}_A$ of point A is 400 ft/sec in the direction shown. Determine the angular velocities ω_4 and ω_5 of the two gears and show the velocity images of the two gears. Determine also the velocity $\mathbf{V}_D$ of point D on gear 5. Velocity equations can be written as follows:

I. $\mathbf{V}_B = \mathbf{V}_A + \mathbf{V}_{BA}$

where

$$\begin{aligned} V_B &= \text{direction parallel to pitch line of rack, magnitude unknown} \\ V_A &= \text{400 ft/sec (given), direction perpendicular to } O_2A \\ V_{BA} &= \text{direction perpendicular to } BA\text{, magnitude unknown} \end{aligned}$$

Measured on polygon of Fig. 10.20*b* $V_B = 342$ ft/sec and $V_{BA} = 382$ ft/sec

II. $\mathbf{V}_C = \mathbf{V}_B + \mathbf{V}_{CB}$

where

$$\begin{aligned} V_C &= \text{direction perpendicular to } O_6C\text{, magnitude unknown} \\ V_{CB} &= \text{direction perpendicular to } CB\text{, magnitude unknown} \end{aligned}$$

Measured on polygon of Fig. 10.20*b* $V_C = 120$ ft/sec and $V_{CB} = 366$ ft/sec
Measured on polygon of Fig. 10.20*c*

$$V_{BP_4} = V_B = 342 \text{ ft/sec}, \qquad V_{CM_5} = 680 \text{ ft/sec}$$

$$V_{M_5} = V_{M_4} = 675 \text{ ft/sec}, \qquad V_D = 705 \text{ ft/sec}$$

$$\omega_4 = \frac{V_{BP_4}}{BP} = \frac{342}{\frac{4}{12}} = 1026 \text{ rad/sec} \quad \text{(cw)}$$

$$\omega_5 = \frac{V_{CM_5}}{CM} = \frac{680}{\frac{2}{12}} = 4080 \text{ rad/sec} \quad \text{(ccw)}$$

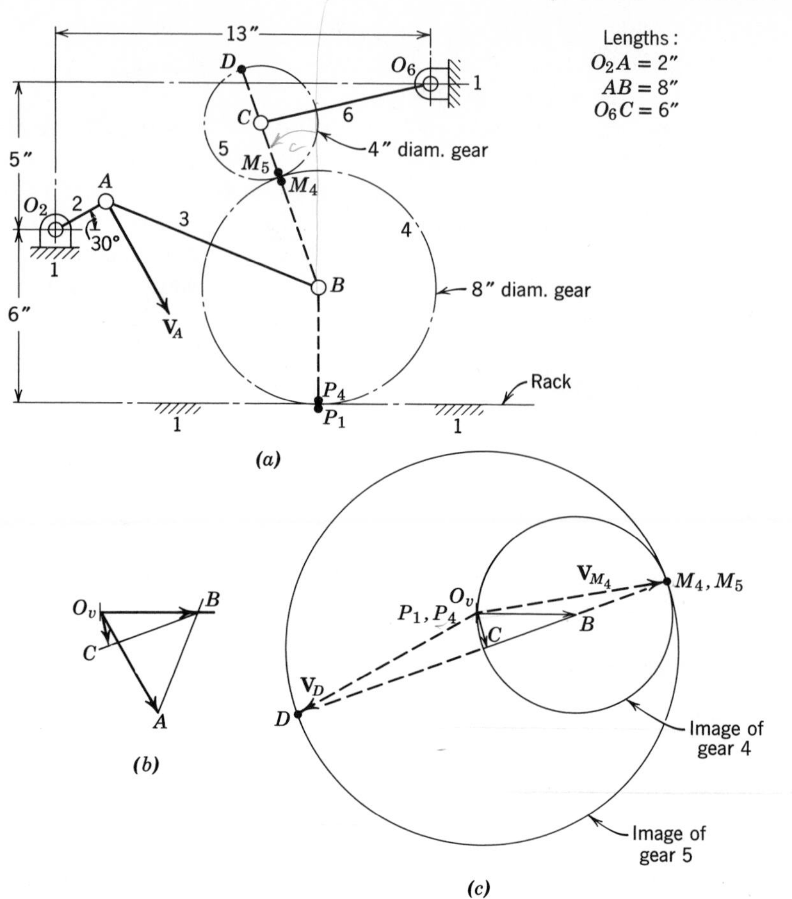

Figure 10.20

Solution. Since the distance BC between the centers of the two gears is constant in all phases of the mechanism, an equivalent link joining the two centers may be visualized. Therefore, a five-bar linkage is first analyzed to determine the velocities $\mathbf{V}_B$ and $\mathbf{V}_C$ of the centers of the gears. The velocity polygon of Fig. 10.20*b* shows the determination of $\mathbf{V}_B$ and $\mathbf{V}_{BA}$ from Eq. I. In a similar manner $\mathbf{V}_C$ and $\mathbf{V}_{CB}$ are determined from Eq. II.

In Fig. 10.20*c*, the velocity vectors $\mathbf{V}_B$ and $\mathbf{V}_C$ of Fig. 10.20*b* are redrawn for the construction of the velocity images of gears 4 and 5. Because the velocity $\mathbf{V}_{P_1}$ of point P_1 is zero and $\mathbf{V}_{P_4} = \mathbf{V}_{P_1}(\mathbf{V}_{P_4P_1} = 0)$ the image of both points P_1 and P_4 is at the pole point O_v as shown. With point P_4 located on the polygon, the velocity image of gear 4 is drawn with B as a center and radius BP_4. The image of point M_4 on the circle is determined by drawing a line through B on the polygon perpendicular to the line M_4B on the configuration diagram. The image of point M_5 is the same as that of point M_4 because $\mathbf{V}_{M_5} = \mathbf{V}_{M_4}$. The image of gear 5 is, therefore, drawn with C as a center and radius CM_5. The image of point D is located on a diameter of the circle opposite point M_5.

The magnitudes and senses of ω_4 and ω_5 can now be determined from $\mathbf{V}_{BP_4}$ and $\mathbf{V}_{CM_5}$, respectively, as shown.

10.12 Instantaneous Centers of Velocity. In the foregoing paragraphs and examples, the velocity analyses of linkages were made from an understanding of relative velocity and the influence of motion constraint on relative velocity. In the following, another concept is utilized to determine the linear velocity of particles in mechanisms, namely, the concept of the instantaneous center of velocity. This concept is based on the fact that at a given instant a pair of coincident points on two links in motion will have identical velocities relative to a fixed link and, therefore, will have zero velocity relative to each other. At this instant either link will have pure rotation relative to the other link about the coincident points. A special case of this is where one link is moving and the other is fixed. A pair of coincident points on these two links will then have zero absolute velocity, and the moving link at this instant will be rotating relative to the fixed link about the coincident points. In both cases the coincident set of points is referred to as an *instantaneous center of velocity*. From the foregoing, it can be seen that an instantaneous center is (*a*) a point in both bodies, (*b*) a point at which the two bodies have no relative velocity, and (*c*) a point about which one body may be considered to rotate relative to the other body at a given instant. It is easily seen that when two links, either both moving or one moving and one fixed, are directly connected together, the center of the connecting joint is an instantaneous center for the two links. When two links, either both moving or one moving and one fixed, are not directly connected, however, an instant center for the two links will also exist for a given phase of the linkage as will be shown in the following section.

In the four-bar linkage of Fig. 10.21, it is obvious that relative to the fixed link, points O_2 and O_4 are locations of particles on links 2 and 4, respectively, which are at zero velocity. It is less obvious that on link 3, which has both translating and angular motion, a particle is also at zero velocity relative to the fixed link. Referring to the velocity polygon shown in Fig. 10.21, the velocity image of link 3 appears as the line AB and none of the particles on this line is at zero velocity. However, if link 3 is visualized large enough in extent as a rigid body to include O_v of the polygon, a particle of zero velocity is then included in the image. To determine the location of O_v, the instantaneous center of link 3 relative to link 1, on the mechanism, a triangle

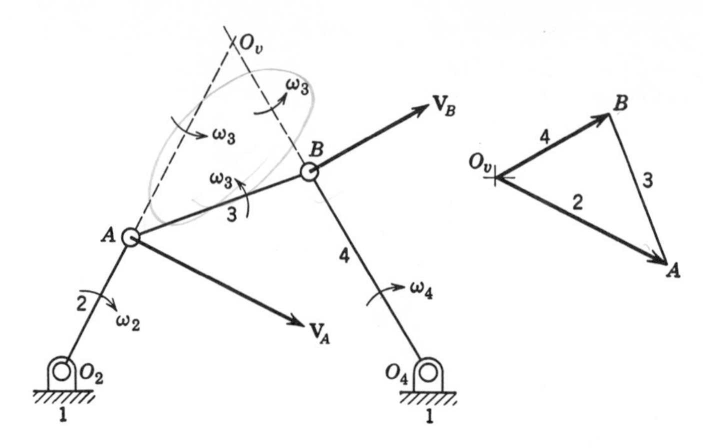

Figure 10.21

similar to O_vBA of the polygon is constructed on the mechanism so that the sides of the two similar triangles are mutually perpendicular. It is important to note that for the particles on link 3 at A and at B, the fixed vectors $\mathbf{V}_A$ and $\mathbf{V}_B$ on link 3 are normal to the lines drawn from the instantaneous center O_v to A and B.

Since A and the instantaneous center O_v are particles on a common rigid link, the magnitude of $\mathbf{V}_A$ may be determined from $V_A = \omega_3(O_vA)$. Similarly, $V_B = \omega_3(O_vB)$. The magnitude of the velocity of any particle on link 3 may be determined from the product of ω_3 and the radial distance from the instantaneous center to the particle, and the direction of the velocity vector is normal to the radial line.

It may also be seen that the instantaneous center of link 3 relative to link 1 changes position with respect to time because of the changes in the shape of the velocity polygon as the mechanism passes through a cycle of phases. However, for links in pure rotation the instantaneous centers are fixed centers, such as O_2 and O_4 of links 2 and 4 of Fig. 10.21.

The determination of velocities by instantaneous centers does not require the velocity polygon of free vectors and is judged by many to be the quicker method. By the method of instantaneous centers, the velocity vectors are shown directly as fixed vectors.

In the solution of a problem, such as in Fig. 10.22, the locations of the instantaneous centers of the moving links relative to the fixed link are generally determined first. For links 2 and 4, O_2 and O_4 are obviously points of zero velocity. For links such as link 3, only the *directions* of the velocities of two particles on the link need to be known since the intersection of the normals to the velocity direction lines determines the instantaneous center.

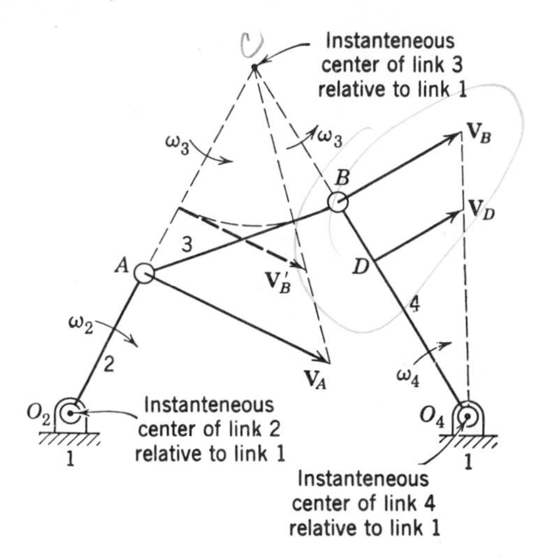

Figure 10.22

Fixed velocity vectors may be determined almost entirely by graphical construction. In Fig. 10.22, assuming ω_2 is the only information given, V_A may be computed from $\omega_2(O_2A)$ and $\mathbf{V}_A$ drawn normal to O_2A using the instantaneous center of link 2 relative to link 1. Considering particles A and B as part of link 3, the magnitude of $\mathbf{V}_B$ may be determined from similar triangles, as shown by the graphical construction, since V_A and V_B are proportional to the distance of A and B from the instantaneous center of link 3 relative to link 1. The equation which justifies the use of similar triangles in determining $\mathbf{V}_B$ may be written as $\omega_3 = V_A/(CA) = V_B/(CB)$. The velocity of any particle on link 4 such as D may be determined graphically from similar triangles as shown using the instantaneous center of link 4 relative to link 1.

For links which are in pure translation, such as the slider in a slider-crank mechanism, the direction lines of the velocities of all of its particles are parallel, and the normals, also being parallel, intersect at infinity. Thus, the instantaneous center of a link in translation is at an infinite distance from the link in a direction normal to the path of translation.

10.13 Instantaneous Center Notation. In the foregoing, instantaneous centers of velocity were determined for each of the moving links relative to the fixed link. The system of labeling these points is shown in Fig. 10.23 where the instantaneous center of link 3 relative to the fixed link is labeled 31 to indicate the motion of "3 relative to 1". Link 1 has the same instantaneous center relative to link 3 when link 3 is considered the fixed

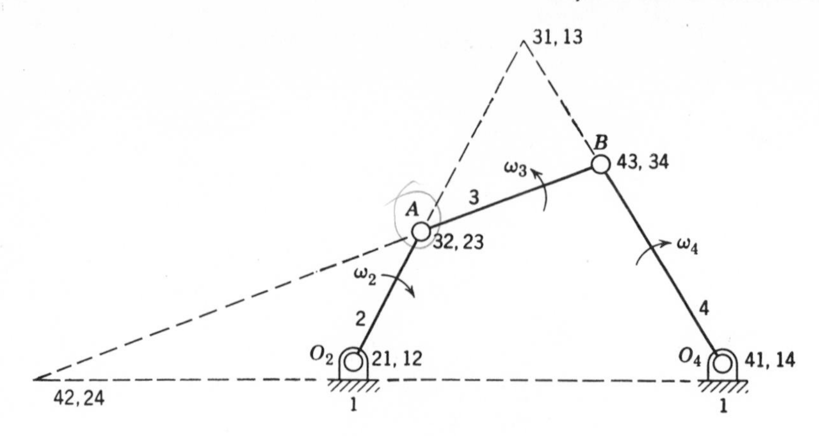

Figure 10.23

link, in which case link 1 appears to be rotating in the opposite sense ($\omega_{13} = -\omega_{31}$) relative to link 3. Since points 31 and 13 are the same point, either designation is acceptable although the simpler notation 13 is preferred. The instantaneous center of link 2 relative to link 1 is labeled 21 or 12, and that of link 4 relative to link 1 is labeled 41 or 14 as shown.

Also of interest is the instantaneous center of one link relative to another where both links are moving relative to the fixed link. Such a center is shown at point A in Fig. 10.23 where both A_2 and A_3 have a common absolute velocity $\mathbf{V}_A$ because of the pinned joint so that the relative velocities $\mathbf{V}_{A_3A_2}$ and $\mathbf{V}_{A_2A_3}$ are zero. It is obvious that point A is the instantaneous center 32 about which link 3 is rotating relative to link 2 at an angular velocity ω_{32}. Point A is also the instantaneous center 23. In a similar manner point B is the instantaneous center 43 or 34. The instantaneous center 42 or 24 is also shown in Fig. 10.23. However, the method of determining its location will not be presented until the next section.

10.14 Kennedy's Theorem. For three independent bodies in general plane motion, Kennedy's theorem states that the three instantaneous centers lie on a common straight line. In Fig. 10.24, three independent links (1, 2, and 3) are shown in motion relative to each other. There are three instantaneous centers (12, 13, and 23), whose instantaneous locations are to be determined.

If link 1 is regarded as a fixed link, or datum link, the velocities of particles A_2 and B_2 on link 2 and the velocities of D_3 and E_3 on link 3 may be regarded as absolute velocities relative to link 1. The instantaneous center 12 may be located from the intersection of the normals to the velocity direction lines

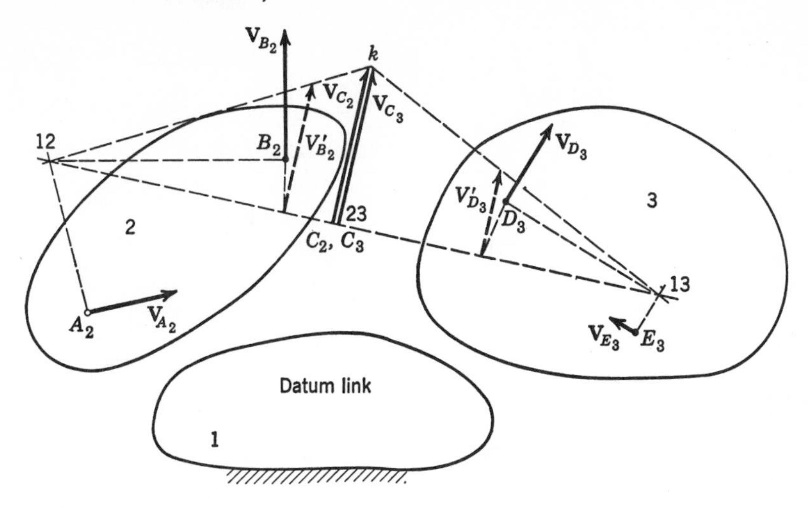

Figure 10.24

drawn from A_2 and B_2. Similarly, the center 13 is located from normals drawn from particles D_3 and E_3. The instantaneous centers 12 and 13 are relative to link 1.

It remains to determine the third instantaneous center 23. On a line drawn through the centers 12 and 13, there exists a particle C_2 on link 2 at an absolute velocity $\mathbf{V}_{C_2}$ having the same direction as the absolute velocity $\mathbf{V}_{C_3}$ of a particle C_3 on link 3. Since $\mathbf{V}_{C_2}$ is proportional to the distance of C_2 from 12, the magnitude of $\mathbf{V}_{C_2}$ is determined from the graphical construction shown and $\mathbf{V}_{C_3}$ is determined in a similar manner. From the intersection of the construction lines at k, a common location of C_2 and C_3 is determined such that the absolute velocities $\mathbf{V}_{C_2}$ and $\mathbf{V}_{C_3}$ are identical. This location is the instantaneous center 23, since the absolute velocities of the coincident particles are common and the relative velocities $\mathbf{V}_{C_2C_3}$ and $\mathbf{V}_{C_3C_2}$ are zero. It should be obvious that 23 is on a straight line with 12 and 13 in order that the directions of $\mathbf{V}_{C_2}$ and $\mathbf{V}_{C_3}$ are common.

Kennedy's theorem is extremely useful in determining the locations of instantaneous centers in mechanisms having a large number of links, many of which are in general plane motion.

10.15 Determination of Instantaneous Centers by Kennedy's Theorem. In a mechanism consisting of n links, there are $n - 1$ instantaneous centers relative to any given link. For n number of links, there is a total of $n(n - 1)$ instantaneous centers. However, since for each location of instantaneous centers there are two centers, the total number N of locations is given by

$$N = \frac{n(n-1)}{2}$$

The number of locations of centers increases rapidly with numbers of links as shown below.

n Links	N Centers
4	6
5	10
6	15
7	21

Example 10.6

For the Whitworth mechanism shown in Fig. 10.25, determine the fifteen locations of instantaneous centers of zero velocity.

Solution. Because of the large number of locations to be determined, it is desirable to use a system of accounting for the centers as they are determined. The circle diagram shown in Fig. 10.25 is one of the simplest means of accounting. The numbers of the links are designated on the periphery of the circle, and the chord linking any two numbers represents an instantaneous center. In the upper circle are shown eight centers which may be determined by inspection. Five of the centers (12, 14, 23, 45, and 56) are at pin-jointed connections as shown. Two centers (16 and 34) are at infinity, since link 6 is in translation relative to link 1, and link 3 is in translation relative to link 4. Because the absolute velocity directions of points B and C of link 5 are known, the intersection of the normals locates 15. Thus eight centers are located by inspection, and the solid lines of the circle diagram indicate that these centers are located.

For centers less obviously determined, Kennedy's theorem may be used. In the upper circle, to locate center 13 a dashed line is drawn such that it closes two triangles. The triangle 1–2–3 represents the three centers (12, 23, and 13) of links 1, 2, and 3, which according to Kennedy's theorem lie on a straight line. Similarly, triangle 1–3–4 represents the centers 13, 34, and 14, which also lie on a straight line. The intersection of the two lines on the mechanism locates the center 13, which must lie on both lines. The dashed line may be made solid to indicate that the unknown center has been located. The lower circle shows the next step in which the center 24 is located using triangles 2–3–4 and 1–2–4. It may be seen that 24 is the logical center to determine rather than 25 or 26, which cannot be drawn as common to two triangles until other centers have been determined.

In Fig. 10.25, ten of the fifteen centers are shown. Figure 10.26 shows the same mechanism with all fifteen centers located.

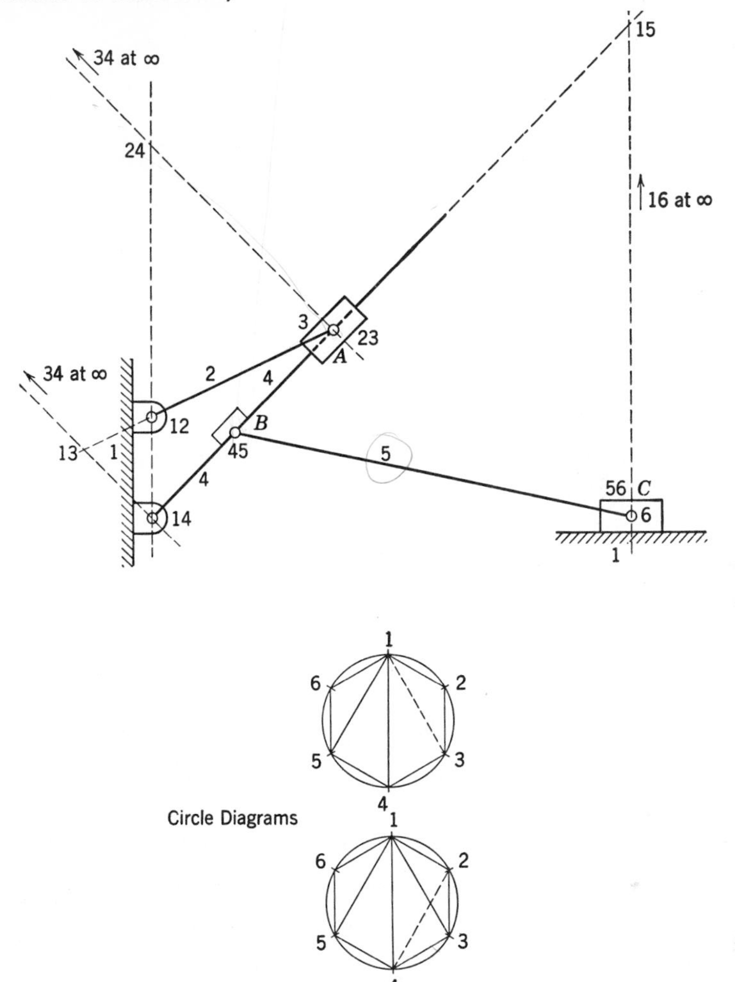

Figure 10.25

10.16 Determination of Velocity by Instantaneous Centers. Kennedy's theorem may be used to great advantage in determining directly the absolute velocity of any given particle of a mechanism without necessarily determining the velocities of intermediate particles as required by the vector polygon method. In connection with the Whitworth mechanism of Fig. 10.25, for example, the velocity of the tool support (link 6) may be determined from the known speed of the driving link 2 without first determining the velocities of points on the connecting links 3, 4, and 5.

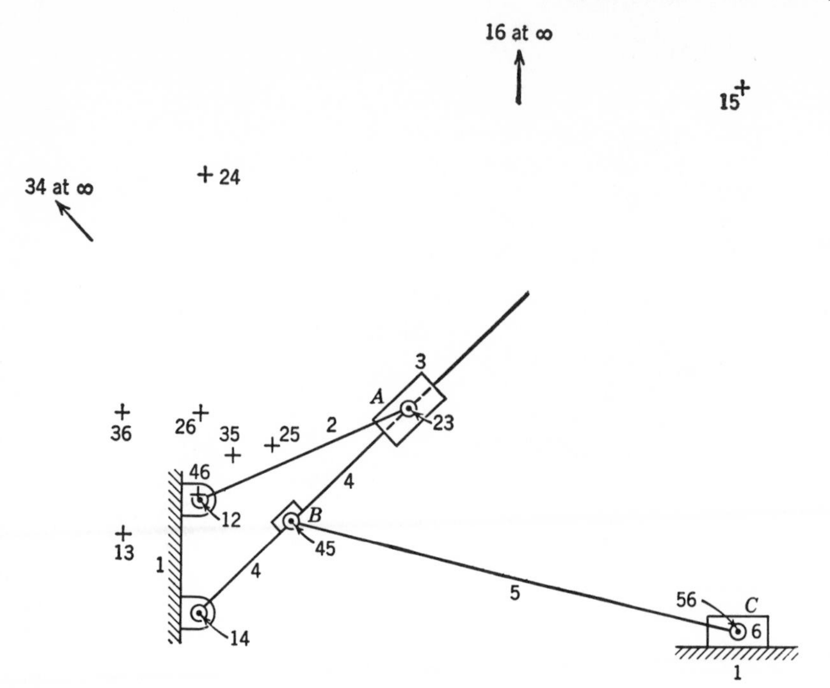

Figure 10.26

Example 10.7

For the Whitworth mechanism shown in Fig. 10.27, determine the absolute velocity $\mathbf{V}_C$ of the tool support when the driving link 2 rotates at a speed such that $\mathbf{V}_A = 30$ ft/sec as shown.

Solution. Two solutions for $\mathbf{V}_C$ are shown in Fig. 10.27. In the first of these (Fig. 10.27*a*), links 1, 3, and 5 are involved such that instantaneous centers 13, 15, and 35 are used. $\mathbf{V}_A$ is the known absolute velocity of a particle on link 3 relative to link 1; thus, links 3 and 1 are involved. The absolute velocity $\mathbf{V}_C$ is to be determined for a particle on link 5 also relative to link 1, thus involving links 5 and 1. According to Kennedy's theorem, the instantaneous centers 13, 15, and 35 are on a common straight line as shown in Fig. 10.27*a*. Using center 13, the absolute velocity $\mathbf{V}_{P_3}$ for a particle P_3 located at 35 on link 3 may be determined graphically from similar triangles by swinging $\mathbf{V}_A$ to position $\mathbf{V}'_A$ using center 13 as a pivot point. Point 35 represents the location of coincident particles P_3 on link 3 and P_5 on link 5, for which the absolute velocities are common (see Fig. 10.24). Thus, $\mathbf{V}_{P_3}$ is also the absolute velocity $\mathbf{V}_{P_5}$ of a particle on link 5. Since both P_5 and C are points on link 5, the absolute velocity $\mathbf{V}_C$ may be determined from similar triangles by swinging $\mathbf{V}_{P_5}$ to position $\mathbf{V}'_{P_5}$ using center 15 as a pivot point. The length of $\mathbf{V}_C$ is measured to determine magnitude of velocity.

In the above solution, the centers 13 and 15 relative to the fixed link are *pivot points*, and the center 35 of the moving links is the *transfer point*. By

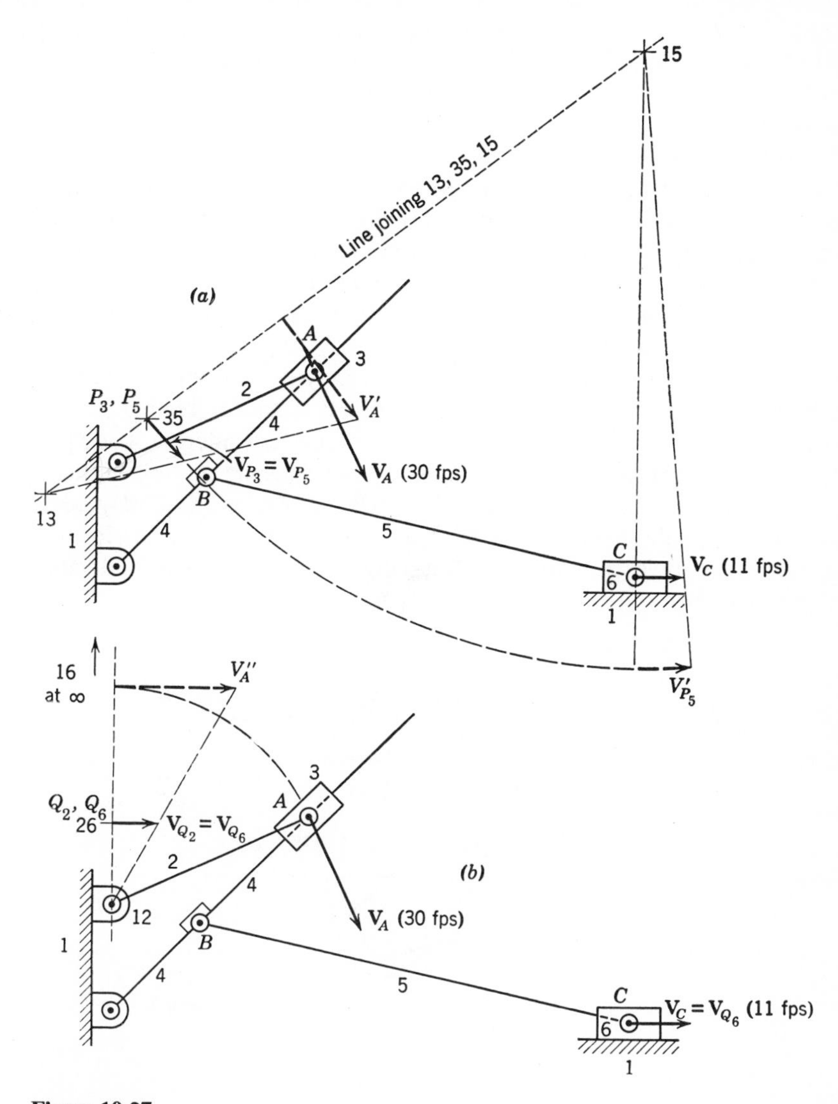

Figure 10.27

properly identifying these points, the determination of velocities becomes systematic.

The second solution (Fig. 10.27*b*) for $\mathbf{V}_C$ is similar to the first, in which pivot points 12 and 16 are used because $\mathbf{V}_A$ represents the absolute velocity of a particle on link 2 and $\mathbf{V}_C$ is the absolute velocity of a particle on link 6. Center 26 is the transfer point representing the location of coincident particles Q_2 and Q_6 on links 2 and 6, for which the absolute velocities $\mathbf{V}_{Q_2}$ and $\mathbf{V}_{Q_6}$ are common. $\mathbf{V}_{Q_2}$ is determined graphically from $\mathbf{V}_A$ using center 12 as a pivot point. Since pivot point 16 is at infinity, link 6 is in pure translation relative to link 1 so that $\mathbf{V}_C$ is the same in magnitude and direction as $\mathbf{V}_{Q_2}$ and $\mathbf{V}_{Q_6}$, as shown.

10.17 Rolling Elements. The method of instantaneous centers is frequently applied to mechanisms consisting of rolling elements as in epicyclic gear trains (Fig. 10.28). As shown previously, the relative velocity of the coincident particles at the point of contact of two rolling links is zero. Thus, an instantaneous center exists at the point of contact.

For the reduction drive shown in Fig. 10.28, the instantaneous centers are as shown. The speed reduction ratio ω_{31}/ω_{41} (the internal gear speed to carrier speed when the sun gear is fixed) may be determined from linear velocities of particles as shown. Assuming that the absolute angular velocity ω_{41} of the carrier is known, $\mathbf{V}_A$ may be determined considering A as a particle on link 4. $\mathbf{V}_A$ is also the absolute velocity of a particle on link 2; therefore, using the center 12, the absolute velocity $\mathbf{V}_{P_2}$ of P_2 on link 2 may be determined graphically from similar triangles. Since center 23 is the location of coincident particles on links 2 and 3 having a common absolute velocity, ω_{31} may be calculated from $\mathbf{V}_{P_3}$.

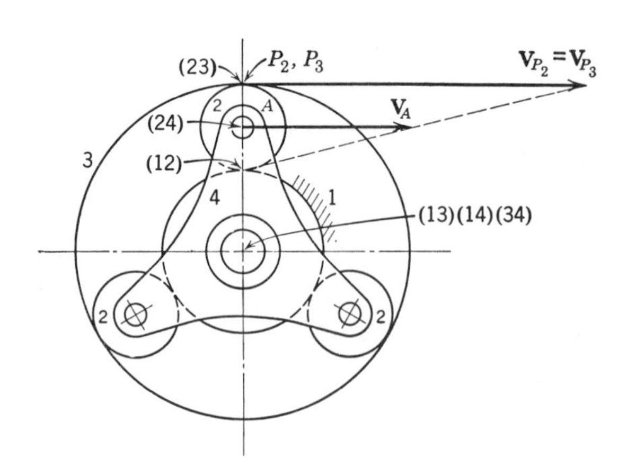

Figure 10.28

10.18 Graphical Determination of Acceleration in Mechanisms. As in the determination of velocities of particles in a mechanism, the linear accelerations of particles may also be determined by graphical construction of acceleration polygons and acceleration images. It is important that the relative acceleration of pairs of particles be understood.

10.19 Relative Acceleration of Particles in Mechanisms. If the acceleration $\mathbf{A}_Q$ of a particle Q is known, the acceleration of another particle $\mathbf{A}_P$ may be determined by adding the relative acceleration vector $\mathbf{A}_{PQ}$ as shown in the following vector equation:

$$\mathbf{A}_P = \mathbf{A}_Q + \mathbf{A}_{PQ} \tag{10.26}$$

As discussed in the sections on relative velocity, it is shown that the relative velocity of a pair of particles depends on the type of constraint used in a given mechanism. Similarly, the relative acceleration $\mathbf{A}_{PQ}$ in mechanisms depends on the type of built-in constraint.

10.20 Relative Acceleration of Particles in a Common Link. As shown in Fig. 10.29*a*, when two particles P and Q in the same rigid link are considered, the fixed distance PQ constrains particle P to move on a circular arc relative to Q regardless of the absolute linear motion of Q. Therefore, since the path of P relative to Q is circular, the acceleration vector $\mathbf{A}_{PQ}$ may be represented by the perpendicular components of acceleration $\mathbf{A}^n_{PQ}$ and $\mathbf{A}^t_{PQ}$, respectively normal and tangent to the relative path at P. Regardless of the linear absolute acceleration of Q, the angular motions of the link relative to Q are the same as relative to the fixed link because a particle such as Q has no angular motion. For the circular path of P relative to Q, the angular velocity ω_r of the radius of curvature PQ is the same as the absolute angular velocity ω_3 of the link. Also, the angular acceleration α_r of the radius of curvature is the same as the absolute angular acceleration α_3 of the link.

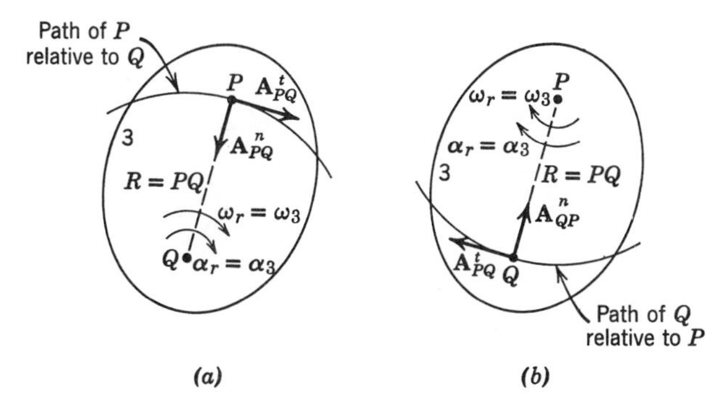

Figure 10.29

The magnitude of the normal relative acceleration $\mathbf{A}_{PQ}^n$ may be determined from Eq. 10.2*a*.

$$|\mathbf{A}_{PQ}^n| = (PQ)\omega_3^2 = \frac{V_{PQ}^2}{PQ} \tag{10.27}$$

The magnitude of the tangential relative acceleration $\mathbf{A}_{PQ}^t$ may be determined from Eq. 10.3*a*.

$$|\mathbf{A}_{PQ}^t| = (PQ)\alpha_3 \tag{10.28}$$

Because the relative path is circular, dR/dt is zero.

It is to be observed that the direction of $\mathbf{A}_{PQ}^n$ is normal to the relative path and that its sense is toward the center of curvature Q so that the vector is directed from P toward Q as shown in Fig. 10.29*a*. The direction of $\mathbf{A}_{PQ}^t$ is tangent to the relative path (normal to line PQ), and the sense of the vector depends on the sense of α_r. In Fig. 10.29*b* the relative acceleration vectors $\mathbf{A}_{QP}^n$ and $\mathbf{A}_{QP}^t$ of Q relative to P are shown where the magnitudes and senses of ω_3 and α_3 are the same as in Fig. 10.29*a*. The relative path shown is that of Q observed at P. It is to be noted that $\mathbf{A}_{QP}^n = -\mathbf{A}_{PQ}^n$ and $\mathbf{A}_{QP}^t = -\mathbf{A}_{PQ}^t$, where the minus signs indicate "opposite in sense."

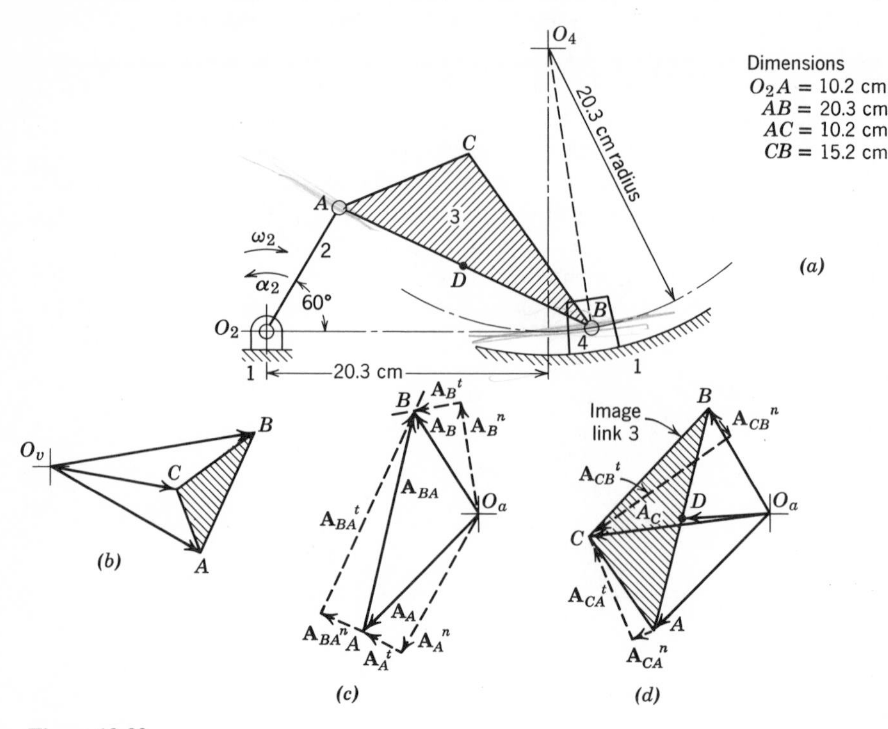

Figure 10.30

Example 10.8

When the mechanism is in the phase shown in Fig. 10.30*a*, link 2 rotates with an angular velocity ω_2 of 30 rad/sec and an angular acceleration α_2 of 240 rad/sec^2 in the directions given. Determine the acceleration $\mathbf{A}_B$ of point B, the acceleration $\mathbf{A}_C$ of point C, the angular acceleration α_3 of link 3, the angular acceleration α_4 of link 4, and the relative acceleration α_{34}. Velocity and acceleration equations can be written as follows:

I. $\quad \mathbf{V}_B = \mathbf{V}_A + \mathbf{V}_{BA}$
II. $\quad \mathbf{V}_C = \mathbf{V}_A + \mathbf{V}_{CA}$
III. $\quad \mathbf{V}_C = \mathbf{V}_B + \mathbf{V}_{CB}$

where

V_B = direction perpendicular to O_4B, magnitude unknown
$V_A = (O_2A)\omega_2 = (10.2)30 = 306$ cm/sec, direction perpendicular to O_2A
V_{BA} = direction perpendicular to BA, magnitude unknown
V_C = direction unknown, magnitude unknown
V_{CA} = direction perpendicular to CA, magnitude unknown
V_{CB} = direction perpendicular to CB, magnitude unknown

Measured on polygon of Fig. 10.30*b*

$$V_B = 366 \text{ cm/sec}$$
$$V_{BA} = 230 \text{ cm/sec}$$
$$V_{CA} = 113 \text{ cm/sec}$$
$$V_{CB} = 175 \text{ cm/sec}$$

IV. $\quad \mathbf{A}_B = \mathbf{A}_A + \mathbf{A}_{BA}$

$$\mathbf{A}_B^n + \mathbf{A}_B^t = \mathbf{A}_A^n + \mathbf{A}_A^t + \mathbf{A}_{BA}^n + \mathbf{A}_{BA}^t$$

where

$A_B^n = \dfrac{V_B^2}{O_4B} = \dfrac{366^2}{20.3} = 6598$ cm/sec^2, direction from B towards O_4
A_B^t = direction perpendicular to $\mathbf{A}_B^n$, magnitude unknown
$A_A^n = \dfrac{V_A^2}{O_2A} = \dfrac{306^2}{10.2} = 9180$ cm/sec^2, direction from A toward O_2
$A_A^t = (O_2A)\alpha_2 = (10.2)240 = 2448$ cm/sec^2, direction perpendicular to $\mathbf{A}_A^n$
$A_{BA}^n = \dfrac{V_{BA}^2}{BA} = \dfrac{230^2}{20.3} = 2605$ cm/sec^2, direction from B toward A
A_{BA}^t = direction perpendicular to $\mathbf{A}_{BA}^n$, magnitude unknown

Measured on polygon of Fig. 10.30c

$A_B = 7040$ cm/sec^2, $A_B^t = 2470$ cm/sec^2, and $A_{BA}^t = 12{,}900$ cm/sec^2

$$\alpha_3 = \frac{A_{BA}^t}{BA} = \frac{12{,}900}{20.3} = 635 \text{ rad/sec}^2 \quad \text{(ccw)}$$

$$\alpha_4 = \frac{A_B^t}{O_4B} = \frac{2470}{20.3} = 122 \text{ rad/sec}^2 \quad \text{(cw)}$$

$$\alpha_{34} = \alpha_3 - \alpha_4 = 635 - (-122) = 757 \text{ rad/sec}^2 \quad \text{(ccw)}$$

V. $\mathbf{A}_C = \mathbf{A}_A + \mathbf{A}_{CA}^n + \mathbf{A}_{CA}^t$

VI. $\mathbf{A}_C = \mathbf{A}_B + \mathbf{A}_{CB}^n + \mathbf{A}_{CB}^t$

where

A_C = direction unknown, magnitude unknown

$$A_{CA}^n = \frac{V_{CA}^2}{CA} = \frac{113^2}{10.2} = 1252 \text{ cm/sec}^2, \text{ direction from } C \text{ toward } A$$

A_{CA}^t = direction perpendicular to $\mathbf{A}_{CA}^n$, magnitude unknown

$$A_{CB}^n = \frac{V_{CB}^2}{CB} = \frac{175^2}{15.2} = 2014 \text{ cm/sec}^2, \text{ direction from } C \text{ toward } B$$

A_{CB}^t = direction perpendicular to $\mathbf{A}_{CB}^n$, magnitude unknown

Measured on polygon of Fig. 10.30d, $A_C = 10{,}400$ cm/sec^2

Solution. The velocity polygon of Fig. 10.30b shows the determination of $\mathbf{V}_B$ and $\mathbf{V}_{BA}$ from Eq. I. In a similar manner $\mathbf{V}_C$, $\mathbf{V}_{CA}$, and $\mathbf{V}_{CB}$ are determined from Eqs. II and III. The shaded triangle ABC of the velocity polygon is the velocity image of link 3.

Equation IV expresses $\mathbf{A}_B$ in terms of $\mathbf{A}_A$ and $\mathbf{A}_{BA}$, and all of the components of this equation are known as indicated in magnitude, sense, and direction or in direction. In constructing the acceleration polygon Fig. 10.30c starting with the right side of Eq. IV, the vector $\mathbf{A}_A^n$ is drawn from pole O_a to which is added $\mathbf{A}'_A$. This gives the vector $\mathbf{A}_A$ whose tip is labeled "A". Next, add the vector $\mathbf{A}_{BA}^n$ starting at point A, and to it add the direction of $\mathbf{A}_{BA}^t$. As can be seen, it is impossible to complete the solution using only the components on the right side of Eq. IV. Therefore, consider the left side of the equation and draw vector $\mathbf{A}_B^n$ from O_a and to it add the direction of $\mathbf{A}_B^t$. The intersection of the direction of $\mathbf{A}_{BA}^t$ and the direction of $\mathbf{A}_B^t$ completes the polygon. Arrowheads are now added to the vectors $\mathbf{A}_{BA}^t$ and $\mathbf{A}_B^t$ so that the addition of the vectors of the polygon agrees with the addition of the terms of Eq. IV. The resultant of the vectors $\mathbf{A}_B^n$ and $\mathbf{A}_B^t$ gives $\mathbf{A}_B$ whose tip is labeled "B". The resultant of $\mathbf{A}_{BA}^n$ and $\mathbf{A}_{BA}^t$ is also shown on the polygon.

The magnitudes and senses of α_3 and α_4 can now be determined from $\mathbf{A}_{BA}^t$ and $\mathbf{A}_B^t$, respectively, as shown.

To determine $\mathbf{A}_C$ it is necessary to use Eqs. V and VI, which give the relations between $\mathbf{A}_C$ and $\mathbf{A}_A$ and $\mathbf{A}_B$. The components of these equations are known as indicated. The acceleration vectors $\mathbf{A}_A$ and $\mathbf{A}_B$ are redrawn in Fig. 10.30*d* from Fig. 10.30*c* without their normal and tangential components for clarity. Use Eq. V, and draw the vector $\mathbf{A}_{CA}^n$ from point A in Fig. 10.30*d* and to it add the direction of $\mathbf{A}_{CA}^t$. Consider next Eq. VI and draw the vector $\mathbf{A}_{CB}^n$ from point B and to it add the direction of $\mathbf{A}_{CB}^t$. The intersection of the direction of $\mathbf{A}_{CA}^t$ and the direction of $\mathbf{A}_{CB}^t$ completes the polygon. This intersection is point C, which gives $\mathbf{A}_C$. Arrowheads are now added to the vectors $\mathbf{A}_{CA}^t$ and $\mathbf{A}_{CB}^t$ so that the vector addition checks with Eqs. V and VI. The shaded triangle ABC of Fig. 10.30*d* is the acceleration image of link 3.

The acceleration of any point D as shown on link 3 can be determined by locating its corresponding position on the acceleration image of link 3. The vector from O_a to D is $\mathbf{A}_D$ as shown in Fig. 10.30*d*.

10.21 Relative Acceleration of Coincident Particles on Separate Links. Coriolis Component of Acceleration. The next mechanism to be considered is one in which there is relative sliding between two links, as between links 3 and 4 as shown in Fig. 10.31, and it is required to determine ω_4 and α_4 given ω_2 and α_2. In this mechanism points A_2 and A_3 are the same point, and point A_4 is their projection on link 4. In order to find ω_4 and α_4, the velocity and acceleration of the two coincident points A_2 and A_4, each on separate links, must be analyzed.[2]

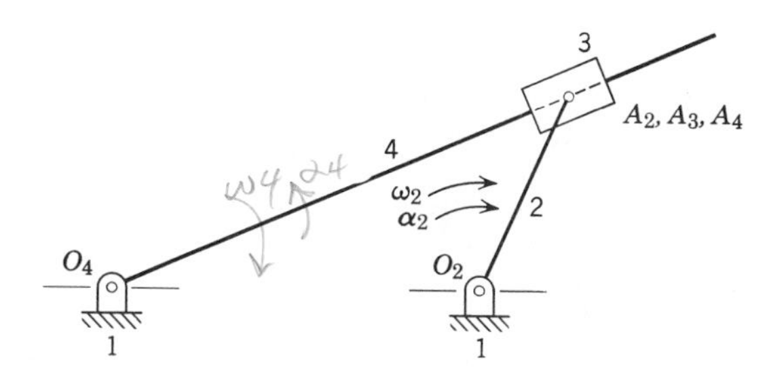

Figure 10.31

[2] Point A_3 could have been used instead of A_2 as the point coincident with A_4. However, point A_2 is generally preferred because it is on a link directly connected to the ground and its motion can be easily visualized.

The equation for the velocity of point A_4 can be written as follows:

$$\mathbf{V}_{A_4} = \mathbf{V}_{A_2} + \mathbf{V}_{A_4A_2} \tag{10.29}$$

In this equation $\mathbf{V}_{A_2}$ is known in magnitude, sense, and direction and $\mathbf{V}_{A_4}$ and $\mathbf{V}_{A_4A_2}$ are known in direction. The velocity polygon can easily be drawn, and $\mathbf{V}_{A_4}$ determined from which ω_4 can be calculated.

The acceleration of point A_4 can be determined from the following equation

$$\mathbf{A}_{A_4} = \mathbf{A}_{A_2} + \mathbf{A}_{A_4A_2} \tag{10.30}$$

which can be expanded as follows:

$$\mathbf{A}^n_{A_4} + \mathbf{A}^t_{A_4} = \mathbf{A}^n_{A_2} + \mathbf{A}^t_{A_2} + \mathbf{A}^n_{A_4A_2} + \mathbf{A}^t_{A_4A_2} + 2\boldsymbol{\omega}_2 \times \mathbf{V}_{A_4A_2} \tag{10.31}$$

In going from Eq. 10.30 to Eq. 10.31, the following substitution was made

$$\mathbf{A}_{A_4A_2} = \mathbf{A}^n_{A_4A_2} + \mathbf{A}^t_{A_4A_2} + 2\boldsymbol{\omega}_2 \times \mathbf{V}_{A_4A_2}$$

To determine the relative acceleration between two moving coincident points, it is necessary to add a third component as shown. This component is known as *Coriolis component* which was developed using vector mathematics in Section 10.6. Also because points A_4 and A_2 are coincident, the terms $\mathbf{A}^n_{A_4A_2}$ and $\mathbf{A}^t_{A_4A_2}$ do not represent the usual normal and tangential components of two points on the same rigid body as previously considered. For this reason they often appear in the literature written with a capital script $\mathscr{A}$. The magnitude of $\mathbf{A}^n_{A_4A_2}$ can be calculated from the relation

$$|\mathbf{A}^n_{A_4A_2}| = \frac{V^2_{A_4A_2}}{R} \tag{10.32}$$

where R is the radius of curvature of the path of point A_4 relative to point A_2. This component is directed from the coincident points along the radius toward the center of curvature. The tangential component $\mathbf{A}^t_{A_4A_2}$ is known in direction and is tangent to the path of $\mathbf{A}_4$ relative to $\mathbf{A}_2$ at the coincident points. The magnitude of the Coriolis component $2\boldsymbol{\omega}_2 \times \mathbf{V}_{A_4A_2}$ is easily calculated because $\boldsymbol{\omega}_2$ is given data and $\mathbf{V}_{A_4A_2}$ can be determined from the velocity polygon. The direction of this component is normal to the path of A_4 relative to A_2, and its sense is the same as that of $\mathbf{V}_{A_4A_2}$ rotated about its origin 90° in the direction of ω_2. An example of this method of determining the direction will be given in a later section.

In Eq. 10.31 all of the components can easily be determined in magnitude, sense, and direction or in direction except $\mathbf{A}^n_{A_4A_2}$. This component calculated from $V^2_{A_4A_2}/R$ can only be determined if the radius of curvature R of the path of A_4 relative to A_2 is known. Unfortunately, this path is not easily determined for the mechanism shown in Fig. 10.31 so it is necessary to rewrite

Eq. 10.31 in the following form:

$$\mathbf{A}^n_{A_2} + \mathbf{A}^t_{A_2} = \mathbf{A}^n_{A_4} + \mathbf{A}^t_{A_4} + \mathbf{A}^n_{A_2A_4} + \mathbf{A}^t_{A_2A_4} + 2\boldsymbol{\omega}_4 \times \mathbf{V}_{A_2A_4} \quad (10.33)$$

With Eq. 10.31 written in this form, $\mathbf{A}^n_{A_2A_4}$ can easily be evaluated as zero because the path of A_2 relative to A_4 is a straight line and R is infinite. The acceleration polygon can now be drawn and $\mathbf{A}^t_{A_4}$ determined from which α_4 is calculated.

While it is easy to see in Fig. 10.31 that the path of point A_2 relative to point A_4 is a straight line by inverting the mechanism and letting link 4 be the fixed link, it is very difficult to visualize the path of A_4 relative to A_2. As a means of determining this path, consider Fig. 10.32 where link 2 is now the fixed link. In this figure link 1 is placed in a number of angular positions relative to link 2, and the relative position of A_4 is determined for each position of link 1. It may be seen that the position of link 4 is always in a

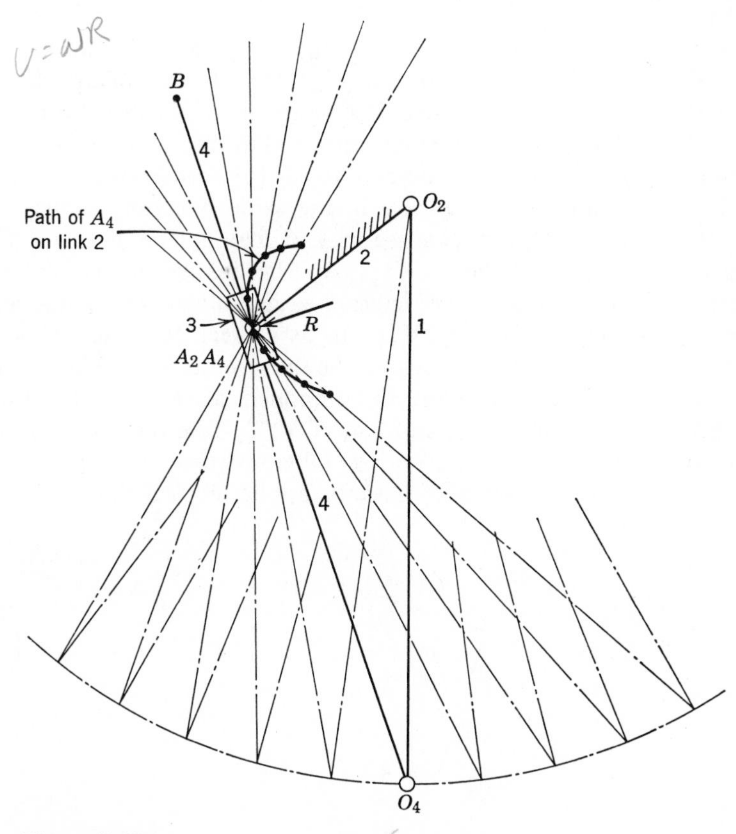

Figure 10.32

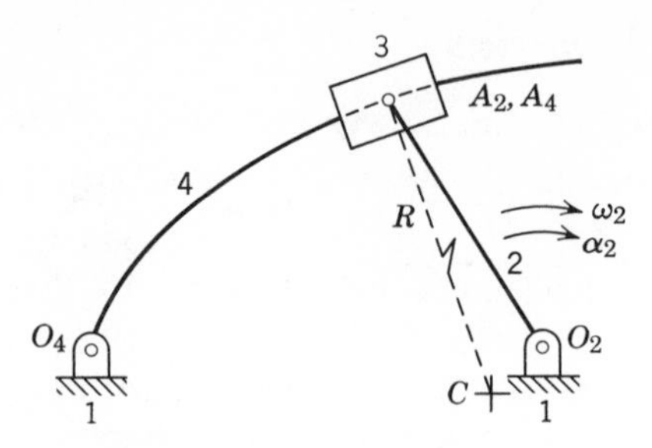

Figure 10.33

direction from O_4 through A_2 and that A_4 is a fixed distance from O_4. As shown, the path of A_4 on link 2 is curvilinear and tangent to link 4 at point A_2. Unfortunately, the path is not circular so the radius of curvature is difficult to determine.

Consider next the case where link 4 of Fig. 10.31 has been replaced by a curved link of circular form as shown in Fig. 10.33. In this linkage the path of A_2 relative to A_4 is a circular arc of known radius and center of curvature. The magnitude of $\mathbf{A}^n_{A_2A_4}$ is therefore not zero, and the vector representing this component will be directed from point A toward the center of curvature C.

The Coriolis component is always in the same direction as the $\mathbf{A}^n_{A_2A_4}$ component, if one exists, but its sense may or may not be the same. Considering the Coriolis term $2\boldsymbol{\omega}_4 \times \mathbf{V}_{A_2A_4}$ for the linkage of Fig. 10.33, its direction and sense can easily be determined as follows. Draw the vector representing the relative velocity $\mathbf{V}_{A_2A_4}$ in its correct direction and sense. Rotate this vector 90° about its origin in the same sense as ω_4. This will give the direction and sense of the Coriolis component as shown in Fig. 10.34. As can be seen, the terms $\mathbf{A}^n_{A_2A_4}$ and $2\boldsymbol{\omega}_4 \times \mathbf{V}_{A_2A_4}$ have the same sense for this case and will therefore add together. Obviously, this method of determining the direction and sense of Coriolis applies even if the $\mathbf{A}^n_{A_2A_4}$ component is zero.

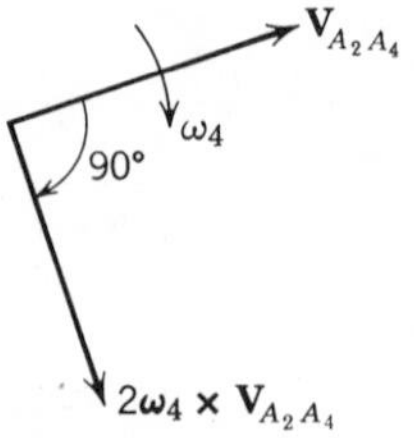

Figure 10.34

Example 10.9

In the crank-shaper mechanism shown in Fig. 10.35*a* link 2 rotates at a constant angular velocity ω_2 of 10 rad/sec. Determine the acceleration $\mathbf{A}_{A_4}$ of point A_4 on link 4 and the angular acceleration α_4 when the mechanism is in the phase shown. Velocity and acceleration equations can be written as follows:

I. $\mathbf{V}_{A_4} = \mathbf{V}_{A_2} + \mathbf{V}_{A_4A_2}$

where

V_{A_4} = direction perpendicular to O_4A_4, magnitude unknown
$V_{A_2} = (O_2A_2)\omega_2 = (4)10 = 40$ in./sec, direction perpendicular to O_2A_2
$V_{A_4A_2}$ = direction parallel to O_4A_4, magnitude unknown

Measured on polygon of Fig. 10.35*b*

$$V_{A_4} = 13 \text{ in./sec} \quad \text{and} \quad V_{A_4A_2} = 38 \text{ in./sec}$$

$$\omega_4 = \frac{V_{A_4}}{O_4A_4} = \frac{13}{10} = 1.3 \text{ rad/sec} \quad \text{(ccw)}$$

II. $\mathbf{A}_{A_4} = \mathbf{A}_{A_2} + \mathbf{A}_{A_4A_2}$

III. $\mathbf{A}_{A_2} = \mathbf{A}_{A_4} + \mathbf{A}_{A_2A_4}$

$$\mathbf{A}^n_{A_2} + \mathbf{A}^t_{A_2} = \mathbf{A}^n_{A_4} + \mathbf{A}^t_{A_4} + \mathbf{A}^n_{A_2A_4} + \mathbf{A}^t_{A_2A_4} + 2\boldsymbol{\omega}_4 \times \mathbf{V}_{A_2A_4}$$

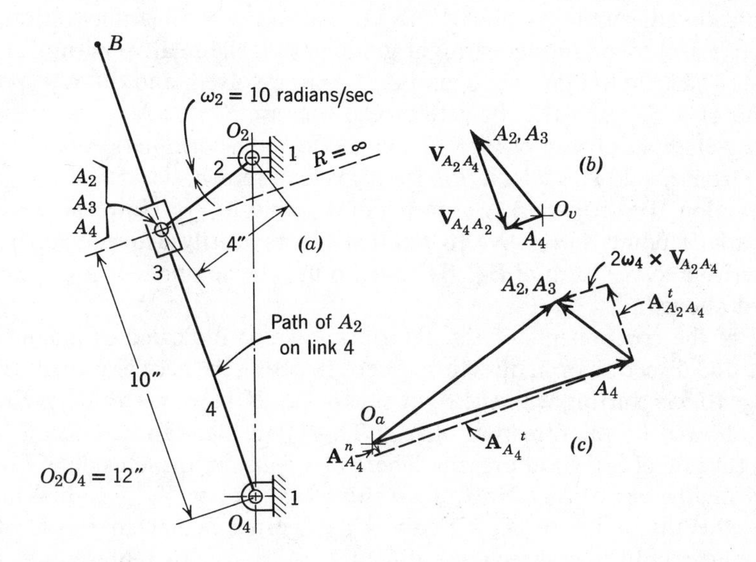

Figure 10.35

where

$$A^n_{A_2} = \frac{V^2_{A_2}}{O_2A_2} = \frac{40^2}{4} = 400 \text{ in./sec}^2\text{, direction from } A_2 \text{ toward } O_2$$

$$A^t_{A_2} = 0(\alpha_2 = 0)$$

$$A^n_{A_4} = \frac{V^2_{A_4}}{O_4A_4} = \frac{13^2}{10} = 16.9 \text{ in./sec}^2\text{, direction from } A_4 \text{ toward } O_4$$

$$A^t_{A_4} \doteq \text{direction perpendicular to } \mathbf{A}^n_{A_4}\text{, magnitude unknown}$$

$$A^n_{A_2A_4} = \frac{V^2_{A_2A_4}}{R} = 0 \qquad (R = \infty)$$

$$2\omega_4 V_{A_2A_4} = 2(1.3)38 = 98.8 \text{ in./sec, direction perpendicular to } \mathbf{V}_{A_2A_4}$$

$$A^t_{A_2A_4} = \text{direction perpendicular to } 2\boldsymbol{\omega}_4 \times \mathbf{V}_{A_2A_4}\text{, magnitude unknown}$$

Measured on polygon of Fig. 10.35*c*

$$A_{A_4} = 475 \text{ in./sec}^2 \quad \text{and} \quad A^t_{A_4} = 474 \text{ in./sec}^2$$

$$\alpha_4 = \frac{A^t_{A_4}}{O_4A_4} = \frac{474}{10} = 47.4 \text{ rad/sec}^2 \quad \text{(cw)}$$

Solution. Link 4 is a guide link which constrains points A_2 and A_3 to follow a straight line path on link 4. Two pairs of coincident points may be considered, either A_2 and A_4 or A_3 and A_4. For this illustration, A_2 and A_4 are chosen, and the straight guide path is the relative path of A_2 on link 4. Thus, the vectors $\mathbf{V}_{A_2A_4}$ and $\mathbf{A}_{A_2A_4}$ are involved, and the $\mathbf{A}^n_{A_2A_4}$ component of $\mathbf{A}_{A_2A_4}$ can easily be determined because $R = \infty$.

The velocity polygon of Fig. 10.35*b* shows the determination of $\mathbf{V}_{A_4}$ and $\mathbf{V}_{A_4A_2}$ from Eq. I. The calculation for ω_4 is also shown.

Equation II expresses $\mathbf{A}_{A_4}$ in terms of $\mathbf{A}_{A_2}$ and $\mathbf{A}_{A_4A_2}$. However, because the path of point A_4 relative to point A_2 is not easily determined, Eq. II is rewritten in the form of Eq. III so as to use the component $\mathbf{A}_{A_2A_4}$ as discussed above.

All of the components of Eq. III are known as indicated in magnitude, sense, and direction or in direction. In constructing the acceleration polygon of Fig. 10.35*c* starting with the right side of Eq. III, the vector $\mathbf{A}^n_{A_4}$ is drawn first followed by the direction of $\mathbf{A}^t_{A_4}$. This is all that can be laid off from the right side of Eq. III at present. Therefore, consider the left side of Eq. III and draw the vector $\mathbf{A}_{A_2}$. Next, draw the vector $2\boldsymbol{\omega}_4 \times \mathbf{V}_{A_2A_4}$ so that its tip meets the tip of vector $\mathbf{A}_{A_2}$. Draw $\mathbf{A}^t_{A_2A_4}$ perpendicular to the Coriolis component until it intersects the direction of the vector representing $\mathbf{A}^t_{A_4}$; this completes the polygon. Arrowheads are now added to the vectors $\mathbf{A}^t_{A_4}$

and $\mathbf{A}^t_{A_2A_4}$ so that the addition of the vectors of the polygon agrees with the addition of the terms of Eq. III. The magnitude and sense of α_4 can now be determined from $\mathbf{A}^t_{A_4}$ as shown.

Example 10.10

In the mechanism shown in Fig. 10.36*a*, link 2 drives link 3 through a pin at point *B*. Link 2 rotates at a uniform angular velocity ω_2 of 50 rad/sec, and the radius of curvature *R* of the slot in link 3 is 12 in. Determine the acceleration $\mathbf{A}_{B_3}$ of point B_3 on link 3 and the angular acceleration α_3 for the position shown. Velocity and acceleration equations can be written as follows:

I. $\quad \mathbf{V}_{B_3} = \mathbf{V}_{B_2} + \mathbf{V}_{B_3B_2}$

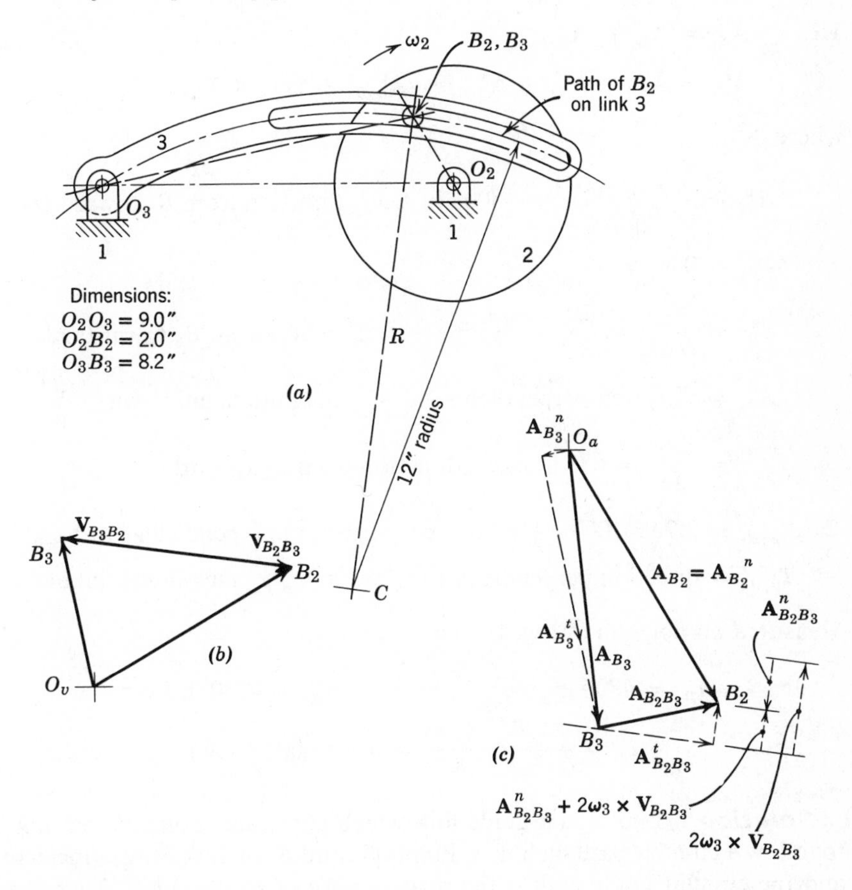

Figure 10.36

where

$$V_{B_3} = \text{direction perpendicular to } O_3B_3\text{, magnitude unknown}$$

$$V_{B_2} = (O_2B_2)\omega_2 = (2)50 = 100 \text{ in./sec, direction perpendicular to } O_2B_2$$

$$V_{B_3B_2} = \text{direction perpendicular to } \mathbf{R}\text{, magnitude unknown}$$

Measured on polygon of Fig. 10.36*b*

$$V_{B_3} = 65 \text{ in./sec} \qquad \text{and} \qquad V_{B_3B_2} = 100 \text{ in./sec}$$

$$\omega_3 = \frac{V_{B_3}}{O_3B_3} = \frac{65}{8.2} = 7.93 \text{ rad/sec (ccw)}$$

II. $\mathbf{A}_{B_3} = \mathbf{A}_{B_2} + \mathbf{A}_{B_3B_2}$

III. $\mathbf{A}_{B_2} = \mathbf{A}_{B_3} + \mathbf{A}_{B_2B_3}$

$$\mathbf{A}^n_{B_2} + \mathbf{A}^t_{B_2} = \mathbf{A}^n_{B_3} + \mathbf{A}^t_{B_3} + \mathbf{A}^n_{B_2B_3} + \mathbf{A}^t_{B_2B_3} + 2\boldsymbol{\omega}_3 \times \mathbf{V}_{B_2B_3}$$

where

$$A^n_{B_2} = \frac{V^2_{B_2}}{O_2B_2} = \frac{100^2}{2} = 5000 \text{ in./sec}^2\text{, direction from } B_2 \text{ toward } O_2$$

$$A^t_{B_2} = 0 \ (\alpha_2 = 0)$$

$$A^n_{B_3} = \frac{V^2_{B_3}}{O_3B_3} = \frac{65^2}{8.2} = 515 \text{ in./sec}^2\text{, direction from } B_3 \text{ toward } O_3$$

$$A^t_{B_3} = \text{direction perpendicular to } \mathbf{A}^n_{B_3}\text{, magnitude unknown}$$

$$\frac{V^2_{B_2B_3}}{R} = \frac{100^2}{12} = 833 \text{ in./sec}^2\text{, direction from } B_2 \text{ toward } C$$

$$2\omega_3V_{B_2B_3} = 2(7.93)100 = 1585 \text{ in./sec}^2\text{, direction perpendicular to } \mathbf{V}_{B_2B_3}$$

$$A^t_{B_2B_3} = \text{direction perpendicular to } 2\boldsymbol{\omega}_3 \times \mathbf{V}_{B_2B_3}\text{, magnitude unknown}$$

Measured on polygon of Fig. 10.36*c*

$$A_{B_3} = 4800 \text{ in./sec}^2 \qquad \text{and} \qquad A^t_{B_3} = 4730 \text{ in./sec}^2$$

$$\alpha_3 = \frac{A^t_{B_3}}{O_3B_3} = \frac{4730}{8.2} = 577 \text{ rad/sec}^2 \text{ (cw)}$$

Solution. Link 3 is a guide link which constrains point B_2 on link 2 to follow a circular path on link 3. Points B_2 and B_3 on link 3 are coincident, and the circular guide path is the relative path of B_2 on link 3. Therefore, the vectors $\mathbf{V}_{B_2B_3}$ and $\mathbf{A}_{B_2B_3}$ are involved in the analysis.

The velocity polygon of Fig. 10.36*b* shows the determination of $\mathbf{V}_{B_3}$ and $\mathbf{V}_{B_3B_2}$ from Eq. I. The calculation for ω_3 is also shown.

Equation II gives $\mathbf{A}_{B_3}$ in terms of $\mathbf{A}_{B_2}$ and $\mathbf{A}_{B_3B_2}$. Because the path of B_2 relative to B_3 is known to be a circular arc, and the path of B_3 relative to B_2 is not easily determined, Eq. II is rewritten in the form of Eq. III so as to use the component $\mathbf{A}_{B_2B_3}$.

All of the components of Eq. III are known as indicated in magnitude, sense, and direction, or in direction. The acceleration polygon of Fig. 10.36*c* is started with the right side of Eq. III by drawing the vector $\mathbf{A}^n_{B_3}$ followed by the direction of $\mathbf{A}^t_{B_3}$. This is all that can be laid off on the right side of Eq. III at the moment so consider the left side of the equation and draw the vector $\mathbf{A}_{B_2}$. The vectors $\mathbf{A}^n_{B_2B_3}$ and $2\boldsymbol{\omega}_3 \times \mathbf{V}_{B_2B_3}$ have opposite sense. Determine the resultant of these two vectors, and add it to the polygon so that its tip meets the tip of vector $\mathbf{A}_{B_2}$. Draw $\mathbf{A}^t_{B_2B_3}$ perpendicular to $\mathbf{A}^n_{B_2B_3}$, until it intersects the direction of the vector representing $\mathbf{A}^t_{B_3}$; this completes the polygon. Arrowheads are now added to the vectors $\mathbf{A}^t_{B_3}$ and $\mathbf{A}^t_{B_2B_3}$ so that the addition of the vectors of the polygon agrees with the addition of the terms of Eq. III. The magnitude and sense of α_3 can now be determined from $\mathbf{A}^t_{B_3}$ as shown.

10.22 Relative Acceleration of Coincident Particles at the Point of Contact of Rolling Elements. An important type of constraint in mechanisms is that which occurs because one link is constrained to roll on another link without relative sliding of the two surfaces at the point of contact. In Fig. 10.37 are shown the rolling pitch circles of a pair of gears in mesh with particles P_3 on link 3 and P_2 on link 2 coincident

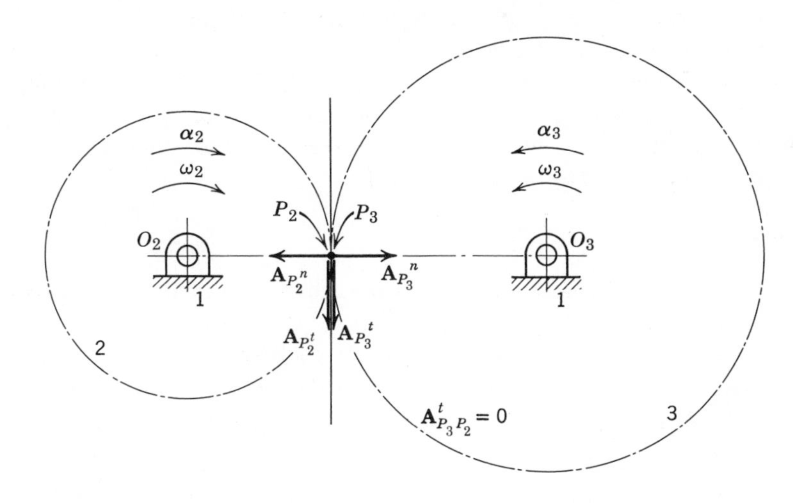

Figure 10.37

in position at the point of contact of the rolling circles. As concluded in an earlier paragraph, the relative velocity $\mathbf{V}_{P_3P_2}$ of the coincident particles is zero, and the absolute velocities $\mathbf{V}_{P_3}$ and $\mathbf{V}_{P_2}$ are identical.

The relative acceleration $\mathbf{A}_{P_3P_2}$ of the coincident particles may be represented by component accelerations, a component $\mathbf{A}^t_{P_3P_2}$ in the t–t direction of the common tangent to the surfaces at the point of contact, and a component $\mathbf{A}^n_{P_3P_2}$ in a direction normal to the surfaces at the point of contact. The tangential component of relative acceleration $\mathbf{A}^t_{P_3P_2}$ is the vector difference of the absolute tangential accelerations $\mathbf{A}^t_{P_3}$ and $\mathbf{A}^t_{P_2}$ shown in Fig. 10.37. Like the tangential velocities $\mathbf{V}_{P_3}$ and $\mathbf{V}_{P_2}$, the tangential accelerations $\mathbf{A}^t_{P_3}$ and $\mathbf{A}^t_{P_2}$ are identical because of the condition of no slipping of the surfaces at the point of contact. No slipping requires that there be no relative motion of the two particles in the direction of possible sliding, which is the tangent direction. Thus, because $\mathbf{A}^t_{P_3}$ and $\mathbf{A}^t_{P_2}$ are identical, the tangential component of acceleration of P_3 relative to P_2 is zero.

The normal component of relative acceleration $\mathbf{A}^n_{P_3P_2}$ is the vector difference of the absolute accelerations $\mathbf{A}^n_{P_3}$ and $\mathbf{A}^n_{P_2}$ shown in Fig. 10.37 in the normal direction. It may be seen that the absolute normal acceleration of P_3 is toward O_3 and that of P_2 is toward O_2. These are parallel vectors, but the senses of the vectors are opposite so that the magnitude of $\mathbf{A}^n_{P_3P_2}$ is the sum of the magnitudes of $\mathbf{A}^n_{P_3}$ and $\mathbf{A}^n_{P_2}$. Thus, it is important to observe that a normal relative acceleration $\mathbf{A}^n_{P_3P_2}$ exists although the tangential relative acceleration is zero.

In a mechanism such as is shown in Fig. 10.37 where the centers of the gears are fixed, it is not necessary to draw an acceleration polygon to determine $\mathbf{A}_{P_3}$ and α_3. The angular acceleration α_3 can easily be determined from α_2 and the ratio of the gear radii using the fact that $\mathbf{A}^t_{P_3} = \mathbf{A}^t_{P_2}$. After α_3 and ω_3 have been found, the components $\mathbf{A}^n_{P_3}$ and $\mathbf{A}^t_{P_3}$ can be calculated and combined to give $\mathbf{A}_{P_3}$. In more complex cases where gear centers are in motion, as in the following example, it is recommended that solutions be undertaken using polygon construction.

Example 10.11

In the mechanism shown in Fig. 10.38*a*, gear 2 rotates about O_2 with a constant angular velocity ω_2 of 10 rad/sec, and gear 3 rolls on gear 2. Determine the acceleration $\mathbf{A}_{P_3}$ of point P_3 on gear 3 and the velocity and acceleration images of gears 2 and 3. Velocity and acceleration equations can be written as follows:

I. $\mathbf{V}_B = \mathbf{V}_A + \mathbf{V}_{BA}$

II. $\mathbf{V}_{P_2} = \mathbf{V}_A + \mathbf{V}_{P_2A}$

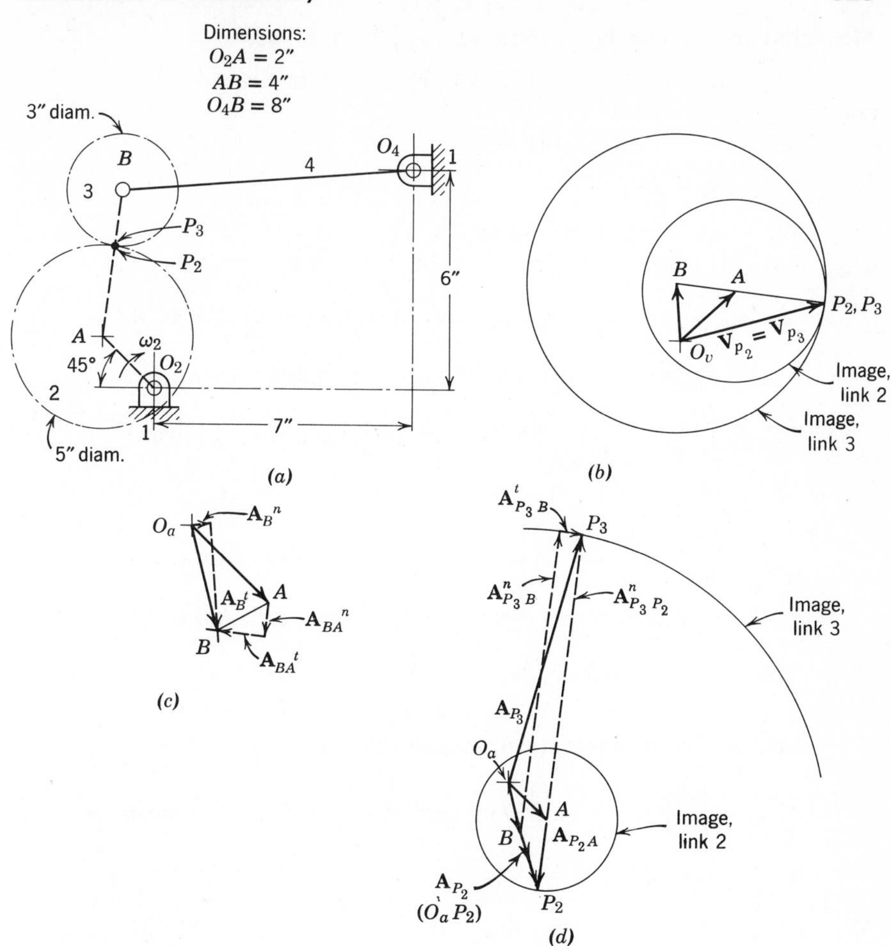

Figure 10.38

where

V_B = direction perpendicular to O_4B, magnitude unknown

$V_A = (O_2A)\omega_2 = (2)10 = 20$ in./sec, direction perpendicular to O_2A

V_{BA} = direction perpendicular to line joining points B and A, magnitude unknown

V_{P_2} = direction perpendicular to O_2P_2, magnitude unknown

V_{P_2A} = direction perpendicular to P_2A, magnitude unknown

Measured on polygon $V_B = 16$ in./sec, $V_{BA} = 16$ in./sec

$$V_{P_2} = 41 \text{ in./sec}, \; V_{P_2A} = 25 \text{ in./sec}$$

and

$$V_{P_3B} = 41 \text{ in./sec}$$

III. $$\mathbf{A}_B = \mathbf{A}_A + \mathbf{A}_{BA}$$

$$\mathbf{A}_B^n + \mathbf{A}_B^t = \mathbf{A}_A^n + \mathbf{A}_A^t + \mathbf{A}_{BA}^n + \mathbf{A}_{BA}^t$$

where

$$\mathbf{A}_B^n = \frac{V_B^2}{O_4B} = \frac{16^2}{8} = 32 \text{ in./sec}^2\text{, direction from } B \text{ toward } O_4$$

$$A_B = \text{direction perpendicular to } \mathbf{A}_B^n\text{, magnitude unknown}$$

$$A_A^n = \frac{V_A^2}{O_2A} = \frac{20^2}{2} = 200 \text{ in./sec}^2\text{, direction from } A \text{ toward } O_2$$

$$A_A^t = 0(\alpha_2 = 0)$$

$$A_{BA}^n = \frac{V_{BA}^2}{BA} = \frac{16^2}{4} = 64 \text{ in./sec}^2\text{, direction from } B \text{ toward } A$$

$$A_{BA}^t = \text{direction perpendicular to } \mathbf{A}_{BA}^n\text{, magnitude unknown}$$

IV. $$\mathbf{A}_{P_2} = \mathbf{A}_A + \mathbf{A}_{P_2A}^n + \mathbf{A}_{P_2A}^t$$

where

$$A_{P_2} = \text{direction unknown, magnitude unknown}$$

$$A_{P_2A}^n = \frac{V_{P_2A}^2}{P_2A} = \frac{25^2}{2.5} = 250 \text{ in./sec}^2\text{, direction from } P_2 \text{ toward } A$$

$$A_{P_2A}^t = 0(\alpha_2 = 0)$$

V. $$\mathbf{A}_{P_3} = \mathbf{A}_{P_2} + \mathbf{A}_{P_3P_2}^n + \mathbf{A}_{P_3P_2}^t$$

where

$$A_{P_3} = \text{direction unknown, magnitude unknown}$$

$$A_{P_3P_2}^n = \text{direction parallel to line } AB\text{, magnitude unknown}$$

$$A_{P_3P_2}^t = 0$$

VI. $$\mathbf{A}_{P_3} = \mathbf{A}_B + \mathbf{A}_{P_3B}^n + \mathbf{A}_{P_3B}^t$$

where

$$A_{P_3} = \text{direction unknown, magnitude unknown}$$

$$A_{P_3B}^n = \frac{V_{P_3B}^2}{P_3B} = \frac{41^2}{1.5} = 1120 \text{ in./sec}^2\text{, direction from } P_3 \text{ toward } B$$

$$\mathbf{A}_{P_3B}^t = \text{direction perpendicular to } \mathbf{A}_{P_3B}^n\text{, magnitude unknown}$$

Measured on polygon,

$$A_{P_3} = 965 \text{ in./sec}^2$$

Solution. It may be seen that the motions of the centers of the gears at A and B are the same as the pins of an equivalent four-bar linkage connecting points O_2, A, B, and O_4. The velocity polygon in Fig. 10.38*b* shows the determination of $\mathbf{V}_B$ and $\mathbf{V}_{BA}$ from Eq. I. In a similar manner $\mathbf{V}_{P_2}$ and $\mathbf{V}_{P_2A}$ are determined from Eq. II. The point P_3 is also known because $\mathbf{V}_{P_2} = \mathbf{V}_{P_3}$. The velocity image of link 2 is a circle with point A as the center and a radius AP_2. The velocity image of link 3 is a circle with point B as the center and a radius BP_3.

The acceleration polygon of Fig. 10.38*c* shows the determination of $\mathbf{A}_B$ and $\mathbf{A}_{BA}$ from Eq. III whose components are known as indicated.

In order to have a clearer diagram for determining $\mathbf{A}_{P_2}$ and $\mathbf{A}_{P_3}$, the acceleration vectors $\mathbf{A}_A$ and $\mathbf{A}_B$ from Fig. 10.38*c* are redrawn to a different scale in Fig. 10.38*d*. The acceleration polygon shows the determination of $\mathbf{A}_{P_2}$ from Eq. IV knowing $\mathbf{A}_A$ and $\mathbf{A}^n_{P_2A}(\mathbf{A}^t_{P_2A} = 0)$. The acceleration image of link 2 is a circle with point A as the center and a radius AP_2.

To determine $\mathbf{A}_{P_3}$ it is necessary to use Eqs. V and VI. The polygon shows the vector $\mathbf{A}_{P_2}$ to which is added the direction of $\mathbf{A}^n_{P_3P_2}$ from Eq. V ($\mathbf{A}^t_{P_3P_2} = 0$). Next, from Eq. VI the vector $\mathbf{A}^n_{P_3B}$ and the direction of $\mathbf{A}^t_{P_3B}$ are added to the vector $\mathbf{A}_B$. The intersection of the direction of $\mathbf{A}^n_{P_3P_2}$ and the direction of $\mathbf{A}^t_{P_3B}$ closes the polygon and determines point P_3. The image of link 3 is a circle with B as a center and a radius BP_3.

10.23 Analytical Solution of Relative Velocity and Acceleration Equations by Unit Vectors. Another method of velocity and acceleration analysis is to use the equations of relative motion but to express the components of these equations in unit vector form. By doing this, an analytical solution can be developed in place of a graphical one which uses velocity and acceleration polygons.

Consider the four-bar linkage shown in Fig. 10.39. The velocity and

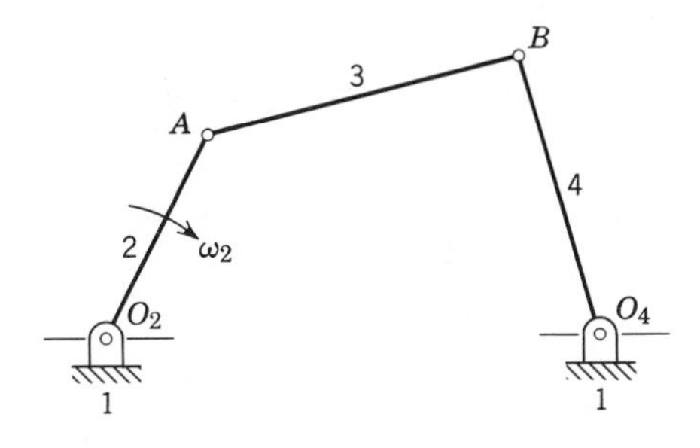

Figure 10.39

acceleration equation for points A and B can be expressed as follows:

$$\mathbf{V}_B = \mathbf{V}_A + \mathbf{V}_{BA}$$

$$\mathbf{A}_B = \mathbf{A}_A + \mathbf{A}_{BA}$$

$$\mathbf{A}_B^n + \mathbf{A}_B^t = \mathbf{A}_A^n + \mathbf{A}_A^t + \mathbf{A}_{BA}^n + \mathbf{A}_{BA}^t$$

The basic equations from which the magnitudes of the above components can readily be calculated are

$$V = r\omega \tag{10.34}$$

$$A^n = V\omega \tag{10.35}$$

$$A^t = r\alpha \tag{10.36}$$

It is obvious that these equations cannot give direction or sense. By writing them as cross products, however, directions as well as magnitudes can easily be determined. Equations 10.34, 10.35, and 10.36 may be rewritten as cross products as follows:

$$\mathbf{V} = \boldsymbol{\omega} \times \mathbf{r} \tag{10.37}$$

$$\mathbf{A}^n = \boldsymbol{\omega} \times \mathbf{V} = \boldsymbol{\omega} \times (\boldsymbol{\omega} \times \mathbf{r}) \tag{10.38}$$

$$\mathbf{A}^t = \boldsymbol{\alpha} \times \mathbf{r} = \dot{\boldsymbol{\omega}} \times \mathbf{r} \tag{10.39}$$

To illustrate this method in general terms, consider Eq. 10.37 and let

$$\boldsymbol{\omega} = \omega\mathbf{k}$$

and

$$\mathbf{r} = x_A\mathbf{i} + y_A\mathbf{j}$$

Keeping in mind that

$$\mathbf{i} \times \mathbf{i} = \mathbf{j} \times \mathbf{j} = \mathbf{k} \times \mathbf{k} = 0$$

$$\mathbf{i} \times \mathbf{j} = -\mathbf{j} \times \mathbf{i} = \mathbf{k}$$

$$\mathbf{j} \times \mathbf{k} = -\mathbf{k} \times \mathbf{j} = \mathbf{i}$$

$$\mathbf{k} \times \mathbf{i} = -\mathbf{i} \times \mathbf{k} = \mathbf{j}$$

Eq. 10.37 can be expanded to give

$$\mathbf{V} = \boldsymbol{\omega} \times \mathbf{r} = -\omega y_A\mathbf{i} + \omega x_A\mathbf{j}$$

from which the magnitude, direction, and sense of the velocity of the point in question can easily be determined. It is easier, however, to solve Eq. 10.37 if it is written as a determinant; thus,

$$\mathbf{V} = \boldsymbol{\omega} \times \mathbf{r} = \begin{vmatrix} \mathbf{i} & \mathbf{j} & \mathbf{k} \\ 0 & 0 & \omega \\ x_A & y_A & 0 \end{vmatrix} = -\omega y_A\mathbf{i} + \omega x_A\mathbf{j}$$

Consider next Eq. 10.38 written as a determinant.

$$\mathbf{A}^n = \boldsymbol{\omega} \times (\boldsymbol{\omega} \times \mathbf{r}) = \omega\mathbf{k} \times \begin{vmatrix} \mathbf{i} & \mathbf{j} & \mathbf{k} \\ 0 & 0 & \omega \\ x_A & y_A & 0 \end{vmatrix}$$

$$= \begin{vmatrix} \mathbf{i} & \mathbf{j} & \mathbf{k} \\ 0 & 0 & \omega \\ -\omega y_A & \omega x_A & 0 \end{vmatrix} = -\omega^2 x_A \mathbf{i} - \omega^2 y_A \mathbf{j}$$

From the above discussion, illustrated with Eqs. 10.37 and 10.38, it can be seen that it is an easy matter to express the components of the relative motion equations in unit vector form. A complete solution is then obtained by substituting in the equations of relative motion, and summing the **i** and **j** components. A complete solution is illustrated in the following example which analyzes the same mechanism as Example 10.2.

Example 10.12

Consider the Geneva Mechanism as analyzed in Example 10.2, and let it now be required to determine ω_3 and α_3 using the equations of relative motion with components expressed in unit vector form. A skeleton diagram of the mechanism is shown in Fig. 10.40.

The velocity and acceleration equations are written as follows considering the velocity and acceleration of point A_2 relative to A_3 because the path of point A_2 is known to be a straight line relative to A_3.

I. $\quad \mathbf{V}_{A_2} = \mathbf{V}_{A_3} + \mathbf{V}_{A_2A_3}$

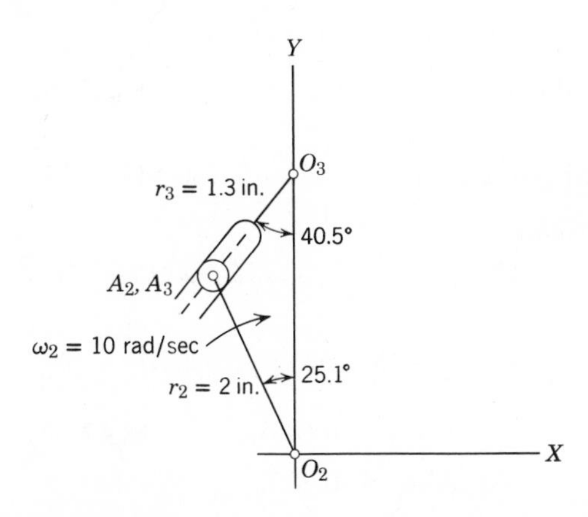

Figure 10.40

where

$$\mathbf{V}_{A_2} = \boldsymbol{\omega}_2 \times \mathbf{r}_2$$

$$\mathbf{V}_{A_3} = \boldsymbol{\omega}_3 \times \mathbf{r}_3$$

$$\begin{aligned}\mathbf{V}_{A_2A_3} &= V_{A_2A_3}(\sin 40.5\mathbf{i} + \cos 40.5\mathbf{j}) \\ &= V_{A_2A_3}(0.649\mathbf{i} + 0.760\mathbf{j}) \\ &= 0.649V_{A_2A_3}\mathbf{i} + 0.760V_{A_2A_3}\mathbf{j}\end{aligned}$$

$$\boldsymbol{\omega}_2 = -10\mathbf{k} \text{ rad/sec}$$

$$\boldsymbol{\omega}_3 = \omega_3\mathbf{k}$$

$$\begin{aligned}\mathbf{r}_2 &= r_2(-\sin 25.1\mathbf{i} + \cos 25.1\mathbf{j}) \\ &= 2.0(-0.424\mathbf{i} + 0.906\mathbf{j}) \\ &= -0.848\mathbf{i} + 1.812\mathbf{j}\end{aligned}$$

$$\begin{aligned}\mathbf{r}_3 &= r_3(-\sin 40.5\mathbf{i} - \cos 40.5\mathbf{j}) \\ &= 1.3(-0.649\mathbf{i} - 0.760\mathbf{j}) \\ &= -0.843\mathbf{i} - 0.987\mathbf{j}\end{aligned}$$

Substituting the values of $\boldsymbol{\omega}_2$, $\mathbf{r}_2$ and $\boldsymbol{\omega}_3$, $\mathbf{r}_3$ into the equations for $\mathbf{V}_{A_2}$ and $\mathbf{V}_{A_3}$, respectively, gives

$$\mathbf{V}_{A_2} = \boldsymbol{\omega}_2 \times \mathbf{r}_2 = \begin{vmatrix} \mathbf{i} & \mathbf{j} & \mathbf{k} \\ 0 & 0 & -10 \\ -0.848 & 1.812 & 0 \end{vmatrix} = 18.12\mathbf{i} + 8.48\mathbf{j}$$

$$\mathbf{V}_{A_3} = \boldsymbol{\omega}_3 \times \mathbf{r}_3 = \begin{vmatrix} \mathbf{i} & \mathbf{j} & \mathbf{k} \\ 0 & 0 & \omega_3 \\ -0.843 & -0.987 & 0 \end{vmatrix} = 0.987\omega_3\mathbf{i} - 0.843\omega_3\mathbf{j}$$

Substituting the above values for $\mathbf{V}_{A_2}$, $\mathbf{V}_{A_3}$, and $\mathbf{V}_{A_2A_3}$ into Eq. I, gives the following:

$$18.12\mathbf{i} + 8.48\mathbf{j} = 0.987\omega_3\mathbf{i} - 0.843\omega_3\mathbf{j} + 0.649V_{A_2A_3}\mathbf{i} + 0.760V_{A_2A_3}\mathbf{j}$$

Summing **i** *components*,

$$18.12\mathbf{i} = 0.987\omega_3\mathbf{i} + 0.649V_{A_2A_3}\mathbf{i}$$

Summing **j** *components*,

$$8.48\mathbf{j} = -0.843\omega_3\mathbf{j} + 0.760V_{A_2A_3}\mathbf{j}$$

Therefore,

$$0.987\omega_3 + 0.649V_{A_2A_3} = 18.12$$

$$-0.843\omega_3 + 0.760V_{A_2A_3} = 8.48$$

Multiplying the second equation by 0.987/0.843 and adding the two equations gives

$$1.539 V_{A_2A_3} = 28.06$$
$$V_{A_2A_3} = 18.2 \text{ in./sec}$$

Therefore,

$$\mathbf{V}_{A_2A_3} = 18.2(0.649\mathbf{i} + 0.760\mathbf{j}) = 11.88\mathbf{i} + 13.9\mathbf{j}$$

and

$$\omega_3 = \frac{18.12 - (0.649)18.2}{0.987} = 6.35 \text{ rad/sec} \quad \text{(ccw)}$$

II. $\mathbf{A}_{A_2} = \mathbf{A}_{A_3} + \mathbf{A}_{A_2A_3}$

$$\mathbf{A}^n_{A_2} + \mathbf{A}^t_{A_2} = \mathbf{A}^n_{A_3} + \mathbf{A}^t_{A_3} + \mathbf{A}^n_{A_2A_3} + 2\boldsymbol{\omega}_3 \times \mathbf{V}_{A_2A_3} + \mathbf{A}^t_{A_2A_3}$$

where

$$\mathbf{A}^n_{A_2} = \boldsymbol{\omega}_2 \times \mathbf{V}_{A_2} = \boldsymbol{\omega}_2 \times (\boldsymbol{\omega}_2 \times \mathbf{r}_2)$$

$$= -10\mathbf{k} \times \begin{vmatrix} \mathbf{i} & \mathbf{j} & \mathbf{k} \\ 0 & 0 & -10 \\ -0.848 & 1.812 & 0 \end{vmatrix} = \begin{vmatrix} \mathbf{i} & \mathbf{j} & \mathbf{k} \\ 0 & 0 & -10 \\ 18.12 & 8.48 & 0 \end{vmatrix}$$

$$\mathbf{A}^n_{A_2} = 84.8\mathbf{i} - 181.2\mathbf{j}$$

$$\mathbf{A}^t_{A_2} = 0(\boldsymbol{\alpha}_2 = 0)$$

$$\mathbf{A}^n_{A_3} = \boldsymbol{\omega}_3 \times \mathbf{V}_{A_3} = \boldsymbol{\omega}_3 \times (\boldsymbol{\omega}_3 \times \mathbf{r}_3)$$

$$= 6.35\mathbf{k} \times \begin{vmatrix} \mathbf{i} & \mathbf{j} & \mathbf{k} \\ 0 & 0 & 6.35 \\ -0.843 & -0.987 & 0 \end{vmatrix} = \begin{vmatrix} \mathbf{i} & \mathbf{j} & \mathbf{k} \\ 0 & 0 & 6.35 \\ 6.25 & -5.33 & 0 \end{vmatrix}$$

$$= 33.7\mathbf{i} + 39.5\mathbf{j}$$

$$\mathbf{A}^t_{A_3} = \dot{\boldsymbol{\omega}}_3 \times \mathbf{r}_3 = \begin{vmatrix} \mathbf{i} & \mathbf{j} & \mathbf{k} \\ 0 & 0 & \dot{\omega}_3 \\ -0.843 & -0.987 & 0 \end{vmatrix} = 0.987\dot{\omega}_3\mathbf{i} - 0.843\dot{\omega}_3\mathbf{j}$$

$$\mathbf{A}^n_{A_2A_3} = 0(\mathbf{R} = \infty)$$

$$2\boldsymbol{\omega}_3 \times \mathbf{V}_{A_2A_3} = \begin{vmatrix} \mathbf{i} & \mathbf{j} & \mathbf{k} \\ 0 & 0 & 12.70 \\ 11.88 & 13.9 & 0 \end{vmatrix}$$

$$= -176\mathbf{i} + 150\mathbf{j}$$

$$\mathbf{A}^t_{A_2A_3} = A^t_{A_2A_3}(0.649\mathbf{i} + 0.760\mathbf{j})$$

$$= 0.649 A^t_{A_2A_3}\mathbf{i} + 0.760 A^t_{A_2A_3}\mathbf{j}$$

Substituting the above values into the component form of Eq. II gives

$$84.8\mathbf{i} - 181.2\mathbf{j} = 33.7\mathbf{i} + 39.5\mathbf{j} + 0.987\dot{\omega}_3\mathbf{i} - 0.843\dot{\omega}_3\mathbf{j}$$
$$-176\mathbf{i} + 150\mathbf{j} + 0.649A^t_{A_2A_3}\mathbf{i} + 0.760A^t_{A_2A_3}\mathbf{j}$$

Summing **i** *components,*

$$84.8\mathbf{i} = 33.7\mathbf{i} + 0.987\dot{\omega}_3\mathbf{i} - 176\mathbf{i} + 0.649A^t_{A_2A_3}\mathbf{i}$$

Summing **j** *components,*

$$-181.2\mathbf{j} = 39.5\mathbf{j} - 0.843\dot{\omega}_3\mathbf{j} + 150\mathbf{j} + 0.760A^t_{A_2A_3}\mathbf{j}$$

Therefore,

$$0.987\dot{\omega}_3 + 0.649A^t_{A_2A_3} = 227.1$$
$$-0.843\dot{\omega}_3 + 0.760A^t_{A_2A_3} = -370.7$$

Multiplying the second equation by 0.987/0.843 and adding the two equations gives

$$1.539A^t_{A_2A_3} = -206.9$$
$$A^t_{A_2A_3} = -135.1 \text{ in./sec}^2$$

Therefore,

$$\mathbf{A}^t_{A_2A_3} = -135.1(0.649\mathbf{i} + 0.760\mathbf{j}) = -87.68\mathbf{i} - 102.7\mathbf{j}$$

and

$$\dot{\omega}_3 = \frac{227.1 - (0.649)(-135.1)}{0.987} = 318 \text{ rad/sec}^2 \quad \text{(ccw)}$$

$$\mathbf{A}^t_{A_3} = 0.987\dot{\omega}_3\mathbf{i} - 0.843\dot{\omega}_3\mathbf{j}$$
$$= 318(0.987\mathbf{i} - 0.843\mathbf{j})$$
$$\mathbf{A}^t_{A_3} = 314\mathbf{i} - 269\mathbf{j}$$
$$|\mathbf{A}^t_{A_3}| = \sqrt{314^2 + 269^2} = 413 \text{ in./sec}^2$$
$$\mathbf{A}^n_{A_3} = 33.7\mathbf{i} + 39.5\mathbf{j}$$
$$|\mathbf{A}^n_{A_3}| = \sqrt{33.7^2 + 39.5^2} = 52.5 \text{ in./sec}^2$$
$$\mathbf{A}_{A_3} = \mathbf{A}^n_{A_3} + \mathbf{A}^t_{A_3}$$
$$|\mathbf{A}_{A_3}| = \sqrt{52.5^2 + 413^2}$$
$$= 416 \text{ in./sec}^2$$

10.24 Graphical Differentiation. A kinematic method which should not be overlooked is the method of graphical differentiation which may be used successfully regardless of the complexity of the arrangement of the links of a mechanism. The disadvantage of the method is its lack of accuracy. Graphical differentiation depends on a graphical determination of the slope of a curve to determine derivatives. The method is illustrated in Fig. 10.41 for a linkage in which the driving link 2 rotates at constant angular velocity and the driven link 4 oscillates as shown. Twelve phases of the mechanism are shown to scale K_s for equal increments of time as given by the equal angular displacements of link 2. The velocity and acceleration of point B are desired. Curves are shown for the coordinate displacements X and Y of the point B as it traverses its curvilinear path. See Fig. 10.41*b*.

The abscissa of the displacement curve is a line of arbitrary length L divided into 12 equal parts to represent equal time intervals in one revolution of link 2. Since the time for one revolution of link 2 is $1/n$ min or $60/n$ sec (n = r/min), the time scale for the abscissa is $K_t = 60/nL$ sec/in. The displacements X, Y of point B are shown on the ordinate of the displacement curve to the same scale K_s as used in the layout of the mechanism.

Graphical differentiation is accomplished by drawing a tangent to the

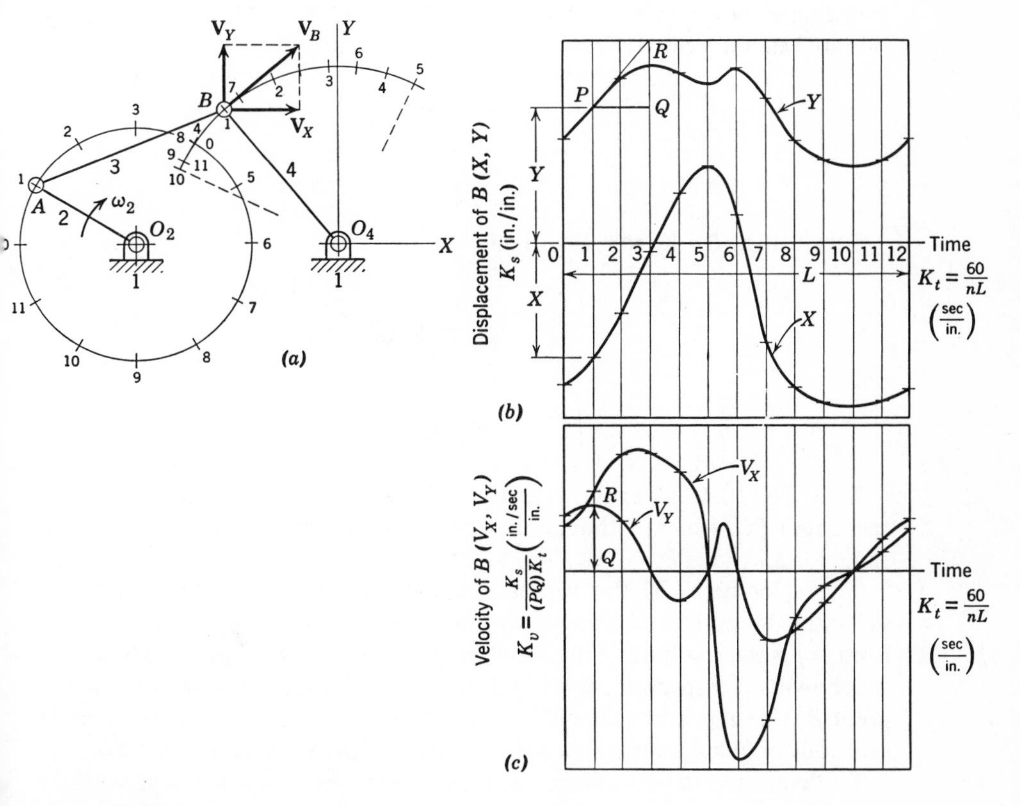

Figure 10.41

displacement curve at some point (such as for position 1 in Fig. 10.41*b*) and determining the slope of the curve from the triangle *PRQ*. The slope represents velocity or the derivative of displacement with respect to time.

$$V = \frac{(QR)K_s}{(PQ)K_t} \tag{10.40}$$

In Eq. 10.40 the lengths *QR* and *PQ* are inches on the paper, and the scales K_s and K_t are required to convert slope to units of velocity. If K_s is in inches per inch and K_t in seconds per inch, velocity is then in inches per second.

To plot a curve of velocity against time as shown in Fig. 10.41*c*, slopes at the incremental points on the displacement curve are evaluated graphically. If *PQ* is taken as the same length for all triangles drawn to determine slope, then the distance *QR* is the variable showing the variations in velocity. *QR* may be transferred from the triangle of the displacement curve to the velocity curve as the ordinate. As shown in Fig. 10.41*c*, the velocity curves for the coordinate velocities $\mathbf{V}_X$ and $\mathbf{V}_Y$ of point *B* are plotted. However, since *QR* is in inches, the velocity scale K_v for the curves must be determined.

$$V = (QR)K_v$$

$$K_v = \frac{V}{QR}$$

Substituting for *V* from Eq. 10.40,

$$K_v = \frac{(QR)K_s}{(QR)(PQ)K_t}$$

$$K_v = \frac{K_s}{(PQ)K_t} \tag{10.41}$$

Thus Eq. 10.41 gives the velocity scale in terms of the other scales and the length *PQ*, which, although an arbitrarily chosen length in inches, is the same for all triangles.

The velocity $\mathbf{V}_B$ is the vector resultant of the component coordinate velocities $\mathbf{V}_X$ and $\mathbf{V}_Y$. As shown in Fig. 10.41*a*, for position 1 of the mechanism, $\mathbf{V}_B$ is the resultant of its components and should be normal to line O_4B. As shown by inspection of the velocity curves, the maximum velocity of point *B* is near positions 6 and 7. Also, the curves show that at the extreme positions of link 4, namely, positions 5 and 10, the velocity of *B* is zero.

To determine the coordinate accelerations $\mathbf{A}_X$ and $\mathbf{A}_Y$ of *B*, the velocity curves may be differentiated graphically in a similar manner and curves may be shown of acceleration against time. The acceleration scale may be calculated from the following expression:

$$K_a = \frac{K_v}{(P'Q')K_t} \tag{10.42}$$

where K_a is the acceleration scale and $P'Q'$ is an arbitrary length similar to PQ.

The accuracy of differentiating graphically depends on the care taken in drawing tangents and on the number of increments into which the abscissa of the displacement curve is divided. Accuracy increases as the number of increments is increased and the individual increments are made smaller.

As has been shown above, graphical differentiation is a very simple method of determining velocity and acceleration curves from a displacement–time curve when a complete cycle of a mechanism is to be analyzed. The method is rapid in plotting one curve from another but unfortunately, the accuracy is limited. It is obvious that in a case where the equations for displacement, velocity, and acceleration are readily available, as in the slider-crank mechanism, it is easier to calculate the values and plot the curves if desired than to resort to graphical differentiation. In other mechanisms, however, such as that shown in Fig. 10.41, graphical differentiation is much quicker than analytical methods providing that sufficient accuracy can be obtained.

The accuracy of this method can be greatly improved by using a digital computer to perform the differentiation instead of doing it graphically. This can easily be done if the displacement–time values, or the equation from which they can be calculated, are available. The example which follows shows a comparison of the values of velocity for the piston in a slider-crank mechanism found by computer differentiation and by formula.

Example 10.13

A slider-crank mechanism with a crank of 2 in. and a connecting rod of 8 in. operates at a crank speed of 3300 r/min. Determine the piston velocity (ft/sec) for 90° of crank rotation starting from top dead center in increments of 1° by the following methods.

1. Numerical differentiation of the displacement–time values calculated from the equation

$$x = R(1 - \cos\theta) + \frac{R^2}{2L}\sin^2\theta$$

from Chapter 2 using finite difference methods.

2. Direct calculation of velocity from the equation

$$V = R\omega\left[\sin\theta + \frac{R}{2L}\sin 2\theta\right]$$

Show the improvement in accuracy for numerical differentiation by also taking increments of 0.1° and 0.01°.

Solution.

$$\omega = \frac{2\pi n}{60} = 345.40 \text{ rad/sec}$$

$$\text{Time for stroke } (180°) = 0.00909 \text{ sec}$$

$$\text{Time for } 1° \text{ of crank rotation} = 0.0000505 \text{ sec}$$

$$K_v = \frac{K_s}{(PQ)K_t} = \frac{\frac{1}{12}}{0.0000505} = 1650.1 \text{ ft/sec in.}^{-1}$$

The space scale is taken full size and converted to ft/in. The term $(PQ)K_t$ is the value of the increment in seconds and changes if the increment is changed.

After the value of piston displacement x has been calculated for each angular increment, the change in displacement Δx between increments is determined. The value of Δx is proportional to the velocity for the particular point under consideration, and the product of Δx and K_v gives the velocity in ft/sec. An illustration of this is shown in Fig. 10.42.

Values of velocities determined by formula and by numerical differentiation for increments of 1°, 0.1°, and 0.01° are shown in the table which follows. It is interesting to note how closely the velocities by differentiation with 0.01° increments match the velocities calculated by formula.

Although not included in this example, piston accelerations can be determined in a similar manner from velocities.

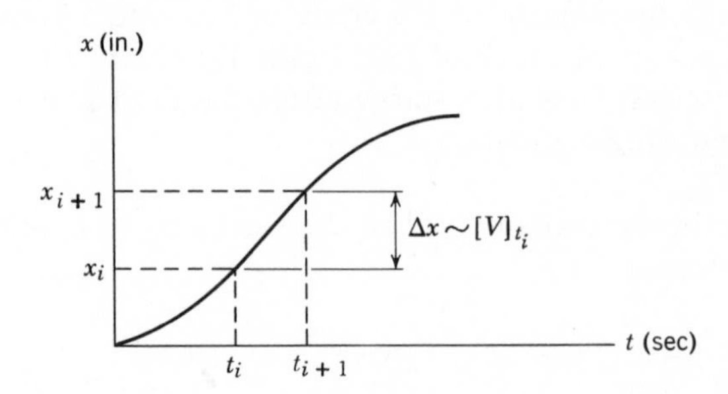

Figure 10.42

Numerical Differentiation

θ (degrees)	V(ft/sec) formula	V(ft/sec) 1° increment	V(ft/sec) 0.1° increment	V(ft/sec) 0.01° increment
1	1.26	0.63	1.19	1.25
2	2.51	1.88	2.45	2.51
3	3.77	3.14	3.70	3.76
4	5.02	4.39	4.96	5.01
5	6.27	5.65	6.21	6.26
6	7.52	6.89	7.45	7.49
7	8.76	8.14	8.70	8.76
8	10.00	9.38	9.94	10.01
9	11.23	10.62	11.17	11.23
10	12.46	11.85	12.40	12.45
20	24.33	23.76	24.27	24.31
30	35.03	34.53	34.98	35.04
40	44.11	43.70	44.07	44.11
50	51.21	50.91	51.18	51.22
60	56.11	55.92	56.09	56.02
70	58.75	58.67	58.74	58.69
80	59.18	59.21	59.19	59.17
90	57.59	57.72	57.61	57.59

It should be mentioned that the values of V with 1° increments will more nearly match those of V by formula if the latter are calculated at the midpoints of the intervals, that is, at 0.5°, 1.5°, 2.5°, and so on. This can be seen in the following tabulation for the first ten single-degree increments.

θ (degrees)	V(ft/sec) formula
0.5	0.628
1.5	1.883
2.5	3.138
3.5	4.391
4.5	5.642
5.5	6.890
6.5	8.135
7.5	9.376
8.5	10.613
9.5	11.844

The reason for this is that V by numerical differentiation more nearly represents the velocity at the midpoint instead of at the end of the interval. As the increments become smaller, this difference will decrease until it becomes negligible as in the case of the 0.01° increments.

10.25 *Kinematic Analysis by Complex Numbers.* In addition to the methods of velocity and acceleration analysis already presented, analytical solutions with kinematic vectors in complex form are often used.

A simple kinematic case is shown in Fig. 10.43*a* in which link 2 rotates about a fixed axis O_2. It is desired to determine the velocity and acceleration vectors $\mathbf{V}_P$ and $\mathbf{A}_P$ of particle P when the link is in the phase given by θ_2 and the known instantaneous angular velocity and angular acceleration are ω_2 and α_2.

The position of particle P may be represented by the vector $\mathbf{r}_P$ shown in Fig. 10.43*b*. By establishing real and imaginary axes as shown, $\mathbf{r}_P$ may be

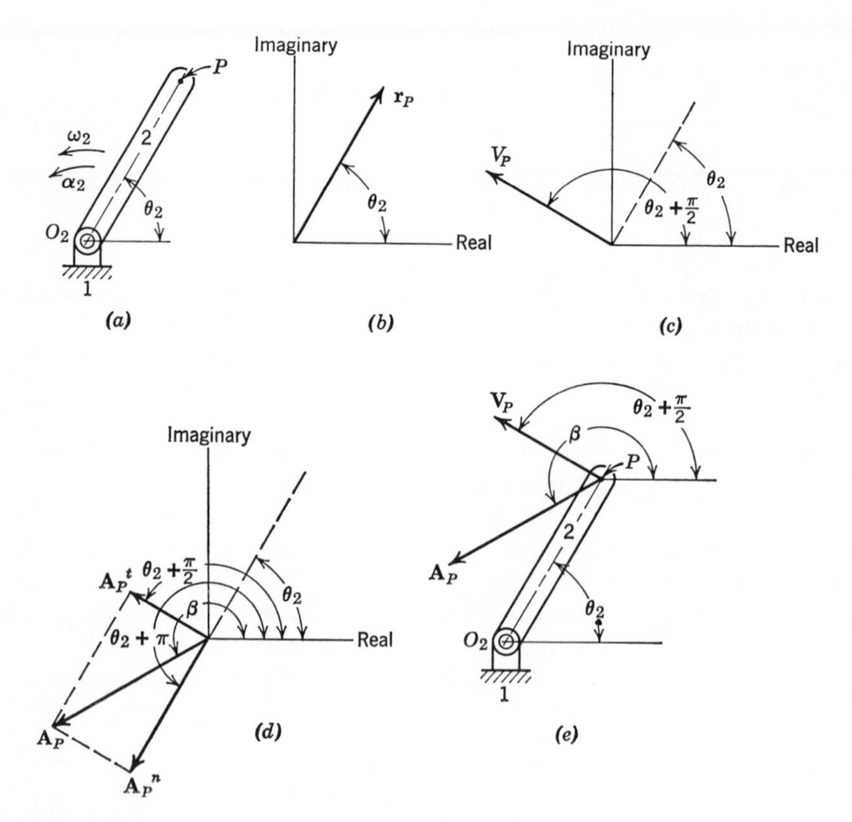

Figure 10.43

expressed by a complex number in any of the following equivalent forms:

$$\mathbf{r}_P = a + ib$$
$$\mathbf{r}_P = r_P(\cos\theta_2 + i\sin\theta_2)$$
$$\mathbf{r}_P = r_P e^{i\theta_2} \tag{10.43}$$

Although all forms of the complex number are useful ones, the simplest form for differentiation is the exponential form in which r_P is the magnitude of the position vector and $e^{i\theta_2}$ represents a vector of unit length at a counterclockwise angular position θ_2. Differentiation of Eq. 10.43 yields the velocity vector $\mathbf{V}_P$.

$$\mathbf{V}_P = \dot{\mathbf{r}}_P = r_P\dot{\theta}_2 ie^{i\theta_2}$$
$$\mathbf{V}_P = r_P\omega_2(ie^{i\theta_2}) \tag{10.44}$$

where $\dot{\mathbf{r}}_P = d\mathbf{r}_P/dt$ and $\dot{\theta}_2 = d\theta_2/dt = \omega_2$. The term in parentheses in Eq. 10.44 is the unit vector multiplied by i and is equivalent to $i(\cos\theta_2 + i\sin\theta_2)$. Using trigonometric relationships, it may be shown that $i(\cos\theta_2 + i\sin\theta_2)$ is equal to $\cos[\theta_2 + \pi/2] + i\sin[\theta_2 + \pi/2]$ so that $ie^{i\theta_2} = e^{i(\theta_2 + \pi/2)}$. Thus

$$\mathbf{V}_P = r_P\omega_2 e^{i(\theta_2 + \pi/2)} \tag{10.45}$$

As shown in Fig. 10.43c, the direction of the velocity vector $\mathbf{V}_P$ is given by the angle $(\theta_2 + \pi/2)$ and is shown to be at an angle 90° greater than the angle of $\mathbf{r}_P$. Thus, multiplication of the unit vector by i rotates the vector 90° in the counterclockwise sense. Also, each subsequent multiplication of the unit vector by i rotates the vector an additional 90° increment in the counterclockwise sense.

Differentiation of the velocity Eq. 10.44 gives the acceleration vector $\mathbf{A}_P$ as follows:

$$\mathbf{A}_P = \ddot{\mathbf{r}}_P = r_P\omega_2^2 i^2 e^{i\theta_2} + r_P\dot{\omega}_2 ie^{i\theta_2}$$
$$= r_P\omega_2^2(i^2 e^{i\theta_2}) + r_P\alpha_2(ie^{i\theta_2}) \tag{10.46}$$

where $\alpha_2 = d\omega_2/dt = \dot{\omega}_2$. The first right-hand term of Eq. 10.46 represents the normal component of acceleration $\mathbf{A}_P^n$, in which $r_P\omega_2^2$ is the magnitude and i^2 indicates that the direction is 180° greater than θ_2 as shown in Fig. 10.43d. The second term is the tangential component of acceleration $\mathbf{A}_P^t$ of magnitude $r_P\alpha_2$ and direction 90° greater than θ_2 as indicated by i. To designate the directions of the component accelerations, Eq. 10.46 may be rewritten as follows:

$$\mathbf{A}_P = r_P\omega_2^2 e^{i(\theta_2 + \pi)} + r_P\alpha_2 e^{i(\theta_2 + \pi/2)} \tag{10.47}$$

Equations 10.46 and 10.47 show that the acceleration vector $\mathbf{A}_P$ is the resultant of two perpendicular vectors. To determine the magnitude of the resultant vector and its angular position, the following algebraic steps may be made beginning with Eq. 10.46.

$$\begin{aligned}\mathbf{A}_P &= -r_P\omega_2^2(\cos\theta_2 + i\sin\theta_2) + r_P\alpha_2(i\cos\theta_2 - \sin\theta_2)\\ &= -(r_P\omega_2^2\cos\theta_2 + r_P\alpha_2\sin\theta_2) + i(-r_P\omega_2^2\sin\theta_2 + r_P\alpha_2\cos\theta_2)\\ &= a + ib \qquad (10.48)\end{aligned}$$

As Eq. 10.48 shows, the acceleration $\mathbf{A}_P$ may also be expressed as the resultant of two component vectors in which "a" is the real component and "b" is the perpendicular imaginary component. The magnitude of $\mathbf{A}_P$ may be determined as follows:

$$\begin{aligned}\mathbf{A}_P &= \sqrt{a^2 + b^2}\\ &= \sqrt{(r_P\omega_2^2\cos\theta_2 + r_P\alpha_2\sin\theta_2)^2 + (-r_P\omega_2^2\sin\theta_2 + r_P\alpha_2\cos\theta_2)^2}\\ &= \sqrt{(r_P\omega_2^2)^2 + (r_P\alpha_2)^2} \qquad (10.49)\end{aligned}$$

The direction of $\mathbf{A}_P$ is given by the angle β in Fig. 10.43d, and this angle may be determined as follows:

$$\tan\beta = \frac{b}{a} = \frac{(-\omega_2^2\sin\theta_2 + \alpha_2\cos\theta_2)}{-(\omega_2^2\cos\theta_2 + \alpha_2\sin\theta_2)} \qquad (10.50)$$

Using the angle β, the acceleration vector $\mathbf{A}_P$ may be expressed as a single

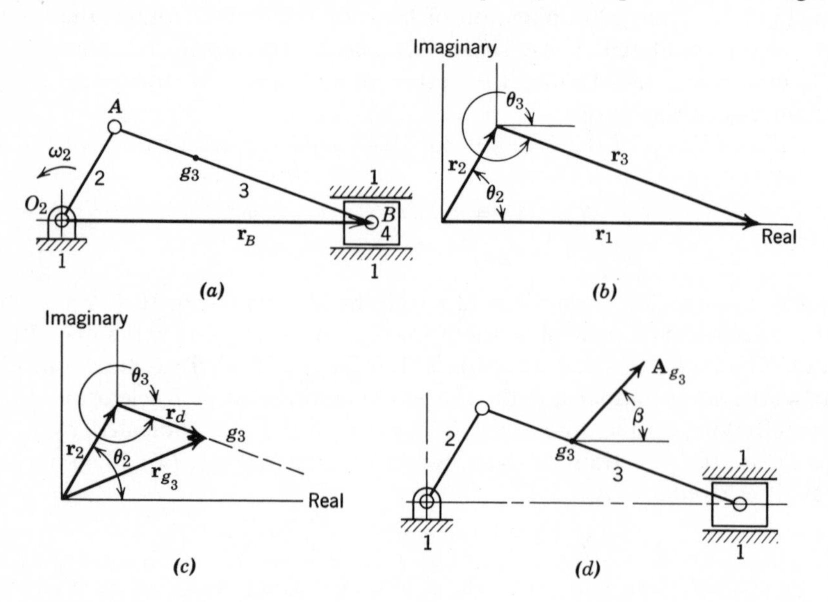

Figure 10.44

vector instead of two vectors as follows:

$$\mathbf{A}_P = A_P e^{i\beta} \tag{10.51}$$

In Fig. 10.43*e*, the velocity and acceleration vectors $\mathbf{V}_P$ and $\mathbf{A}_P$ are shown as fixed vectors at the particle P on the link.

It is important to note that the preceding relationships are based on the assumption that ω_2 and α_2 are known quantities for all phases θ_2 of the link. In many problems related to machinery the link may rotate at constant angular velocity so that ω_2 is constant and α_2 is zero. If, for example, α_2 is not zero but is a constant, then ω_2 is a function of time or θ_2. Considering the case where α_2 = constant = k, and ω_2 is zero at the initial condition $\theta_2 = 0$, the dependency of ω_2 on θ_2 may be determined as follows:

$$\frac{d\omega_2}{dt} = \alpha_2 = k$$

Expressing the derivative $d\omega_2/dt$ as $(d\omega_2/d\theta_2)(d\theta_2/dt) = \omega_2(d\omega_2/d\theta_2)$,

$$\omega_2\left(\frac{d\omega_2}{d\theta_2}\right) = k$$

$$\int \omega_2 \, d\omega_2 = k \int d\theta_2$$

$$\omega_2^2 = 2k\theta_2 + C_1 \tag{10.52}$$

C_1 is the constant of integration and is equal to zero for $\omega_2 = 0$ at $\theta_2 = 0$.

10.26 Kinematic Analysis of the Slider-Crank by Complex Numbers. In the slider-crank mechanism of Fig. 10.44*a*, the crank rotates at constant angular velocity ω_2, and the velocity $\mathbf{V}_B$ and acceleration $\mathbf{A}_B$ of the slider are to be determined. As shown, the position of particle B relative to the fixed point O_2 is given by the vector $\mathbf{r}_B$. Referring to Fig. 10.44*b*, it may be seen that two independent vector equations may be written for $\mathbf{r}_B$, namely, $\mathbf{r}_B = \mathbf{r}_1$ and $\mathbf{r}_B = \mathbf{r}_2 + \mathbf{r}_3$. The obvious result of combining the equations for $\mathbf{r}_B$ is the following vector equation:

$$\mathbf{r}_1 = \mathbf{r}_2 + \mathbf{r}_3 \tag{10.53}$$

If the vectors are to be represented by complex numbers, real and imaginary axes may be shown as in Fig. 10.44*b*, and Eq. 10.53 may be written as follows:

$$r_1 e^{i\theta_1} = r_2 e^{i\theta_2} + r_3 e^{i\theta_3} \tag{10.54}$$

where r_2 and r_3 are the fixed lengths of links 2 and 3, and r_1 is a variable length giving the position of the slider. The angle θ_1 of r_1 is fixed at $\theta_1 = 0$ so that $e^{i\theta_1} = 1$. Thus

$$\mathbf{r}_1 = r_2 e^{i\theta_2} + r_3 e^{i\theta_3} \tag{10.55}$$

Two successive differentiations of Eq. 10.55 yield expressions giving the velocity $\mathbf{V}_B$ and acceleration $\mathbf{A}_B$ as follows:

$$\mathbf{V}_B = \dot{\mathbf{r}}_1 = r_2\dot{\omega}_2(ie^{i\theta_2}) + r_3\omega_3(ie^{i\theta_3}) \tag{10.56}$$

$$\mathbf{A}_B = \ddot{\mathbf{r}}_1 = r_2(i\alpha_2 - \omega_2^2)(e^{i\theta_2}) + r_3(i\alpha_3 - \omega_3^2)(e^{i\theta_3}) \tag{10.57}$$

It may be seen from an inspection of Eqs. 10.55, 10.56, and 10.57 that, although the differentiations are made to determine the kinematic values of particle B, the equations also involve the angular velocities and accelerations of links 2 and 3 as well as their angular positions. In these equations, r_2, r_3, θ_2, ω_2, and α_2 are the known quantities, and the unknown quantities to be determined are six in number, namely, r_1, $\dot{r}_1$, $\ddot{r}_1$, θ_3, ω_3, and α_3.

Two of the unknowns, r_1 and θ_3, may be determined from Eq. 10.55 by separately equating the real and imaginary parts of the equation as follows:

$$\mathbf{r}_1 = r_2(\cos\theta_2 + i\sin\theta_2) + r_3(\cos\theta_3 + i\sin\theta_3)$$

$$r_1 = r_2\cos\theta_2 + r_3\cos\theta_3 \qquad \text{(Real)} \tag{10.58}$$

$$0 = r_2\sin\theta_2 + r_3\sin\theta_3 \qquad \text{(Imaginary)} \tag{10.59}$$

Equation 10.59 may be solved to determine θ_3.

$$\theta_3 = \sin^{-1}\left[\frac{-r_2}{r_3}\sin\theta_2\right] \tag{10.60}$$

and Eq. 10.58 may then be used to determine r_1.

In a similar manner, the unknowns $\dot{r}_1$ and ω_3 may be obtained from Eq. 10.56 by separately equating the real and imaginary parts of the equation.

$$\dot{\mathbf{r}}_1 = r_2\omega_2(i\cos\theta_2 - \sin\theta_2) + r_3\omega_3(i\cos\theta_3 - \sin\theta_3)$$

$$\dot{r}_1 = -r_2\omega_2\sin\theta_2 - r_3\omega_3\sin\theta_3 \qquad \text{(Real)} \tag{10.61}$$

$$0 = r_2\omega_2\cos\theta_2 + r_3\omega_3\cos\theta_3 \qquad \text{(Imaginary)} \tag{10.62}$$

Equation 10.62 allows the determination of ω_3.

$$\omega_3 = -\omega_2\frac{r_2\cos\theta_2}{r_3\cos\theta_3} \tag{10.63}$$

and Eq. 10.61 may then be used to determine $\dot{r}_1 = V_B$.

The remaining unknowns, $\ddot{r}_1$ and α_3, are determined from the real and imaginary parts of Eq. 10.57.

$$\ddot{r}_1 = -r_2(\omega_2^2\cos\theta_2 + \alpha_2\sin\theta_2) - r_3(\omega_3^2\cos\theta_3 + \alpha_3\sin\theta_3) \qquad \text{(Real)} \tag{10.64}$$

$$0 = r_2(\alpha_2\cos\theta_2 - \omega_2^2\sin\theta_2) + r_3(\alpha_3\cos\theta_3 - \omega_3^2\sin\theta_3) \qquad \text{(Imaginary)} \tag{10.65}$$

From Eq. 10.65, the unknown α_3 may be determined

$$\alpha_3 = \frac{r_2(\omega_2^2 \sin\theta_2 - \alpha_2 \cos\theta_2)}{r_3 \cos\theta_3} + \frac{\omega_3^2 \sin\theta_3}{\cos\theta_3} \tag{10.66}$$

and $\ddot{r}_1 = A_B$ may then be determined from Eq. 10.64. For constant angular velocity of the crank, the angular acceleration α_2 is zero, so that Eqs. 10.64 and 10.66 giving A_B and α_3 are somewhat simplified.

Those kinematic quantities of engineering interest, such as the velocity $\mathbf{V}_B$ and acceleration $\mathbf{A}_B$ of the slider and ω_3 and α_3 of the connecting rod, may be determined numerically from the preceding equations for all phases θ_2 of the crank and for arbitrary values of the crank speed ω_2 and L/R ratio (r_3/r_2). Although the calculations to be undertaken involve voluminous arithmetical operations, such operations may be assigned to the digital computer with the advantage that a great number of variations of the problem may be solved to optimize a design.

The velocity and acceleration of other particles of the mechanism may also be of engineering interest. For example, as will be discussed in Chapter 11, the accelerations of the mass centers of the individual links are important because they are related to the forces acting on the links. In considering the acceleration $\mathbf{A}_{g_3}$ of the mass center g_3 of link 3 in Fig. 10.44*a*, the following equations follow from the vector addition shown in Fig. 10.44*c*.

$$\mathbf{r}_{g_3} = \mathbf{r}_2 + \mathbf{r}_d = r_2 e^{i\theta_2} + r_d e^{i\theta_3} \tag{10.67}$$

$$\mathbf{V}_{g_3} = \dot{\mathbf{r}}_{g_3} = r_2\omega_2(ie^{i\theta_2}) + r_d\omega_3(ie^{i\theta_3}) \tag{10.68}$$

$$\mathbf{A}_{g_3} = \ddot{\mathbf{r}}_{g_3} = r_2(i\alpha_2 - \omega_2^2)(e^{i\theta_2}) + r_d(i\alpha_3 - \omega_3^2)(e^{i\theta_3}) \tag{10.69}$$

For constant angular speed of the crank, $\alpha_2 = 0$, so that

$$\begin{aligned}
\mathbf{A}_{g_3} &= -r_2\omega_2^2(e^{i\theta_2}) + r_d(i\alpha_3 - \omega_3^2)(e^{i\theta_3}) \\
&= (-r_2\omega_2^2 \cos\theta_2 - r_d\alpha_3 \sin\theta_3 - r_d\omega_3^2 \cos\theta_3) \\
&\quad + i(-r_2\omega_2^2 \sin\theta_2 + r_d\alpha_3 \cos\theta_3 - r_d\omega_3^2 \sin\theta_3) \\
&= a_{g_3} + ib_{g_3}
\end{aligned} \tag{10.70}$$

The magnitude of A_{g_3} may be determined from $A_{g_3} = \sqrt{a_{g_3}^2 + b_{g_3}^2}$, and the angle β which $\mathbf{A}_{g_3}$ makes with the real axis may be determined from $\tan\beta = b_{g_3}/a_{g_3}$. The vector $\mathbf{A}_{g_3}$ is shown as a fixed vector in Fig. 10.44*d*.

Example 10.14

The slider-crank of an internal combustion engine (Fig. 10.44*a*) includes a crank of 2.0 in. length and a connecting rod of 8.0 in. length. The crank speed of the engine is constant at 3000 r/min (314 rad/sec). Determine the acceleration of the mass center $\mathbf{A}_{g_3}$ of the connecting rod when the crank angle is $\theta_2 = 30°$. The mass center g_3 is located 2.0 in. from the crank pin at A. In addition, determine curves showing (1) the magnitude of $\mathbf{A}_{g_3}$ versus θ_2 and (2) the angle β which $\mathbf{A}_{g_3}$ makes with the real axis versus θ_2.

Solution. The calculation of the acceleration A_{g_3} may be made using Eq. 10.70 and the following given data: $r_2 = 2.0$ in., $r_3 = 8.0$ in., $r_d = 2.0$ in., $\omega_2 = 314$ rad/sec, and $\theta_2 = 30°$. However, before the calculation can be undertaken, the unknowns θ_3, ω_3, and α_3 must first be determined.

The connecting-rod angle θ_3 may be determined from Eq. 10.60 as follows:

$$\sin\theta_3 = -\frac{r_2}{r_3}\sin\theta_2 = -\frac{2.0}{8.0}\sin 30°$$

$$= -0.125$$

$$\theta_3 = -7.18° \text{ or } 352.82°$$

$$\cos\theta_3 = 0.992$$

It may be seen that for $\sin\theta_3 = -0.125$ there are two positions of the connecting rod, either $\theta_3 = 352.82°$ or $187.18°$, depending on whether the slider is to the right or to the left of the crank center O_2.

The angular velocity ω_3 and the angular acceleration α_3 of the connecting rod may be determined from Eqs. 10.63 and 10.66, respectively.

$$\omega_3 = -\omega_2\frac{r_2\cos\theta_2}{r_3\cos\theta_3} = -(314)\frac{2.0}{8.0}\frac{\cos 30°}{\cos(352.82°)}$$

$$= -68.56 \text{ rad/sec}$$

$$\alpha_3 = \frac{r_2}{r_3}\frac{(\omega_2^2\sin\theta_2 - \alpha_2\cos\theta_2)}{\cos\theta_3} + \frac{\omega_3^2\sin\theta_3}{\cos\theta_3}$$

$$= \frac{2.0}{8.0}\frac{(314^2\sin 30° - 0)}{\cos(352.82°)} + \frac{(-68.56)^2\sin(352.82°)}{\cos(352.82°)}$$

$$= 11{,}840 \text{ rad/sec}^2$$

With the preceding quantities determined, the real and imaginary components of the acceleration $\mathbf{A}_{g_3}$ may be determined by evaluating a_{g_3} and b_{g_3} in Eq. 10.70 as follows:

$$a_{g_3} = -r_2\omega_2^2\cos\theta_2 - r_d\alpha_3\sin\theta_3 - r_d\omega_3^2\cos\theta_3$$

$$= -2.0(314)^2(0.866) - 2.0(11{,}840)(-0.125)$$

$$-2.0(-68.56)^2(0.992)$$

$$= -177{,}300 \text{ in./sec}^2$$

$$\begin{aligned} b_{g3} &= -r_2\omega_2^2 \sin\theta_2 + r_d\alpha_3 \cos\theta_3 - r_d\omega_3^2 \sin\theta_3 \\ &= -2.0(314)^2(0.500) + 2.0(11{,}840)(0.992) \\ &\quad -2.0(-68.56)^2(-0.125) \\ &= -74{,}020 \text{ in./sec}^2 \end{aligned}$$

The magnitude of $\mathbf{A}_{g3}$ is the vector sum of the components:

$$\begin{aligned} A_{g3} &= \sqrt{a_{g3}^2 + b_{g3}^2} = \sqrt{(-177{,}300)^2 + (-74{,}020)^2} \\ &= 192{,}200 \text{ in./sec}^2 \end{aligned}$$

and the angle of $\mathbf{A}_{g3}$ with the real or horizontal axis is given by β.

$$\tan\beta = \frac{b_{g3}}{a_{g3}} = \frac{-74{,}020}{-177{,}300} = 0.417$$

$$\beta = 202.67°$$

Calculations similar to those illustrated for increasing crank angle θ_2 in 10° increments make possible the plotting of the curves of A_{g3} and β for one cycle of the crank as shown in Fig. 10.45.

10.27 Inversion of the Slider-Crank. Of the inversions of the slider-crank, the crank shaper (Fig. 10.46*a*) is interesting to analyze by complex numbers because the Coriolis component of acceleration is involved. In Fig. 10.46*b* are shown the vectors giving the position $\mathbf{r}_{B_2}$ of particle B_2 on the crank at the pin connection to the slider. Two independent vector equations for the position of B_2, namely, $\mathbf{r}_{B_2} = \mathbf{r}_4$ and $\mathbf{r}_{B_2} = \mathbf{r}_1 + \mathbf{r}_2$, may

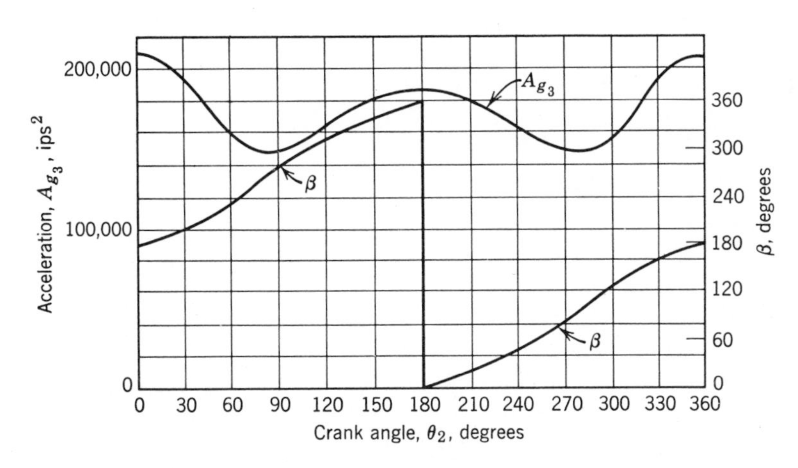

Figure 10.45

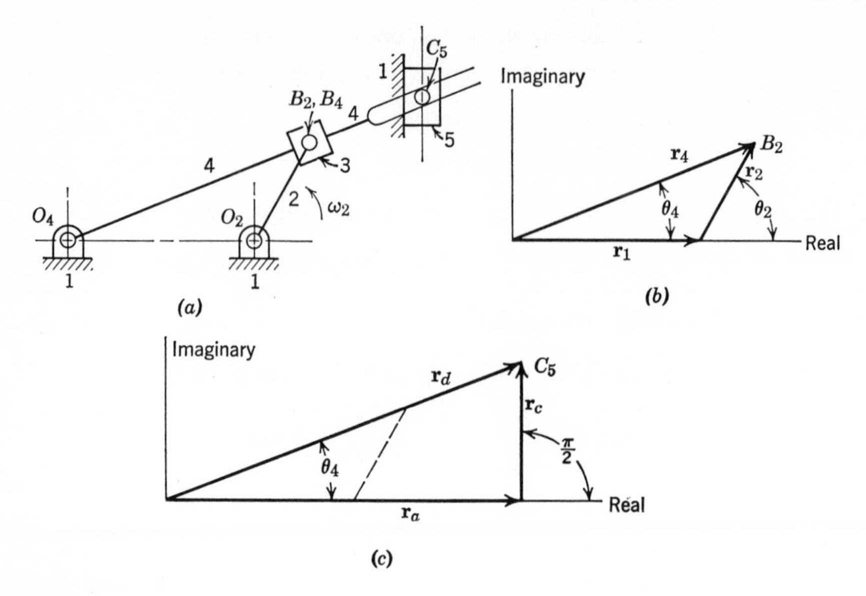

Figure 10.46

be combined to give the following vector equation:

$$\mathbf{r}_4 = \mathbf{r}_1 + \mathbf{r}_2$$

$$r_4(e^{i\theta_4}) = r_1 + r_2(e^{i\theta_2}) \tag{10.71}$$

Differentiating Eq. 10.71 yields the following velocity equation:

$$r_4\omega_4(ie^{i\theta_4}) + \dot{r}_4(e^{i\theta_4}) = r_2\omega_2(ie^{i\theta_2}) \tag{10.72}$$

From an inspection of Eq. 10.72 term by term, it may be seen that the equation is another form of the equation $\mathbf{V}_{B_4} + \mathbf{V}_{B_2B_4} = \mathbf{V}_{B_2}$ for the coincident particles B_4 and B_2. Differentiating Eq. 10.72 yields the following acceleration equation:

$$r_4\omega_4^2(i^2e^{i\theta_4}) + r_4\alpha_4(ie^{i\theta_4}) + 2\dot{r}_4\omega_4(ie^{i\theta_4}) + \ddot{r}_4(e^{i\theta_4})$$
$$= r_2\omega_2^2(i^2e^{i\theta_2}) + r_2\alpha_2(ie^{i\theta_2}) \tag{10.73}$$

Inspection of Eq. 10.73 term by term shows that the equation is an alternate form of the equation $\mathbf{A}^n_{B_4} + \mathbf{A}^t_{B_4} + 2\boldsymbol{\omega}_4 \times \mathbf{V}_{B_2B_4} + \mathbf{A}^t_{B_2B_4} = \mathbf{A}^n_{B_2} + \mathbf{A}^t_{B_2}$.

In the crank-shaper mechanism, link 2 is the driving link usually rotating at a known constant angular velocity ω_2 with α_2 equal to zero. Thus, referring to Eqs. 10.71, 10.72, and 10.73, it may be seen that r_1, r_2, θ_2, ω_2, and α_2 are the known quantities, and the six unknowns to be determined are θ_4, ω_4, α_4, r_4, $\dot{r}_4$, and $\ddot{r}_4$. By equating the real and imaginary parts of each of the Eqs. 10.71, 10.72, and 10.73, six equations are obtained which make possible the determination of the six unknowns.

After θ_4, ω_4, and α_4 have been determined for a known value of ω_2 and an arbitrary value of θ_2, it becomes possible to determine numerically the velocity and acceleration of other particles of the mechanism. For example, since the crank shaper is a quick-return mechanism, it is of interest to determine the velocity $\mathbf{V}_{C_5}$ of the tool-holding slider (link 5) of Fig. 10.46*a* for comparison of the magnitudes of the slider velocity during the working and return strokes of the mechanism. In Fig. 10.46*c* is shown the vector polygon which includes the position vector $\mathbf{r}_c$ of particle C_5. From the polygon,

$$\mathbf{r}_d = \mathbf{r}_a + \mathbf{r}_c$$

$$\mathbf{r}_c = \mathbf{r}_d - \mathbf{r}_a$$

$$r_c(e^{i\pi/2}) = r_d(e^{i\theta_4}) - r_a \tag{10.74}$$

Differentiating Eq. 10.74 gives the following velocity expression:

$$\mathbf{V}_{C_5} = \dot{r}_c(e^{i\pi/2}) = r_d\omega_4(ie^{i\theta_4}) + \dot{r}_d(e^{i\theta_4}) \tag{10.75}$$

In Eqs. 10.74 and 10.75, $\mathbf{r}_a$ is a known fixed length, and θ_4 and ω_4 are known from previously developed equations. Equating the real and imaginary parts of each of Eqs. 10.74 and 10.75, four equations become available for the determination of the four unknowns r_c, $\dot{r}_c$, r_d, and $\dot{r}_d$, of which $\dot{r}_c$ is the velocity magnitude V_{C_5} of the slider.

10.28 Analysis of the Four-Bar Linkage. In the analysis of the four-bar linkage of Fig. 10.47*a* by complex numbers, the trigonometric relationships are somewhat more involved than those for the slider-crank or the crank shaper because the basic vector polygon consists of four sides instead of three.

Referring to Fig. 10.47*a*, the position vector $\mathbf{r}_B$ for point *B* of the linkage is the resultant in two independent vector equations: $\mathbf{r}_B = \mathbf{r}_2 + \mathbf{r}_3$ and $\mathbf{r}_B = \mathbf{r}_1 + \mathbf{r}_4$. These equations may be combined as shown in the following,

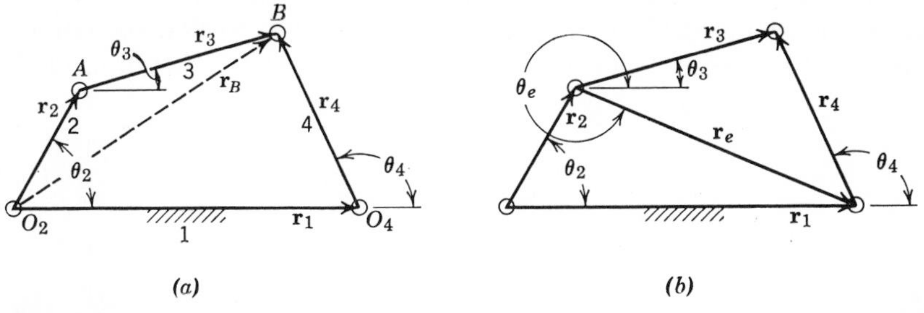

Figure 10.47

and derivatives may be taken to determine the velocity $\mathbf{V}_B$ and the acceleration $\mathbf{A}_B$ of point B.

$$\mathbf{r}_B = r_2(e^{i\theta_2}) + r_3(e^{i\theta_3}) = r_1 + r_4(e^{i\theta_4}) \tag{10.76}$$

$$\mathbf{V}_B = r_2\omega_2(ie^{i\theta_2}) + r_3\omega_3(ie^{i\theta_3}) = r_4\omega_4(ie^{i\theta_4}) \tag{10.77}$$

$$\begin{aligned}\mathbf{A}_B &= r_2(i\alpha_2 - \omega_2^2)(e^{i\theta_2}) + r_3(i\alpha_3 - \omega_3^2)(e^{i\theta_3}) \\ &= r_4(i\alpha_4 - \omega_4^2)(e^{i\theta_4})\end{aligned} \tag{10.78}$$

In the preceding equations, the fixed lengths r_1, r_2, r_3, and r_4 are known as is the angular position $\theta_1 = 0$ of the fixed link 1. If link 2 is the driving link, then θ_2, ω_2, and α_2 are also known quantities. It may be seen that the six unknown quantities to be determined from the three equations are θ_3, θ_4, ω_3, ω_4, α_3, and α_4.

The angular velocities ω_3 and ω_4 may be determined by equating the real and imaginary parts of the velocity Eq. 10.77 as follows:

$$r_2\omega_2 \sin\theta_2 + r_3\omega_3 \sin\theta_3 = r_4\omega_4 \sin\theta_4$$

$$r_2\omega_2 \cos\theta_2 + r_3\omega_3 \cos\theta_3 = r_4\omega_4 \cos\theta_4$$

Multiplying the upper equation by $\cos\theta_4$ and the lower by $\sin\theta_4$,

$$r_2\omega_2 \sin\theta_2 \cos\theta_4 + r_3\omega_3 \sin\theta_3 \cos\theta_4 = r_4\omega_4 \sin\theta_4 \cos\theta_4$$

$$r_2\omega_2 \cos\theta_2 \sin\theta_4 + r_3\omega_3 \cos\theta_3 \sin\theta_4 = r_4\omega_4 \sin\theta_4 \cos\theta_4$$

Subtracting the upper equation from the lower,

$$r_2\omega_2(\cos\theta_2 \sin\theta_4 - \sin\theta_2 \cos\theta_4) + r_3\omega_3(\cos\theta_3 \sin\theta_4 - \sin\theta_3 \cos\theta_4) = 0$$

$$r_2\omega_2 \sin(\theta_4 - \theta_2) + r_3\omega_3 \sin(\theta_4 - \theta_3) = 0$$

$$\omega_3 = -\frac{r_2}{r_3}\omega_2 \frac{\sin(\theta_4 - \theta_2)}{\sin(\theta_4 - \theta_3)} \tag{10.79}$$

In a similar manner,

$$\omega_4 = \frac{r_2}{r_4}\omega_2 \frac{\sin(\theta_3 - \theta_2)}{\sin(\theta_3 - \theta_4)} \tag{10.80}$$

The angular accelerations α_3 and α_4 may be determined by solving simultaneously the two equations obtained from Eq. 10.78 on equating real and imaginary parts. The expressions so determined are the following:

$$\alpha_3 = \frac{\left\{\begin{aligned}&-r_2\alpha_2 \sin(\theta_4 - \theta_2) + r_2\omega_2^2 \cos(\theta_4 - \theta_2) \\ &\qquad + r_3\omega_3^2 \cos(\theta_4 - \theta_3) - r_4\omega_4^2\end{aligned}\right\}}{r_3 \sin(\theta_4 - \theta_3)} \tag{10.81}$$

$$\alpha_4 = \frac{\left\{\begin{aligned}&r_2\alpha_2 \sin(\theta_3 - \theta_2) - r_2\omega_2^2 \cos(\theta_3 - \theta_2) \\ &\qquad + r_4\omega_4^2 \cos(\theta_3 - \theta_4) - r_3\omega_3^2\end{aligned}\right\}}{r_4 \sin(\theta_3 - \theta_4)} \tag{10.82}$$

Before numerical evaluations can be made of the angular values ω_3, ω_4, α_3, and α_4 by Eqs. 10.79 through 10.82, it is necessary that the angles θ_3 and θ_4 be determined as functions of θ_2. The determination of these angles is the more difficult problem of the four-bar linkage. Although it would appear that the two unknown angles may be determined from position equation 10.76 by equating real and imaginary components, the difficulty encountered is that the trigonometric relationships are algebraically complicated. An alternate approach to the solution of these angles is one which includes the auxiliary position vector $\mathbf{r}_e$ shown in Fig. 10.47*b*. The variable length r_e and the angle θ_e may be determined as functions of θ_2 as follows:

$$\mathbf{r}_e = \mathbf{r}_1 - \mathbf{r}_2$$

$$r_e(e^{i\theta_e}) = r_1 - r_2(e^{i\theta_2})$$

$$r_e \cos \theta_e = r_1 - r_2 \cos \theta_2 \tag{10.83}$$

$$r_e \sin \theta_e = -r_2 \sin \theta_2 \tag{10.84}$$

Squaring both Eqs. 10.83 and 10.84 and adding them as simultaneous equations yields the length r_e.

$$r_e^2 = r_1^2 + r_2^2 - 2r_1r_2 \cos \theta_2 \tag{10.85}$$

and the angle θ_e may then be determined from Eq. 10.84.

$$\sin \theta_e = -\frac{r_2}{r_e} \sin \theta_2 \tag{10.86}$$

The auxiliary vector $\mathbf{r}_e$ may also be used in the following equation in order to evolve a relationship of θ_3 and θ_4 to r_e and θ_e:

$$\mathbf{r}_3 = \mathbf{r}_e + \mathbf{r}_4$$

$$r_3 \cos \theta_3 = r_e \cos \theta_e + r_4 \cos \theta_4 \tag{10.87}$$

$$r_3 \sin \theta_3 = r_e \sin \theta_e + r_4 \sin \theta_4 \tag{10.88}$$

Squaring Eqs. 10.87 and 10.88 and adding them,

$$r_3^2 = r_e^2 + r_4^2 + 2r_er_4(\cos \theta_e \cos \theta_4 + \sin \theta_e \sin \theta_4)$$

$$= r_e^2 + r_4^2 + 2r_er_4 \cos (\theta_e - \theta_4)$$

$$\cos (\theta_e - \theta_4) = \frac{r_3^2 - r_e^2 - r_4^2}{2r_er_4} \tag{10.89}$$

Rewriting Eqs. 10.87 and 10.88 as follows:

$$r_4 \cos \theta_4 = r_3 \cos \theta_3 - r_e \cos \theta_e$$

$$r_4 \sin \theta_4 = r_3 \sin \theta_3 - r_e \sin \theta_e$$

and squaring and adding leads to the following:

$$\cos(\theta_3 - \theta_e) = \frac{r_3^2 + r_e^2 - r_4^2}{2r_3r_e} \tag{10.90}$$

Since θ_e and r_e are known, Eqs. 10.89 and 10.90 yield the angles θ_4 and θ_3.

The equations which determine angular positions are critical ones. It may be seen that Eq. 10.86 determines two values of θ_e and that Eqs. 10.89 and 10.90 give two values of $(\theta_e - \theta_4)$ and $(\theta_3 - \theta_e)$. Care must be taken to select the realistic values of these angles for the linkage under investigation; otherwise data may be obtained for other linkages. For example, Fig. 10.48 shows a four-bar linkage in which the driver (link 2) rotates a full 360°, with the follower (link 4) oscillating through an angle less than 90°. The lengths of the links are given, and the linkage is shown to scale in the phase $\theta_2 = 60°$. Substitution of the given numerical data in Eq. 10.85 gives $r_e =$ 6.08 in. According to Eq. 10.86,

$$\sin\theta_e = -0.427$$

$$\theta_e = -25.28° \qquad \text{or} \qquad -154.72°$$

Obviously, $\theta_e = -25.28°$ is the correct value, and the alternate value is for a linkage where the pivot O_4 is to the left of the crank center O_2. Equation

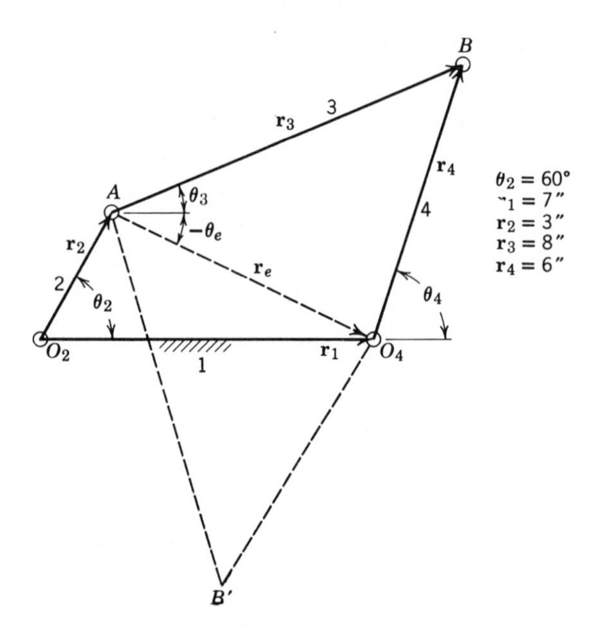

Figure 10.48

10.89 yields the following:

$$\cos(\theta_e - \theta_4) = -0.123$$

$$\theta_e - \theta_4 = 97.08° \quad \text{or} \quad -97.08°$$

Again, two angular values are given. Using $\theta_e = -25.28°$, one obtains the respective angles

$$\theta_4 = -122.37° \quad \text{or} \quad 71.80°$$

Referring to Fig. 10.48, $\theta_4 = 71.80°$ is the correct value for the linkage under investigation, and the alternate angle gives the angular position of link 4 in a different linkage ($O_2AB'O_4$), with the follower oscillating below the pivot O_4 rather than above O_4. According to Eq. 10.90,

$$\cos(\theta_3 - \theta_e) = 0.667$$

$$\theta_3 - \theta_e = 48.10° \quad \text{or} \quad -48.10°$$

$$\theta_3 = 22.82° \quad \text{or} \quad -73.38°$$

The correct value is $\theta_3 = 22.82°$, and the alternate angle is for the position of link 3 given by AB' in Fig. 10.48.

After the correct angles have been determined, the calculations of ω_3, ω_4, α_3, and α_4 may follow from Eqs. 10.79 through 10.82. The linear velocity and linear acceleration of any particle of the linkage, such as the mass center of any of the links, may then be determined in a manner such as that illustrated in Example 10.14 for the slider crank.

Interesting examinations may be made of the preceding equations of the four-bar linkage, particularly with regard to the angular motions given by Eqs. 10.79 through 10.82. It will be observed that most of the terms of these equations involve the differences of two angles which in some phase become zero when the angular positions of two links are equal. For example, the equations show that ω_3, ω_4, α_3, and α_4 become infinite when the angular position of the connecting link (link 3) is equal to that of the follower (link 4). It is interesting to ponder the possible linkage configuration and phase which would result in infinite angular motions.

10.29 Complex Mechanisms. The addition of extra links to the basic four-bar mechanism increases the complexity of the kinematic analysis of the mechanism. In Fig. 10.49 is shown a six-bar linkage in which the addition of links 5 and 6 to the basic four-bar linkage (1, 2, 3, 4) forms a second four-bar linkage consisting of links 3, 4, 5, and 6. By the addition of the two links, the number of kinematic unknowns is increased by six (θ_5, θ_6, ω_5, ω_6, α_5, and α_6) so that the total number of unknowns is twelve. Thus, twelve independent equations are required to determine the unknowns.

Combining the independent vector equations for the position of point B,

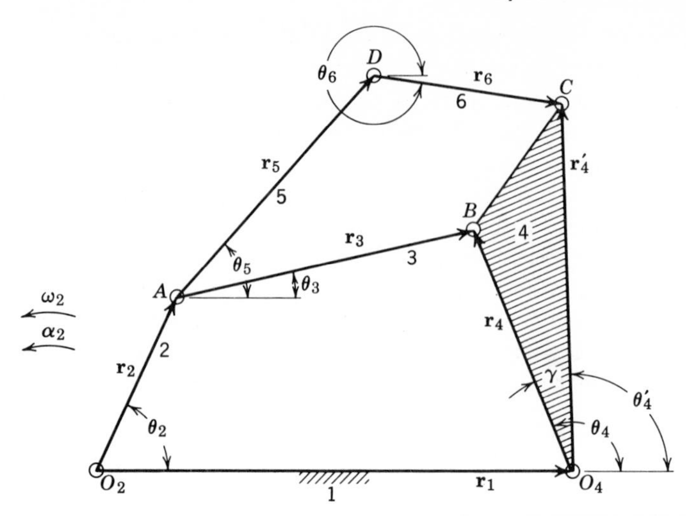

Figure 10.49

the following equation is obtained to include the kinematic quantities of the links in the loop formed by the lower four-bar linkage.

$$\mathbf{r}_2 + \mathbf{r}_3 = \mathbf{r}_1 + \mathbf{r}_4 \tag{10.91}$$

Similarly, a second combined vector equation may be written for the position of point C such as to involve a loop of links including links 5 and 6.

$$\mathbf{r}_2 + \mathbf{r}_5 + \mathbf{r}_6 = \mathbf{r}_1 + \mathbf{r}'_4 \tag{10.92}$$

The independent equations 10.91 and 10.92 may be written in complex number form.

$$r_2(e^{i\theta_2}) + r_3(e^{i\theta_3}) = r_1 + r_4(e^{i\theta_4}) \tag{10.93}$$

$$r_2(e^{i\theta_2}) + r_5(e^{i\theta_5}) + r_6(e^{i\theta_6}) = r_1 + r'_4(e^{i\theta'_4}) \tag{10.94}$$

By equating the real and imaginary components of the preceding equations, four independent equations are obtained from which the four unknown angles θ_3, θ_4, θ_5, and θ_6 may be determined as functions of θ_2. The angle θ'_4 may be given as θ_4 minus the fixed angle γ of link 4 shown in Fig. 10.49. The determination of the unknown angles is complicated trigonometrically and requires the determination of auxiliary lengths and angles as illustrated in the discussion of the four-bar linkage.

Differentiation of Eqs. 10.93 and 10.94 and equating the real and imaginary components result in four additional independent equations which may be used in the determination of the four angular velocities ω_3, ω_4, ω_5, and ω_6

as functions of ω_2. The solution of the numerous simultaneous equations is best accomplished by the use of determinants. Further differentiation and equating of real and imaginary parts yield four equations for the determination of the four unknown angular accelerations.

The preceding method of analysis may be applied to plane mechanisms of higher order of complexity. If two more links are added to the linkage of Fig. 10.49, making an eight-link mechanism, a third independent vector equation enclosing another independent loop of links makes available the additional equations required for solution.

10.30 Spatial Linkages. Velocity and acceleration analysis of three-dimensional linkages can be more easily accomplished by using vectors expressed in complex form than by any other method. Figure 10.50 shows a vector in space, and it may be designated as follows:

$$\mathbf{r}_p = (r_p \sin \phi)e^{i\theta} + (r_p \cos \phi)j = r_p(e^{i\theta} \sin \phi + j \cos \phi)$$

where θ is the angle between the real axis and the projection of the vector on the complex plane, and ϕ is the angle between the j-axis and the vector.

If a vector $\mathbf{r}_p = (x_1, y_1, z_1)$ is given in rectangular coordinates, it may be converted to the above system by means of the following relations:

$$\theta = \tan^{-1}\left(\frac{y_1}{x_1}\right)$$

$$\phi = \tan^{-1}\frac{\sqrt{x_1^2 + y_1^2}}{z_1}$$

$$r_p = \sqrt{x_1^2 + y_1^2 + z_1^2}$$

For the linkage shown in Fig. 10.51 link 2 is the driver which rotates in

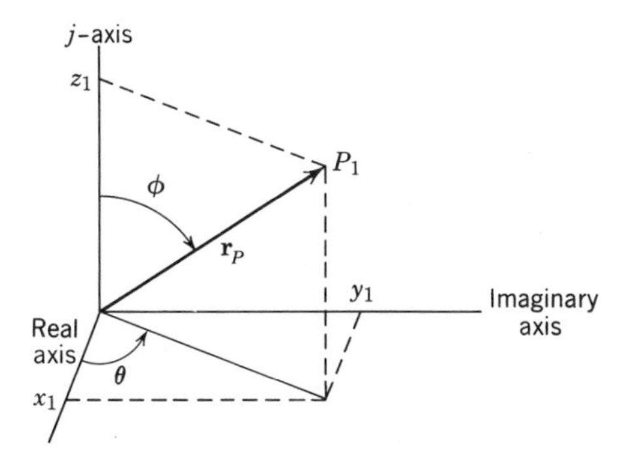

Figure 10.50

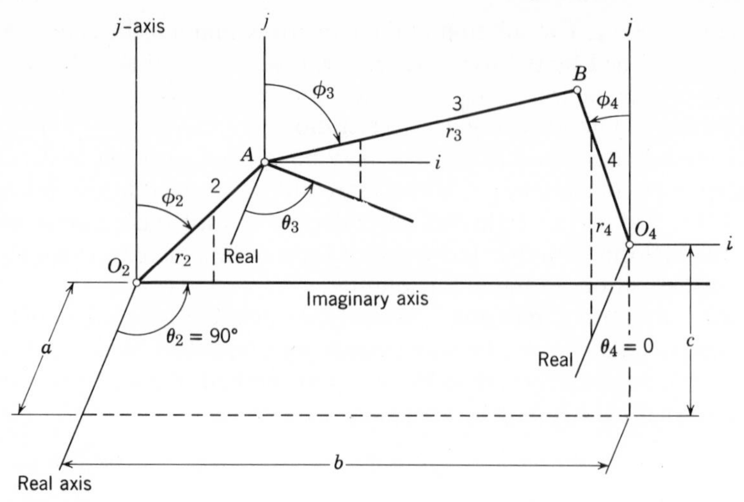

Figure 10.51

the plane defined by the imaginary axis and the j-axis while link 4 is constrained to rotate in the plane of the real axis and the j-axis. Given ϕ_2, $\dot{\phi}_2$, $\ddot{\phi}_2$, and the proportions of the linkage, equations can be developed[3] for the angular position, velocity, and acceleration of link 4.

The equation for the links in complex form may be written as follows:

$$r_2(e^{i\theta_2} \sin \phi_2 + j \cos \phi_2) + r_3(e^{i\theta_3} \sin \phi_3 + j \cos \phi_3)$$
$$= a + ib + jc + r_4(e^{i\theta_4} \sin \phi_4 + j \cos \phi_4) \quad (10.95)$$

By separating the preceding equation into real, imaginary, and j-components three algebraic equations may be obtained from which any of the three unknown terms θ_3, ϕ_3, or ϕ_4 may be determined.

Differentiating the above equation with respect to time

$$r_2[(\dot{\phi}_2 \cos \phi_2 + i\dot{\theta}_2 \sin \phi_2)e^{i\theta_2} - j\dot{\phi}_2 \sin \phi_2]$$
$$+ r_3[(\dot{\phi}_3 \cos \phi_3 + i\dot{\theta}_3 \sin \phi_3)e^{i\theta_3} - j\dot{\phi}_3 \sin \phi_3]$$
$$= r_4[(\dot{\phi}_4 \cos \phi_4 + i\dot{\phi}_4 \sin \phi_4)e^{i\theta_4} - j\dot{\phi}_4 \sin \phi_4)] \quad (10.96)$$

Separating the above into real, imaginary, and j-components and noting that $\dot{\theta}_2 = \dot{\theta}_4 = 0$, the equation for $\dot{\phi}_4$ can be derived as

$$\dot{\phi}_4 = \frac{r_2\dot{\phi}_2}{r_4}\left[\frac{\sin \phi_2 \cos \phi_3 - \cos \phi_2 \sin \phi_3 \cos (\theta_2 - \theta_3)}{\sin \phi_4 \cos \phi_3 - \cos \phi_4 \sin \phi_3 \cos (\theta_4 - \theta_3)}\right] \quad (10.97)$$

[3] F. H. Raven, "Position, Velocity, and Acceleration Analysis, and Kinematic Synthesis of Plane and Space Mechanisms by a Generalized Procedure Called the Method of Independent Position Equations," L.C. Card 58-5833, University Microfilms, Ann Arbor, Michigan.

By means of a second differentiation and separating into real, imaginary, and *j*-components, the equation for $\ddot{\phi}_4$ can be shown to be

$$\dot{\phi}_4 = \frac{r_2(\ddot{\phi}_2 \sin \phi_2 + \dot{\phi}_2^2 \cos \phi_2) + r_3(\ddot{\phi}_3 \sin \phi_3 + \dot{\phi}_3^2 \cos \phi_3) - r_4\dot{\phi}_4^2 \cos \phi_4}{r_4 \sin \phi_4} \tag{10.98}$$

Other types of constraints for spatial four-bar linkages may be found in a paper by Harrisberger.

Reference

Harrisberger, L. "A Number Synthesis Survey of Three-Dimensional Mechanisms," *Trans. ASME, J. Eng. Ind., Series B*, Vol. 87, No. 2, May 1965.

Problems

10.1 A turbine operates at 15,000 r/min. Calculate the velocity and acceleration of the tip of the rotor blade which is 10 in. from the center of rotation.

10.2 The tip of a turbine blade has a linear velocity of 600 m/sec. Calculate the angular velocity in revolutions per minute for the following wheel diameters: 7, 40, 75, and 90 cm.

10.3 In an automotive engine, the maximum piston acceleration is 1000g (g = 32.2 ft/sec^2) at a given constant crank speed. The crank radius is $2\frac{1}{2}$ in., and the connecting rod length is 10 in. Determine the crank speed in r/min and the linear speed of the crank-pin center in feet per second and feet per minute. Determine also the maximum piston velocity in fps and fpm at this crank speed. See Eqs. 2.3 and 2.4 of Chapter 2.

10.4 The particle Q of the body shown in Fig. 10.52 is in motion along a curvilinear path. The radius of curvature of the path, its angular velocity and acceleration, and

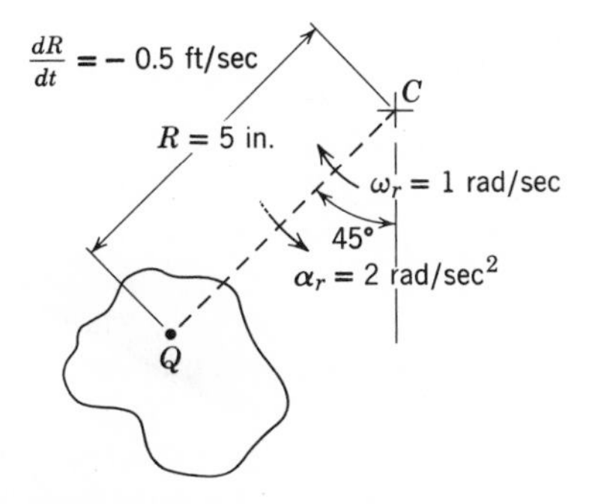

Figure 10.52

its rate of change with time are as shown. Determine the magnitudes of $\mathbf{A}_Q^n$, $\mathbf{A}_Q^t$, and $\mathbf{A}_Q$ and show as vectors on a sketch of the figure.

10.5 A point mass P travels a curvilinear path about point A as shown in Fig. 10.53. If $A_P = 64$ cm/sec^2, determine ω_R.

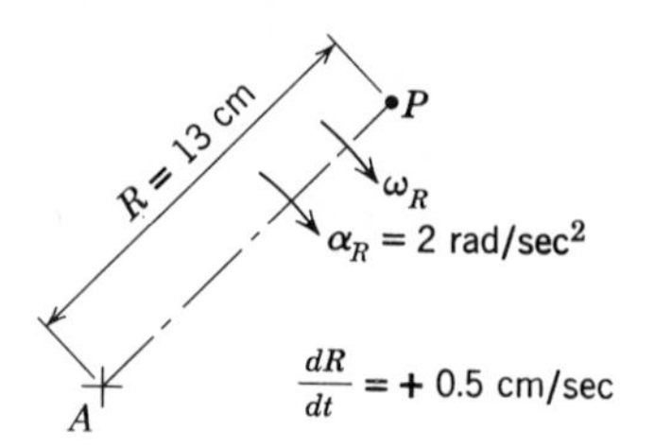

Figure 10.53

10.6 In Fig. 10.54, link 2 and disk 3 rotate about the same fixed axis 0. $\omega_2 =$ 15 rad/sec (cw) and $\alpha_2 = 0$. $\omega_3 = 10$ rad/sec (cw) and $\alpha_3 = 30$ rad/sec^2 (ccw). Point P_2 in link 2 is coincident with point P_3 in disk 3. Determine $\mathbf{V}_{P_2P_3}$, $\mathbf{A}_{P_2P_3}$, ω_{23}, and α_{23}.

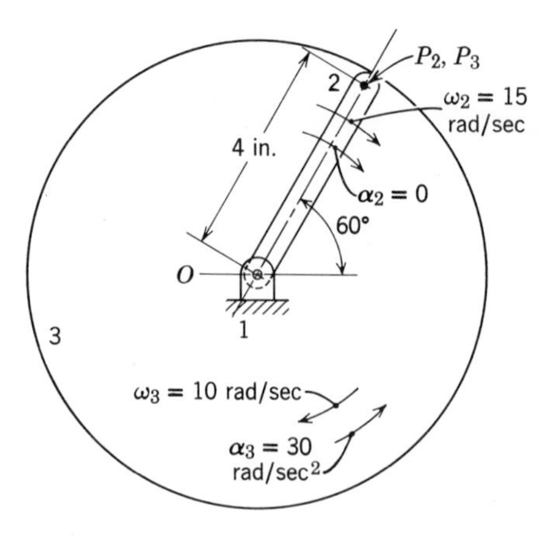

Figure 10.54

10.7 Given points A and B in a common link as shown in Fig. 10.55 calculate the velocity and acceleration of point B relative to point A.

10.8 The wheel in Fig. 10.56 rolls without slipping. The velocity and acceleration of P are as shown. For each phase make a sketch of the wheel and graphically determine the values of $\mathbf{A}_P^n$ and $\mathbf{A}_P^t$ using convenient scales. Calculate R and ω_r of the path of P and locate the center of curvature C. What information is needed to determine α_r?

10.9 The centrifugal blower shown in Fig. 10.57 rotates at an angular velocity of 900 rad/sec. The velocity of a particle of gas P relative to the blade is $\mathbf{V}_{PB} = 685$ cm/sec. Determine the velocity of the blade tip and the absolute velocity of particle P.

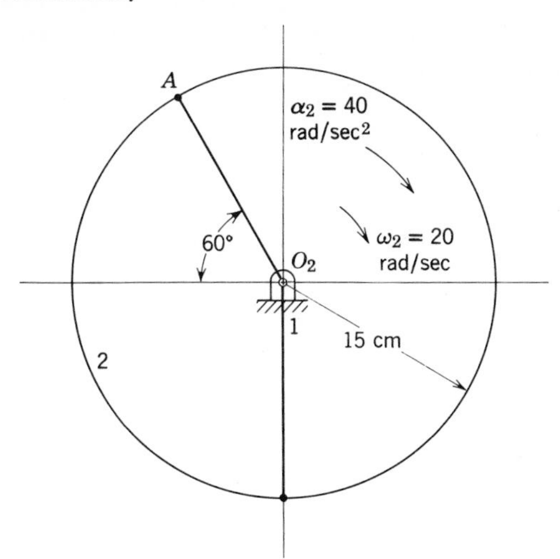

Figure 10.55

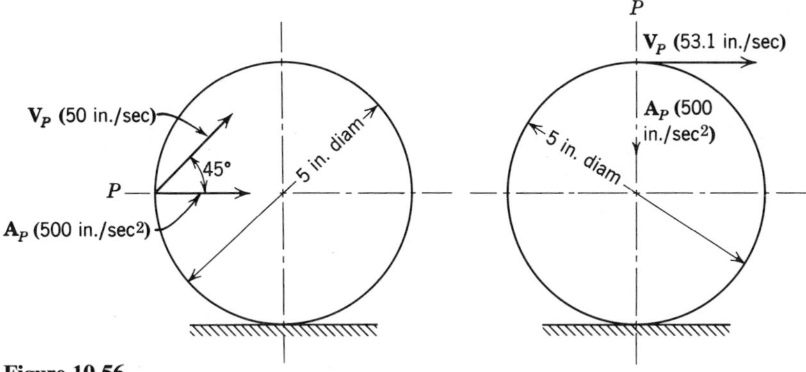

Figure 10.56

10.10 In Fig. 10.58 is shown a gas particle P leaving the passage of a turbine wheel at a velocity $\mathbf{V}_{PB}$ relative to the blades. The angle α is the departure angle of the particle relative to the passage and is measured from the plane of rotation. If the wheel speed is 10,000 r/min, determine angle α so that the absolute velocity of P is 100 m/sec in a direction parallel to the exit-duct axis. What is the sense of rotation of the wheel when viewed from the right? The radius from the axis of rotation to particle P is 30 cm.

10.11 For the linkage shown in Fig. 10.59 determine ω_4 and α_4 using unit vectors. Link 2 rotates at a constant angular velocity.

10.12 Calculate $\mathbf{V}_B$ and $\mathbf{A}_B$ for the linkage shown in Fig. 10.60 using unit vectors. ω_2 is constant.

10.13 In Fig. 10.61 link 2 rotates at a constant angular velocity. Determine $\mathbf{V}_C$ and $\mathbf{A}_C$ using unit vectors.

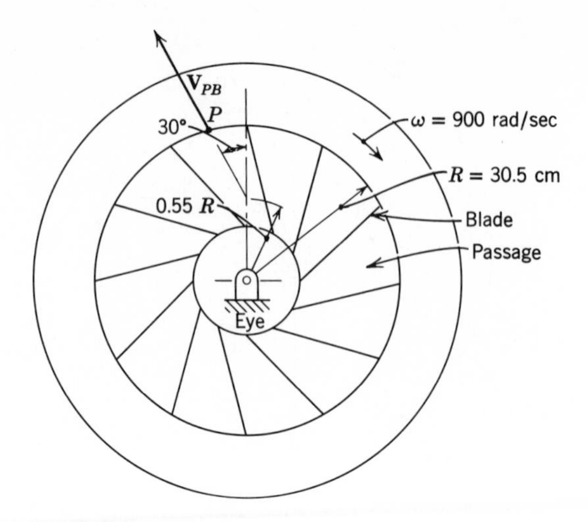

Figure 10.57

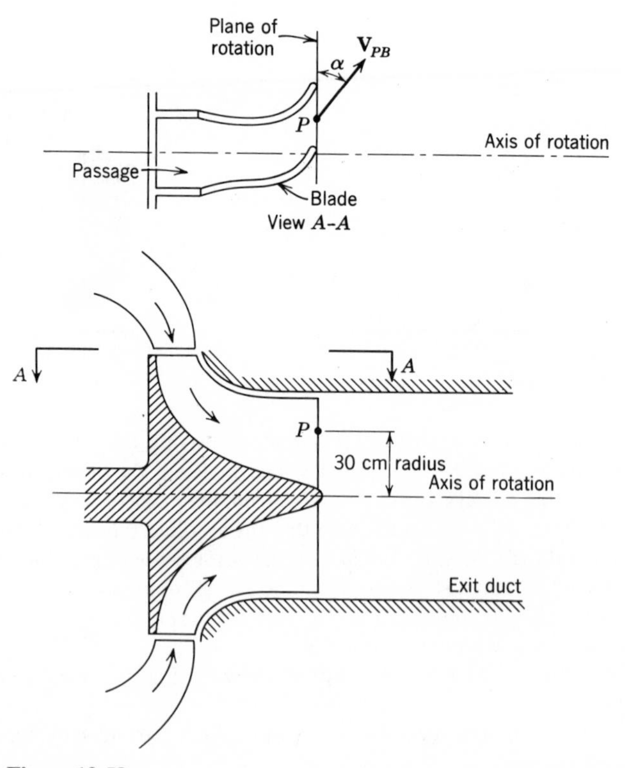

Figure 10.58

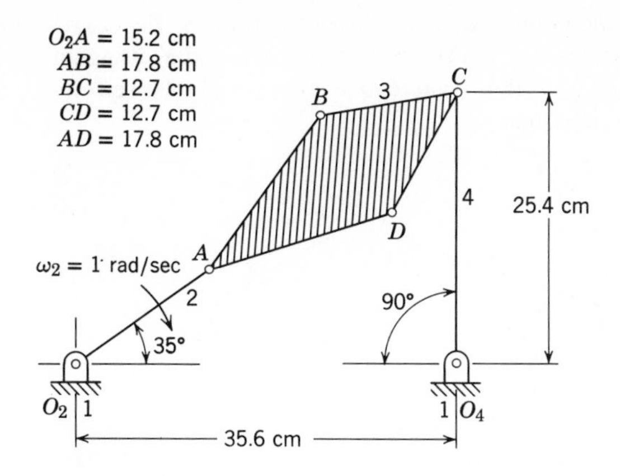

Figure 10.59

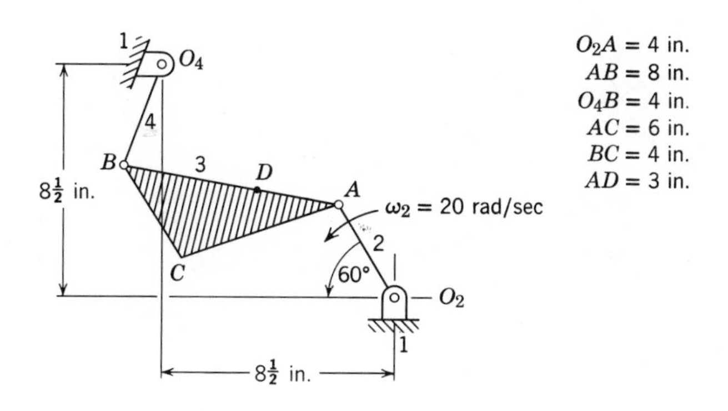

Figure 10.60

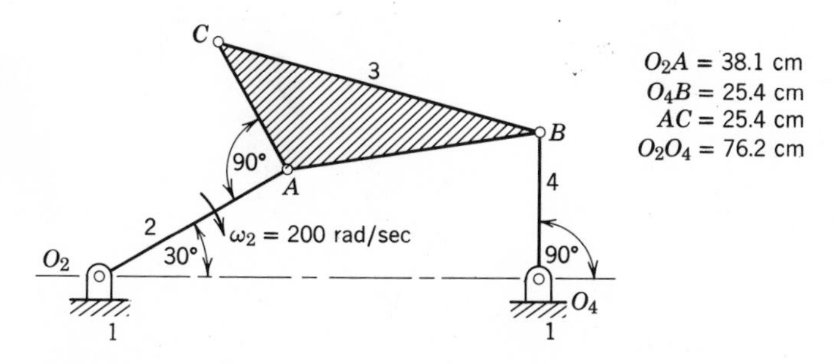

Figure 10.61

10.14 For the crank-shaper mechanism shown in Fig. 10.62, link 2 rotates at a constant angular velocity. Calculate $\mathbf{V}_{A_4}$, $\mathbf{A}_{A_4}$, ω_4, and α_4 using unit vectors.

10.15 In Fig. 10.63 link 2 rotates at a constant angular velocity as shown. Calculate $\mathbf{V}_B$ and $\mathbf{A}_B$ using unit vectors.

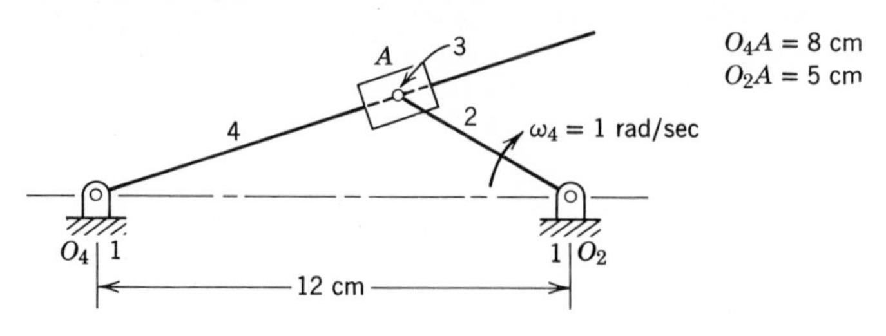

Figure 10.62

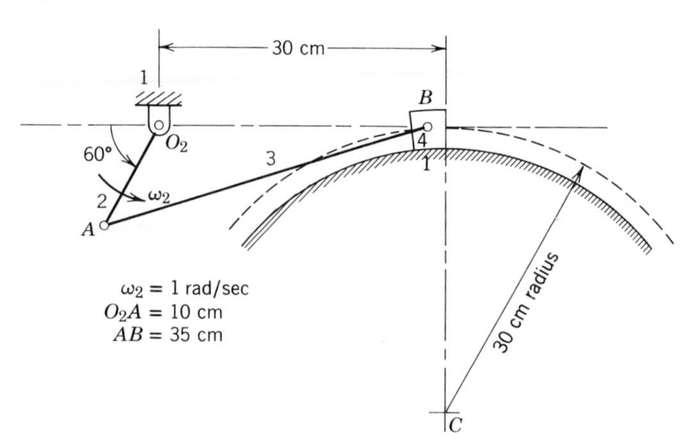

Figure 10.63

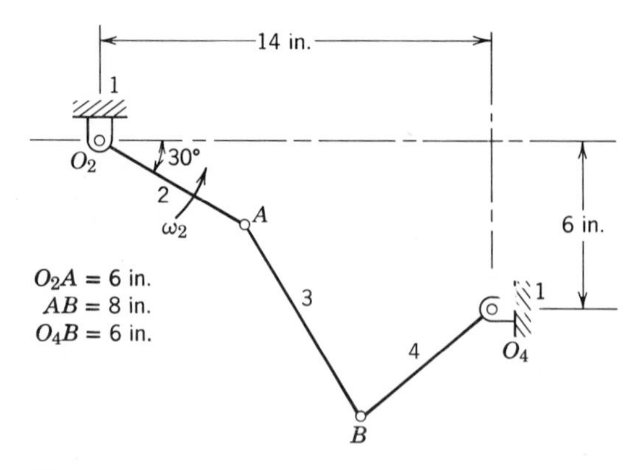

Figure 10.64

10.16 For the linkage shown in Fig. 10.64 calculate $\mathbf{V}_B$, $\mathbf{A}_B$, ω_4, and α_4 using unit vectors. Link 2 rotates at a constant angular velocity of 160 rad/sec.

10.17 In the mechanism shown in Fig. 10.65, link 4 rotates at a constant angular velocity and $\mathbf{V}_B = 24.4$ m/sec. Calculate α_2 using unit vectors.

10.18 Determine by means of unit vectors α_2 of the linkage given in Fig. 10.66. Link 4 rotates at a constant angular velocity as shown.

10.19 Using unit vectors, determine the acceleration of the follower for the phase shown for the cam mechanism of Fig. 10.67.

10.20 For the crank-shaper mechanism shown in Fig. 10.68, link 2 rotates at a constant angular velocity. Determine $\mathbf{A}_{A_4}$, ω_4, and α_4 using unit vectors.

10.21 The driving link 2 of Fig. 10.69 rotates at a constant angular velocity. Determine by unit vectors the velocity and acceleration of link 6.

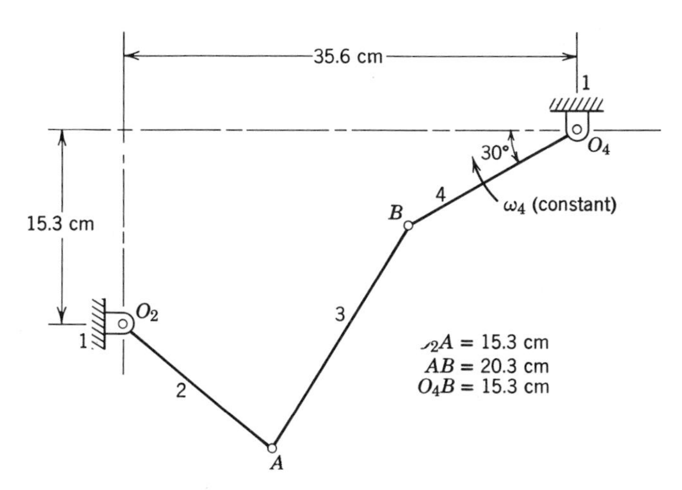

Figure 10.65

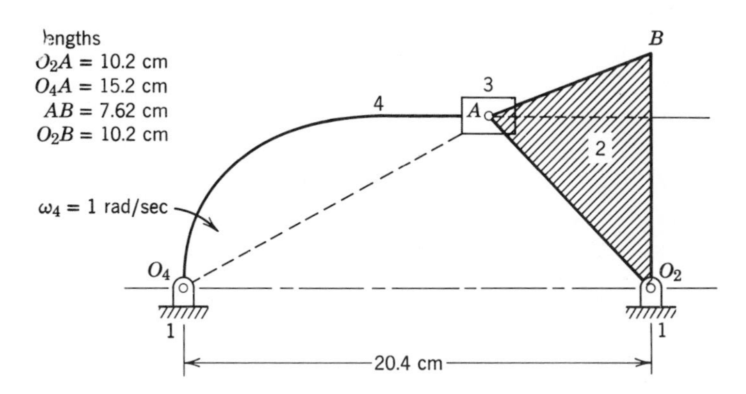

Figure 10.66

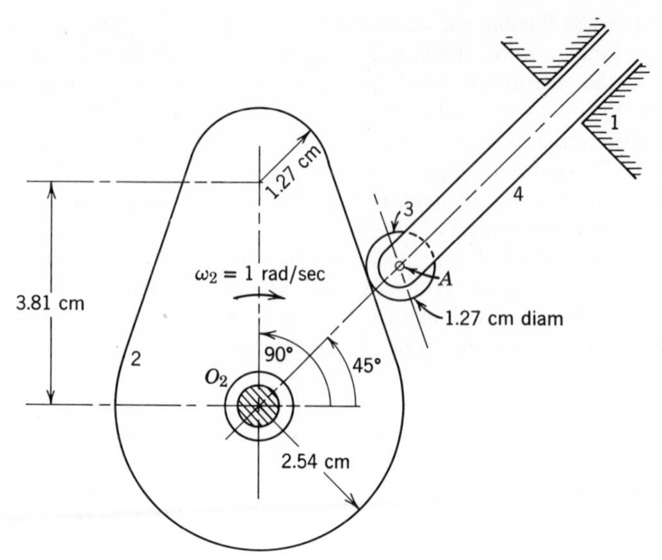

Figure 10.67

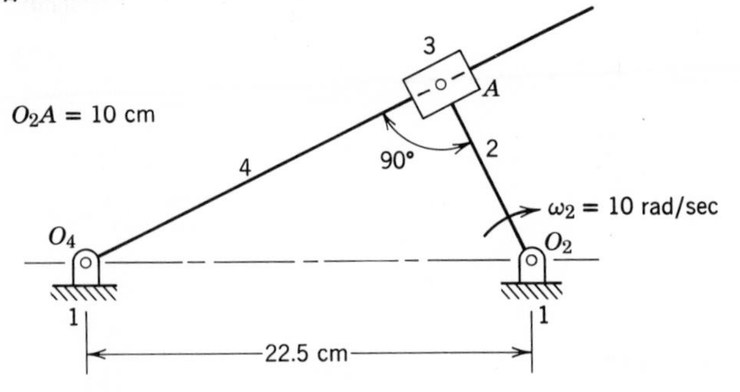

Figure 10.68

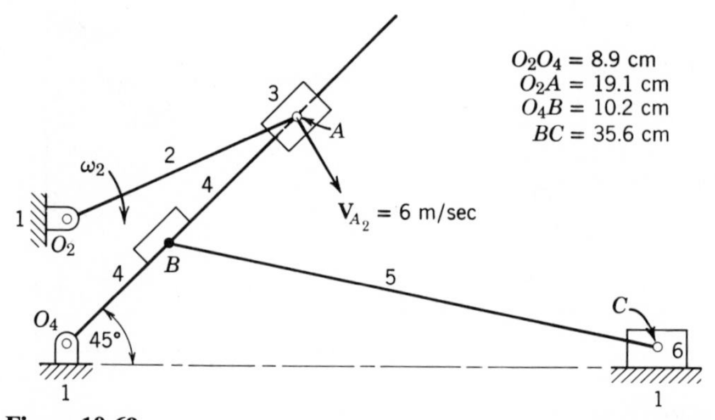

Figure 10.69

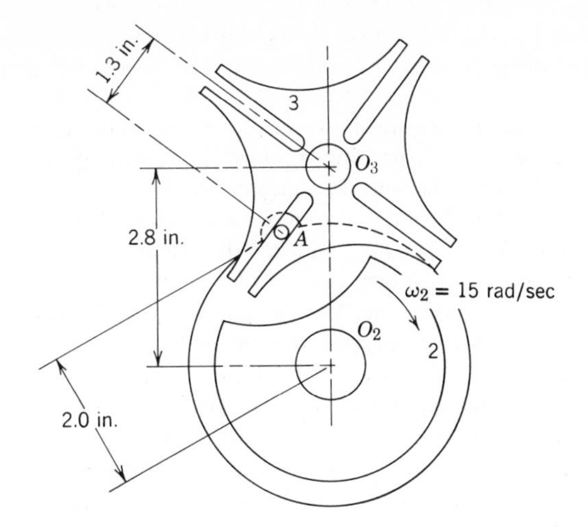

Figure 10.70

10.22 Given the Geneva mechanism as shown in Fig. 10.70, determine and plot ω_3 and α_3 for one-quarter revolution of link 3. Let θ_2 be taken in increments of 5° starting where the pin A of link 2 first engages the slot in link 3. Link 2 rotates at a constant angular velocity in the direction shown. Use the equations of relative motion written in terms of unit vectors and solve by determinants and the computer.

10.23 Given the cam and follower of Fig. 10.71 determine $\mathbf{V}_Q$, $\mathbf{A}_Q$, ω_2, and α_2 for the curved follower by unit vectors.

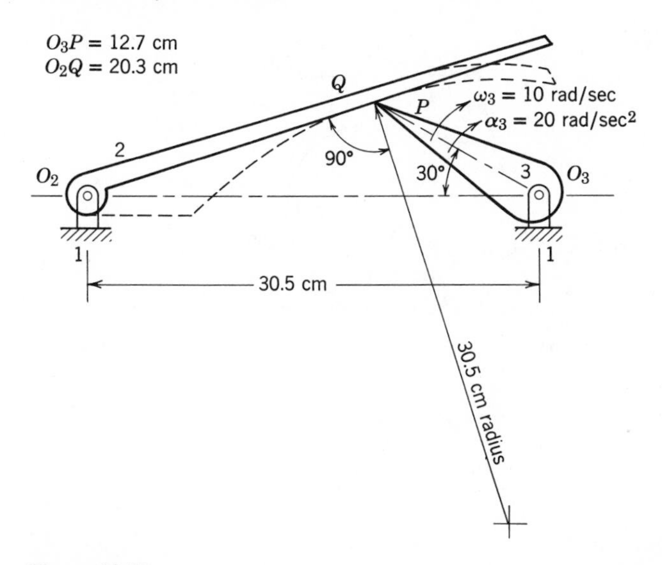

Figure 10.71

10.24 In Fig. 10.72 cam 2 rotates in the direction shown at a constant angular velocity. Determine $\mathbf{V}_{B_4}$, $\mathbf{A}_{B_4}$, ω_4, and α_4 using unit vectors.

10.25 For the linkage shown in Fig. 10.59, draw the velocity polygon and determine $\mathbf{V}_B$ and $\mathbf{V}_D$.

10.26 Link 2 of the mechanism shown in Fig. 10.60 rotates at a constant angular velocity of 20 rad/sec in the direction shown. Draw the velocity polygon and determine $\mathbf{V}_C$, $\mathbf{V}_D$, ω_3, and ω_4. Indicate the velocity image of link 3.

10.27 In Fig. 10.61 link 2 rotates at a constant angular velocity. Draw the velocity polygon and determine $\mathbf{V}_C$, ω_3, and ω_4. Indicate the velocity image of each link.

10.28 For the mechanism shown in Fig. 10.62, determine the relative velocities $\mathbf{V}_{A_2A_3}$ and $\mathbf{V}_{A_3A_4}$ in cm/sec and in./sec. Also calculate ω_{24}, ω_{34}, and ω_{32}.

10.29 (*a*) Using a velocity polygon, determine $\mathbf{V}_B$ and ω_4 of the slider-crank mechanism shown in Fig. 10.63 for $\omega_2 = 1$ rad/sec. Determine $\mathbf{V}_B$ for a crank speed of 3000 r/min without redrawing the velocity polygon. (*b*) Draw an acceleration polygon and determine $\mathbf{A}_B$ and α_4 for $\omega_2 = 1$ rad/sec and $\alpha_2 = 0$. Determine $\mathbf{A}_B$ for a crank speed of 3000 r/min without redrawing the acceleration polygon.

10.30 In the mechanism shown in Fig. 10.64, link 2 rotates at a constant angular velocity of 160 rad/sec. Construct the velocity and the acceleration polygon, and determine ω_4 and α_4.

10.31 Given the offset slider-crank mechanism as shown in Fig. 10.73, determine $\mathbf{V}_B$, $\mathbf{A}_B$, ω_3, and α_3 by drawing velocity and acceleration polygons.

10.32 In the mechanism shown in Fig. 10.65, link 4 is the driver and rotates at a constant angular velocity. If $\mathbf{V}_B = 24.4$ m/sec, draw the velocity and acceleration polygons and determine α_2.

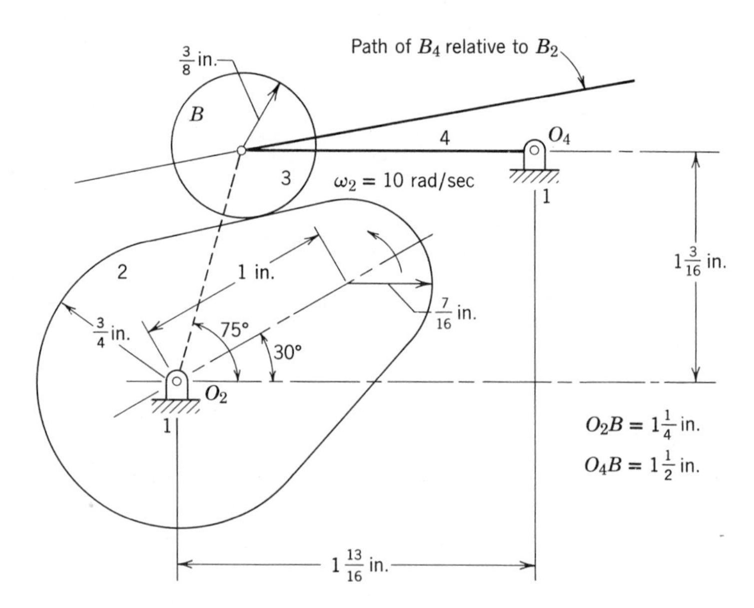

Figure 10.72

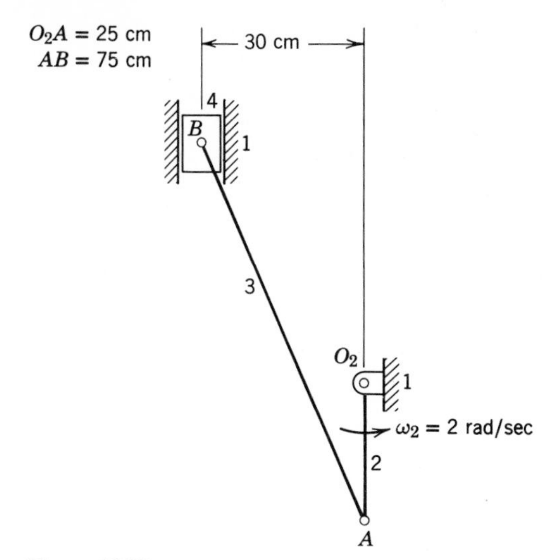

Figure 10.73

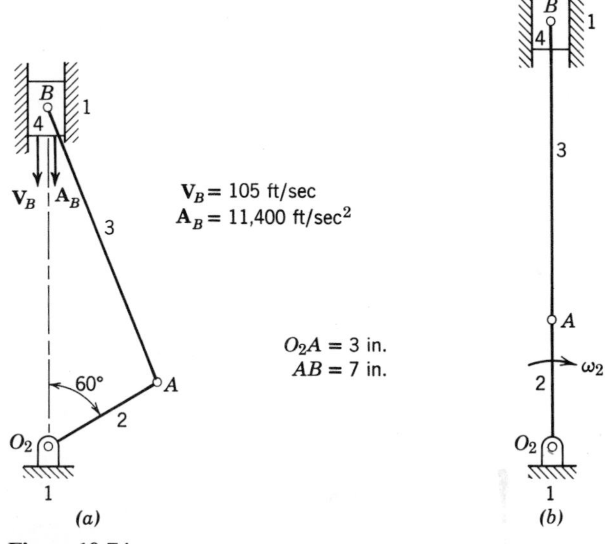

Figure 10.74

10.33 (*a*) In the slider-crank mechanism of Fig. 10.74*a* the velocity and acceleration of the slider are given. Using a vector polygon, determine $\mathbf{V}_A$ and calculate ω_2 and ω_3. If ω_2 is constant, draw the acceleration polygon and calculate α_3. (*b*) Using the value of ω_2 from part *a*, determine the acceleration of the slider in Fig. 10.74*b* and α_3.

10.34 Construct the velocity and acceleration polygons for the linkage shown in Fig. 10.75. Link 2 rotates at a constant angular velocity.

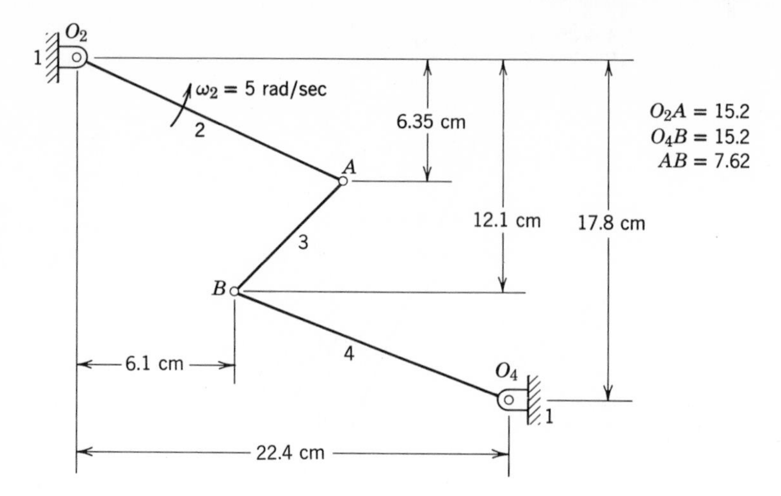

Figure 10.75

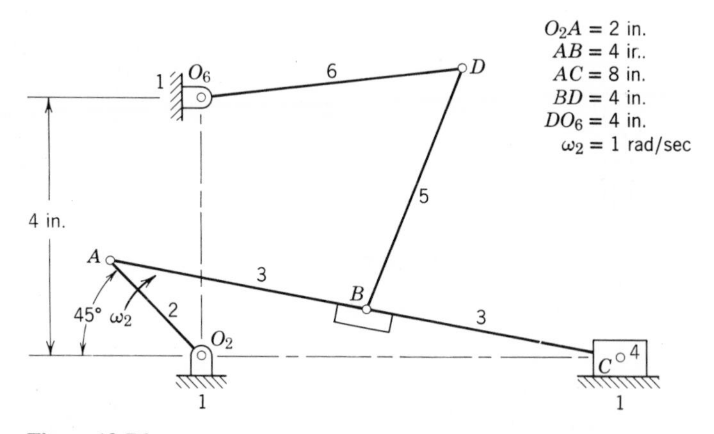

Figure 10.76

10.35 (*a*) For the mechanism shown in Fig. 10.76, link 2 rotates at constant angular velocity. Construct the velocity polygon and determine the velocity of point D. (*b*) Using the results of part *a*, construct the acceleration polygon and determine $\mathbf{A}_D$. (*c*) By proportion calculate the velocity and acceleration of point D if the angular velocity of the driving link is increased to 1200 r/min.

10.36 (*a*) Determine the velocity of point F in Fig. 10.77 for a speed of 3600 r/min of link 2. Construct the velocity polygon for a unit speed of $\omega_2 = 1$ rad/sec. (*b*) Referring to part *a* determine the acceleration of point F for a uniform speed of 3600 r/min of link 2. Construct the acceleration polygon for a unit speed of $\omega_2 = 1$ rad/sec and $\alpha_2 = 0$. What difficulties arise if $\alpha_2 = 1$ rad/sec^2 is used?

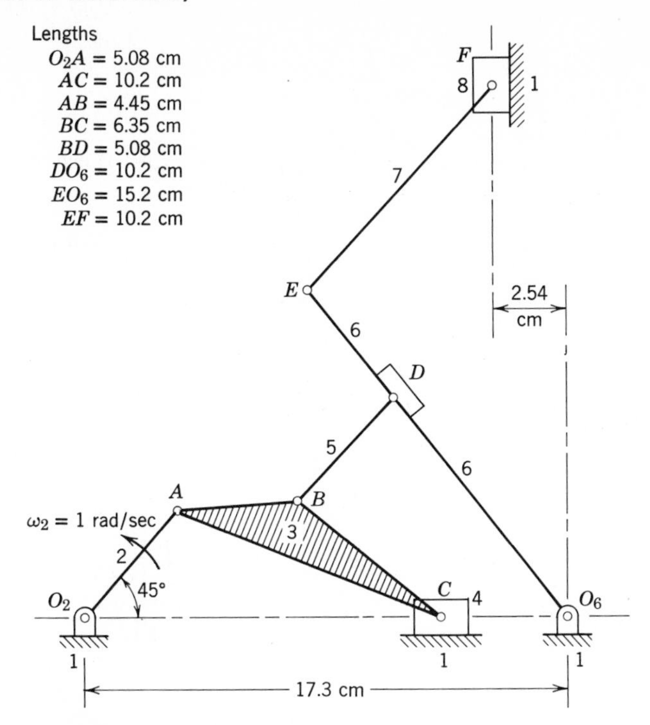

Figure 10.77

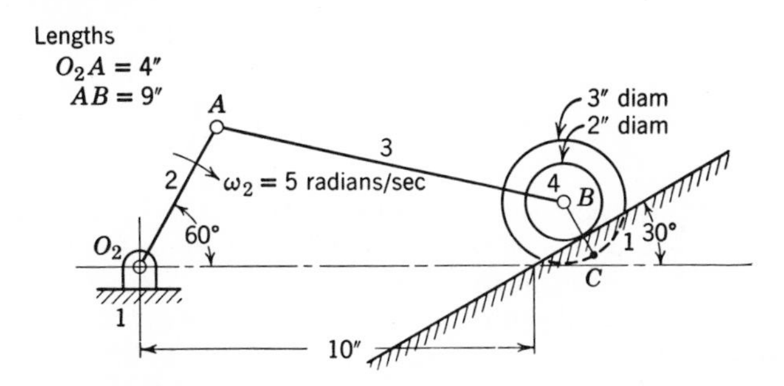

Figure 10.78

10.37 (*a*) The wheel of the mechanism of Fig. 10.78 rolls without slipping. Construct the velocity image of the wheel and determine $\mathbf{V}_C$ and ω_4. (*b*) Referring to part *a*, construct the acceleration image of the wheel and determine $\mathbf{A}_C$ and α_4.

10.38 (*a*) For the locomotive linkage shown in Fig. 10.79, determine the velocities of points *C*, *R*, and *S* for a locomotive speed of 60 mi/h. Calculate ω_2 and ω_3. (*b*) Referring to part *a*, determine the accelerations of points *C*, *R*, and *S* for a locomotive speed of 60 mi/h. Calculate α_2 and α_3.

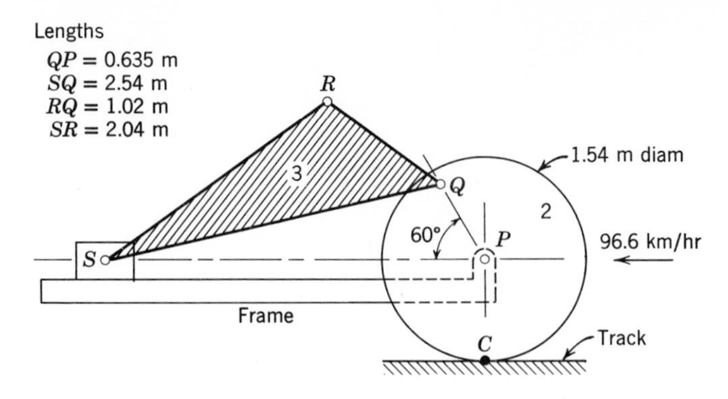

Figure 10.79

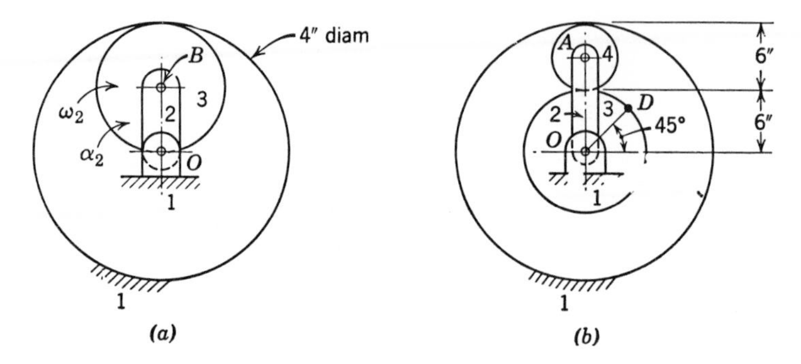

Figure 10.80

10.39 (*a*) In the epicyclic gear train of Fig. 10.80*a*, the carrier (link 2) rotates such that $\omega_2 = 12$ rad/sec and $\alpha_2 = 48$ rad/sec^2 at the instant shown. Construct the velocity polygon and show the velocity image of the planet (link 3). Determine ω_3. (*b*) Referring to part *a*, construct the acceleration polygon and show the acceleration image of the planet. Determine α_3.

10.40 (*a*) The carrier (link 2) of the epicyclic gear train of Fig. 10.80*b* rotates clockwise at a uniform speed of 10 rad/sec. Determine the velocity of point D_3. (*b*) Referring to part *a*, determine the acceleration of point D_3.

10.41 (*a*) Determine the velocity of the piston of the mechanism of Fig. 10.81. Calculate ω_{32}. (*b*) Determine the acceleration of the piston of the mechanism of Fig. 10.81. Calculate α_{32} ($\alpha_2 = 0$).

10.42 (*a*) In Fig. 10.82, links 4 and 5 are gears in mesh. Construct the velocity images of links 4 and 5 when (1) $\omega_5 = 0$ and (2) $\omega_5 = 5$ rad/sec. (*b*) Referring to part *a*, construct the acceleration images of links 4 and 5 when (1) $\omega_5 = 0$ and (2) $\omega_5 = 5$ rad/sec ($\alpha_2 = 0$).

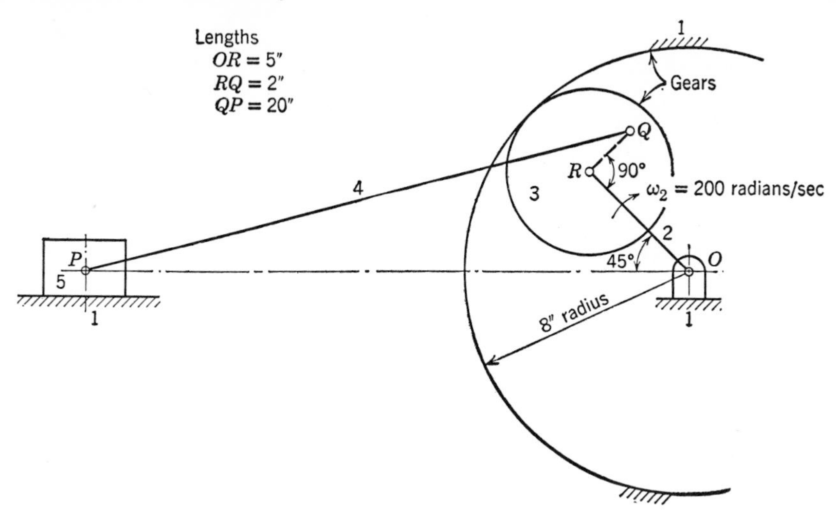

Figure 10.81

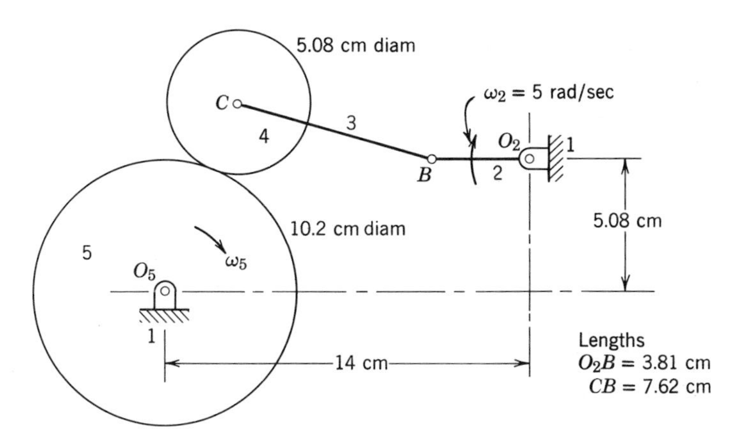

Figure 10.82

10.43 For the mechanism shown in Fig. 10.83, link 2 rotates at a constant angular velocity of 2 rad/sec and slider 5 moves with a constant linear velocity of 10 in./sec. Draw the velocity and acceleration polygons and determine $\mathbf{V}_C$ and $\mathbf{A}_C$.

10.44 In Fig. 10.84 is shown a double slider-crank mechanism with the cranks rotating at constant angular velocities. Draw the velocity and acceleration polygons and determine the velocity and acceleration of point D relative to point B.

10.45 A double slider-crank mechanism is shown in Fig. 10.85. Draw the velocity and acceleration polygons and determine tne velocity and acceleration of each slider. The crank 2 rotaies at a constant angular velocity.

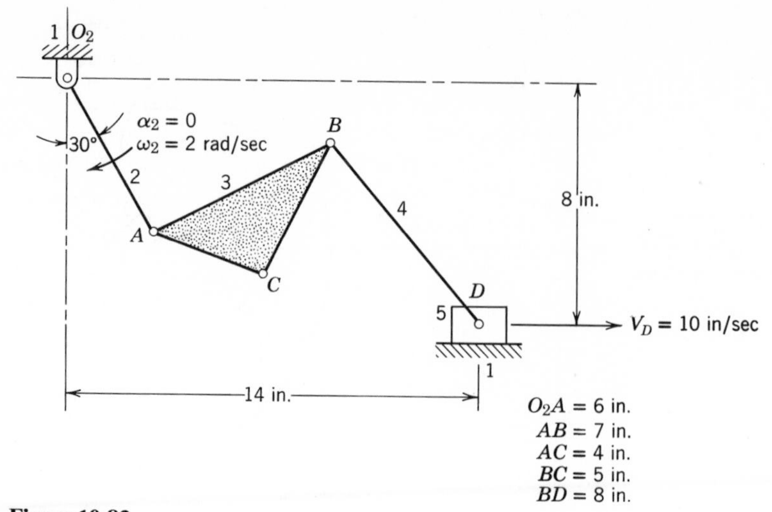

Figure 10.83

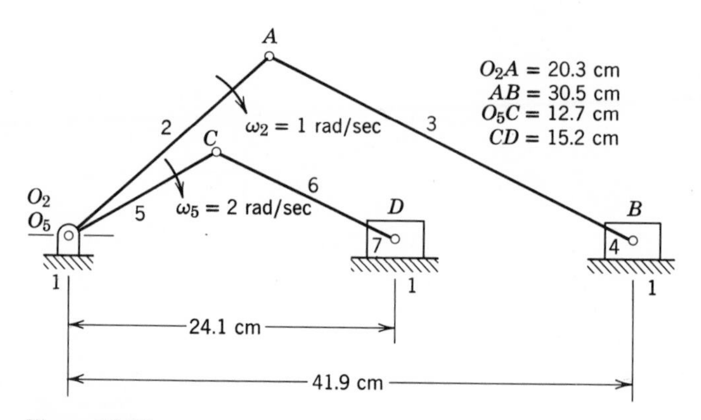

Figure 10.84

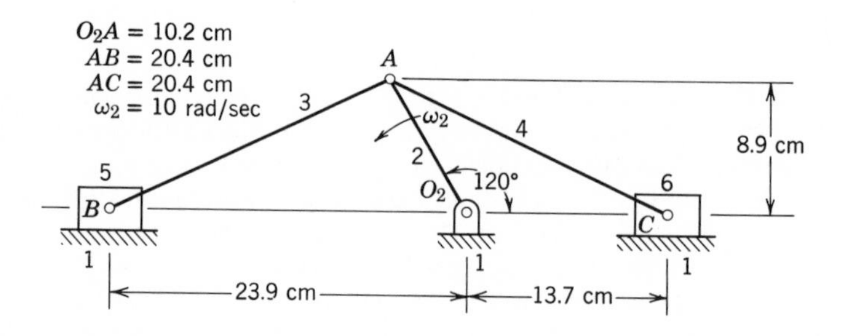

Figure 10.85

10.46 A diagram of a three-cylinder radial engine is shown in Fig. 10.86. If the crank rotates at a constant angular velocity of 1 rad/sec, determine the velocities and accelerations of the pistons by polygon methods.

10.47 A toggle mechanism is shown in Fig. 10.87 with link 2 rotating at a constant angular velocity. Construct the velocity and acceleration polygons and determine $\mathbf{V}_C$ and $\mathbf{A}_C$.

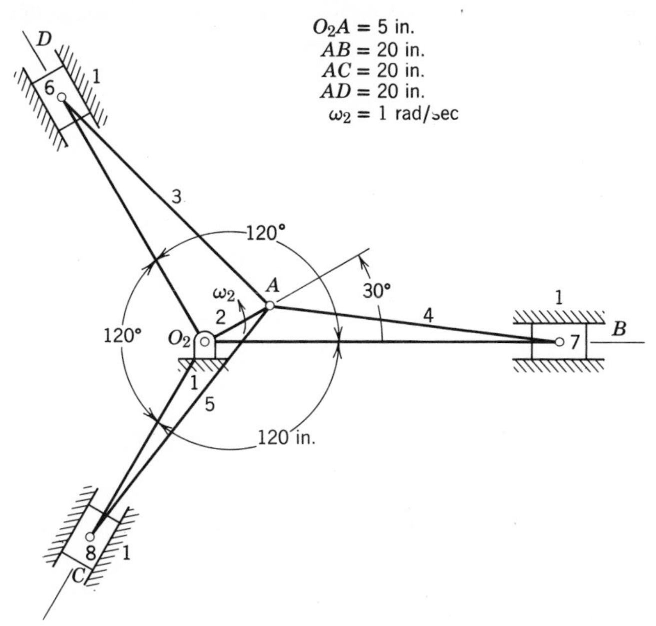

Figure 10.86

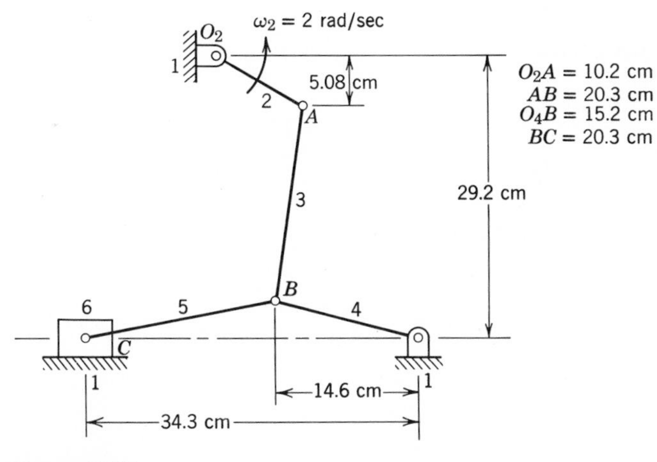

Figure 10.87

10.48 (*a*) Given the velocity of point C in the mechanism shown in Fig. 10.88, draw the velocity polygon and determine $\mathbf{V}_{AC}$ and ω_{34}. (*b*) Construct the acceleration polygon if $\mathbf{A}_C = 0$ and determine $\mathbf{A}_{AC}$ and α_{34}.

10.49 (*a*) For the mechanism shown in Fig. 10.89, draw the velocity polygon and determine the velocity image of each link. Determine $\mathbf{V}_{g_3}$ of the center of gravity g_3 of the connecting rod from its velocity image. Determine the velocity of sliding $\mathbf{V}_{C_5C_4}$. (*b*) Draw the acceleration polygon and determine the acceleration image of each link. Determine $\mathbf{A}_{g_3}$ from the image of the connecting rod.

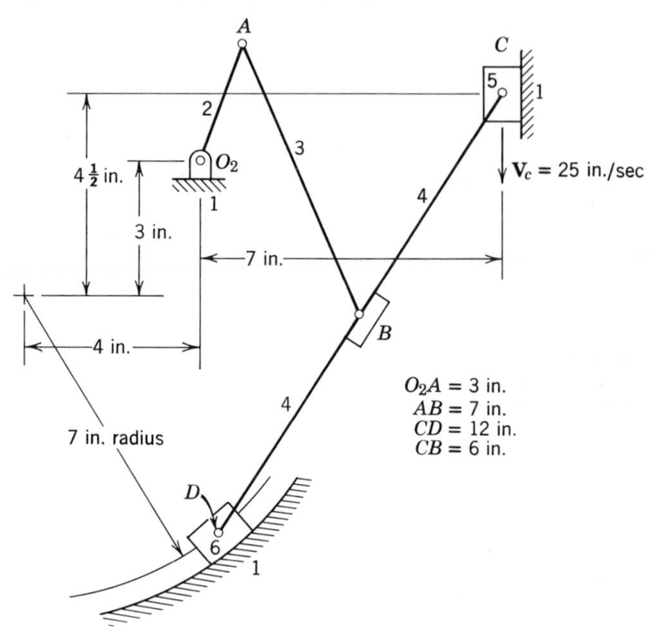

Figure 10.88

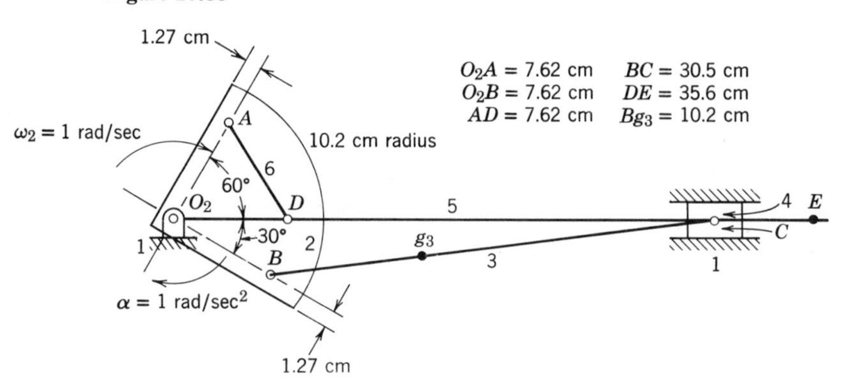

Figure 10.89

10.50 Determine the angular acceleration of link 2 of Fig. 10.66.

10.51 For the radial roller follower of Fig. 10.67 driven by the disk cam, determine the follower acceleration for the phase shown. Also, determine the follower acceleration for the phase in which the cam is rotated 45° from the phase shown. For convenience in drawing, rotate the follower relative to the cam.

10.52 A crank-shaper mechanism is shown in Fig. 10.68 with link 4 perpendicular to link 2. Determine $\mathbf{V}_{A_4A_2}$, $\mathbf{A}_{A_4}$, ω_4, and α_4 using polygons.

10.53 For the Whitworth quick-return mechanism shown in Fig. 10.69, determine the velocity and acceleration of the tool holder (link 6) by velocity and acceleration polygons. Give the results in units of m/sec and m/sec^2.

10.54 Link 2 of the Geneva mechanism shown in Fig. 10.70 rotates clockwise at a constant angular velocity of 15 rad/sec. Determine the angular velocity and angular acceleration of link 3.

10.55 Draw the velocity and acceleration polygons for the cam and followers shown in Fig. 10.71. Point P is on body 3 and point Q on body 2. Determine $\mathbf{V}_Q$, $\mathbf{A}_Q$, ω_2, and α_2 for the flat-faced follower and for the curved-faced follower. Give values of velocity and acceleration in units of centimeters per second and centimeters per second^2.

10.56 For the slider-crank mechanism shown in Fig. 10.90, (*a*) determine all instantaneous centers, (*b*) determine velocity of point B by instantaneous center methods, (*c*) check $\mathbf{V}_B$ found in part *b* by drawing a velocity polygon.

10.57 For the offset slider-crank shown in Fig. 10.91, determine the velocity of the slider in cm/sec using instantaneous centers if $\omega_2 = 2$ rad/sec.

10.58 Given the mechanism shown in Fig. 10.92, locate all instantaneous centers.

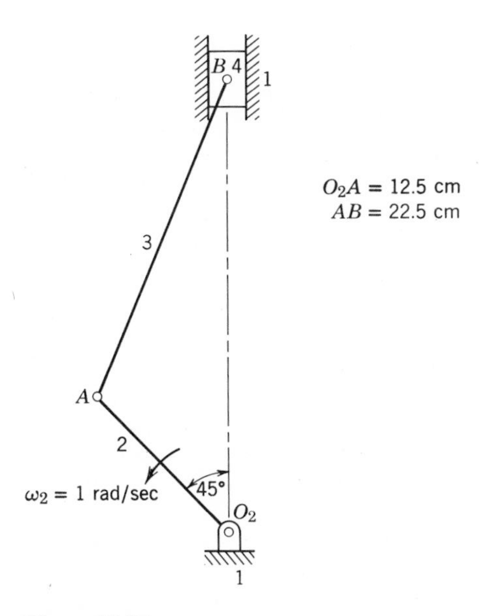

Figure 10.90

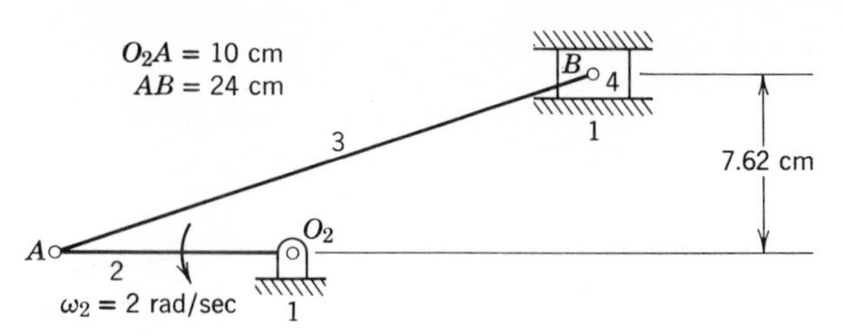

Figure 10.91

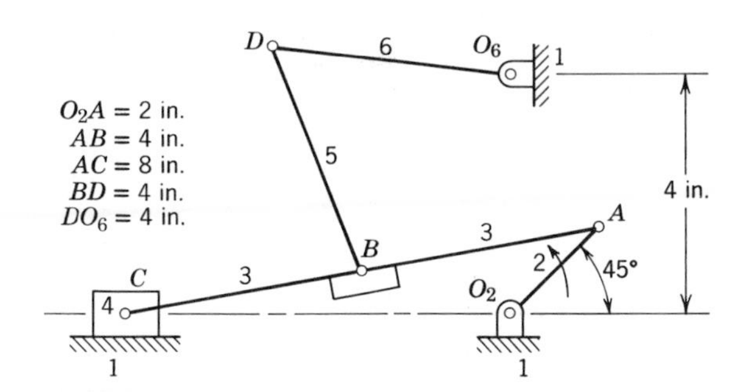

Figure 10.92

10.59 Without determining other instantaneous centers, locate the instantaneous centers 13 and 17 in Fig. 10.77. On what line is 37 located?

10.60 Locate the six instantaneous centers of Fig. 10.78.

10.61 Determine all instantaneous center locations for the mechanism of Fig. 10.81. $\mathbf{V}_R$ is 83.3 ft/sec and $\mathbf{V}_P$ is to be determined. Determine $\mathbf{V}_P$ using instantaneous centers 12, 15, 25. Check your answer by determining $\mathbf{V}_P$ using instantaneous centers 13, 14, 34, and also 13, 15, 35.

10.62 In Fig. 10.82, $\omega_2 = 5$ rad/sec and $\omega_5 = 5$ rad/sec. Determine ω_4 using the instantaneous centers 15, 14, 45.

10.63 For the crank-shaper mechanism shown in Fig. 10.93, determine (*a*) all instantaneous center locations, and (*b*) the velocity of the tool holder (link 5) using the known velocity $\mathbf{V}_{A_2} = 40$ ft/sec.

10.64 (*a*) Given the mechanism shown in Fig. 10.94, locate all instantaneous centers. (*b*) Determine $\mathbf{V}_D$ by instantaneous center methods if $\mathbf{V}_A = 25$ in./sec with ω_2 turning counterclockwise.

10.65 The claw mechanism shown in Fig. 10.95 is used to shift items to the left with intermittent motion. Gears 2 and 3 are in mesh at P, and the velocity of P is given by a vector 1 in. long. Using instantaneous centers, determine the vector representing the velocity of C of the claw.

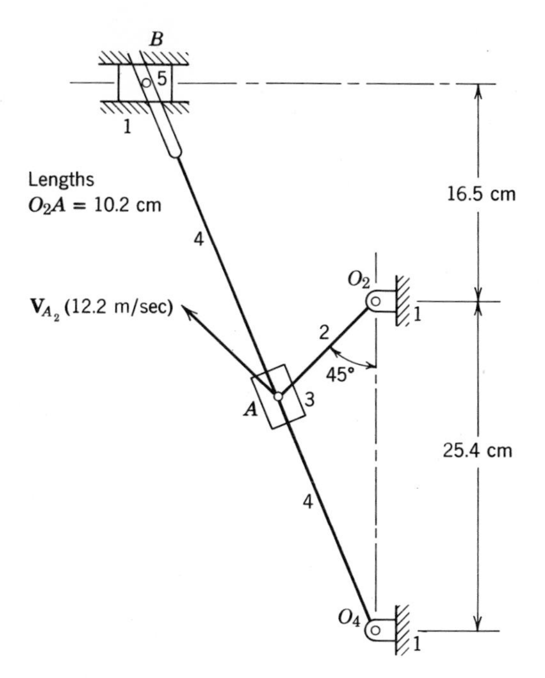

Figure 10.93

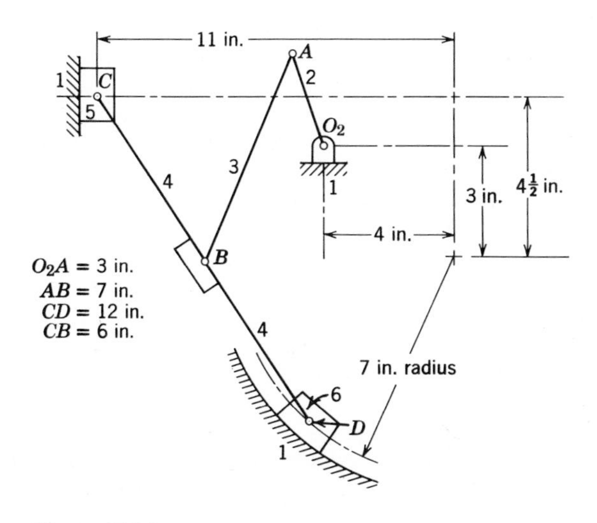

Figure 10.94

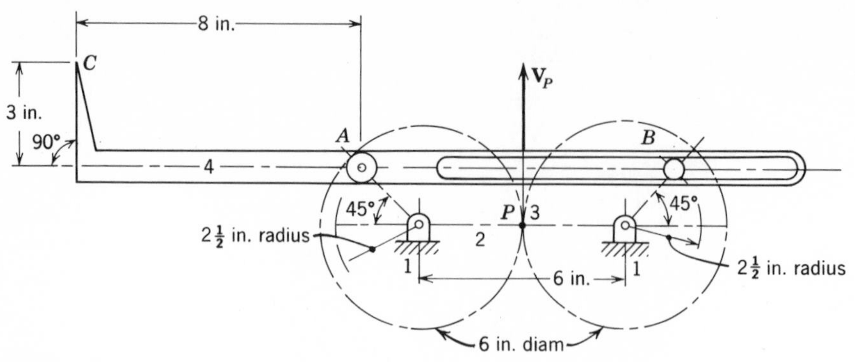

Figure 10.95

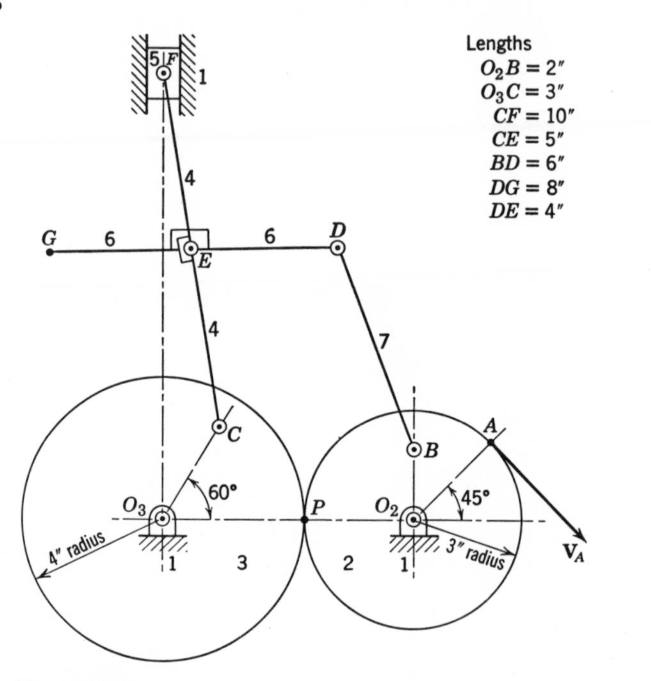

Figure 10.96

10.66 In the mechanism shown in Fig. 10.96, gears 2 and 3 are in mesh. The velocity of point A is given by a vector 1 in. long. By using instantaneous centers, determine the velocity of the slider (link 5). Also determine ω_4.

10.67 In the epicyclic gear train of Fig. 10.97 the sun gear (link 3) and the internal gear (link 5) have the same sense of rotation. $\mathbf{V}_{R_5}$ is one-half the length of $\mathbf{V}_{P_3}$. Determine the location of instantaneous center 14. Determine the angular velocity ratios ω_4/ω_3 and ω_2/ω_3.

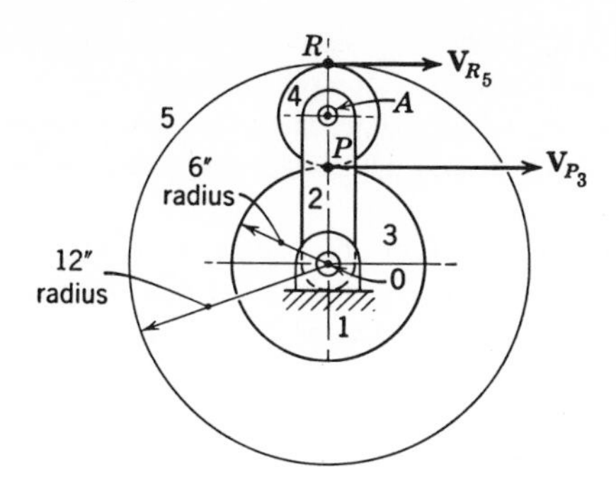

Figure 10.97

10.68 Using real and imaginary axes, sketch to scale the vectors given by the following complex numbers: $\mathbf{r} = 8e^{i\pi/3}$, $\mathbf{r} = -4e^{i\pi}$, $\mathbf{r} = 8(\cos 60° + i \sin 60°)$, $\mathbf{r} = 10 - 40i$, $\mathbf{r} = -4 - 8i$, $\dot{\mathbf{r}} = i(\cos 120° + i \sin 120°)$, $\ddot{\mathbf{r}} = i^2(\cos 120° + i \sin 120°)$.

10.69 The link shown in Fig. 10.43a is made to rotate counterclockwise about O_2 at a uniform angular acceleration α_2 beginning from rest at $\theta_2 = 0$. From the equations of section 10.25, determine expressions for the magnitude $\mathbf{A}_P$ and angle β in terms of α_2 rather than ω_2. Numerically evaluate $\mathbf{A}_P$ and β for $\alpha_2 = 10$ rad/sec^2, $O_2P = 4$ in., and $\theta_2 = 120°$.

10.70 In Example 10.14 are shown the calculations of $\mathbf{A}_{g_3}$ and β of the acceleration $\mathbf{A}_{g_3}$ of the connecting rod in the slider crank in Fig. 10.44. Determine $\mathbf{A}_{g_3}$ and β when the crank angle θ_2 is 120° instead of 30°.

10.71 Using the equations in Section 10.26 continue the calculations of Example 10.14 to determine $\mathbf{V}_B$ and $\mathbf{A}_B$ of the slider for $\theta_2 = 30°$.

10.72 The crank of the Scotch yoke in Fig. 10.98 rotates at constant ω_2. Beginning

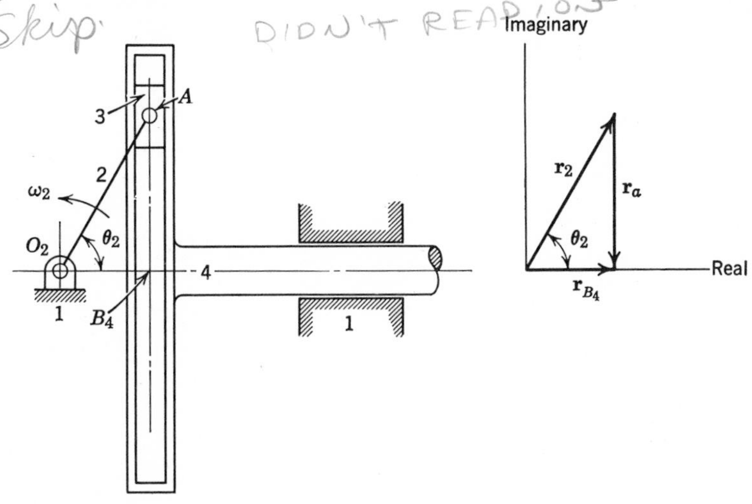

Figure 10.98

with the vector equation $\mathbf{r}_{B_4} = \mathbf{r}_2 + \mathbf{r}_a$, derive expressions for the magnitudes $\mathbf{V}_{B_4}$ and $\mathbf{A}_{B_4}$ of the yoke using the equations of complex numbers.

10.73 In the mechanism of Fig. 10.99 the crank drives the linkage at constant ω_2. Using complex number equations, determine expressions giving θ_4, ω_4, and α_4 in terms of ω_2 and known lengths.

10.74 In Fig. 10.100 is shown an offset slider crank for which the instantaneous values of ω_2 and α_2 are known at the phase θ_2. Beginning with the vector equation $\mathbf{r}_a + \mathbf{r}_d = \mathbf{r}_2 + \mathbf{r}_3$ and using the equations of complex numbers, determine expressions for θ_3, ω_3, α_3, r_a, $\dot{r}_a$, and $\ddot{r}_a$.

10.75 Referring to the four-bar linkage of Fig. 10.48, the angular positions of the links are numerically determined in Section 10.28 for the data given in the figure and for $\theta_2 = 60°$. Evaluate numerically the values of ω_3, ω_4, α_3, and α_4 when $\theta_2 = 60°$, $\omega_2 = 1$ rad/sec, and $\alpha_2 = 0$. Evaluate also the magnitude and angle of the acceleration vector of point B.

10.76 Equations 10.79 and 10.80 for the four-bar linkage show that ω_3 and ω_4 may be zero when $(\theta_4 - \theta_2)$ or $(\theta_3 - \theta_2)$ are zero, respectively. By sketches show

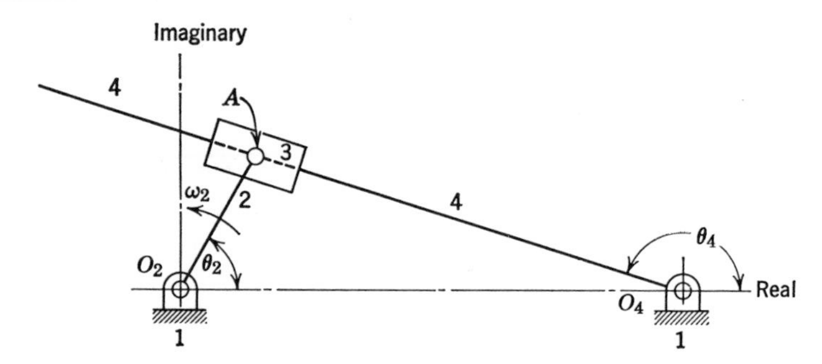

Figure 10.99

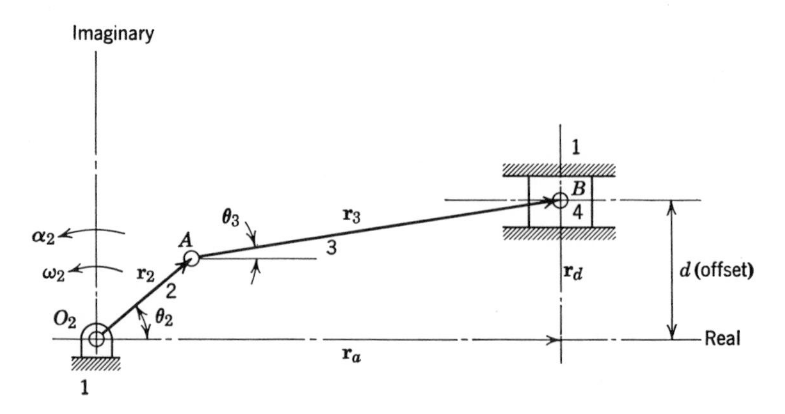

Figure 10.100

possible positions of the linkage for (*a*) $\omega_3 = 0$ and (*b*) $\omega_4 = 0$. The same equations show that ω_3 and ω_4 may be infinite when link angles are such as to make the denominators of the equations zero. Sketch possible positions of the linkage for (*a*) $\omega_3 = \infty$ and (*b*) $\omega_4 = \infty$. Are these positions practical?

10.77 The spatial linkage shown in Fig. 10.51 in the text has the following dimensions:

$$r_2 = 5 \text{ in.} \qquad a = 4 \text{ in.}$$

$$r_4 = 8 \text{ in.} \qquad b = 16 \text{ in.}$$

$$c = 4 \text{ in.}$$

$$\theta_2 = 90^\circ \qquad \phi_2 = 30^\circ \qquad \dot{\phi}_2 = 10 \text{ rad/sec}$$

$$\theta_4 = 0^\circ \qquad \phi_4 = 45^\circ$$

Calculate $\dot{\phi}_4$ and give its direction.

10.78 Derive Eq. 10.98 for $\ddot{\phi}_4$ from the equation for $\dot{\phi}_4$.

11

Force Analysis of Machinery

11.1 Introduction. In order to design the parts of a machine or mechanism for strength, it is necessary to determine the forces and torques acting on the individual links. Each component of a complete machine, however small, should be carefully analyzed for its role in transmitting force. A four-bar mechanism, for example, in reality consists of eight links if the pins or bearings connecting the primary members are included. Bearings, pins, screws, and other fasteners are often critical elements in machinery because of the concentration of force at these elements. Mechanisms which transmit force by direct surface contact on small areas of contact, such as cams, gears, Geneva wheel pins, are also important in this respect.

In machines doing useful work, the forces associated with the principal function of the machine are usually known or assumed. For example, in a piston-type engine or piston-type compressor, the gas force on the piston is known or assumed; in a quick-return mechanism such as the crank shaper or the Whitworth machine, the resistance of the cutting tool is assumed. Such forces are termed *static* forces since in a machine analysis they are classed differently from inertia forces, which are expressed in terms of the accelerated motion of the individual links.

In machines doing useful work, the forces associated with the principal which produce the accelerated motion of the link are often greater than the static forces related to the primary function of the machine. In many rotary machines, such as bladed compressor and turbine wheels, precautions are necessary to avoid runaway conditions in which speeds may exceed structurally safe design speeds.

11.2 Centrifugal Force in Rotor Blades. In rotors, inertia force, the product of mass and acceleration, is known as *centrifugal force*. In high-speed rotors with blades (such as compressor and turbine wheels, supercharger wheels, fans, and propellers), centrifugal forces tend to separate the blades from the rotor. Figure 11.1 shows a simple type of bladed rotor. To determine the centrifugal force producing a resisting centripetal force at the base (section a–a) of any given blade, an integration is required since the acceleration is a function of R. Assuming that the rotor is at constant angular velocity ω, the inertia force dF acting on the element of the blade shown is the product of the mass of the element dM and the centripetal acceleration $A^n = \omega^2 R$ from Eq. 10.2a. Therefore,

$$dF = (dM)A^n = \omega^2 R \, dM \tag{11.1}$$

It is recalled from the study of mechanics that inertia force is opposite in sense to the centripetal acceleration, hence the term *centrifugal* force. The mass of the element is the product of mass density w/g (w is weight density in pounds per cubic inch and g is 386 in./sec^2) and the volume of the element $bt(dR)$; b, t, R are in inches.

$$dF = \omega^2 R \frac{w}{g} bt \, dR$$

$$F = bt \frac{w}{g} \omega^2 \int_{R=R_i}^{R=R_o} R \, dR \tag{11.2}$$

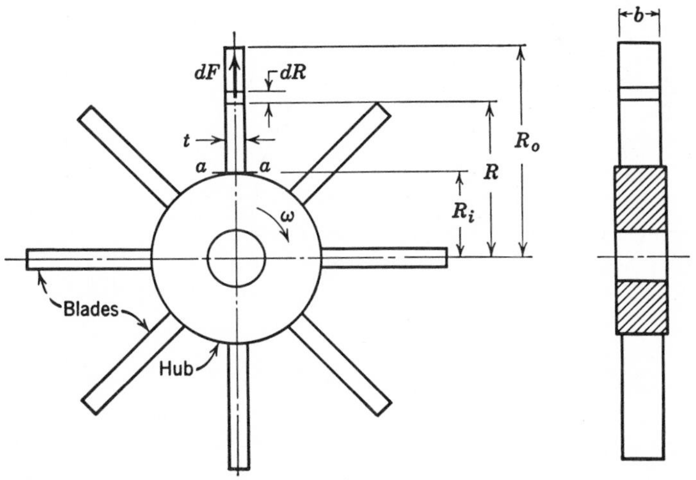

Figure 11.1

The average tensile stress s_b at the base of the blade due to the inertia force is P/A, in which $P = F$ and $A = bt$.

$$s_b = \frac{F}{bt} = \frac{w}{g}\omega^2 \int_{R=R_i}^{R=R_o} R\, dR \tag{11.3}$$

Equation 11.3 shows that the stress at the base of the blade is independent of the cross-sectional area $A = bt$ but depends on rotor speed, mass density, and the inner and outer radii of the blades.

Rotor blades such as the wide blades of a fan are idealized in Fig. 11.2. The fan has the form of a disk with slots between the blades. The element of inertia force dF is the same as given by Eq. 11.1, in which the mass of the element is

$$dM = \frac{w}{g} tR\, d\phi\, dR$$

and

$$dF = \frac{w}{g} t\omega^2 R^2\, dR\, d\phi$$

Therefore,

$$F = \frac{w}{g} t\omega^2 \int_{\phi=0}^{\phi=2\pi/N} \int_{R=R_i}^{R=R_o} R^2\, dR\, d\phi$$

$$F = \frac{2\pi}{N}\frac{w}{g} t\omega^2 \int_{R=R_i}^{R=R_o} R^2\, dR \tag{11.4}$$

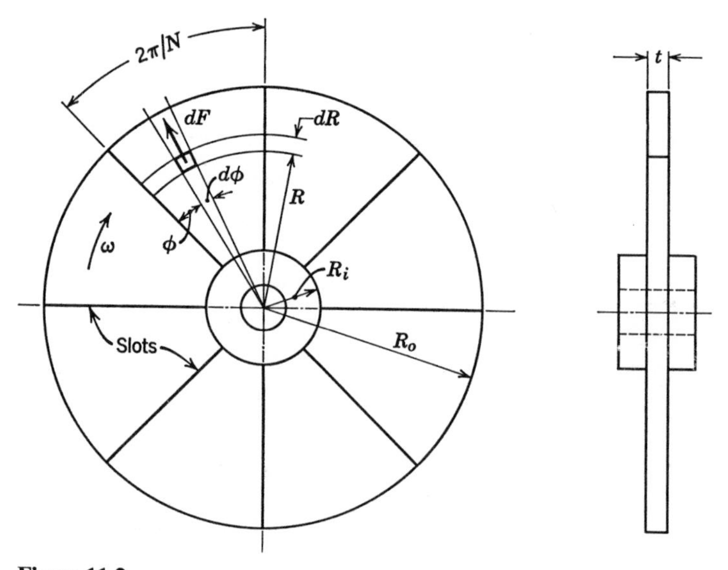

Figure 11.2

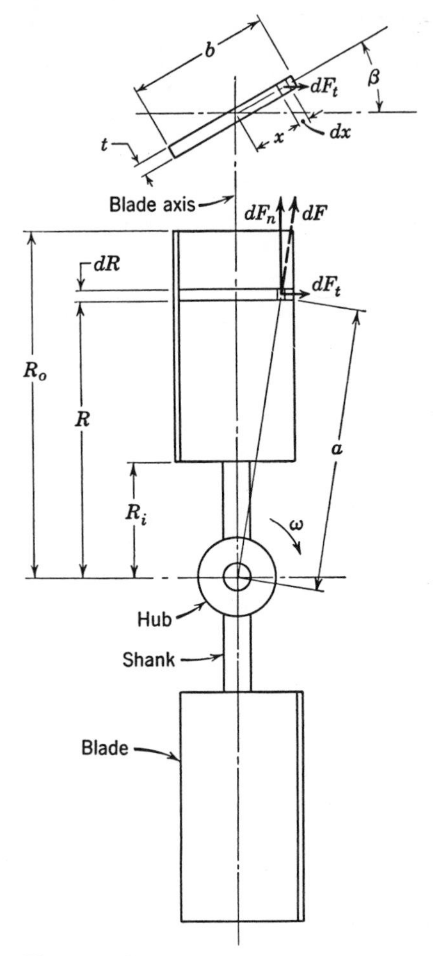

Figure 11.3

where N is the number of blades and F is the force tending to separate the blade from the hub. The average stress at the base of the blade with cross-sectional area $A = 2\pi R_i t/N$ is

$$s_b = \frac{w}{g}\frac{\omega^2}{R_i}\int_{R=R_i}^{R=R_o} R^2\, dR \tag{11.5}$$

In aircraft propellers, the blades are set at a blade angle as shown in Fig. 11.3. In such cases, inertia forces produce a twisting moment on the blade. Referring to Fig. 11.3, the inertia force dF on an element $t(dx)(dR)$ is

$$dF = \omega^2 a\, dM$$

in which $dM = (w/g)t\,dx\,dR$. The inertia force dF which is due to A'' of the element of mass may be shown as the components dF_n and dF_t, in which dF_n produces a tensile force on the blade parallel to the blade axis and dF_t produces a twisting moment dM_t about the axis of the blade because of the moment arm $x \sin \beta$.

$$dF_n = \frac{R}{a}\,dF = \omega^2 R\,dM$$

$$dF_n = \frac{w}{g}\,t\omega^2 R\,dx\,dR \tag{11.6}$$

$$dF_t = \frac{x \cos \beta}{a}\,dF = \omega^2 x \cos \beta\,dM$$

$$dF_t = \frac{w}{g}\,t\omega^2 x \cos \beta\,dx\,dR \tag{11.7}$$

$$dM_t = x \sin \beta\,dF_t$$

$$dM_t = \frac{w}{g}\,t \cos \beta \sin \beta \omega^2 x^2\,dx\,dR \tag{11.8}$$

The total inertia force of the blade producing tension in the shank parallel to the axis of the blade is

$$F_n = \frac{w}{g}\,t\omega^2 2 \int_{x=0}^{x=b/2} \int_{R=R_i}^{R=R_o} R\,dx\,dR \tag{11.9}$$

and the total twisting moment on the shank is

$$M_t = \frac{w}{g}\,t\omega^2 \cos \beta \sin \beta 2 \int_{x=0}^{x=b/2} \int_{R=R_i}^{R=R_o} x^2\,dx\,dR \tag{11.10}$$

11.3 Inertia Force. Inertia Torque. From the study of mechanics, it is known that the following equations of motion apply for a rigid body in plane motion.

$$\Sigma \mathbf{F} = M\mathbf{A}_g \tag{11.11}$$

$$\Sigma T = I\alpha \tag{11.12}$$

in which $\Sigma \mathbf{F}$ is the vector sum, or the resultant $\mathbf{R}$, of a system of forces acting on the body in the plane of motion; M is the mass of the body; and $\mathbf{A}_g$ is the acceleration of the mass center g (center of gravity) of the body. ΣT is the sum of the moments of the forces and torques about an axis through the mass center normal to the plane of motion; I is the moment of inertia of the body about the same axis through the mass center; and α is the angular acceleration of the body in the plane of motion. The unit of mass M commonly used is the slug (lb-sec^2/ft) and the unit of moment of inertia I is slug-ft^2 (lb-sec^2-ft). For the International System of Units the unit of mass is kilogram (kg) and the unit of moment of inertia is kg-m^2. See Appendix 3 for conversion factors.

Figure 11.4 shows a rigid body in plane motion acted upon by forces for which the resultant **R** is determined from the polygon of free force vectors shown. Since **R** represents Σ **F**, Eq. 11.11 may be rewritten:

$$\mathbf{R} = M\mathbf{A}_g \tag{11.13}$$

For the case in which the forces are known, the acceleration $\mathbf{A}_g$ of the body may be calculated from Eq. 11.13 provided the mass is also known. The direction of $\mathbf{A}_g$ is parallel to **R** and in the same sense as **R**.

The line of action of **R** is determined as shown in Fig. 11.4, and, from the principle of moments, Re is equal to Σ T. Eq. 11.12 may be rewritten

$$Re = I\alpha \tag{11.14}$$

The angular acceleration α of the body may be determined from Eq. 11.14 if the forces and the moment of inertia I of the body are known. α is in the same angular sense as the moment Re.

Equations of motion in the forms of Eqs. 11.11 through 11.14 are useful when accelerations are to be determined including magnitude, direction, and sense. However, for mechanisms with constrained motion, accelerations are usually known from a kinematic analysis as discussed in Chapter 10, and the forces and moments which produce the accelerations are to be determined.

When $\mathbf{A}_g$ of a given link is known and $M\mathbf{A}_g$ is calculable, a simplification in concept results if $M\mathbf{A}_g$, expressed in units of force, is regarded as a force vector $\mathbf{F}_o$ and is shown as the equilibrant of **R** on the free-body diagram of the

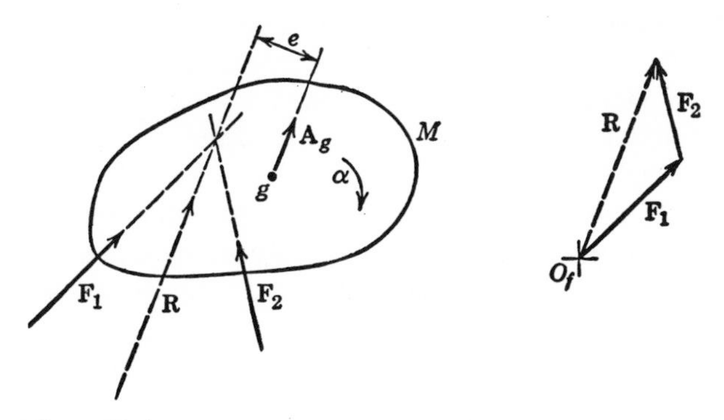

Figure 11.4

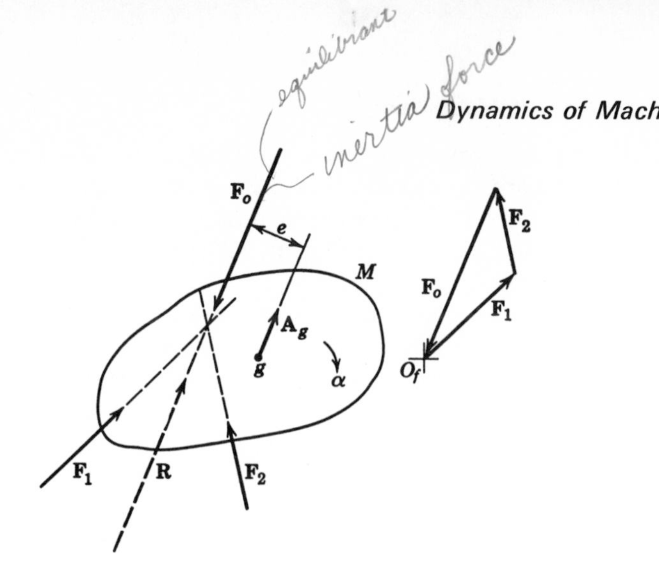

Figure 11.5

link. In Fig. 11.5, the body of Fig. 11.4 is shown with $\mathbf{F}_o$ as an equilibrant. As a vector, $\mathbf{F}_o$ is shown parallel in direction to $\mathbf{A}_g$, which is also parallel to $\mathbf{R}$ and is equal in magnitude to $\mathbf{R}$ from Eq. 11.13. To be the equilibrant of $\mathbf{R}$, however, $\mathbf{F}_o$ must be shown in a sense opposite to $\mathbf{A}_g$. Also, the line of action of $\mathbf{F}_o$ must be such that its moment about the mass center is equal and *opposite* to the moment of R. Equation 11.14 may be used to determine the distance e of the line of action of $\mathbf{F}_o$.

$$e = \frac{I\alpha}{R}$$

$$e = \frac{I\alpha}{F_o} = \frac{I\alpha}{MA_g} \tag{11.15}$$

It should be noted that the moment of $\mathbf{F}_o$ about the mass center is opposite in sense to α. By showing $\mathbf{F}_o$ opposite in sense to $\mathbf{A}_g$ and the moment of $\mathbf{F}_o$ opposite in sense to α, it appears to represent a resistance to the accelerated motion of the link and in a sense is a measure of the inertia of the link. Thus $\mathbf{F}_o$ is termed an *inertia force*.

In Fig. 11.5, $\mathbf{F}_o$ is shown as the equilibrant of $\mathbf{R}$. An alternative representation as in Fig. 11.6 is to show $\mathbf{F}_o$ at the mass center g and to add an *inertia torque* or *inertia couple* T_o in a sense opposite to α. The magnitudes of $\mathbf{F}_o$ and T_o are given by the following equations:

$$F_o = MA_g \tag{11.16}$$

$$T_o = I\alpha \tag{11.17}$$

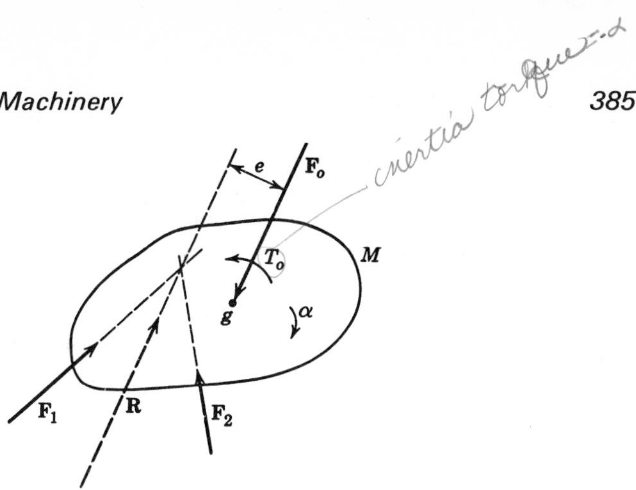

Figure 11.6

When $\mathbf{A}_g$ is zero and α is of a value other than zero, only the inertia couple T_o remains.

It may be seen from Fig. 11.5 that, by showing known mass-acceleration effects of a link as an inertia force, the equations of motion (11.11 and 11.12) may be interpreted as equations of static equilibrium and may be written as such:

$$\Sigma \mathbf{F} = 0$$

$$\Sigma T = 0$$

in which $\Sigma \mathbf{F}$ includes $\mathbf{F}_o$, and ΣT includes T_o. In Fig. 11.5, the polygon of free force vectors, including $\mathbf{F}_o$, closes as required for static equilibrium.

The inertia force method is a simple and useful method since kinetic problems involving rigid-body linkages in plane motion are reduced to problems of static equilibrium. Because of the constrained motion of linkages, accelerations and inertia forces and couples of the individual links may be determined first, and the forces producing accelerated motion may then be determined from the laws of static equilibrium.

11.4 Force Determination. In the force analysis of a complete mechanism, a free-body diagram of each link should usually be made to indicate the forces acting on the link. In determining the directions of these forces, the following laws from the study of statics will be recalled.

1. A rigid body acted on by two forces is in static equilibrium only if the two forces are colinear and equal in magnitude but opposite in sense. If only the points of application of the two forces are known, such as A and B of Fig. 11.7, the directions of the two forces may be determined from the direction of the line joining A and B.

2. For a rigid body acted upon by three forces in static equilibrium, the

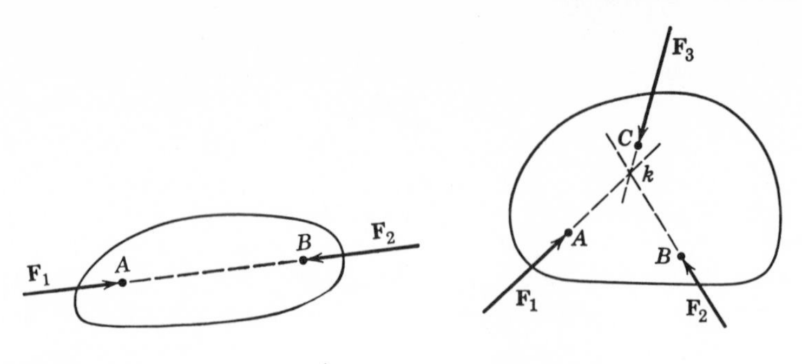

Figure 11.7 Figure 11.8

lines of action of the three forces are concurrent at some point such as k in Fig. 11.8. Thus, if the lines of action of two of the forces are known, the line of action of the third force must pass through its point of application and the point of concurrency k. In some cases a greater number than three forces on a body may be reduced to three by determining the resultant of the known forces.

3. A rigid body acted upon by a couple is in static equilibrium only if acted upon by another coplanar couple equal in magnitude and opposite in sense as shown in Fig. 11.9.

In the case of a static force analysis, the vector sum of the forces on each link must equal zero for equilibrium. This must also be true for a dynamic analysis when inertia forces are used. To use the concept of inertia forces is, therefore, an advantage because both static and dynamic cases can be treated in the same manner. In both types of analyses, the vector equations can be solved analytically or graphically to determine the unknown forces.

Factors which determine whether an analytical or a graphical solution should be made are the type of mechanism and the number of positions to be analyzed. For relatively simple mechanisms such as cams and gears, an analytical solution is generally used. For the analysis of a linkage at only one

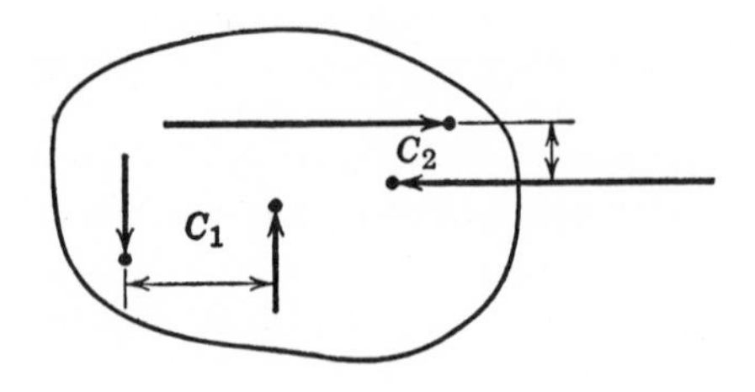

Figure 11.9

position, a graphical solution is much quicker than an analytical one. However, if several positions or a complete cycle are to be studied, analytical methods should be selected. This is especially true if computer facilities are available or pocket or desk calculators with vector-resolver capabilities. It should be mentioned, however, that even when an analytical solution is used, it is often desirable to check the results at one position by graphical means.

11.5 Methods of Linkage Force Analysis. There are two general methods of performing a complete force analysis on a linkage (*a*) by superposition, and (*b*) by the use of transverse and radial components. These methods are discussed in the following sections.

Superposition. The principle of superposition may be used in the force analysis of a rigid body in static equilibrium. This principle states that a resultant effect may be determined from the summation of several effects which are equivalent to the total effect. By this method, a linkage with several forces acting on it can easily be analyzed by determining the effect of these forces taken one at a time. The results of the several single-force analyses are then added together to give the total forces acting on each joint in the linkage. Superposition can also be used advantageously to combine the results of a static and an inertia-force analysis that were made independently.

Although this method is easy to use, it has the disadvantage that the linkage must be analyzed several times which often becomes tedious. Another disadvantage is that an accurate analysis cannot be made if friction forces are to be considered. In linkages with turning pairs this problem does not usually arise because friction forces are generally small enough to be neglected. However, with sliding pairs such as with the piston and cylinder in the slider-crank mechanism, superposition would not be a suitable method of analysis if friction is to be considered between the piston and cylinder. Errors would occur because of a change in direction of the force between the piston and cylinder in the separate solutions required for superposition. An example of this is given by Hirschhorn to which the reader is referred.[1]

An analytical solution of a force analysis by superposition is given in Example 11.1, and a graphical solution is given in Example 11.2.

Example 11.1

In Fig. 11.10*a* is shown the linkage from Fig. 10.7 of Example 10.1 for which a velocity and acceleration analysis has been made. It is required to determine the bearing forces on each link and the shaft torque T_s at O_2 by superposition using unit vectors.

[1] J. Hirschhorn, *Kinematics and Dynamics of Plane Mechanisms*, McGraw-Hill Book Company, 1962.

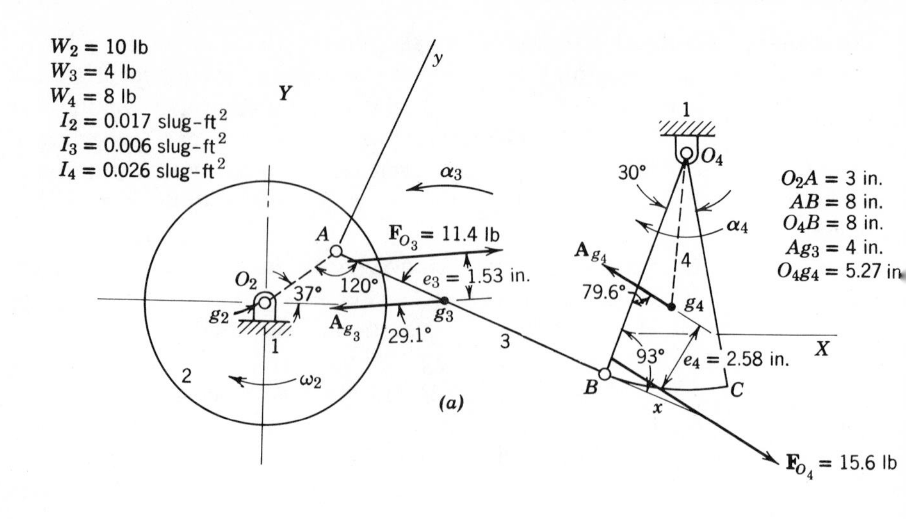

$W_2 = 10$ lb
$W_3 = 4$ lb
$W_4 = 8$ lb
$I_2 = 0.017$ slug-ft^2
$I_3 = 0.006$ slug-ft^2
$I_4 = 0.026$ slug-ft^2
$O_2A = 3$ in.
$AB = 8$ in.
$O_4B = 8$ in.
$Ag_3 = 4$ in.
$O_4g_4 = 5.27$ in.
$F_{O_3} = 11.4$ lb
$e_3 = 1.53$ in.
$e_4 = 2.58$ in.
$F_{O_4} = 15.6$ lb
(a)

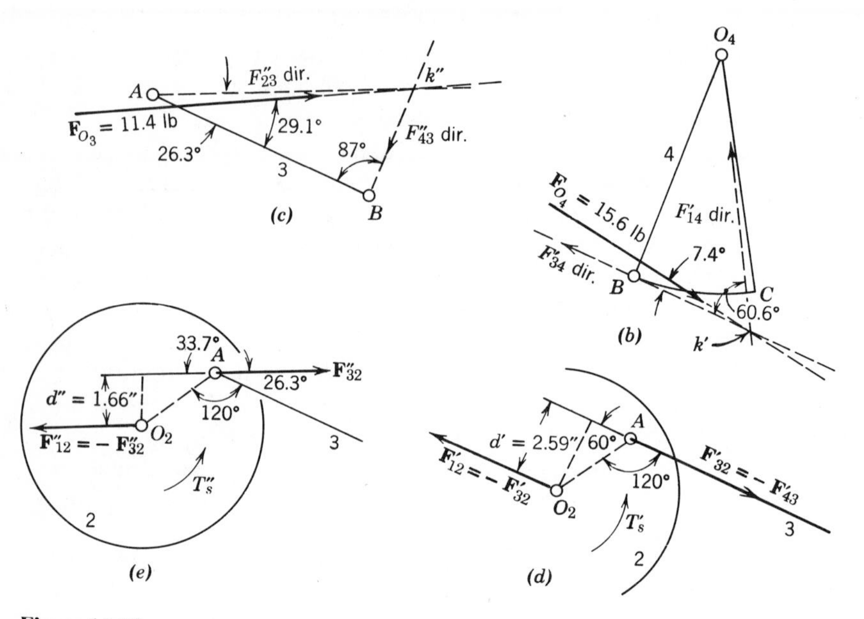

F''_{23} dir.
F''_{43} dir.
$F_{O_3} = 11.4$ lb
(c)
$F_{O_4} = 15.6$ lb
F'_{14} dir.
F'_{34} dir.
(b)
$d'' = 1.66''$
$F''_{12} = -F''_{32}$
(e)
$d' = 2.59''$
$F'_{12} = -F'_{32}$
$F'_{32} = -F'_{43}$
(d)

Figure 11.10

Solution. From the solution of Example 10.1

$$\omega_3 = 4.91 \text{ rad/sec (ccw)}$$
$$\omega_4 = 7.82 \text{ rad/sec (ccw)}$$
$$\alpha_3 = 241 \text{ rad/sec}^2 \text{ (ccw)}$$
$$\alpha_4 = +129 \text{ rad/sec}^2 \text{ (cw)}$$
$$\mathbf{A}_A = -72\mathbf{i} - 124.8\mathbf{j}$$
$$|\mathbf{A}_A| = 144 \text{ ft/sec}^2$$
$$\mathbf{A}_B = -88.1\mathbf{i} + 35.8\mathbf{j}$$
$$|\mathbf{A}_B| = 95.1 \text{ ft/sec}^2$$

Using the above values, the following were calculated and their directions determined.

$$|\mathbf{A}_{g_3}| = 91.6 \text{ ft/sec}^2$$
$$|\mathbf{A}_{g_4}| = 62.7 \text{ ft/sec}^2$$

The angles at which the vector $\mathbf{A}_{g_3}$ and $\mathbf{A}_{g_4}$ act are shown in Fig. 11.10*a*.

The magnitudes of the inertia forces and the offset distances can be calculated as follows:

$$F_{o_2} = 0(A_{g_2} = 0)$$

$$F_{o_3} = M_3 A_{g_3} = \frac{4 \times 91.6}{32.2} = 11.4 \text{ lb}$$

$$F_{o_4} = M_4 A_{g_4} = \frac{8 \times 62.7}{32.2} = 15.6 \text{ lb}$$

$$e_3 = \frac{I_3 \alpha_3}{F_{o_3}} = \frac{0.006 \times 241}{11.4} = 0.127 \text{ ft} = 1.53 \text{ in.}$$

$$e_4 = \frac{I_4 \alpha_4}{F_{o_4}} = \frac{0.026 \times 129}{15.6} = 0.215 \text{ ft} = 2.58 \text{ in.}$$

The vectors $\mathbf{F}_{o_3}$ and $\mathbf{F}_{o_4}$ are shown in Fig. 11.10*a* in their correct positions relative to their respective acceleration vectors, that is, parallel to acceleration vectors, opposite in sense, and offset distance e so that $F_o \times e$ gives a torque whose sense is opposite to that of α.

The solution of this problem will be carried out using superposition considering (*a*) only $\mathbf{F}_{o_4}$ acting, (*b*) only $\mathbf{F}_{o_3}$ acting, and (*c*) addition of the two analyses. All components have been taken relative to the xy axes.

FORCE ANALYSIS WITH ONLY $\mathbf{F}_{o_4}$ ACTING. Figure 11.10*b* shows a free-body diagram of link 4 upon which act forces $\mathbf{F}_{o_4}$, $\mathbf{F}'_{34}$, and $\mathbf{F}'_{14}$. The direction of

$\mathbf{F}_{o_4}$ is known and that of $\mathbf{F}'_{34}$ is along link 3 because link 3 becomes a two-force member when $\mathbf{F}_{o_3}$ is omitted; the direction lines of these two vectors intersect at point k'. Link 4 is in equilibrium under the action of these three forces with no couple acting upon it so that the direction of vector $\mathbf{F}'_{14}$ must pass through points k' and O_4. From the angles given on the free body diagram and the magnitude of $\mathbf{F}_{o_4}$, the forces $\mathbf{F}'_{34}$ and $\mathbf{F}'_{14}$ can be determined as follows:

$$\mathbf{F}_{o_4} = 15.6(\cos 7.4°\mathbf{i} + 7.4°\mathbf{j})$$
$$= 15.5\mathbf{i} - 2.01\mathbf{j}$$
$$\mathbf{F}'_{14} = F'_{14}(\cos 60.6°\mathbf{i} + \sin 60.6°\mathbf{j})$$
$$= -0.491F'_{14}\mathbf{i} + 0.871F'_{14}\mathbf{j}$$
$$\mathbf{F}'_{34} = -F'_{34}\mathbf{i}$$

Because link 4 is in equilibrium under the action of the three forces given above,

$$\mathbf{F}_{o_4} + \mathbf{F}'_{14} + \mathbf{F}'_{34} = 0$$

Substituting the above relations in the equation of equilibrium,

$$15.5\mathbf{i} - 2.01\mathbf{j} - 0.491F'_{14}\mathbf{i} + 0.871F'_{14}\mathbf{j} - F'_{34}\mathbf{i} = 0$$

Summing **i** *components,*

$$15.5\mathbf{i} - 0.491F'_{14}\mathbf{i} - F'_{34}\mathbf{i} = 0$$

Summing **j** *components,*

$$-2.01\mathbf{j} + 0.871F'_{14}\mathbf{j} = 0$$
$$F'_{14} = 2.30 \text{ lb}$$

Also $15.5\mathbf{i} - 0.491 \times 2.30\mathbf{i} - F'_{34}\mathbf{i} = 0$

$$F'_{34} = 14.3 \text{ lb}$$

Therefore,

$$\mathbf{F}'_{14} = -0.491 \times 2.30\mathbf{i} + 0.871 \times 2.50\mathbf{j} = -1.13\mathbf{i} + 2.01\mathbf{j}$$
$$\mathbf{F}'_{34} = -14.3\mathbf{i}$$

To calculate the shaft torque T'_s necessary to hold link 2 in equilibrium under the action of the couple produced by $\mathbf{F}'_{32}$ and $\mathbf{F}'_{12}$, refer to Fig. 11.10*d* where

$$F'_{32} = F'_{43} = 14.3 \text{ lb}$$

and

$$d' = 2.59 \text{ in.}$$

Therefore,

$$T'_s = F'_{32}\,d' = 14.3 \times 2.59$$
$$= 37.2 \text{ lb-in. (ccw)}$$

Torque T'_s could also have been easily determined using vector equations for F'_{32} and d' and the relation

$$\mathbf{T}'_s = -(\mathbf{F}_{32} \times \mathbf{d}')$$

FORCE ANALYSIS WITH ONLY $\mathbf{F}_{o_3}$ ACTING. Figure 11.10*c* shows a free-body diagram of link 3 under the action of three forces $\mathbf{F}_{o_3}$, $\mathbf{F}''_{23}$, and $\mathbf{F}''_{43}$. The direction of $\mathbf{F}_{o_3}$ is known and that of $\mathbf{F}''_{43}$ is along line O_4B because link 4 becomes a two-force member when $\mathbf{F}_{o_4}$ is omitted. The intersection of the known directions of $\mathbf{F}_{o_3}$ and $\mathbf{F}''_{43}$ gives point k''. The direction of $\mathbf{F}''_{23}$ must pass through points k'' and A because link 3 is in equilibrium under the action of these three forces with no couple acting upon it. From the information given on the free-body diagram, the forces $\mathbf{F}''_{23}$ and $\mathbf{F}''_{43}$ can be determined as follows:

$$\mathbf{F}_{o_3} = 11.4(\cos 29.1°\mathbf{i} + \sin 29.1°\mathbf{j})$$
$$= 9.94\mathbf{i} + 5.53\mathbf{j}$$
$$\mathbf{F}''_{23} = F''_{23}(-\cos 26.3°\mathbf{i} - \sin 26.3°\mathbf{j})$$
$$= -0.897F''_{23}\mathbf{i} - 0.443F''_{23}\mathbf{j}$$
$$\mathbf{F}''_{43} = F''_{43}(\cos 87°\mathbf{i} - \sin 87°\mathbf{j})$$
$$= 0.052F''_{43}\mathbf{i} - 0.999F''_{43}\mathbf{j}$$

Because link 3 is in equilibrium under the action of the three forces given above,

$$\mathbf{F}_{o_3} + \mathbf{F}''_{23} + \mathbf{F}''_{43} = 0$$

Substituting the above relations in the equation of equilibrium

$$9.94\mathbf{i} + 5.53\mathbf{j} - 0.897F''_{23}\mathbf{i} - 0.443F''_{23}\mathbf{j} + 0.052F''_{43}\mathbf{i} - 0.999F''_{43}\mathbf{j} = 0$$

Summing **i** *components,*

$$9.94\mathbf{i} - 0.897F''_{23}\mathbf{i} + 0.052F''_{43}\mathbf{i} = 0$$

Summing **j** *components,*

$$5.53\mathbf{j} - 0.443F''_{23}\mathbf{j} - 0.999F''_{43}\mathbf{j} = 0$$

Solving the equations for **i** and **j** components simultaneously

$$F''_{23} = 11.1 \text{ lb}$$
$$F''_{43} = 0.605 \text{ lb}$$

Therefore,

$$\mathbf{F}''_{23} = -0.897 \times 11.1\mathbf{i} - 0.443 \times 11.1\mathbf{j}$$
$$= -9.98\mathbf{i} - 4.93\mathbf{j}$$
$$\mathbf{F}''_{43} = 0.052 \times 0.605\mathbf{i} - 0.999 \times 0.605\mathbf{j}$$
$$= 0.032\mathbf{i} - 0.604\mathbf{j}$$

The shaft torque T''_s necessary to hold link 2 in equilibrium under the action of the couple produced by $\mathbf{F}''_{32}$ and $\mathbf{F}''_{12}$ can be calculated from Fig. 11.10*e* where

$$\mathbf{F}''_{32} = 11.1 \text{ lb}$$

and

$$\mathbf{d}'' = 1.66 \text{ in.}$$

Therefore,

$$T''_s = F''_{32}\, d'' = 11.1 \times 1.66$$
$$= 18.5 \text{ lb-in. (ccw)}$$

If desirable, T''_s can be calculated using the following vector equations

$$\mathbf{F}''_{32} = -\mathbf{F}''_{23} = 9.98\mathbf{i} + 4.93\mathbf{j}$$
$$\mathbf{d}'' = 0.736\mathbf{i} - 1.49\mathbf{j}$$

and

$$\mathbf{T}''_s = -(\mathbf{F}''_{32} \times \mathbf{d}'')$$
$$= 18.5\mathbf{k} \text{ lb-in. (ccw)}$$

TOTAL FORCES:

$$\mathbf{F}_{32} = \mathbf{F}'_{32} + \mathbf{F}''_{32} = \mathbf{F}'_{43} + \mathbf{F}''_{32}$$
$$= 14.3\mathbf{i} + 9.98\mathbf{i} + 4.93\mathbf{j}$$
$$= 24.3\mathbf{i} + 4.93\mathbf{j}$$

and

$$|\mathbf{F}_{32}| = 24.8 \text{ lb}$$
$$\mathbf{F}_{43} = \mathbf{F}'_{43} + \mathbf{F}''_{43}$$
$$= 14.3\mathbf{i} + 0.032\mathbf{i} - 0.604\mathbf{j}$$
$$= 14.3\mathbf{i} - 0.604\mathbf{j}$$

and

$$|\mathbf{F}_{43}| = 14.4 \text{ lb}$$
$$\mathbf{F}_{14} = \mathbf{F}'_{14} + \mathbf{F}''_{14} = \mathbf{F}'_{14} + \mathbf{F}''_{43}$$
$$= -1.13\mathbf{i} + 2.01\mathbf{j} + 0.032\mathbf{i} - 0.604\mathbf{j}$$
$$= -1.10\mathbf{i} - 1.41\mathbf{j}$$

and

$$|\mathbf{F}_{14}| = 1.78 \text{ lb}$$

$$T_s = T_s' + T_s''$$

$$= 37.2 + 18.5$$

$$= 55.7 \text{ lb-in. (ccw)}$$

Therefore,

$$F_{14} = 1.78 \text{ lb}$$

$$F_{43} = 14.4 \text{ lb}$$

$$F_{32} = 24.8 \text{ lb}$$

$$F_{12} = 24.8 \text{ lb}$$

$$T_s = 55.7 \text{ lb-in. (ccw)}$$

Example 11.2

A graphical analysis of the linkage of Example 11.1 will now be made by superposition using the International System of Units.

Solution. The following values were determined from an acceleration polygon (not shown) for the linkage.

$$A_{g_2} = 0$$

$$A_{g_3} = 27.92 \text{ m/sec}^2$$

$$A_{g_4} = 19.11 \text{ m/sec}^2$$

$$\alpha_3 = 241 \text{ rad/sec}^2 \text{ (ccw)}$$

$$\alpha_4 = 129 \text{ rad/sec}^2 \text{ (cw)}$$

Vectors representing the above values are shown on the configuration diagram (Fig. 11.11*a*).

The magnitudes of the inertia forces and offset distances can be calculated as follows:

$$F_{o_2} = 0(A_{g_2} = 0)$$

$$F_{o_3} = M_3 A_{g_3} = \frac{17.79 \times 27.92}{9.807} = 50.65 \text{ N}$$

$$F_{o_4} = M_4 A_{g_4} = \frac{35.58 \times 19.11}{9.807} = 69.33 \text{ N}$$

$$e_3 = \frac{I_3 \alpha_3}{F_{o_3}} = \frac{0.008 \times 241}{50.65} = 0.0381 \text{ m} = 3.81 \text{ cm}$$

$$e_4 = \frac{I_4 \alpha_4}{F_{o_4}} = \frac{0.035 \times 129}{69.33} = 0.0651 \text{ m} = 6.51 \text{ cm}$$

FORCE ANALYSIS WITH ONLY $\mathbf{F}_{o4}$ ACTING (Figs. 11.11*b*, *c*, *d*, *e*).

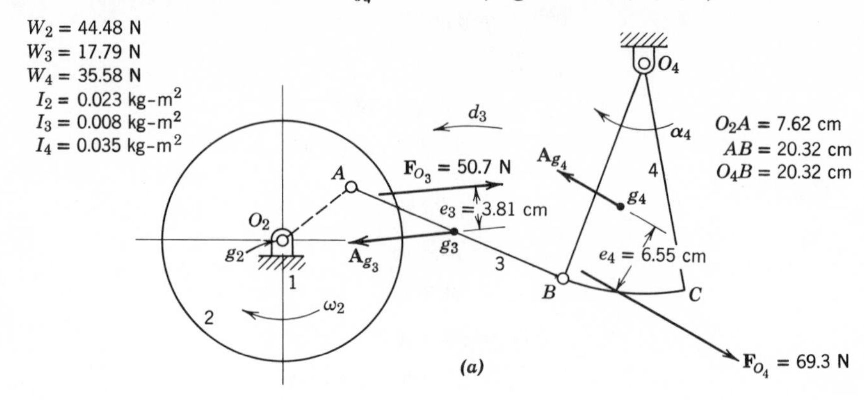

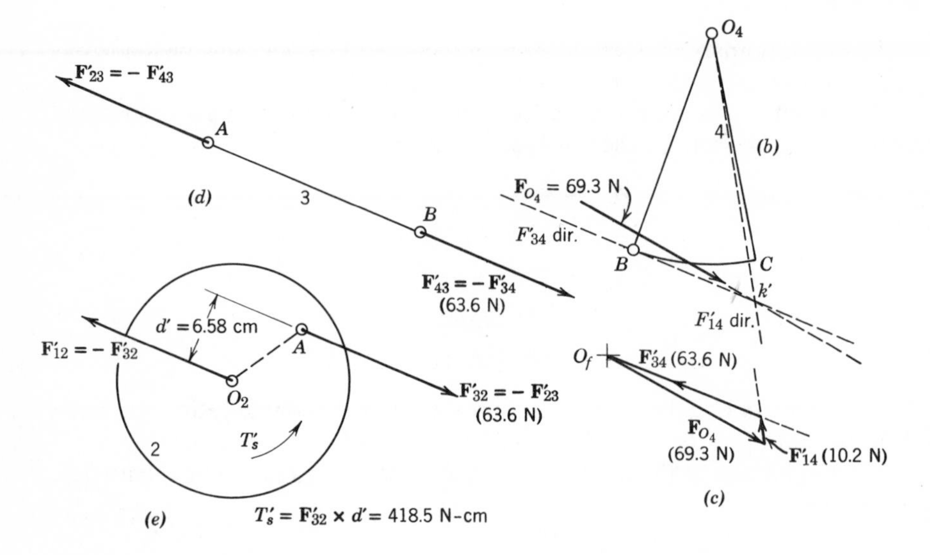

Figure 11.11

The vectors $\mathbf{F}_{o_3}$ and $\mathbf{F}_{o_4}$ are shown on the configuration diagram (Fig. 11.11*a*) in their correct positions relative to their respective acceleration vectors, that is parallel to acceleration vector, opposite in sense, and offset a distance e so that $F_o \times e$ gives a torque whose sense is opposite to that of α.

The solution of this problem will be carried out using superposition considering (*a*) only $\mathbf{F}_{o4}$ acting, (*b*) only $\mathbf{F}_{o3}$ acting, and (*c*) addition of the two analyses.

FORCE ANALYSIS WITH ONLY $\mathbf{F}_{o4}$ ACTING. Figure 11.11*b* shows a free-body

FORCE ANALYSIS WITH ONLY $\mathbf{F}_{o_3}$ ACTING (Figs. 11.11*f*, *g*, *h*, *i*).
RESULTANT FORCE ANALYSIS (Figs. 11.11*j*, *k*, *l*).

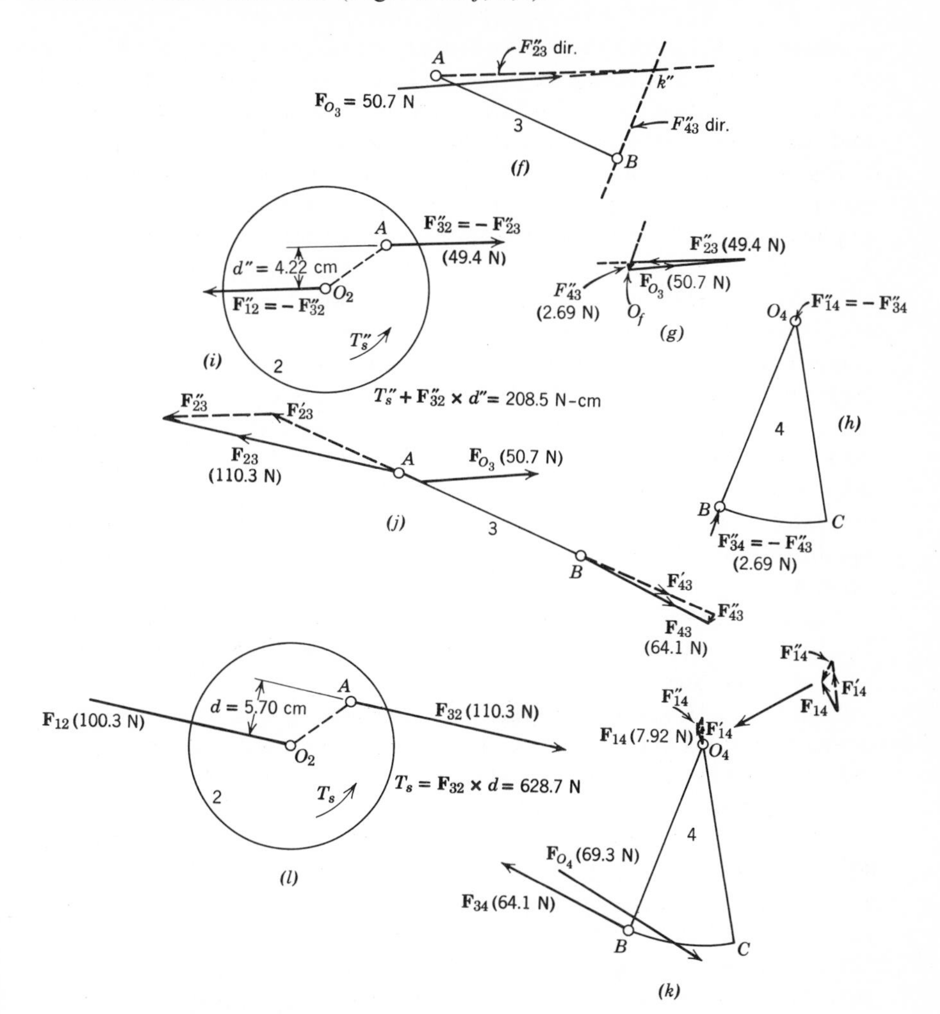

Figure 11.11 (continued).

diagram of link 4 upon which act $\mathbf{F}_{o_4}$, $\mathbf{F}'_{34}$, and $\mathbf{F}'_{14}$. The direction of $\mathbf{F}_{o_4}$ is known and that of $\mathbf{F}'_{34}$ is along link 3 because link 3 becomes a two-force member when $\mathbf{F}_{o_3}$ is omitted from the analysis. The direction lines of these two vectors intersect at point k'. Link 4 is in equilibrium under the action of these three forces with no couple acting upon it; therefore, the direction of vector $\mathbf{F}'_{14}$ must pass through points k' and O_4.

Figure 11.11*c* shows the force polygon for the determination of the magnitudes and senses of the vectors $\mathbf{F}'_{34}$ and $\mathbf{F}'_{14}$. Because link 4 is in equilibrium, $\mathbf{F}_{o_4} + \mathbf{F}'_{14} + \mathbf{F}'_{34} = 0$ and the polygon closes as shown.

Figure 11.11*d* shows the free-body diagram of link 3 with force $\mathbf{F}'_{23}$ acting at point A and $\mathbf{F}'_{43}$ acting at point B. Link 3 is in equilibrium under the action of these two forces. The directions of these forces are along the link, and their magnitudes are equal to the magnitude of $\mathbf{F}'_{34}$ as shown.

Figure 11.11*e* shows the free-body diagram of link 2. Force $\mathbf{F}'_{32}$ acts at point A as shown and is balanced by the equal and opposite force $\mathbf{F}'_{12}$ applied at point O_2. Because these forces are not colinear and form a couple $\mathbf{F}'_{32} \times d'$, it is necessary to apply a torque T'_s with a sense opposite to that of the couple to hold link 2 in equilibrium. Torque T'_s is applied to the shaft upon which link 2 is mounted.

FORCE ANALYSIS WITH ONLY $\mathbf{F}_{o_3}$ ACTING. Figure 11.11*f* shows a free-body diagram of link 3 under the action of three forces $\mathbf{F}_{o_3}$, $\mathbf{F}''_{23}$, and $\mathbf{F}''_{43}$. The direction of $\mathbf{F}_{o_3}$ is known and that of $\mathbf{F}''_{43}$ is along line O_4B because link 4 becomes a two-force member when $\mathbf{F}_{o_4}$ is omitted from the analysis. The intersection of the known directions of $\mathbf{F}_{o_3}$ and $\mathbf{F}''_{43}$ gives point k''. The direction of $\mathbf{F}''_{23}$ must pass through points k'' and A because link 3 is in equilibrium under the action of these three forces with no couple acting upon it.

Figure 11.11*g* shows the force polygon for the determination of vectors $\mathbf{F}''_{23}$ and $\mathbf{F}''_{43}$. Because link 3 is in equilibrium, $\mathbf{F}_{o_3} + \mathbf{F}''_{23} + \mathbf{F}''_{43} = 0$ and the polygon closes as shown.

Figure 11.11*h* shows the free-body diagram of link 4 with force $\mathbf{F}''_{34}$ acting at point B and $\mathbf{F}''_{14}$ acting at point O_4. Link 4 is in equilibrium under the action of these two forces so that they act along the line O_4B and their magnitudes are equal to that of $\mathbf{F}''_{43}$ as shown.

Figure 11.11*i* shows the free-body diagram of link 2. Forces $\mathbf{F}''_{32}$ and $\mathbf{F}''_{12}$ act at points A and O_2, respectively. Because a couple is formed by $\mathbf{F}''_{32} \times d''$, it is necessary to apply torque T''_s with a sense opposite to that of the couple to hold link 2 in equilibrium.

TOTAL FORCES. Figure 11.11*j* shows the free-body diagram of link 3 in equilibrium with forces $\mathbf{F}_{o_3}$, $\mathbf{F}_{23}$, and $\mathbf{F}_{43}$ where by superposition $\mathbf{F}_{23} = \mathbf{F}'_{23} + \mathbf{F}''_{23}$ and $\mathbf{F}_{43} = \mathbf{F}'_{43} + \mathbf{F}''_{43}$. Figure 11.11*k* shows the free-body diagram of link 4 in equilibrium under the action of forces $\mathbf{F}_{o_4}$, $\mathbf{F}_{34}$, and $\mathbf{F}_{14}$. Force $\mathbf{F}_{14} = \mathbf{F}'_{14} + \mathbf{F}''_{14}$ as shown. Figure 11.11*l* shows link 2 in equilibrium with forces $\mathbf{F}_{32}$ and $\mathbf{F}_{12}$ and torque $T_s = T'_s + T''_s$.

TRANSVERSE AND RADIAL COMPONENTS. In the force analysis of linkages, the problem often arises of determining the force between two connecting links when each link has a force acting upon it such as an inertia force. Although such a problem may generally be solved using superposition, it is often easier to work with the two forces simultaneously in determining their effect on the joint connecting the two links. To illustrate this method, refer to Fig. 11.12*a* which shows a four-bar linkage with force $\mathbf{F}_3$ acting on link 3 at point *S* and $\mathbf{F}_4$ acting on link 4 at point *T*, and $\mathbf{F}_{34}$ is to be determined. Considering first link 3, $\mathbf{F}_3$ is resolved into components perpendicular and parallel to link 3. These are known as transverse and radial components, respectively, and are labeled $\mathbf{F}_3^{TA}$ and $\mathbf{F}_3^{RA}$. By considering moments about point *A*, the effect of the transverse component $\mathbf{F}_3^{TA}$ acting on link 3 is to produce a transverse force $\mathbf{F}_{34}^{TA}$ acting on link 4 at point *B*. The moments of these transverse forces about point *A* are equal (moments due to radial forces are zero) and can be expressed as

$$\mathbf{F}_{34}^{TA} \times AB = \mathbf{F}_3^{TA} \times AS$$

from which

$$\mathbf{F}_{34}^{TA} = \mathbf{F}_3^{TA} \times \frac{AS}{AB}$$

A graphical method for finding $\mathbf{F}_{34}^{TA}$ is shown in Fig. 11.12*a* using similar triangles.

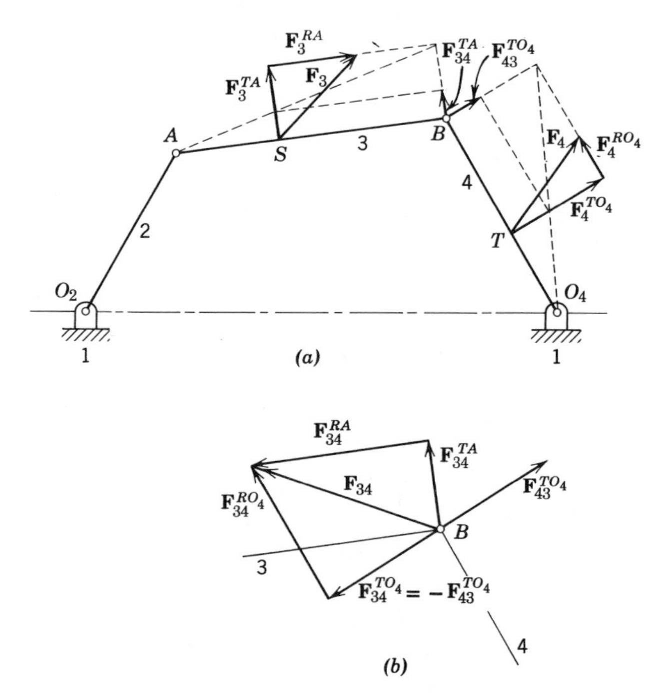

Figure 11.12

Next, consider $\mathbf{F}_4$ resolved into transverse and radial components $\mathbf{F}_4^{TO_4}$ and $\mathbf{F}_4^{RO_4}$, respectively. By considering moments about point O_4, component $\mathbf{F}_4^{TO_4}$ produces a transverse force $\mathbf{F}_{43}^{TO_4}$ acting on link 3 at point B. The moments of these forces about point O_4 are equal and can be expressed as

$$\mathbf{F}_{43}^{TO_4} \times O_4B = \mathbf{F}_4^{TO_4} \times O_4T$$

from which

$$\mathbf{F}_{43}^{TO_4} = \mathbf{F}_4^{TO_4} \times \frac{O_4T}{O_4B}$$

The graphical method for finding $\mathbf{F}_{43}^{TO_4}$ is also shown on Fig. 11.12*a*. Although the directions of the radial components $\mathbf{F}_{34}^{RA}$ and $\mathbf{F}_{43}^{RO_4}$ are known, they have been omitted from Fig. 11.12*a* for clarity.

Figure 11.12*b* shows the forces $\mathbf{F}_{34}^{TA}$ and $\mathbf{F}_{43}^{TO_4}$ acting at point B redrawn to a larger scale. Also shown in $\mathbf{F}_{34}^{TO_4} = -\mathbf{F}_{43}^{TO_4}$. The directions of $\mathbf{F}_{34}^{RA}$ and $\mathbf{F}_{34}^{RO_4}$ are shown so that $\mathbf{F}_{34}$ can be determined from the following equations

$$\mathbf{F}_{34} = \mathbf{F}_{34}^{TA} + \mathbf{F}_{34}^{RA}$$

and

$$\mathbf{F}_{34} = \mathbf{F}_{34}^{TO_4} + \mathbf{F}_{34}^{RO_4}$$

An analytical and a graphical solution of the force analysis of the linkage of Example 11.1 by this method is given in Examples 11.3 and 11.4, respectively.

Example 11.3

An analysis of the linkage of Example 11.1 will now be made using transverse and radial components. Figure 11.10*a* has been redrawn in Fig. 11.13*a*. In the solution all components have been taken relative to the *xy* axes.

Solution. With reference to Fig. 11.13*b*

$$\mathbf{F}_{o_3} = 9.94\mathbf{i} + 5.53\mathbf{j} \text{ from Example 11.1}$$

and

$$\mathbf{F}_{o_3}^{TA} = 5.53\mathbf{j}$$

Taking moments about point A

$$\mathbf{F}_{34}^{TA} \times AB = \mathbf{F}_{o_3}^{TA} \times AS$$

$$\mathbf{F}_{34}^{TA} = \mathbf{F}_{o_3}^{TA} \times \frac{AS}{AB} = (5.53\mathbf{j}) \times \frac{0.864}{8.00}$$

$$= 0.598\mathbf{j}$$

From Fig. 11.13*c*,

$$\mathbf{F}_{o_4} = 15.5\mathbf{i} - 2.01\mathbf{j} \text{ from Example 11.1}$$

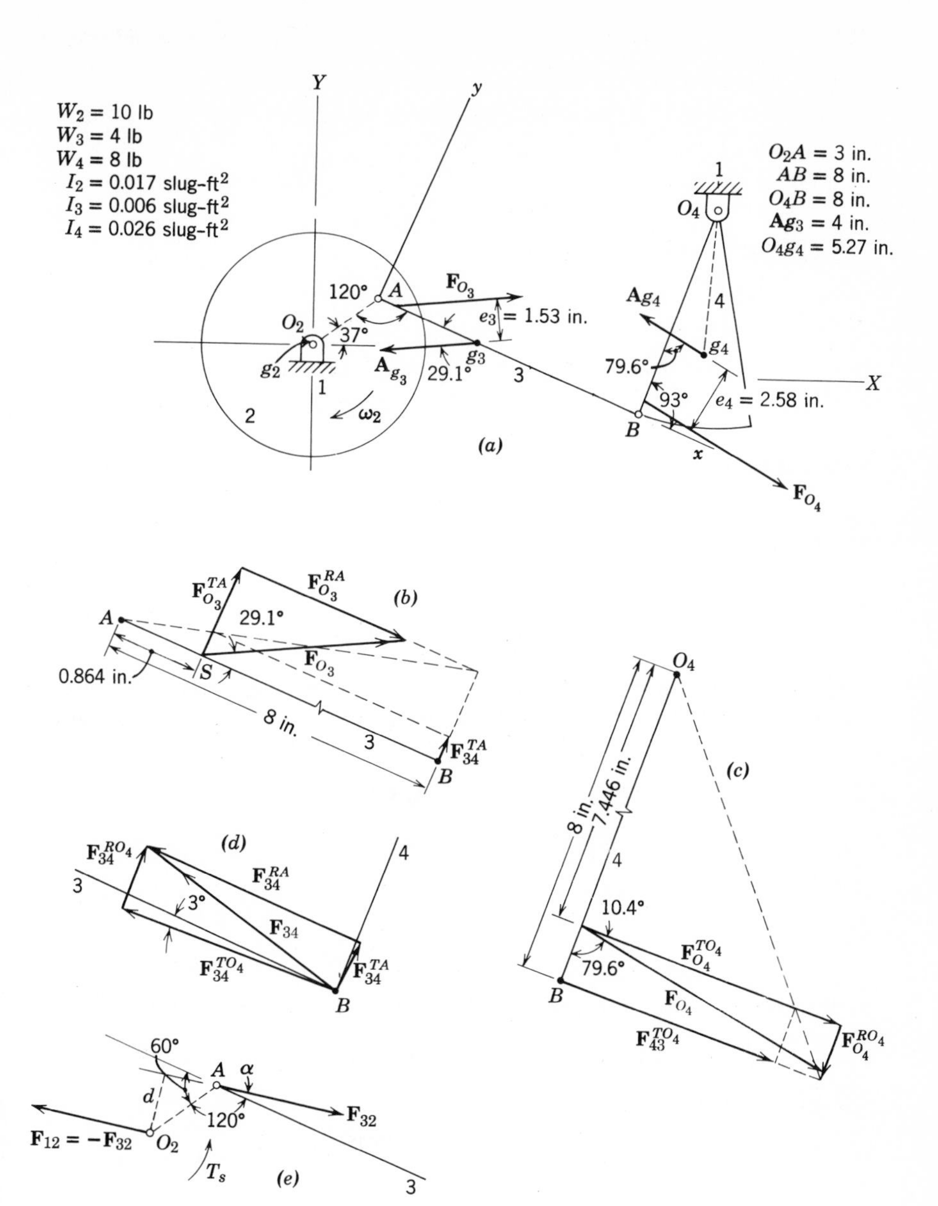

$W_2 = 10$ lb
$W_3 = 4$ lb
$W_4 = 8$ lb
$I_2 = 0.017$ slug-ft^2
$I_3 = 0.006$ slug-ft^2
$I_4 = 0.026$ slug-ft^2
$O_2A = 3$ in.
$AB = 8$ in.
$O_4B = 8$ in.
$Ag_3 = 4$ in.
$O_4g_4 = 5.27$ in.
$e_3 = 1.53$ in.
$e_4 = 2.58$ in.
(a)
(b)
0.864 in.
8 in.
(c)
8 in.
7.446 in.
(d)
(e)

Figure 11.13

and

$$\mathbf{F}_{o_4}^{TO_4} = \mathbf{F}_{o_4} \cos 10.4^\circ$$
$$= (15.5\mathbf{i} - 2.01\mathbf{j}) \cos 10.4^\circ$$
$$= 15.2\mathbf{i} + 1.97\mathbf{j}$$

Taking moments about point O_4,

$$\mathbf{F}_{43}^{TO_4} \times O_4B = \mathbf{F}_{o_4}^{TO_4} \times O_4T$$

$$\mathbf{F}_{43}^{TO_4} = \mathbf{F}_{o_4}^{TO_4} \times \frac{O_4T}{O_4B} = (15.2\mathbf{i} + 1.97\mathbf{j}) \times \frac{7.466}{8.00}$$

$$= 14.2\mathbf{i} + 1.84\mathbf{j}$$

Therefore,

$$\mathbf{F}_{34}^{TO_4} = -\mathbf{F}_{43}^{TO_4} = -14.2\mathbf{i} - 1.84\mathbf{j}$$

From Fig. 11.13*d*,

$$\mathbf{F}_{34} = \mathbf{F}_{34}^{TA} + \mathbf{F}_{34}^{RA}$$

also

$$\mathbf{F}_{34} = \mathbf{F}_{34}^{TO_4} + \mathbf{F}_{34}^{RO_4}$$

In these two equations, $\mathbf{F}_{34}^{TA}$ and $\mathbf{F}_{34}^{TO_4}$ are known in magnitude and direction, and $\mathbf{F}_{34}^{RA}$ and $\mathbf{F}_{34}^{RO_4}$ are known in direction which can be expressed as follows:

$$\mathbf{F}_{34}^{RA} = -F_{34}^{RA}\mathbf{i}$$
$$\mathbf{F}_{34}^{RO_4} = F_{34}^{RO_4}(-\cos 87^\circ\mathbf{i} + \sin 87^\circ\mathbf{j})$$
$$= -0.0523F_{34}^{RO_4}\mathbf{i} + 0.999F_{34}^{RO_4}\mathbf{j}$$

Therefore,

$$\mathbf{F}_{34} = -F_{34}^{RA}\mathbf{i} + 0.598\mathbf{j}$$

also

$$\mathbf{F}_{34} = -14.2\mathbf{i} - 1.84\mathbf{j} - 0.0523F_{34}^{RO_4}\mathbf{i} + 0.999F_{34}^{RO_4}\mathbf{j}$$

Setting the two equations for $\mathbf{F}_{34}$ equal to each other, one obtains

$$-F_{34}^{RA}\mathbf{i} + 0.598\mathbf{j} = -14.2\mathbf{i} - 1.84\mathbf{j} - 0.0523F_{34}^{RO_4}\mathbf{i} + 0.999F_{34}^{RO_4}\mathbf{j}$$

Summing **i** *components*,

$$-F_{34}^{RA}\mathbf{i} = -14.2\mathbf{i} - 0.0523F_{34}^{RO_4}\mathbf{i}$$

Summing **j** *components*,

$$0.598\mathbf{j} = -1.84\mathbf{j} + 0.999F_{34}^{RO_4}\mathbf{j}$$
$$F_{34}^{RO_4} = 2.44 \text{ lb}$$

Also

$$-F_{34}^{RA}\mathbf{i} = -14.2\mathbf{i} - 0.0523 \times 2.44\mathbf{i}$$
$$F_{34}^{RA} = 14.3 \text{ lb}$$

Therefore,

$$\mathbf{F}_{34} = -14.3\mathbf{i} + 0.598\mathbf{j}$$
$$|\mathbf{F}_{34}| = 14.3 \text{ lb}$$

Because link 3 is in equilibrium under the action of $\mathbf{F}_{23}$, $\mathbf{F}_{o_3}$, and $\mathbf{F}_{43}$

$$\mathbf{F}_{23} + \mathbf{F}_{o_3} + \mathbf{F}_{43} = 0$$

Substituting the equations for $\mathbf{F}_{o_3}$ and $\mathbf{F}_{43}$ into the above, one obtains

$$\mathbf{F}_{23} + 9.94\mathbf{i} + 5.53\mathbf{j} + 14.3\mathbf{i} - 0.598\mathbf{j} = 0$$

Therefore,

$$\mathbf{F}_{23} = -24.3\mathbf{i} - 4.94\mathbf{j}$$
$$|\mathbf{F}_{23}| = 24.8 \text{ lb}$$

Link 4 is in equilibrium under the action of $\mathbf{F}_{14}$, $\mathbf{F}_{o_4}$, and $\mathbf{F}_{34}$, therefore

$$\mathbf{F}_{14} + \mathbf{F}_{o_4} + \mathbf{F}_{34} = 0$$

Substituting the equations for $\mathbf{F}_{o_4}$ and $\mathbf{F}_{34}$ into the preceding equation, one obtains

$$\mathbf{F}_{14} + 15.5\mathbf{i} - 2.01\mathbf{j} - 14.3\mathbf{i} + 0.598\mathbf{j} = 0$$

Therefore,

$$\mathbf{F}_{14} = -1.14\mathbf{i} + 1.41\mathbf{j}$$
$$|\mathbf{F}_{14}| = 1.81 \text{ lb}$$

The shaft torque T_s necessary to hold link 2 in equilibrium under the action of the couple produced by $\mathbf{F}_{32}$ and $\mathbf{F}_{12}$ can be calculated by reference to Fig. 11.13*e* where

$$\mathbf{F}_{32} = -\mathbf{F}_{23} = 24.3\mathbf{i} + 4.94\mathbf{j}$$
$$|\mathbf{F}_{32}| = 24.8 \text{ lb}$$
$$\alpha = \tan^{-1}\left(\frac{4.94}{24.3}\right) = 11.5^\circ$$

and

$$d = (O_2A) \sin 48.5^\circ = 3 \sin 48.5^\circ = 2.25 \text{ in.}$$

Therefore,

$$T_s = F_{32}\, d = 24.8 \times 2.25$$
$$= 55.7 \text{ lb-in. (ccw)}$$

The same result can be obtained using the equation for $\mathbf{F}_{32}$ given above and the following:

$$\mathbf{d} = 0.448\mathbf{i} - 2.20\mathbf{j}$$
$$\mathbf{T}_s = -(\mathbf{F}_{32} \times \mathbf{d})$$
$$= 55.7\mathbf{k} \text{ lb-in. (ccw)}$$

Therefore,

$$F_{14} = 1.81 \text{ lb}$$
$$F_{43} = 14.3 \text{ lb}$$
$$F_{32} = 24.8 \text{ lb}$$
$$F_{12} = 24.8 \text{ lb}$$
$$T_s = 55.7 \text{ lb-in. (ccw)}$$

Example 11.4

A graphical analysis of the linkage of Example 11.1 will now be made using transverse and radial components. Figure 11.13*a* has been redrawn in Fig. 11.14*a* to an enlarged scale.

Solution. Inertia forces $\mathbf{F}_{o_3}$ and $\mathbf{F}_{o_4}$ were determined in Example 11.1 and their magnitudes found to be

$$F_{o_3} = 11.4 \text{ lb}$$
$$F_{o_4} = 15.6 \text{ lb}$$

Their directions and locations relative to their centers of gravity are shown in Fig. 11.14*a*. Link 4 is a three-force member under the action of the forces $\mathbf{F}_{14}$, $\mathbf{F}_{o_4}$, and $\mathbf{F}_{34}$. Link 3 is also a three-force member under the action of $\mathbf{F}_{23}$, $\mathbf{F}_{o_3}$, and $\mathbf{F}_{43}$. No couple acts on either link 3 or link 4 so that both links are in equilibrium under the action of their respective forces. The force $\mathbf{F}_{34}$ is determined from $\mathbf{F}_{o_3}$ and $\mathbf{F}_{o_4}$ by using transverse and radial components of these forces and taking moments about points A and O_4. As shown in Fig. 11.14*a*, $\mathbf{F}_{o_3}$ is resolved into transverse and radial components $\mathbf{F}_{o_3}^{TA}$ and $\mathbf{F}_{o_3}^{RA}$ at point S, and $\mathbf{F}_{o_4}$ is resolved into components $\mathbf{F}_{o_4}^{TO_4}$ and $\mathbf{F}_{o_4}^{RO_4}$ at point T. Taking moments about point A, $\mathbf{F}_{34}^{TA}$ is determined from the relation

$$\mathbf{F}_{34}^{TA} \times AB = \mathbf{F}_{o_3}^{TA} \times AS$$

where $\mathbf{F}_{o_3}^{TA}$ and distances AB and AS are known. In a similar manner, $\mathbf{F}_{43}^{TO_4}$ is found by taking moments about point O_4 which gives the relation

$$\mathbf{F}_{43}^{TO_4} \times O_4B = \mathbf{F}_{o_4}^{TO_4} \times O_4T$$

where $\mathbf{F}_{o_4}^{TO_4}$ and distances O_4B and O_4T are known. The solution of these equations by similar triangles is shown in Fig. 11.14*a* to give $\mathbf{F}_{34}^{TA}$ and $\mathbf{F}_{43}^{TO_4}$.

The force $\mathbf{F}_{34}$ can now be found from the equations

$$\mathbf{F}_{34} = \mathbf{F}_{34}^{TA} + \mathbf{F}_{34}^{RA}$$
$$\mathbf{F}_{34} = \mathbf{F}_{34}^{TO_4} + \mathbf{F}_{34}^{RO_4}$$

where

$$\mathbf{F}_{34}^{TO_4} = -\mathbf{F}_{43}^{TO_4}$$

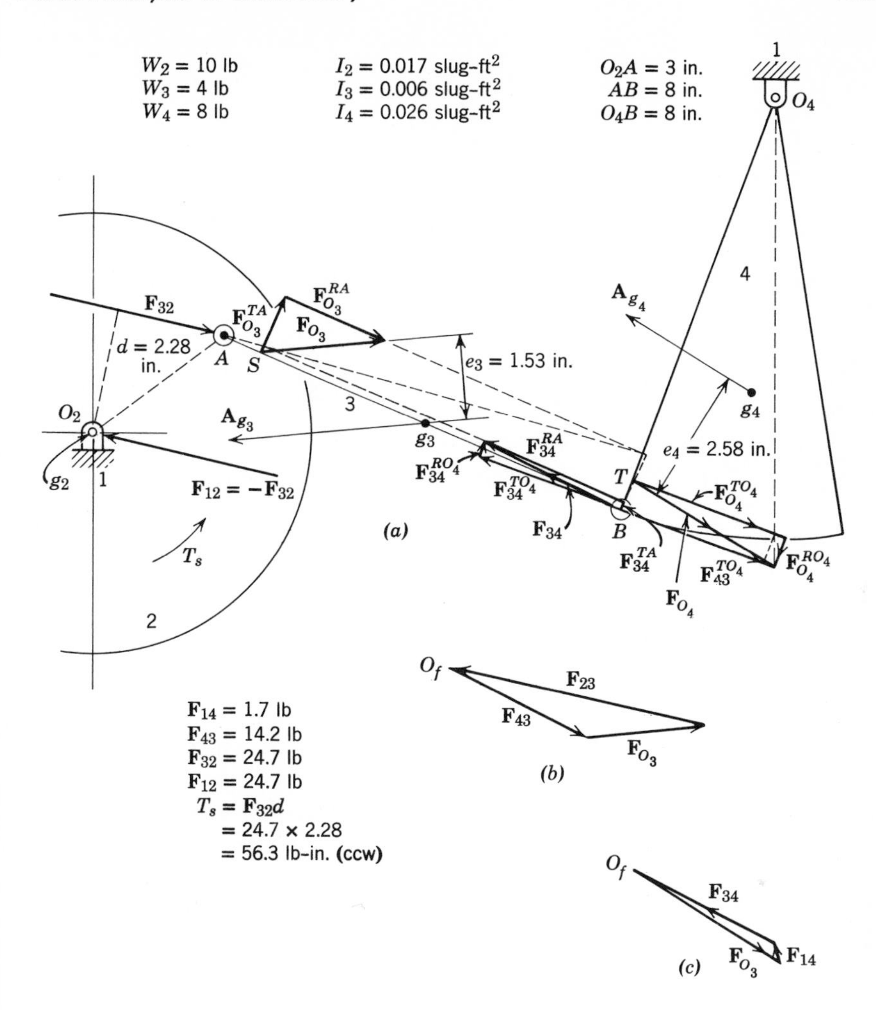

Figure 11.14

In these two equations $\mathbf{F}_{34}^{TA}$ and $\mathbf{F}_{34}^{TO_4}$ are known in magnitude, sense, and direction, and $\mathbf{F}_{34}^{RA}$ and $\mathbf{F}_{34}^{RO_4}$ are known in direction. The graphical solution of these equations to give $\mathbf{F}_{34}$ is shown in Fig. 11.14*a*.

Figure 11.14*b* shows the force polygon for the determination of $\mathbf{F}_{23}$ from the equation of equilibrium

$$\mathbf{F}_{43} + \mathbf{F}_{o_3} + \mathbf{F}_{23} = 0$$

where $\mathbf{F}_{43}$ and $\mathbf{F}_{o_3}$ are known. Figure 11.14*c* shows the polygon for determining $\mathbf{F}_{14}$ from the equation

$$\mathbf{F}_{o_4} + \mathbf{F}_{14} + \mathbf{F}_{34} = 0$$

where $\mathbf{F}_{o_4}$ and $\mathbf{F}_{34}$ are known.

The direction of the shaft torque T_s necessary to hold link 2 in equilibrium under action of the couple produced by $\mathbf{F}_{32}$ and $\mathbf{F}_{12}$ is shown in Fig. 11.14*a*. The magnitude of T_s is determined from the relation $T_s = F_{32}d$ as shown below Fig. 11.14*a*.

11.6 Linkage Analysis by the Method of Virtual Work. The methods of force analysis presented so far have been based on the principle of equilibrium of forces. Another method applicable for the analyses of linkages is that of *virtual work*, and it often results in much simpler solutions. This method is based on the principle that if a rigid body is in equilibrium under the action of external forces, the total work done by these forces is zero for a small displacement of the body.

To review the concept of work, consider Fig. 11.15, which shows a force **F** acting on a particle at point A. If the particle moves from point A to A' through a small distance $\boldsymbol{\delta}\mathbf{s}$, the work of force **F** during the displacement $\boldsymbol{\delta}\mathbf{s}$ is

$$\delta U = (F)(\delta s) \cos \theta$$

As can be seen from the equation, the work done is the scaler product of the displacement and the component of the force in the direction of the displacement, and can be positive, negative, or zero depending on angle θ. If θ is less than 90°, the force component and the displacement have the same sense and work is positive. If $\theta = 90°$, work is zero, and for θ greater than 90° but less than 270°, work is negative. If the preceeding equation is

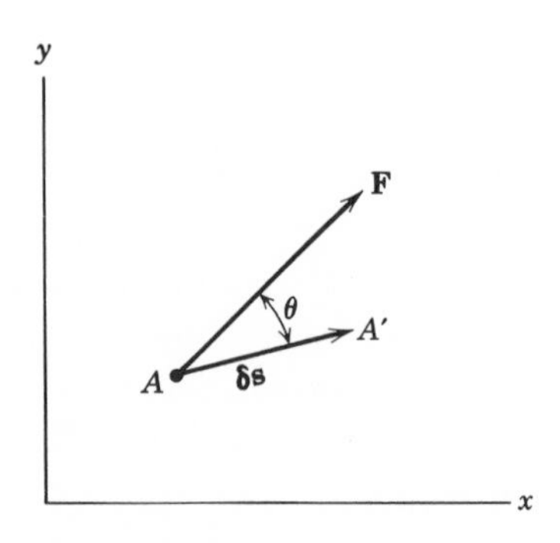

Figure 11.15

compared with the vector equation

$$\mathbf{A} \cdot \mathbf{B} = AB \cos \theta$$

it can be seen that the equation for work can be written as the dot product of the force and displacement vectors as follows

$$\delta U = \mathbf{F} \cdot \boldsymbol{\delta}\mathbf{s} \tag{11.18}$$

The term *virtual work* is used for this method of analysis to indicate work which results from an infinitesimal displacement which is imaginary. Such a displacement is called a *virtual displacement* and labeled δs to distinguish it from an actual displacement ds. Although virtual displacements are imaginary, they must be consistent with the constraints of the mechanism under consideration. A virtual displacement may also be a measure of rotation and is labeled $\boldsymbol{\delta\theta}$. The virtual work done by a torque T is, therefore, $\delta U = \mathbf{T} \cdot \boldsymbol{\delta\theta}$.

From the definition of virtual work it follows that if a system which is in equilibrium under the action of external forces and torques is given a virtual displacement, the total virtual work must be zero. This concept may be expressed mathematically as follows:

$$\delta U = \Sigma\, \mathbf{F}_n \cdot \boldsymbol{\delta}\mathbf{s}_n + \Sigma\, \mathbf{T}_n \cdot \boldsymbol{\delta\theta}_n = 0 \tag{11.19}$$

In the application of this equation it must be remembered that the virtual displacements $\boldsymbol{\delta}\mathbf{s}_n$ and $\boldsymbol{\delta\theta}_n$ must be consistent with the constraints of the mechanism. As an example of this, consider Fig. 11.16 in which a four-bar linkage is acted upon by forces $\mathbf{F}_3$ and $\mathbf{F}_4$ applied at points C and D, respectively, and it is required to determine torque T_2 necessary to maintain static equilibrium. If link 2 is given a virtual displacement $\boldsymbol{\delta\theta}_2$, the equations

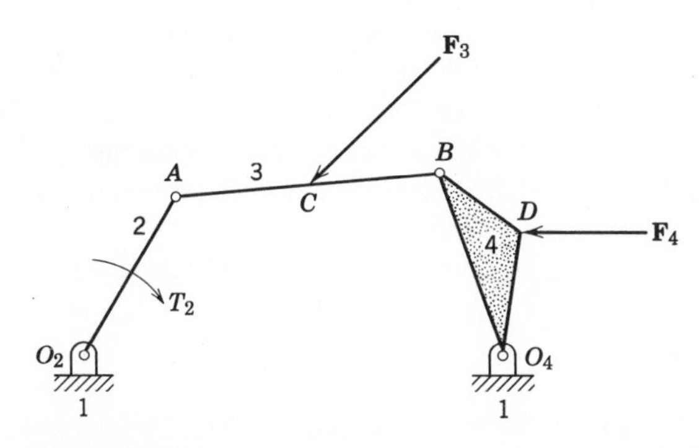

Figure 11.16

for $\delta \mathbf{s}_C$ and $\delta \mathbf{s}_D$ must be expressed as a function of $\delta \boldsymbol{\theta}_2$ in order to solve Eq. 11.19 for T_2.

The method of virtual work can also be applied to dynamic analyses if inertia forces and inertia torques are considered as applied forces and torques. Equation 11.19 can be modified for the dynamic case by dividing each term by dt. This is permissible because each virtual displacement takes place in the same interval of time. Making this change, one obtains

$$\Sigma\, \mathbf{F}_n \cdot \frac{\delta \mathbf{s}_n}{dt} + \Sigma\, \mathbf{T}_n \cdot \frac{\delta \boldsymbol{\theta}_n}{dt} = 0$$

and

$$\Sigma\, \mathbf{F}_n \cdot \mathbf{V}_n + \Sigma\, \mathbf{T}_n \cdot \boldsymbol{\omega}_n = 0 \tag{11.20}$$

Therefore, the virtual work of the external forces and torques is proportional to the velocity of the points of application of the forces on the links. Equation 11.20 can be expanded to give terms for applied forces and torques and for inertia forces and torques as follows:

$$\Sigma\, \mathbf{T}_n \cdot \boldsymbol{\omega}_n + \Sigma\, \mathbf{F}_n \cdot \mathbf{V}_n + \Sigma\, \mathbf{F}_{o_n} \cdot \mathbf{V}_{g_n} + \Sigma\, \mathbf{T}_{o_n} \cdot \boldsymbol{\omega}_n = 0 \tag{11.21}$$

where

$$\mathbf{F}_{o_n} \cdot \mathbf{V}_{g_n} = -\frac{W_n}{g} \mathbf{A}_{g_n} \cdot \mathbf{V}_{g_n} \quad \text{(inertia force)}$$

$$\mathbf{T}_{o_n} \cdot \boldsymbol{\omega}_n = -I_n \boldsymbol{\alpha}_n \cdot \boldsymbol{\omega}_n \quad \text{(inertia torque)}$$

After a velocity and acceleration analysis has been made, Eq. 11.21 can easily be solved for one unknown which would usually be the torque required on the driving link to hold the mechanism in equilibrium.

In Eq. 11.21, only the virtual work done by the external forces and torques on a linkage appears. The internal forces between connecting links occur in pairs. These are equal in magnitude but opposite in sense so that their net work during any displacement is zero. For this reason, Eq. 11.21 cannot be used to evaluate bearing forces between connecting links.

Although analyses by the method of virtual work can be made graphically, an analytical solution is easier. In a graphical solution if the forces and velocities are not both in the same direction, components of the forces must be used. In Eq. 11.21 with the terms expressed as dot products, the problem of directions is automatically taken care of.

An analytical solution of a force analysis by the method of virtual work is given in Example 11.5.

Example 11.5

In Fig. 11.17 is shown the linkage of Fig. 11.10*a*. Use the method of virtual work to determine the torque T_2 necessary to hold the linkage in equilibrium.

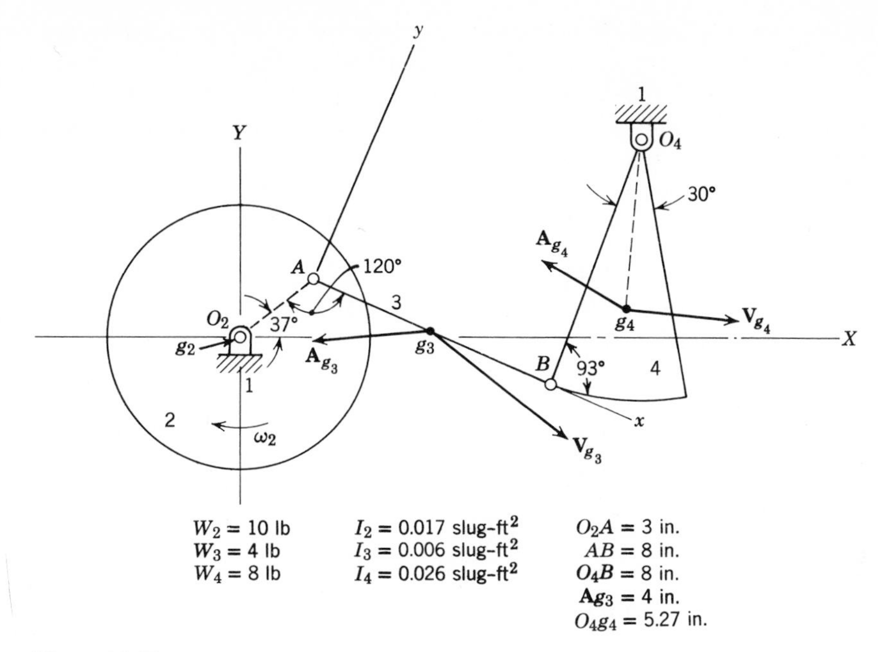

Figure 11.17

In the solution all components have been taken relative to the xy axes.

Solution. From the solution of Example 10.1

$$\mathbf{V}_A = 5.2\mathbf{i} - 3.0\mathbf{j} \qquad \boldsymbol{\omega}_2 = -24\mathbf{k}\,\text{rad/sec}$$

$$\mathbf{V}_B = 5.2\mathbf{i} + 0.27\mathbf{j} \qquad \boldsymbol{\omega}_3 = 4.91\mathbf{k}\,\text{rad/sec}$$

$$\mathbf{A}_A = -72\mathbf{i} - 124.8\mathbf{j} \qquad \boldsymbol{\omega}_4 = 7.82\mathbf{k}\,\text{rad/sec}$$

$$\mathbf{A}_B = -88.1\mathbf{i} + 35.8\mathbf{j} \qquad \boldsymbol{\alpha}_3 = 241\mathbf{k}\,\text{rad/sec}^2$$

$$\boldsymbol{\alpha}_4 = -129\mathbf{k}\,\text{rad/sec}^2$$

Using the above values, the following velocities and accelerations of the centers of gravity of links 3 and 4 were determined. They are shown on the linkage in Fig. 11.17.

$$\mathbf{V}_{g_3} = 5.2\mathbf{i} - 1.37\mathbf{j} \qquad \mathbf{A}_{g_3} = -80.01\mathbf{i} - 44.53\mathbf{j}$$

$$\mathbf{V}_{g_4} = 3.27\mathbf{i} + 1.06\mathbf{j} \qquad \mathbf{A}_{g_4} = -62.17\mathbf{i} + 8.08\mathbf{j}$$

Writing Eq. 11.21 for the linkage of Fig. 11.17,

$$\mathbf{T}_2 \cdot \boldsymbol{\omega}_2 + \mathbf{F}_{o_3} \cdot \mathbf{V}_{g_3} + \mathbf{F}_{o_4} \cdot \mathbf{V}_{g_4} + \mathbf{T}_{o_3} \cdot \boldsymbol{\omega}_3 + \mathbf{T}_{o_4} \cdot \boldsymbol{\omega}_4 = 0$$

where

$$\mathbf{T}_2 \cdot \boldsymbol{\omega}_2 = (T_2\mathbf{k}) \cdot (-24\mathbf{k}) = -24T_2$$

$$\mathbf{F}_{o_3} \cdot \mathbf{V}_{g_3} = \left[\frac{-W_3}{g}\mathbf{A}_{g_3}\right] \cdot \mathbf{V}_{g_3}$$

$$= \frac{-4}{32.2}(-80.01\mathbf{i} - 44.53\mathbf{j})(5.2\mathbf{i} - 1.37\mathbf{j})$$

$$= 44.13 \text{ ft-lb/sec}$$

$$\mathbf{F}_{o_4} \cdot \mathbf{V}_{g_4} = \left[\frac{-W_4}{g}\mathbf{A}_{g_4}\right] \cdot \mathbf{V}_{g_4}$$

$$= \frac{-8}{32.2}(-62.17\mathbf{i} + 8.08\mathbf{j})(3.27\mathbf{i} + 1.06\mathbf{j})$$

$$= 48.32 \text{ ft-lb/sec}$$

$$\mathbf{T}_{o_3} \cdot \boldsymbol{\omega}_3 = (-I_3\boldsymbol{\alpha}_3) \cdot \boldsymbol{\omega}_3$$

$$= -0.006(241\mathbf{k}) \cdot (4.91\mathbf{k}) = -7.099 \text{ ft-lb/sec}$$

$$\mathbf{T}_{o_4} \cdot \boldsymbol{\omega}_4 = (-I_4\boldsymbol{\alpha}_4) \cdot \boldsymbol{\omega}_4$$

$$= -0.026(-129\mathbf{k}) \cdot (7.82\mathbf{k}) = 26.23 \text{ ft-lb/sec}$$

Therefore,

$$-24T_2 + 44.13 + 48.32 - 7.099 + 26.23 = 0$$

and

$$T_2 = 4.649 \text{ lb-ft}$$

$$= 55.79 \text{ lb-in. (ccw)}$$

The value for T_2 agrees closely with the value found in Example 11.1.

11.7 Linkage Motion Analysis from Dynamic Characteristics. In the preceding sections, linkages were considered which operated continuously with the driving link rotating at a known, generally uniform, angular velocity. From this it was possible to make a velocity, acceleration, and inertia force analysis of the mechanism. Combining these forces with the static forces acting on the linkage, one was able to complete the analysis and determine the bearing loads.

Occasionally, the problem arises of having to determine the velocity and acceleration characteristics of a mechanism where the driving force is produced by a rapid release of energy as from a spring, solenoid, or air cylinder. A circuit breaker is such a mechanism with its design depending upon the time required for a motion cycle, and the applied force necessary to open the breaker.

In the solution of such problems, energy methods are generally preferred because of the ease with which they can be applied. Of the several methods that have been developed, the one by Quinn[2] will be presented here. His method is based on the distribution of kinetic energy in a mechanism and is referred to as Quinn's *energy-distribution theorem* which states that "the percent of the total kinetic energy which the link of a mechanism contains will remain the same in any given position regardless of the speed." This theorem applies to mechanisms in which there is no change in the mass or moment of inertia of the links with speed, and in which a linear relationship exists between the velocities of the various links in a given position.

In applying this method, a convenient value is assumed for the velocity of the input link of a mechanism, and a velocity analysis made including the determination of the velocities of the centers of gravity of the various links. From this, the kinetic energy of each link can be calculated, and the ratio of the kinetic energy of any link to the total for all the links expressed as

$$\varepsilon = \frac{KE}{\Sigma\, KE} \tag{11.22}$$

This ratio is known as an *energy-contribution coefficient*[3] and is a constant for any link in a particular phase regardless of its velocity. Values of ε can be calculated for various links over the range of several phases, and curves of ε versus crank angle or slider displacement plotted if desired.

In addition to the coefficient ε which can easily be calculated for any link as shown above, it is necessary in making a dynamic analysis to know the variation of the external forces acting on the mechanism as well as the input velocity in some reference phase. The reference phase is usually the starting phase where the input velocity is zero. With this information, the actual velocity of the input link in a particular phase can be determined as follows:

1. Knowing the external forces in relation to the phase positions, calculate the work input to the mechanism between the starting phase and the phase under consideration.
2. Because a change in kinetic energy is equal to the work done, the actual change in kinetic energy of the mechanism from the starting phase to the phase in question is now known from item 1.
3. For the starting phase, the input velocity is zero so that the kinetic energy is also zero. Therefore, the kinetic energy of the mechanism for the phase in question is known from item 2.

[2] B. E. Quinn, "Energy Method for Determining Dynamic Characteristics of Mechanisms," *ASME Journal of Applied Mechanics*, Vol. 16, No. 3, pp. 283–288, September 1949.

[3] J. Hirschorn, *Kinematics and Dynamics of Plane Mechanisms*, p. 201, McGraw-Hill Book Company, 1962.

4. After calculating the value of ε for the input link for the phase in question, determine the kinetic energy for the input link from item 3.

5. From item 4, calculate the velocity of the input link from the relation $KE = \frac{1}{2}I\omega^2$ for a rotating input link or from $KE = \frac{1}{2}MV^2$ if the input is a slider.

A sample problem using this method is given in Example 11.6.

From the above procedure, the velocity of the input link can be calculated for as many phases as necessary and a graph of input velocity versus displacement of the input link drawn. If many phases are to be analyzed, the calculations can easily be made and the graph plotted using a digital computer.

Example 11.6

For the slider-crank mechanism shown in Fig. 11.18, determine the angular velocity ω_2 of link 2 when the mechanism is in phase IV. The starting position is phase I where $\omega_2 = 0$. The torque T_2 on link 2 is constant at 90 lb-in. (ccw). The force $\mathbf{P}$ on piston 4 varies uniformly from 161 lb in phase I to 36 lb in phase IV.

Solution. The angular velocity ω_2' of link 2 is assumed to be 1 rad/sec (cw), where the prime denotes an *assumed* value. From a velocity analysis of the linkage with $\omega_2' = 1$ rad/sec, the velocities at phase IV are as follows:

$$V_A' = 2 \text{ in./sec} \qquad V_{g3}' = 2.00 \text{ in./sec}$$

$$V_B' = 2.05 \text{ in./sec} \qquad \omega_3 = 0.049 \text{ rad/sec (ccw)}$$

$$V_{BA}' = 0.39 \text{ in./sec}$$

Calculating the kinetic energy of the links based on the assumed value of $\omega_2' = 1$ rad/sec,

$$\begin{aligned} KE_2' &= \tfrac{1}{2}I_2(\omega_2')^2 = \tfrac{1}{2} \times 0.024 \times 1 \\ &= 0.012 \text{ lb-in.} \end{aligned}$$

$$\begin{aligned} KE_3' &= \tfrac{1}{2}I_3(\omega_3')^2 + \tfrac{1}{2}M_3(V_{g3}')^2 \\ &= \tfrac{1}{2} \times 0.09 \times (0.049)^2 + \tfrac{1}{2} \times \tfrac{3}{386} \times (2.00)^2 \\ &= 0.000108 + 0.01554 \\ &= 0.01565 \text{ lb-in.} \end{aligned}$$

$$\begin{aligned} KE_4' &= \tfrac{1}{2}M_4(V_B')^2 = \tfrac{1}{2} \times \tfrac{2}{386} \times (2.05)^2 \\ &= 0.01089 \text{ lb-in.} \end{aligned}$$

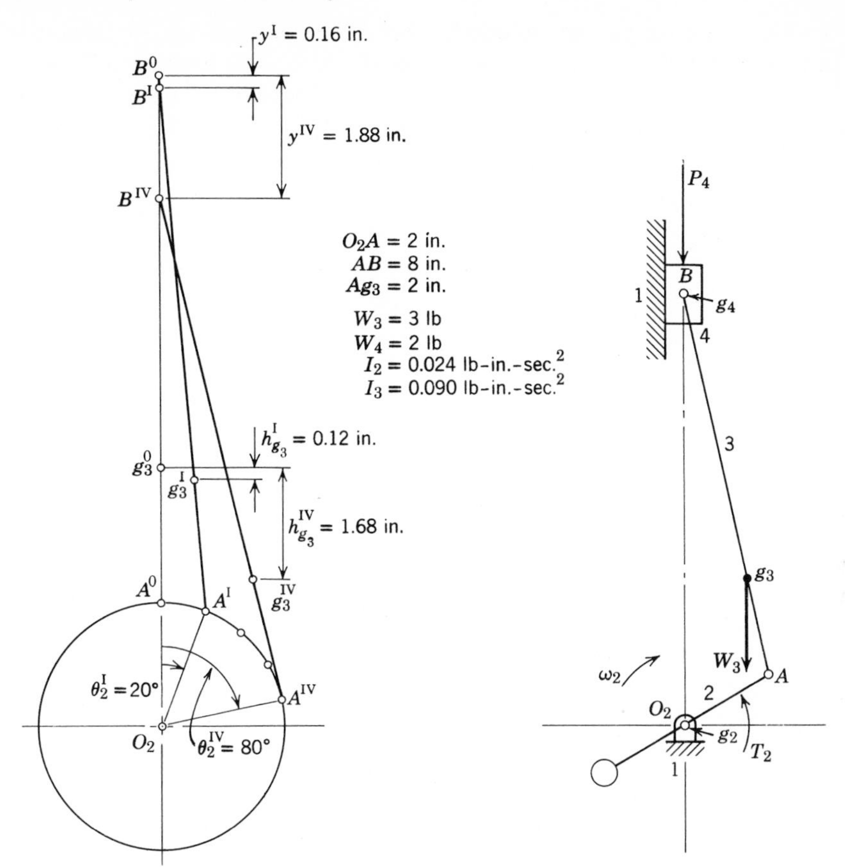

Figure 11.18

Therefore,

$$\Sigma\, KE' = 0.03854 \text{ lb-in.}$$

From Eq. 11.22,

$$\varepsilon_2 = \frac{KE'_2}{\Sigma\, KE'} = \frac{0.012}{0.03854} = 0.3114$$

The external work input to the mechanism by forces $\mathbf{P}_4$ and $\mathbf{W}_3$ and torque T_2 between phases I and IV is calculated as follows:

$$\begin{aligned} Wk_{P_4} &= \tfrac{1}{2}(P_4^{IV} + P_4^{I})(y^{IV} - y^{I}) \\ &= \tfrac{1}{2}(36 + 161)(1.88 - 0.16) \\ &= 169.42 \text{ lb-in.} \end{aligned}$$

$$Wk_{W_3} = W_3(h_{g3}^{IV} - h_{g3}^{I})$$
$$= 3(1.68 - 0.12)$$
$$= 4.68 \text{ lb-in.}$$
$$Wk_{T_2} = -T_2(\theta_2^{IV} - \theta_2^{I})$$
$$= -90(80 - 20) \times \frac{\pi}{180}$$
$$= -94.25 \text{ lb-in.}$$

Therefore,

$$\Sigma\, Wk = 79.85 \text{ lb-in.}$$

From the fact that the change in kinetic energy must equal the work done

$$\Sigma\, KE^{IV} - \Sigma\, KE^{I} = 79.85 \text{ lb-in.}$$

but

$$\Sigma\, KE^{I} = 0$$

Therefore,

$$\Sigma\, KE^{IV} = 79.85 \text{ lb-in.}$$

The actual value of ω_2 in phase IV is calculated from the relation

$$KE_2^{IV} = \tfrac{1}{2} I_2 (\omega_2^{IV})^2$$

where

$$KE_2^{IV} = (\varepsilon_2^{IV})(\Sigma\, KE^{IV})$$
$$= 0.3114 \times 79.85$$
$$= 24.87 \text{ lb-in.}$$

Therefore,

$$\omega_2^{IV} = \left[\frac{2KE_2^{IV}}{I_2}\right]^{1/2} = \left[\frac{2 \times 24.87}{0.024}\right]^{1/2}$$
$$= 45.52 \text{ rad/sec}$$

The angular acceleration of link 2 can be determined for any phase position from a plot of ω_2 versus θ_2 using the subnormal construction with the appropriate scale factors for ω_2 and θ_2. The angular acceleration α_2 can also be found using the rate-of-change-of-energy method developed by Hirschorn.[4]

11.8 Linkage Force Analysis by Complex Numbers. Another analytical method of force analysis is by expressing vectors in complex form. This method is especially applicable when a complete cycle of a linkage is to be analyzed and computer facilities are available.

[4] J. Hirschorn, "Dynamic Acceleration Analysis," *Machine Design*, Vol. 32, No. 4, pp. 151–155, 1960.

In Fig. 11.19*a* is shown a typical four-bar linkage in a given phase of the motion cycle. A shaft torque T_s acts on the drive link (link 2) at O_2. The accelerations A_g of the mass centers and the angular accelerations α of the moving links may be determined numerically by complex numbers as demonstrated in Chapter 10. The three inertia forces $\mathbf{F}_o$, which are related to the accelerations, represent the dynamic loading of the mechanism. The objective of the analysis is the determination of the bearing forces and the shaft torque which produce the dynamic loading.

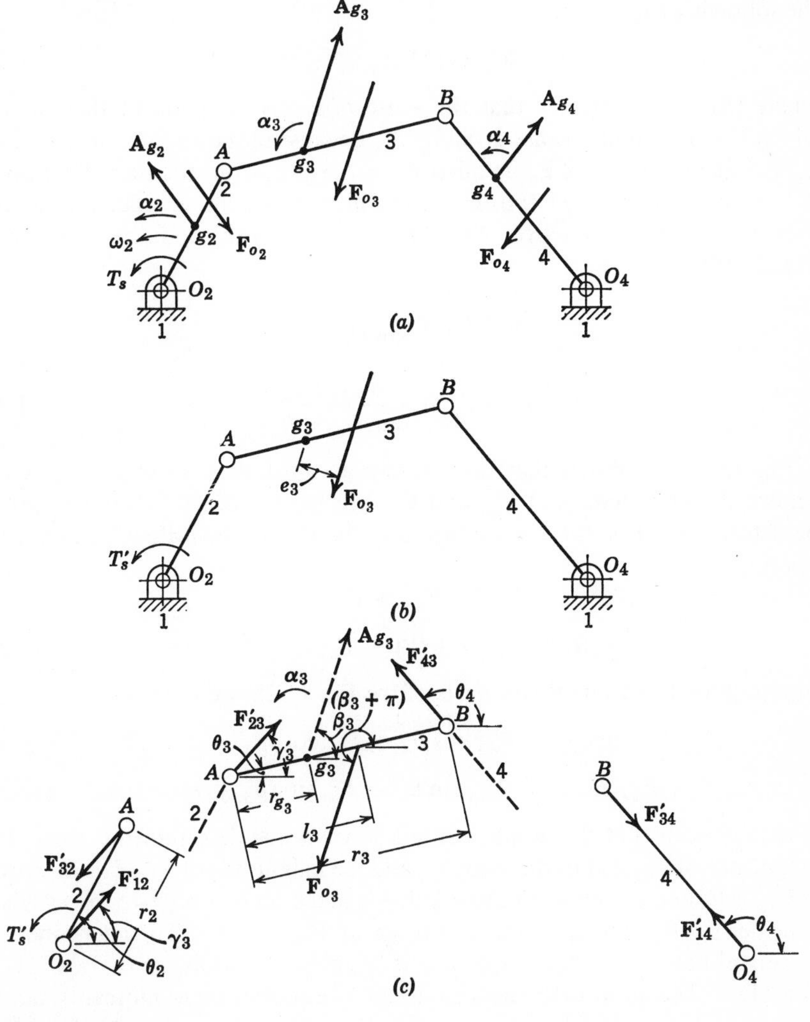

Figure 11.19

Figure 11.19*b* shows the four-bar linkage with inertia force $\mathbf{F}_{o_3}$ as the only load vector acting so that the bearing forces and shaft torque to be determined are those related to $\mathbf{F}_{o_3}$ alone. Similar independent force analyses with $\mathbf{F}_{o_2}$ and $\mathbf{F}_{o_4}$ acting alone may be made, and eventually the resultant bearing forces and shaft torque may be obtained by superposition.

In the analysis with $\mathbf{F}_{o_3}$ acting alone, the free body to be considered first is that of link 3 shown in Fig. 11.19*c*. Assuming that the acceleration A_{g_3} (expressed as $A_{g_3}e^{i\beta_3}$) and the angular acceleration α_3 have been determined as in Chapter 10, the inertia force vector $\mathbf{F}_{o_3}$ may be determined from the following expression:

$$\mathbf{F}_{o_3} = (M_3 A_{g_3})e^{i(\beta_3 + \pi)} \tag{11.23}$$

where $(\beta_3 + \pi)$ indicates that the sense of $\mathbf{F}_{o_3}$ is opposite to that of $\mathbf{A}_{g_3}$, which has the angular sense given by β_3. Because of the angular acceleration α_3, the line of action of $\mathbf{F}_{o_3}$ is offset $e_3 = I_3\alpha_3/F_{o_3}$ from the line of action of $\mathbf{A}_{g_3}$ as shown in Fig. 11.19*b*. For convenience in making calculations, the location of the line of action of $\mathbf{F}_{o_3}$ may be given by the distance l_3 shown in Fig. 11.19*c*.

$$l_3 = r_{g_3} + \frac{e_3}{\sin(\beta_3 - \theta_3)}$$

$$l_3 = r_{g_3} + \frac{I_3\alpha_3/F_{o_3}}{\sin(\beta_3 - \theta_3)} \tag{11.24}$$

Figure 11.19*c* shows that three forces act on link 3, of which $\mathbf{F}_{o_3}$ is the known dynamic load and $\mathbf{F}'_{23}$ and $\mathbf{F}'_{43}$ are the unknown bearing forces to be determined. For static equilibrium of the forces, the following equations apply:

$$\mathbf{F}'_{23} + \mathbf{F}'_{43} + \mathbf{F}_{o_3} = 0$$

$$F'_{23}(e^{i\gamma'_3}) + F'_{43}(e^{i\theta_4}) + F_{o_3}(e^{i(\beta_3 + \pi)}) = 0 \tag{11.25}$$

Equating real and imaginary parts of the Eq. 11.25, we obtain

$$F'_{23}\cos\gamma'_3 + F'_{43}\cos\theta_4 + F_{o_3}\cos(\beta_3 + \pi) = 0 \tag{11.26}$$

$$F'_{23}\sin\gamma'_3 + F'_{43}\sin\theta_4 + F_{o_3}\sin(\beta_3 + \pi) = 0 \tag{11.27}$$

It may be seen that the number of unknowns to be determined is three, the magnitude of $\mathbf{F}'_{23}$ and its direction γ'_3, and the magnitude of $\mathbf{F}'_{43}$. The direction of $\mathbf{F}'_{43}$ is θ_4 and is known because link 4 is acted on by only two forces when considering $\mathbf{F}_{o_3}$ acting alone as shown in Fig. 11.19*c*. To determine the three unknowns, another equation is required in addition to Eqs. 11.26 and 11.27. The additional equation is one of equilibrium of moments, about either point *A* or *B*. Choosing point *A*, the sum of the moments about this point is required to be zero as follows:

$$F'_{43} r_3 \sin(\theta_4 - \theta_3) - F_{o_3} l_3 \sin(\beta_3 - \theta_3) = 0$$

$$F'_{43} = F_{o_3} \frac{l_3 \sin(\beta_3 - \theta_3)}{r_3 \sin(\theta_4 - \theta_3)} \tag{11.28}$$

On determining F'_{43} from Eq. 11.28, the real and imaginary components of F'_{23} may then be determined from Eqs. 11.26 and 11.27 as follows:

$$\mathscr{R}F'_{23} = F'_{23} \cos \gamma'_3 = -F'_{43} \cos \theta_4 - F_{o_3} \cos(\beta_3 + \pi) \tag{11.29}$$

$$\mathscr{I}F'_{23} = F'_{23} \sin \gamma'_3 = -F'_{43} \sin \theta_4 - F_{o_3} \sin(\beta_3 + \pi) \tag{11.30}$$

The symbols $\mathscr{R}$ and $\mathscr{I}$ indicate *real* and *imaginary* components of the vector $\mathbf{F}'_{23}$. The resultant of these components is the vector $\mathbf{F}'_{23}$, the magnitude of which may be determined as follows:

$$F'_{23} = \sqrt{(\mathscr{R}F'_{23})^2 + (\mathscr{I}F'_{23})^2} \tag{11.31}$$

The direction of $\mathbf{F}'_{23}$ is the angle γ'_3 which may be determined from the following expression:

$$\tan \gamma'_3 = \frac{\mathscr{I}F'_{23}}{\mathscr{R}F'_{23}} \tag{11.32}$$

Thus, the magnitudes and directions of the bearing forces at A and B may be calculated from the preceding equations. From the free body of link 4 shown in Fig. 11.19*c*, it is to be observed that the bearing force $\mathbf{F}'_{14}$ at O_4 is identical to the force $\mathbf{F}'_{43}$ because only two forces act on link 4. Similarly, since there are only two forces on link 2 as shown in the free body, the bearing force $\mathbf{F}'_{12}$ at O_2 is identical to $\mathbf{F}'_{23}$.

The final step of determining the shaft torque T'_s may be realized from the static equilibrium of couples acting on link 2.

$$\begin{aligned} T'_s &= -F'_{12} r_2 \sin(\theta_2 - \gamma'_3) \\ &= -F'_{23} r_2 \sin(\theta_2 - \gamma'_3) \end{aligned} \tag{11.33}$$

The preceding analysis has led to the determination of equations giving bearing forces and shaft torque due to the load $\mathbf{F}_{o_3}$. A similar analysis with only $\mathbf{F}_{o_4}$ acting would yield another set of equations giving the influence of $\mathbf{F}_{o_4}$ on the bearing forces and shaft torque, and a third analysis would give the influence of $\mathbf{F}_{o_2}$. The resultant force at each of the bearings would be determined by superposition by summing the real and imaginary components calculated in the individual analyses. At bearing A, for example, the superposed resultant real and imaginary components of the bearing force are the sums $\Sigma\, \mathscr{R}F_{23}$ and $\Sigma\, \mathscr{I}F_{23}$, and the resultant bearing force $\mathbf{F}_{23}$ may be determined from

$$F_{23} = \sqrt{(\Sigma \mathcal{R}F_{23})^2 + (\Sigma \mathcal{I}F_{23})^2} \tag{11.34}$$

and the angle γ_3 of the resultant force from

$$\tan \gamma_3 = \frac{\Sigma \mathcal{I}F_{23}}{\Sigma \mathcal{R}F_{23}} \tag{11.35}$$

Example 11.7

The drive link of the four-bar linkage shown in Fig. 11.20*a* rotates about O_2 at a constant angular velocity $\omega_2 = 100$ rad/sec. Using the data given in Fig. 11.20*a*, determine the bearing forces and the shaft torque at O_2 due to the dynamic load $\mathbf{F}_{o_3}$ when the mechanism is in the phase where $\theta_2 = 60°$. The angular positions θ_3 and θ_4 of links 3 and 4 in this phase are the same as those determined in Chapter 10, Section 10.28, which refers to Fig. 10.48.

Solution. Preliminary to determining the inertia load $\mathbf{F}_{o_3}$, the following kinematic quantities are evaluated from Eqs. 10.79, 10.80, and 10.81.

Angular velocities of links 3 and 4:

$$\omega_3 = -\frac{r_2}{r_3}\omega_2 \frac{\sin(\theta_4 - \theta_2)}{\sin(\theta_4 - \theta_3)}; \qquad \omega_4 = \frac{r_2}{r_4}\omega_2 \frac{\sin(\theta_3 - \theta_2)}{\sin(\theta_3 - \theta_4)}$$

Angular acceleration of link 3:

$$\alpha_3 = \frac{\left\{\begin{array}{c} -r_2\alpha_2 \sin(\theta_4 - \theta_2) + r_2\omega_2^2 \cos(\theta_4 - \theta_2) \\ + r_3\omega_3^2 \cos(\theta_4 - \theta_3) - r_4\omega_4^2 \end{array}\right\}}{r_3 \sin(\theta_4 - \theta_3)}$$

Substitution of the data given in Fig. 11.20*a* in the preceding equations in the order they are written yields the following numerical values:

$$\omega_3 = -10.16 \text{ rad/sec (cw)}$$

$$\omega_4 = 40.05 \text{ rad/sec (ccw)}$$

$$\alpha_3 = 3361 \text{ rad/sec}^2 \text{ (ccw)}$$

The acceleration $\mathbf{A}_{g_3}$ of the mass center of link 3 may be determined as the sum of two acceleration vectors, Eq. 10.69, as follows:

$$\mathbf{A}_{g_3} = r_2(i\alpha_2 - \omega_2^2)(e^{i\theta_2}) + r_{g_3}(i\alpha_3 - \omega_3^2)(e^{i\theta_3})$$

Noting that $\alpha_2 = 0$, this equation may be expanded as follows:

$$\mathbf{A}_{g_3} = (-r_2\omega_2^2 \cos\theta_2 - r_{g_3}\alpha_3 \sin\theta_3 - r_{g_3}\omega_3^2 \cos\theta_3) + i(-r_2\omega_2^2 \sin\theta_2 + r_{g_3}\alpha_3 \cos\theta_3 - r_{g_3}\omega_3^2 \sin\theta_3)$$

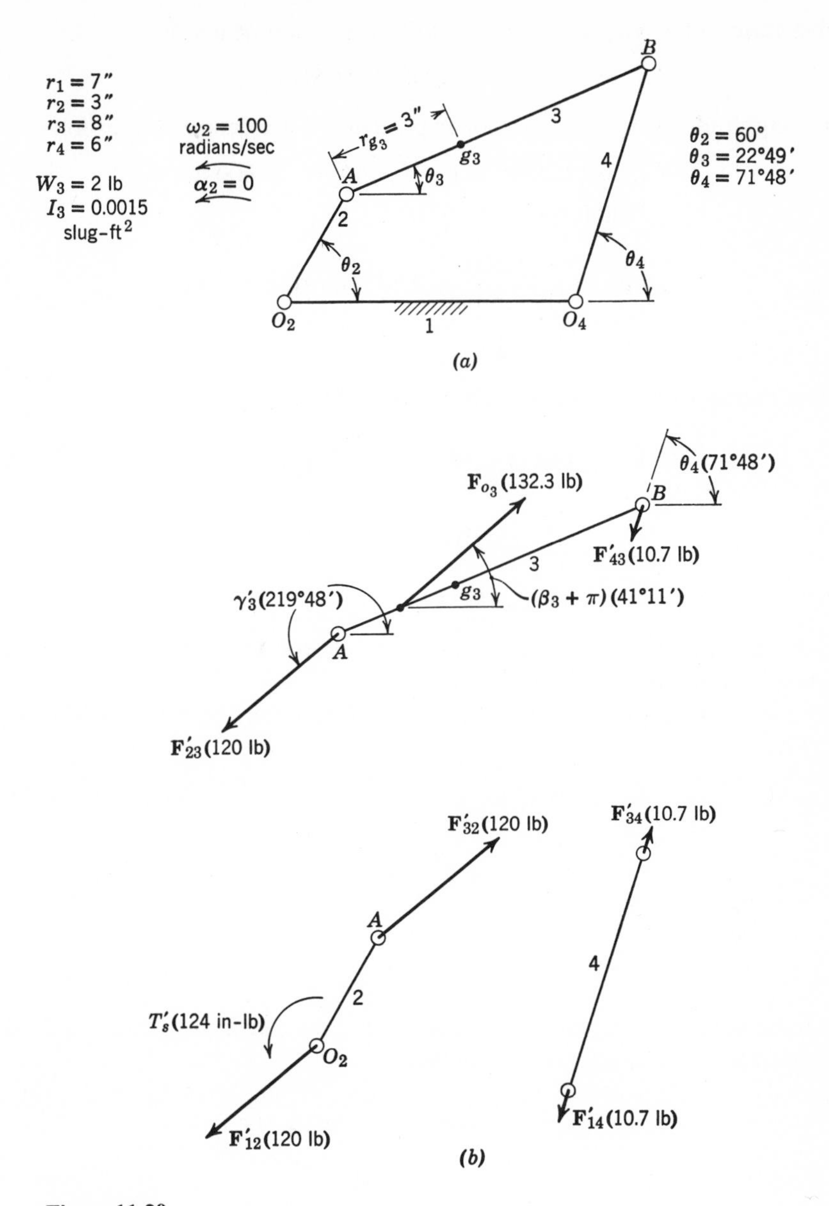

$r_1 = 7''$
$r_2 = 3''$
$r_3 = 8''$
$r_4 = 6''$
$W_3 = 2$ lb
$I_3 = 0.0015$ slug-ft^2
$\omega_2 = 100$ radians/sec
$\alpha_2 = 0$
$r_{g_3} = 3''$
$\theta_2 = 60°$
$\theta_3 = 22°49'$
$\theta_4 = 71°48'$
A
B
g_3
θ_3
θ_2
θ_4
O_2
O_4
1
2
3
4
(a)
$\mathbf{F}_{o_3}$ (132.3 lb)
θ_4 (71°48′)
$\mathbf{F}'_{43}$ (10.7 lb)
γ'_3 (219°48′)
$(\beta_3 + \pi)$ (41°11′)
$\mathbf{F}'_{23}$ (120 lb)
$\mathbf{F}'_{32}$ (120 lb)
$\mathbf{F}'_{34}$ (10.7 lb)
T'_s (124 in-lb)
$\mathbf{F}'_{12}$ (120 lb)
$\mathbf{F}'_{14}$ (10.7 lb)
(b)

Figure 11.20

Substitution of given data yields the following numerical values:

$$\mathbf{A}_{g_3} = -19{,}200 - 16{,}800i$$

The magnitude of $\mathbf{A}_{g_3}$ and its angular position β_3 are determined as follows:

$$|\mathbf{A}_{g_3}| = \sqrt{(-19{,}200)^2 + (-16{,}800)^2} = 25{,}600 \text{ in./sec}^2$$
$$= 2130 \text{ ft/sec}^2$$

$$\tan \beta_3 = \frac{-16{,}800}{-19{,}200} = 0.875 \text{ (third quadrant)}$$

$$\beta_3 = 221.18°$$

The inertia force $\mathbf{F}_{o_3}$ may now be evaluated from Eq. 11.23.

$$\mathbf{F}_{o_3} = (M_3 A_{g_3})e^{i(\beta_3 + \pi)} = \frac{W_3}{g} A_{g_3} e^{i(\beta_3 + \pi)}$$

$$= \frac{2}{32.2}(2130)e^{i(221.18° + 180°)}$$

$$= 132.3e^{i(41.18°)}$$

which indicates that the magnitude of the vector $\mathbf{F}_{o_3}$ is 132.3 lb, and its angular position is 41.18°.

The location of the line of action of $\mathbf{F}_{o_3}$ is l_3 and may be determined from Eq. 11.24.

$$l_3 = r_{g_3} + \frac{I_3 \alpha_3}{F_{o_3} \sin (\beta_3 - \theta_3)}$$

$$= 3.00 + \frac{0.0015(3361)(12)}{132.3 \sin 198.37°}$$

$$= 3.00 - 1.45$$

$$= 1.55 \text{ in.}$$

It is to be noted that sin 198.37° is negative and that the factor (12) is required for consistency of length units in inches.

Equation 11.28 gives the magnitude of the bearing force $\mathbf{F}'_{43}$ at B. Substitution of numerical values in this equation yields the following:

$$F'_{43} = -10.7 \text{ lb}$$

and substituting further in Eqs. 11.29 and 11.30 gives the numerical values of $\mathscr{R}F'_{23}$ and $\mathscr{I}F'_{23}$, from which the magnitude of the resultant $\mathbf{F}'_{23}$ is determined from Eq. 11.31 and its angle γ'_3 from Eq. 11.32.

$$F'_{23} = 120 \text{ lb} \qquad \gamma'_3 = 219.80°$$

Finally, the shaft torque T_s' at O_2 is determined by substituting in Eq. 11.33.

$$T_s' = 124 \text{ in.-lb}$$

In Fig. 11.20*b* are shown the forces on the several free bodies of the four-bar linkage, and these are shown to scale according to the preceding determined numerical values.

11.9 Engine Force Analysis. In Fig. 11.21 is shown the slider-crank mechanism of a typical single-cylinder, four-stroke cycle, internal combustion engine. Shown also are the vectors which represent the principal loads on the mechanism: (*a*) the static gas load **P** on the piston, and (*b*) the dynamic loads $\mathbf{F}_{o4}$ and $\mathbf{F}_{o3}$ acting on the piston and connecting rod, respectively. The inertia force $\mathbf{F}_{o2}$ of the crank is zero since it is usual to counterweight the crankshaft such that the mass center is at the axis of rotation O_2. Thus, the crankshaft itself is nominally balanced so that $\mathbf{A}_{g2}$ is zero. If the analysis is made for constant rotative speed of the crank ($\alpha_2 = 0$), the inertia couple of the crankshaft is also zero. Gravity forces also act on the mechanism, but because the weights of the moving parts are small compared to the principal loads, these are usually neglected.

Also shown in Fig. 11.21 is a typical curve showing the variation of combustion chamber gas pressure in the four-stroke cycle corresponding with two revolutions of the crankshaft. Magnitudes of gas pressure are determined from a thermodynamic analysis or from experimental measurements of combustion chamber pressure. The gas force **P** on the piston is the product of gas pressure and piston head area.

Two force analyses of the engine mechanism will be made using (*a*) superposition solved graphically (Example 11.8), and (*b*) an analytical solution with unit vectors (Example 11.9). In the second case it will not be necessary to use superposition because of certain simplifications that can be made and which will be discussed in a later section.

Example 11.8

The crankshaft speed of the slider-crank engine shown in Fig. 11.21 is 3000 r/min. Using the data given, determine the loads on the mechanism when the crank is in the phase $\theta_2 = 60°$. From a force analysis of the mechanism, determine the forces transmitted through the wrist-pin bearing, the crank-pin bearing, and the main bearings. Determine also the crankshaft torque T_s.

Solution. As shown in Fig. 11.21, for the phase $\theta_2 = 60°$ the mechanism is in the expansion (power) stroke and the gas pressure is 200 psig. The corresponding gas load on the piston is $P = 1410$ lb. The inertia force $\mathbf{F}_{o4}$ (360 lb) also acts on the piston, and its magnitude is determined from the product $M_4 A_B$. Inertia force $\mathbf{F}_{o3}$ (1230 lb) of the connecting rod has the

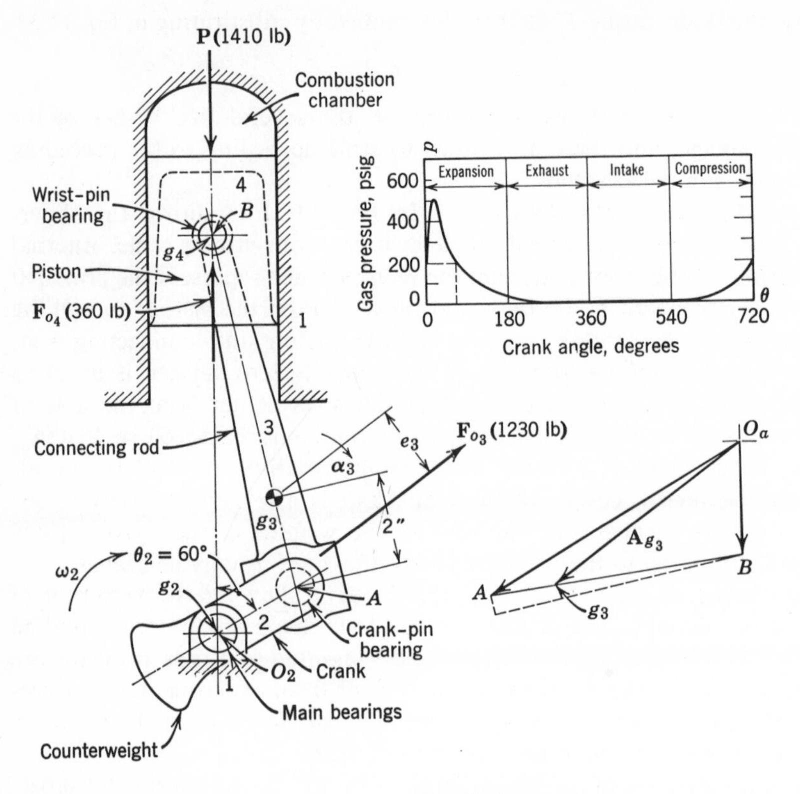

Data:

- Crankshaft speed, 3000 r/min
- Stroke, 4 in.
- Crank length, 2 in.
- Piston weight, 2 lb
- Piston area, 7.05 sq in.
- Connecting-rod weight, 3 lb
- Connecting-rod length, 8 in.
- Connecting-rod moment of inertia, $I_3 = 0.0075$ slug-ft^2

From acceleration polygon:

$$A_A = 16{,}500 \text{ fps}^2$$
$$A_{g_3} = 13{,}200 \text{ fps}^2$$
$$A_B = 5{,}800 \text{ fps}^2$$
$$\alpha_3 = 21{,}900 \text{ rad/sec}^2$$

Inertia forces:

$$F_{o_4} = M_4 A_B = \frac{W_4}{g} A_B = \frac{2}{32.2}(5800) = 360 \text{ lb}$$

$$F_{o_3} = \frac{W_3}{g} A_{g_3} = \frac{3}{32.2}(13{,}200) = 1230 \text{ lb}$$

$$e_3 = \frac{I_3 \alpha_3}{F_{o_3}} = \frac{0.0075(21{,}900)}{1230} = 0.133 \text{ ft} = 1.60 \text{ in.}$$

Gas force:

$$P = pA_p = 200(7.05) = 1410 \text{ lb}$$

Figure 11.21

magnitude M_3A_{g3}, the direction of the acceleration $\mathbf{A}_{g3}$ of the mass center of the rod, and a line of action offset e_3 because of the angular acceleration α_3.

In Fig. 11.22 is shown the force analysis of the mechanism in which $\mathbf{P}$, $\mathbf{F}_{o4}$, and $\mathbf{F}_{o3}$ are the known loads on the linkage. Superposition of forces is used to determine the unknown forces. In Fig. 11.22*a* forces in the linkage due to the loads $\mathbf{P}$ and $\mathbf{F}_{o4}$ on the piston are determined, and in Fig. 11.22*b* the forces due to $\mathbf{F}_{o3}$ on the connecting rod are determined. Finally, resultant forces are determined by superposition as shown in Fig. 11.22*c*. It is to be noted that friction forces are not included in the analysis; because of the pressure lubrication of the bearings and the cylinder wall, friction is assumed to be small and negligible.

Referring to Fig. 11.22*a*, $\mathbf{F}_4$ (1050 lb) is the resultant of the colinear forces $\mathbf{P}$ and $\mathbf{F}_{o4}$. Beginning with the free body of the piston (link 4), three forces are shown concurrent at B. The direction of the connecting rod force $\mathbf{F}'_{34}$ on the piston is known since only two forces act on link 3. The direction of the cylinder wall force $\mathbf{F}'_{14}$ on the side of the piston is normal to the wall in the absence of friction, and the line of action of $\mathbf{F}'_{14}$ is through the point of concurrency at B. Beginning with the known force $\mathbf{F}_4$, the equilibrium force polygon shown is constructed to determine the magnitudes and senses of $\mathbf{F}'_{34}$ and $\mathbf{F}'_{14}$. The two colinear forces $\mathbf{F}'_{43}$ and $\mathbf{F}'_{23}$ on the free body of the connecting rod (link 3) are equal in magnitude to $\mathbf{F}'_{34}$ of the polygon. Also, the two parallel but noncolinear forces $\mathbf{F}'_{32}$ and $\mathbf{F}'_{12}$ on the free body of the crank (link 2) are equal in magnitude to $\mathbf{F}'_{34}$. Thus all the unknown forces are determined from one force polygon. As shown on the free body of the crank, the shaft torque T'_s on the crankshaft at O_2 is the counterclockwise equilibrant of the couple formed by $\mathbf{F}'_{32}$ and $\mathbf{F}'_{12}$.

Referring to Fig. 11.22*b*, the known force on the linkage is $\mathbf{F}_{o3}$ of the connecting rod. Isolating the connecting rod as a free body as shown, it may be seen that three concurrent forces act. The direction of the piston force $\mathbf{F}''_{43}$ on the rod is known since only two forces act on the piston, one of which must be normal to the piston side. The direction of the crank force $\mathbf{F}''_{23}$ on the rod is through the point k of concurrency determined from the intersection of the lines of action of $\mathbf{F}_{o3}$ and $\mathbf{F}''_{43}$. Construction of the equilibrium force polygon shown determines the magnitudes and sense of $\mathbf{F}''_{43}$ and $\mathbf{F}''_{23}$. The two noncolinear forces $\mathbf{F}''_{32}$ and $\mathbf{F}''_{12}$ on the free body of the crank are equal in magnitude to $\mathbf{F}''_{23}$. The shaft torque T''_s is the equilibrant of the couple of the forces on the crank. It is to be observed that the moment arm of the couple is small because the inertia force $\mathbf{F}_{o3}$ very nearly acts through the crank-pin center at A.

Although this analysis is primarily graphical, it is interesting to note that the polygon of Fig. 11.22*a* involving forces $\mathbf{F}_4$, $\mathbf{F}'_{14}$, and $\mathbf{F}'_{34}$ could have just as easily been treated analytically because the directions of the three forces

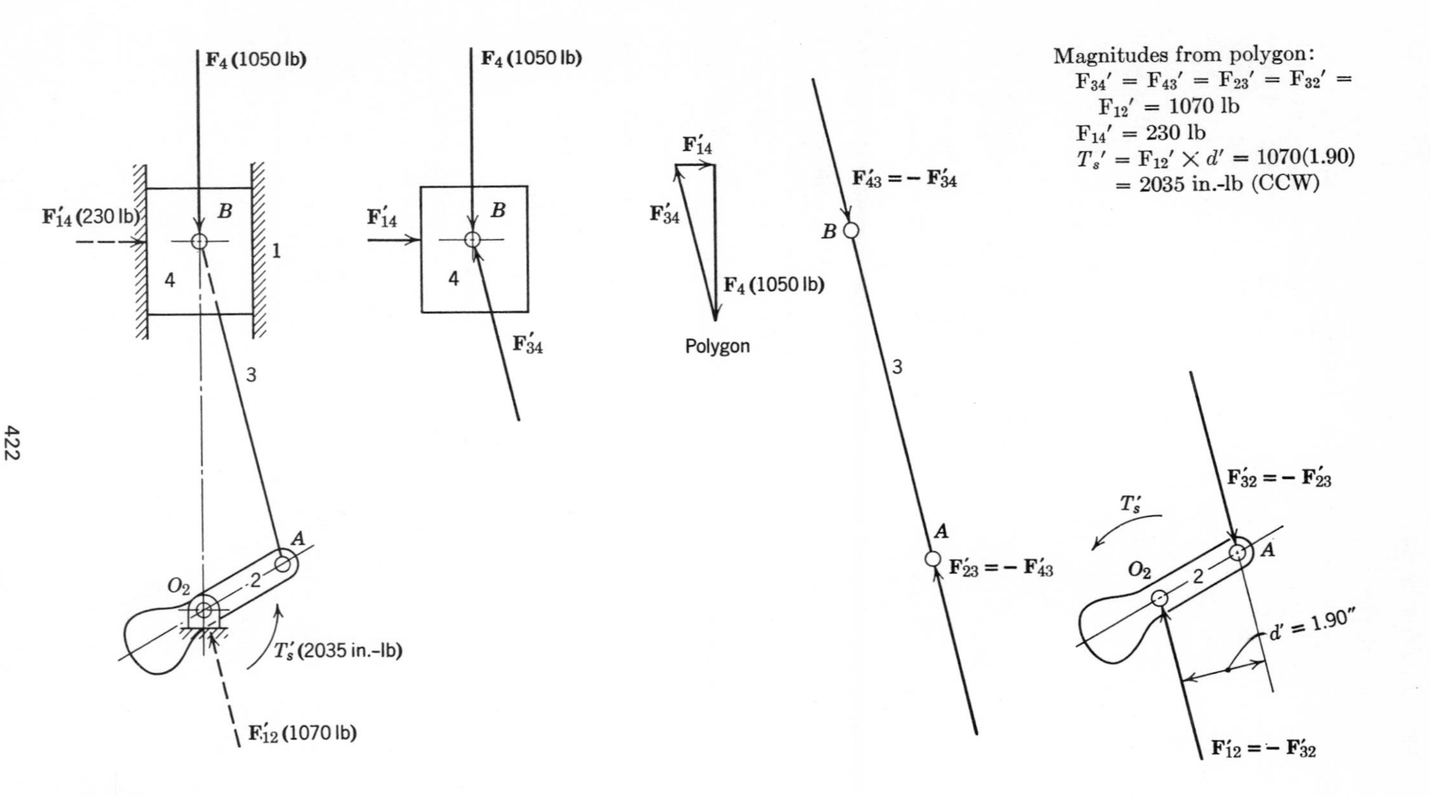

Figure 11.22a

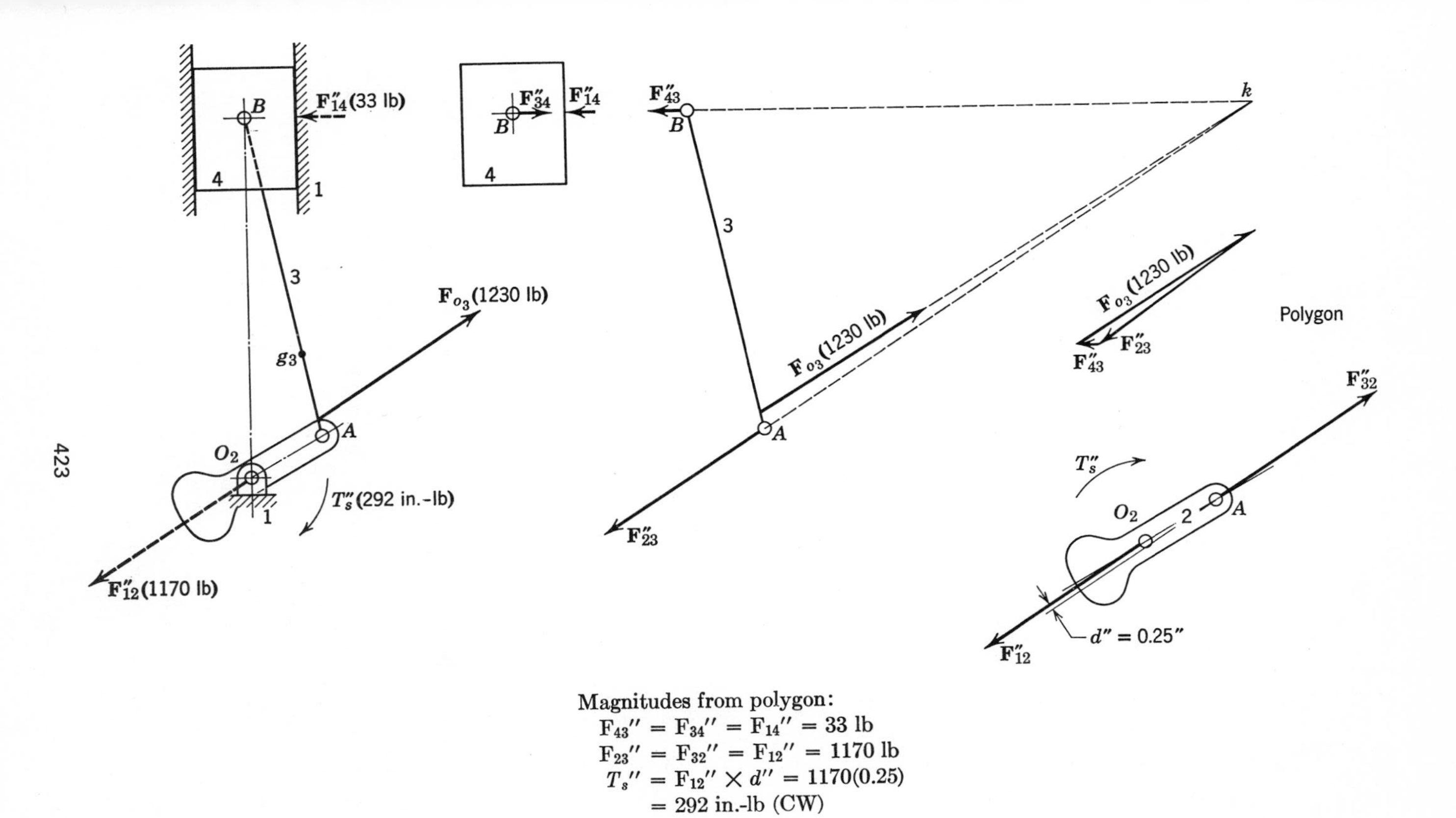

Figure 11.22b

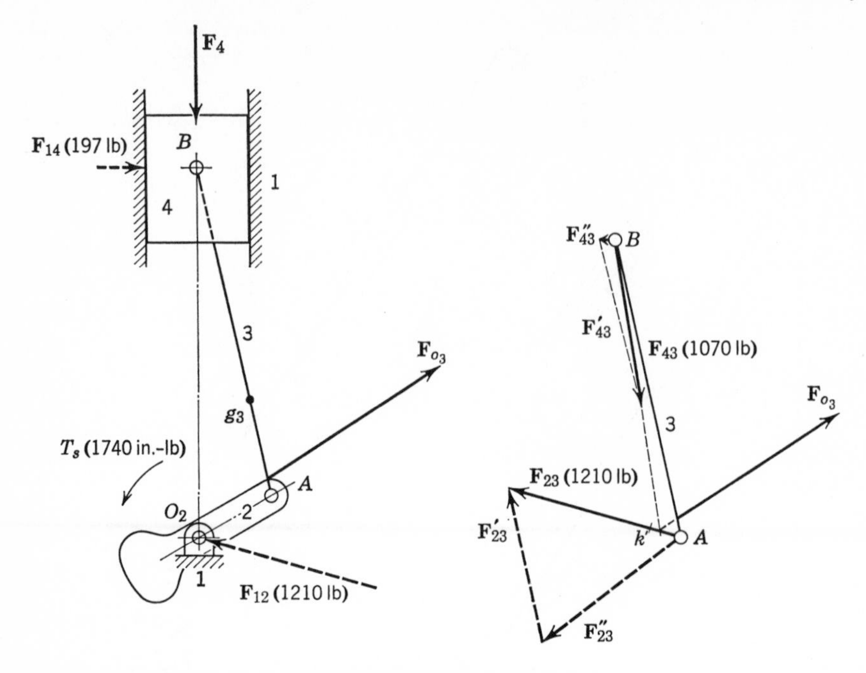

Bearing forces:
Wrist-pin, 1070 lb ($\mathbf{F}_{43}$)
Crank-pin, 1210 lb ($\mathbf{F}_{23}$)
Mains, 1210 lb ($\mathbf{F}_{12}$)
Crankshaft torque:
T_s = 1740 in.-lb (CCW)

Figure 11.22c

can easily be written in unit vector form. The polygon of Fig. 11.22*b* with forces $\mathbf{F}_{o_3}$, $\mathbf{F}''_{23}$, and $\mathbf{F}''_{43}$ is more quickly solved graphically, however, because the directions of $\mathbf{F}_{o_3}$ and $\mathbf{F}''_{23}$ are not readily known in unit vector form without further calculations to determine angles.

The resultant forces obtained by superposition are shown in Fig. 11.22*c*. The free body of the connecting rod shows the resultant forces acting at the pin-connected ends of the rod. At the upper end of the rod is shown the resultant force $\mathbf{F}_{43}$ transmitted through the wrist-pin bearing. $\mathbf{F}_{43}$ is the vector sum of $\mathbf{F}'_{43}$ and $\mathbf{F}''_{43}$. Similarly, the resultant force $\mathbf{F}_{23}$ transmitted through the crank-pin bearing at *A* is the vector sum of $\mathbf{F}'_{23}$ and $\mathbf{F}''_{23}$. It is to be observed that the connecting rod is in equilibrium under the action of these resultant forces and the inertia force $\mathbf{F}_{o_3}$ and that the three forces intersect at a common point k'. The resultant force through the main bearings is $\mathbf{F}_{12}$, which is identical to the force $\mathbf{F}_{23}$ through the crank-pin bearing. The crankshaft torque T_s is the algebraic sum of T_s' and T_s''.

11.10 Dynamically Equivalent Masses. Any rigid link in plane motion, having a mass M and a moment of inertia I, may be represented by an equivalent system of two point masses such that the inertia of the two masses is kinetically equivalent to the inertia of the link. In Fig. 11.23 is shown the inertia force $\mathbf{F}_o$ of a link, displaced e from the mass center g by virtue of its angular acceleration α. Also shown are the two point masses M_P and M_Q, which are to be the equivalent of the link in order that the resultant of the inertia forces $\mathbf{F}_P = M_P\mathbf{A}_P$ and $\mathbf{F}_Q = M_Q\mathbf{A}_Q$ is equal to $\mathbf{F}_o = M\mathbf{A}_g$. Therefore,

$$\mathbf{F}_P + \mathbf{F}_Q = \mathbf{F}_o \tag{11.36}$$

Although the proof is not undertaken here, it may be shown that to satisfy Eq. 11.36 the following three equivalents must be met:

1. Equivalence of mass. The sum of the point masses must be equal to the mass M of the link.

$$M_P + M_Q = M \tag{11.37}$$

2. Equivalence of mass center. The mass center of the system of the two point masses must be at the mass center of the link. This requires that the point masses lie on a common link through g. This also requires that the sum of the moments of the point masses about g be zero.

$$M_P l_P - M_Q l_Q = 0 \tag{11.38}$$

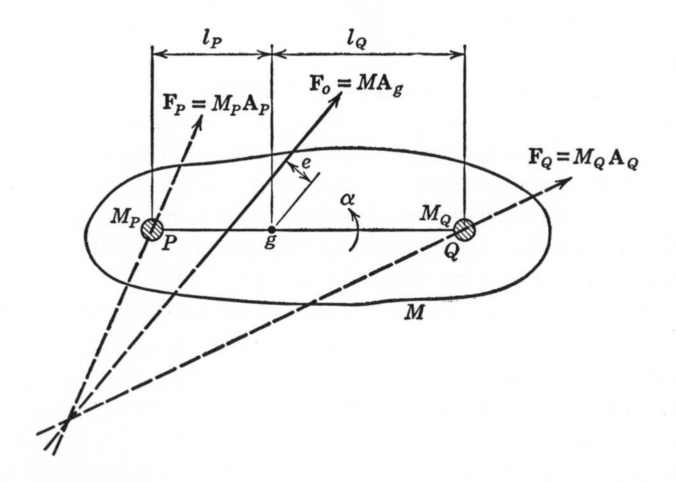

Figure 11.23

3. Equivalence of moment of inertia. The sum of the moments of inertia of the point masses about g must be equal to the moment of inertia I of the link.

$$M_P l_P^2 + M_Q l_Q^2 = I \tag{11.39}$$

If a link is to be replaced by the equivalent system of two point masses, it is necessary to determine the four quantities of the system, the magnitudes of the two masses M_P and M_Q and the two distances l_P and l_Q. The last three equations given are not in the best form for determining these quantities. A more useful form of these equations may be derived as follows: By solving the first two equations 11.37 and 11.38 simultaneously for M_P and M_Q, the following equations are obtained:

$$M_P = M \frac{l_Q}{l_P + l_Q} \tag{11.40}$$

$$M_Q = M \frac{l_P}{l_P + l_Q} \tag{11.41}$$

Then, substituting these equations in Eq. 11.39, the following is obtained:

$$M l_P l_Q = I$$

$$l_P l_Q = \frac{I}{M} \tag{11.42}$$

Since there are only three equations, and four quantities are to be determined, it may be seen that one of the quantities must be arbitrarily chosen. Usually, one of the distances l_P or l_Q is so chosen, and the other is then calculated from Eq. 11.42. With the distances determined, the magnitudes of M_P and M_Q may then be calculated from Eqs. 11.40 and 11.41.

11.11 Application of Equivalent Masses. Dynamically equivalent systems of two masses are most widely used in the analysis of automotive and aircraft piston engines, particularly with regard to connecting rods. Although applications of the method are made with approximations of small error, simplification in engine analysis is the primary advantage. Also, the method has influenced the design of counterweights on the crankshaft to reduce shaking of the engine.

Figure 11.24 shows a typical automotive connecting rod for which the center of gravity, weight, and moment of inertia are given. By arbitrarily locating one of the equivalent masses M_B at the wrist-pin bearing B, one of the inertia forces is determined from the piston acceleration. The location of the second mass M_P is as shown and is calculated from Eq. 11.42 by solving for l_P. Because of the shape of the connecting rod, the center of gravity is near the crank-pin center A as shown. Because of the nearness of P to A,

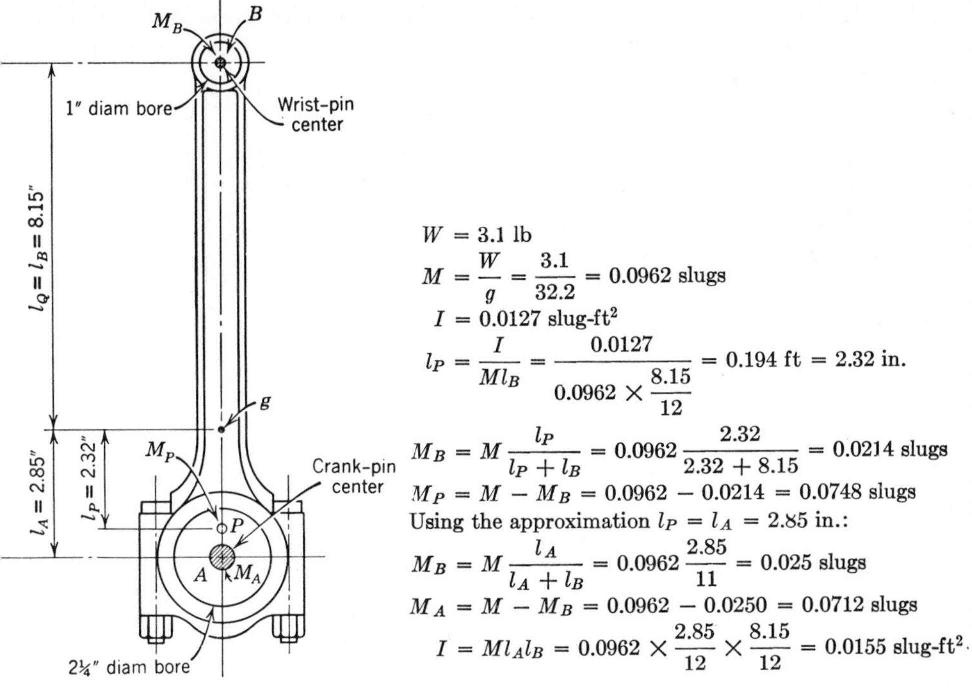

$$W = 3.1 \text{ lb}$$

$$M = \frac{W}{g} = \frac{3.1}{32.2} = 0.0962 \text{ slugs}$$

$$I = 0.0127 \text{ slug-ft}^2$$

$$l_P = \frac{I}{Ml_B} = \frac{0.0127}{0.0962 \times \frac{8.15}{12}} = 0.194 \text{ ft} = 2.32 \text{ in.}$$

$$M_B = M\frac{l_P}{l_P + l_B} = 0.0962\frac{2.32}{2.32 + 8.15} = 0.0214 \text{ slugs}$$

$$M_P = M - M_B = 0.0962 - 0.0214 = 0.0748 \text{ slugs}$$

Using the approximation $l_P = l_A = 2.85$ in.:

$$M_B = M\frac{l_A}{l_A + l_B} = 0.0962\frac{2.85}{11} = 0.025 \text{ slugs}$$

$$M_A = M - M_B = 0.0962 - 0.0250 = 0.0712 \text{ slugs}$$

$$I = Ml_Al_B = 0.0962 \times \frac{2.85}{12} \times \frac{8.15}{12} = 0.0155 \text{ slug-ft}^2.$$

Figure 11.24

the approximation may be made with little error that $l_P = l_A$. Thus, the second mass is at the crank-pin center, and the inertia force may be determined from the acceleration of the crank pin, which is constant in all phases of the mechanism when operating at constant crankshaft speed. As shown in Fig. 11.24, the approximation results in a moment of inertia of the equivalent system slightly larger than the true moment of inertia of the link since Ml_Al_B is greater than Ml_Pl_B. The magnitudes of the masses M_A and M_B are calculated from Eqs. 11.40 and 11.41 using l_A and l_B.

Since M_A is a mass rotating about the crankshaft axis with constant centrifugal force (at constant crankshaft speed), counterweights attached to the crankshaft may be of such mass as to counterbalance M_A of the connecting rod as well as the mass of the crank.

11.12 Engine Force Analysis Using Point Masses. In Fig. 11.25*a* is shown the internal combustion engine mechanism with approximate kinetically equivalent point masses replacing the connecting rod. One of the point masses, M_{B_3}, is located at the wrist-pin axis and the other, M_{A_3}, at the crank-pin axis. Thus, the dynamic loading of the connecting

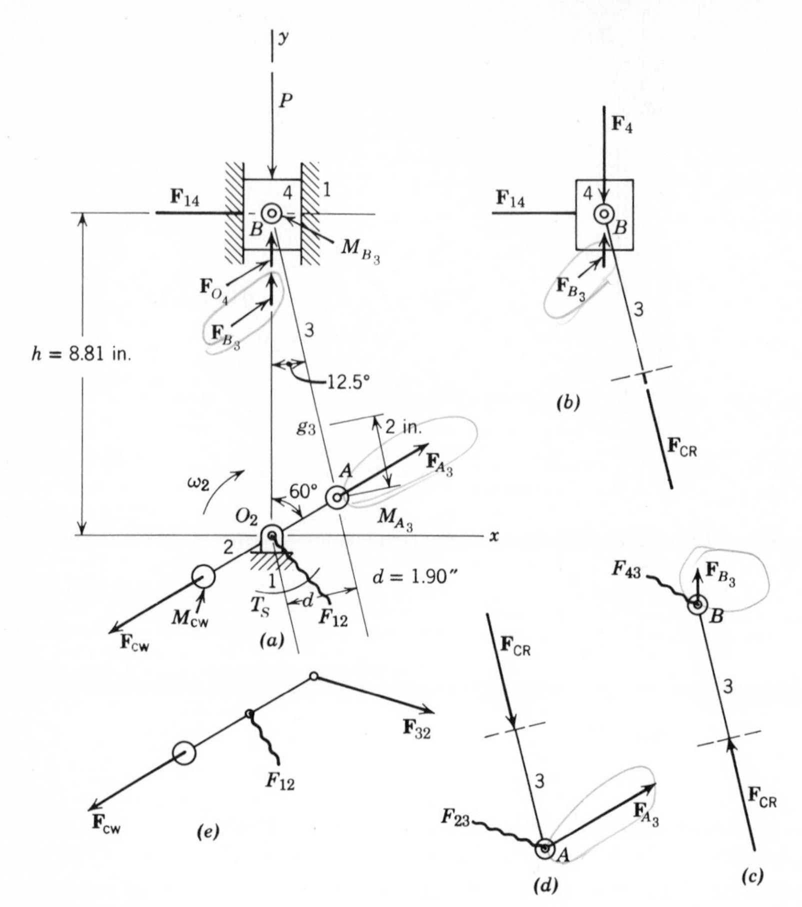

Figure 11.25

rod is represented by the inertia force vectors $\mathbf{F}_{B_3}$ and $\mathbf{F}_{A_3}$, the magnitudes of which are $F_{B_3} = M_{B_3}A_B$ and $F_{A_3} = M_{A_3}A_A$. For all phases of the mechanism, the line of action of $\mathbf{F}_{B_3}$ is on the line of reciprocation of the wrist pin, and the force $\mathbf{F}_{A_3}$ is always radially outward on the crank line O_2A when the speed of the crank is uniform.

As shown in Fig. 11.25*a*, it is usual practice to add a mass M_{cw} to the counterweight of the crank so that an inertia force $\mathbf{F}_{cw}$ is induced to balance the inertia force $\mathbf{F}_{A_3}$ of the connecting-rod mass. By counterweighting in this manner, the masses rotating with the crank (crankshaft mass plus M_{A_3}) are balanced to put the mass center of the combination at O_2 so that no force from these masses acts on the main bearings.

It is to be observed that all the forces acting on the connecting rod, the inertia forces and the bearing forces, act at the ends of the rod at A and B. There are no transverse components of force between the ends of the rod to bend or shear the link, and therefore the member is in axial tension or compression. This is a result of assuming that M_{A_3} may be placed at the crank-pin center A rather than at the correct point slightly removed from A. The fact that the connecting-rod force is axial in direction makes it possible to undertake the force analysis of the engine without superposition as illustrated in the following example.

Example 11.9

An analysis of the engine mechanism of Example 11.8 will now be made using approximate kinetically equivalent point masses for the connecting rod. In the solution all components have been taken relative to the xy axes.

Solution. From the data and solution of Example 11.8

$$n = 3000 \text{ r/min}$$

$$W_3 = 3 \text{ lb} \qquad A_A = 16{,}500 \text{ ft/sec}^2$$

$$l_3 = 8 \text{ in.} \qquad A_B = 5800 \text{ ft/sec}^2$$

$$l_A = 2 \text{ in.} \qquad F_{o_4} = 360 \text{ lb}$$

$$l_B = 6 \text{ in.} \qquad P = 1410 \text{ lb}$$

With the data given above, the approximate kinetically equivalent masses of the connecting rod can be calculated as follows:

$$M_3 = \frac{W_3}{g} = \frac{3}{32.2} = 0.0933 \text{ slugs}$$

$$M_{B_3} = M_3\left(\frac{l_A}{l_3}\right) = 0.0933(\tfrac{2}{8}) = 0.0233 \text{ slugs}$$

$$M_{A_3} = M_3 - M_{B_3} = 0.0933 - 0.0233 = 0.0700 \text{ slugs}$$

The inertia forces of the two ends of the connecting rod can also be calculated.

$$F_{B_3} = M_{B_3}A_B = 0.0233(5800) = 135 \text{ lb}$$

$$F_{A_3} = M_{A_3}A_A = 0.0700(16{,}500) = 1155 \text{ lb}$$

Figure 11.25a shows the forces acting on the mechanism. Of these forces $\mathbf{P}$, $\mathbf{F}_{o_4}$, $\mathbf{F}_{B_3}$, $\mathbf{F}_{A_3}$, and $\mathbf{F}_{cw}$ are known in magnitude, sense, and direction; force $\mathbf{F}_{14}$ is known in direction only. Nothing is known about force $\mathbf{F}_{12}$ except that it acts at point O_2. The equations for these forces, except $\mathbf{F}_{12}$, can be written as follows:

$$\mathbf{F}_4 = \mathbf{P} + \mathbf{F}_{o_4} = -1410\mathbf{j} + 360\mathbf{j} = -1050\mathbf{j}$$

$$\mathbf{F}_{B_3} = 135\mathbf{j}$$

$$\mathbf{F}_{A_3} = 1155(\cos 30°\mathbf{i} + \sin 30°\mathbf{j})$$

$$= 1000.26\mathbf{i} + 577.50\mathbf{j}$$

$$\mathbf{F}_{cw} = -1000.26\mathbf{i} - 577.50\mathbf{j}$$

$$\mathbf{F}_{14} = F_{14}\mathbf{i}$$

In Fig. 11.25*b* is shown a freebody-diagram of the piston and the upper part of the connecting rod. These members are acted upon by forces $\mathbf{F}_4$, $\mathbf{F}_{B_3}$, $\mathbf{F}_{14}$, and $\mathbf{F}_{CR}$. The force $\mathbf{F}_{CR}$ is the force of the lower part of the connecting rod acting upon the upper part. Its direction is along the rod because the rod becomes a two-force member when equivalent masses are placed at points *A* and *B*. The equation for $\mathbf{F}_{CR}$ can be written as follows:

$$\mathbf{F}_{CR} = F_{CR}(-\sin 12.5°\mathbf{i} + \cos 12.5°\mathbf{j})$$

$$= -0.2164F_{CR}\mathbf{i} + 0.9763F_{CR}\mathbf{j}$$

Because the piston and the upper part of the rod are in equilibrium under the action of the four forces

$$\mathbf{F}_4 + \mathbf{F}_{B_3} + \mathbf{F}_{14} + \mathbf{F}_{CR} = 0$$

Substituting the relations previously determined in the equation of equilibrium, one obtains

$$-1050\mathbf{j} + 135\mathbf{j} + F_{14}\mathbf{i} - 0.2164F_{CR}\mathbf{i} + 0.9763F_{CR}\mathbf{j} = 0$$

Summing **i** *components,*

$$F_{14}\mathbf{i} - 0.2164F_{CR}\mathbf{i} = 0$$

Summing **j** *components,*

$$-1050\mathbf{j} + 135\mathbf{j} + 0.9763F_{CR}\mathbf{j} = 0$$

$$F_{CR} = 937.2 \text{ lb}$$

Also,

$$F_{14}\mathbf{i} - 0.2164 \times 937.2\mathbf{i} = 0$$

$$F_{14} = 202.8 \text{ lb}$$

Therefore,

$$\mathbf{F}_{CR} = -0.2164 \times 937.2\mathbf{i} + 0.9763 \times 937.2\mathbf{j}$$

$$= -202.81\mathbf{i} + 915\mathbf{j}$$

$$\mathbf{F}_{14} = 202.8\mathbf{i}$$

Consider next Fig. 11.25*c*, which shows the upper part of the connecting rod acted upon by forces $\mathbf{F}_{B_3}$, $\mathbf{F}_{CR}$, and $\mathbf{F}_{43}$. The following equation of equilibrium can be written:

$$\mathbf{F}_{B_3} + \mathbf{F}_{CR} + \mathbf{F}_{43} = 0$$

where

$$\mathbf{F}_{43} = F_{43}(\lambda_x \mathbf{i} + \lambda_y \mathbf{j}) \qquad \text{(direction unknown; } \boldsymbol{\lambda} \text{ is a unit vector in the direction of } \mathbf{F}_{43})$$

Substituting in the equation of equilibrium,

$$135\mathbf{j} - 202.81\mathbf{i} + 915\mathbf{j} + \lambda_x F_{43}\mathbf{i} + \lambda_y F_{43}\mathbf{j} = 0$$

Summing **i** *components*,

$$-202.81\mathbf{i} + \lambda_x F_{43}\mathbf{i} = 0$$

$$\lambda_x F_{43}\mathbf{i} = 202.81\mathbf{i}$$

Summing **j** *components*,

$$135\mathbf{j} + 915\mathbf{j} + \lambda_y F_{43}\mathbf{j} = 0$$

$$\lambda_y F_{43}\mathbf{j} = -1050\mathbf{j}$$

Therefore,

$$\mathbf{F}_{43} = 202.81\mathbf{i} - 1050\mathbf{j}$$

$$|\mathbf{F}_{43}| = 1069.4 \text{ lb}$$

Figure 11.25*d* shows a freebody-diagram of the lower part of the connecting rod under action of forces $\mathbf{F}_{CR}$, $\mathbf{F}_{A_3}$, and $\mathbf{F}_{23}$. The following equation of equilibrium can be written:

$$\mathbf{F}_{CR} + \mathbf{F}_{A_3} + \mathbf{F}_{23} = 0$$

where

$$\mathbf{F}_{23} = F_{23}(\lambda_x \mathbf{i} + \lambda_y \mathbf{j}) \qquad \text{(direction unknown)}$$

and

$$\mathbf{F}_{CR} = 202.81\mathbf{i} - 915\mathbf{j} \qquad \text{(sense opposite to that of force } \mathbf{F}_{CR} \text{ acting on upper part of rod)}$$

Substituting in the equation of equilibrium,

$$202.81\mathbf{i} - 915\mathbf{j} + 1000.26\mathbf{i} + 577.50\mathbf{j} + \lambda_x F_{23}\mathbf{i} + \lambda_y F_{23}\mathbf{j} = 0$$

Summing **i** *components*,

$$202.81\mathbf{i} + 1000.26\mathbf{i} + \lambda_x F_{23}\mathbf{i} = 0$$

$$\lambda_x F_{23}\mathbf{i} = -1203.07\mathbf{i}$$

Summing **j** *components,*

$$-915\mathbf{j} + 577.50\mathbf{j} + \lambda_y F_{23}\mathbf{j} = 0$$

$$\lambda_y F_{23}\mathbf{j} = 337.5\mathbf{j}$$

Therefore,

$$\mathbf{F}_{23} = -1203.07\mathbf{i} + 337.5\mathbf{j}$$

$$|\mathbf{F}_{23}| = 1249.5 \text{ lb}$$

Figure 11.25*e* shows the crank and counterweight under action of forces $\mathbf{F}_{32}$, $\mathbf{F}_{cw}$, and $\mathbf{F}_{12}$. The equation of equilibrium is

$$\mathbf{F}_{32} + \mathbf{F}_{cw} + \mathbf{F}_{12} = 0$$

where

$$\mathbf{F}_{32} = -\mathbf{F}_{23} = 1203.07\mathbf{i} - 337.5\mathbf{j}$$

and

$$\mathbf{F}_{12} = F_{12}(\lambda_x\mathbf{i} + \lambda_y\mathbf{j}) \qquad \text{(direction unknown)}$$

Substituting in the above equation of equilibrium

$$1203.07\mathbf{i} - 337.5\mathbf{j} - 1000.26\mathbf{i} - 577.50\mathbf{j} + \lambda_x F_{12}\mathbf{i} + \lambda_y F_{12}\mathbf{j} = 0$$

Summing **i** *components,*

$$1203.07\mathbf{i} - 1000.26\mathbf{i} + \lambda_x F_{12}\mathbf{i} = 0$$

$$\lambda_x F_{12}\mathbf{i} = -202.81\mathbf{i}$$

Summing **j** *components,*

$$-337.5\mathbf{j} - 577.50\mathbf{j} + \lambda_y F_{12}\mathbf{j} = 0$$

$$\lambda_y F_{12}\mathbf{j} = 915\mathbf{j}$$

Therefore,

$$\mathbf{F}_{12} = -202.81\mathbf{i} + 915\mathbf{j}$$

$$|\mathbf{F}_{12}| = 937.2 \text{ lb}$$

Comparing the equation for $\mathbf{F}_{12}$ with that for $\mathbf{F}_{CR}$ acting on the upper part of the rod, one can see that the equations are identical so that the two vectors are parallel and have the same sense and magnitude.

The shaft torque T_s necessary to hold link 2 in equilibrium can easily be calculated from the relation

$$T_s = F_{14}h = 202.8 \times 8.81 = 1786.8 \text{ lb-in. (ccw)}$$

Torque T_s can also be obtained from the relation $F_{12}d$.

The results of the analysis are as follows:

$$F_{14} = 203 \text{ lb}$$
$$F_{43} = 1069 \text{ lb (wrist-pin force)}$$
$$F_{23} = 1250 \text{ lb (crank-pin force)}$$
$$F_{12} = 937 \text{ lb (main-bearing force)}$$
$$T_s = 1787 \text{ lb-in. (ccw)}$$

If these values are compared with those from Example 11.8 in which equivalent point masses were not used for the connecting rod, it will be seen that there is close agreement with the exception of the force $\mathbf{F}_{12}$ at the main bearing. In this analysis the magnitude and direction of $\mathbf{F}_{12}$ are quite different from those determined in Example 11.8. This difference, however, is not due to the use of equivalent masses but rather to the use of additional counterweight on the crank to balance inertia force $\mathbf{F}_{A_3}$; this partly unloads the main bearing.

It is interesting to note how easily the shaft torque T_s can be determined using the *method of virtual work.* This is illustrated in the following:

$$\mathbf{T}_2 \cdot \boldsymbol{\omega}_2 + \mathbf{F}_4 \cdot \mathbf{V}_B + \mathbf{F}_{B_3} \cdot \mathbf{V}_B + \mathbf{F}_{A_3} \cdot \mathbf{V}_A + \mathbf{F}_{\text{cw}} \cdot \mathbf{V}_{\text{cw}} = 0$$

$$\omega_2 = \frac{2\pi n}{60} = \frac{2\pi \times 3000}{60} = 314.16 \text{ rad/sec (cw)}$$

$$|\mathbf{V}_B| = R\omega\left[\sin\theta + \frac{R}{2L}\sin 2\theta\right]$$

$$= \tfrac{2}{12} \times 314.16\left[\sin 60^\circ + \frac{2}{2 \times 8}\sin 120^\circ\right]$$

$$= 51.01 \text{ ft/sec}$$

Therefore,

$$\mathbf{V}_B = -51.01\mathbf{j}$$
$$\mathbf{F}_4 = \mathbf{P} + \mathbf{F}_{o_4} = -1050\mathbf{j}$$
$$\mathbf{F}_{B_3} = 135\mathbf{j}$$
$$\mathbf{T}_2 \cdot \boldsymbol{\omega}_2 = (T_2\mathbf{k}) \cdot (-314.16\mathbf{k}) = -314.16T_2 \text{ ft-lb/sec}$$
$$\mathbf{F}_4 \cdot \mathbf{V}_B = (-1050\mathbf{j}) \cdot (-51.01\mathbf{j}) = 53{,}563.7 \text{ ft-lb/sec}$$
$$\mathbf{F}_{B_3} \cdot \mathbf{V}_B = (135\mathbf{j}) \cdot (-51.01\mathbf{j}) = -6886.8 \text{ ft-lb/sec}$$
$$\mathbf{F}_{A_3} \cdot \mathbf{V}_A = 0 \text{ (force and velocity vectors at right angles)}$$
$$\mathbf{F}_{\text{cw}} \cdot \mathbf{V}_{\text{cw}} = 0 \text{ (force and velocity vectors at right angles)}$$

Therefore,

$$-314.16T_2 + 53{,}563.7 - 6886.8 = 0$$

and

$$T_2 = 148.58 \text{ lb-ft}$$

$$= 1783 \text{ lb-in. (ccw)}$$

This value agrees closely with the value found previously.

11.13 Engine Block. In the foregoing discussion of the piston engine linkage, the frame of the engine, or the engine block, is considered the fixed member. However, in automotive installations, the engine block is supported on flexible mountings in order that a minimum of the unbalanced resultant force of the engine is transmitted to the engine supports. Figure 11.26 shows a free-body diagram of the engine block (link 1) fastened to the supporting link 0. The supporting link is shown rigid to illustrate the nature of the forces and moments which are transmitted to the supports. With flexible supports, the force system becomes one involving nonrigid members and a vibration analysis is required.

The effects of the slider-crank mechanism of Fig. 11.25 is shown on the free-body diagram of the block in Fig. 11.26*a*. $\mathbf{F}_{41}$ is the reactive force of the piston on the cylinder wall, and $\mathbf{F}_{21}$ is the force of the crankshaft on the main bearings fixed to the engine block. The force $\mathbf{P}$ by the gas pressure acts on the head of the block. These are known forces from the analysis made of the slider-crank mechanism. The reactive forces $\mathbf{R}_1$ and $\mathbf{R}_2$ of the engine supports are the unknowns to be determined.

The free-body diagram of Fig. 11.26*b* shows the forces acting on the combination of engine block with slider crank. Forces such as $\mathbf{F}_{41}$ and $\mathbf{F}_{14}$, as well as $\mathbf{F}_{21}$ and $\mathbf{F}_{12}$, are internal forces and are therefore not shown. The gas forces $\mathbf{P}$ are colinear, equal, and opposite, and therefore do not effect $\mathbf{R}_1$ and $\mathbf{R}_2$. $\mathbf{F}_{cw}$ and $\mathbf{F}_{A_3}$ are also colinear, equal, and opposite. Those vectors shown on the free body which affect the forces $\mathbf{R}_1$ and $\mathbf{R}_2$ are the inertia forces $\mathbf{F}_{o4}$ and $\mathbf{F}_{B_3}$ and the shaft torque T_s.

For static equilibrium of vertical forces, the resultant reactive force $\mathbf{R}$ (which is equal to $\mathbf{R}_1 + \mathbf{R}_2$) of the supports should be the same for both free bodies. The equilibrium force polygons for the two free bodies are shown in Fig. 11.26, and $\mathbf{R}$ is the same in both. From the polygon of Fig. 11.26*b* in particular, it may be seen that the resultant force $\mathbf{R}$ is equal in magnitude to the sum of the inertia forces $\mathbf{F}_{o4}$ and $\mathbf{F}_{B_3}$ of the masses reciprocating with the wrist-pin center. $\mathbf{R}$ is the equilibrant of the inertia forces. The opposite vector $\mathbf{S}$, which is the resultant of the inertia forces, is termed the *shaking force* since, if flexible supports are used, the engine block will be raised from its supports when the inertia forces are directed upward as shown, and will be pressed

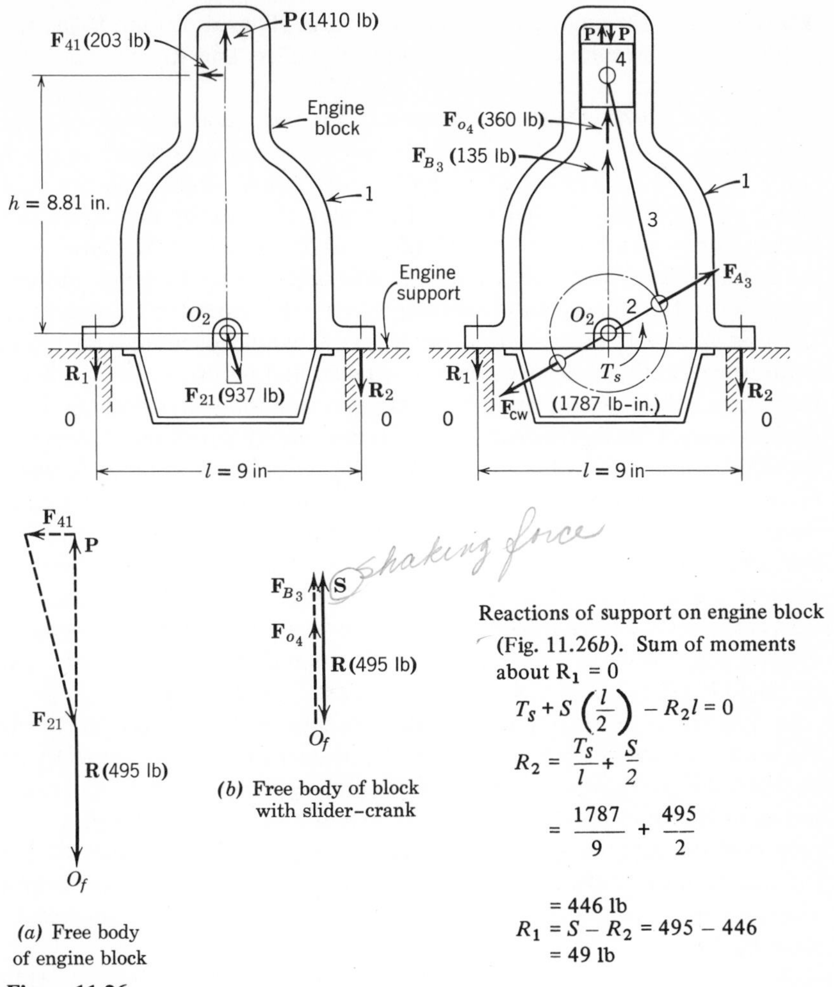

(*a*) Free body of engine block

(*b*) Free body of block with slider–crank

Figure 11.26

against the supports when the inertia forces are directed downward in other phases of the cycle of the mechanism.

Thus, the reciprocating masses at the wrist-pin center cause a vertical vibration or vertical shaking of the engine. It is to be observed that no resultant horizontal force acts on the engine, so there is no vibrational excitation in this direction. However, because of the shaft torque T_s, a couple acts on the engine, which, if mounted on flexible supports, excites an angular oscillation of the engine as the shaft torque changes in magnitude and sense during the engine cycle. Thus, T_s is a *shaking couple* which is also transmitted to the engine supports and makes $\mathbf{R}_2$ greater than $\mathbf{R}_1$ as shown in Fig. 11.26. Calculations of the magnitudes of $\mathbf{R}_1$ and $\mathbf{R}_2$ are shown below Fig. 11.26 based on the free body of Fig. 11.26*b*.

11.14 Engine Output Torque. Of particular interest in engines is the shaft torque variation in the engine cycle corresponding to the 720° crank cycle. A plot of shaft torque versus crank angle θ shows a large variation in magnitude and sense of the torque, since by inspection of free bodies it may be seen that in some phases the torque is in the same sense as the crank motion and in other phases it is opposite to the crank motion. It would seem that the assumption of constant crank speed in the engine force analysis is invalid since a variation in torque would produce a variation in crank speed in the cycle. However, it is usual and necessary to fix a flywheel to the crankshaft. As shown in a following discussion, a flywheel of relatively small moment of inertia will reduce crank speed variations to negligibly small values (1 or 2% of crank speed). Because of its importance to flywheel design, an analytical method of evaluating output torque T as a function of crank angle θ is developed below. Output torque T and shaft torque T_s are the same in magnitude but of opposite sense. T_s is the torque on the mechanism as a free body and is the resisting torque of the load; T is the torque delivered to the flywheel and to the vehicle or load which the engine is driving. As shown in Example 11.9, the magnitude of the torque T_s may be calculated from either of two expressions: $T_s = F_{12}d$ or $T_s = F_{14}h$.

In Fig. 11.27 is shown the engine mechanism of Fig. 11.25*a* in which the main bearing force $\mathbf{F}_{12}$ is known to be parallel to the connecting rod axis because of the method of counterweighting the crankshaft. Considering the equilibrium of forces on the entire mechanism as a free body, it may be seen that the simple polygon of forces in Fig. 11.27 determines both $\mathbf{F}_{12}$ and $\mathbf{F}_{14}$ from the known colinear forces $\mathbf{P}$, $\mathbf{F}_{o4}$, and $\mathbf{F}_{B_3}$ acting at the wrist-pin center. The known forces may be combined and shown as the resultant force $\mathbf{F}_B$.[5]

$$\mathbf{F}_B = \mathbf{P} - (\mathbf{F}_{o4} + \mathbf{F}_{B_3}) \tag{11.43}$$

From the force polygon,

$$F_{12} = \frac{F_B}{\cos\phi} \tag{11.44}$$

$$F_{14} = F_B \tan\phi \tag{11.45}$$

Noting in Fig. 11.27 that $d = h \sin\phi$, we may evaluate the output torque

[5] For cases where the sense of $(\mathbf{F}_{o4} + \mathbf{F}_{B_3})$ is reversed, this equation becomes $\mathbf{F}_B = \mathbf{P} + (\mathbf{F}_{o4} + \mathbf{F}_{B_3})$.

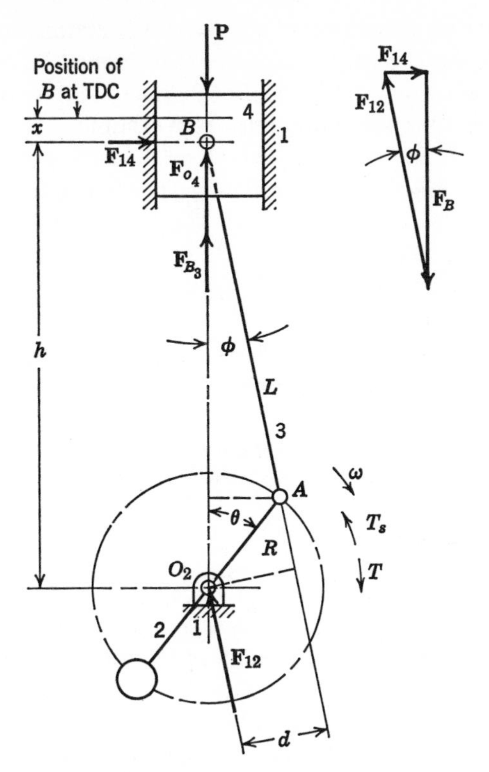

Figure 11.27

T from either of the two expressions as follows:

$$T = F_{12}d = F_B \frac{h \sin \phi}{\cos \phi}$$
$$= F_B h \tan \phi \tag{11.46}$$

or

$$T = F_{14}h$$
$$= F_B h \tan \phi \tag{11.47}$$

Although the output torque is evaluated from different equations, Eqs. 11.46 and 11.47 show that they are identical. To determine torque as a function of crank angle θ from these equations, each of the right-hand factors (F_B, h, and ϕ) must be expressed as functions of θ. The first of these factors F_B depends on the several forces indicated in Eq. 11.43, each of which must also be expressed as functions of θ. An equation for the gas force P cannot be written directly as a function of θ since P is determined experimentally on an indicator diagram and is shown either versus θ as in Fig. 11.21 or versus piston position x in Fig. 11.27. Piston position is expressed as a function of θ in the following equations from Chapter 2:

$$x = R\left[1 - \cos\theta + \frac{L}{R}(1 - \cos\phi)\right]$$

$$\frac{x}{2R} = \frac{1}{2}\left[1 - \cos\theta + \frac{L}{R}(1 - \cos\phi)\right] \tag{11.48}$$

R is crank length, and L is connecting-rod length. The angle ϕ may be determined in terms of θ from the following relationship given by triangles in Fig. 11.27.

$$L\sin\phi = R\sin\theta$$

$$\phi = \sin^{-1}\left(\frac{R}{L}\sin\theta\right) \tag{11.49}$$

In Eq. 11.48, x is positive downward from the top-dead-center (T.D.C.) position of the piston and is expressed as a fraction of the stroke $2R$. The fraction gives the location on the abcissa of the indicator diagram at which the gas pressure is read for a given θ. Crank angle θ is positive clockwise in the sense of crank motion as shown in Fig. 11.27. For convenience, values of $x/2R$ for given angles θ have been calculated for a number of L/R ratios and are given in Table 11.1.

The sum of the inertia forces $\mathbf{F}_{o4} + \mathbf{F}_{B_3}$ in Eq. 11.43 may be calculated from $(M_4 + M_{B_3})A_B$, in which A_B is the piston acceleration. As shown in Chapter 2, the piston acceleration is a function of θ in the following expression:

$$A_B = R\omega^2\left(\cos\theta + \frac{R}{L}\cos 2\theta\right)$$

$$\frac{A_B}{R\omega^2} = \left(\cos\theta + \frac{R}{L}\cos 2\theta\right) \tag{11.50}$$

Values of $A_B/R\omega^2$ from Eq. 11.50 are tabulated in Table 11.1 for a number of L/R ratios as a function of θ. At the beginning of the stroke when θ is small, A_B is positive downward in the direction toward the crankshaft axis. However, since inertia force is opposite in sense to acceleration, $\mathbf{F}_{o4}$ and $\mathbf{F}_{B_3}$ are given a minus sign in Eq. 11.43 and are shown opposite to $\mathbf{P}$ in Fig. 11.27.

Figure 11.28 shows the variation in one engine cycle of the resultant force or combined force $\mathbf{F}_B$. Also shown are the individual curves for gas force $\mathbf{P}$

Table 11.1 SLIDER-CRANK FUNCTIONS FOR OUTPUT TORQUE, PISTON ACCELERATION, AND PISTON POSITION

	$\frac{L}{R} = 4.0$			$\frac{L}{R} = 4.5$			$\frac{L}{R} = 5.0$		
θ	$\frac{h}{R}\tan\phi$	$\frac{A_B}{R\omega^2}$	$\frac{x}{2R}$	$\frac{h}{R}\tan\phi$	$\frac{A_B}{R\omega^2}$	$\frac{x}{2R}$	$\frac{h}{R}\tan\phi$	$\frac{A_B}{R\omega^2}$	$\frac{x}{2R}$
0	0	1.2500	0	0	1.2222	0	0	1.2000	0
15	0.3214	1.1824	0.0212	0.3144	1.1583	0.0208	0.3088	1.1391	0.0204
30	0.6091	0.9910	0.0827	0.5968	0.9771	0.0809	0.5870	0.9660	0.0795
45	0.8341	0.7071	0.1779	0.8196	0.7071	0.1774	0.8081	0.7071	0.1716
60	0.9769	0.3750	0.2974	0.9642	0.3889	0.2921	0.9540	0.4000	0.2878
75	1.0303	0.0423	0.4298	1.0227	0.0666	0.4230	1.0169	0.0856	0.4177
90	1.0000	−0.2500	0.5635	1.0000	−0.2222	0.5563	1.0000	−0.2000	0.5505
105	0.9015	−0.4753	0.6886	0.9091	−0.4512	0.6819	0.9149	−0.4320	0.6765
120	0.7551	−0.6250	0.7974	0.7680	−0.6111	0.7921	0.7781	−0.6000	0.7878
135	0.5801	−0.7071	0.8851	0.5946	−0.7071	0.8815	0.6061	−0.7071	0.8787
150	0.3909	−0.7410	0.9487	0.4032	−0.7549	0.9469	0.4130	−0.7660	0.9455
165	0.1962	−0.7494	0.9872	0.2032	−0.7735	0.9867	0.2087	−0.7927	0.9863
180	0	−0.7500	1.0000	0	−0.7778	1.0000	0	−0.8000	1.0000

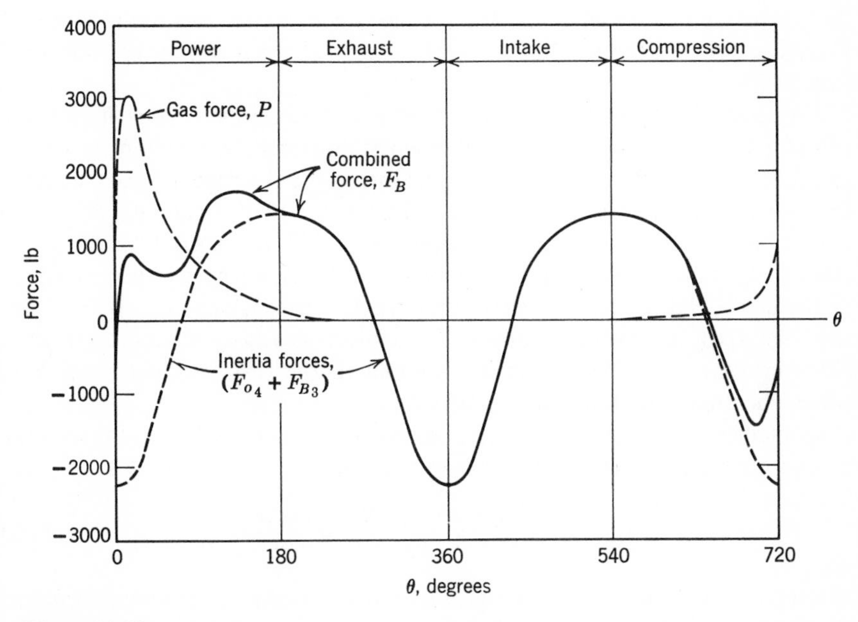

Figure 11.28

and the inertia forces $\mathbf{F}_{o_4} + \mathbf{F}_{B_3}$. It may be seen from Table 11.1 that accelerations are positive near top dead center and are negative near the bottom of the stroke. Inertia forces are negative near the top of the stroke and positive near the bottom of the stroke as shown in Fig. 11.28.

The product $h \tan \phi$ in Eq. 11.47 may be expressed in terms of θ as follows. From triangles in Fig. 11.27,

$$h \tan \phi = (R \cos \theta + L \cos \phi) \tan \phi$$

$$= R \tan \phi \left(\cos \theta + \frac{L}{R} \cos \phi \right)$$

$$\frac{h}{R} \tan \phi = \tan \phi \left(\cos \theta + \frac{L}{R} \cos \phi \right) \tag{11.51}$$

As shown in Table 11.1, values of $(h/R) \tan \phi$ are tabulated as a function of θ for various L/R ratios from Eqs. 11.51 and 11.49. It is to be observed that for downward strokes ($0° < \theta < 180°$), $(h/R) \tan \phi$ values are positive. Although they are not shown in the table, for upward strokes ($180° < \theta < 360°$), these values are negative because $\tan \phi$ is negative.

A typical curve of output torque of a single-cylinder engine for one complete cycle is shown in Fig. 11.29. Positive values of output torque are those which are in the same sense as the crank motion. It may be seen that the torque changes sign where the torque is zero either because $(h/R) \tan \phi$ is zero or because the resultant force $\mathbf{F}_B$ is zero at the locations of θ shown in Fig. 11.28.

The dashed line of Fig. 11.29 represents the average torque T_{av} of one cylinder for one complete cycle. The work done on the mechanism by the gas force during the power stroke produces the average torque; without this work, the average torque would be zero, and the changes of torque would be due to the inertia forces alone. Although the output torque in the crankshaft is greatly variable as shown in Fig. 11.29, with a flywheel attached to the shaft, the torque delivered after the flywheel is very nearly constant and equal to the average torque T_{av}. Under steady-state operation at a given crank speed, the torque T_{av} is equal to the resisting load torque T_L which the engine with flywheel is driving.

Horsepower output for one cylinder may be determined from the average torque output and the speed of the crankshaft.

$$\text{hp} = \frac{T_{av}\omega}{550} = \frac{T_{av}}{550} \frac{2\pi n}{60} = \frac{T_{av} n}{5250} \tag{11.52}$$

in which T_{av} is in foot-pounds and n is in revolutions per minute. If friction is neglected in the torque analysis, the power given by Eq. 11.52 is very

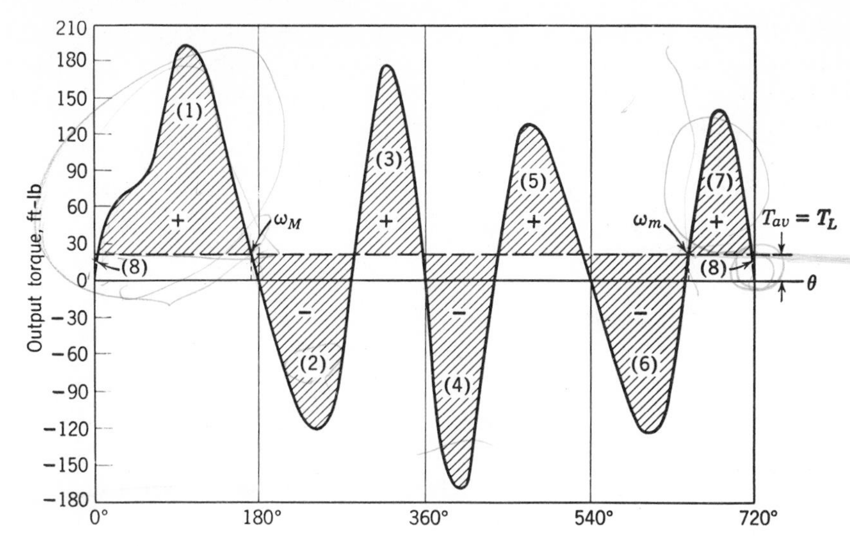

Figure 11.29

nearly equal to the indicated horsepower (ihp) as determined from the indicator diagram of gas pressure and stroke.

11.15 Flywheel Size. As shown in Fig. 11.29, the output torque of the slider-crank mechanism is greater than the load torque for some portions of the engine cycle and is less in other parts of the cycle. Since the curve of Fig. 11.29 is a plot of torque versus θ, the shaded area represents work which either increases or decreases the kinetic energy of the system by causing an increase or decrease in crankshaft speed. The degree to which crank speed is increased or decreased depends on the inertia of the system since kinetic energy involves both mass, or moment of inertia, and speed. Control of the crank-speed fluctuations is obtained primarily from a flywheel for which the moment of inertia may be calculated.

Figure 11.30 shows a single-cylinder engine with a flywheel. The free-body diagram of the flywheel shows the unbalance of torques acting on the flywheel to accelerate its angular motion. For output torque T of the slider-crank greater than the load torque T_L, the equation of motion may be written:

$$T - T_L = I\alpha \tag{11.53}$$

in which I is the moment of inertia of the flywheel about the crank axis, and α is in the sense of the resultant torque. Since $\alpha = (d\omega/dt)(d\theta/d\theta) = \omega(d\omega/d\theta)$, Eq. 11.53 may be rewritten:

$$T - T_L = I\omega\frac{d\omega}{d\theta}$$

$$(T - T_L)\,d\theta = I\omega\,d\omega$$

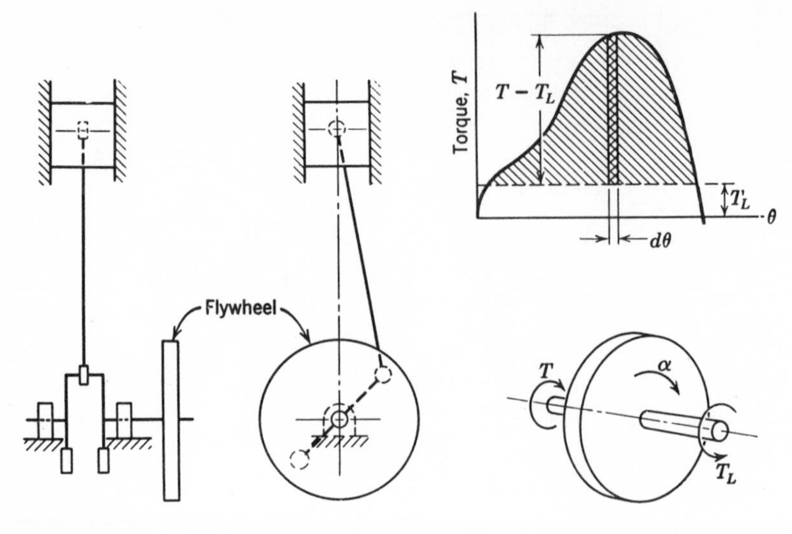

Figure 11.30

Integrating,

$$\int_{\theta \text{ at } \omega_m}^{\theta \text{ at } \omega_M} (T - T_L)\, d\theta = I \int_{\omega_m}^{\omega_M} \omega\, d\omega$$

$$= \tfrac{1}{2} I(\omega_M^2 - \omega_m^2) \qquad (11.54)$$

In Eq. 11.54, the left-hand term is the work done on the flywheel and is represented by the shaded area of the torque diagrams of Figs. 11.29 and 11.30; the right-hand term is the corresponding change in kinetic energy of the flywheel due to its change in speed. Positive shaded areas of the torque diagram in Fig. 11.29 represent regions in the engine cycle where work is done to increase flywheel speed, and negative areas, the work to decrease speed. Limits of θ on the integral of Eq. 11.54 are found by inspection so as to determine the greatest change in speed of the flywheel in the engine cycle where ω_M is the maximum and ω_m is the minimum angular velocity of the flywheel. The shaded loop in Fig. 11.29 having the greatest area would appear to represent the region of the greatest speed change. As shown, for a single-cylinder engine, the largest loop is in the power stroke, as would be expected because of the work done by the expanding gas to speed up the engine. Thus, ω_M corresponds with θ at the end of the first loop. However, ω_m is not at the beginning of the first loop (1), but rather at the beginning of the seventh loop (7) since this loop is also positive and is nearly adjacent to the first loop except for the small negative loop (8) between the positive areas.

The locations of maximum and minimum crank speeds ω_M and ω_m on the torque diagram are not always easily determined by inspection. In such cases, a systematic arithmetic method may be used. For example, in Fig. 11.31, a torque diagram is shown with positive and negative areas above and below the average torque line. The relative magnitudes of the areas are given for the loops. If at the beginning of the first loop the speed is the datum value ω_0, then the speed at the end of the first loop is greater than ω_0 because of the positive area $A_1 = 7$. At the end of the second loop, which is negative, the speed is lower than at the end of the first loop but greater than ω_0 because the algebraic sum of the first two areas is positive, $A_1 + A_2 = 7 - 4 = 3$. At the end of each loop in Fig. 11.31 is shown the sum of the areas from the beginning of the first loop $(A_1 + A_2 + \cdots + A_n)$. The sum of the areas of all of the loops must equal zero since the average torque line is established at such a position that the sum of positive areas above the average torque line equals the sum of the negative areas below the average torque line. The maximum value of the sums gives the location of ω_M, which is the maximum speed greater than ω_0. As shown in Fig. 11.31, the location of ω_M is at the end of the first loop where the greatest sum is +7. Similarly, the minimum value of the sums (−2 at the end of the fourth loop) gives the location of ω_m, which is the speed most below ω_0. The algebraic sum of the areas between the locations of ω_m and ω_M represents the work done by torque to change the flywheel kinetic energy from a minimum to a maximum value.

The integral expression of Eq. 11.54 may be represented by the area A.

$$A = \int_{\theta \text{ at } \omega_m}^{\theta \text{ at } \omega_M} (T - T_L)\, d\theta \tag{11.55}$$

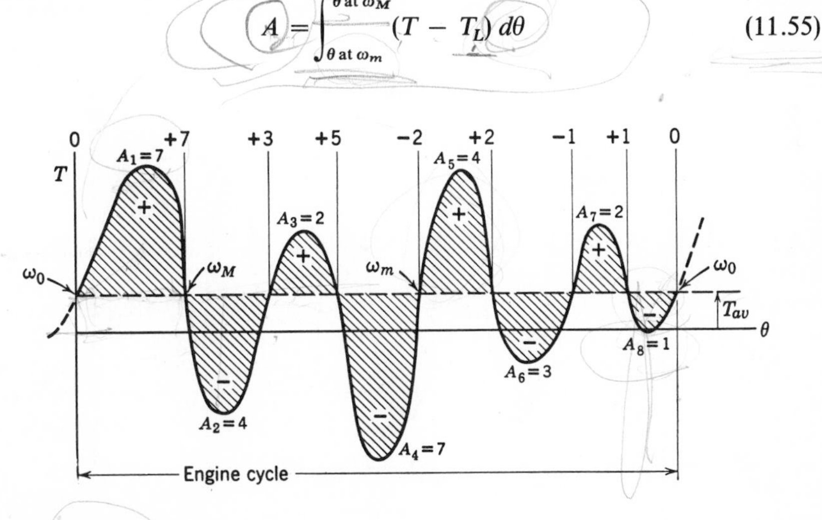

Figure 11.31

which is the algebraic summation of areas of loops in the cycle giving the greatest change in flywheel speed. The net positive area in square inches may be obtained by use of a planimeter and converted to foot-pounds of work A with the proper use of scales for the diagram. Substitution in Eq. 11.54 gives

$$A = \tfrac{1}{2}I(\omega_M^2 - \omega_m^2) \tag{11.56}$$

In multicylinder engines, the firing order and crank arrangement are such that the pulses of torque from the power strokes of the individual cylinders are uniformly distributed throughout the engine cycle of 720°. In a six-cylinder engine, for example, the cranks are spaced at 120° of crank angle (720/6) so that a power stroke begins every 120° of crankshaft rotation. The resultant torque curve, obtained by the superposition of the torque curves of the individual cylinders in proper phase relationship, is shown in Fig. 11.32. As shown, the loops of the torque curve are uniform in the sense that the loops are of the same form every 120°. The dashed line of average torque is located to make each positive loop equal in area to a negative loop. Therefore, in Eq. 11.55, A is determined from the area of any individual loop. The minimum speed ω_M is at the end of every negative loop, and ω_M is at the end of every positive loop.

In order to determine the required moment of inertia I of the flywheel, a coefficient of fluctuation K is assigned so that the fluctuation, or difference of maximum and minimum speeds, is a small fraction of the average design speed ω_{av} of the engine.

$$K = \frac{\omega_M - \omega_m}{\omega_{av}} \tag{11.57}$$

The average speed ω_{av} is

$$\omega_{av} = \frac{\omega_M + \omega_m}{2} \tag{11.58}$$

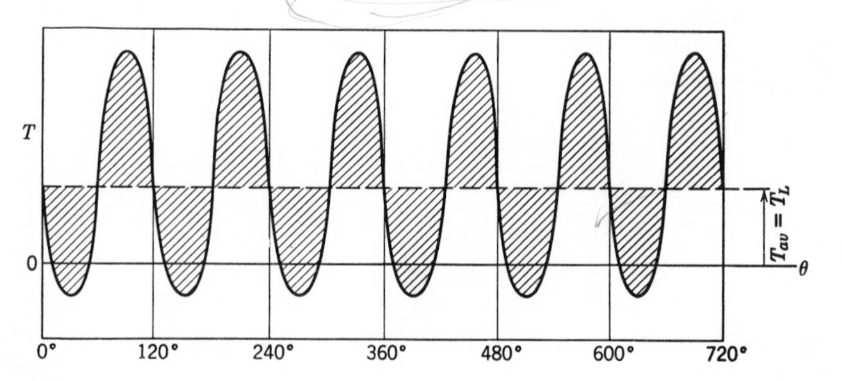

Figure 11.32

Equation 11.56 may be rewritten

$$A = \tfrac{1}{2}I(\omega_M + \omega_m)(\omega_M - \omega_m)$$
$$= \tfrac{1}{2}I(2\omega_{av})(K\omega_{av})$$
$$A = IK\omega_{av}^2$$
$$= IK\frac{4\pi^2 n^2}{(60)^2}$$

Solving for I gives

$$I = 91\frac{A}{Kn^2} \tag{11.59}$$

Equation 11.59 gives the required moment of inertia of the flywheel for the average speed n in revolutions per minute at which the torque analysis is made; the units of I are slug-feet^2 (lb-ft-sec^2) and those of A are foot-pounds.

Example 11.10

Determine the required moment of inertia of a flywheel for a single-cylinder engine for which Fig. 11.29 is the torque diagram at a speed of 3300 r/min. The maximum allowable fluctuation in speed in the engine cycle is 40 r/min.

Solution. From Fig. 11.29,

Area of first loop	=	+0.762 in.^{2}
Area of seventh loop	=	+0.265
		1.027
Negative area of eighth loop	=	0.007
		1.020 in.^{2}

In Fig. 11.29, the torque scale is 120 ft-lb/in. and the angular scale is $(\frac{8}{9})\pi$ rad/in.; therefore each square inch of the torque diagram represents 335 ft-lb of work.

$$A = 1.020(335) = 342 \text{ ft-lb}$$

$$K = \tfrac{40}{3300} = 0.01212$$

$$I = 91\frac{A}{Kn^2} = \frac{91(342)}{0.01212 \times (3300)^2}$$
$$= 0.236 \text{ slug-ft}^2$$

Flywheels are usually disks in automotive engines and rim type in steam engines or punch presses (Fig. 11.33). In automotive installations, a thin disk of large diameter gives the lowest weight for the moment of inertia required;

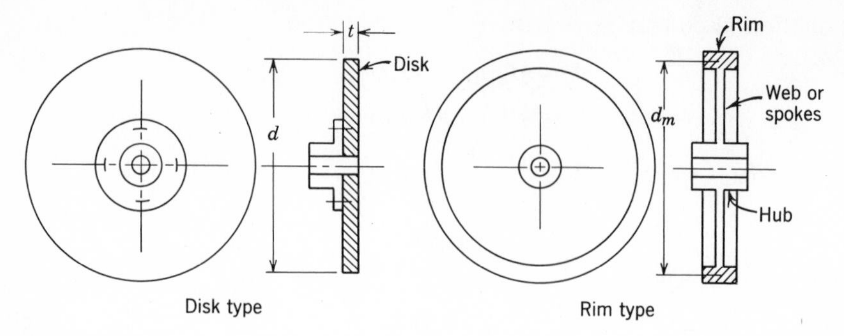

Figure 11.33

however, a compromise as to diameter must be made because of the stresses in the material caused by centrifugal force and because of space and road clearance requirements. Large flywheels of the rim type in steam engines and punch presses are limited in diameter primarily by allowable stresses due to centrifugal force.

For disk-type flywheels, the moment of inertia and weight W are related as follows:

$$I = \frac{Mr^2}{2} = \frac{Wd^2}{8g} \tag{11.60}$$

in which r and d are radius and diameter respectively of the disk. The diameter d in Eq. 11.60 is normally given in dimensions of feet.

Example 11.11

Determine the weight W and thickness t of a steel disk-type flywheel 12 in. in diameter to give the moment of inertia required in Example 11.10. The weight density of steel is $w = 490$ lb/ft³.

Solution.

$$I = \frac{Wd^2}{8g}$$

$$W = \frac{8gI}{d^2} = \frac{8 \times 32.2 \times 0.236}{1} = 61 \text{ lb}$$

$$W = (\text{vol of disk})\ w = \frac{\pi d^2}{4} tw$$

$$t = \frac{4W}{\pi d^2 w} = \frac{4 \times 61}{\pi (1)^2 490} = 0.158 \text{ ft}$$

$$= 1.90 \text{ in.}$$

It may be seen that a small speed fluctuation may be obtained with a reasonable weight of flywheel. However, as Eq. 11.59 indicates, a larger flywheel is required at low speeds, although A may be much less at low speeds.

For rim-type flywheels $I = Mk^2$, in which k is the radius of gyration. It is sufficiently accurate to assume that the mean radius r_m of the rim is equal to k.

$$I = Mr_m^2 = \frac{W}{4g}d_m^2 \tag{11.61}$$

The solution of Eq. 11.61 for W gives only the weight of the rim. The weights of the hub, web, or spokes also contribute a small amount to the moment of inertia of the flywheel with the result that the speed fluctuation is somewhat less than the assigned value.

Components of an engine installation other than the flywheel may contribute flywheel effect. The crankshaft and the equivalent mass of the connecting rod at the crank pin act as a flywheel. In automotive installations during road operation with clutch engaged, the rotating parts of the driving system as well as the car itself serve to reduce engine speed fluctuation to the degree that almost no flywheel is required. However, the idling condition at low speed in automotive engines is the prime condition for automotive flywheel design. Flywheel effect is important also in maintaining motor spin during starting. Aircraft piston engines are normally without a flywheel because of the large flywheel effect of propellers and propeller reduction gears. In the design of all reciprocating machinery such as diesel-electric systems, compressors, steam winches operated by donkey engines, quick-return mechanisms, motorcycles, outboard motors, and punch presses, flywheel effect is required, but the degree to which moment of inertia must be added in a flywheel depends on the requirements of the installation.

11.16 Forces on Gear Teeth. For gears in mesh, the line of transmission of force is along the line of action, which is always normal to the contacting tooth surfaces as the teeth traverse the arc of action. As shown in Fig. 11.34, the line of action of the tooth force $\mathbf{F}$ is at the pressure angle ϕ to the tangent of the pitch circles. The tooth of the driver shown in Fig. 11.34 is in contact with a tooth of the driven gear at the pitch point. In this position the teeth are in the state of pure rolling, and no friction due to relative sliding exists. At other positions in the arc of action, relative sliding exists and the resultant force on the gear tooth is inclined to the line of action by the angle of friction. In a force analysis of mechanisms with gears, the friction angle may be neglected with little error in the determination of the magnitude of tooth force.

If two sets of gear teeth are in contact, the transmitted force is divided

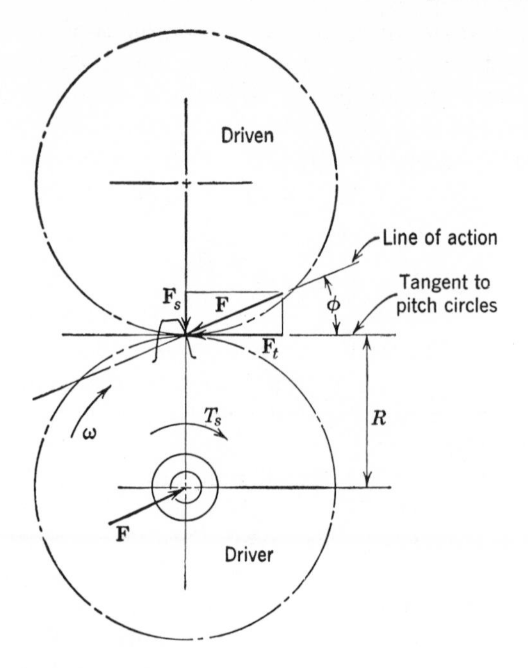

Figure 11.34

between the two sets of teeth. The free-body diagram of the driving gear, for example, would show two tooth forces, both of which act along the line of action. The resultant of the two forces, equal to the transmitted force, also acts along the line of action. The proportion of transmitted force carried by each tooth depends on the accuracy with which the gear teeth are in mesh, which in turn depends on the accuracy of manufacture of the tooth forms. Since one tooth is likely to carry more force than the other, it is usual to assume that one tooth carries the full transmitted force.

As shown in Fig. 11.34, the tooth force **F** is represented by the components $\mathbf{F}_t$ and $\mathbf{F}_s$, of which $\mathbf{F}_t$ is called the tangential force and $\mathbf{F}_s$ the separating force. $F_t = F \cos \phi$ and $F_s = F \sin \phi$. In many problems, the shaft torque T_s applied to the driving gear is known and F_t may be determined from the equilibrium of moments about the shaft axis

$$T_s = F_t R \tag{11.62}$$

in which R is the pitch radius of the gear. The transmitted force **F** may be determined from

$$F = \frac{F_t}{\cos \phi} = \frac{T_s}{R \cos \phi} \tag{11.63}$$

Equation 11.63 shows that, for a given torque applied to the gear, the tooth force **F** increases with pressure angle. The separating force $\mathbf{F}_s$ also increases with pressure angle. It may be seen that $\mathbf{F}_t$ acts to shear and bend the tooth and that $\mathbf{F}_s$ acts to compress the tooth. The transmitted force **F** causes high local stresses in the material in the vicinity of contact on the tooth face.

Example 11.12

It is desired to determine the tooth forces acting on the several gears of the planetary gear train shown in Fig. 11.35*a*. One hundred horsepower is

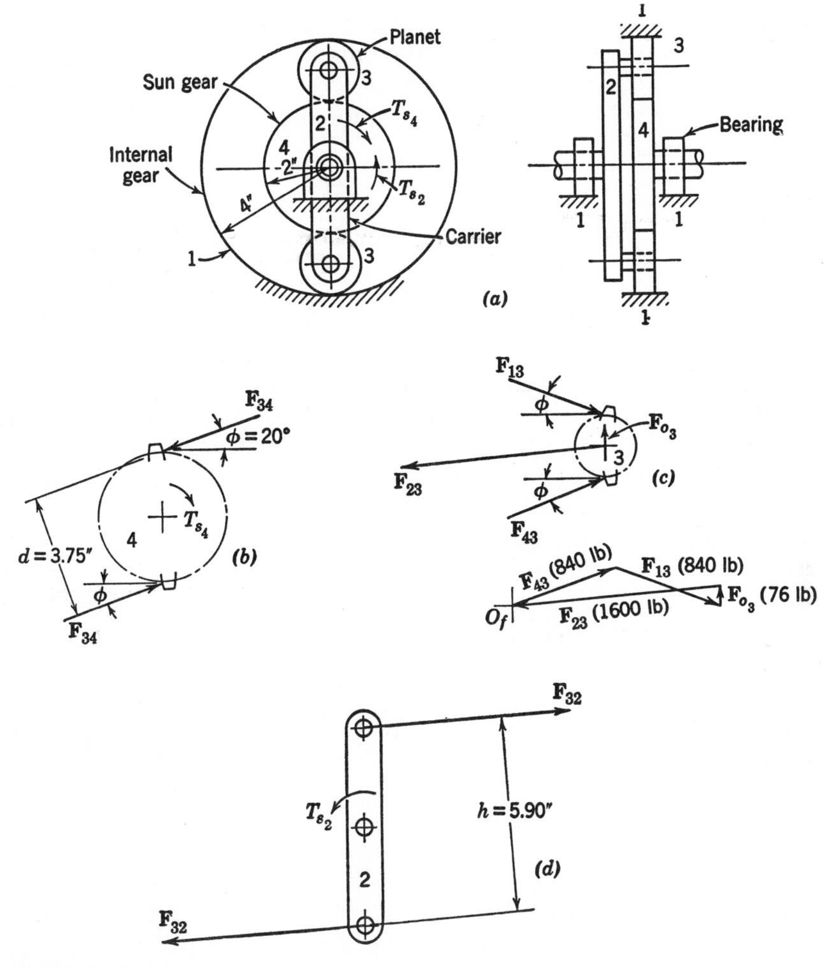

Figure 11.35

transmitted by the gear train at constant speed. The sun gear (link 4) rotating clockwise at $n_4 = 2000$ r/min is the input side of the train, and the carrier (link 2) rotating clockwise at $n_2 = 667$ r/min is the output side. The shaft torque T_{s4} acting on the sun gear is the driving torque, and T_{s2} acting on the carrier is the resisting shaft torque of the load. Spur gears with involute shaped teeth at 20° pressure angle are used in the gear train.

Solution. Carefully drawn free-body diagrams of the individual links as shown in Fig. 11.35 aid in the determination of the forces acting on each link for static equilibrium. Inertia forces are zero for the sun gear and carrier as well as for the internal gear since the accelerations of the mass centers of these elements are zero; the inertia torques are also zero since the train operates at constant angular velocity and zero angular acceleration. Insofar as the planets are concerned, centrifugal inertia forces act because of the centripetal acceleration of the centers of the planets. Assuming a planet weight of the order of 2 lb, the centrifugal force acting is

$$F_{o3} = M_3 A_{g3} = \frac{2}{32.2}\frac{3}{12}\left(\frac{2\pi 667}{60}\right)^2$$

$$= 76 \text{ lb}$$

The free-body diagram of the sun gear in Fig. 11.35*b* shows the driving torque T_{s4} and two tooth forces $\mathbf{F}_{34}$, which are shown along the line of action for 20° pressure angle. Since the power transmitted and the speed of the gear are known, T_{s4} may be calculated as follows:

$$\text{hp} = \frac{T_{s4}\omega_4}{550}$$

$$T_{s4} = \frac{550 \text{ hp}}{\omega_4} = \frac{550 \text{ hp}}{2\pi n_4/60}$$

$$= \frac{550(100)}{2\pi 2000/60}$$

$$= 262 \text{ ft-lb} \quad \text{(cw)}$$

Since two couples in equilibrium act on link 4,

$$F_{34}d = T_{s4}$$

$$F_{34} = \frac{T_{s4}}{d} = \frac{262}{3.75/12} = 840 \text{ lb}$$

The free-body diagram of the planet in Fig. 11.35*c* shows that four forces act on the link, of which the forces $\mathbf{F}_{43}$ and $\mathbf{F}_{o3}$ are known. The direction and sense of the tooth force $\mathbf{F}_{13}$ may be ascertained by considering the moment equilibrium of forces acting about the planet center. If $\mathbf{F}_{13}$ also acts at the pressure angle of 20°, then $\mathbf{F}_{13}$ and $\mathbf{F}_{43}$ are equal in magnitude to satisfy equilibrium of moments about the planet center. The force $\mathbf{F}_{23}$ of the carrier acting on the planet is the remaining unknown, which may be determined from the force polygon in Fig. 11.35*c* or analytically as follows:

The equation of equilibrium can be written from the free-body diagram of the planet as

$$\mathbf{F}_{13} + \mathbf{F}_{o_3} + \mathbf{F}_{43} + \mathbf{F}_{23} = 0$$

where

$$\begin{aligned}\mathbf{F}_{13} &= F_{13}(\cos 20°\mathbf{i} - \sin 20°\mathbf{j})\\ &= 0.9397F_{13}\mathbf{i} - 0.3420F_{13}\mathbf{j}\\ \mathbf{F}_{o_3} &= 76.0\mathbf{j}\\ \mathbf{F}_{43} &= F_{43}(\cos 20°\mathbf{i} + \sin 20°\mathbf{j})\\ &= 840 \times 0.9397\mathbf{i} + 840 \times 0.3420\mathbf{j}\\ &= 789.35\mathbf{i} + 287.28\mathbf{j}\end{aligned}$$

But,

$$|\mathbf{F}_{13}| = |\mathbf{F}_{43}|$$

Therefore,

$$\mathbf{F}_{13} = 789.35\mathbf{i} - 287.28\mathbf{j}$$

Also,

$$\mathbf{F}_{23} = F_{23}(\lambda_x\mathbf{i} + \lambda_y\mathbf{j}) \text{ (direction unknown)}$$

Substituting in the equation of equilibrium

$$789.35\mathbf{i} - 287.28\mathbf{j} + 76.0\mathbf{j} + 789.35\mathbf{i} + 287.28\mathbf{j} + \lambda_x F_{23}\mathbf{i} + \lambda_y F_{23}\mathbf{j} = 0$$

Summing **i** *components,*

$$\begin{aligned}789.35\mathbf{i} + 789.35\mathbf{i} + \lambda_x F_{23}\mathbf{i} &= 0\\ \lambda_x F_{23}\mathbf{i} &= -1578.70\mathbf{i}\end{aligned}$$

Summing **j** *components,*

$$\begin{aligned}76.0\mathbf{j} + \lambda_y F_{23}\mathbf{j} &= 0\\ \lambda_y F_{23}\mathbf{j} &= -76.0\mathbf{j}\end{aligned}$$

Therefore,

$$\begin{aligned}\mathbf{F}_{23} &= -1578.70\mathbf{i} - 76.0\mathbf{j}\\ |\mathbf{F}_{23}| &= 1580.53 \text{ lb}\end{aligned}$$

From the free-body diagram of the carrier in Fig. 11.35*d*, the shaft torque

T_{s_2} may be determined from equilibrium of moments about the carrier axis.

$$T_{s_2} = F_{32}h$$
$$= 1600 \times \frac{5.9}{12}$$
$$= 787 \text{ ft-lb (ccw)}$$

T_{s_2} may also be determined from the transmitted horsepower and the speed of the carrier.

$$\text{hp} = \frac{T_{s_2}\omega_2}{550}$$

$$T_{s_2} = \frac{550 \text{ hp}}{\omega_2} = \frac{550 \text{ hp}}{2\pi n_2/60}$$

$$T_{s_2} = \frac{550(100)}{2\pi 667/60}$$
$$= 787 \text{ ft-lb (ccw)}$$

The foregoing solution indicates that the tooth force to be expected is 840 lb. However, a subtlety exists in interpreting the effect of the inertia force $\mathbf{F}_{o_3}$ on forces. In the solution above, it was assumed that the carrier constrained the planet center to remain at the meshing pressure angle of 20° at the pitch point by providing the reaction to the planet inertia force $\mathbf{F}_{o_3}$. However, if a large clearance exists for the pin connecting the planet to the carrier, then the carrier cannot provide the reaction to $\mathbf{F}_{o_3}$, in which case the planet will displace slightly toward the internal gear in such a way that it meshes with the internal gear at a pressure angle slightly less than 20° and with the sun gear at a pressure angle slightly more than 20°. The result is that $\mathbf{F}_{43}$ will be somewhat greater than 840 lb and $\mathbf{F}_{13}$ somewhat less.

An interesting extension of this example is to consider the effect of replacing the spur gears in the planetary drive of Fig. 11.35 with helical gears while keeping the center distances, gear ratios, and power transmitted the same. If the helical gears have a normal pressure angle ϕ_n of 20° and a helix angle ψ of 30°, the following relations can be developed for planet 3 assuming that the normal force in the plane of rotation $\mathbf{F}_{13}$ remains at 840 lb. Referring to Fig. 11.36,

$$\tan\phi = \frac{\tan\phi_n}{\cos\psi} = \frac{\tan 20}{\cos 30}$$

$$\phi = 22.8°$$

$$\mathbf{F}_{13} = F_{13}(\cos\phi\mathbf{i} - \sin\phi\mathbf{j})$$
$$= 840(\cos 22.8°\mathbf{i} - \sin 22.8°\mathbf{j})$$
$$= 774.39\mathbf{i} - 325.45\mathbf{j}$$

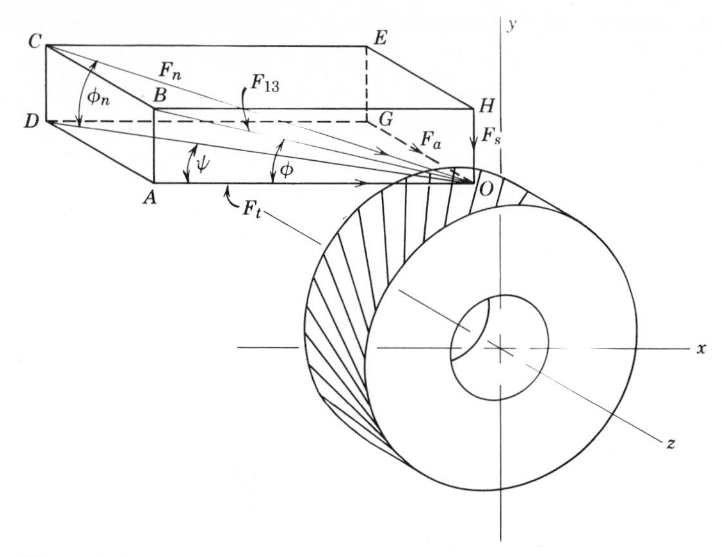

Figure 11.36

Therefore,

$$\mathbf{F}_t = 774.39\mathbf{i}$$

$$\mathbf{F}_s = -325.45\mathbf{j}$$

and

$$|\mathbf{F}_a| = F_t \tan \psi = 774.39 \tan 30°$$

$$= 447.09 \text{ lb}$$

$$\mathbf{F}_a = 447.09\mathbf{k}$$

Therefore,

$$\mathbf{F}_n = 774.39\mathbf{i} - 325.45\mathbf{j} + 447.09\mathbf{k}$$

$$|\mathbf{F}_n| = 951.57 \text{ lb}$$

It is left to the reader to determine how this change will affect the bearing force $\mathbf{F}_{23}$.

11.17 Cam Forces. At high cam speeds, the force transmitted at the contact of cam and follower is high and may cause serious wear of the contacting surfaces. Figure 11.37 shows a disk cam with radial roller follower. Two phases of the cam are shown as it rotates counterclockwise at a uniform speed n_c of 8550 r/min. In Fig. 11.37*a*, the cam is in such a phase that the acceleration $\mathbf{A}_f$ of the follower is away from the cam. In this phase, the inertia force $\mathbf{F}_f$ of the follower is such that even without the force $\mathbf{S}$ of the compressed spring, the follower is held in contact with the cam. However, in Fig. 11.37*b*, the phase of the cam is such that a high downward acceleration $\mathbf{A}_f$ of the follower is present. In this instance the follower inertia force $\mathbf{F}_f$ is great enough to cause the follower to leave the cam unless a force **S** equal to $\mathbf{F}_f$ is applied by the spring. Assuming that the weight of the follower including roller, stem, and spring is 1 lb, then

$$F_f = M_f A_f = \frac{1}{32.2}(101{,}000) = 3140 \text{ lb}$$

and the required spring force is $S = 3140$ lb. Since **S** and $\mathbf{F}_f$ are equal forces but opposite in sense, the force **F** acting at the contacting surfaces of cam and roller is zero. Since the spring is compressed $\delta = \frac{1}{2}$ in. corresponding to the lift of the cam, the required spring constant k is

$$k = \frac{S}{\delta} = \frac{3140}{\frac{1}{2}} = 6280 \text{ lb/in.}$$

Returning to the phase of the cam shown in Fig. 11.37*a*, it may be observed that the force **F** at the contacting surfaces will be high because it represents a reaction to both the inertia force $\mathbf{F}_f$ and the spring force **S** as shown in the free-body diagram of the follower in Fig. 11.37*c*. **F** is normal to the surfaces of contact, and the angle which **F** makes with the direction of motion of the follower is the pressure angle ϕ. For static equilibrium, the summation of forces in the direction of follower displacement is zero.

$$F \cos \phi - F_f - S = 0$$

$$F = \frac{F_f + S}{\cos \phi} \tag{11.64}$$

Inertia force $\mathbf{F}_f$ is

$$F_f = M_f A_f = \frac{1}{32.2}(111{,}000) = 3450 \text{ lb}$$

The spring is compressed $\frac{1}{8}$ in.

$$S = k\delta = 6280 \times \tfrac{1}{8} = 785 \text{ lb}$$

The pressure angle $\phi = 25°$ ($\cos 25° = 0.907$).

From Eq. 11.64,

$$F = \frac{3450 + 785}{0.907} = 4670 \text{ lb}$$

A surface force $F = 4670$ lb and spring constant $k = 6280$ lb/in. are high for a cam of the size shown. A stress analysis would show that the speed

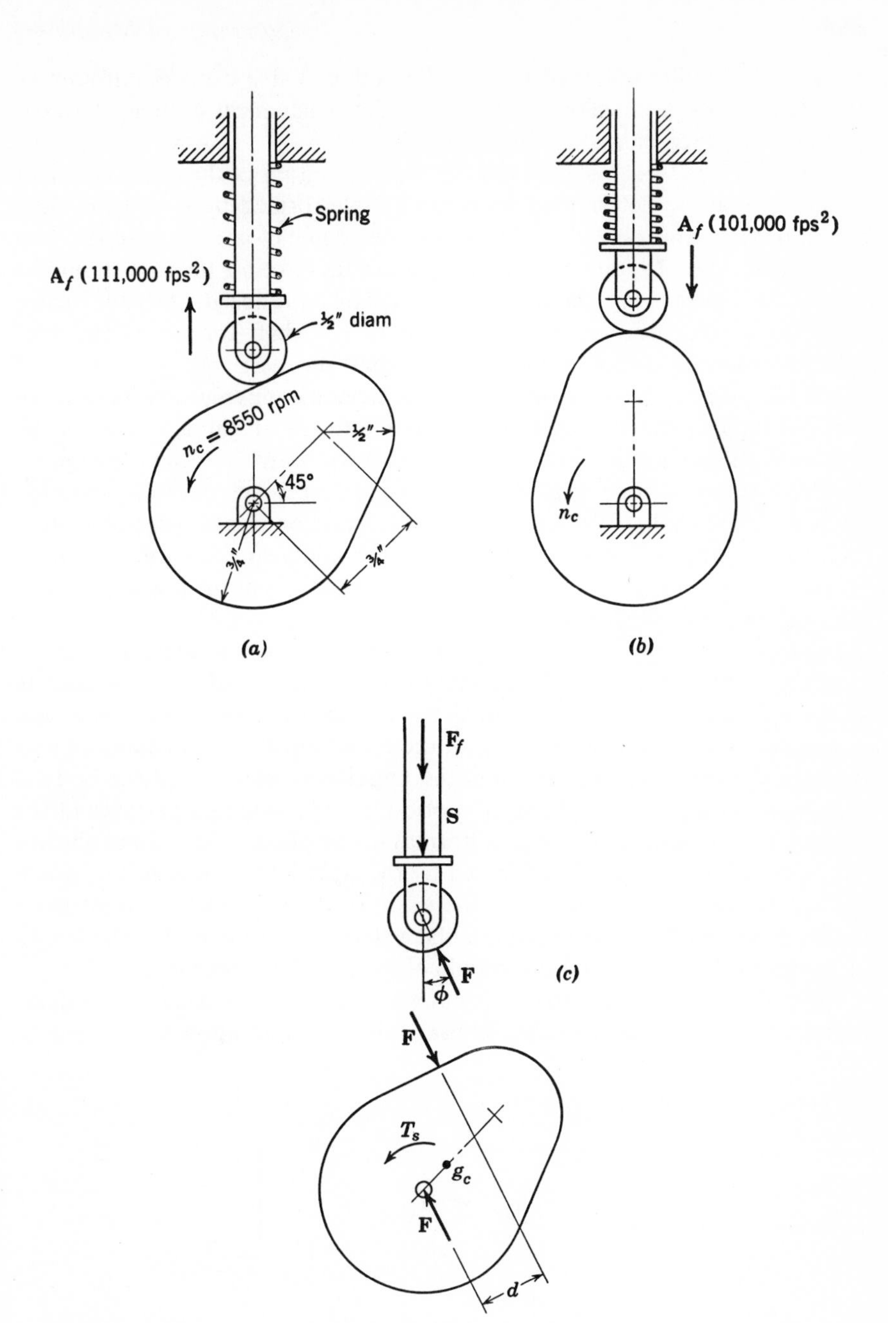
Spring
A_f (111,000 fps²)
½" diam
n_c = 8550 rpm
½"
45°
¾"
¾"
(a)
A_f (101,000 fps²)
n_c
(b)
F_f
S
F
φ
(c)
F
T_s
g_c
F
d

Figure 11.37

of the cam should be limited to a smaller value. Valve cams in automotive installations are a constant challenge to the mechanical designer because of the demand for increased engine speeds.

In Fig. 11.37*c* is also shown the free-body diagram of the cam. The shaft torque T_s may be determined from the couple formed by **F** and the shaft reaction equal and opposite to **F**. Since the center of gravity g_c of the cam is not at the axis of rotation, a centrifugal inertia force for the cam should be shown. However, since the cam rotates at constant speed, the cam inertia force does not influence the calculation of the shaft torque T_s but does enter into the calculation of the resultant shaft reaction.

11.18 Gyroscopic Forces. In vehicles having engines with rotating parts of high moment of inertia, gyroscopic forces are in action when the vehicle is changing direction of motion. In automotive vehicles undergoing roadway turns at high velocity, gyroscopic forces act on such spinning parts as crankshaft, flywheel, clutch, transmission gears, propeller shaft and wheels. Engine parts as well as propeller and gear reduction system of an airplane are under the action of gyroscopic effects in turns and pullouts. Locomotives and ships are similarly affected.

Figure 11.38 shows a rigid body spinning at a constant angular velocity ω about a spin axis through the mass center. The angular momentum **H** of the spinning body is represented by a vector whose magnitude is $I\omega$, in which I is the moment of inertia of the body about the spin axis and axis through the mass center. Although the angular momentum of the body is in a plane parallel to the planes of motion of the individual particles of the body, it is represented by a vector normal to the plane of motion as shown. The sense of the vector is determined by the right-hand screw rule in which the arrowhead of the vector is in the sense of the advance of a right-hand screw turned in the sense of the angular velocity ω of the body. The length of the vector represents the magnitude of the angular momentum.

From the study of mechanics it is known that the rate of change of angular momentum with respect to time is proportional to an applied torque T as

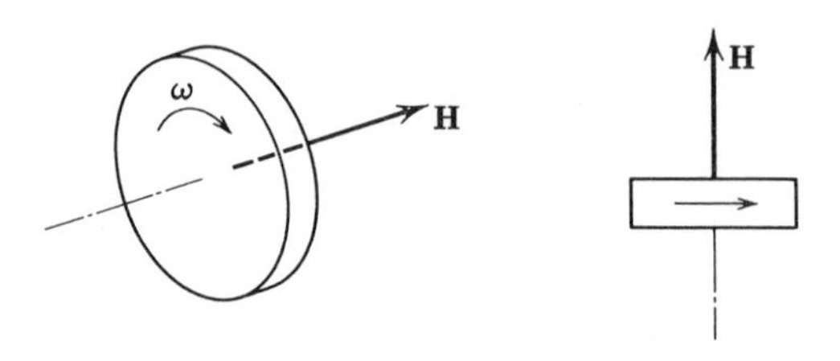

Figure 11.38

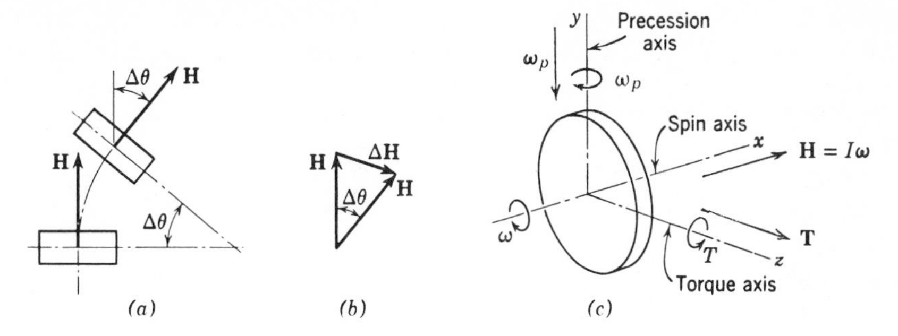

Figure 11.39

determined from the following equation of motion.

$$T = I\alpha = I\frac{d\omega}{dt} = \frac{d}{dt}(I\omega)$$

Also

$$H = I\omega$$

Therefore,

$$T = \frac{dH}{dt} \tag{11.65}$$

In the case shown in Fig. 11.38, a torque applied in the plane of motion of the spinning body in the sense of ω increases the angular momentum at a given rate which may be shown as an increase in the length of the vector.

In the foregoing discussion the spin axis was considered fixed. If the spin axis is made to change angular position as in a vehicle traversing a plane curved path as shown in Fig. 11.39*a* gyroscopic action results. For constant ω, the magnitude of the angular momentum remains constant for an angular displacement $\Delta\theta$ of the spin axis as shown by the vectors. However, a change in angular momentum exists because of the change in direction of the momentum as shown by the polygon of free vectors in Fig. 11.39*b*. For a small value of $\Delta\theta$, the magnitude of the change in angular momentum $\mathbf{\Delta H}$ is

$$\Delta H = (I\omega)\,\Delta\theta$$

The rate of change of angular momentum with respect to time is

$$\frac{dH}{dt} = \lim_{\Delta t \to 0} \frac{\Delta H}{\Delta t} = \lim_{\Delta t \to 0} (I\omega)\frac{\Delta\theta}{\Delta t} = I\omega\frac{d\theta}{dt}$$

Therefore,

$$\frac{dH}{dt} = I\omega\omega_p$$

and

$$T = I\omega\omega_p \tag{11.66}$$

in which $\omega_p = d\theta/dt$ is the angular velocity of precession of the spin axis, or the rate at which the spin axis displaces angularly.

The magnitude of the torque **T** which is associated with the precession of the spin axis can easily be determined from Eq. 11.66, and it is now necessary to determine its direction. Referring again to Fig. 11.39*b*, one can see that as Δt approaches zero, the vector $\Delta\mathbf{H}$ becomes normal to the vector **H** which has the same direction as the spin axis. Therefore, the torque vector **T** will also be normal to the spin axis in the plane of **H** and $\Delta\mathbf{H}$. Figure 11.39*c* shows the x axis as the spin axis and the y axis as the precession axis. The z axis becomes the torque axis because of the fact that the direction of torque **T** is normal to the spin axis and lies in the xz plane. From the orientation of the spin, precession, and torque axes, it can be seen that Eq. 11.66 can be written in the following vector form:

$$\mathbf{T} = \boldsymbol{\omega}_p \times I\boldsymbol{\omega} \tag{11.67}$$

The applied torque **T** in Eq. 11.67 is a couple referred to as the *gyroscopic couple*. Because this couple has the same direction as $\Delta\mathbf{H}$, the couple lies in the xy plane and represents a torque applied to the body about the z or torque axis. Thus, it may be seen that to cause precession of a spinning body a torque must be applied to the body in a plane normal to the plane in which the spin axis is precessing.

The flywheel of an automotive engine is an example of a spinning body which is subject to the gyroscopic couple in roadway turns at high vehicular speed. As shown in Fig. 11.40, the flywheel of a single-cylinder engine is

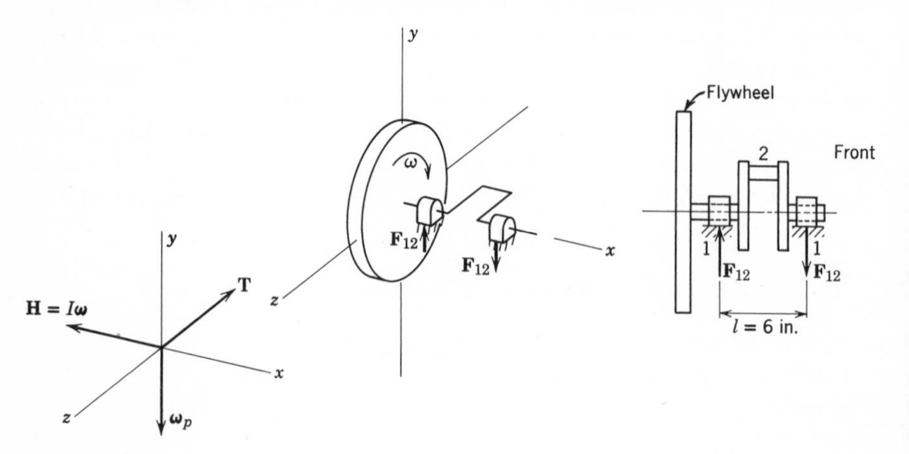

Figure 11.40

fixed to the crankshaft, which in turn is supported by the main bearings. The crankshaft and the equivalent mass of the connecting rod are also spinning masses which may be considered as part of the flywheel. Forces $\mathbf{F}_{12}$, which represent the gyroscopic couple, are applied to the crankshaft by the bearings. These forces are superposed on the forces produced by the operation of the slider-crank mechanism. Other bearing forces which are induced by the turning vehicular motion are those resulting from centrifugal force while the vehicle is in the curved path.

Example 11.13

For the single-cylinder engine of Fig. 11.40, determine the bearing forces $\mathbf{F}_{12}$ caused by the gyroscopic action of the flywheel of Example 11.10 as the engine vehicle traverses a 1000-ft-radius curve at 60 mi/h (88 ft/sec) in a turn to the right. The engine speed is 3300 r/min and is turning clockwise viewed from the front of the engine.

Solution.

$$I \text{ of flywheel} = 0.236 \text{ slug-ft}^2$$

$$\omega = \frac{2\pi n}{60} = \frac{2\pi 3300}{60} = 346 \text{ rad/sec}$$

$$\omega_p = \frac{V}{R} = \frac{88}{1000} = 0.088 \text{ rad/sec}$$

$$\mathbf{T} = \boldsymbol{\omega}_p \times I\boldsymbol{\omega}$$

where

$$\boldsymbol{\omega}_p = -0.088\mathbf{j}$$

$$I\boldsymbol{\omega} = (0.236)(-346\mathbf{i}) = -81.66\mathbf{i}$$

Therefore,

$$\mathbf{T} = (-0.088\mathbf{j}) \times (-81.66\mathbf{i})$$

$$= -7.18\mathbf{k} \text{ ft-lb (cw)}$$

and

$$F_{12} = \frac{T}{l} = \frac{7.18}{\frac{6}{12}} = 14.4 \text{ lb}$$

With the direction of torque $\mathbf{T}$ clockwise, the sense of force $\mathbf{F}_{12}$ applied to the crankshaft by the bearing will be up at the rear bearing and down at the front bearing as shown in Fig. 11.40.

As may be seen from Example 11.13, the gyroscopic forces on the bearings are small compared to those due to slider-crank action. These forces are greater in sharper roadway turns. Gyroscopic forces on bearings supporting clutch, transmission gears, and drive shaft are also small because of the low moment of inertia of the parts. However, the moment of inertia of the front wheel may be great enough to apply appreciable gyroscopic forces to the ball joints and the steering mechanism. The gyroscopic forces of the spinning parts of the engine that are transmitted to the car body have the effect of raising the front end of the car on its suspension as it traverses a curve to the right.

The magnitudes of gyroscopic forces of heavy flywheels of shipboard engines (several tons in some cases) may be of large magnitude. Metal airplane propellers of large diameter cause high gyroscopic couples in some maneuvers as shown in the following example.

Example 11.14

Determine the gyroscopic couple of a 10-ft-diameter solid aluminum alloy four-bladed propeller in which each blade weighs 40 lb. The test maneuver of the airplane is a power-on flat spin in which the propeller speed is 1500 r/min, and the rotation of the flat spin is 1 rad/sec. The radius of gyration k of the propeller with respect to the propeller axis is approximately one-half of the propeller radius.

Solution.

$$k = \tfrac{1}{2}(5) = 2.5 \text{ ft}$$

$$I = Mk^2 = \frac{4(40)}{32.2}(2.5)^2$$

$$= 31.0 \text{ slug-ft}^2$$

$$\omega = \frac{2\pi n}{60} = \frac{2\pi(1500)}{60} = 157 \text{ rad/sec}$$

$$\omega_p = 1 \text{ rad/sec}$$

$$T = 31(157)(1)$$

$$= 4860 \text{ ft-lb}$$

The effect of the couple is to impose a large load on bearings supporting the propeller shaft as well as to impose large bending moments on the individual blades near the propeller hub. The gyroscopic effect is great enough to affect the maneuver by raising or lowering the nose of the airplane depending on the senses of the propeller spin and the precession.

It is a characteristic of the gyroscope that a gyroscopic couple must be applied in order to cause precession. In many instrument applications such as the gyrocompass and artificial horizon used in aircraft, precession is undesirable and great care is taken to reduce the gyroscopic couple to a minimum as the vehicle in which it is mounted undergoes turns that would cause precession. The resistance of a gyroscope to precession is greater as $I\omega$ is greater; a high moment of inertia and high spin velocity give it the characteristic of "rigidity" against precessing in space. Rigidity is the desired characteristic in the gyrocompass, which provides a fixed datum required for navigational purposes. Although the gyroscope is mounted in low-friction bearings in such a manner that the vehicle's turning transmits a minimum of gyroscopic couple, some torque is nevertheless applied by friction and the gyroscope must periodically be reset to the desired datum position. The rigidity characteristic in gyroscopes is also utilized in control equipment. In naval gun directors, the gyroscope provides a datum during pitching and rolling of the ship and an electrical signal may be transmitted to machinery which holds gun positions relative to the gyroscope rather than relative to the ship. Gyroscopes in automatic pilots control the flight position of aircraft by transmitting signals to the control surfaces as wind currents and other disturbances cause the aircraft to yaw, pitch, and roll.

11.19 Moment of Inertia Determination. In the foregoing discussions of force analyses, the moments of inertia I of the individual links were known or assumed. The designer or analyst of a machine is often confronted with the need for determining moment of inertia. Formulas are available in handbooks and textbooks on mechanics for the determination of moment of inertia of bodies having simple geometric form such as cylinders, disks, and bars and tubes of round and rectangular cross section. Many machine elements such as gears, pulleys, flywheels, gyroscopes, rotors, and shafts are simple enough in form that determination of moments of inertia by formula is quite accurate. Although calculations of I for links of more complex forms, such as connecting rods, crankshafts, planet carriers, and odd-shaped cams, may be made by considering the complicated forms as composities of simpler forms, the determinations are less accurate. If parts are available, moments of inertia may be determined experimentally in most cases. One of the most useful experimental methods is to mount the part in such a way that it may oscillate as a pendulum and to observe the period of oscillation, which is a function of the moment of inertia of the pendulum.

Figure 11.41 shows a pendulum suspended from the knife-edge at O so that O is the axis of rotation about which the pendulum oscillates from θ_1 to $-\theta_1$. The mass center g is at a distance l_O from O. Two forces act on the pendulum: the force of gravity W, and the supporting force of the knife-edge. The following equation of motion is written using the moment center O:

$$\Sigma T_O = I_O \alpha$$

$$-W l_O \sin\theta = I_O \alpha = I_O \frac{d^2\theta}{dt^2} \tag{11.68}$$

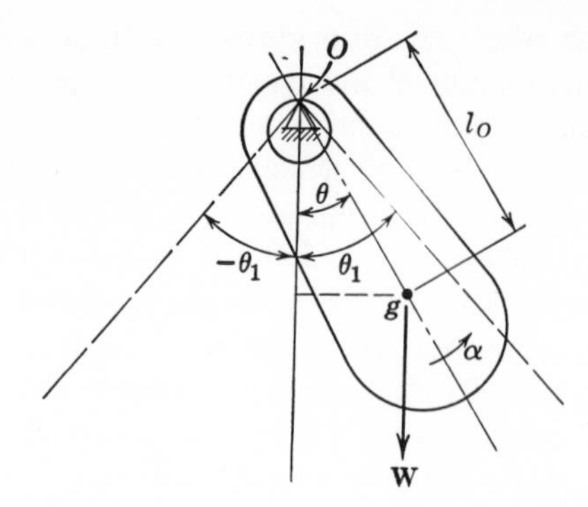

Figure 11.41

I_o is the moment of inertia of the pendulum about the axis through O. The moment T_o depends upon the position θ of the pendulum from vertical. Since α is in the same sense as increasing values of θ, the minus sign of Eq. 11.68 indicates that T_o is in the opposite sense to α. For small oscillations of the pendulum, $\theta = \sin\theta$ may be assumed with little error. Thus

$$\frac{d^2\theta}{dt^2} = -\frac{Wl_o}{I_o}\theta \tag{11.69}$$

Equation 11.69 is a differential equation, which on double integration yields an equation relating time t and θ.

$$t = \sqrt{(I_o/Wl_o)}\cos^{-1}\left(\frac{\theta}{\theta_1}\right) \tag{11.70}$$

The two constants of integration are evaluated for the conditions: $\omega = d\theta/dt = 0$ at $t = 0$, and $\theta = \theta_1$ at $t = 0$. Since the time of the oscillation is measured from the starting position $\theta = \theta_1$, the time to reach the vertical position is determined by substituting $\theta = 0$.

$$t = \frac{\pi}{2}\sqrt{I_o/Wl_o} \tag{11.71}$$

The period τ of the pendulum, or the time for one complete oscillation is four times the time given by Eq. 11.71.

$$\tau = 2\pi\sqrt{I_o/Wl_o}$$

$$I_o = \frac{\tau^2}{4\pi^2}Wl_o \tag{11.72}$$

I_o is the moment of inertia about the axis through O. Usually, the moment of inertia I about the axis through the mass center is wanted and may be determined from the parallel axis theorem.

$$I_O = I + Ml_O^2$$

$$I = I_O - Ml_O^2$$

$$= \frac{\tau^2}{4\pi^2} Wl_O - \frac{W}{g} l_O^2$$

$$I = Wl_O\left(\frac{\tau^2}{4\pi^2} - \frac{l_O}{g}\right) \tag{11.73}$$

Thus I may be determined from Eq. 11.73 by experimentally noting the time for a large number of oscillations of a part suspended as a pendulum. A connecting rod, for example, may be suspended on a knife-edge from either the wrist-pin bore or the crank-pin bore. The quantity in parentheses in Eq. 11.73 approaches zero as l_O becomes large because the two terms are nearly equal in magnitude. Under these conditions, the accuracy of determining I depends on the accuracy of measuring both l_O and τ. Accuracy is greatly increased by making l_O a small measurable value other than zero. Thus, accuracy is better for the case in which the connecting rod is suspended from the end closest to the center of gravity. It should be mentioned, however, that it is often difficult to obtain an accurate time of oscillation if the point about which the body is swung is too close to the center of gravity.

The moment of inertia of a part may also be determined experimentally by mounting the part on a pendulum made of a lightweight platform suspended by chords as shown in Fig. 11.42. To determine the moment of inertia of the part about the centroidal axis g–g, the part is oriented such that g–g is directly below and parallel to the suspension axis O–O. The period for small oscillations is observed by counting oscillations for a time of several minutes. The following equation, which determines I of the part,

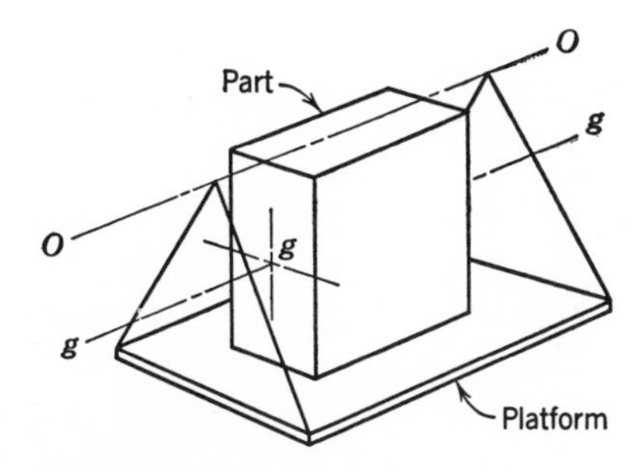

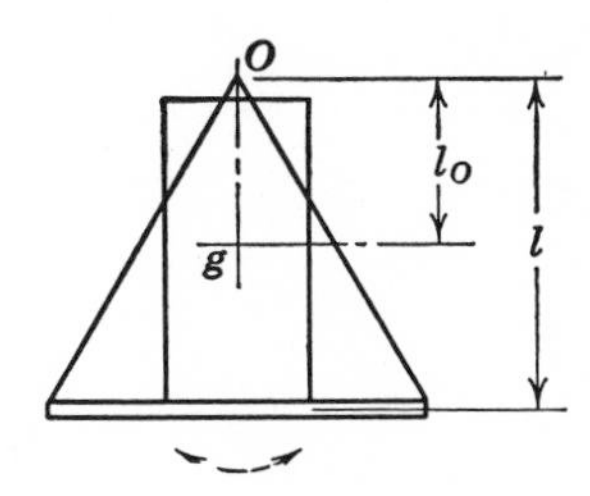

Figure 11.42

is a modification of Eq. 11.73, in which the second term accounts for the effect of the platform. τ_b represents the period of the platform without the part.

$$I = Wl_O\left(\frac{\tau^2}{4\pi^2} - \frac{l_O}{g}\right) + \frac{W_b l}{4\pi^2}(\tau^2 - \tau_b^2) \tag{11.74}$$

where

W = weight of part
W_b = weight of platform
l_O = distance from O to center of gravity of part
l = distance from O to center of gravity of platform
τ = period of platform with part
τ_b = period of platform alone

To allow for accuracy in determining I when using Eq. 11.74, the length of the suspension should be such that l_O is as small as possible but accurately measurable.

A third method for determining I is to orient the part on an equilateral triangular platform suspended as shown in Fig. 11.43 and to observe the period of the apparatus as a torsional pendulum oscillating about axis g–g. To determine the moment of inertia of the part about the centroidal axis g–g, the part is oriented such that g–g is parallel to the three vertical suspension chords and is equidistant r from the three chords.

The moment of inertia I of the part is determined from the following equation in which the second term accounts for the effect of the platform.

$$I = \frac{Wr^2\tau^2}{4\pi^2 l} + \frac{W_b r^2}{4\pi^2 l}(\tau^2 - \tau_b^2) \tag{11.75}$$

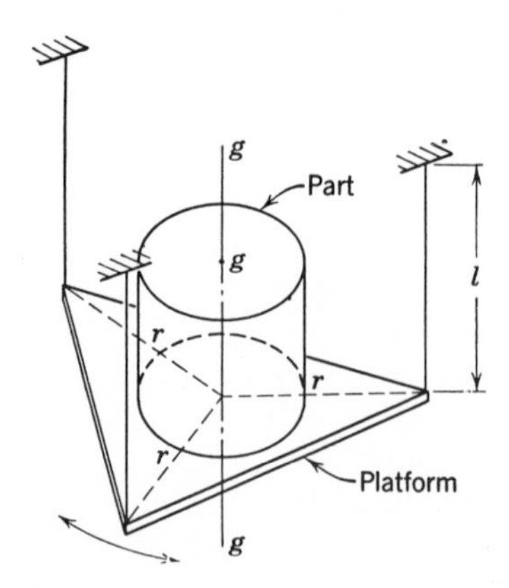

Figure 11.43

where

$$W = \text{weight of part}$$
$$W_b = \text{weight of platform}$$
$$\tau = \text{torsional period of platform with part}$$
$$\tau_b = \text{torsional period of platform alone}$$

Problems

11.1 The rotor of a jet compressor has blades 10 cm long mounted on a 90 cm diameter hub. Assuming the configuration of Fig. 11.1 and that $b = 7$ cm, $t = 0.6$ cm, and $w = 0.0272\ \text{N/cm}^3$, determine the allowable rotative speed of the rotor for an allowable maximum blade stress of 6900 N/cm^2. The blades are aluminum.

11.2 The blades of one of the stages of a jet engine compressor are 4 in. long and are mounted on a rotor hub 36 in. in diameter. Assuming that the blades have the configuration of Fig. 11.1, that $b = 3$ in., and $t = \frac{1}{4}$ in., and that the rotor speed is 8000 r/min, determine the force exerted on the hub by the centrifugal force of an individual blade and the corresponding stress at the base of the blade for (*a*) a steel blade ($w = 0.285\ \text{lb/in.}^3$) and for (*b*) an aluminum alloy blade ($w = 0.10\ \text{lb/in.}^3$).

11.3 Consider a blade of the type shown in Fig. 11.1 in which the cross-sectional area of the base is A_b and that of the tip is A_t. Assuming a uniform taper of cross-sectional area with radius, derive an expression for the stress s_b due to centrifugal force at the base of the blade in terms of the taper ratio $k = A_t/A_b$.

11.4 Assuming that a fan is simulated by the configuration of Fig. 11.2, determine the maximum blade stress (N/cm^2) due to centrifugal force for a 26-cm-diameter fan. The fan has eight blades of 0.107 cm thickness mounted on a hub of 6-cm diameter rotating at 3600 r/min ($w = 0.0769\ \text{N/cm}^3$).

11.5 Assume that a 15-ft-diameter solid aluminum alloy propeller rotating at 1200 r/min has the uniform blade shape shown in Fig. 11.3. The blade length is 5 ft, $b = 8$ in., and $t = 1$ in., and the blade angle is 20°. Determine the tensile force on the shank by the centrifugal force on the blade and the corresponding twisting moment on the shank ($w = 0.10\ \text{lb/in.}^3$).

11.6 In Fig. 11.44, link 2 rotates about a fixed axis at O_2. For the data given, determine the inertia force vector $\mathbf{F}_o$ and show it in its proper position on a scale drawing of link 2. Show also the resultant force vector $\mathbf{R}$ representing the forces which produce the angular motions shown.

11.7 For the mechanism shown in Fig. 11.45, determine the magnitudes, directions, senses, and locations of the inertia forces acting on links 2, 3, and 4. Show the results on a scale drawing of the mechanism. Draw the given acceleration polygon to scale for use in determining unknown accelerations.

11.8 In the four-bar mechanism of Fig. 11.46, the center of gravity of link 3 is coincident with the centroid of the rectangle shown. From the given information, determine the inertia force of link 3, and show it as a vector in its correct relationship to the mechanism in the phase shown.

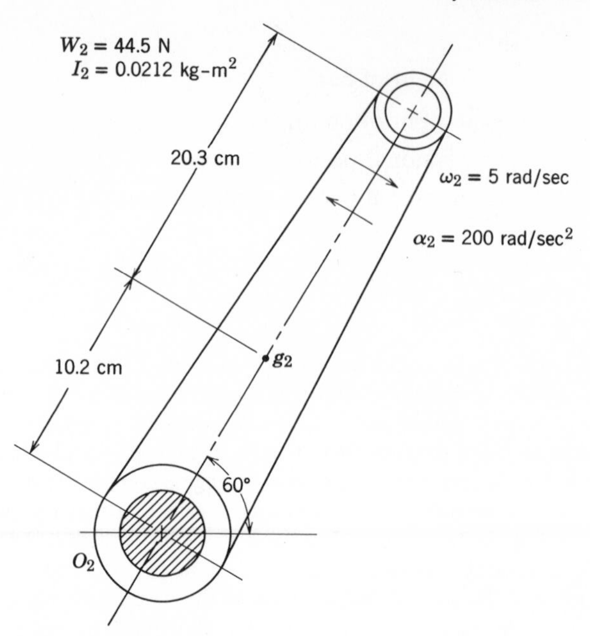

Figure 11.44

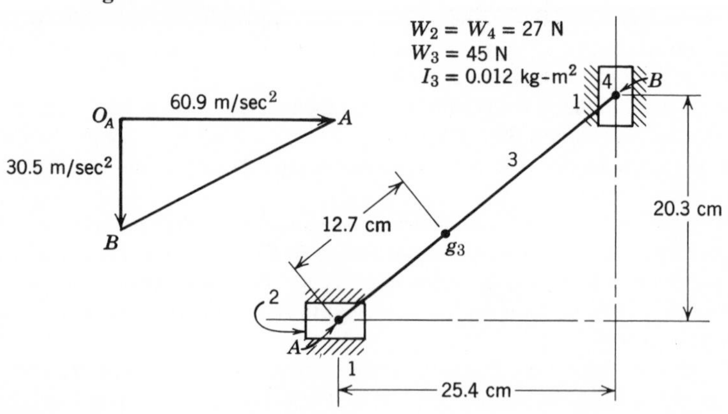

Figure 11.45

11.9 The link in Fig. 11.47, rotating about the fixed center O_2, is in motion in such a way that the center of gravity is accelerating in the direction shown and its normal component $\mathbf{A}^n_{g_2} = 2000$ ft/sec^2. Using vector polygons, determine the force $\mathbf{F}_A$ and the reactive force at O_2 that are producing the motion of the link.

11.10 For the mechanism of Fig. 11.48, determine the forces $\mathbf{F}_{14}$ and $\mathbf{F}_{12}$ due to the action of the inertia force $\mathbf{F}_{o4}$. Also determine the shaft torque T_s applied to link 2 at O_2. Draw the mechanism to scale and show the answers in their proper locations.

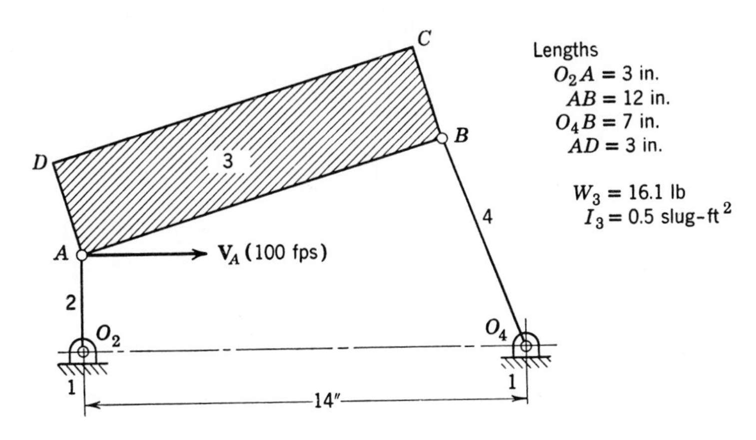

Figure 11.46

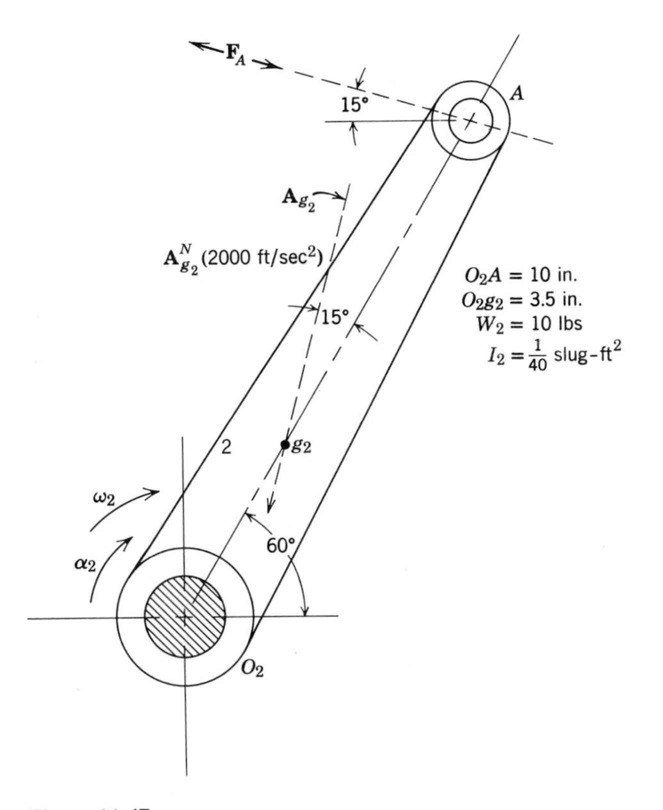

Figure 11.47

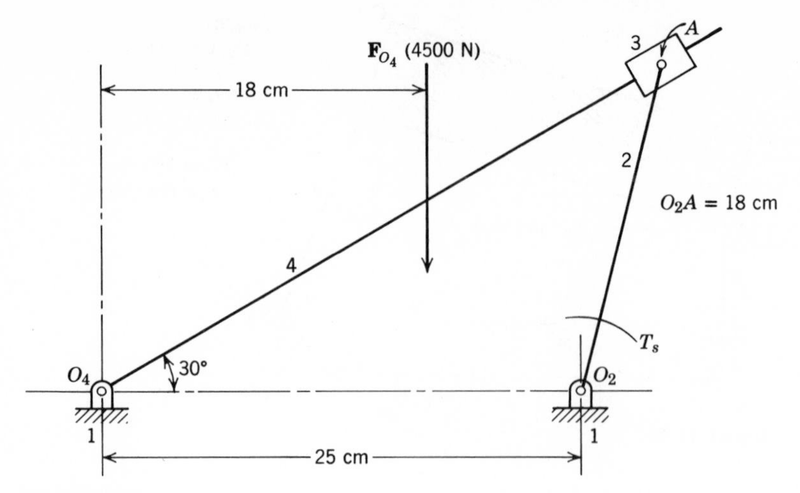

Figure 11.48

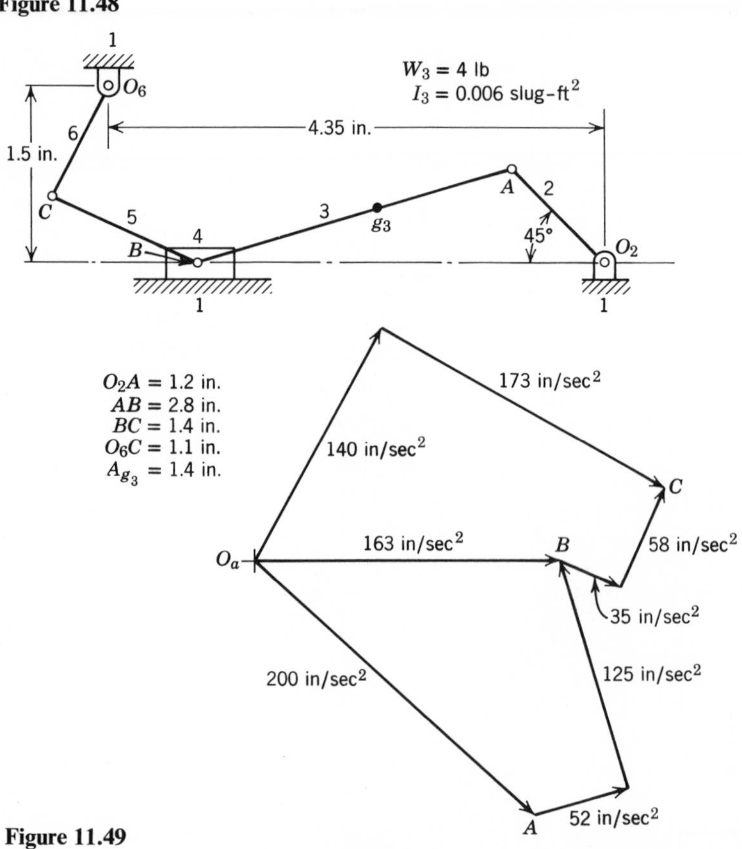

Figure 11.49

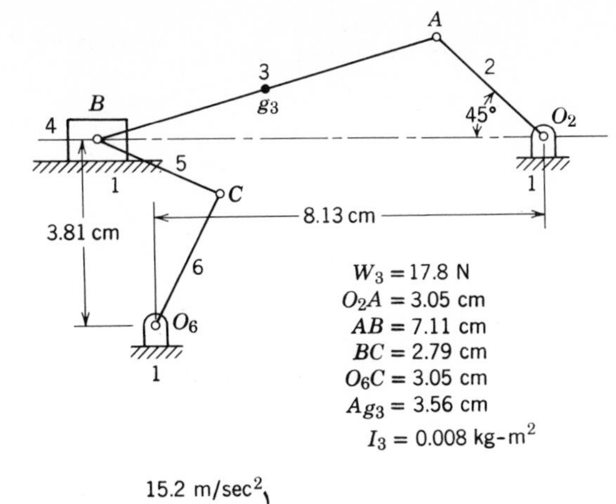

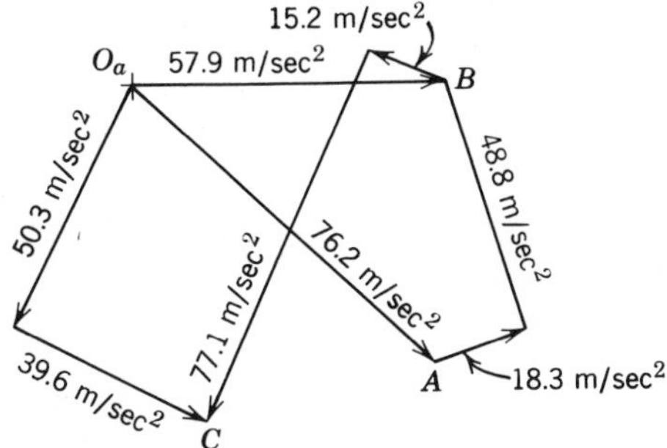

Figure 11.50

11.11 Given the mechanism shown in Fig. 11.49 and its acceleration polygon, calculate $\mathbf{F}_{o_3}$, and show it on the configuration diagram in its correct location.

11.12 For the linkage shown in Fig. 11.50 with its acceleration polygon, calculate $\mathbf{F}_{o_3}$ and show it on the configuration diagram in its correct location.

11.13 In the crank-shaper mechanism of Fig. 11.51, the tool holder 6 is acted upon by a static force of 100 lb as the tool cuts the work. Using force polygons, determine the forces acting on the bearings at A, B, C, O_2, and O_4 due to the tool force. Also determine the shaft torque T_s applied to link 2 at O_2. Draw the free-body diagram of each link (except link 1), and show the forces acting to scale. Also show T_s on link 2.

11.14 A 75-lb cutting force acts on the Whitworth mechanism as shown in Fig. 11.52. From a static force analysis of the mechanism, determine the forces acting on the bearings due to $\mathbf{F}_6$ and the shaft torque T_s applied to link 2 at O_2.

11.15 For the mechanism shown in Fig. 11.53, determine the forces acting on the bearings due to $\mathbf{F}_6$ and the torque T_s on the shaft at O_2.

11.16 Refer to Fig. 11.54. Using the method of superposition, construct force polygons and determine the forces on the bearings at A, B, C, and O_2 to maintain static equilibrium. Determine also the shaft torque T_s at O_2 of the driving link.

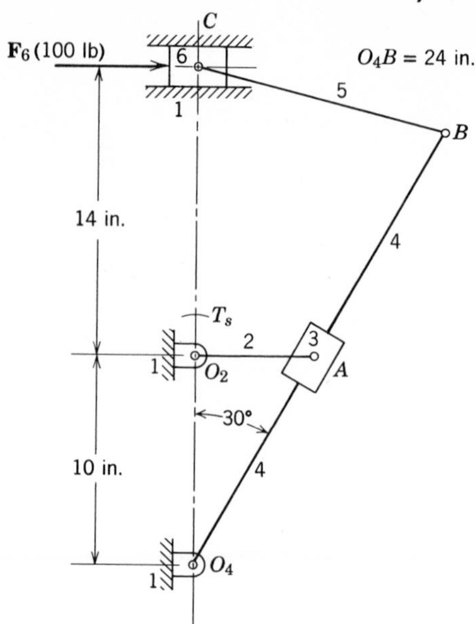

Figure 11.51

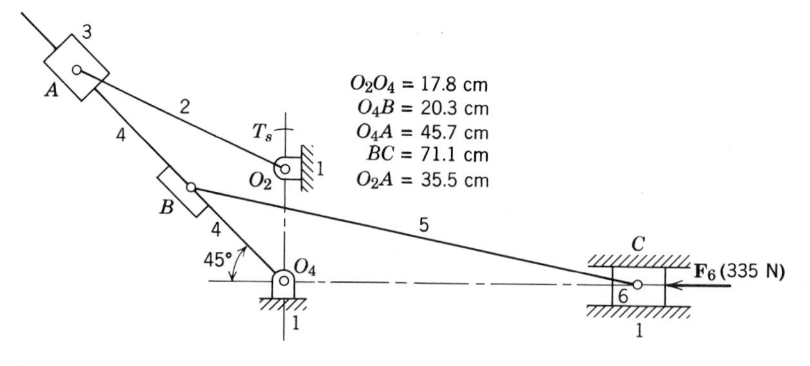

Figure 11.52

11.17 The flyball governor of Fig. 11.55 rotates about the Y–Y axis at a constant angular velocity. The spring exerts a force of 100 lb to balance the inertial force on the balls. Determine the rotation speed (r/min) of the governor in position shown. Each ball weighs 3.22 lb.

11.18 The tensioning mechanism of Fig. 11.56 is shown in both the open and closed position. **P** is the force applied to the handle, and **Q** represents the tension in the cable. By eye, sketch the force polygons for both the open and the closed positions, and show that the ratio of Q/P becomes infinite when the points A, B, and C are on a straight line.

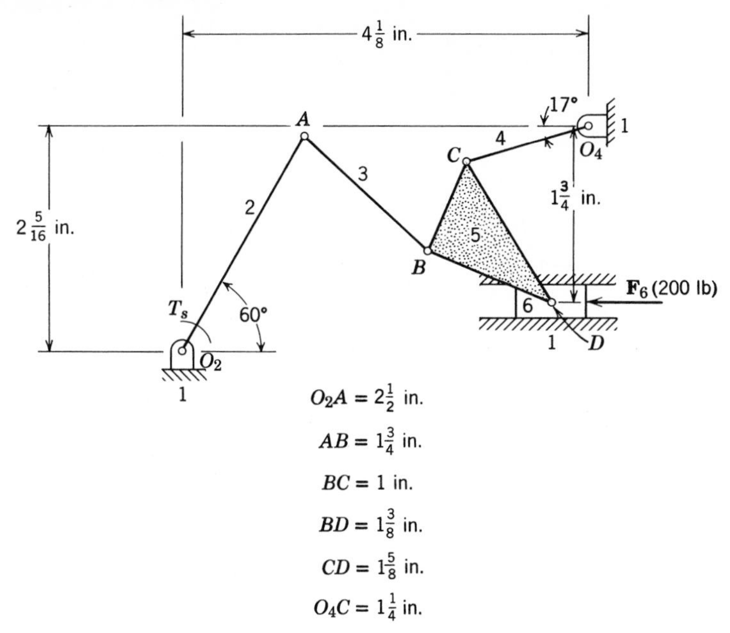

Figure 11.53

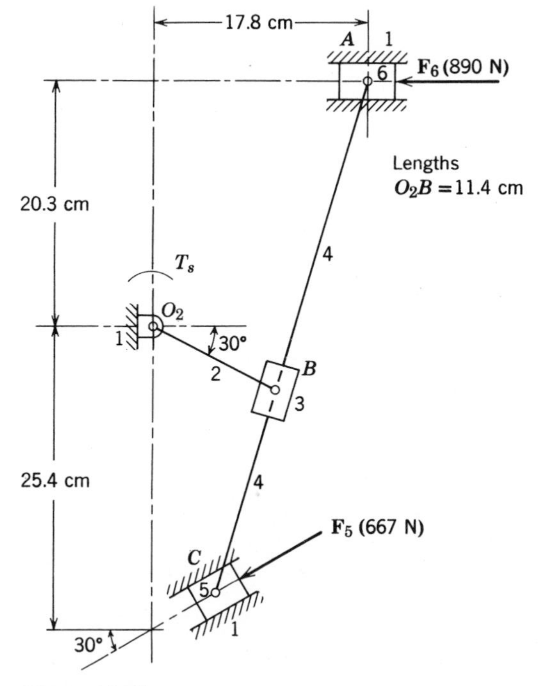

Figure 11.54

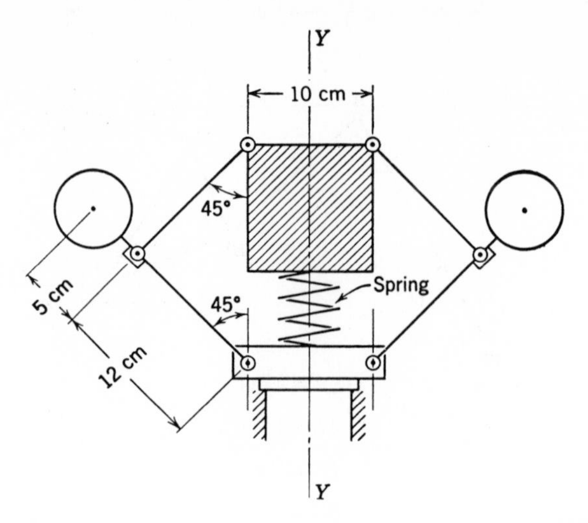

Figure 11.55

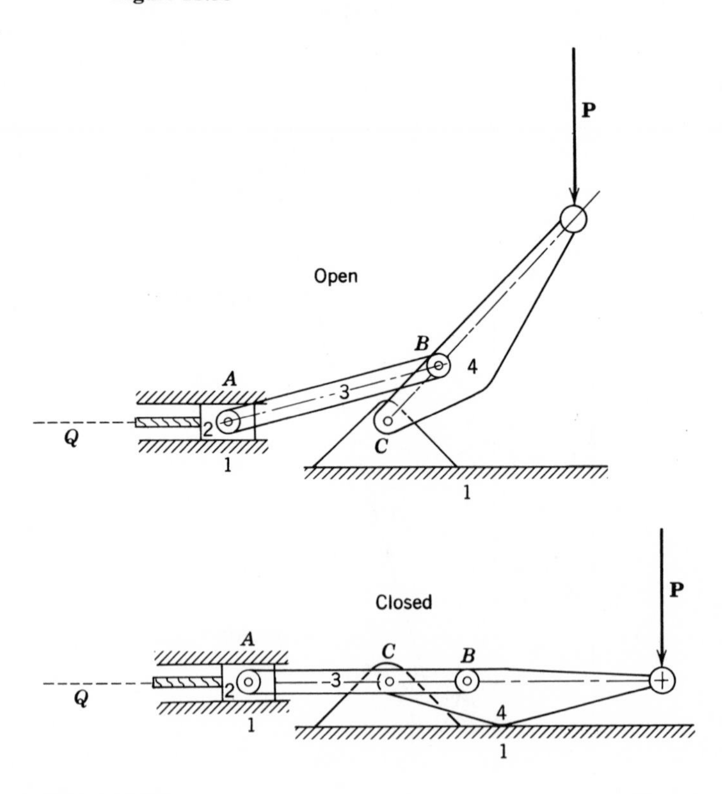

Figure 11.56

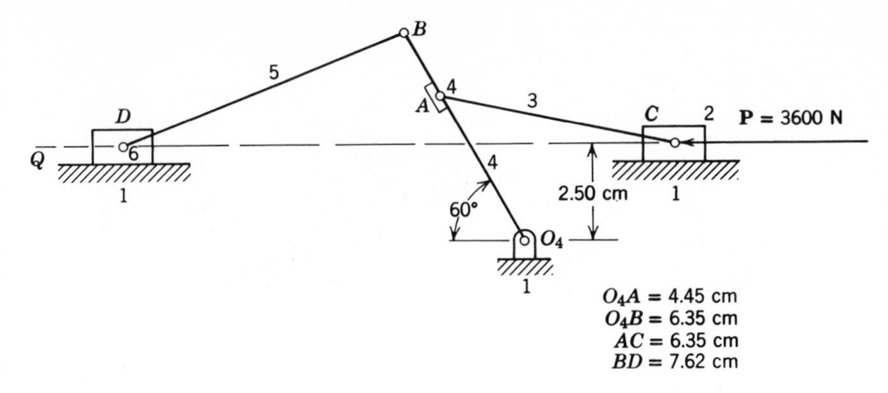

Figure 11.57

11.19 Determine force **Q** which must be applied to link 6 in Fig. 11.57 to maintain static equilibrium of the system under the action of force **P**.

11.20 For the mechanism of Fig. 11.58, determine the force **Q** necessary to maintain static equilibrium of the system under the action of the force $P = 1000$ lb. Solve using transverse and radial components.

11.21 Refer to Fig. 11.59. Given the resistance $\mathbf{P} = 5338$ N, determine the force which must be applied at m in the direction shown to maintain static equilibrium.

11.22 The four-bar mechanism of Fig. 11.60 is driven at O_2 at a constant angular velocity of 500 rad/sec. From the data given, make a complete dynamic analysis including a kinematic analysis, inertia force determinations, and a force analysis. Solve using transverse and radial components.

11.23 From the slider-crank data given in Fig. 11.61, make a complete dynamic analysis including a kinematic analysis, inertia force determinations, and a force analysis.

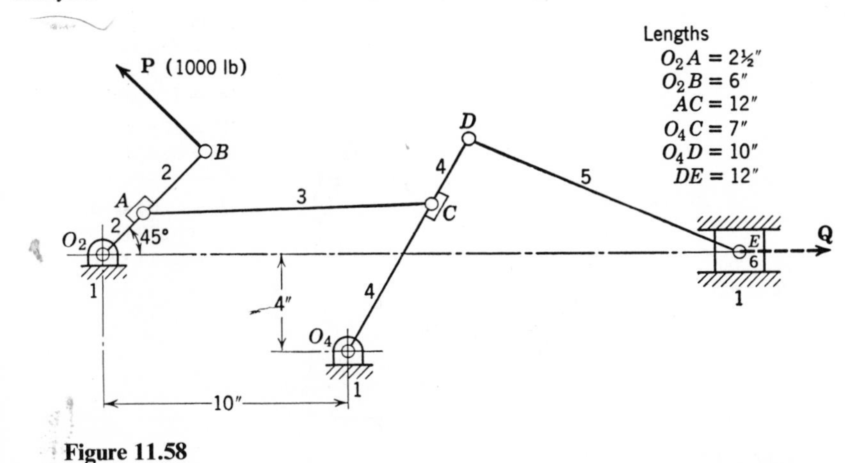

Figure 11.58

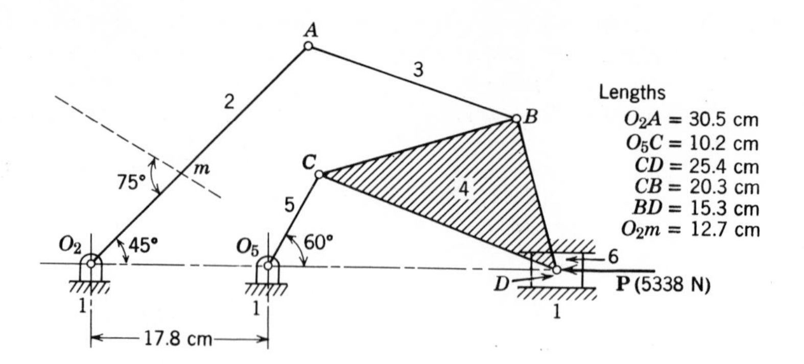

Figure 11.59

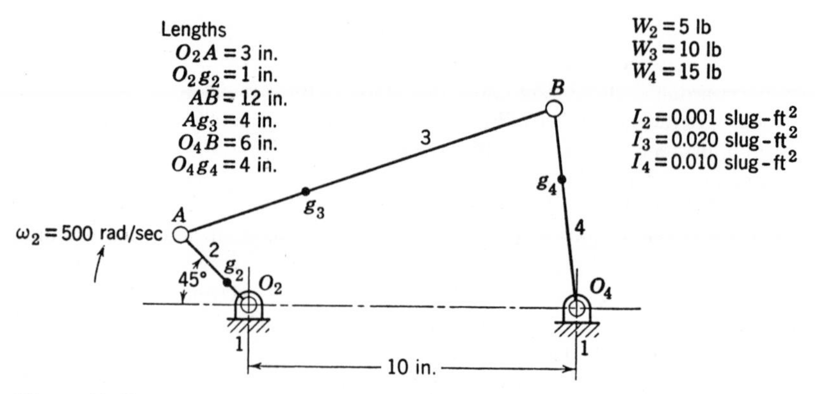

Figure 11.60

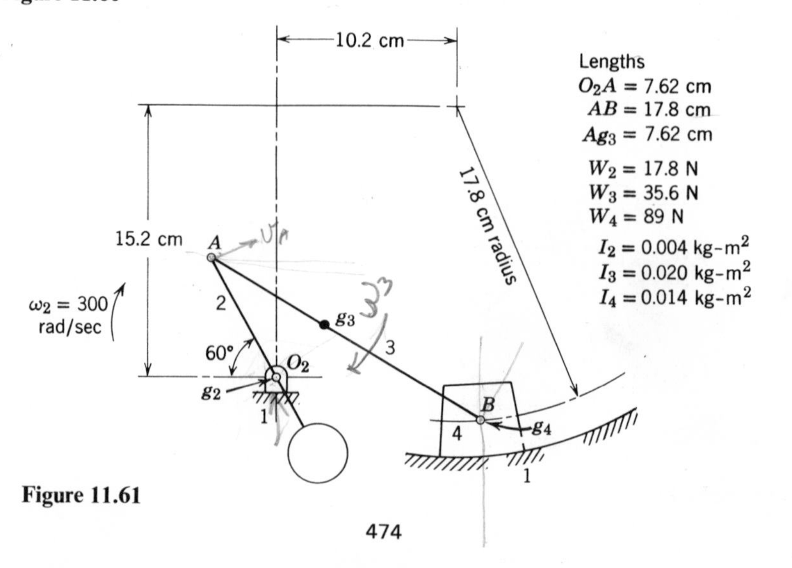

Figure 11.61

11.24 The Scotch yoke mechanism is often utilized in actuating small vibration tables as shown in Fig. 11.62. The motion is simple harmonic. Determine the maximum force on the bearing when the crank of length $e = \frac{1}{8}$ in. rotates at 6000 r/min giving a vibration frequency of 6000 cycles/min. Include the inertia effects of all parts inducing force on the bearing.

11.25 From the data given for the donkey engine of Fig. 11.63, make a force analysis and determine the forces on the bearings at O_2, A, and B.

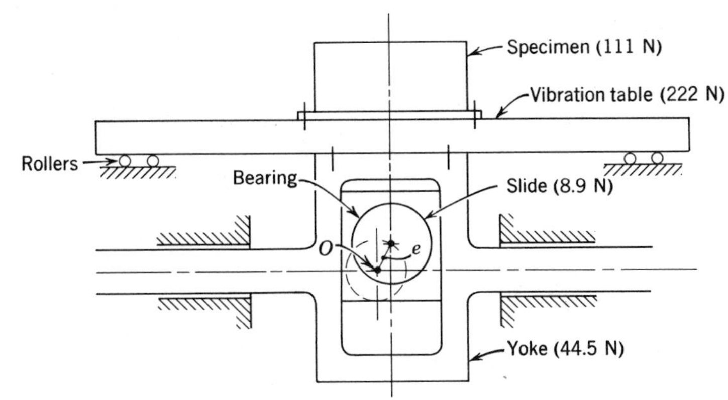

Figure 11.62

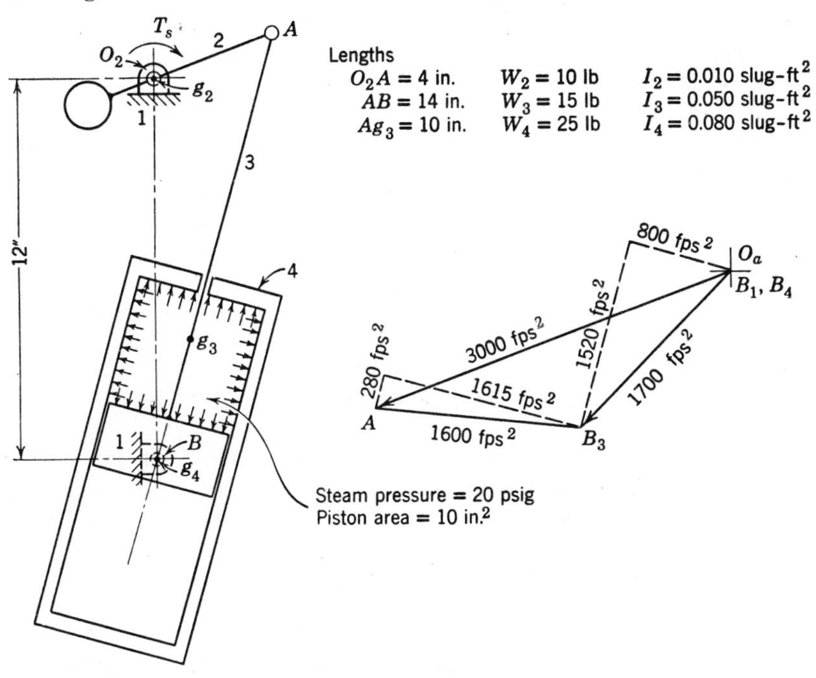

Figure 11.63

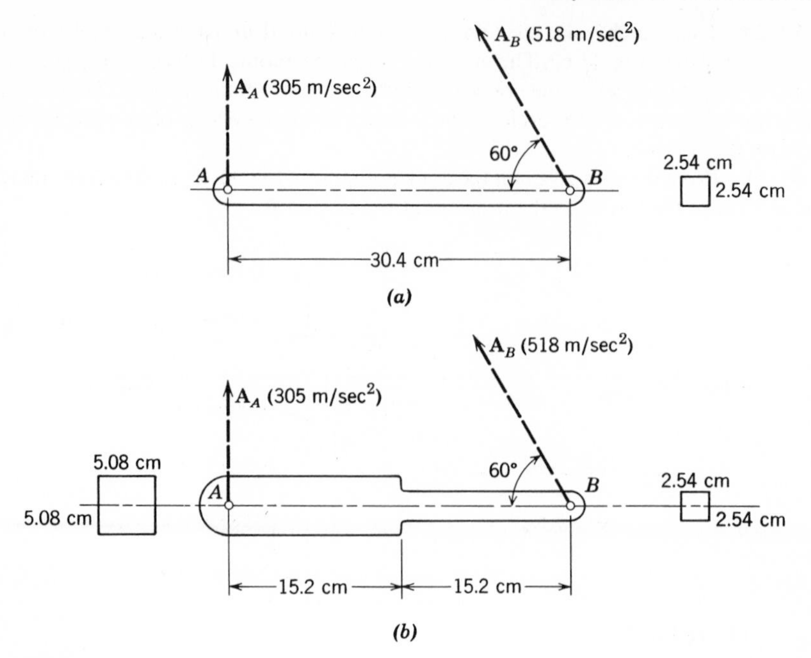

Figure 11.64

11.26 In Fig. 11.64*a*, the uniform steel bar is in motion and the accelerations of points *A* and *B* are as shown. Determine the following, giving magnitude, direction, and sense: (*a*) Transverse and longitudinal distribution of acceleration. (*b*) Transverse and longitudinal distribution of inertia force (for steel $w = 0.285$ lb/in.^{3}). (*c*) Transverse and longitudinal resultant inertia forces. Show line of action. (*d*) Resultant inertia force. Show line of action. Determine the same quantities for the nonuniform steel bar of Fig. 11.64*b*.

11.27 Determine the shaft torque T_s that must be applied to link 2 in Fig. 11.60 to hold the linkage in equilibrium. Use the method of virtual work.

11.28 Using the method of virtual work, calculate the crankshaft torque T_s necessary to hold the linkage in equilibrium for the slider-crank mechanism of Fig. 11.65. The gas force may be assumed zero.

11.29 For the mechanism of Fig. 11.61, determine the crankshaft torque T_s necessary to hold the linkage in equilibrium. Use the method of virtual work.

11.30 Calculate the shaft torque T_s which must be applied to link 2 of Fig. 11.63 to hold the linkage in equilibrium. Use the method of virtual work.

11.31 Use the method of virtual work to determine the crankshaft torque T_s necessary to hold the linkage in equilibrium for the two-cylinder engine shown in Fig. 11.66.

11.32 For the slider-crank mechanism shown in Fig. 11.18, a spring provides the driving force on piston 4 with a spring rate of 100 lb/in. of deflection. Assume that the

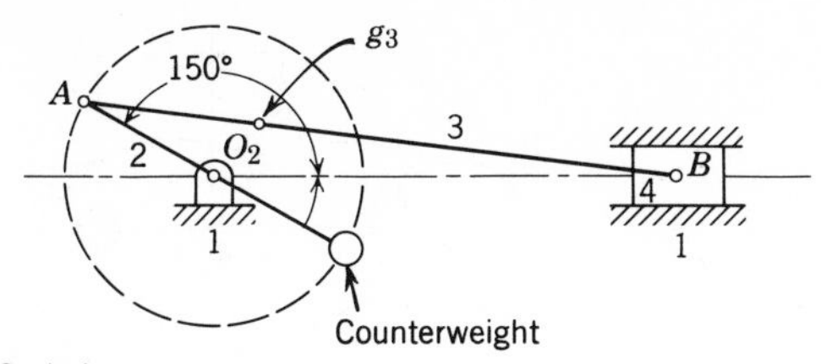

Crank speed, 4000 r/min
Piston weight, 12.2 N
Connecting rod weight, 8.9 N
Stroke, 10.2 cm
Connecting rod length, 20.3 cm
g_3 at 5.08 cm from A
$I_3 = 0.007$ kg-m^2

Figure 11.65

spring is compressed 1.75 in. from its free length when piston 4 is at its starting position at phase I and that ω_2 is zero at this position. Determine ω_2 at position III if T_2 is constant at 50 lb-in. (ccw) and $\theta_2^{III} = 60°$.

11.33 For the link shown in Fig. 11.67 M_Q, l_Q, and l_P are known. (*a*) Determine an equation for M_P in terms of the above values. (*b*) Calculate the moment of inertia of the link if $l_Q = 7.6$ cm, $l_P = 5.1$ cm, and $M_Q = 0.073$ kg.

11.34 For the link shown in Fig. 11.68, the two point masses at A and B are intended to be kinetically equivalent. Determine whether they are kinetically equivalent.

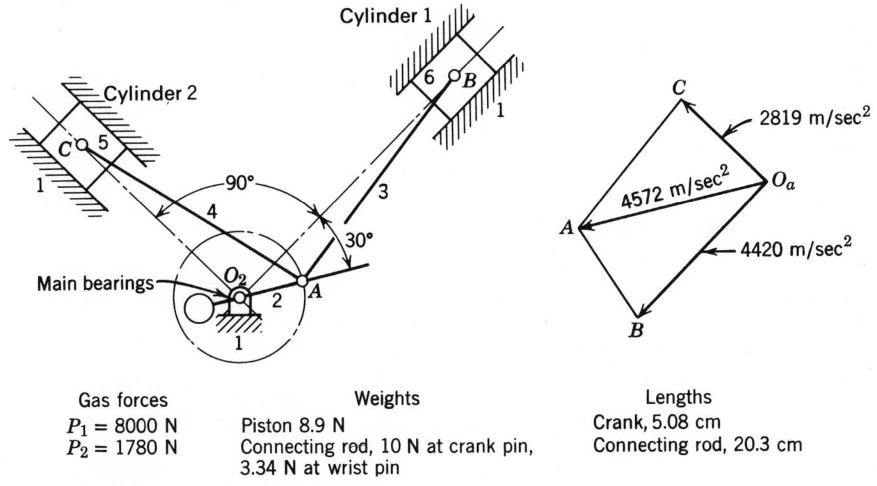

Gas forces
$P_1 = 8000$ N
$P_2 = 1780$ N

Weights
Piston 8.9 N
Connecting rod, 10 N at crank pin, 3.34 N at wrist pin

Lengths
Crank, 5.08 cm
Connecting rod, 20.3 cm

Moment of inertia of connecting rods, = 0.010 kg-m^2
Distance from crank center to center of gravity of connecting rods, = 5.08 cm

Figure 11.66

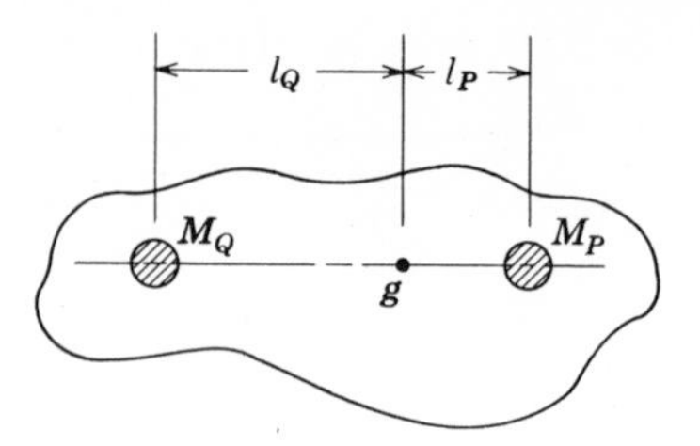

Figure 11.67

11.35 A link is shown in Fig. 11.69 that has its mass divided between points A and B in the manner shown. Determine whether the two masses are kinetically equivalent.

11.36 For the link shown in Fig. 11.70, determine the location of W_2 and the magnitudes of W_1 and W_2 so that they are kinetically equivalent. Determine also the magnitudes, directions, and senses of the inertia forces of the point masses and show them on a drawing of the link to scale.

11.37 The single-cylinder four-stroke engine of Fig. 11.65 is shown in the intake phase in which the gas force may be assumed zero. From the data given, determine the following: (*a*) The velocity and acceleration polygons. (*b*) The true kinetically equivalent masses of the connecting rod locating one of them at point B. (*c*) The approximate kinetically equivalent masses locating one at B and the other at A. (*d*) The inertia forces $\mathbf{F}_{o_4}$, $\mathbf{F}_{B_3}$, $\mathbf{F}_{A_3}$ and show on the diagram to scale using the masses of c. (*e*) Using the free-body diagram of the complete mechanism (excluding link 1), determine $\mathbf{F}_{14}$ and $\mathbf{F}_{12}$

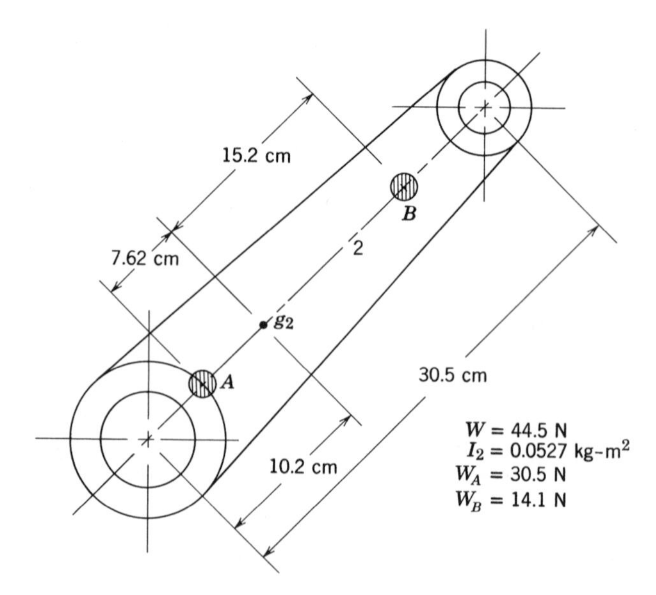

Figure 11.68

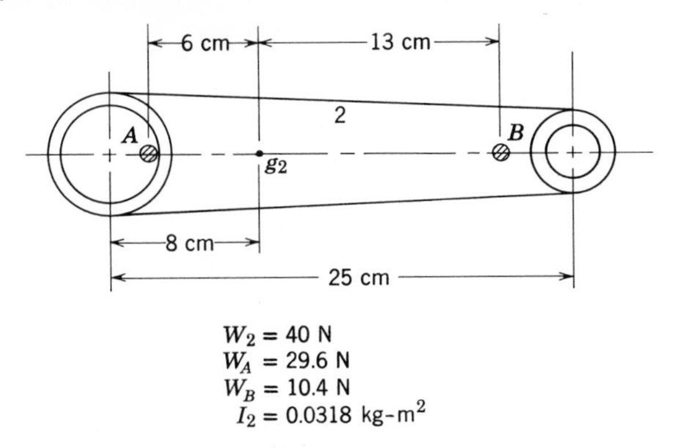

Figure 11.69

due to the inertia forces. Determine the crankshaft torque T_s. Assume that the counterweight balances the crank and the equivalent masses at A.

11.38 For the two-cylinder 90° V engine shown in Fig. 11.66, determine the resultant force on the main bearings due to the gas forces and inertia forces. The center of gravity of the crankshaft is at O_2. However, there is no counterweight to counterbalance the connecting rod equivalent masses at the crank pin. Determine the crankshaft torque T_s. Show your answers on the layout of the mechanism.

11.39 In Fig. 11.71, two free-body diagrams are shown of a single-cylinder engine in which the rotating masses are counterbalanced. Figure 11.71*a* is a free-body diagram of the moving parts of the slider crank, and Fig. 11.71*b* is a free-body diagram of the engine block together with the slider crank. Show vectors of the forces and torques acting on the free bodies when the engine operates at constant crank speed and is under gas pressure during the power stroke. Explain each vector.

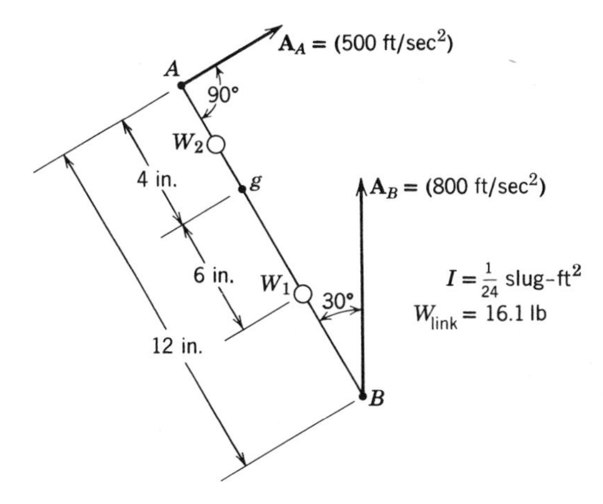

Figure 11.70

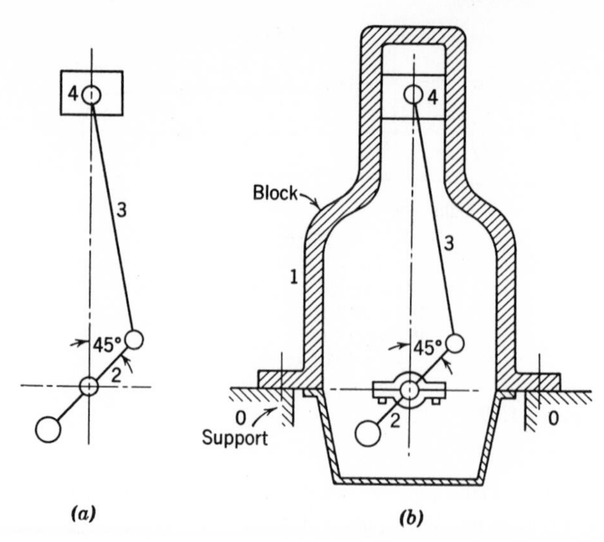

Figure 11.71

11.40 Sketch a free-body diagram of the engine block of the two-cylinder 90° V engine of Fig. 11.66 and show vectors of the forces acting. Explain each vector.

11.41 Using Table 11.1, determine the instantaneous output torque of the single-cylinder engine shown in Fig. 11.65. Zero gas pressure may be assumed since the phase shown is in the intake stroke. Inertia forces act.

11.42 By the analytical method and Table 11.1, determine the instantaneous output torque of the two-cylinder 90° V engine of Fig. 11.66 for a crank speed of 4000 r/min (acceleration polygon of Fig. 11.66 does not apply).

11.43 In Fig. 11.72 is shown a two-cylinder, four-cycle engine with cranks 90° apart. Using Table 11.1, determine and plot the output torque for one cylinder for each 15° crank angle of the 720° cycle. Plot the output torque of the engine versus crank angle by superposing the torque curves of the two cylinders with a phase angle of 90°. Assume that the gas pressure during the power stroke varies as shown in Fig. 11.72 and that the gas pressure is zero for the other strokes. Assume also that all masses rotating with the crank are counterbalanced.

11.44 The torque output diagram of a single-cylinder engine is shown in Fig. 11.73. Determine: (*a*) The average output torque and the kilowatt output for the engine which operates at 3500 r/min, (*b*) The locations of the crank angles at which the crankshaft speed is a maximum and a minimum during the engine cycle, and (*c*) Calculate the work done to change the speed from the minimum to the maximum.

11.45 Assume that the torque output diagram of Fig. 11.74 is for the first cylinder of a two-cylinder in-line engine with cranks at 180°. On this diagram, superimpose the same diagram for the second cylinder. Determine the locations of maximum and minimum crankshaft speeds in terms of the crank angle of the first cylinder.

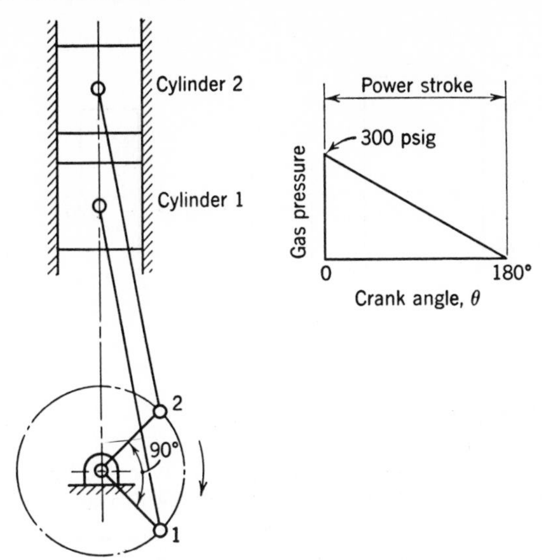

Figure 11.72 Data: crank speed, 2800 r/min; stroke, 6 in.; connecting-rod length, 12 in.; piston weight, 4 lb; weight of connecting rod at crank pin, 3 lb, at wrist pin 1 lb; piston diameter, 5 in.

11.46 If each square inch of a torque diagram represents 375 ft-lb of work, the area between the points of ω_M and ω_m is 1.20 in.^{2}, the engine speed is 3500 r/min, and the maximum allowable fluctuation of speed in the engine cycle is 35 r/min, determine the following: (*a*) The moment of inertia of a steel disk-type flywheel. (*b*) The weight and thickness of the flywheel if the diameter is 15 in. ($w = 490$ lb/ft^3).

11.47 A single-cylinder engine has a 23 cm mean diameter rim-type flywheel weighing 200 N. The engine operates at 3000 r/min and has an allowable fluctuation of speed in the engine cycle of 30 r/min. Determine the total work output in Newton-metres.

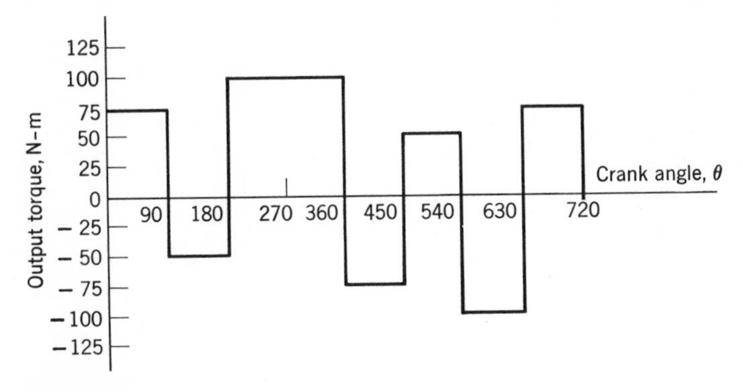

Figure 11.73

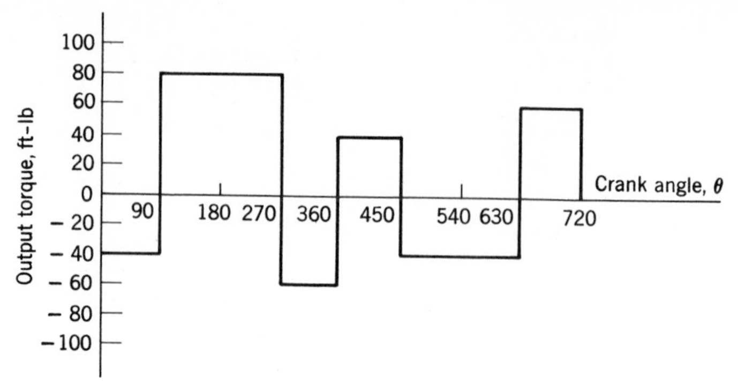

Figure 11.74

11.48 The torque output diagram of Fig. 11.73 is for a single-cylinder engine at 3000 r/min. Determine the weight of a steel disk-type flywheel required to limit the crank speed to 10 r/min above and 10 r/min below the average speed of 3000 r/min. The outside diameter of the flywheel is 25 cm. Determine also the weight of the rim of a rim-type steel flywheel of 25 cm mean diameter for the same allowable fluctuation in speed.

11.49 In the punch press shown in Fig. 11.75*a*, a slider-crank mechanism with flywheel is used to punch holes in steel plates. A hole is punched for each revolution of

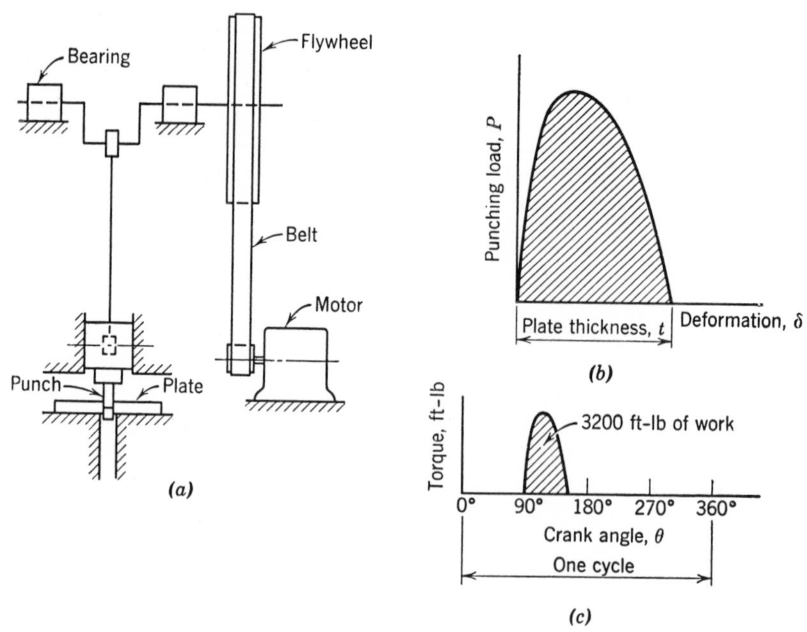

Figure 11.75

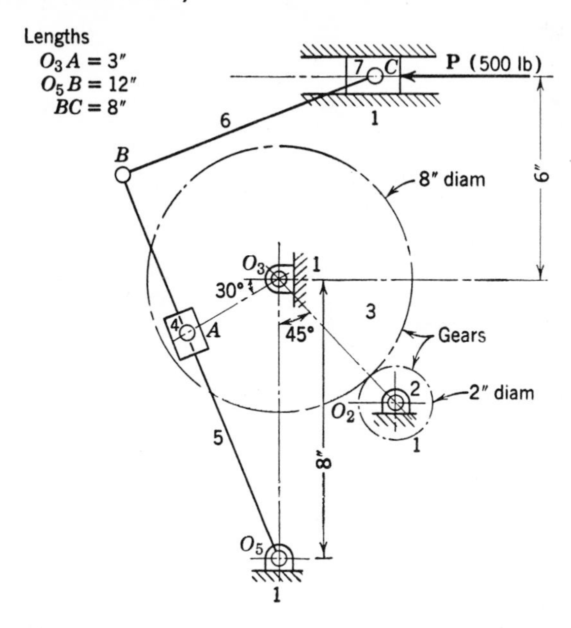

Figure 11.76

the flywheel, which operates at an average speed of 300 r/min. The load P on the punch during punching is the force necessary to shear the plate, and the variation of punching force with shearing deformation of the plate is shown in Fig. 11.75*b*. For the maximum size of plate and hole to be punched, it is estimated that 3200 ft-lb of work is required to punch the hole and that the punching is accomplished in one sixth of a revolution of the flywheel. Figure 11.75*c* shows the torque diagram for one cycle of punching. Determine the following: (*a*) The average crank torque for one cycle. (*b*) The horsepower required of the motor. (*c*) The required moment of inertia of flywheel for a minimum speed of 280 r/min just after punching.

11.50 Gears are normally used to drive crank-shaper machinery as shown in Fig. 11.76. Gears 2 and 3 are in mesh on standard pitch circles and have 20° stub teeth. Determine the tooth force on gears 2 and 3 and the shaft torque at O_2 to maintain static equilibrium of the mechanism acted upon by the known cutting tool force of 500 lb.

11.51 For the mechanism of Fig. 11.77, determine the force at the pitch point of the meshed gears due to the inertia force of link 4. The pressure angle of the gear teeth is 20°. The mechanism is driven by a shaft at O_2.

11.52 The gear and rack of Fig. 11.78 are in static equilibrium under the static force **P**. Note that since there is no bearing support of the gear at O, the force **P** causes a tooth of the gear to bear against two teeth on the rack. The resultant of the two forces on the gear tooth must be equal, opposite, and colinear with **P** for static equilibrium of the gear. If the gears are of $14\frac{1}{2}°$ tooth form, determine the force of contact on one of the rack teeth.

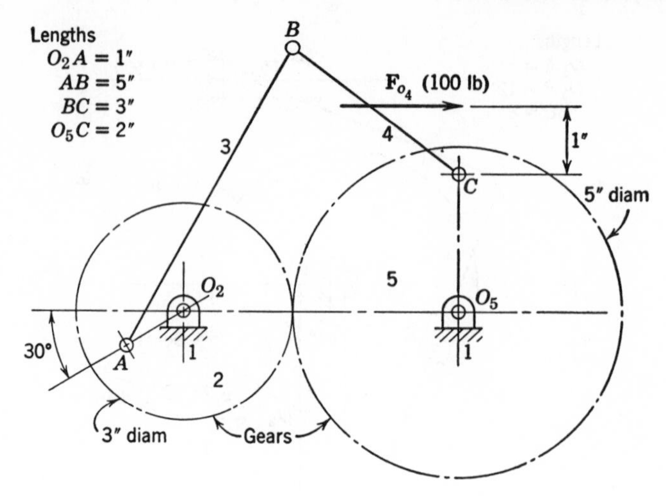

Figure 11.77

11.53 Make a force analysis of the mechanism of Fig. 11.79, which is in static equilibrium under the action of the force **P**. The shaft at O_2 is capable of producing a resisting torque. Is it necessary to know the pressure angle of the gear teeth to determine T_s? Why? Sketch an enlarged view of the contact of the teeth, and show force vectors.

11.54 A disk cam, rotating at 200 r/min, lifts a radial roller follower with simple harmonic motion through a maximum displacement of 2.5 in. while the cam rotates through 90°. There is then a dwell of the follower for 180° of cam rotation, followed by a return of the follower in simple harmonic motion in the remaining 90° of cam rotation. For a follower weight of 8 lb, determine the inertia force of the follower for each 15° of cam rotation and plot the results. Determine and plot the resultant of the weight force and inertia force. Determine the required spring constant of a spring which will maintain contact of the follower with the cam throughout the cam rotation.

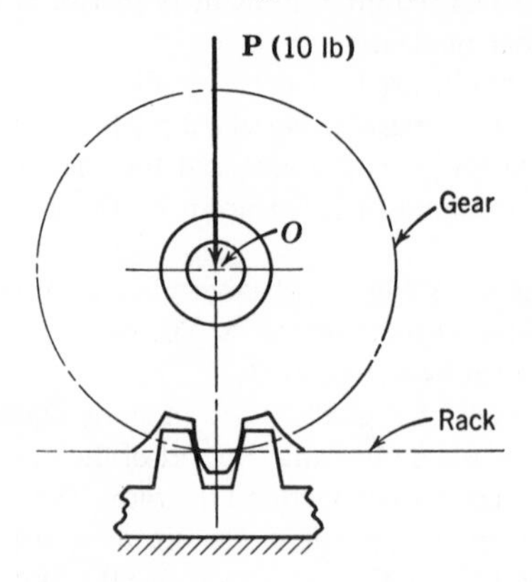

Figure 11.78

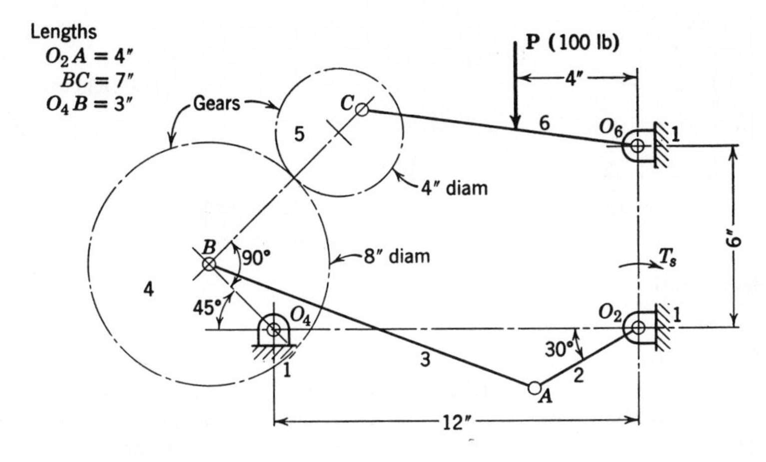

Figure 11.79

11.55 A spring-load radial roller follower of a disk cam is lifted through a distance of 5 cm with simple harmonic motion. The weight of the follower including the weight of the spring is 45 N. Because of its very low spring constant, the spring is assumed to apply a constant force of 25 N. Determine the maximum speed of the cam so that the follower does not lose contact with the cam.

11.56 In a Geneva mechanism, the pin of the driving wheel exerts a contacting force against the slot of the driven wheel. Determine this force for the following data: driver speed, 400 r/min (constant); center distance of the wheels, 4 in.; pin location, on 3-in. radius; number of slots, four; phase, pin radius at 30° from line of wheel centers; weight of driven wheel, 1 lb; radius of gyration of driven wheel, 1 in. Determine the pin force due to inertia of the driven wheel, and determine the torque on the driving shaft.

11.57 The gyroscope as shown in Fig. 11.80 is often used to demonstrate gyroscopic precession due to the action of the gravitational force. Determine the angular velocity of precession of a gyroscope of 16.1 lb weight spinning at 6000 r/min. The radius r is 4 in., and a is 17.9 in.

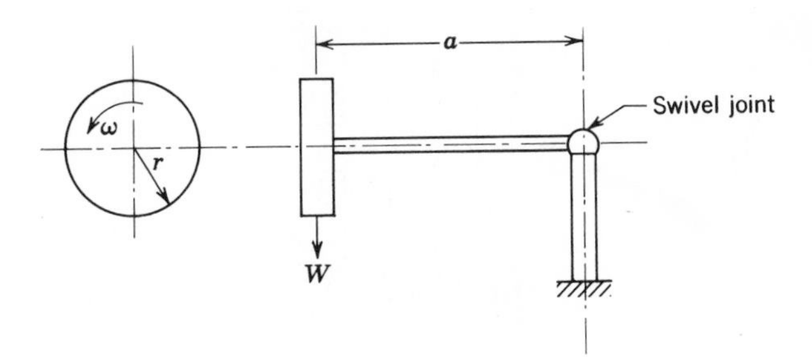

Figure 11.80

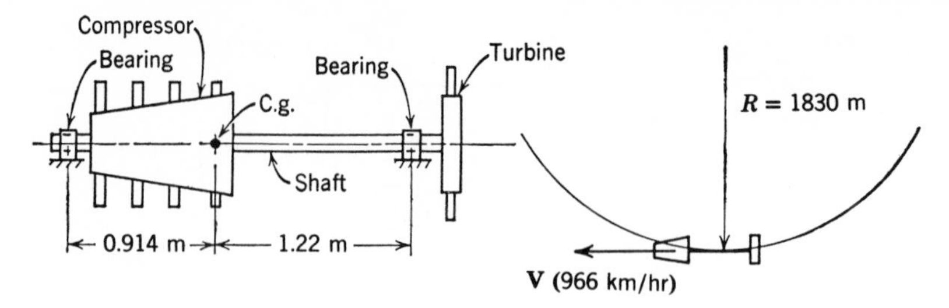

Figure 11.81

11.58 The rotor of a jet airplane engine is supported by two bearings as shown in Fig. 11.81. The rotor assembly including compressor, turbine, and shaft is 1500 lb in weight and has a radius of gyration of 9 in. Determine the maximum bearing force as the airplane undergoes a pullout on a 6000-ft-radius curve at a constant airplane speed of 600 mi/h and an engine rotor speed of 10,000 r/min. Include the effect of centrifugal force due to the pullout as well as the gyroscopic effect.

11.59 In the bevel gear planetary gear train of Fig. 11.82, the carrier (link 4) rotates at 1200 r/min. Determine the force on the bearings of planet (link 3) produced by gyroscopic action. I about the spin axis of the planet is 0.060 lb-sec^2-in.

11.60 In Fig. 11.83 is shown the gimbal mounting of a gyroscope used in instrument applications to maintain a fixed axis in space. Low-friction bearings are used to minimize the precession of the gyro. Through which bearings must friction torque be applied to cause precession of the x axis in the xz plane? If the gyro spins at 10,000 r/min and I of the gyro is 0.001 slug-ft^2, how much friction torque must be applied continuously to cause a precession at 1°/h?

11.61 To determine the moment of inertia I about an axis through the mass center of the connecting rod of Fig. 11.84*a*, the rod is suspended as a pendulum and the period of small oscillations is observed. (*a*) When suspended on a knife-edge at O_1, 59 oscillations

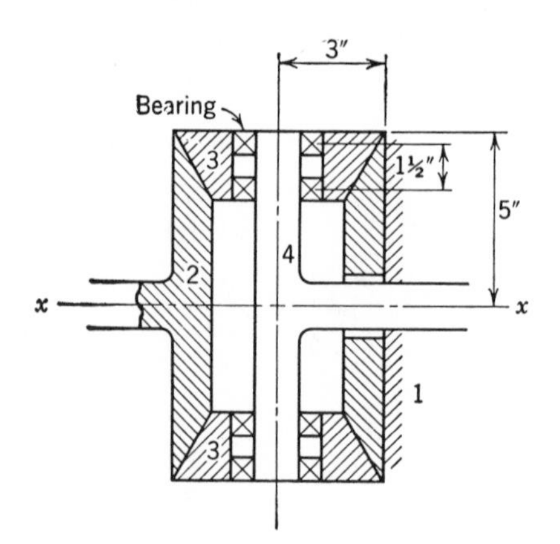

Figure 11.82

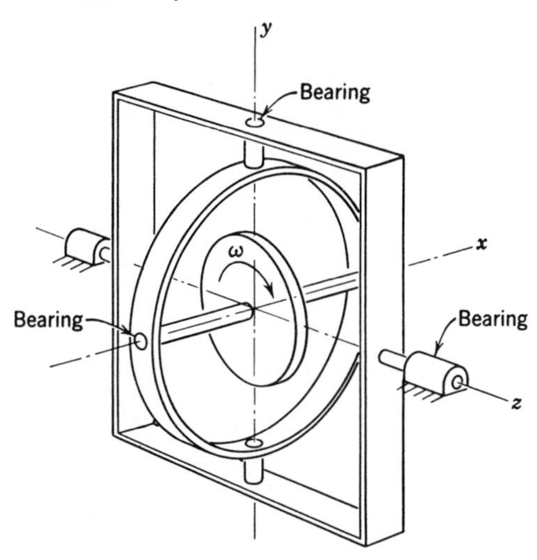

Figure 11.83

are counted in 60 sec. Determine I. Calculate the percent difference in I if 58 oscillations are counted in 60 sec. (*b*) When suspended at O_2, 66 oscillations are counted in 60 sec. Determine I, and calculate the percent difference in I if only 65 oscillations are counted in 60 sec. Which suspension gives the greater accuracy in determining I? Why?

11.62 In an experiment, the ring of Fig. 11.84*b* makes 107 oscillations in 1 min when supported as shown. The ring weighs 1.203 lb. Determine the moment of inertia I

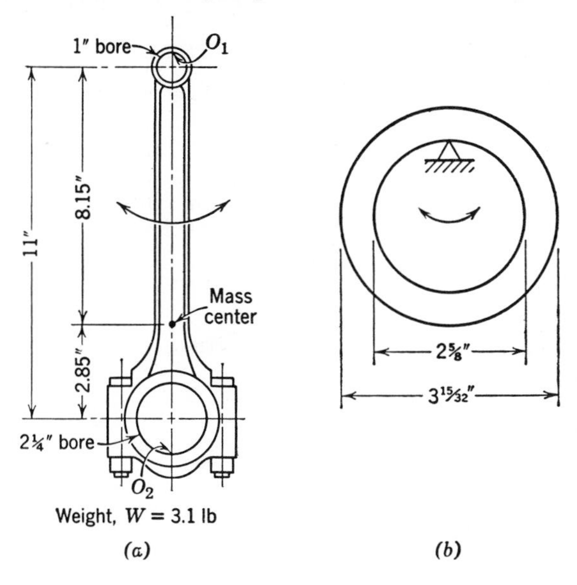

Figure 11.84

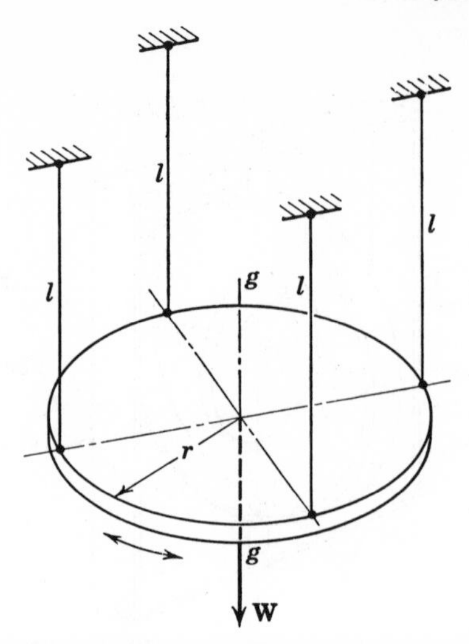

Figure 11.85

of the ring about an axis through its center of gravity (*a*) theoretically and (*b*) from the experimental data. Compute the percent error in I based on the theoretical value.

11.63 In Fig 11.85 is shown a solid thin disk suspended as a torsional pendulum by four weightless strings. Derive an expression for the determination of I of the disk about the g–g axis in terms of the disk weight W, l, r, and the torsional period τ of small oscillations.

11.64 The slider crank of Fig. 11.86 is operated at a uniform crank speed of 200 rad/sec. The lengths, the centers of gravity, the weights, and the moments of inertia

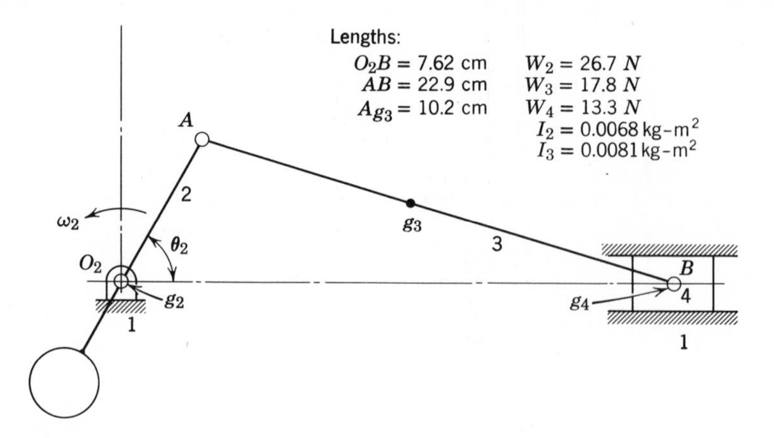

Figure 11.86

I of the links are given. Using the analytical method of complex numbers, determine the numerical values of the following for the phase $\theta_2 = 60°$: (*a*) Magnitude and angle of the inertia forces $\mathbf{F}_{o_3}$ and $\mathbf{F}_{o_4}$. Refer to Section 10.26 for kinematic equations. (*b*) Magnitude and angle of the bearing force $\mathbf{F}'_{34}$ due to inertia force $\mathbf{F}_{o_4}$. (*c*) Magnitude and angle of the bearing force $\mathbf{F}''_{34}$ due to inertia force $\mathbf{F}_{o_3}$. (*d*) Magnitude and angle of the resultant force $\mathbf{F}_{34}$.

12

Balance of Machinery

12.1 Introduction. As discussed in Chapter 11, the inertia forces of the slider-crank mechanism in an engine cause shaking of the engine block. Shaking forces in machines due to inertia forces may be minimized by balancing inertia forces in opposition to each other in such a way that little or no force is transmitted to the machine supports.

In Fig. 12.1, for example, the rotating mass M without counterbalance induces a shaking force equal to the inertia force $\mathbf{F}$ that is transmitted to the bearings and the supports. Because of the rotation, the shaking force has the characteristics of forced vibration at a circular frequency ω. The degree to which the forced vibration is undesirable depends upon the frequency of the forced vibration and the natural frequency of the flexible members through which force is transmitted such as the shaft or the supports. If conditions are near resonance, amplitudes of vibration may become large enough to cause discomfort as in an automobile or may cause failure of the shaft, the bearings, or the supports. As shown in Fig. 12.1, the shaking force may be minimized by counterbalancing in such a manner that the resultant of the inertia forces of the mass M and the counterbalance is zero.

In the subsequent discussions, methods are shown for determining the requirements of balance in (*a*) systems of masses rotating about a common axis, and (*b*) systems of reciprocating masses. In piston engines, both systems are present, the crankshaft being a system of rotating masses and the pistons being one of reciprocation.

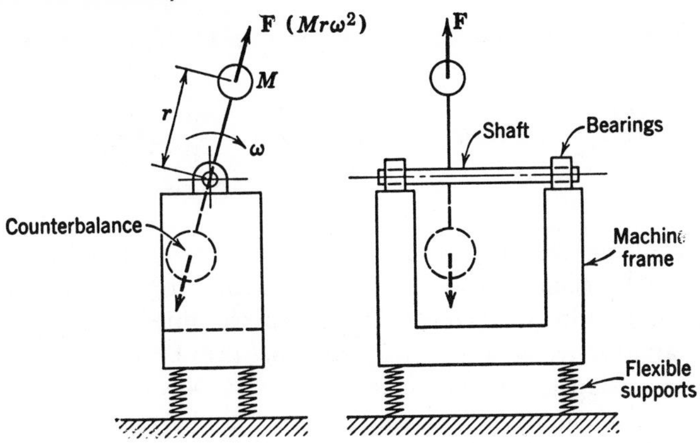

Figure 12.1

12.2 Balance of Rotors. Figure 12.2 shows a rigid rotor consisting of a system of three masses rotating in a common *transverse* plane about the axis O–O. A fourth mass is to be added to the system so that the sum of the inertia forces (shaking force) is zero and balance is achieved. For constant ω, the inertia force for any given mass M is $F = Mr\omega^2$ with direction and sense

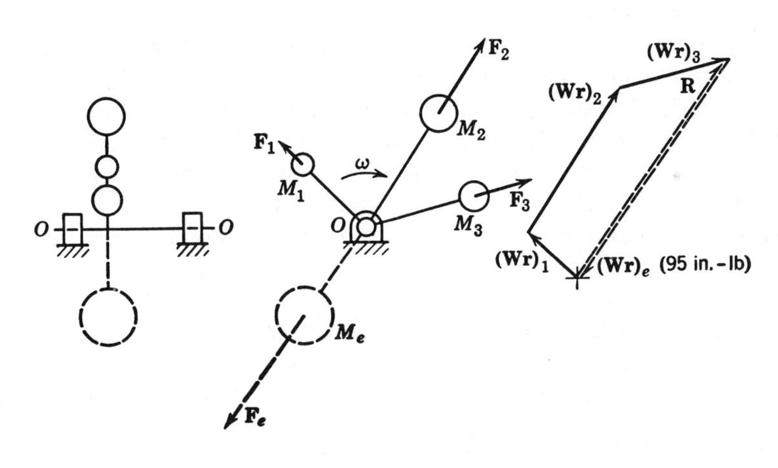

Number	W, lb	r, in.	Wr, in.-lb
1	3	8	24
2	5	12	60
3	4	10	40

Figure 12.2

radially outward. For balance, the vector sum of the inertia forces of the system is zero.

$$\Sigma \mathbf{F} = \Sigma (\mathbf{M}\mathbf{r}\omega^2) = \Sigma \left(\frac{\mathbf{W}}{\mathbf{g}} \mathbf{r}\omega^2 \right) = \frac{\omega^2}{g} \Sigma (\mathbf{W}\mathbf{r}) = 0$$

$$\Sigma (\mathbf{W}\mathbf{r}) = 0 \tag{12.1}$$

Since for all masses ω^2/g is constant, balance is achieved if Eq. 12.1 is satisfied. $(\mathbf{Wr})$ for each mass is a vector in the same direction and sense as the inertia force. In Fig. 12.2, values of Wr for the three known masses are tabulated, and the Wr value of the fourth mass is to be determined to satisfy Eq. 12.1 for balance. As shown in the vector polygon, the resultant $\mathbf{R}$ represents the unbalance of the three masses, and $(Wr)_e = 95$ in.-lb is the magnitude of the equilibrant required for balance. In Fig. 12.2, the balancing weight W_e is a 10-lb weight at $r_e = 9.5$ in., although any arbitrary value of W_e or r_e may be selected. Without the balancing mass, the resultant force in the rotating system is $R\omega^2/g$, which causes a bending of the shaft and exerts forces on the bearings supporting the shaft; in Fig. 12.2, the left bearing would carry a greater part of the unbalanced load. With the balancing mass added, the shaft bending and bearing loads are reduced to a minimum. Any number of masses rotating in a common radial plane may be balanced with a single mass.

For the case in which the rotating masses of a rigid rotor lie in a common *axial* plane as in Fig. 12.3, the inertia forces are parallel vectors. Balance of inertia forces is achieved in this case as in the previous case by satisfying Eq. 12.1. However, balance of the *moments* of inertia forces is also required. In the system of Fig. 12.2, moment equilibrium is inherent since the inertia force vectors are concurrent. In Fig. 12.3, however, the inertia forces are not concurrent when viewed in the axial plane. Thus, for balance of moments, the moments of the inertia forces about an arbitrarily chosen axis normal to the axial plane must be zero.

$$\Sigma Fa = \Sigma \frac{W}{g} r\omega^2 a = \frac{\omega^2}{g} \Sigma Wra = 0$$

$$\Sigma Wra = 0 \tag{12.2}$$

in which a is the moment arm of any given inertia force.

The magnitude of the resultant force $\mathbf{R}$ of the three unbalanced masses in Fig. 12.3 is the algebraic sum as well as the vector sum of $(\mathbf{Wr})$ terms of the three masses since the inertia force vectors are parallel. As shown in the table of Fig. 12.3, upward Wr values are taken as positive. The line of action of $\mathbf{R}$ is determined using the principle of moments in which moments are taken about the moment center O. The distance a_R from the moment center

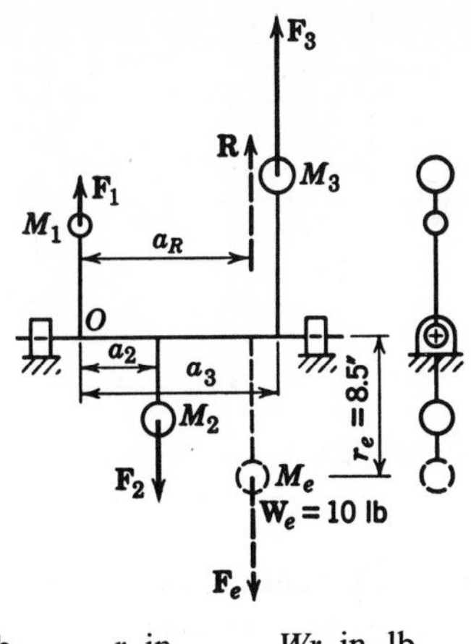

Number	W, lb	r, in.	Wr, in.-lb	a, in.	Wra
1	5	7	+35	0	0
2	10	5	−50	5	−250
3	10	10	+100	12	+1200
			+85		+950

$$R = \Sigma\, Wr = 85$$

$$W_e r_e = -R = -85 \text{ in.-lb}$$

$$a_R = \frac{\Sigma\, Wra}{\Sigma\, Wr} = \frac{950}{85} = 11.2 \text{ in.}$$

Figure 12.3

O locates the line of action of **R**. As shown in the table of Fig. 12.3, counterclockwise values of *Wra* are positive. To satisfy Eqs. 12.1 and 12.2 for balance, the equilibrant $(\mathbf{Wr})_e$ is equal, opposite, and colinear with **R**. In Fig. 12.3, a 10-lb weight at $r_e = 8.5$ in. is shown as the balancing weight.

In some instances, as shown in Fig. 12.4, the resultant of the system of masses to be balanced is a couple. The resultant force **R** for the two equal masses in Fig. 12.4 is zero. However, because the inertia forces of the two masses are not colinear, an unbalanced couple exists. To meet the requirements of moment balance, *two* additional masses are needed to provide a balancing couple.

In the foregoing cases, balancing requirements are met by determining the minimum number of additional masses to achieve balance. Often, more than the minimum number are used. For example, in Fig. 12.3 the single counterbalancing mass is added to reduce shaking forces to zero and to remove load

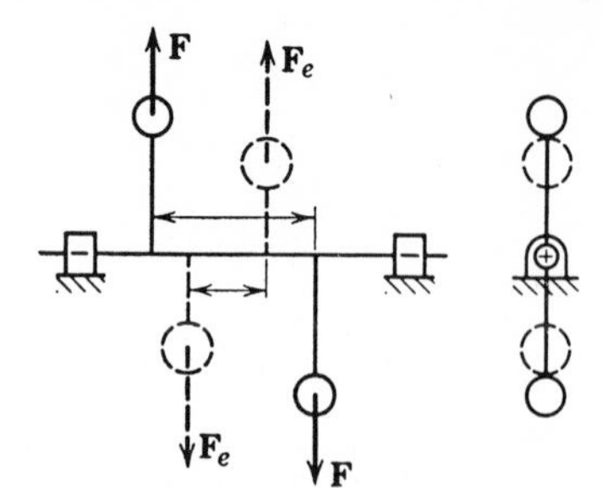

Figure 12.4

from the bearings supporting the shaft. However, the shaft is under the action of bending, which in some cases may be severe. Balance may also be achieved by providing a counterbalance opposite each mass, a total of three counterbalances, with the advantage that shaft bending is reduced to near zero. As shown in Fig. 12.5*a*, crankshafts are frequently balanced by counterbalancing each crank separately to reduce shaft bending. Greater total weight is a disadvantage in utilizing large numbers of counterweights. As shown for the crankshaft in Fig. 12.5*b*, the symmetrical distribution of cranks provides balance without the addition of counterweights, but to reduce shaft bending intermediate main bearings are added.

The most general case of distribution of rotating masses on a rigid rotor is that in which the masses lie in various transverse and axial planes as in Fig. 12.6. As in the foregoing cases, Eq. 12.1 must be satisfied for balance of

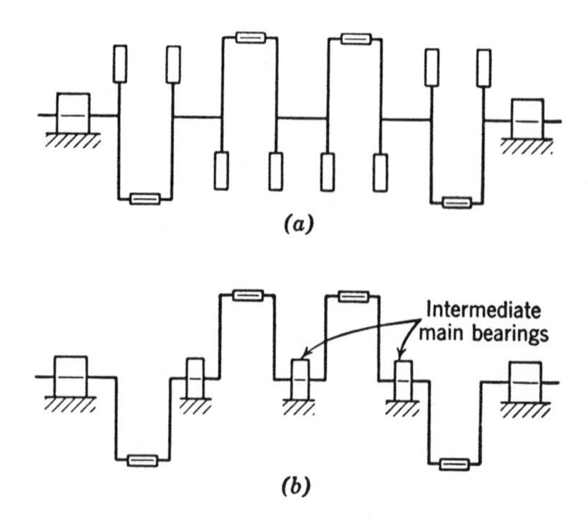

Figure 12.5

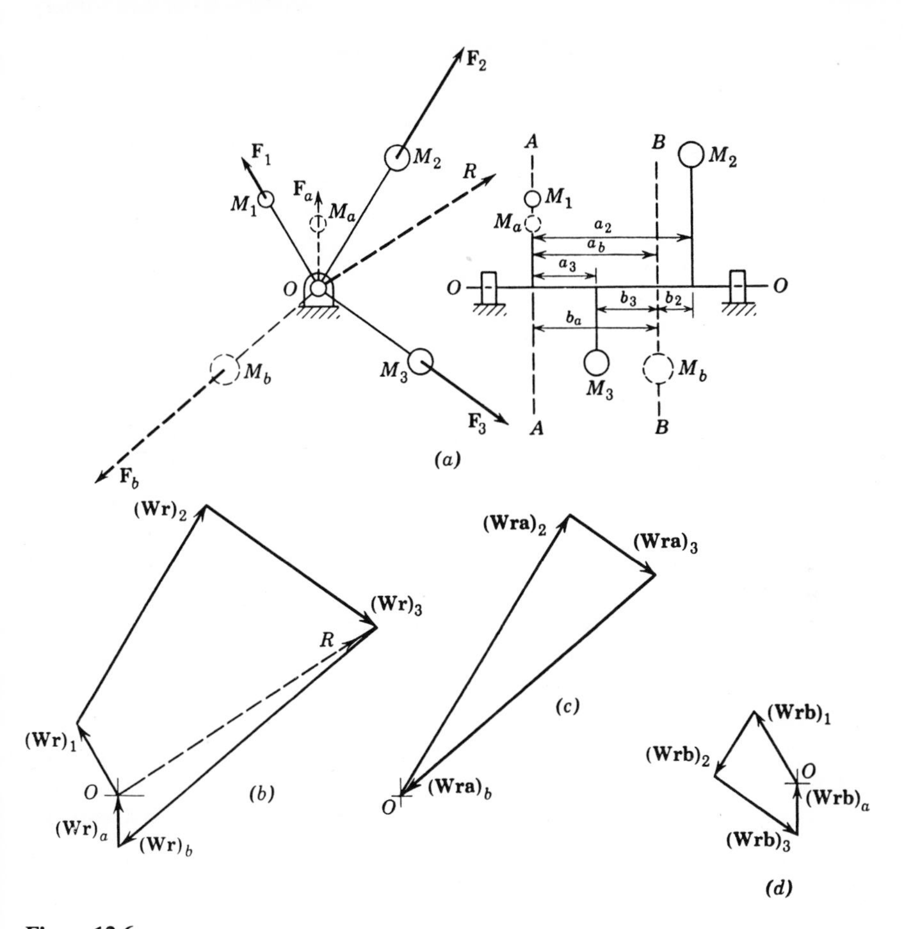

Figure 12.6

Number	W, lb	r, in.	Wr	a, in.	Wra	b, in.	Wrb
1	5	8	40	0	0	10	400
2	10	12	120	13	1560	−3	−360
3	10	10	100	5	500	5	500
a	5*	5	25*	0	0	10	250*
b	16.3*	10	163*	10	1630*	0	0

* From polygons and calculations.

$$(Wr)_b = \frac{(Wra)_b}{a_b} = \frac{1630}{10} = 163 \text{ in.-lb}$$

$$(Wr)_a = \frac{(Wrb)_a}{b_a} = \frac{250}{10} = 25 \text{ in.-lb}$$

inertia forces. As shown in Fig. 12.6*b*, the resultant **R** of the three unbalanced masses of Fig. 12.6*a* is obtained from a vector polygon. Although it would appear that a single balancing mass would satisfy Eq. 12.1, a consideration of moment balance shows that a minimum of two balancing masses is required.

In Fig. 12.6*a*, the transverse plane *A–A* is arbitrarily chosen about which moments of the inertia forces are evaluated. It may be seen that the moments of the various individual forces are in different axial planes. For moment balance, the *vector* sum of the moments of the forces must be zero.

$$\Sigma\,(\mathbf{Wra}) = 0 \tag{12.3}$$

Equation 12.3 is similar to Eq. 12.2 except that a vector sum is indicated rather than an algebraic sum. Since in the general case, the resultant unbalanced moment is in a different axial plane from the resultant **R** of unbalanced forces, a single balancing mass does not satisfy both Eqs. 12.1 and 12.3.

In Fig. 12.6*c* is shown the vector polygon of moments taken about the transverse plane *A–A*. Plane *B–B* is chosen as a transverse plane in which a balancing mass M_b is to be placed to achieve balance of moments. Magnitudes of the moment vectors are tabulated as shown. Although moment vectors are usually represented in direction and sense according to the right-hand screw rule, in Fig. 12.6*c* they are shown in the same direction and sense as the inertia forces. In Fig. 12.6*c*, the known moment vectors $(\mathbf{Wra})_2$ and $(\mathbf{Wra})_3$ are laid off first, and the closing side $(\mathbf{Wra})_b$ determines the required moment vector for balance. The direction of $(\mathbf{Wra})_b$ shows the axial plane in which M_b is to be placed. As shown, the magnitude of the force vector $(\mathbf{Wr})_b$ is calculated from $(Wra)_b/a_b$ and laid off on the force polygon of Fig. 12.6*b*. For balance of forces, a second mass M_a is required to close the force polygon as indicated by $(\mathbf{Wr})_a$. $(\mathbf{Wr})_a$ and $(\mathbf{Wr})_b$ form the equilibrant of **R**. By placing M_a in plane *A–A* such that it has zero moment about plane *A–A*, the moment vector polygon (Fig. 12.6*c*) for balance is unchanged. Thus, both Eqs. 12.1 and 12.3 are satisfied. As indicated in the table in Fig. 12.6, a 5-lb weight at $r_a = 5$ in. in plane *A–A* and a 16.3-lb weight at $r_b = 10$ in. in plane *B–B* are used for balance. Figure 12.6*a* shows the axial planes of the balancing masses as determined from the directions of $(\mathbf{Wr})_a$ and $(\mathbf{Wr})_b$ in Fig. 12.6*b*.

In Fig. 12.6*d* is shown the moment vector polygon in which moments are taken about plane *B–B* to determine the moment vector $(\mathbf{Wrb})_a$ due to M_a in plane *A–A*. The vector $(\mathbf{Wr})_a$ obtained from this polygon is the same as in the previous solution. As shown in the table and in Fig. 12.6*d*, the sense of $(\mathbf{Wrb})_2$ is negative since M_2 is on the opposite side of plane *B–B* from M_1 and M_3.

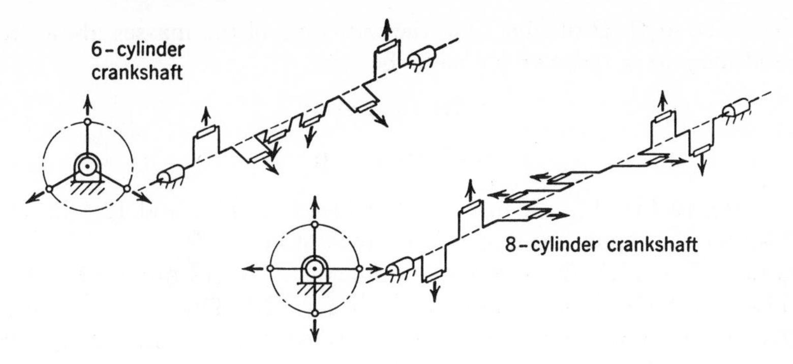

Figure 12.7

The problem of Fig. 12.6 may be solved using any two of the three vector polygons shown. Also, for the general case of Fig. 12.6 any number of masses may be balanced by a minimum of two masses placed in any two arbitrarily selected transverse planes such as A–A and B–B.

As shown in Fig. 12.7, the cranks of six- and eight-cylinder in-line engines are arranged so that crankshaft balance is obtained by symmetry, although the individual crank masses (including equivalent connecting rod masses) are in different axial planes.

12.3 Dynamic and Static Balance. The requirements for balance of rigid rotors as illustrated in Section 12.2 are those for *dynamic* balance or balance due to the action of inertia forces. *Static* balance is a balance of forces due to the action of gravity. Figure 12.8 shows a rigid rotor with the shaft laid on horizontal parallel ways. Under the action of gravity, the rotor will not roll if it is in static balance regardless of the angular position of the rotor. The requirement for static balance is that the center of gravity of the system of masses be at the axis O–O of rotation. For the center of

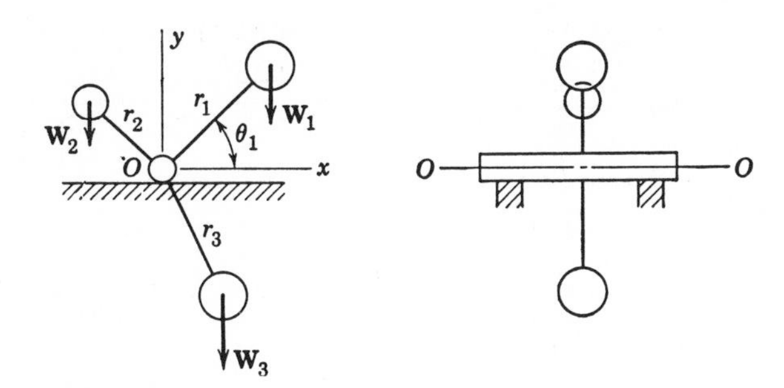

Figure 12.8

gravity to be at O–O of Fig. 12.8, the moments of the masses about the x axis and the y axis, respectively, must be zero.

$$\Sigma\, Wr \sin\theta = 0 \tag{12.4}$$

$$\Sigma\, Wr \cos\theta = 0 \tag{12.5}$$

Referring to Fig. 12.2, it may be seen that Eqs. 12.4 and 12.5 for static equilibrium also apply for dynamic balance of inertia forces. In the vector polygon of Fig. 12.2, the vertical components of forces are represented in Eq. 12.4 and the horizontal components in Eq. 12.5. Thus, if the conditions for dynamic balance are met, the conditions for static balance are also met. This is true also for the rotors of Figs. 12.3, 12.4, and 12.6. However, it is not true that if a rotor is statically balanced, it is also dynamically balanced. The rotor of Fig. 12.4, for example, is statically balanced without the balancing masses, but it is not dynamically balanced because of the unbalance of moments in the axial plane. Thus, static balance fails to indicate moment balance required for the dynamic case. A static balance is a reliable test of dynamic balance only in the case of Fig. 12.2, where all masses lie in a common transverse plane and a dynamic unbalance of moment is unlikely.

The use of horizontal parallel ways as in Fig. 12.8 is a simple method for "shop" balancing or "production" balancing of rotors having masses in a common radial plane. As shown in Fig. 12.9, airplane propellers are tested for dynamic balance in this manner from a test for static balance. A high degree of balance is achieved by adding washers to the propeller hub as shown. Also, as shown, washers are added to the blade shank in balancing the individual propeller blades against a standard balancing moment. Balancing by removing metal by drilling is avoided in propeller balancing because of the stress concentration caused by holes. Rotors having the general shape of a

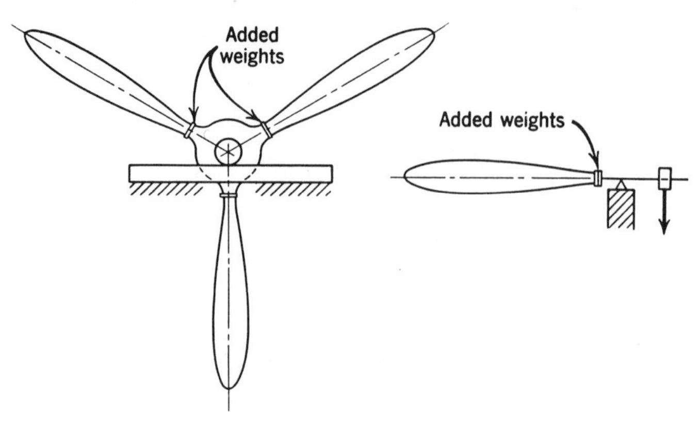

Figure 12.9

thin disk such as gears, pulleys, wheels, cams, fans, flywheels, and impellers are often balanced statically.

12.4 Balancing Machines. Although the dynamic balance of a rotor is properly met in the design of the rotor, some unbalance, however small, results from the manufacture of the rotor. Carefully machined parts are likely to be in better balance than cast parts. In many instances, it is more economical to allow an unbalance caused by manufacture and to balance the part by adding or removing material as indicated by a balancing machine. Commercial balancing machines are available which enable balancing of parts at mass production rates of manufacture.

The degree to which a rotor is to be dynamically balanced depends upon the speed at which the rotor is to operate. A small unbalance of mass may be tolerable at low speed since the inertia force representing unbalance may be small, but because the unbalanced force increases as the square of the speed, the unbalance transmitted to the bearings may be large at high speed. The rotor of a jet engine operating at greater than 10,000 r/min, for example, must be balanced to a high degree. For such rotors, the individual compressor and turbine blades are balanced in pairs at opposite locations on the rotor in such a way that, if one blade is damaged, the pair is replaced to restore balance.

The principle on which dynamic balancing machines are based is shown in Fig. 12.10. The rotor to be dynamically balanced is supported on flexible springs and rotated at the speed the rotor is to be used. As shown in Fig. 12.10, the springs permit only a lateral oscillation of the rotor under the action of the unbalanced force **F**. If there is also an unbalanced moment in the rotor, the amplitudes of oscillation of the two springs will be different and in some

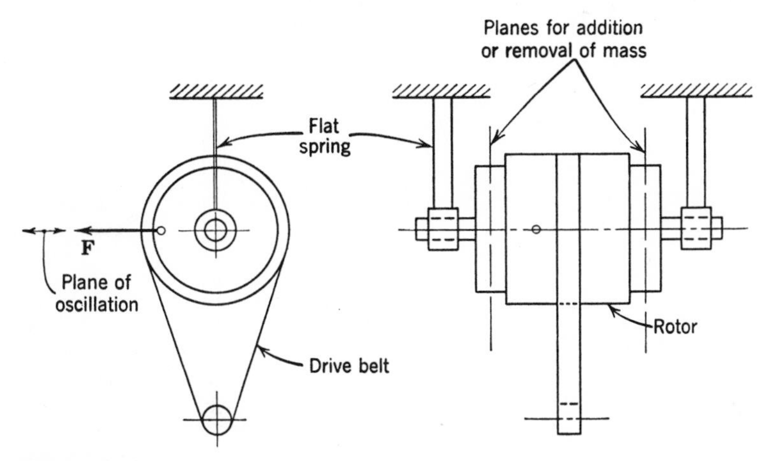

Figure 12.10

cases opposite in sign. The amplitudes of oscillation of each spring are measured with a highly sensitive electronic pickup which is calibrated to show the amount of unbalance. The machine also indicates the angular position of the unbalance on the rotor by imparting a signal at the instant the rotating force vector is horizontal and the amplitude is maximum. After the amount and angular position of the unbalance is read, the rotor is removed from the machine and material is added by soldering masses to it or is removed by drilling holes. As shown in Fig. 12.10, material is usually added or removed at two specific locations where it is not injurious to the rotor surface. In electric motor armatures, for example, it is not always possible to add or remove material in the region of the electrical windings. Long rotors such as armatures, crankshafts, and jet engine rotors are balanced in machines of this kind.

12.5 Balance of Reciprocating Masses. As shown in Fig. 11.26, the shaking of a piston engine is due primarily to the inertia forces of the reciprocating masses located at the wrist pin. The masses rotating with the crankshaft are normally balanced and they do not transmit a shaking force to the engine block. As shown in the slider-crank free body of Fig. 12.11, the effect of the inertia force **F** of the reciprocating masses is the transmission of force to the engine block at the cylinder wall and at the main bearings. The vertical component of the main bearing force $\mathbf{F}_{12}^{y}$ and the inertia force **F** are equal, opposite, and colinear. The horizontal component of the bearing force $\mathbf{F}_{12}^{x}$ and the cylinder wall force $\mathbf{F}_{14}$ are equal and opposite and form a

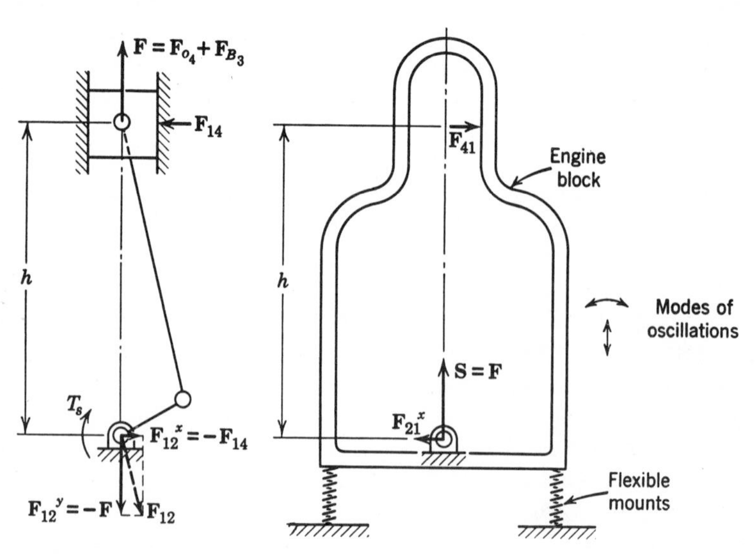

Figure 12.11

couple $\mathbf{F}_{14}h$ since they are not colinear. The effect of the reciprocating masses on the engine block, as shown in the free body of the block, is a shaking force $\mathbf{S} = \mathbf{F}$ and a shaking couple $\mathbf{F}_{41}h$. Since both the shaking force and shaking couple change in magnitude and sense during the engine cycle, forced vibrations are imposed on the engine block. If the engine block is mounted on flexible supports, the mode of oscillation of the block imposed by $\mathbf{S}$ is an up and down mode; the shaking couple produces a rotational oscillation or lateral rocking.

In Chapter 11, it was shown that the gas forces do not contribute to the vertical shaking but that they do produce a shaking couple as do the inertia forces of the reciprocating masses. Since output torque and shaking couple are equal (except in sense), the output torque diagrams of Chapter 11 show the variations of the shaking couple in the engine cycle due both to gas force and to inertia forces of the reciprocating masses.

In the following discussion, it is shown that the resultant shaking force on an engine block may be reduced to zero in some instances by combining several slider-cranks to form a multicylinder engine in which the individual shaking forces balance one another. The resultant shaking couple of a multicylinder engine, however, is not reduced to zero as shown by the torque diagram of the six-cylinder engine in Fig. 11.32. However, by the proper design of the flexible supports connecting the engine block to the supporting frame, the oscillations due to shaking couple may be isolated from the frame for certain shaking couple frequencies.

Figure 12.12 shows a typical arrangement of cranks in a six-cylinder in-line engine. In this engine, the cranks are fixed at 120° to each other as shown and all slider-crank parts are the same as to size, shape, and weight. As shown in the table of Fig. 12.12, the inertia force $\mathbf{F}$ of the individual reciprocating masses are calculated from the following equation:

$$F = MA_B = MR\omega^2\left[\cos\theta + \frac{R}{L}\cos 2\theta\right]$$

in which M is the combined masses M_4 and M_{B_3} for a single cylinder, R is the crank length, L is the connecting-rod length, ω is the crank angular velocity, and θ is the crank angle from TDC. (top dead center).

As the table shows, the arrangement of cranks for a six-cylinder engine is such that the resultant of the six inertia forces is zero for the position of the crankshaft given by $\theta_1 = 20°$. It may be shown that the resultant is zero for all positions of the crankshaft. Thus no shaking force is transmitted to the main bearings supporting the crankshaft or to the engine block. The six-cylinder engine is well known for its inherent balance of reciprocating masses. The single-cylinder engine, the two-, three-, and four-cylinder engines of the

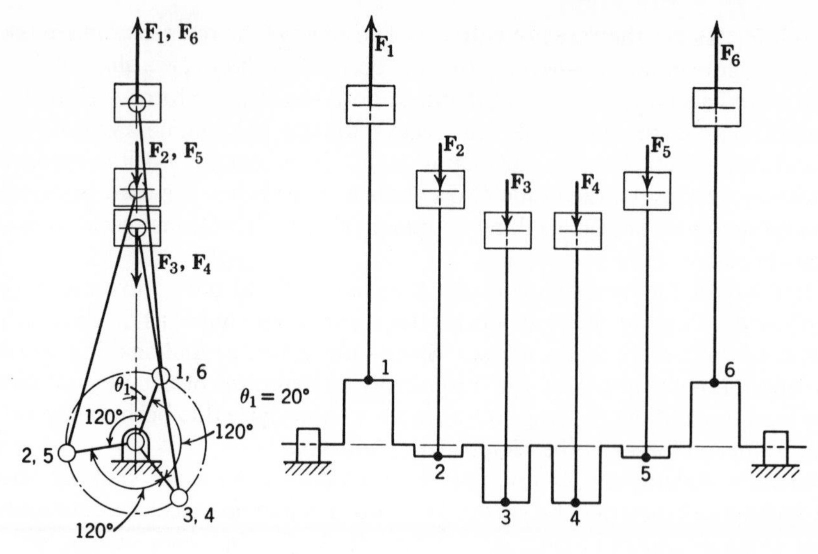

Figure 12.12. Data: $M = 2.5/32.2 = 0.0777$ slugs; $R = 2$ in.; $n = 3000$ rpm. $R/L = \frac{1}{4}$ $\quad MR\omega^2 = 1280$ lb

Number	$\theta°$	$\cos\theta$	$\cos 2\theta$	$R/L \cos 2\theta$	$(\cos\theta + R/L \cos 2\theta)$	F, lb
1	20	+0.904	+0.766	+0.191	+1.131	+1450
2	260	−0.174	−0.940	−0.235	−0.409	−525
3	140	−0.766	+0.174	+0.044	−0.722	−925
4	140	−0.766	+0.174	+0.044	−0.722	−925
5	260	−0.174	−0.940	−0.235	−0.409	−525
6	20	+0.940	+0.766	+0.191	+1.131	+1450
		$\Sigma = 0$	$\Sigma = 0$	$\Sigma = 0$	$\Sigma = 0$	$\Sigma = 0$

in-line type are not inherently balanced against shaking by reciprocating masses as is the six-cylinder engine.

12.6 Analytical Determination of Unbalance. Analytical methods are available for the determination of the unbalance or the shaking force of a multicylinder engine. The method leads to simple algebraic expressions which give the magnitude and sense of unbalance as a function of crank position θ_1. Crank position of a multicylinder engine in the engine cycle is given by the crank angle θ_1 of the first cylinder as shown in Fig. 12.13. In automotive engines, the first cylinder is at the front end, and θ_1 is measured clockwise in the direction of rotation when viewed from the front end.

The following analytical derivation applies only to in-line types of engines in which the cylinders are in line on the same side of the crankshaft. The reciprocating mass M and the R/L ratio is the same for each cylinder.

As shown in Fig. 12.13, θ_1 of the first crank locates the clockwise position of the crankshaft in the engine cycle, ϕ_2 and ϕ_3 are the fixed angles of cranks 2 and 3, respectively measured clockwise from crank 1. Although three cylinders are shown in Fig. 12.13, any number of cylinders may be considered. The inertia force F of any given cylinder at θ is

$$F = MR\omega^2\left[\cos\theta + \frac{R}{L}\cos 2\theta\right]$$

$$= MR\omega^2 \cos\theta + M\frac{R^2}{L}\omega^2 \cos 2\theta \qquad (12.6)$$

The two right-hand terms in Eq. 12.6 are the first two terms of a series, the remaining terms of which are usually considered negligible. The first term (first harmonic) is known as the primary force F_p, and the second term (second harmonic), as the secondary force F_s. Thus,

$$F = F_p + F_s \qquad (12.7)$$

in which $F_p = MR\omega^2 \cos\theta$, and $F_s = M(R^2/L)\omega^2 \cos 2\theta$.

The summation of the inertia forces of a multicylinder engine is the

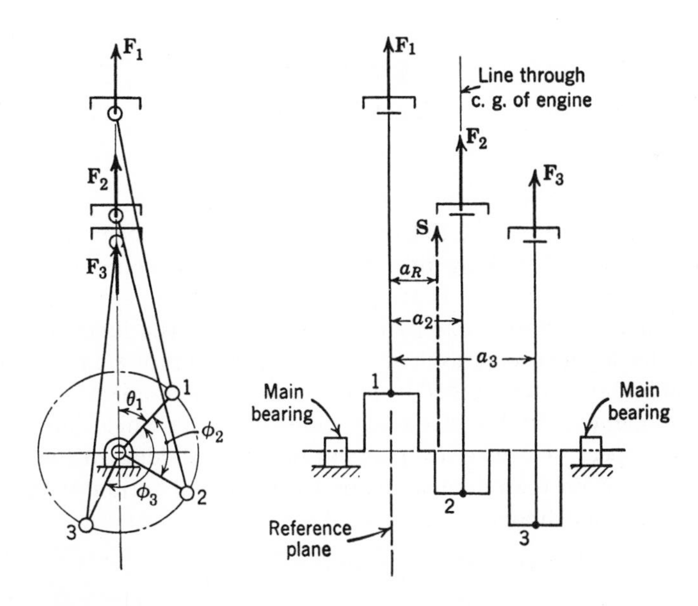

Figure 12.13

resultant force or shaking force S which represents the unbalance.

$$S = \Sigma F = \Sigma F_p + \Sigma F_s \tag{12.8}$$

In some engines, the primary forces may be balanced although the secondary forces may not. The reverse may also be the case. The equation of unbalance of primary forces is developed as follows, in which $\theta = \theta_1 + \phi$.

$$\begin{aligned}\Sigma F_p &= \Sigma MR\omega^2 \cos\theta \\ &= MR\omega^2 \Sigma \cos\theta \\ &= MR\omega^2 \Sigma \cos(\theta_1 + \phi) \\ &= MR\omega^2 \Sigma [\cos\theta_1 \cos\phi - \sin\theta_1 \sin\phi]\end{aligned}$$

Since $\cos\theta_1$ and $\sin\theta_1$ are constant for all terms of the summation,

$$\Sigma F_p = MR\omega^2[\cos\theta_1 \Sigma \cos\phi - \sin\theta_1 \Sigma \sin\phi] \tag{12.9}$$

The equation of unbalance of secondary forces is similar in form.

$$\Sigma F_s = M\frac{R^2}{L}\omega^2[\cos 2\theta_1 \Sigma \cos 2\phi - \sin 2\theta_1 \Sigma \sin 2\phi] \tag{12.10}$$

It may be seen from Eqs. 12.9 and 12.10 that, for any given arrangement of cranks in a multicylinder engine, the angles ϕ are known such that $\Sigma \cos\phi$, $\Sigma \sin\phi$, $\Sigma \cos 2\phi$, and $\Sigma \sin 2\phi$ may be evaluated, and the equations of unbalance become functions only of θ_1. It may also be seen that for balance, or zero shaking force, the following summations must all be zero.

$$\begin{aligned}\Sigma \cos\phi &= 0 \\ \Sigma \sin\phi &= 0 \\ \Sigma \cos 2\phi &= 0 \\ \Sigma \sin 2\phi &= 0\end{aligned}$$

Another mode of shaking must be considered for multicylinder engines. Viewing the engine of Fig. 12.13 from the side, it may be seen that the line of action of the resultant shaking force in the axial plane may not lie on a line of symmetry between the main bearings. Moreover, the line of action of the resultant **S** may be shifting axially in the axial plane as a function of θ_1. In this event, the engine oscillates in an end over end rotational mode. The line of action of **S** may be determined from the principle of moments in terms of a primary moment C_p and a secondary moment C_s, in which moments are taken with respect to a reference plane at the first cylinder. In Fig. 12.13, a is the distance from the reference plane to the line of action of the inertia force of any given cylinder.

$$\begin{aligned}C_p &= \Sigma F_p a = MR\omega^2 \Sigma a \cos\theta \\ &= MR\omega^2[\cos\theta_1 \Sigma a \cos\phi - \sin\theta_1 \Sigma a \sin\phi]\end{aligned} \tag{12.11}$$

and

$$C_s = \frac{MR^2\omega^2}{L}[\cos 2\theta_1 \Sigma a \cos 2\phi - \sin 2\theta_1 \Sigma a \sin 2\phi] \quad (12.12)$$

$$C = C_p + C_s \quad (12.13)$$

The distance a_R of the line of action of the shaking force **S** may be determined from the resultant moment C about the reference plane as follows:

$$a_R = \frac{C}{S} \quad (12.14)$$

In certain cases, the shaking force **S** is zero, indicating a balance of inertia forces, but the resultant moment C is not. In this case, the resultant is a couple C in the axial plane, which produces an end over end axial shaking couple. In some cases, a_R is not a function of θ_1 but is constant. In this case, if constant a_R places the line of action of **S** other than through the center of gravity of the engine, an end over end shaking couple exists.

Example 12.1

Determine the unbalance S of the reciprocating masses of the conventional four-cylinder engine shown in Fig. 12.14, in which the cranks are at 180°. Determine also the unbalance of axial shaking couple.

Solution. The fixed angles ϕ are shown in Fig. 12.14. It should be noted that, although ϕ_1 and ϕ_4 are zero, their cosine functions are unity and must be taken into account in the equations that determine unbalance.

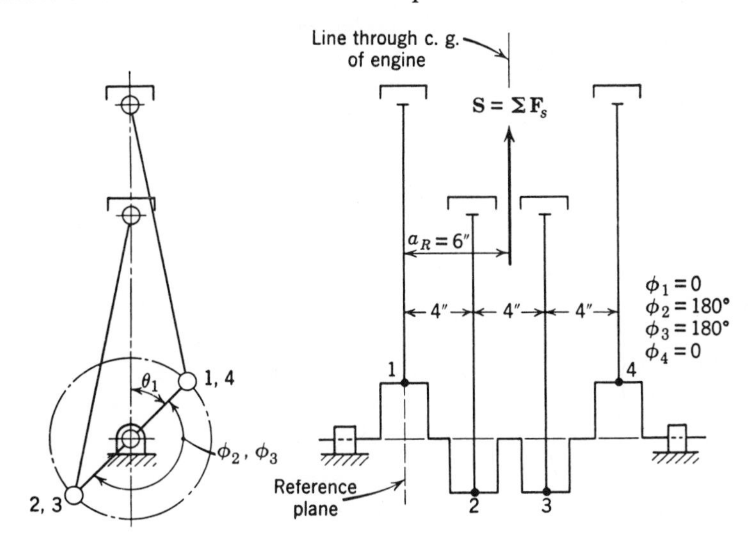

Figure 12.14

The following summations are made to determine the constants which appear in the equations of unbalance.

$$\Sigma \cos \phi = 1 - 1 - 1 + 1 = 0$$
$$\Sigma \sin \phi = 0 + 0 + 0 + 0 = 0$$
$$\Sigma \cos 2\phi = 1 + 1 + 1 + 1 = 4$$
$$\Sigma \sin 2\phi = 0 + 0 + 0 + 0 = 0$$

Referring to Eqs. 12.9 and 12.10, it may be seen that the primary forces are balanced and that the secondary forces are not.

$$\Sigma F_p = MR\omega^2[\cos \theta_1(0) - \sin \theta_1(0)] = 0$$

$$\Sigma F_s = \frac{MR^2\omega^2}{L}[\cos 2\theta_1(4) - \sin 2\theta_1(0)]$$

$$\Sigma F_s = 4\frac{MR^2\omega^2}{L}\cos 2\theta_1$$

$$S = \Sigma F_p + \Sigma F_s$$

$$S = 4\frac{MR^2\omega^2}{L}\cos 2\theta_1 = \frac{MR^2}{L}(2\omega)^2 \cos 2\theta_1 \tag{12.15}$$

Equation 12.15, which gives the shaking force of the conventional four-cylinder engine as a function of θ_1, is shown plotted in Fig. 12.15. It may be seen that the shaking force curve is a simple harmonic curve whose circular frequency (2ω) is twice the speed of the crankshaft.

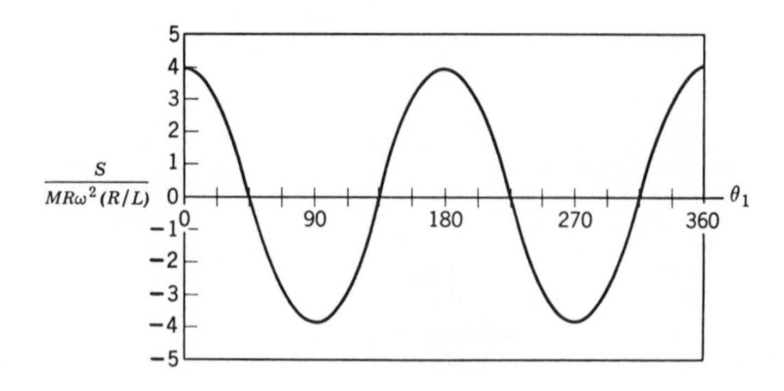

Figure 12.15

The following summations give the constants which apply in the moment equations 12.11 and 12.12.

$$\Sigma\, a \cos \phi = 0(1) + 4(-1) + 8(-1) + 12(1) = 0$$

$$\Sigma\, a \sin \phi = 0(0) + 4(0) + 8(0) + 12(0) = 0$$

$$\Sigma\, a \cos 2\phi = 0(1) + 4(1) + 8(1) + 12(1) = 24$$

$$\Sigma\, a \sin 2\phi = 0(0) + 4(0) + 8(0) + 12(0) = 0$$

Referring to Eqs. 12.11 and 12.12, it may be seen that a secondary moment C_s exists about the reference plane and that the primary moments are zero.

$$C = C_p + C_s = 24 \frac{MR^2\omega^2}{L} \cos 2\theta_1$$

The line of action of the shaking force **S** is determined as follows:

$$a_R = \frac{C}{S} = \frac{24(MR^2\omega^2/L) \cos 2\theta_1}{4(MR^2\omega^2/L) \cos 2\theta_1}$$

$$a_R = 6 \text{ in.}$$

The line of action of the shaking force is constant since a_R is not a function of θ_1. Also, assuming that the line of action of **S** is through the center of gravity of the engine at $a_R = 6$ in., no axial shaking couple exists.

As shown, the only unbalance is a shaking force due to the secondary forces which tend to cause an up and down vibration of the engine. A common device used to balance the secondary forces of a four-cylinder engine is the Lanchester balancer shown in Fig. 12.16. The balancer consists of two meshed gears with eccentric masses as shown. The pitch point of the

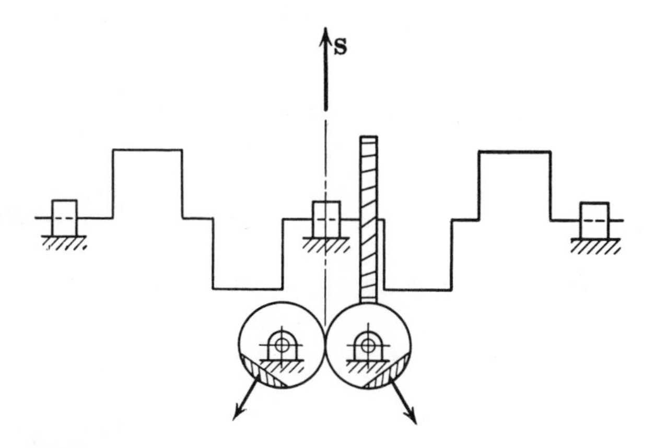

Figure 12.16

meshing gears is directly under the center line of the engine such that the resultant inertia force of the rotating masses counterbalances the shaking force **S**. A crossed helical gear on the crankshaft drives the balancing gears at twice crankshaft speed in order that the balancing forces are of the same circular frequency as the unbalanced secondary forces. In cases where unbalance of primary forces only exists, the balancing gears rotate at crankshaft speed.

Example 12.2

Show that the conventional six-cylinder engine of Fig. 12.12 is in balance according to Eqs. 12.9, 12.10, 12.11, and 12.12. Assume that cylinders are b in. apart.

Solution. To show that the six-cylinder engine is in balance, it is necessary to show that the following summations are zero. From Fig. 12.12, the fixed angles ϕ are

$$\phi_1 = \phi_6 = 0$$

$$\phi_2 = \phi_5 = 240°$$

$$\phi_3 = \phi_4 = 120°$$

$$\Sigma \cos \phi = 1 - \tfrac{1}{2} - \tfrac{1}{2} - \tfrac{1}{2} - \tfrac{1}{2} + 1 = 0$$

$$\Sigma \sin \phi \overset{\Sigma}{=} 0 - \frac{\sqrt{3}}{2} + \frac{\sqrt{3}}{2} + \frac{\sqrt{3}}{2} - \frac{\sqrt{3}}{2} + 0 = 0$$

$$\Sigma \cos 2\phi = 1 - \tfrac{1}{2} - \tfrac{1}{2} - \tfrac{1}{2} - \tfrac{1}{2} + 1 = 0$$

$$\Sigma \sin 2\phi = 0 + \frac{\sqrt{3}}{2} - \frac{\sqrt{3}}{2} - \frac{\sqrt{3}}{2} + \frac{\sqrt{3}}{2} + 0 = 0$$

$$\Sigma a \cos \phi = 0(1) + b(-\tfrac{1}{2}) + 2b(-\tfrac{1}{2}) + 3b(-\tfrac{1}{2}) + 4b(-\tfrac{1}{2}) + 5b(1) = 0$$

$$\Sigma a \sin \phi = 0(0) + b\left(-\frac{\sqrt{3}}{2}\right) + 2b\left(\frac{\sqrt{3}}{2}\right) + 3b\left(\frac{\sqrt{3}}{2}\right) + 4b\left(-\frac{\sqrt{3}}{2}\right) + 5b(0) = 0$$

$$\Sigma a \cos 2\phi = 0(1) + b(-\tfrac{1}{2}) + 2b(-\tfrac{1}{2}) + 3b(-\tfrac{1}{2}) + 4b(-\tfrac{1}{2}) + 5b(1) = 0$$

$$\Sigma a \sin 2\phi = 0(0) + b\left(\frac{\sqrt{3}}{2}\right) + 2b\left(-\frac{\sqrt{3}}{2}\right) + 3b\left(-\frac{\sqrt{3}}{2}\right) + 4b\left(\frac{\sqrt{3}}{2}\right) + 5b(0) = 0$$

Substitution of the preceding summations in Eqs. 12.9, 12.10, 12.11, and 12.12 shows that there is no resultant shaking force and no resultant axial moment, and thus signifies a balance of inertia forces of the six reciprocating masses.

The usual straight-edge engine consists of a combination of two four-cylinder engines at 90° crank angle as shown in Fig. 12.17. One of the four-cylinder engines is split with two cylinders at the front end and two at the rear, and the second four-cylinder engine is in the center. As shown in Example 12.1, the four-cylinder engine is unbalanced as to secondary forces. For the split four-cylinder engine the unbalance is

$$S_1 = 4\frac{MR^2\omega^2}{L}\cos 2\theta_1$$

The shaking force of the middle set of four cylinders in terms of θ_3 of the first cylinder in the middle set is

$$S_2 = 4\frac{MR^2\omega^2}{L}\cos 2\theta_3$$

However, since $\theta_3 = \theta_1 + 270°$,

$$S_2 = 4\frac{MR^2\omega^2}{L}\cos 2(\theta_1 + 270°)$$
$$= -4\frac{MR^2\omega^2}{L}\cos 2\theta_1$$

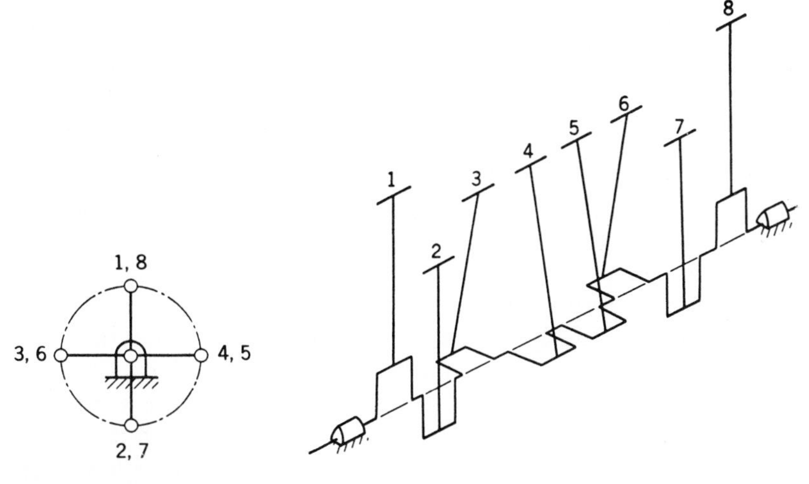

Figure 12.17

Since $S_1 = -S_2$, the resultant shaking force is zero. Also, since the lines of action of $\mathbf{S}_1$ and $\mathbf{S}_2$ are coincident at the center of the engine, there is no axial shaking couple. Thus, like the six-cylinder engine, the straight-eight of Fig. 12.17 is a balanced engine.

Small engines with one, two, or three cylinders are used in a multitude of applications such as outboard motors, mowers, and garden machinery. Air compressors and compressors for spraying are reciprocating machines of one, two, and three cylinders. The balance of reciprocating masses in these machines is poor, as the equations of unbalance would show. The three-cylinder machine with cranks at 120° is in balance as to shaking force but an axial shaking couple exists. Because comfort is unimportant in small engine installations, except possibly with outboard motors, unbalance is tolerable. If the expense is warranted, small engines may be mounted on springs or rubber mounts to isolate the machine vibrations from the frame on which the engine is supported.

12.7 Firing Order. In multicylinder engines, the crank arrangements are such that there is a smooth distribution of torque in the engine cycle as well as a balance of inertia forces of the reciprocating masses. For example, in the four-cylinder engine, a power stroke begins every 180° of crank angle in the following order of cylinder numbers: 1–3–4–2. In the six-cylinder engine, a power stroke begins every 120° of crank angle with a 1–5–3–6–2–4 firing order. The eight-cylinder engine fires every 90° of crank angle. In this discussion of firing order, only the four-stroke cycle engine is considered where one power stroke per cylinder occurs for every 720° of crank rotation. Four events take place in the 720° cycle which are intake, compression, power or expansion, and exhaust.

12.8 V Engines. As shown in Fig. 12.18, the V engine consists of two in-line engines in which the crankshaft is common to both engines. The axial planes in which the two sets of pistons reciprocate intersect at the crankshaft axis and form a V of angle β. In automotive installations, V-8 and V-12 engines are common in which β is either 60° or 90°. Small engines and compressors are often V-2 or V-4.

In Fig. 12.18 is shown a common arrangement of cylinders used in the V-8 engine in which the cranks are at 90° and β is 90°. The engine consists of two four-cylinder in-line engines or two "banks" of four cylinders each. As shown, the connecting rods of each pair of cylinders, one from each bank, are side by side on a common crank or "throw." It may be seen that the side by side arrangement introduces a small axial couple. In some instances, this couple is minimized by reversing the side by side position of some pairs of cylinders from those of other pairs. In other instances, the connecting rods are offset, with the cylinders in the same transverse plane but with the crank-pin ends of the connecting rods set side by side. In the following analysis of

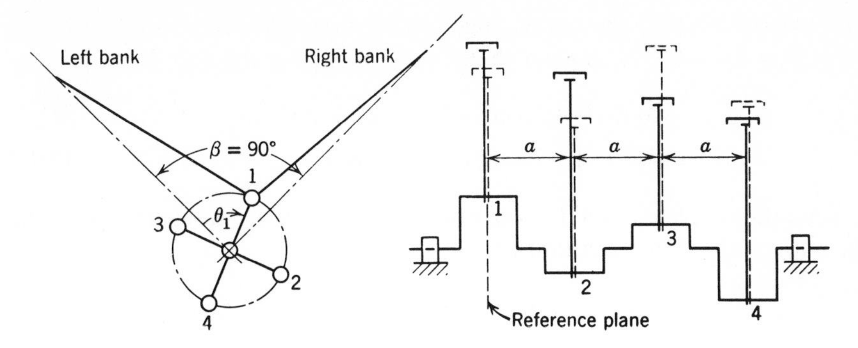

Figure 12.18

balance of the V-8 engine, the effect of the side by side arrangement on balance is neglected.

Since Eqs. 12.9, 12.10, 12.11, and 12.12 apply to in-line engines only, each bank of the V-8 engine may be analyzed separately for balance or unbalance. Any unbalanced force or couple of a given bank is in the axial plane in which the cylinders reciprocate. The resultant unbalance of the complete engine is determined from the vector sum of the unbalance of the two banks.

The following summations apply to either bank since the fixed angles ϕ are the same for both banks.

$$\phi_1 = 0 \qquad \phi_2 = 90^\circ \qquad \phi_3 = 270^\circ \qquad \phi_4 = 180^\circ$$

$$\Sigma \cos \phi = 0 \qquad \Sigma \sin \phi = 0$$

$$\Sigma \cos 2\phi = 0 \qquad \Sigma \sin 2\phi = 0$$

$$\Sigma\, a \cos \phi = -3a \qquad \Sigma\, a \sin \phi = -a$$

$$\Sigma\, a \cos 2\phi = 0 \qquad \Sigma\, a \sin 2\phi = 0$$

As may be seen from the summations, there is no resultant shaking force **S** for either bank since the primary and secondary forces are balanced. However, as shown by the summations, a resultant moment due to primary forces exists, and since **S** is zero the resultant moment is manifest as an axial shaking couple. For the left bank, the axial shaking couple C_L may be evaluated from Eq. 12.11 in terms of θ_1 measured from TDC of the left bank.

$$C_L = MR\omega^2[-3a \cos \theta_1 + a \sin \theta_1] \tag{12.16}$$

For the right bank, the crank angle of the first cylinder is $-(\beta - \theta_1) = \theta_1 - \beta = \theta_1 - 90°$. The axial shaking couple C_R for the right bank is

$$C_R = MR\omega^2[-3a\cos(\theta_1 - 90) + a\sin(\theta_1 - 90)]$$
$$= MR\omega^2[-3a\sin\theta_1 - a\cos\theta_1] \tag{12.17}$$

Since the couples $\mathbf{C}_L$ and $\mathbf{C}_R$ are in axial planes at 90° as shown in Fig. 12.19, the magnitude of the resultant couple C is

$$C = \sqrt{C_L^2 + C_R^2} \tag{12.18}$$

Substitution of the values of C_L and C_R from Eqs. 12.16 and 12.17 gives

$$C = \sqrt{10}\, MR\omega^2 a \tag{12.19}$$

It may be seen that the resultant unbalanced couple of the V-8 engine is independent of θ_1 and is therefore constant in magnitude for all angular positions of the crankshaft. The axial plane in which the resultant couple $\mathbf{C}$ lies is given by the angle α measured clockwise from the plane of the left bank as shown in Fig. 12.19.

$$\tan\alpha = \frac{C_R}{C_L} \tag{12.20}$$

It may be seen that α is a function of θ_1 and that the vector $\mathbf{C}$ rotates with the engine crankshaft. The angle that $\mathbf{C}$ makes with the first crank is γ, which may

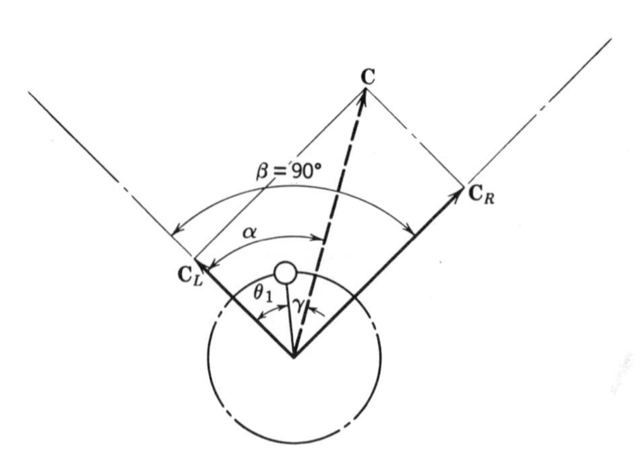

Figure 12.19

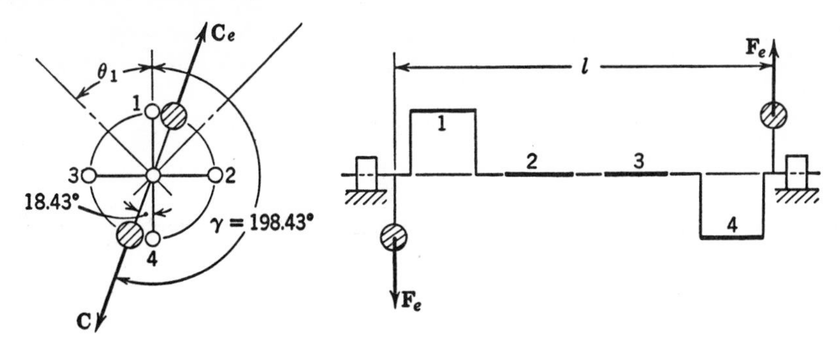

Figure 12.20

be determined as follows, since $\alpha = \theta_1 + \gamma$,

$$\tan(\theta_1 + \gamma) = \frac{C_R}{C_L}$$

$$\frac{\tan\theta_1 + \tan\gamma}{1 - \tan\theta_1 \tan\gamma} = \frac{C_R}{C_L}$$

$$\tan\gamma = \frac{C_R - C_L \tan\theta_1}{C_L + C_R \tan\theta_1} \tag{12.21}$$

Substitution of the values of C_L and C_R from Eqs. 12.16 and 12.17 gives

$$\tan\gamma = \frac{-1}{-3}$$

$$\gamma = 198.43° \qquad \text{(third quadrant)} \tag{12.22}$$

As the crankshaft turns in the clockwise sense, the resultant axial shaking couple **C** acts in an axial plane which leads the axial plane of the first crank by a constant angle $\gamma = 198.43°$, or leads the fourth crank by 18.43°. Figure 12.20 shows the resultant unbalanced axial couple **C** in its correct position with respect to cranks 1 and 4. As shown, the engine is completely balanced by the introduction of a balancing couple $\mathbf{C}_e$, the equilibrant of **C**, in the form of two counterweights such that

$$C_e = F_e l = -\sqrt{10}\, MR\omega^2 a$$

12.9 Opposed Engines. As shown in Fig. 12.21, the opposed engine consists of two banks of cylinders, or two in-line engines on opposite sides of the crankshaft in a common horizontal plane. The opposed engine is a special case of the V engine in which $\beta = 180°$, and the determination of

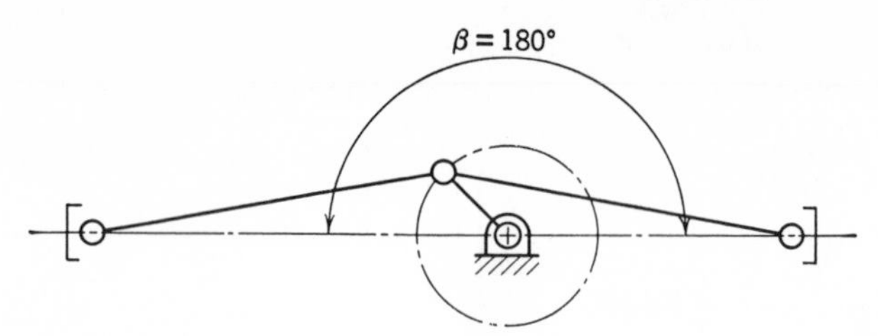

Figure 12.21

balance or unbalance may be made as in V engines. The resultant shaking force S and the resultant unbalanced axial moment C lie in the horizontal plane. Opposed engines of two, four, and six cylinders are commonly used in light aircraft, and six-cylinder installations are used in helicopters and buses.

Problems

12.1 The rigid rotor of Fig. 12.22 shown with three masses is to be balanced by the addition of a fourth mass. Determine the required weight and angular position of the balancing mass, which is to be located at $r_4 = 10$ in. Show answers on a scale drawing of the rotor.

12.2 For the rigid rotor of Fig. 12.23, determine the bearing reactions at A and B for a rotor speed of 2000 r/min.

12.3 Determine the bearing reactions of the rigid rotor of Fig. 12.24 for a rotor speed of 1200 r/min. Determine the mass or masses which should be added to the rotor at a radius of 2 in. in order that the bearing reactions are due only to the weight of the rotor. Show results using one mass and two masses.

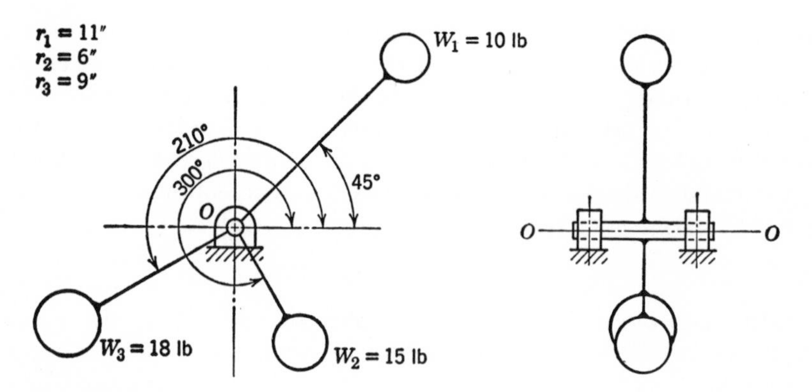

Figure 12.22

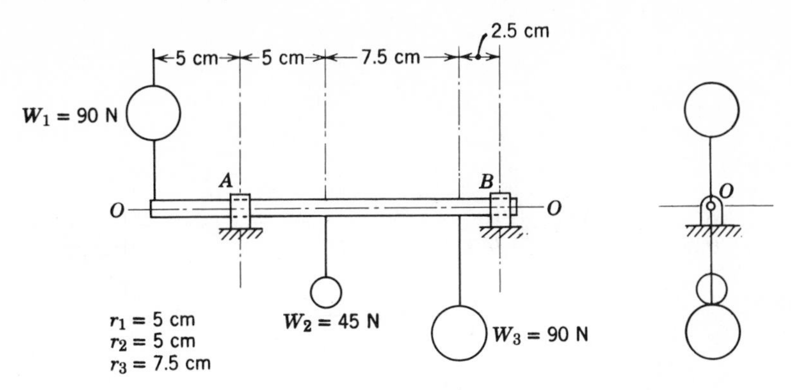

Figure 12.23

12.4 For the rigid rotor of Fig. 12.25 shown with two masses, determine the weights W_A and W_B in planes A–A and B–B respectively which put the rotor in dynamic balance for a rotor speed of 500 r/min. Determine also the angular positions of the balancing weights.

12.5 Weights W_1 and W_2 of the rotor in Fig. 12.26 rotate in the transverse planes shown. Determine the weights W_3 and W_4 in planes 3 and 4, respectively, which give dynamic rotating balance. Shown the correct angular positions of W_3 and W_4.

12.6 The crankshaft of Fig. 12.27 has four equal cranks at 90° and spaced 10 cm apart. Each crank is equivalent to 18 N at a radial distance of 5 cm. Calculate the bearing reactions due to the inertia forces if the shaft is run at 3000 r/min. Balance this system with two weights W_A and W_B in the planes of W_1 and W_3, respectively, and at a radial distance of 5 cm. Determine W_A and W_B, and show their positions.

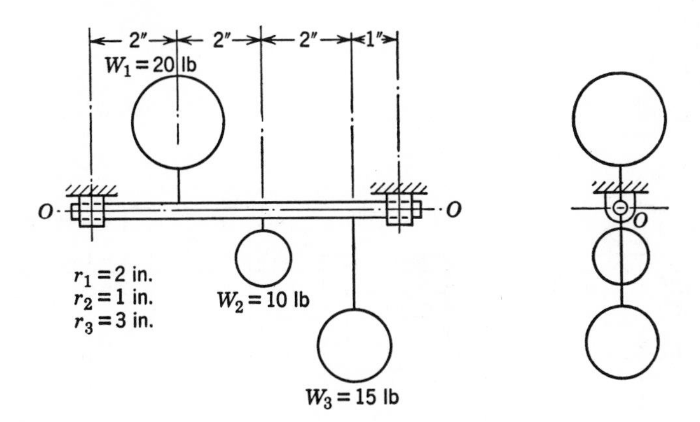

Figure 12.24

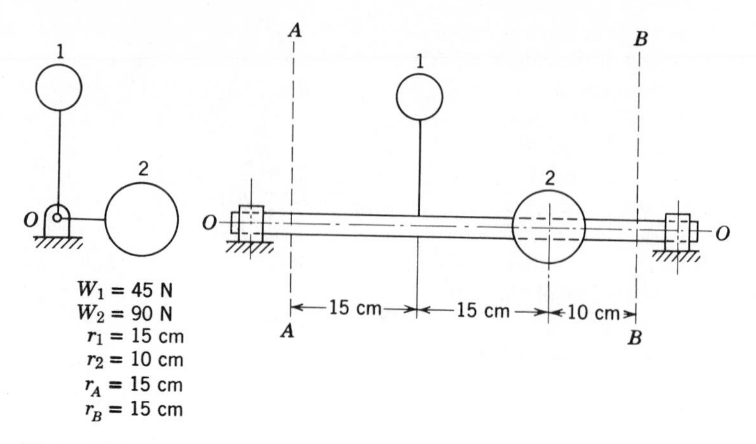

Figure 12.25

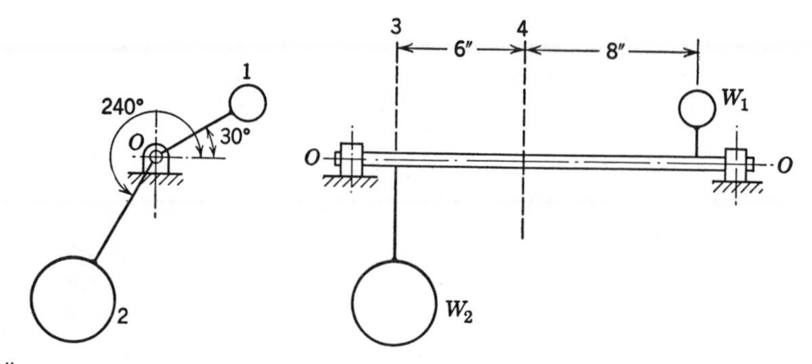

$W_1 = 8$ lb
$W_2 = 20$ lb
$r_1 = 5$ in.
$r_2 = 7\frac{1}{2}$ in.
$r_3 = 4$ in.
$r_4 = 4$ in.

Figure 12.26

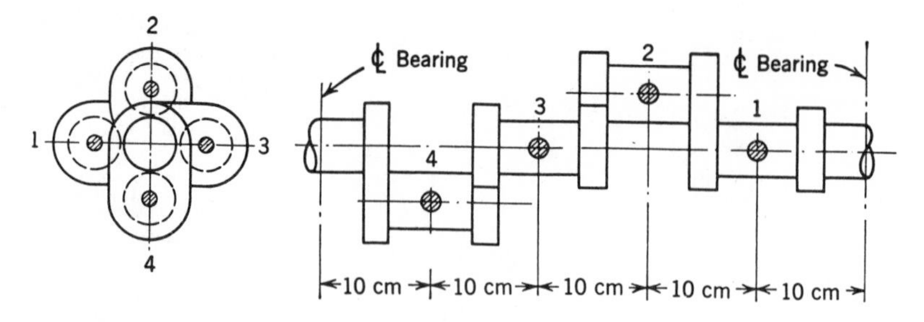

Figure 12.27

12.7 The shaking force produced by a given unbalance in a rotor increases with the rotative speed of the rotor. An unbalance of 1 oz at an eccentricity of 1 in. (Wr = 1 oz-in.) may be small at a low speed and large at a high speed. Calculate the inertia force of 1 oz at 1 in. for speeds in 1000 r/min increments to 10,000 r/min, and plot a curve of inertia force versus speed.

12.8 A jet engine rotor weighs 6700 N. Determine the amount that the center of gravity of the rotor mass may be eccentric from the axis of rotation to produce an inertia force equal to the weight of the rotor at speeds of 1000, 5000, and 10,000 r/min.

12.9 The degree of unbalance permitted in rotors is often specified by limiting the centripetal acceleration of the center of gravity of the rotor to $g/4$. Determine the eccentricity which produces this amount of acceleration at 5000 r/min, and give the permissible amount of unbalance in N-cm (Wr) for a 4500 N rotor.

12.10 The rotor with the steel gears shown in Fig. 12.28 was dynamically balanced in a balancing machine by the addition of the clay masses shown on the periphery of the gears. However, the balancing is to be achieved by drilling holes in the webs of the gears at the diameters shown. Determine the size and location of the holes for dynamic balance.

12.11 In the Scotch yoke mechanism shown in Fig. 12.29, the yoke (link 4) is in simple harmonic motion when the crank of length R rotates at constant angular velocity ω_2. Write a methematical expression for shaking force due to the reciprocating mass M of the yoke.

12.12 For the Scotch yoke mechanism of Fig. 12.29, sketch the supporting frame (link 1) of the mechanism as a free body and show vectors of forces and couples imposed on the frame by the moving parts of the mechanism in approximately the phase shown. Link 2 is driven at constant angular velocity by torque applied at O_2. Designate the shaking force and the shaking couple.

12.13 In Fig. 12.30, the weight of the reciprocating weight at D is 28.6 N, and at C the reciprocating weight is 14.3 N. Determine the resultant unbalanced force due to the reciprocating masses for the phase shown. $R\omega^2 = 305$ m/sec^2.

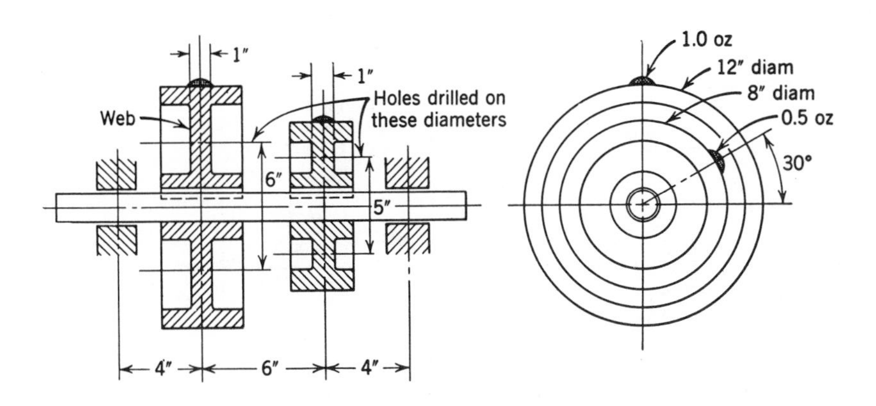

Figure 12.28

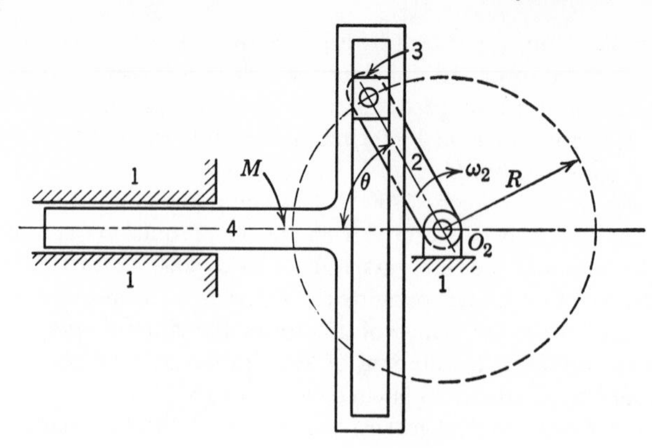

Figure 12.29

12.14 In the three-cylinder radial engine shown in Fig. 12.31, all of the connecting rods are attached to a common crank. The reciprocating mass of each cylinder is M_r, and the equivalent mass of each connecting rod at the crank pin is M_A. M_r and M_A are equal. The mass center of the crank is at O, but there is no counterweight to balance the masses M_A at the crank pin. Calculate inertia forces, and, using force polygons, determine the resultant shaking force **S** on the engine for the phase shown when the crank speed is such that $M_r R\omega^2 = 1000$ lb. $R/L = \frac{1}{4}$. Show **S** as a vector on the drawing of the mechanism.

12.15 Using the data of Problem 12.14 and Eq. 11.47, calculate the shaking couple produced by the reciprocating weights of the three-cylinder engine of Fig. 12.31, when $\theta_1 = 30°$ and $R = 3$ in.

12.16 Draw a sketch of the engine block of the three-cylinder radial engine of Fig. 12.31, and show vectors of forces imposed on the block by the reciprocating inertia forces of the slider-crank mechanisms.

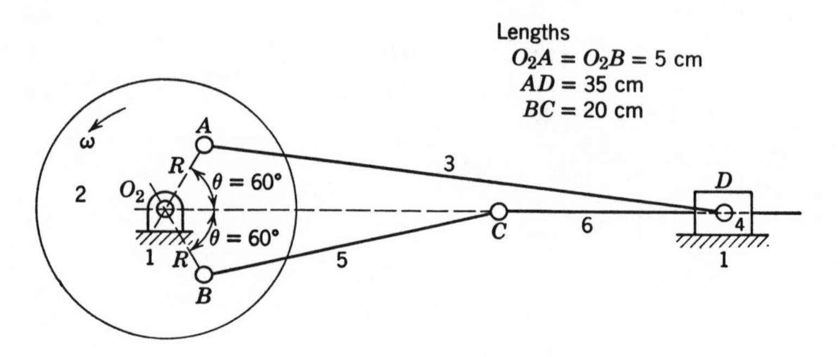

Figure 12.30

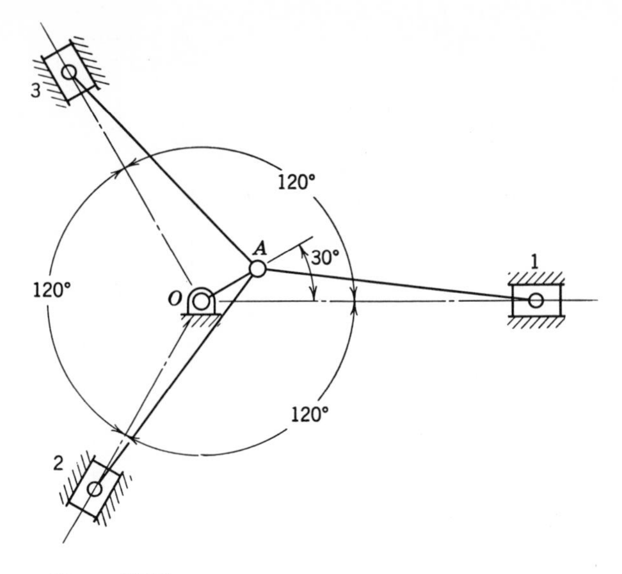

Figure 12.31

12.17 Figure 12.32 shows the four-cylinder mechanism of Fig. 12.14 in the engine block and shows the shaking force **S** at the center line of the four cylinders. M_e is the mass of the complete engine including the block, and the center of gravity is located c distance from the line of action of **S**. The engine is supported by motor mounts having spring constants k. Due to the reciprocating shaking force, the engine vibrates. For the displacements x and ϕ shown, write the equations of motion $\Sigma F_x = M_e A = M_e(d^2x/dt^2)$ and $\Sigma T = I\alpha = I(d^2\phi/dt^2)$.

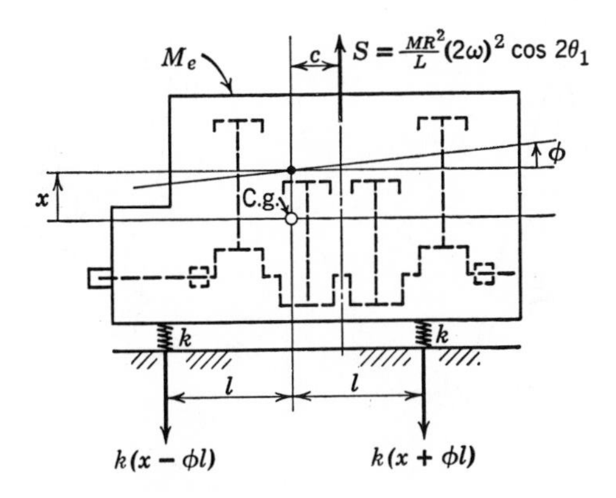

Figure 12.32

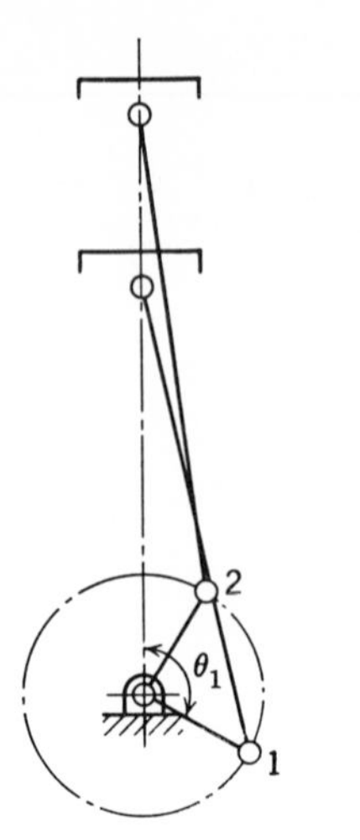

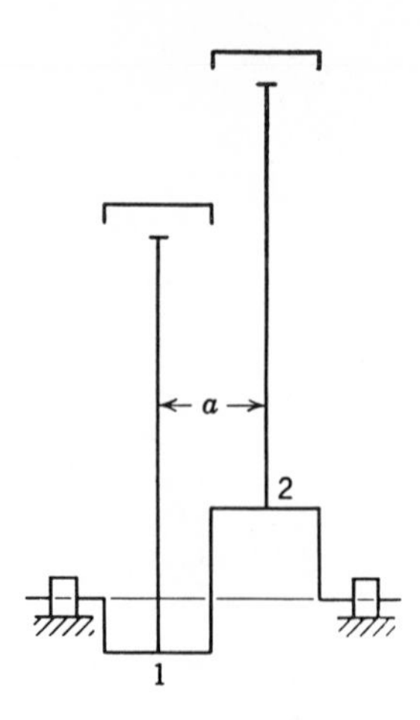

Figure 12.33

12.18 By the analytical method, determine the equations of unbalance of reciprocating masses for the two-cylinder engine of Fig. 12.33 in which the cranks are at 90°. Determine equations for S and a_R in terms of θ_1. Determine S and a_R for $\theta_1 = 30°$, $MR\omega^2 = 8900$ N, $R/L = \frac{1}{4}$, and $a = 10$ cm.

12.19 Work Problem 12.18 for a two-cylinder engine with cranks at 180° instead of 90°.

12.20 For the three-cylinder engine of Fig. 12.34 with cranks as shown, determine the equation of the unbalanced shaking force S of reciprocating masses in terms of θ_1.

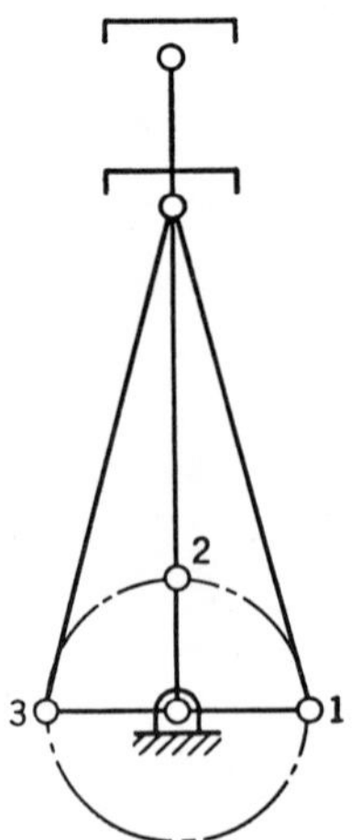

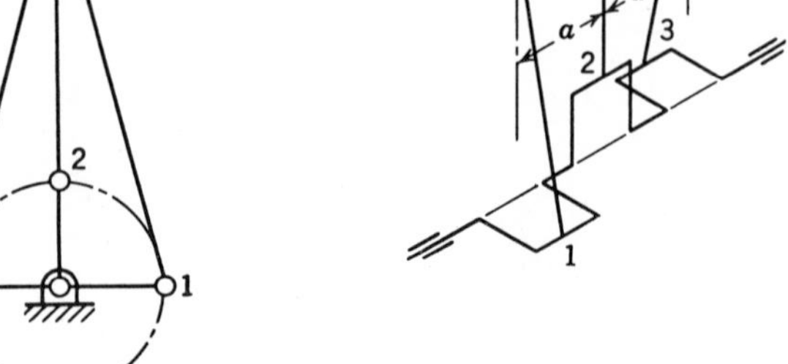

Figure 12.34

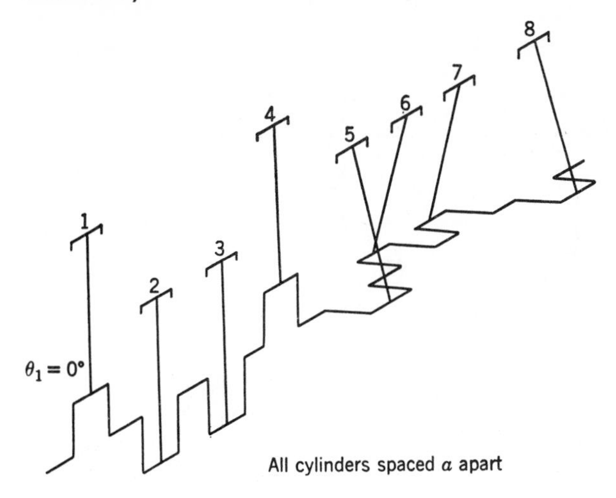

Figure 12.35

Determine also the equation for the distance a_R of the line of action of S from the plane of cylinder 1. Plot curves of S versus θ_1 and a_R versus θ_1 for a complete engine cycle using the following data: Reciprocating weight, 3.22 lb; crank speed, 3000 r/min; stroke, 4 in.; $R/L = \frac{1}{4}$; distance between cylinders, 4 in.

12.21 The equations of unbalance of reciprocating masses for the conventional four-cylinder engine are developed in Example 12.1. The straight-eight-cylinder engine in Fig. 12.35 consists of two four-cylinder engines in tandem with their crank throw planes at 90°. Determine the magnitude and direction of the resultant shaking force or couple for the phase shown ($\theta_1 = 0$).

12.22 For the two-cylinder 90° V engine shown in Fig. 12.36, derive the following equations of unbalance as a function of θ_1: resultant primary force F_p, resultant secondary

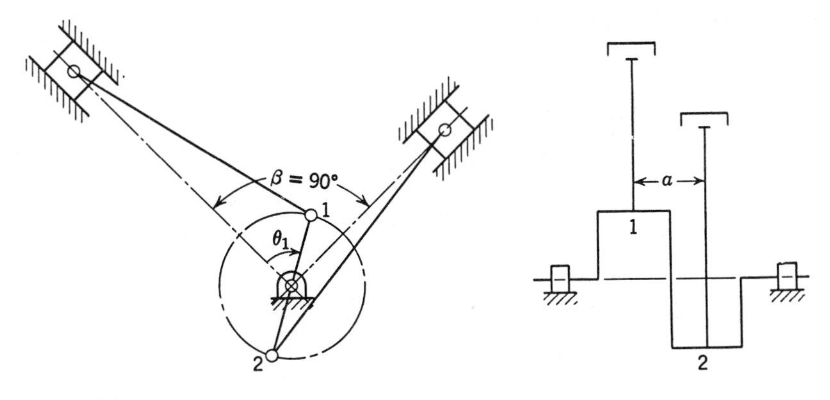

Figure 12.36

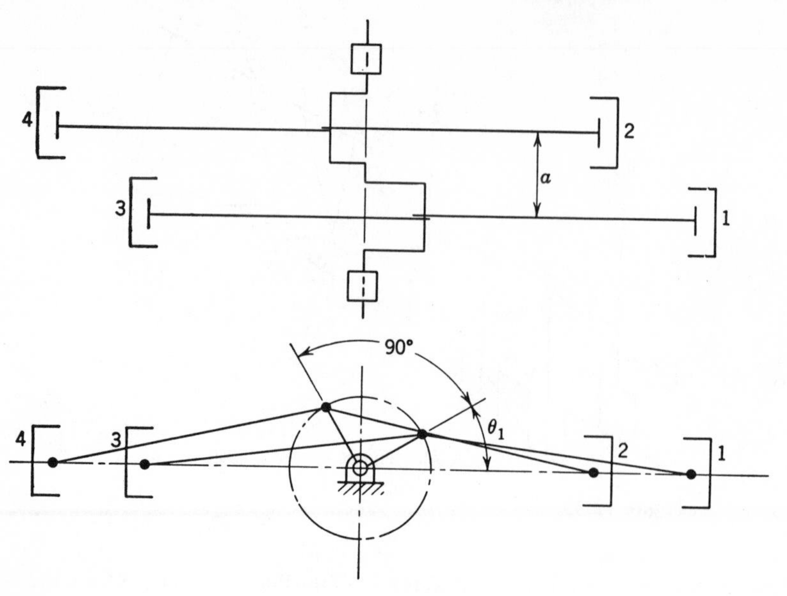

Figure 12.37

force F_s, resultant shaking force S, direction of shaking force, and the distance a_R from the plane of the first cylinder to the line of action of S. Using the equations, determine S and a_R for $\theta_1 = 60°$. $MR\omega^2 = 1$, $R/L = \frac{1}{4}$, $a = 10$ cm.

12.23 For the four-cylinder opposed engine of Fig. 12.37, derive the following equations (in terms of θ_1) of unbalance due to reciprocating masses: primary force F_p, secondary force F_s, shaking force S, and the distance a_R from the plane of cylinder 1 to the line of action of S. Evaluate S and a_R for $\theta_1 = 90°$, assuming $MR\omega^2$ and the distance between sets of cylinders are unity. $R/L = \frac{1}{4}$. For what angle or angles θ_1, if any, will the resultant primary force be zero?

13

Vibration in Machines

13.1 Introduction. Vibration is inherent in all machines because of the motion of the individual parts which rotate, oscillate, or reciprocate. When the forces on an individual part are such that the displacement of the mass center of the part is oscillating or periodically reversing in sense, it is said to be vibrating. The reciprocating motion of the piston of the slider-crank is a simple example of vibration. A more complex motion is the displacement of the mass center of the connecting rod of the slider crank, which periodically reverses sense in two coordinate directions; in addition, the connecting rod undergoes reversing angular displacements. However, as shown in Chapter 12, the connecting rod mass may be considered two equivalent masses having simpler motions. A rotor with mass center at the axis of rotation theoretically does not vibrate. However, if the mass center of the rotor is even slightly eccentric with the axis of rotation, vibration occurs; the mass center moves on a circular path with coordinate displacements having simple harmonic motion.

In the preceding paragraph, vibration is described in terms of the motion of rigid bodies. However, in real machines, the members are elastically flexible, some members being more flexible than others. Members having relatively high elastic flexibility are referred to as *springs*. Consider a rotor consisting of a large rigid mass and a relatively flexible shaft (spring) simply supported by two rigidly held bearings. If the mass center of the large mass is slightly eccentric with the axis of rotation, the inertia force induced by the rotation causes the shaft to flex such that the eccentricity is increased with an accompanying increase in the inertia force. However, the shaft resists bending as the eccentricity increases and the rotor eventually comes to equilibrium at some eccentricity greater than the initial eccentricity. Under certain conditions, bending failure of the shaft results because the eccentricity tends to approach very large values before equilibrium is achieved. As the equilibrium eccentricity becomes large, the forces transmitted through the bearings of the shaft also become large and may affect the displacement of other springy members.

The degree to which vibration is undesirable depends on how highly parts are stressed because of the vibration or on the disturbance caused by the shaking. Disturbance might mean lack of riding comfort, as in vehicles, or lack of steadiness, as in an airplane camera mounting.

Vibration problems of varying complexity are numerous and are treated in many excellent textbooks. It is intended in this chapter to discuss vibration theory in connection with elementary systems and to point out factors which are important in designing machine parts subject to vibration. Much emphasis is placed on the vibration of rotating shafts, since shafts are often the most flexible members in many machines, for example, in jet engines, pumps, electric motors, piston engines, and compressors. Vibration isolation is also treated to illustrate how unavoidable vibration of a machine may be limited in transmission to adjacent supporting frameworks.

13.2 Forced Vibration. Figure 13.1 shows several cases of a mass M forced to vibrate under the action of a periodically reversing force, as in a and b, or under the action of reversing displacement of a support, as in c. In all three cases the mass under consideration is constrained to move in one direction only and therefore has only one degree of freedom. These cases are grouped together because the differential equation of motion and its solution are similar for the three cases.

Figure 13.1a shows a mass M with mass center displaced upward a distance x under the action of the periodically reversing force $F_o \sin \omega t$. The force of the spring S acts downward to resist the upward displacement of the mass. For most elastic springs, the spring force is proportional to the deflection of the spring such that $S = k\delta = kx$ may be written, in which k is the constant of proportionality or spring constant, and δ is the deflection of the spring equal to x. The free-body diagram of the mass shows two forces acting, and the following equation of motion may be written

$$\Sigma F_x = MA_x$$

MA_x is not written as an inertia force since the magnitude and the sense of

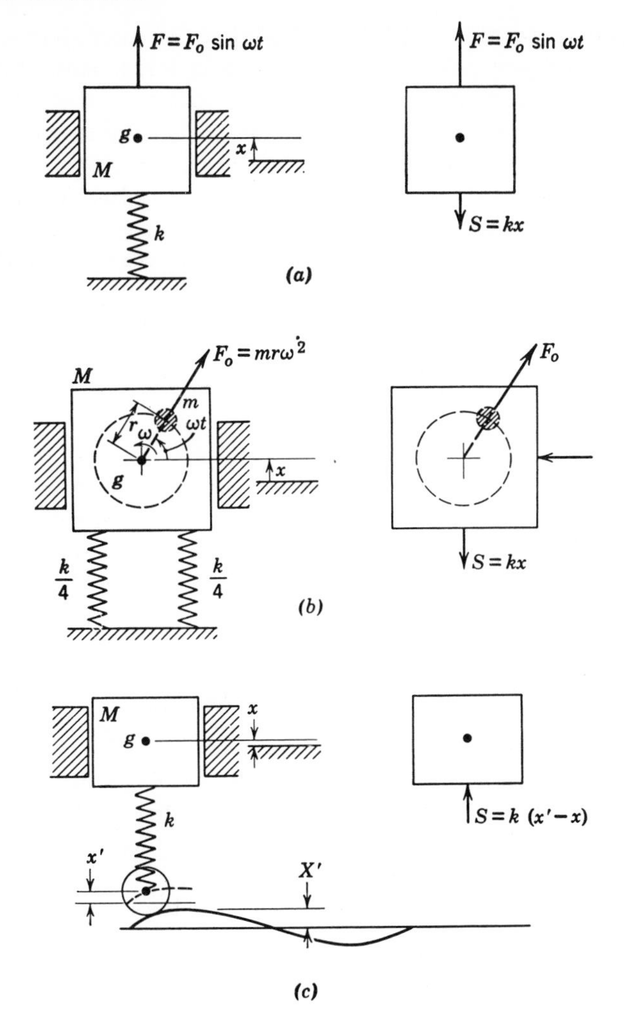

Figure 13.1

A_x are unknown.

$$F_o \sin \omega t - kx = MA_x = M\frac{d^2x}{dt^2}$$

$$\frac{d^2x}{dt^2} + \frac{k}{M}x = \frac{F_o}{M}\sin \omega t \tag{13.1}$$

It should be noted that the gravity force W has been omitted from the free-body diagram of Fig. 13.1*a*. It is left to the reader to prove that Eq. 13.1 would be unchanged if W were included because the spring force due to W is always equal and opposite to W.

In Fig. 13.1*b*, a mass similar to an engine block supported symmetrically by four equal springs is displaced upward a distance x under the action of the inertia force of a mass m rotating with angular velocity ω about a fixed axis in the block. The total mass of the free body shown is M, which includes the rotating mass m. The equation of motion for the free body may be written as follows:

$$\Sigma F_x = MA_x$$

$$F_o \sin \omega t - 4\frac{k}{4}x = MA_x = M\frac{d^2x}{dt^2}$$

$$\frac{d^2x}{dt^2} + \frac{k}{M}x = \frac{F_o}{M}\sin \omega t \tag{13.2}$$

Equation 13.2 is the same in form as Eq. 13.1, in which F_o is the rotating inertia force equal to $mr\omega^2$. Another inertia force acts on m due to d^2x/dt^2, but this is included in $M(d^2x/dt^2)$.

Figure 13.1*c* is a simplified representation of a vehicle of mass M moving on a rough road with a spring suspension between the mass and the wheel. The road is a forcing disturbance which may be represented by a sine wave displacing the wheel upward a distance $x' = X' \sin \omega t$, assuming that the wheel always remains in contact with the road. X' is the amplitude of the road roughness as shown. In this case, the forcing of the displacement of M is accomplished by displacing the wheel. As the wheel rises, the spring is compressed to displace the mass M upward a distance x, which is smaller than x' in order that the spring be in compression. Thus the spring force S acting on M is proportional to the difference of the two displacements or $k(x' - x)$.

$$\Sigma F_x = MA_x$$

$$k(x' - x) = MA_x = M\frac{d^2x}{dt^2}$$

$$kx' - kx = M\frac{d^2x}{dt^2}$$

$$kX' \sin \omega t - kx = M\frac{d^2x}{dt^2}$$

$$\frac{d^2x}{dt^2} + \frac{kx}{M} = \frac{kX'}{M}\sin \omega t$$

$$\frac{d^2x}{dt^2} + \frac{kx}{M} = \frac{F_o}{M}\sin \omega t \tag{13.3}$$

Equation 13.3 is similar to Eqs. 13.1 and 13.2 with F_o representing the force kX', which is a rotating force vector.

In the differential equation 13.3, k, F_o, M, and ω are constants and the variables are x and t. The solution of Eq. 13.3 is $x = C_1 \cos \omega_n t + C_2 \sin \omega_n t + X \sin \omega t$, consisting of three harmonic terms. Two of the terms are harmonic vibrations having a circular frequency $\omega_n = \sqrt{k/M}$, and the third is one having a circular frequency ω of the disturbing force. The first two terms of the solution comprise the general solution which applies for the case in which the right-hand term of Eq. 13.3 is zero. This case is one in which the mass would vibrate freely at a natural frequency ω_n. However, from experience it is known that these vibrations eventually die out due to damping, and in the case of forced vibrations they are transient vibrations that soon disappear. Thus, $x = X \sin \omega t$, which is the particular solution of Eq. 13.3, may be accepted as representing the steady-state vibration. X is the amplitude of forced vibration which may be determined by substitution of $x = X \sin \omega t$ in Eq. 13.3 as follows:

$$\frac{d^2}{dt^2}(X \sin \omega t) + \frac{k}{M} X \sin \omega t = \frac{F_o}{M} \sin \omega t$$

$$-(X\omega^2 \sin \omega t) + \frac{k}{M} X \sin \omega t = \frac{F_o}{M} \sin \omega t \tag{13.4}$$

From Eq. 13.4 it may be seen that $\sin \omega t$ is common in all terms of the equation and that all terms are simple harmonic curves of acceleration as shown in Fig. 13.2. The resultant of the harmonic curves of the left-hand terms of Eq. 13.4 equals the right-hand term if the proper value of X is determined. Canceling $\sin \omega t$ from all terms,

$$-X\omega^2 + \frac{k}{M} X = \frac{F_o}{M}$$

$$X = \frac{F_o/M}{(k/M) - \omega^2} \tag{13.5}$$

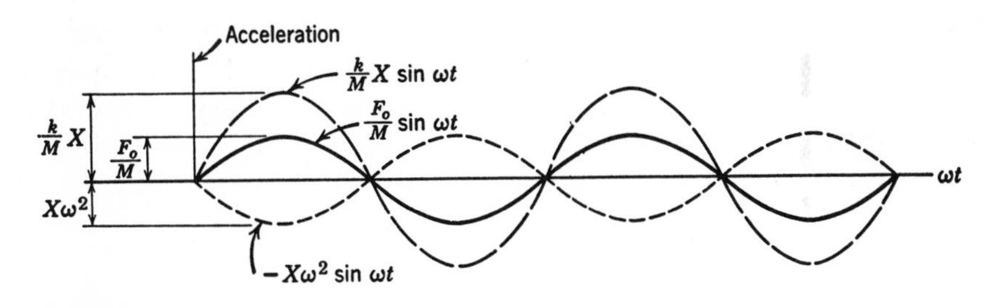

Figure 13.2

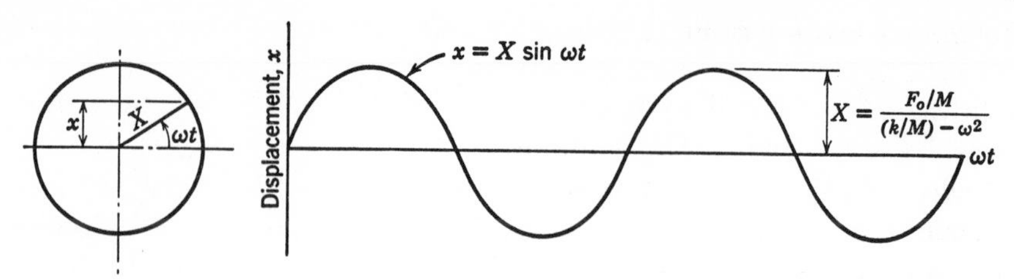

Figure 13.3

Equation 13.5 represents the amplitude of vibration of the resultant simple harmonic curve of displacement given by

$$x = X \sin \omega t = \frac{F_o/M}{(k/M) - \omega^2} \sin \omega t \tag{13.6}$$

as shown in Fig. 13.3. In Eq. 13.6, x is the displacement of M, which appears in Fig. 13.3 as the projection on the vertical axis of a circle having a radius X rotating at the circular frequency ω the same as ω of the rotating vector F_o. Since there are 2π radians in a cycle of vibration, the frequency in cycles per unit time is given by

$$f = \frac{\omega}{2\pi} \tag{13.7}$$

If ω is in radians per second, f is in cycles per second (cycles/sec).[1]

Of primary interest is the maximum displacement of the vibration given by X in Eq. 13.5, which shows that when $\omega^2 = k/M$, the maximum displacement goes to infinite values because the denominator becomes zero. This is a condition to avoid in design where an infinite displacement would cause extreme shaking or breaking stresses.

13.3 Natural Vibration. In the cases of Fig. 13.1, if the forcing function $F_o \sin \omega t$ is absent, a vibration may be induced by displacing the mass M an amplitude X and then releasing the mass, permitting it to vibrate freely. Under these conditions the mass will vibrate naturally between the extremes X and $-X$ with a circular frequency $\omega_n = \sqrt{k/M}$. This may be shown by writing the equation of motion, which is the same as Eq. 13.3, with the right-hand term equal to zero.

$$\frac{d^2x}{dt^2} + \frac{kx}{M} = 0 \tag{13.8}$$

[1] In the International System of Units, 1 cycle/sec = 1 Hz (hertz).

The general solution of Eq. 13.8 is

$$x = C_1 \cos \sqrt{k/M}\, t + C_2 \sin \sqrt{k/M}\, t$$

C_1 and C_2 are constants that may be determined from the boundary conditions of the vibration. If time is started when the mass is at $x = 0$, then the boundary conditions are $x = 0$ at $t = 0$ and $x = X$ at $\sqrt{k/M}\, t = \pi/2$. For these conditions, $C_1 = 0$ and $C_2 = X$, and the solution becomes $x = X \sin \sqrt{k/M}\, t$. Inspection of the solution shows that the displacement of the mass is simple harmonic, in which X is a vector rotating at an angular velocity or circular frequency

$$\omega_n = \sqrt{k/M} \tag{13.9}$$

Thus ω_n is the natural circular frequency depending only on the mass M and the spring constant k. In the absence of damping, the natural free vibration would continue indefinitely. Since an undamped natural vibrating system is a conservative system, the expression of Eq. 13.9 may be derived from energy considerations. In initially displacing the mass X distance, potential energy equal to $kX^2/2$ is stored in the spring. When released, the mass returns to $x = 0$ with a velocity $V = X\omega_n$ such that the kinetic energy $\frac{1}{2}MV^2 = \frac{1}{2}MX^2\omega_n^2$, but the potential energy is zero. Since the system is conservative,

$$\tfrac{1}{2}kX^2 = \tfrac{1}{2}MX^2\omega_n^2$$

$$k = M\omega_n^2$$

$$\omega_n = \sqrt{\frac{k}{M}}$$

In real vibrating systems, however, the energy is not conserved but is eventually dissipated in working against friction.

In machines, forced vibrations rather than natural vibrations occur. Natural vibrations are of interest primarily because the natural circular frequency influences the amplitudes of forced vibrations as indicated in Eq. 13.5. When the circular frequency ω of the forcing function is equal to ω_n, a condition known as *resonance* occurs causing vibration amplitudes to become excessive. In design, ω, k, and M must be chosen to avoid resonant conditions.

In determining ω_n, k is normally given in pounds per inch and M may be given as W/g, where W is weight in pounds and $g = 386$ in./sec^2. ω_n is then in radians per second, and $f_n = \omega_n/2\pi$ is in cycles per second.

13.4 Amplitude of Forced Vibration. Equation 13.5 may be rewritten in the following forms:

$$X = \frac{F_o/M}{(k/M) - \omega^2} = \frac{F_o/M}{(\omega_n^2 - \omega^2)}$$

$$= \frac{F_o/M}{\omega_n^2[1 - (\omega/\omega_n)^2]} = \frac{F_o/M}{(k/M)[1 - (\omega/\omega_n)^2]}$$

$$X = \frac{F_o/k}{1 - (\omega/\omega_n)^2}$$

$$\frac{X}{F_o/k} = \frac{1}{1 - (\omega/\omega_n)^2} \tag{13.10}$$

Equation 13.10 is in nondimensional form showing amplitude as a function of ω and applies for the case of Fig. 13.1*a*. The ratio $X/(F_o/k)$ is known as the magnification factor since it shows the magnitude of the amplitude in terms of the statical displacement caused by F_o. For Fig. 13.1*b*, since F_o is a function of ω in $F_o = mr\omega^2$, Eq. 13.5 may be written in comparable nondimensional form:

$$\frac{XM}{mr} = \frac{(\omega/\omega_n)^2}{1 - (\omega/\omega_n)^2} \tag{13.11}$$

For Fig. 13.1*c*, since $F_o = X'k$

$$\frac{X}{X'} = \frac{1}{1 - (\omega/\omega_n)^2} \tag{13.12}$$

Equations 13.10 and 13.12 are similar functions of ω, whereas Eq. 13.11 is a somewhat different function of ω. These equations are shown plotted in Figs. 13.4*a* and *b*. As may be seen, the equations are somewhat different in form although both curves show a critical condition at $\omega/\omega_n = 1.0$ where amplitude becomes infinite. In Fig. 13.1*a* and *c* the vibrating amplitude becomes smallest at high ratios of ω/ω_n, whereas for Fig. 13.1*b* it is smallest at $\omega/\omega_n = 0$. It should be noted from Eqs. 13.10 through 13.12 that values of the ordinate of Fig. 13.4 are negative for $\omega > \omega_n$ indicating that the displacement X is 180° out of phase with F_o. However, since only magnitude of amplitude is of importance, the curve may also be shown with positive ordinates for values of $\omega/\omega_n > 1.0$.

Example 13.1

The 4448 N vehicle shown in Fig. 13.5 (same as Fig. 13.1*c*) travels along a rough road at 96.6 km/h (26.8 m/sec). The sine curve representing the rough road has an amplitude of 2.54 cm and a wave length of 6.09 m.

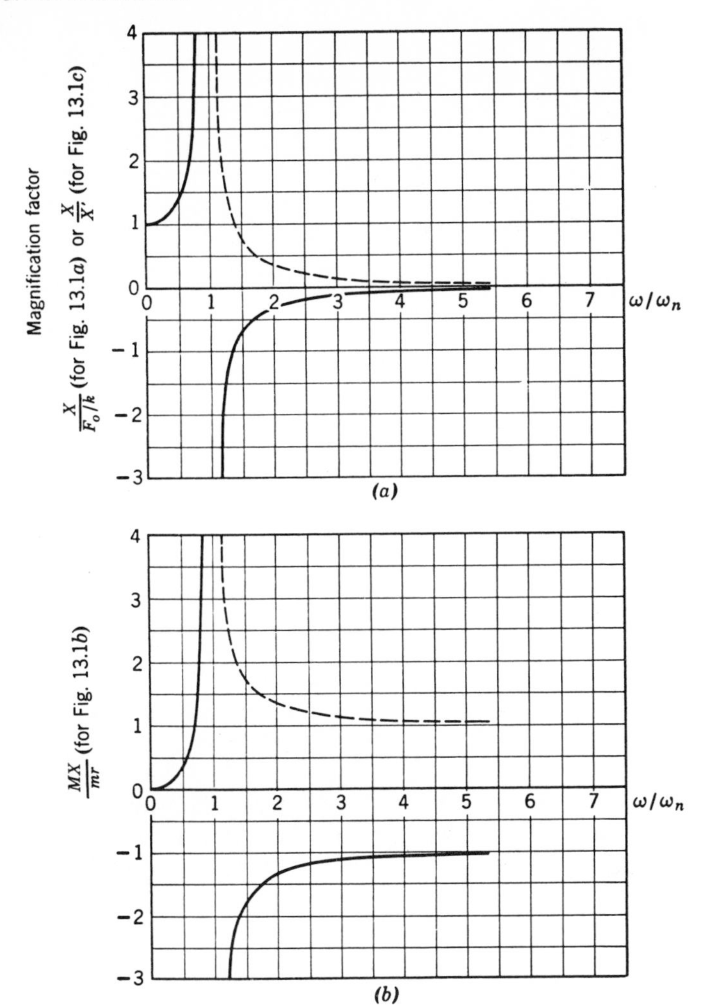

Figure 13.4

Determine (*a*) The spring constant k such that 96.6 km/h is the critical speed at which resonance occurs. (*b*) The spring constant to give a vibration amplitude of 0.636 cm of the vehicle at 96.6 km/h. (*c*) Whether a rider in the vehicle would leave his seat under the conditions of *b*.

Solution. (*a*) Resonance occurs at the condition $\omega = \omega_n$. At 26.8 m/sec the frequency f of the disturbance is 26.8/6.09 = 4.4 Hz. From Eq. 13.7,

$$\omega = 2\pi f$$

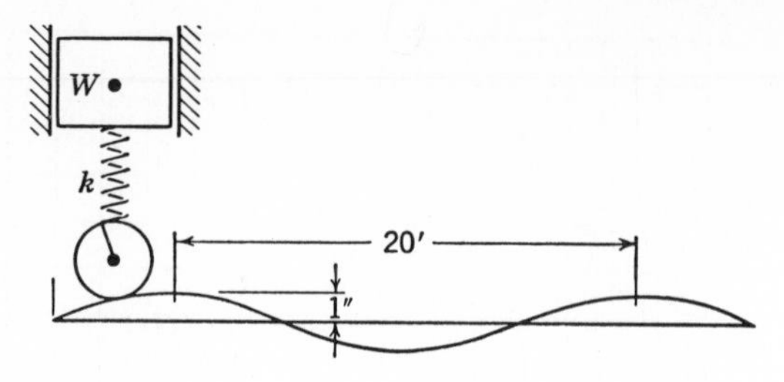

Figure 13.5

The natural frequency ω_n is given by

$$\omega_n^2 = \frac{k}{M} = \frac{kg}{W}$$

For resonance,

$$\omega_n^2 = \omega^2$$

$$\frac{kg}{W} = (2\pi f)^2$$

$$k = (2\pi f)^2 \frac{W}{g} = (2\pi 4.4)^2 \frac{4448 \text{ N}}{980.7 \text{ cm/sec}^2}$$

$$k = 3467 \text{ N/cm}$$

(*b*) $X = 0.636$ cm is the amplitude of the vehicle vibration, and the amplitude of the disturbance is $X' = 2.54$ cm. The ratio X/X' is $\frac{1}{4}$, which from the curve of Fig. 13.4*a* indicates that the ratio ω/ω_n should be 2.2.

$$\frac{\omega}{\omega_n} = 2.2$$

$$\omega_n = \frac{\omega}{2.2}$$

$$\sqrt{\frac{kg}{W}} = \frac{2\pi f}{2.2}$$

$$k = \left(\frac{2\pi f}{2.2}\right)^2 \frac{W}{g} = \left(\frac{2\pi 4.4}{2.2}\right)^2 \times \frac{4448}{980.7}$$

$$k = 717.9 \text{ N/cm}$$

(*c*) The vibration equation for displacement is $x = X \sin \omega t$. The second derivative gives the acceleration $A = -X\omega^2 \sin \omega t$, the maximum absolute value of which is

$$A_{max} = X\omega^2 = X(2\pi f)^2 = 0.636(2\pi 4.4)^2$$
$$A_{max} = 482.6 \text{ cm/sec}^2$$

Since 482.6 cm/sec² is less than g = 980.7 cm/sec², the rider will not leave his seat.

As shown by the calculations, a relatively soft spring (k = 717.9 N/cm compared with k = 3467 N/cm) gives a more comfortable ride. It should be noted that, to realize $\omega/\omega_n = 2.2$ at 96.6 km/h, the vehicle must pass through the critical speed given by $\omega/\omega_n = 1.0$, which is at a speed less than 96.6 km/h. However, with damping devices such as shock absorbers, critical speeds may be passed through with relatively small vehicle vibrations. It should also be noted that the static deflection δ_{st} of the spring due to the vehicle weight is W/k = 4448/717.9 = 6.19 cm and that the spring deflection during vibration at 96.6 km/h varies between 6.19 + 0.636 and 6.19 − 0.636 cm.

13.5 Transmissibility. Figure 13.6 shows the case of Fig. 13.1*b*, in which a vibrating machine is supported by a structure which simulates a building floor. The vibration originating in the machine is transmitted to the floor through the springs S. Thus, the floor is subject to a forced vibration in which the forcing function is the periodically reversing spring force $kx = kX \sin \omega t$. The floor vibration in turn causes discomfort or the shaking of other objects on the floor such as cameras or instruments. The maximum force transmitted through the springs during a cycle of vibration is $F_{tr} = kX$. It is possible to design the springs S such that F_{tr} is near zero, in which case the vibrating machine is said to be isolated.

Since X is given by Eq. 13.5, the transmitted spring force is

$$F_{tr} = kX = \frac{F_o}{1 - (\omega/\omega_n)^2} \tag{13.13}$$

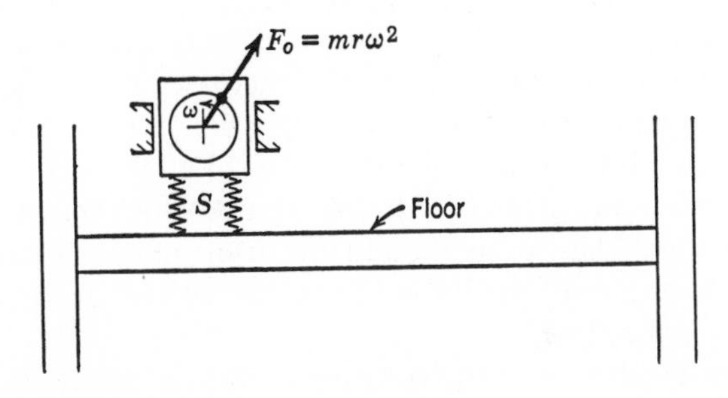

Figure 13.6

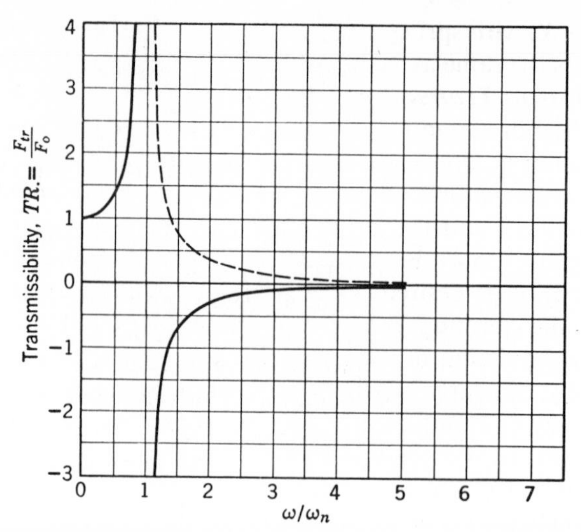

Figure 13.7

If the machine were rigidly fixed to its foundation, the force transmitted to the floor would be F_o. However, with soft springs, the denominator of Eq. 13.13 may be made large so that the transmitted force F_{tr} approaches zero. The ratio of the force F_{tr} with springs to the force F_o for rigid attachment without springs is termed the *transmissibility* TR.

$$\text{TR} = \frac{F_{tr}}{F_o} = \frac{1}{1 - (\omega/\omega_n)^2} \tag{13.14}$$

As shown in Fig. 13.7, the curve of transmissibility versus ω/ω_n given by Eq. 13.14 is exactly the same as that of Fig. 13.4*a*. At $\omega/\omega_n = 1.0$, the transmitted force is infinite, but at values of ω/ω_n greater than 2 there is a considerable reduction in the transmitted force. To make ω/ω_n large, either the design speed ω may be increased or the spring constant may be made small. If ω is low, the spring will have to be very soft for a low TR. This may be impractical. Therefore, ω must be increased so that a harder spring (and higher ω_n) can be used. Even though the transmitted force is low, the amplitude of vibration X as given in Fig. 13.4*b* does not approach zero but approaches a value given by $MX/mr = 1.0$, which may be appreciable. It will be observed that, even though the transmitted vibratory force F_{tr} may be made to approach zero, the spring transmits a steady force equal to the weight of the machine.

Equation 13.14 also gives the transmissibility of the cases of Fig. 13.1*a* and *c* as well as 13.1*b*. For Fig. 13.1*c*, however, the transmitted force is

$F_{tr} = k(X' - X)$ with spring, and $M(d^2x'/dt^2) = -MX'\omega^2$ without spring, but the ratio is the same as in Eq. 13.14.

13.6 Motion Transmissibility. Figure 13.1*c* shows the case in which the wheel or support is vibrating with amplitude X' and transmits a vibrating motion of amplitude X to the machine mass M. The ratio X/X' is termed *motion transmissibility* and is the same as force transmissibility given by Eq. 13.14 and the curve of Fig. 13.7. Thus, the transmissibility curve of Fig. 13.7 may be used in connection with many cases where vibration isolation is desired. Electronic devices and instruments may be isolated from the vibrating motion of ship decks, aircraft floors, and building structures by the design of springs or flexible mounts to give low motion transmissibility. Similarly, engines vibrating from the action of inertia forces may be isolated from automobile frames, aircraft frames, and refrigerator frames by the proper design of isolation springs.

13.7 Damping. In the foregoing, cases of forced vibration are discussed in which the force of friction is neglected. Since friction opposes motion in the sense that it is opposite to velocity, it serves to reduce amplitudes of vibration and is therefore said to damp vibrations. Friction may appear as the viscous resistance of fluids, as the sliding resistance of dry materials in contact, or as internal shearing resistance of the plastic flow of materials as is evident in the hysteresis loop of stress-strain diagrams. Shock absorbers used in automobiles and aircraft are devices utilizing the frictional resistance of fluids to damp vibrations. Commercial spring mounts are made of materials such as rubber, fiber, and cork to utilize the large internal frictional resistance.

Figure 13.8 shows the vibrating mass of Fig. 13.1*a* with a dashpot added to the system to introduce damping by the resistance of a fluid. The force of resistance of the viscous fluid in Fig. 13.8 is proportional to the velocity of the mass M but is opposite in sign to the velocity. Thus, the damping force

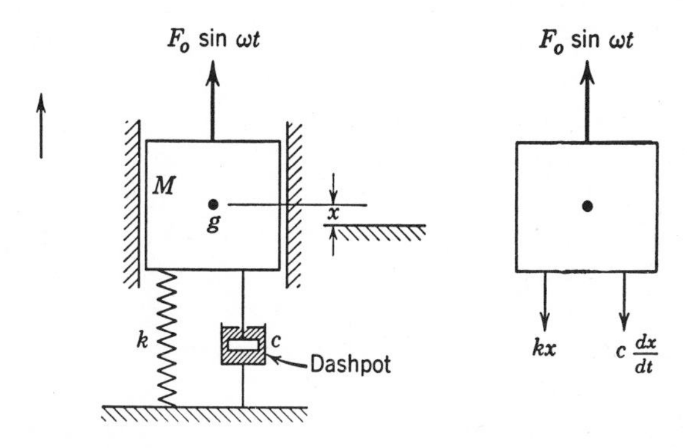

Figure 13.8

is $-cV = -c\,dx/dt$, in which c is the constant of proportionality having the units pound-seconds per inch.

The equation of motion for the free body of Fig. 13.8 may be written as follows:

$$F_o \sin \omega t - cV - kx = MA$$

$$F_o \sin \omega t - c\frac{dx}{dt} - kx = M\frac{d^2x}{dt^2}$$

$$M\frac{d^2x}{dt^2} + c\frac{dx}{dt} + kx = F_o \sin \omega t \tag{13.15}$$

The solution to Eq. 13.15 is $x = X \sin(\omega t - \phi)$, in which the amplitude is

$$X = \frac{F_o/k}{\sqrt{(1 - M\omega^2/k)^2 + (c\omega/k)^2}} \tag{13.16}$$

and the angle ϕ is

$$\tan \phi = \frac{c\omega}{k - M\omega^2} \tag{13.17}$$

ϕ is the angle by which the displacement of the vibration lags the force producing the vibration as shown in Fig. 13.9. Viscosity causes the lag, and as indicated by Eq. 13.17, ϕ is zero in the absence of viscosity.

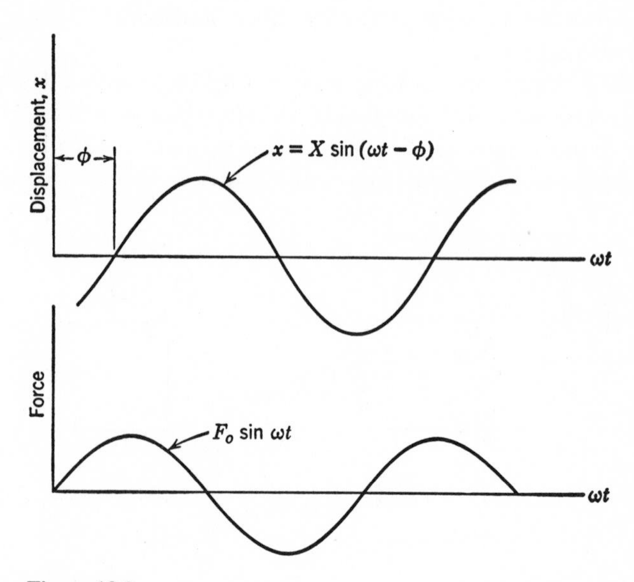

Figure 13.9

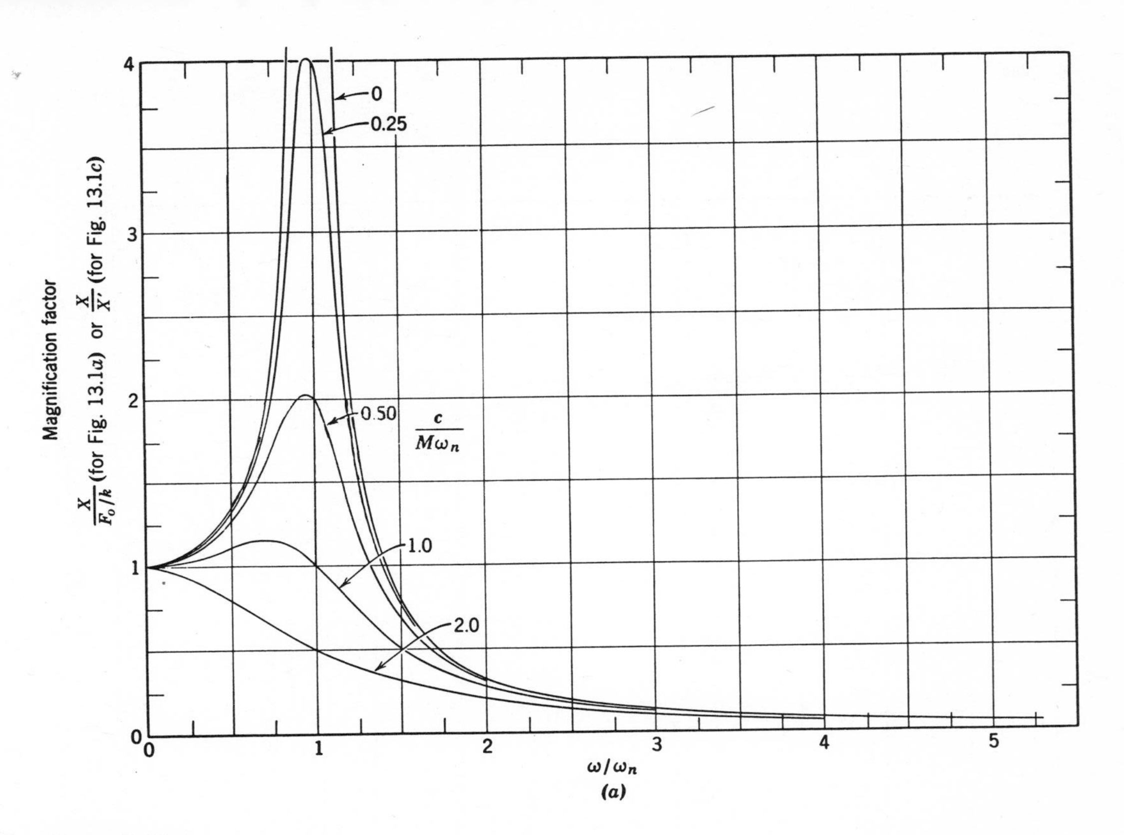
Magnification factor
$\frac{X}{F_o/k}$ (for Fig. 13.1a) or $\frac{X}{X'}$ (for Fig. 13.1c)
0
1
2
3
4
0
0.25
0.50
1.0
2.0
$\frac{c}{M\omega_n}$
0
1
2
3
4
5
ω/ω_n
(a)

Figure 13.10

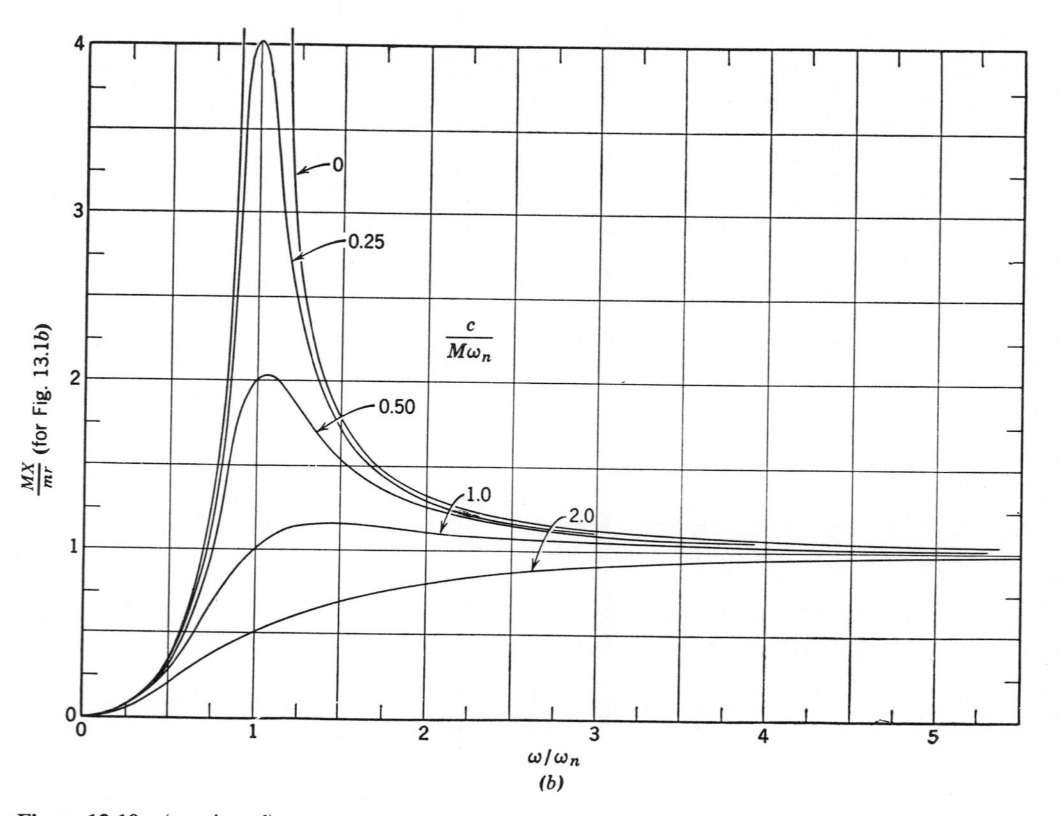

Figure 13.10 (continued)

Of interest is the effect of friction, or damping, on the amplitude of forced vibration given by Eq. 13.16, which may be written in the following non-dimensional form:

$$\frac{X}{F_o/k} = \frac{1}{\sqrt{[1 - (\omega/\omega_n)^2]^2 + [(c/M\omega_n)(\omega/\omega_n)]^2}} \tag{13.18}$$

In Eq. 13.18, the dimensionless $c/M\omega_n$ may be regarded as a damping factor. Figure 13.10*a* shows the effect of damping factor values from 0 to 2 on vibration amplitude given by Eq. 13.18. As shown, damping materially reduces amplitude especially near $\omega/\omega_n = 1.0$. Also, with damping, the curves peak at values of ω/ω_n below one.

For Fig. 13.1*b* with dashpot, in which $F_o = mr\omega^2$, Eq. 13.18 may be written in the following form and plotted as shown in Fig. 13.10*b*.

$$\frac{MX}{mr} = \frac{(\omega/\omega_n)^2}{\sqrt{[1 - (\omega/\omega_n)^2]^2 + [(c/M\omega_n)(\omega/\omega_n)]^2}} \tag{13.19}$$

For Fig. 13.1*c*, Eq. 13.18 may be used to determine X/X' if the dashpot is between the mass M and a fixed datum. For the usual case where the dashpot is between M and the wheel, X/X' is the same as the transmissibility (discussed below) given by Eq. 13.24 or Fig. 13.12.

Transmissibility is also influenced by damping. Referring to Fig. 13.8, since both the spring and the dashpot are connected to the foundation, force by these elements is transmitted to the foundation. The force of the spring is kx and that of the dashpot is $cV = c\,dx/dt$. The resultant of the two forces is the transmitted force F_{tr}.

$$F_{tr} = kx + c\frac{dx}{dt} \tag{13.20}$$

The displacement at any time t is given by $x = X \sin(\omega t - \phi)$, and the velocity is $V = dx/dt = X\omega \cos(\omega t - \phi)$. Thus,

$$F_{tr} = kX \sin(\omega t - \phi) + cX\omega \cos(\omega t - \phi) \tag{13.21}$$

To determine the maximum value of the transmitted force in a cycle of vibration, Eq. 13.21 may be maximized as follows:

$$\frac{d(F_{tr})}{dt} = kX\omega \cos(\omega t - \phi) - cX\omega^2 \sin(\omega t - \phi) = 0$$

$$\frac{\sin(\omega t - \phi)}{\cos(\omega t - \phi)} = \tan(\omega t - \phi) = \frac{kX}{cX\omega} \tag{13.22}$$

Equation 13.22 is represented in the triangle of Fig. 13.11 at the phase for

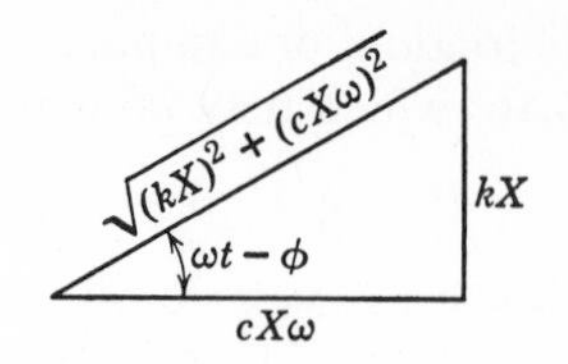

Figure 13.11

maximum F_{tr}. As shown,

$$\sin(\omega t - \phi) = \frac{kX}{\sqrt{(kX)^2 + (cX\omega)^2}}$$

$$\cos(\omega t - \phi) = \frac{cX\omega}{\sqrt{(kX)^2 + (cX\omega)^2}}$$

Substituting in Eq. 13.21,

$$(F_{tr})_{max} = \frac{(kX)^2 + (cX\omega)^2}{\sqrt{(kX)^2 + (cX\omega)^2}} = \sqrt{(kX)^2 + (cX\omega)^2} \quad (13.23)$$

Transmissibility is the ratio of $(F_{tr})_{max}$ to the forcing function F_o. Substituting in Eq. 13.23 the value of X given by Eq. 13.18,

$$(F_{tr})_{max} = \frac{F_o/k\sqrt{k^2 + (c\omega)^2}}{\sqrt{[1 - (\omega/\omega_n)^2]^2 + [(c/M\omega_n)(\omega/\omega_n)]^2}}$$

$$\text{TR} = \frac{(F_{tr})_{max}}{F_o} = \frac{\sqrt{1 + [(c/M\omega_n)(\omega/\omega_n)]^2}}{\sqrt{[1 - (\omega/\omega_n)^2]^2 + [(c/M\omega_n)(\omega/\omega_n)]^2}} \quad (13.24)$$

The effect of damping on transmissibility is shown by the curves of Fig. 13.12 plotted from Eq. 13.24. As shown, high damping reduces the transmitted force for values of $\omega/\omega_n < \sqrt{2}$, but it increases the transmitted force when ω/ω_n exceeds $\sqrt{2}$.

When metallic springs, such as coil or leaf springs, are used to suspend a machine, some damping is present because of the internal friction in the material, but the amount is small and the undamped case may be assumed. As Figs. 13.10 and 13.12 show, k should be chosen such that ω/ω_n at the design speed is greater than $\sqrt{2}$, in which case both the amplitude of vibration and the transmissibility will be low. However, in the absence of much damping in metallic springs, both the amplitude and transmissibility will be

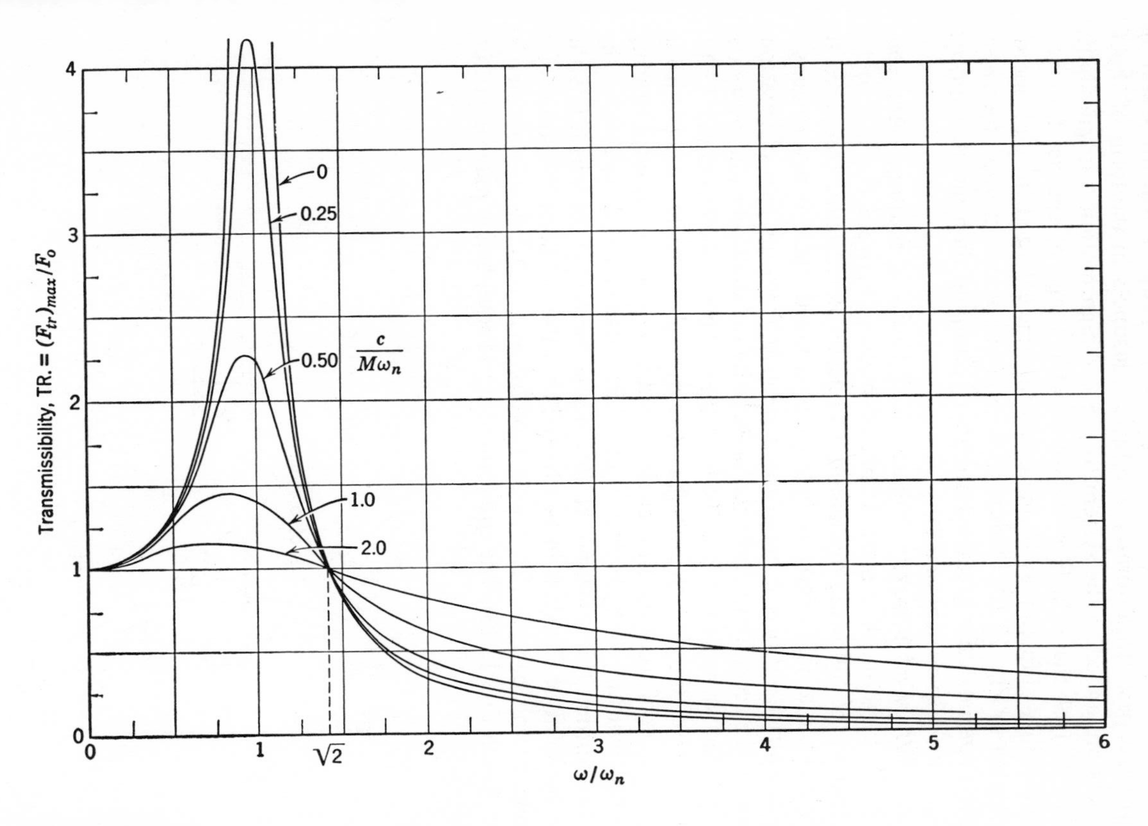

4
3
2
1
0
Transmissibility, TR. = $(F_{tr})_{max}/F_o$
0
0.25
0.50
$\frac{c}{M\omega_n}$
1.0
2.0
0
1
$\sqrt{2}$
2
3
4
5
6
ω/ω_n

Figure 13.12

high at some low critical speed ($\omega/\omega_n = 1.0$) through which the vibrating system must pass. In automotive vehicles, dashpots in the form of shock absorbers are added to the system to prevent high motion transmissibility (Fig. 13.10*a*) of the vehicle body for a wide range of speeds; however, transmissibility of force may be increased to some extent at the high speeds as indicated in Fig. 13.12. Commercial isolators of rubber and other non-metallic materials offer greater amounts of damping than metallic springs ($c/M\omega_n$ is usually not above 0.2).

13.8 Whirl of Shafts. Figure 13.13 shows a rotating rotor consisting of a large disklike mass M placed midway between bearings and a shaft that in the following discussion is considered negligible compared to M. Even when a high degree of balance is attained, there is nevertheless a small eccentricity e of the mass center g of M from the axis of rotation. Because of the eccentricity, centrifugal force by the shaft rotation causes the shaft to deflect an amount r. Viewed from the end of the shaft as in Fig. 13.13, the center O of the deflected shaft at the mass M appears to be whirling about the axis of rotation on a circle of radius r. The inertia force causing the forced whirling is $F_o = M(r + e)\omega^2$. Because of the deflection of the shaft as a spring, the resistance to the inertia force is kr, in which k is the spring constant of the shaft in bending. The sense of the acceleration of g is known in this case so that the MA vector may be shown as an inertia force F_o (as in Fig. 13.13). The equation of static equilibrium may be written

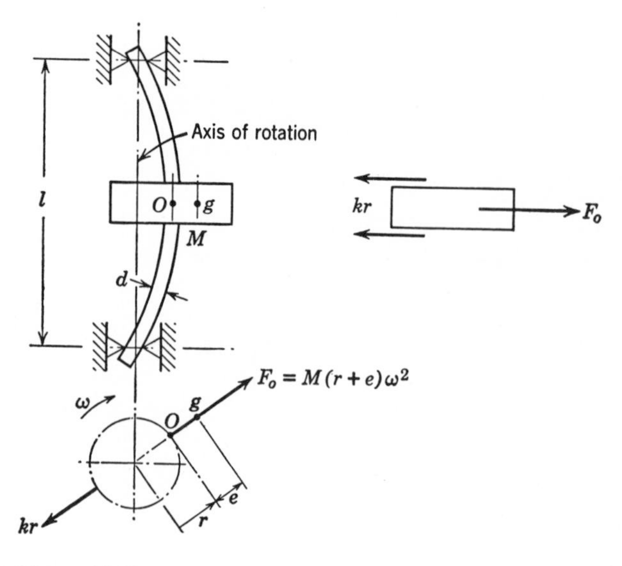

Figure 13.13

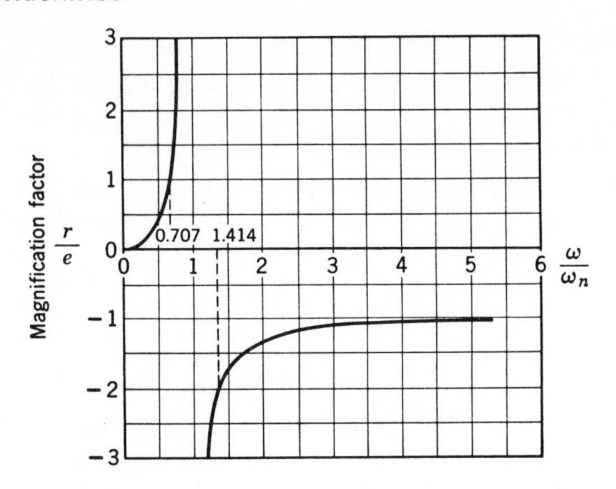

Figure 13.14

as follows:

$$\Sigma F = 0$$

$$M(r + e)\omega^2 - kr = 0 \tag{13.25}$$

To determine the radius of whirl r, Eq. 13.25 may be shown in the following form:

$$r = \frac{e\omega^2}{(k/M) - \omega^2} \tag{13.26}$$

When the speed ω of the shaft is equal to $\sqrt{k/M}$, the denominator of Eq. 13.26 becomes zero, and r becomes intolerably large. The whirling shaft appears as a vibrating beam when viewed from the side where only the projected motion may be observed. Therefore, $\sqrt{k/M}$ of the whirling shaft may be regarded as the natural circular frequency ω_n of the beam when disturbed to vibrate naturally in a bowlike mode. In nondimensional form, Eq. 13.26 may be expressed as follows:

$$\frac{r}{e} = \frac{(\omega/\omega_n)^2}{1 - (\omega/\omega_n)^2} \tag{13.27}$$

Figure 13.14 shows a plot of Eq. 13.27 and indicates that when the speed ω of the shaft is equal to $\omega_n = \sqrt{k/M}$, a critical condition exists because of the very large amplitude of whirl. (The curve of Fig. 13.14 is exactly the same as the curve of Fig. 13.4 representing b of Fig. 13.1.) At the critical condition, ω is designated ω_c and the corresponding shaft speed in revolutions per minute is $n_c = 60\omega_c/2\pi = 60\omega_n/2\pi$, in which $\omega_n = \sqrt{k/M}$ is ordinarily expressed in radians per second. Furthermore,

$$n_c = \frac{60}{2\pi}\omega_n = \frac{60}{2\pi}\sqrt{\frac{k}{M}} = \frac{60}{2\pi}\sqrt{\frac{kg}{W}}$$

$$= \frac{60}{2\pi}\sqrt{\frac{386k}{W}}$$

$$= 188\sqrt{\frac{k}{W}} \tag{13.28}$$

in which n_c is the critical speed in revolutions per minute, k is in pounds per inch, and W, the weight of M, is in pounds. The spring constant k may be evaluated from the static deflection δ_{st} of the shaft due to the weight W. Thus, k equals W/δ_{st} and when substituted in Eq. 13.28 gives the critical speed in the following equation:

$$n_c = 188\sqrt{\frac{1}{\delta_{st}}} \tag{13.29}$$

As shown in textbooks on strength of materials, the static deflection of W placed centrally on a uniform beam may be calculated from $\delta_{st} = Wl^3/48EI_A$. Thus, the critical speed of a shaft with centrally placed mass M may be determined in terms of the shaft dimensions (l is length of shaft between supports, I_A is moment of inertia of area of the shaft cross section equal to $\pi d^4/64$, d is shaft diameter) and the modulus of elasticity E of the shaft material.

$$n_c = 290\sqrt{\frac{Ed^4}{Wl^3}} \tag{13.30}$$

Thus, according to Eq. 13.30, the shaft material and shaft dimensions may be altered, as well as the weight of the mass M, such that the critical speed n_c will be higher or lower than the design speed n at which it is desired to operate. As shown in Fig. 13.14, if n is less than 0.707 or more than 1.414 times n_c, the amplitude of whirl r is less than twice the eccentricity e. For example, if the eccentricity e is 0.001 in., the amplitude of whirl r is 0.002 in. at $n/n_c = \sqrt{2}$.

It is interesting to note from Fig. 13.14 that, at very high speeds above critical ($\omega/\omega_n \gg 1.0$), the value of $r/e = -1$ and $r = -e$, which indicates that the mass center of M is at the axis of rotation. In this case, the mass is not whirling, but the shaft whirls about the mass center of M.

In the preceding discussion, the mass of the shaft is assumed negligible. In the case where the mass of the shaft is large enough to be included, and the shaft is uniform in diameter, 50 percent of the shaft mass m may be included with M to determine the natural circular frequency:

$$\omega_n = \sqrt{\frac{k}{(M + 0.5m)}} \tag{13.31}$$

As shown in Fig. 13.13, the bearings of the shaft are assumed to be rigid. In certain instances, the bearings may be elastically supported, in which case δ_{st} of Eq. 13.29 should include the static deflection of the supports as well as that of the shaft. However, Eq. 13.29 applies only when the flexibility of the supports is the same for all angular positions of the rotor.

13.9 Natural Frequency and Critical Speed. A great many configurations of rotors are conceivable since they may include many masses and many supports, and may have shafts of variable cross section. The behavior of such shafts is similar, at least qualitatively, to the behavior shown by the curve of Fig. 13.14. Of primary interest is the critical speed where the natural frequency and operating speed coincide, a condition which is to be avoided in design of rotating machinery. Actually, all rotors have an infinite number of critical speeds, and curves such as Fig. 13.14 would show many peaks of infinite amplitude forced by unavoidable deviations from the axis of rotation of the centers of gravity of the several rotor masses. Although magnification factor curves may be difficult to obtain mathematically, the critical speeds of shafts are relatively easily determined from calculations of natural frequency. In the following section, several cases are discussed in which critical speed is determined from natural frequency.

13.10 Natural Frequency of a Shaft with Many Masses. In a rotating shaft with several masses as shown in Fig. 13.15*a*, the natural circular frequency of vibration ω_n may be determined for the nonrotating

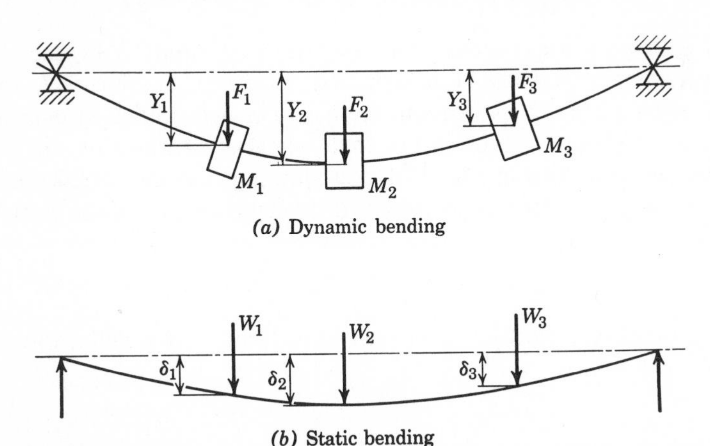

Figure 13.15

condition in which the shaft is deflected and then released to vibrate freely without damping in an up and down mode.

Rayleigh's energy method may be applied in this case. Assuming that the vibrating system is a conservative one, the sum of the potential energy and kinetic energy in any phase of the vibration is constant. Two of these phases are easily analyzed. In the phase in which all of the masses are simultaneously at the maximum displacements Y, the energy stored elastically in the shaft is equal to the potential energy $\Sigma\, FY/2$. In this phase the kinetic energy is zero since all parts of the system are momentarily at zero velocity. Thus, the potential energy is

$$\mathrm{PE} = \frac{F_1 Y_1}{2} + \frac{F_2 Y_2}{2} + \cdots + \frac{F_n Y_n}{2} \tag{13.32}$$

The forces F are those necessary to deflect the shaft as a spring to the shape shown in this phase. Force times displacement determines potential energy. However, since the force is directly proportional to displacement, the average force acting during the displacement Y is $F/2$.

During the vibration, the shaft passes through the undeflected phase in which the potential energy is zero, but the kinetic energy is a maximum since the velocities of the masses are a maximum. Assuming that the motions of the masses are simple harmonic, the velocities are $V = Y\omega_n$ and the kinetic energies are $MV^2/2 = M(Y\omega_n)^2/2$. Thus, the kinetic energy of the system is

$$\mathrm{KE} = \frac{\omega_n^2}{2}\left[M_1 Y_1^2 + M_2 Y_2^2 + \cdots + M_n Y_n^2\right]$$

$$= \frac{\omega_n^2}{2g}\left[W_1 Y_1^2 + W_2 Y_2^2 + \cdots + W_n Y_n^2\right] \tag{13.33}$$

By equating the right-hand terms of Eqs. 13.32 and 13.33, the natural circular frequency ω_n may be determined. However, the forces F and the displacements Y are not known, but these may be determined by considering the statically deflected shape of the shaft when acted upon by the weights of the masses as shown in Fig. 13.15*b*. Assuming that the displacements Y of the vibrating case are proportional to the deflections δ of the static case, then

$$\frac{Y_1}{\delta_1} = \frac{Y_2}{\delta_2} = \cdots = \frac{Y_n}{\delta_n} \tag{13.34}$$

Since the forces to deflect a spring are proportional to the deflections, then

$$\frac{F_1}{W_1} = \frac{Y_1}{\delta_1}, \qquad \frac{F_2}{W_2} = \frac{Y_2}{\delta_2}, \qquad \frac{F_n}{W_n} = \frac{Y_n}{\delta_n} \tag{13.35}$$

Equating the expressions for potential and kinetic energy given by Eqs. 13.32 and 13.33 and using Eqs. 13.34 and 13.35 to eliminate F and Y, the resultant equation giving the natural circular frequency becomes

$$\omega_n^2 = g\frac{[W_1\delta_1 + W_2\delta_2 + \cdots + W_n\delta_n]}{[W_1\delta_1^2 + W_2\delta_2^2 + \cdots + W_n\delta_n^2]}$$

$$\omega_n^2 = g\frac{\Sigma\, W\delta}{\Sigma\, W\delta^2} \tag{13.36}$$

and the critical speed may be determined from $n_c = 60\omega_n/2\pi$.

Rayleigh's equation (Eq. 13.36) is a simple and highly useful expression for determining the fundamental natural frequency of many rotor configurations. The determination of static deflection is the greater part of the effort needed in making calculations as illustrated in the following examples. Beam deflection formulas for many cases are available in textbooks on strength of materials and in handbooks. The area-moment method and other methods may be applied in general cases. Graphical methods, as illustrated in a following paragraph, are also available for determining static deflections of rotors having shafts of varying diameters.

To include shaft mass in the calculations, the shaft may be divided into several lengths, each of which is treated as an additional mass.

Equation 13.36 is not strictly an exact evaluation of natural frequency because the curve of statical deflections is not exactly proportional to the dynamic deflection curve as was assumed. However, the result obtained by the equation is only a percent or two higher than the true fundamental natural frequency. Considering that other factors such as gyroscopic effects during oscillation, press fits of disks on the shaft, and keys alter the critical speed slightly, Eq. 13.36 yields an acceptable answer. The deflection of the supports may be a greater influence on critical speeds and should be added to the deflections of the shaft in Eq. 13.36.

The natural frequency given by Eq. 13.36 is the fundamental, or lowest, frequency of the system of masses. In Fig. 13.15, the shaft is shown deflected in a single half-wave. It is possible for the shaft to vibrate in modes such that the deflected shape consists of several half-waves as shown in Fig. 13.16.

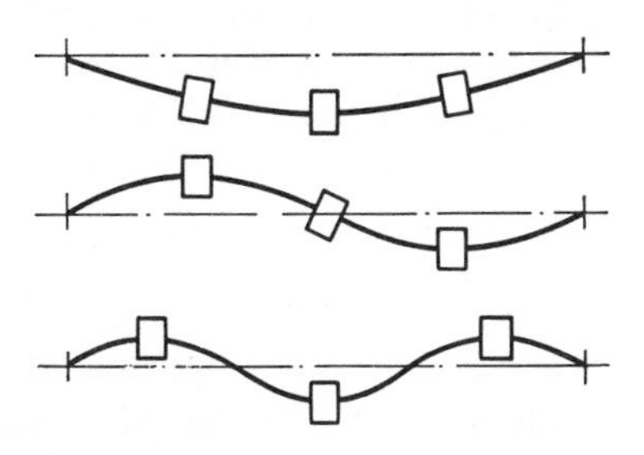

Figure 13.16

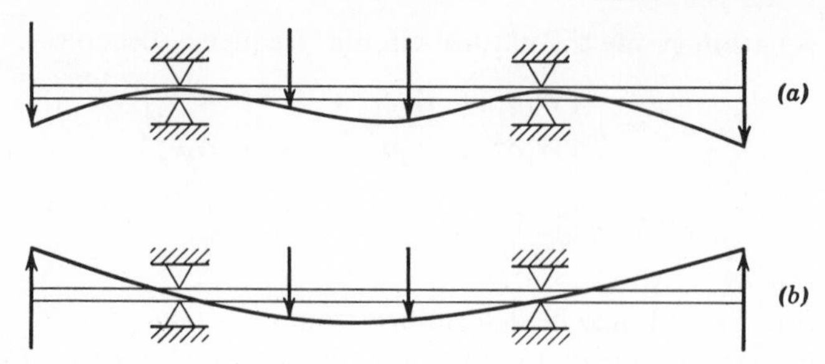

Figure 13.17

There are as many modes as there are masses. Figure 13.16 shows three modes for three masses, although there are an infinite number of modes if the mass of the shaft is included. However, the natural frequency increases as the number of half-waves increases. It is therefore desirable, if possible, to design the proportions of a shaft such that the lowest critical speed is higher than the design speed. However, it is not always possible to do so. In high-speed turbomachinery, the operating speed may be between two critical speeds so that the shaft need not become prohibitively heavy. In this case, it is necessary to pass through the lower critical speed, which may seem dangerous. However, if the rotor is finely balanced and the first critical speed is low, the shaking forces are small in regions near the critical. Also, the amplitude of vibration at critical speed increases to dangerous proportions only if time is allowed for the amplitude to build up; therefore, by acceleration through the critical, amplitudes may be held to acceptable magnitudes. Natural damping of the shaft material, although small, also tends to reduce amplitudes. Many successful machines have been designed to operate between criticals.

Where the shaft is extended outboard of the bearings as in Fig. 13.17*a*, the overhung weights should be reversed in sense as in Fig. 13.17*b* when determining static deflections to be used in Eq. 13.36. It may be seen that the dynamic deflection curve of a single half-wave is simulated in this way to give the lowest natural frequency.

Example 13.2

A 50-lb compressor wheel and a 30-lb turbine wheel are mounted on a common steel shaft as shown in Fig. 13.18*a*. The shaft is to be operated at a design speed of 10,000 r/min. Using Rayleigh's equation (Eq. 13.36)

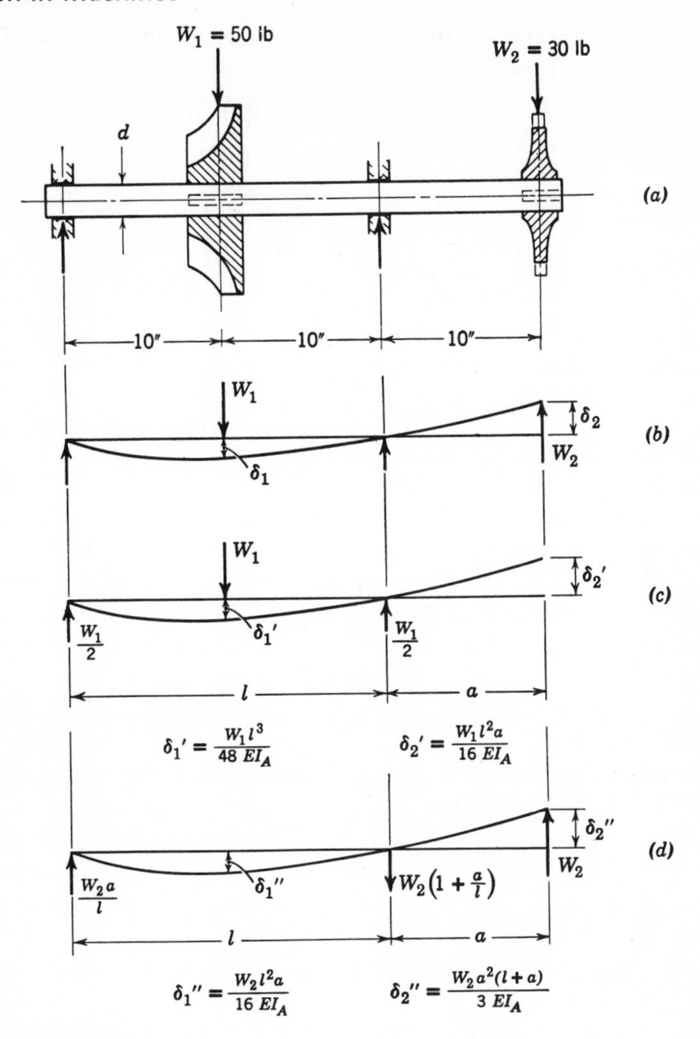

Figure 13.18

determine the diameter of the lightest shaft which can be used for a fundamental critical speed of 12,000 r/min so as to give a margin of safety of 2000 r/min.

Solution. As shown in Fig. 13.18*b*, the overhung load W_2 is reversed in order to obtain a single half-wave deflection curve. Figures 13.18*c* and *d* show the shapes of the deflected beam due to each of the loads acting independently, thus giving two cases for which the static deflection formulas shown are given in texts on strength of materials. By the method of superposition, the deflections δ_1 and δ_2 may be determined as follows:

$$\delta_1 = \delta_1' + \delta_1'' = \frac{W_1 l^3}{48EI_A} + \frac{W_2 l^2 a}{16EI_A}$$

$$= \frac{1}{EI_A}\left[\frac{50(20)^3}{48} + \frac{30(20)^2 10}{16}\right]$$

$$= \frac{15{,}830}{EI_A}$$

$$\delta_2 = \delta_2' + \delta_2'' = \frac{W_1 l^2 a}{16EI_A} + \frac{W_2 a^2(l + a)}{3EI_A}$$

$$= \frac{42{,}500}{EI_A}$$

Using Eq. 13.36,

$$\omega_n^2 = g\left[\frac{W_1\delta_1 + W_2\delta_2}{W_1\delta_1^2 + W_2\delta_2^2}\right]$$

$$= gEI_A\left[\frac{50(15{,}830) + 30(42{,}500)}{50(15{,}830)^2 + 30(42{,}500)^2}\right]$$

For $g = 386$ in./sec^2 and $E = 30 \times 10^6$ psi,

$$\omega_n^2 = 0.359 \times 10^6 I_A$$
$$I_A = 2.79 \times 10^{-6}\omega_n^2$$

For $n_c = 12{,}000$ r/min,

$$\omega_n = \frac{2\pi n_c}{60} = 1260 \text{ rad/sec}$$

Therefore, the required moment of inertia of the shaft cross section is

$$I_A = 2.79 \times 10^{-6}(1260)^2$$
$$= 4.42 \text{ in.}^4$$

Since $I_A = \pi d^4/64$,

$$d^4 = \frac{64}{\pi} I_A = 90 \text{ in.}^4$$

$$d = 3.08 \text{ in.}$$

A 3.125-in.-diameter shaft would be used.

Example 13.3

The supports of the rotor of Example 13.2, Fig. 13.18*a*, were assumed to be rigid. Determine the critical speed of the rotor of Example 13.2 if each of the supports deflects an amount $20{,}000/EI_A$ under the static loading. Use $I_A = 4.42$ in.^{4} and $E = 30 \times 10^6$ psi.

Solution. Because of the flexibility of the supports, the loads W_1 and W_2 will deflect an additional amount. As shown in Fig. 13.19, under the loading the left support deflects downward and the right support upward. As may be seen, there is no effect on the deflection of W_1, but the deflection of W_2 is increased an amount $40{,}000/EI_A$. Therefore, the total static deflections are

$$\delta_1 = \frac{15{,}830}{EI_A}$$

$$\delta_2 = \frac{42{,}500}{EI_A} + \frac{40{,}000}{EI_A} = \frac{82{,}500}{EI_A}$$

Substitution in Eq. 13.36 gives

$$\omega_n^2 = 774{,}000$$

$$\omega_n = 880 \text{ rad/sec}$$

$$n_c = \frac{60}{2\pi}\omega_n = \frac{60}{2\pi}(880)$$

$$= 8400 \text{ r/min}$$

13.11 Shafts with Variable Diameter. Equation 13.36 for critical speed holds for rotor shafts of the type shown in Fig. 13.20*a*, in which the diameter is variable in discrete steps. However, because I_A is variable in such cases, formulas for statical deflections are not simply derived. One of the several graphical methods may be used, such as the one which follows.

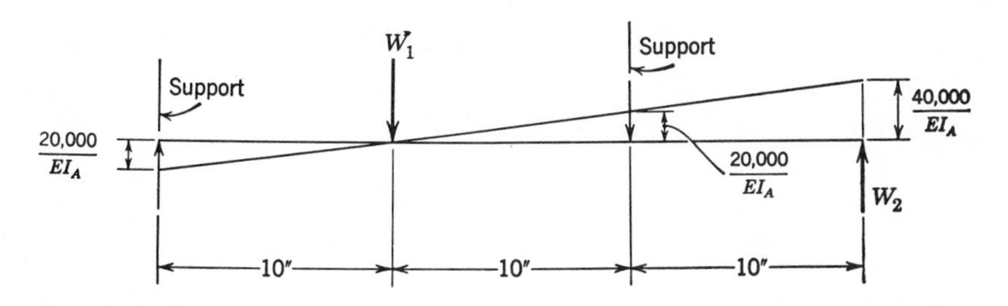

Figure 13.19

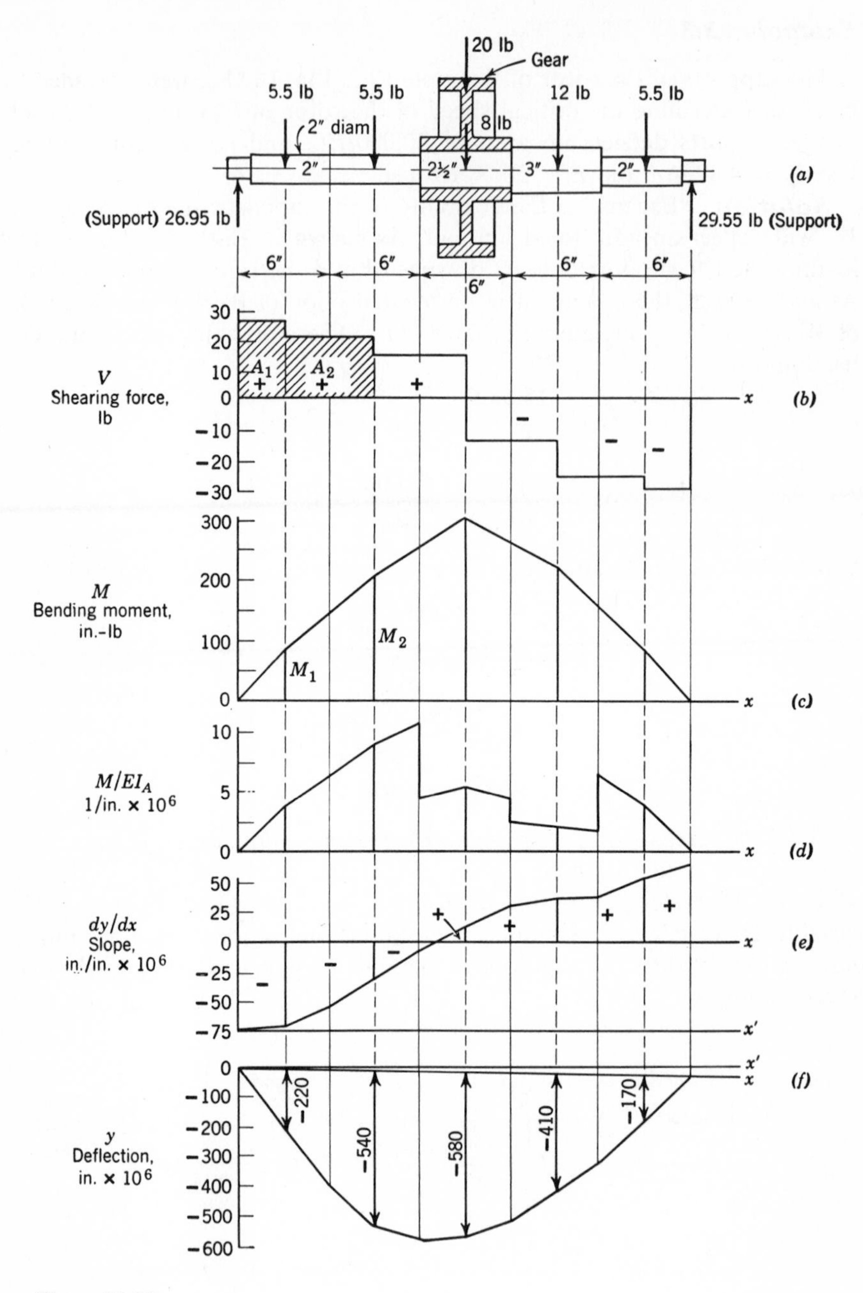
20 lb
Gear
5.5 lb
5.5 lb
12 lb
5.5 lb
2" diam
8 lb
2"
2½"
3"
2"
(a)
(Support) 26.95 lb
29.55 lb (Support)
6"
6"
6"
6"
6"
V
Shearing force, lb
A_1
A_2
(b)
M
Bending moment, in.-lb
M_1
M_2
(c)
M/EI_A
1/in. × 10^6
(d)
dy/dx
Slope, in./in. × 10^6
(e)
y
Deflection, in. × 10^6
−220
−540
−580
−410
−170
(f)

Figure 13.20

It will be recalled from mechanics of materials that the basic differential equation which must be solved to determine statical deflection is the following:

$$\frac{d^2y}{dx^2} = \frac{M}{EI_A} \tag{13.37}$$

in which y is deflection, M is bending moment as a function of x, and I_A is moment of inertia of cross-sectional area as a function of x. Integrated twice, Eq. 13.37 gives beam deflection, of which the first integration is dy/dx, the slope of the elastic curve of the bent beam. Furthermore, starting with the beam loads, two integrations are required to obtain a curve of bending moment. Thus, four integrations are needed to obtain deflections from the known loads.

Since the integration process is one of summing areas under curves, a graphical method of area summing may be used for complex beams having functions with numerous discontinuities. The graphical method will require that curves be drawn to scale in order that areas under the curves may be evaluated by counting squares or by using a planimeter.

In Fig. 13.20*a* is shown a steel rotor with a 20-lb gear and a shaft having three different diameters. The beam is divided into five parts, and weights of the beam are shown at the centers of gravity of the parts. One of the parts includes the weight of the gear. Figure 13.20*a* is a loading diagram from which the shearing force diagram of Fig. 13.20*b* may be determined by conventional methods (the first integration). The bending moment diagram of Fig. 13.20*c* is obtained from the areas of the shearing force diagram (the second integration). For example, the ordinate M_1 is obtained from the area A_1, the ordinate M_2 from the sum of the areas $A_1 + A_2$, and the ordinate M_n from $\Sigma_1^n A$. Due regard is given to the sign of the areas. The areas in square inches must be multiplied by the proper conversion constant obtained from the scales of the shearing force diagram in order that the ordinates of the bending moment diagram be in inch-pounds.

Before further integrations are performed, the bending moment diagram must be converted to an M/EI_A diagram as required by Eq. 13.37. Each ordinate of the bending moment diagram is divided by the appropriate value of EI_A ($E = 30 \times 10^6$ psi for steel and $I_A = \pi d^4/64$) to obtain the M/EI_A ordinates of Fig. 13.20*d*. From the areas of the M/EI_A diagram, the ordinates of Fig. 13.20*e* representing beam slope dy/dx are obtained (the third integration). The ordinates erected from the x' axis are all positive. However, it is known from the expected shape of the bent beam that the slopes are negative near the left end of the beam, positive at the right end, and near the middle of the beam there is a zero slope. Thus, the x axis is arbitrarily drawn such that the negative areas of Fig. 13.20*e* and the positive areas are nearly equal. The fourth integration is made using the areas of Fig. 13.20*e* to obtain the ordinates of static deflection y in Fig. 13.20*f*. It will be observed that the deflection ordinates of static deflection are negative since the areas of the dy/dx curve are negative at the left end where the integration is begun. Although the deflection ordinates are erected from the x' axis, the x axis is drawn as shown since it is known that the deflections of the beam at the supports are zero. Had the arbitrarily drawn x axis of the slope diagram of Fig. 13.20*e* divided the negative and positive areas equally, then the x' axis and the x axis of Fig. 13.20*f* would have coincided.

From the data given in curves a and f, the following values are calculated:

$$\Sigma Wy = 26{,}000 \times 10^{-6} \text{ lb-in.} \qquad \Sigma Wy^2 = 13{,}400{,}000 \times 10^{-12} \text{ lb-in.}^2$$

$$\omega_n^2 = g\frac{\Sigma Wy}{\Sigma Wy^2} = 0.755 \times 10^6$$

$$\omega_n = 0.870 \times 10^3 = 870 \text{ rad/sec}$$

$$n_c = \frac{60(870)}{2\pi} = 8300 \text{ r/min}$$

13.12 Higher-Order Critical Speeds. For rotors having shafts of variable diameters as in the preceding paragraphs, the determination of second- and higher-order critical speeds in bending is relatively more complex than determining the fundamental critical speed from Eq. 13.36. Textbooks by Timoshenko,[2] Den Hartog,[3] and Thomson[4] present methods for rotors with such shafts and for a number of rotors with uniform shafts with and without lumped masses. Figure 13.21 presents cases of uniform simply-supported beams and uniform cantilever beams for which the following formula gives the several natural frequencies:

$$\omega_n = c_n\sqrt{\frac{EI_A g}{Wl^3}} \tag{13.38}$$

c_n is the coefficient in Fig. 13.21 which gives the nth natural frequency, W is the total weight of the beam in pounds, and l is the length of the beam in inches. Although the number of natural frequencies of uniform beams is infinite, only the first three are given since most rotors do not operate at much above the second critical speed. The automotive drive shaft and the

[2] S. P. Timoshenko, D. H. Young, and W. Weaver Jr., *Vibration Problems in Engineering*, John Wiley & Sons, 1974.

[3] J. P. Den Hartog, *Mechanical Vibrations*, McGraw-Hill Book Company, 1956.

[4] W. T. Thomson, *Theory of Vibration with Application*, Prentice-Hall, 1972.

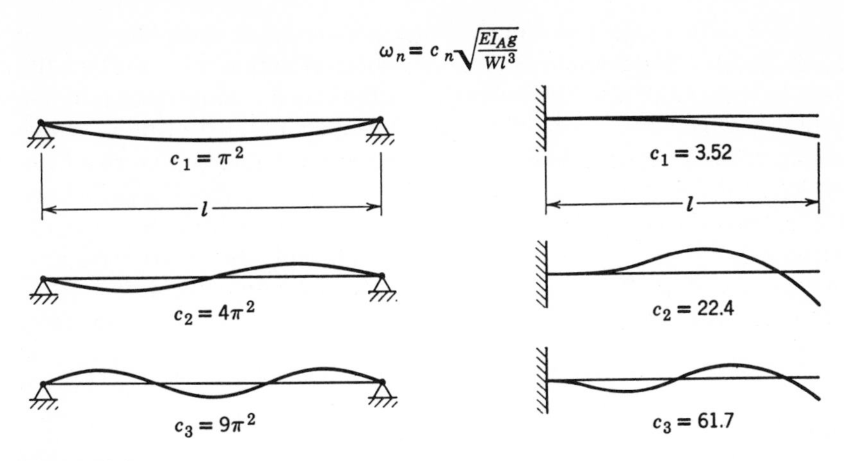

Figure 13.21

quill shaft are examples of uniform simply-supported beams, and turbine and compressor blades are examples of nearly uniform cantilever beams.

Consider the case of the rotor blade shown in Fig. 13.22. The blade is shown as a cantilever beam which undergoes a cycle of disturbance in bending each time it passes a stator blade and causes a change in aerodynamic force. If there are N number of blades on the stator, then the frequency f of the

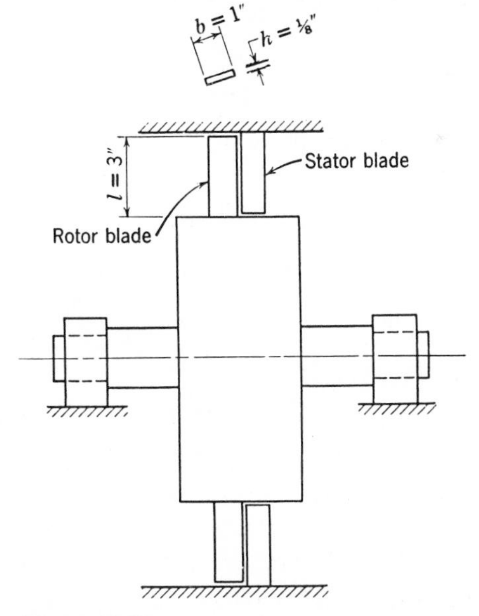

Figure 13.22

disturbance in cycles per minute is the product of N and the rotor speed in revolutions per minute. When the frequency coincides with the natural frequency f_n of the blade in bending, a critical condition exists. For the steel blade shown in Fig. 13.22, the following calculations illustrate the determination of the several critical speeds of the rotor for the case of 30 stator blades.

$$E = 30 \times 10^6 \text{ psi} \qquad g = 386 \text{ in./sec}^2 \qquad l = 3 \text{ in.}$$

$$I_A = \frac{bh^3}{12} = \frac{1 \times (\frac{1}{8})^3}{12} = 0.000163 \text{ in.}^4$$

$$W = \text{volume} \times \text{lb/in.}^3 = (1 \times 3 \times \tfrac{1}{8})0.285 = 0.107 \text{ lb}$$

$$\omega_{n_1} = c_1\sqrt{\frac{EI_A g}{Wl^3}} = 3.52\sqrt{\frac{30 \times 10^6 \times 0.000163 \times 386}{0.107 \times (3)^3}}$$

$$\omega_{n_1} = 2850 \text{ rad/sec}$$

$$f_{n_1} = \frac{60\omega_{n_1}}{2\pi} = \frac{60}{2\pi} \times 2850$$

$$= 27{,}200 \text{ cycles/min}$$

The critical speed of the rotor occurs when $f_{n_1} = Nn_{c_1}$.

$$n_{c_1} = \frac{f_{n_1}}{N} = \frac{27{,}200}{30}$$

$$= 907 \text{ r/min}$$

The second and third critical speeds are

$$n_{c_2} = \frac{c_2}{c_1} n_{c_1} = \frac{22.4}{3.52} \times 907$$

$$= 5780 \text{ r/min}$$

$$n_{c_3} = \frac{c_3}{c_1} n_{c_1} = \frac{61.7}{3.52} \times 907$$

$$= 15{,}900 \text{ r/min}$$

Rotor blades are usually required to be thin and light for high-speed machines and often are required to pass through the first and second critical speeds. Material selection is important in that some materials possess higher damping properties than others, and these properties may mean the difference between success and failure in passing through criticals. Blades are usually curved and are tapered such that the blade thickness is greater at the root than at the tip; this stiffens the blade and raises the criticals somewhat. For other than uniform beams, the coefficients in Eq. 13.38 are different from those in Fig. 13.21.

13.13 Torsional Vibrations. In the foregoing cases of shaft vibration, the modes of oscillation are for shafts acting as springs in bending. Such shafts, when transmitting torque, also act as *torsional* springs. When there are cyclic variations in transmitted torque, torsional modes of vibration are forced which, if coincident in frequency with the natural torsional frequency of the system, will cause resonance. At resonance, the shaft may twist and untwist at amplitudes high enough to produce failure of the shaft in torsion.

The simplest case of natural torsional vibration is shown in Fig. 13.23*a* in which a disk of large moment of inertia I (about the x axis) is made to oscillate about the x axis by an oscillatory twisting of the shaft. The shaft is fixed at one end as shown. To induce the vibration, the disk is displaced the angle A and then released to vibrate freely in an angular or torsional mode. In the following, it is assumed that the moment of inertia of the shaft is negligible compared to that of the lumped mass and that there is no damping.

Considering the disk as a free body, the only torque acting is the spring torque applied by the shaft as shown in Fig. 13.23*b*. At the angular displacement θ of the disk, the spring torque is $-k_t\theta$, where k_t is the spring constant of the shaft in torsion and the minus sign indicates that the torque is opposite in sense to the angular displacement. The equation of angular motion of

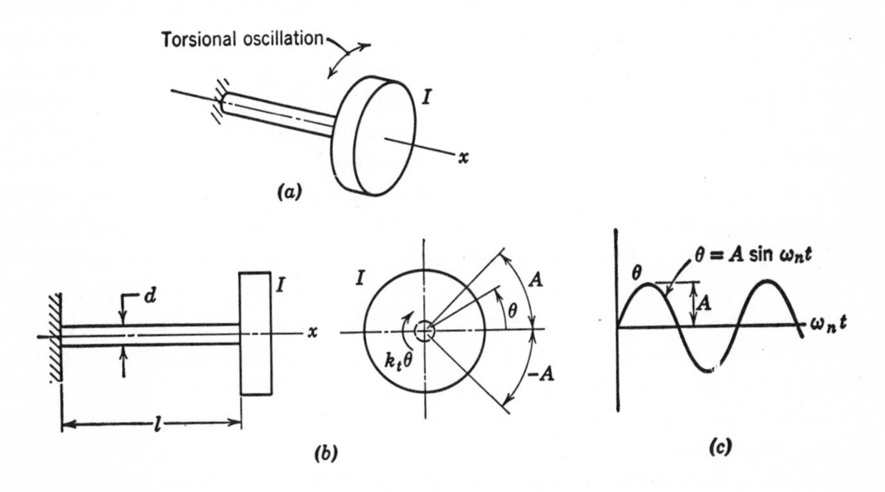

Figure 13.23

the disk about the x axis may be written as follows:

$$\Sigma T_x = I\alpha$$

where I is the moment of inertia of the disk about the x axis and $\alpha = d^2\theta/dt^2$ is the instantaneous angular acceleration of the disk.

$$-k_t\theta = I\alpha = I\frac{d^2\theta}{dt^2}$$

$$\frac{d^2\theta}{dt^2} + \frac{k_t\theta}{I} = 0 \tag{13.39}$$

θ and t are the variables of differential Eq. 13.39, a homogeneous equation for which the solution is $\theta = C_1 \cos \sqrt{k_t/I}\, t + C_2 \sin \sqrt{k_t/I}\, t$. C_1 and C_2 are constants depending on boundary conditions. Assuming that the boundary conditions are $\theta = 0$ at $t = 0$ and $\theta = A$ at $\sqrt{k_t/I}\, t = \pi/2$, then $C_1 = 0$ and $C_2 = A$, and the solution is

$$\theta = A \sin \sqrt{\left(\frac{k_t}{I}\right)}t \tag{13.40}$$

Equation 13.40 is the simple harmonic motion shown by the curve of Fig. 13.23*c* for which the circular frequency is

$$\omega_n = \sqrt{\frac{k_t}{I}} \tag{13.41}$$

It will be observed that ω_n is the natural circular frequency at which the shaft is twisting and untwisting, and that $f_n = \omega_n/2\pi$ is the cyclic frequency. The torsional spring constant k_t is the torque T required to twist one end of the shaft relative to the other a unit angle ϕ. Thus, $k_t = T/\phi$. For shafts of circular cross section, $\phi = Tl/GJ$, and therefore $k_t = GJ/l$. G is the modulus of rigidity (or modulus of elasticity in shear) of the material in pounds per square inch, J is the polar moment of inertia of the cross-sectional area equal to $\pi d^4/32$, and d and l are respectively diameter and length in inches. k_t has the units of inch-pounds per radian. For steel $G =$ 11,500,000 psi, and for aluminum allow $G = 3{,}800{,}000$ psi. I is the moment of inertia of the mass of the disk about the x axis equal to $Mr^2 = (W/g)r^2$, where M is the mass of the disk in lb-sec^2/in., r is radius of gyration of the disk in inches, W is weight in pounds, and $g = 386$ in./sec^2. I is then in lb-sec^2-in., and the ratio k_t/I is in 1/sec^2. ω_n is in radians per second and f_n in cycles per second.

13.14 Torsional Vibration of a Shaft with Two Disks. In most practical situations, a shaft is free to rotate in its support bearings while transmitting torque from one disk to one or more other disks. A common case of a torque-transmitting rotor is shown in Fig. 13.24*a* in which the shaft connects two disks having moments of inertia I_1 and I_2. To determine the natural torsional frequency of the system, the nonrotating condition is first considered. If disk 1 is displaced some angle by an application of torque while disk 2 is held by an equal resisting torque, then on release of the torques the disks will vibrate torsionally at some natural frequency. Experience would show that the frequencies of both disks are the same although the displacements of the disks are opposite in phase, and there is some transverse plane in the shaft between the disks that remains fixed. The fixed plane in the shaft represents the node or the position of a plane having zero motion.

Figure 13.24*b* shows the free bodies of the two disks when the rotor is viewed from the right end. When the phase of disk 1 is the angle θ_1, the phase of disk 2 is θ_2, which lags disk 1 by the angle $\theta_1 - \theta_2$. The torque of the shaft as a torsional spring acting on each of the free bodies is $k_t(\theta_1 - \theta_2)$, in which $\theta_1 - \theta_2$ is the angle through which the shaft is twisted. For torsional equilibrium of the shaft, the torques on the disks are the same in magnitude but opposite in sense as shown. Equations of motion of the individual disks are written in parallel as follows:

Disk 1 Disk 2

$$\Sigma T = I_1\alpha_1 \qquad \Sigma T = I_2\alpha_2$$

$$-k_t(\theta_1 - \theta_2) = I_1\frac{d^2\theta_1}{dt^2} \qquad k_t(\theta_1 - \theta_2) = I_2\frac{d^2\theta_2}{dt^2} \tag{13.42}$$

$$\frac{d^2\theta_1}{dt^2} + \frac{k_t}{I_1}(\theta_1 - \theta_2) = 0 \qquad \frac{d^2\theta_2}{dt^2} - \frac{k_t}{I_2}(\theta_1 - \theta_2) = 0 \tag{13.43}$$

Assuming that the torsional motions of the disks are harmonic, the following solutions apply to Eqs. 13.43:

$$\theta_1 = A_1 \sin \omega_n t \qquad \theta_2 = A_2 \sin \omega_n t \tag{13.44}$$

Substitution of Eqs. 13.44 in Eqs. 13.43 yields

$$\omega_n^2 = \frac{k_t}{I_1}\frac{(A_1 - A_2)}{A_1}, \qquad \omega_n^2 = \frac{k_t}{I_2}\frac{(A_2 - A_1)}{A_2} \tag{13.45}$$

Either of Eqs. 13.45 gives the natural circular frequency of the vibration. Although I_1 and I_2 may be known, the amplitudes of vibration A_1 and A_2 are unknown. However, for equilibrium of torques on the shaft as a free body and as shown by Eqs. 13.42, it may be seen that

$$I_1\frac{d^2\theta_1}{dt^2} = -I_2\frac{d^2\theta_2}{dt^2} \tag{13.46}$$

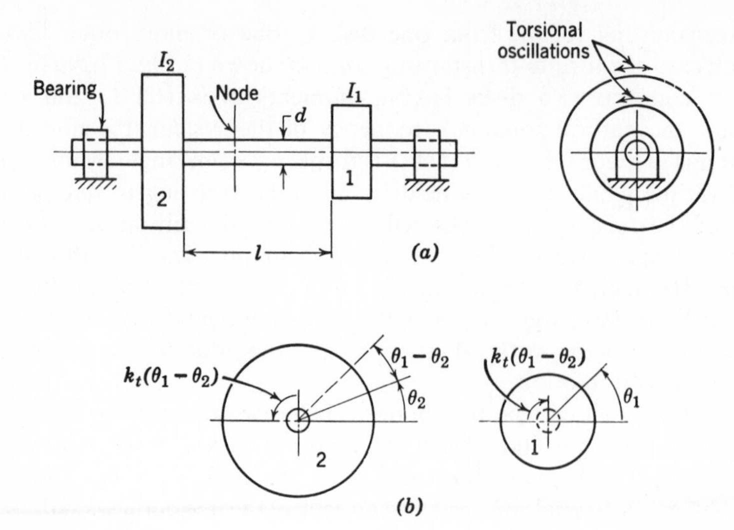

Figure 13.24

Substitution of the solutions given by Eqs. 13.44 in 13.46 gives

$$-I_1 A_1 \omega_n^2 \sin \omega_n t = I_2 A_2 \omega_n^2 \sin \omega_n t$$

$$\frac{A_1}{A_2} = -\frac{I_2}{I_1} \qquad (13.47)$$

Thus, the amplitude terms may be eliminated from Eqs. 13.45, giving the following expression for the natural frequency of the system:

$$\omega_n^2 = \frac{k_t(I_1 + I_2)}{I_1 I_2}$$

$$\omega_n = \sqrt{\frac{k_t(I_1 + I_2)}{I_1 I_2}} \qquad (13.48)$$

$$f_n = \frac{1}{2\pi}\sqrt{\frac{k_t(I_1 + I_2)}{I_1 I_2}} \qquad (13.49)$$

The position of the node between the disks may be determined from the ratio of the amplitudes given in Eq. 13.47. Angles A_1 and A_2 are shown in Fig. 13.25 for the phase in which A_1 is positive and A_2 is negative. In this phase the shaft is twisted the maximum total angle equal to the sum of the absolute values of A_1 and A_2. Since the shaft is of uniform cross section, the distribution of unit angles of twist is distributed linearly along the shaft

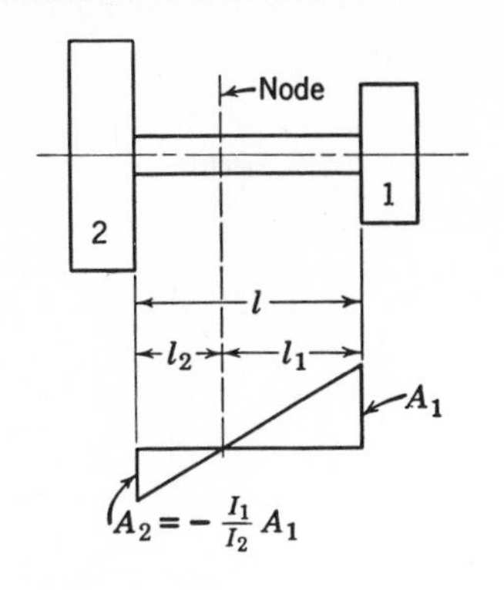

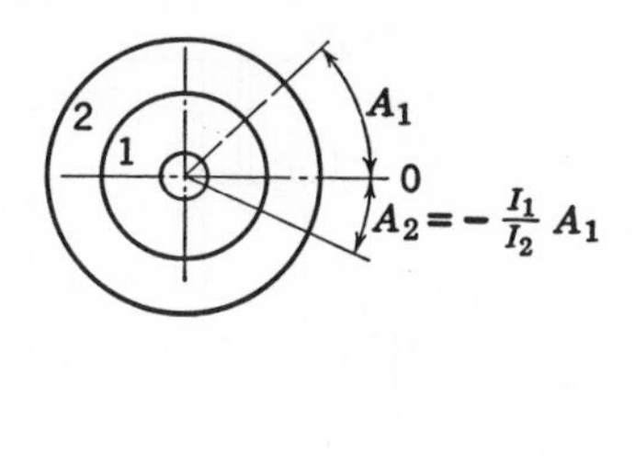

Figure 13.25

such that the position of the node may be determined by the construction shown in Fig. 13.25. From similar triangles,

$$l_1 = \frac{l}{1 + (I_1/I_2)} \tag{13.50}$$

It may be seen from Eq. 13.50 that the node will be nearer to the disk with the greater moment of inertia. When one of the disks is of infinite moment of inertia, the node is then at this disk and the system becomes the same as that of a shaft with one end fixed and a disk at its free end as in Fig. 13.23.

In the foregoing it has been assumed that the two disks oscillate at the frequency given by Eq. 13.49 for the condition where there is no shaft rotation. However, when the shaft is rotating, Eq. 13.49 still applies so long as the *relative* torsional oscillating motion of one disk to the other is the same and the relative oscillatory twisting of the shaft is the same. In this sense the natural torsional frequency is independent of shaft speed. However, the shaft speed becomes of importance if the disturbing or forcing torsional frequency f related to the shaft speed is coincident with the natural torsional frequency f_n to cause resonance. The disturbing frequency depends not only on shaft speed but also on the number of disturbances per shaft revolution. For example, in Fig. 13.26 a bladed disk is shown driven by a disk representing the armature of an electric motor. As a rotor blade passes a stator blade, a disturbance is manifest because of the variation in torque caused by fluid forces. If there are N stator blades, then there are N disturbances per revolution. For a large number of blades, the torque in the shaft is a uniform average torque with a superimposed ripple of N full waves. The forcing frequency in cycles per minute is given by

$$f = Nn \tag{13.51}$$

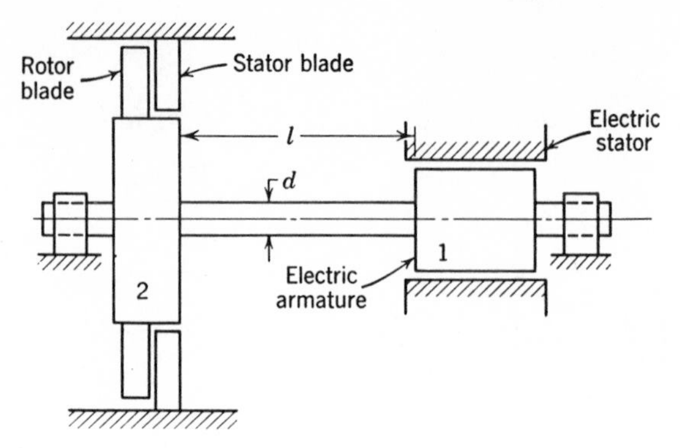

Figure 13.26

in which N is number of blades or disturbances per revolution and n is shaft speed in revolutions per minute. If f coincides with f_n, then n becomes the torsional critical speed n_c of the shaft.

In other applications, torsional disturbances may be caused by meshing gear teeth or by cam action. If the bladed disk of Fig. 13.26 were a gear, the number of disturbances per revolution would be the same as the number of teeth; if it were a cam, the number of disturbances per revolution would be the same as the number of lobes.

Example 13.4

In Fig. 13.26, 15 blades are mounted on disk 2 and there are the same number of stator blades. Disk 2 weighs 88.9 N and has a radius of gyration of 15.24 cm. The electric armature is 44.5 N in weight with a radius of gyration of 7.62 cm. A steel shaft 5.08 cm in diameter and 25.4 cm long transmits the torque from one disk to the other. Determine the torsional critical speed n_c of the rotor based on the disturbance produced by passing blades.

Solution. The natural torsional frequency is given by Eq. 13.49 for which k_t, I_1, and I_2 must be determined.

$$k_t = \frac{GJ}{l}$$

where

$$G = 8{,}000{,}000 \text{ N/cm}^2$$

$$J = \frac{\pi d^4}{32} = \frac{\pi(5.08)^4}{32} = 65.35 \text{ cm}^4$$

$$k_t = \frac{8.0 \times 10^6 \times 65.35}{25.4} = 20.58 \times 10^4 \text{ N-m/rad}$$

$$I_1 = \frac{W_1}{g} r_1^2 = \frac{44.5(7.62)^2}{980.7} = 0.0263 \text{ kg-m}^2$$

$$I_2 = \frac{W_2}{g} r_2^2 = \frac{88.9(15.24)^2}{980.7} = 0.2108 \text{ kg-m}^2$$

$$f_n = \frac{1}{2\pi}\sqrt{\frac{k_t(I_1 + I_2)}{I_1 I_2}} = \frac{1}{2\pi}\sqrt{\frac{20.58 \times 10^4(0.0263 + 0.2108)}{0.0263 \times 0.2108}}$$

$$= 470 \text{ Hz}$$

$$= 28{,}200 \text{ cycles/min}$$

The disturbing frequency for $N = 15$ blades in cycles per minute is

$$f = Nn = 15n$$

Critical speed occurs when $f = f_n$. Then,

$$n = n_c$$

$$15n_c = f_n$$

$$n_c = \frac{f_n}{15} = \frac{28{,}200}{15}$$

$$= 1880 \text{ r/min}$$

13.15 *Torsional Vibration of a Shaft with Many Disks.* For the case of three disks as in Fig. 13.27*a*, there are two lengths of shaft transmitting torque, each of which may have a different torsional spring constant. To determine the natural torsional frequency of this system, the same mathematical steps may be made as in the previous paragraphs for a shaft with two disks. The method requires the writing of three equations of motion, one for each disk, equations relating the amplitudes of the three disks and equations relating amplitudes to moments of inertia. As a result the following quadratic equation must be solved for ω_n.

$$I_1 I_2 I_3(\omega_n^2)^2 - [k_2(I_1 I_2 + I_1 I_3) + k_1(I_2 I_3 + I_1 I_3)]\omega_n^2 + k_1 k_2(I_1 + I_2 + I_3) = 0 \quad (13.52)$$

The solution of Eq. 13.52 gives two values of ω_n, indicating two natural frequencies. The first natural frequency of vibration is one which has a single node as shown in Fig. 13.27*b*, where two adjacent disks are in phase and the third is 180° out of phase. At the second natural frequency, which is higher, there are two nodes as shown in Fig. 13.27*c*.

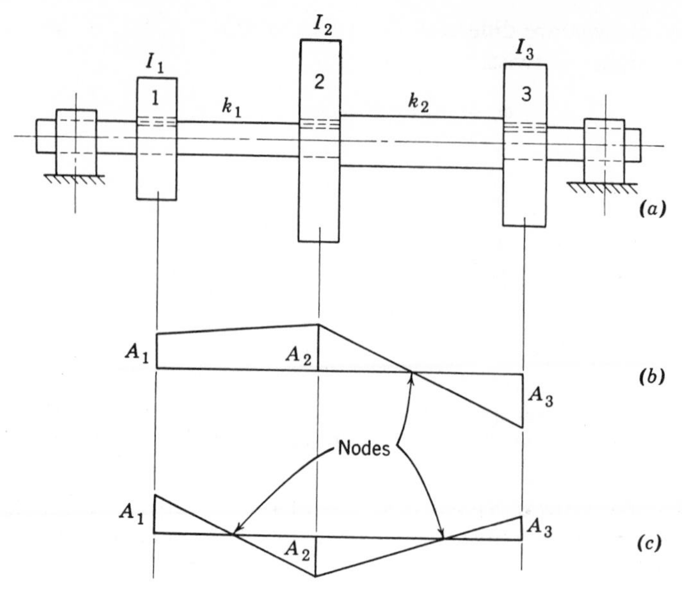

Figure 13.27

For cases in which there are more than three disks, the equations for natural frequency become cubics and equations of higher degree, giving a number of natural frequencies equal to the number of sections of shaft between disks. The higher degree equations become more difficult to solve as the degree increases.

13.16 Stepped Shafts. Where the shaft is variable in diameter as shown for the two-disk rotor of Fig. 13.28, the torsional spring constant is variable. A combined spring constant k_t may be determined in terms of the individual constants k_1, k_2, k_3 $\cdots$ k_n. For springs in series as shown, the instantaneous torque T in each section of shaft is the same. However,

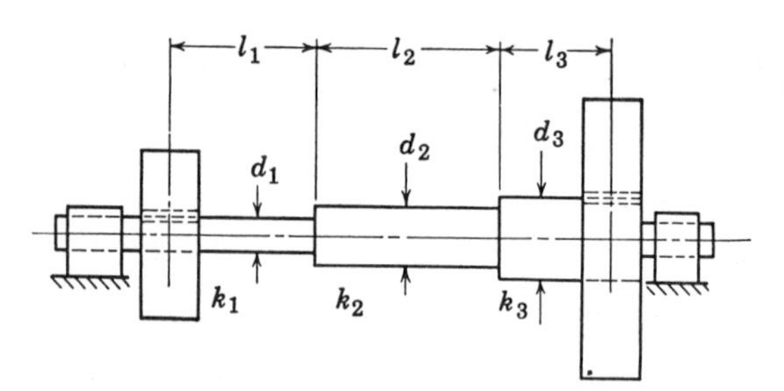

Figure 13.28

the angles of twist are different. The total angle of twist ϕ_t is the sum of the individual angles of twist.

$$\phi_t = \phi_1 + \phi_2 + \phi_3 + \cdots + \phi_n$$

$$\frac{T}{k_t} = \frac{T}{k_1} + \frac{T}{k_2} + \frac{T}{k_3} + \cdots + \frac{T}{k_n}$$

$$\frac{1}{k_t} = \frac{1}{k_1} + \frac{1}{k_2} + \frac{1}{k_3} + \cdots + \frac{1}{k_n}$$

$$\frac{1}{k_t} = \Sigma \frac{1}{k} \tag{13.53}$$

For the rotor with two disks and with a variable diameter shaft, k_t determined by Eq. 13.53 may be substituted in Eq. 13.49 in solving for torsional natural frequency.

13.17 Torsional System with Gears. Figure 13.29*a* shows a torsional system of two disks connected by shafts which transmit torque through gears. When the system oscillates at its natural torsional frequency, the two disks displace relative to each other at amplitudes that depend on the gear ratio as well as the twisting of the two shafts. If the gear ratio is m and the smaller gear is on shaft 2 as shown, then the rotation of gear 2 will

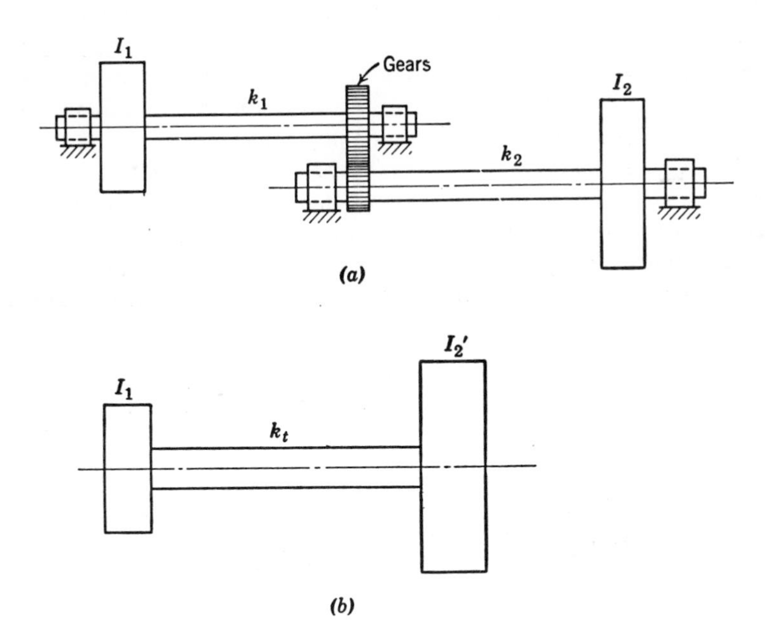

Figure 13.29

be m times greater than the rotation of gear 1 but the torque in shaft 2 at equilibrium conditions will be $1/m$ of the torque in shaft 1. The gears are assumed to have negligible inertia.

To simplify the calculation of natural frequency of the geared system, an equivalent but simpler system as in Fig. 13.29*b* may be solved instead providing certain modifications are made. The equivalent nongeared shaft must have a spring constant and inertias which are the equivalent of the geared system. Consider first the inertias of the two systems. Suppose that both systems have the same stiffness and that in both systems the inertias of disk 1 are I_1. If an acceleration α is applied to shaft 1 in the geared case, then, because of the gears, the acceleration of shaft 2 and disk 2 is $m\alpha$; the torque in shaft 2 is $I_2 m\alpha$, and, because of the gears, the torque T_1 in shaft 1 is m times greater.

$$T_1 = m(I_2 m\alpha)$$

$$T_1 = m^2 I_2 \alpha \tag{13.54}$$

Now, if the same acceleration α is applied to the shaft of the nongeared system, the torque in the shaft is

$$T_1 = I'_2 \alpha \tag{13.55}$$

For the two systems to be equivalent, the torques in both shafts near disk 1 should be the same. Therefore, equating 13.54 and 13.55,

$$I'_2 \alpha = m^2 I_2 \alpha$$

$$I'_2 = m^2 I_2 \tag{13.56}$$

Thus, in the equivalent system, the moment of inertia of disk 2 of the geared system is used but increased m^2 times.

Consider now the spring constants of the two systems. If in the geared case disk 2 is held and a torque T is applied to disk 1, then the torque in shaft 1 is T and in shaft 2 is T/m. The total angular displacement ϕ_t of disk 1 relative to disk 2 consists of the angle of twist ϕ_1 of shaft 1 plus the angular displacement ϕ_2/m of gear 1 due to the angle of twist ϕ_2 of shaft 2. Therefore,

$$\phi_t = \phi_1 + \frac{\phi_2}{m}$$

In terms of torque and spring constants,

$$\phi_t = \frac{T}{k_1} + \frac{T/m}{mk_2}$$

$$= \frac{T}{k_1} + \frac{T}{m^2 k_2} \tag{13.57}$$

For the nongeared case, the angle of twist ϕ_t should be the same as in the geared case for the same torque T.

$$\phi_t = \frac{T}{k_t} \qquad (13.58)$$

Equating Eqs. 13.57 and 13.58,

$$\frac{T}{k_t} = \frac{T}{k_1} + \frac{T}{m^2 k_2}$$

$$k_t = \frac{1}{(1/k_1) + (1/m^2 k_2)} \qquad (13.59)$$

It will be observed that the geared case is reduced to a simpler case by retaining the inertia and spring constant of one disk and shaft and by increasing the inertia of the second disk and spring constant of the second shaft m^2 times. Using Eqs. 13.56 and 13.59 for I'_2 and k_t respectively, the natural torsional frequency may be determined from

$$\omega_n = \sqrt{\frac{k_t(I_1 + I'_2)}{I_1 I'_2}} \qquad (13.60)$$

Problems

13.1 A 44.5 N weight W placed on a spring as in Fig. 13.1*a* deflects the spring $\delta = 1.59$ mm under static conditions. Determine the magnitude of the spring constant k.

13.2 A mass of weight W is suspended from a set of two springs in series as shown in Fig. 13.30*a*. Determine a combined spring constant k in terms of the spring constants k_1 and k_2.

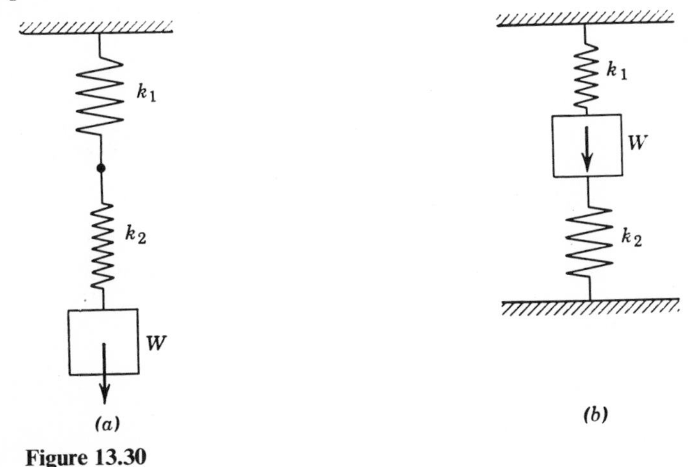

Figure 13.30

13.3 For the spring suspension system of Fig. 13.30*b*, determine the combined spring constant k in terms of k_1 and k_2.

13.4 Add the gravity force vector W to the free-body diagram of Fig. 13.1*a*, and show that the resultant equation of motion is the same as Eq. 13.1. Note that prior to applying the periodic force the spring deflects statically under the action of W.

13.5 A 90 N weight is supported by a spring as in Fig. 13.1*a*. For a spring constant of $k = 350$ N/cm, determine the natural circular frequency of vibration when the weight is disturbed to displace vertically. Determine the natural cyclic frequency.

13.6 A machine mounted on four springs vibrates as the system of Fig. 13.1*b* because of a mass m rotating about an axis in the machine. Determine the amplitude of vibration from Eq. 13.5 using the following data: weight of the rotating mass is 4.45 N, weight of the machine including the rotating mass is 222 N, $r = 5.08$ cm, combined spring constant of the four equal springs $k = 1750$ N/cm, speed of the rotating mass is 3000 r/min. Check your answer using the curve of Fig. 13.4*b*.

13.7 For the machine of Problem 13.6, determine the rotative speed of the rotating mass at which resonance would occur, that is, at which the amplitude of vibration would approach an infinite value.

13.8 Plot a curve of magnification factor versus ω/ω_n from Eq. 13.10 to verify the curve of Fig. 13.4*a*

13.9 A mass of 220 N weight is acted upon by a periodic force $F = 20 \sin \omega t$ as in the case of Fig. 13.1*a*. Determine the amplitude of vibration of the mass at a forcing frequency of 33.3 Hz if the spring constant is 2100 N/cm. Determine the maximum acceleration and maximum velocity during vibration. Determine the maximum kinetic energy of the mass during vibration.

13.10 Counter-rotating eccentric weights as shown in Fig. 13.31 are used to excite the motion of a vibration table mounted on springs. The weights are geared to rotate opposite each other in order that the resultant inertia force of the rotating masses is

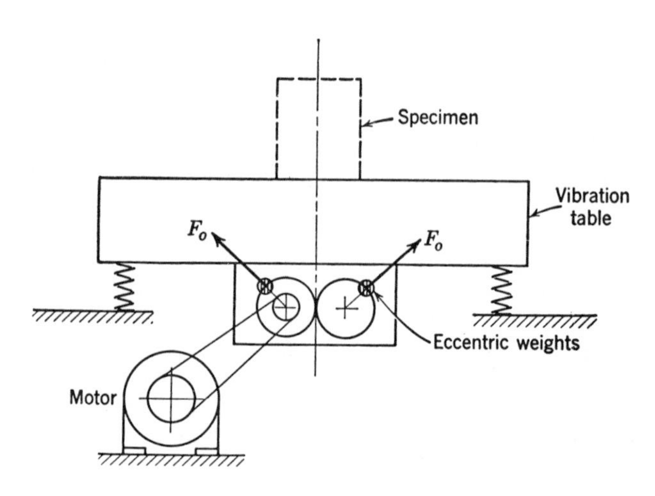

Figure 13.31

always vertical, thereby producing a vertical motion only. The weight of each of the rotating masses is 1 lb at a radius of 3 in. The table weighs 18 lb, and the maximum weight of specimen to be vibrated is 5 lb. Determine the combined spring constant required such that the machine may be operated from 0 to 5000 cycles/min with a MX/mr factor not greater than 1. What is the spring constant of each spring if four springs are used?

13.11 Referring to the vibration table of Problem 13.10, it is desired to test a 2-lb specimen for which the maximum acceleration during vibration is $10g$. Determine the speed in revolutions per minute of the counterweights at which the desired acceleration will occur if the combined spring constant is 35,000 lb/in.

13.12 A 8900 N road vehicle compresses its springs 7.62 cm under static conditions. Assuming the system is that of Fig. 13.1*c*, determine the critical road velocity in km/h when traveling on a road having a sine wave form with an amplitude of 1.27 cm at a wavelength of 15.24 m.

13.13 A 200-lb machine is acted upon by a shaking force equal to $F_o = mr\omega^2$, in which m is a mass of 4 lb weight, $r = 2$ in., and the rotative speed is 3200 r/min. Determine the vibratory force transmitted to the foundation of the machine if the machine is clamped down rigidly. Determine the vibratory force transmitted if the machine is mounted on springs as in Fig. 13.1*b* and the system has a natural frequency of 1000 cycles/min. Determine the amplitude of vibration. If the machine weight were changed from 200 to 100 lb, what would be the effect on transmitted force and amplitude?

13.14 A 350 N electronic instrument is attached to a ship deck which vibrates at an amplitude of 0.16 cm at 50 Hz. The instrument is to be isolated from the deck with four flexible mounts. Determine the required combined spring constant of the mounts such that the amplitude of vibration of the instrument is no more than 0.013 cm. If the springs are equal, what is the spring constant of each mount? What is the static deflection of each mount?

13.15 A mass of 40 lb weight is acted upon by a periodic force $F = 20 \sin \omega t$ at a frequency of 600 cycles/min under the conditions shown in Fig. 13.8, in which a dashpot is in the system. $k = 400$ lb/in., and $c = 4$ lb-sec/in. Determine the amplitude of vibration. Determine the magnitude of the damping force when it is a maximum value during vibration.

13.16 The 4448 N vehicle of Example 13.1 with springs $k = 3467$ N/cm has a critical speed of 96.6 km/h when traveling over a rough road having a sine wave form with an amplitude of 2.54 cm at a wavelength of 6.09 m. Determine the amplitude of vibration at 96.6 km/h if shock absorbers of $c/M\omega_n = 1.0$ are added. Determine c.

13.17 A 200-lb machine (Fig. 13.1*b*) is acted upon by a shaking force $F_o = mr\omega^2$ in which m is a mass of 4-lb weight at a crank radius of $r = 2$ in. It is desired to operate the crank at 6000 r/min. The spring constant is $k = 8200$ lb/in. Determine the amplitude of vibration and the force transmitted to the foundation at 6000 r/min and at the critical speed when (*a*) no damping is present, and (*b*) when damping of $c/M\omega_n = 0.25$ is present.

13.18 A 330 N turbine wheel is mounted on a shaft midway between bearings as in Fig. 13.13. Determine the critical speed in revolutions per minute of the rotor if $k =$ 262,000 N/cm in bending.

13.19 It is desired to operate a 25-lb turbine wheel at 15,000 r/min. The wheel is midway between bearings as in Fig. 13.13, and the bearings are 18 in. apart. Determine the diameter d of steel shaft required in order that the critical speed of the shaft be 40% higher than the desired operating speed. Determine the diameter of the shaft if it is deemed permissible to run through a critical speed of 3000 r/min on the way up to 15,000 r/min.

13.20 A 222 N compressor wheel mounted as in Fig. 13.13 has a critical speed of 3000 r/min. It is known that the unbalance of the rotor is 1.412 N-cm. Determine the eccentricity of the rotor center of gravity with respect to the axis of rotation. Determine the radius of whirl and the force on each bearing for a rotor speed of 12,000 r/min.

13.21 The steel shaft of a rotor is 45 cm long between bearings and 5 cm in diameter. It supports a 130 N impeller at the midspan. Determine the critical speed of the shaft without and with the mass of the shaft included.

13.22 A 75-lb turbine wheel is mounted on a shaft midway between bearings as in Fig. 13.13. Determine the critical speed in revolutions per minute of the rotor if the spring constant of the shaft is 150,000 lb/in. If the spring constant of the two bearing supports together is 50,000 lb/in., determine the change in critical speed because of the flexibility of the supports.

13.23 The rotor shown in Fig. 13.32 has three disks each of different weight. The deflection of the shaft due to the weights has been determined as shown. Determine the fundamental natural frequency of the rotor using Eq. 13.36. What is the critical speed of the rotor in revolutions per minute?

13.24 In Fig. 13.33a is shown a rotor having a compressor wheel and a turbine wheel mounted on a steel shaft of uniform diameter. Determine the diameter d of the shaft which will give a fundamental critical speed 25 percent higher than the operating speed of 8000 r/min. Formulas from which static deflection may be determined are given in Fig. 13.33b.

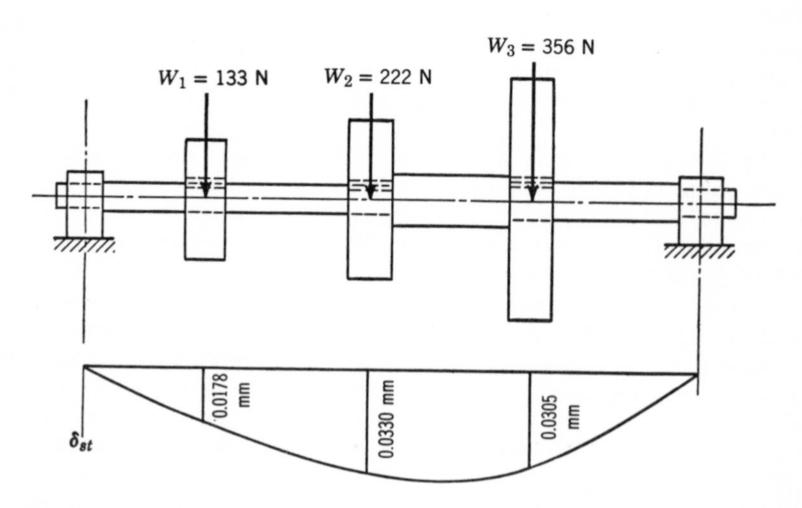

Figure 13.32

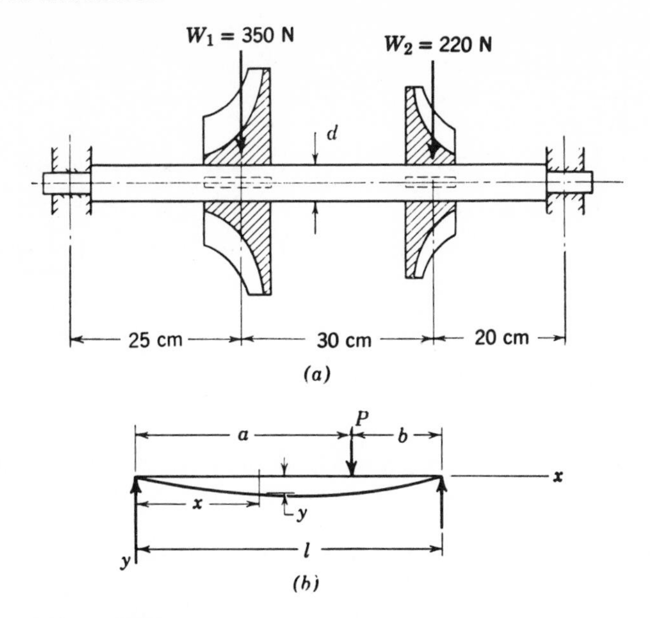

Figure 13.33

$$y = \frac{Pbx}{6EI_A l}(l^2 - b^2 - x^2) \qquad \text{for } x \leqq a$$

$$y = \frac{Pbx}{6EI_A l}(l^2 - b^2 - x^2) + \frac{P(x-a)^3}{6EI_A} \qquad \text{for } x \geqq a$$

13.25 In Fig. 13.34*a* the 50-lb wheel is placed at the midspan of a steel shaft, and in Fig. 13.34*b* the same wheel is shown moved toward the right bearing. Determine the effect of the position of the wheel on the shaft's fundamental critical speed. Static deflections may be calculated from the formulas given in Fig. 13.33*b*.

13.26 Determine the fundamental critical speed of the rotor with steel shaft shown in Fig. 13.35. Consult a textbook on strength of materials to calculate static deflections.

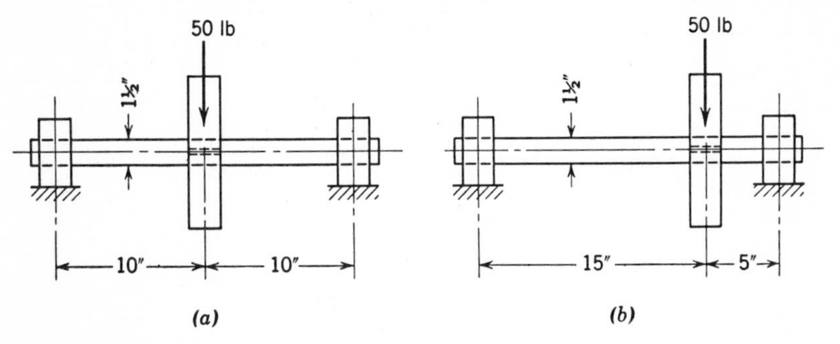

Figure 13.34

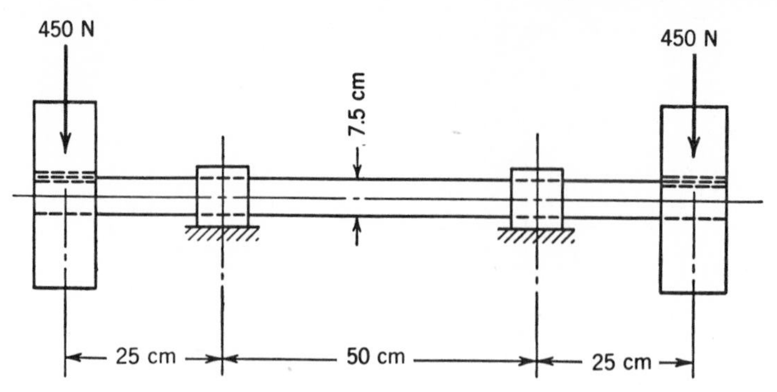

Figure 13.35

13.27 Using Rayleigh's energy method, determine the critical speed of the steel rotor shown in Fig. 13.36. Determine static deflection using the graphical method. Neglect the mass of the shaft.

13.28 Determine the critical speed of the rotor with steel shaft shown in Fig. 13.37. Neglect the shaft mass.

13.29 A steel shaft of variable diameter and with a 89 N disk is shown in Fig. 13.38. Calculate the fundamental critical speed of the rotor, and include the mass of the shaft in the calculations. Divide the shaft into five equal lengths in using the graphical method to determine static deflections.

13.30 The automotive drive shaft shown in Fig. 13.39 is a steel, thin-walled tube with universal couplings at the ends. The mean diameter of the tube is 3 in., and the wall thickness is t. The maximum expected speed of the shaft is 4500 r/min. Determine the value of t or D to insure that the first critical speed in bending is 50% higher than 4500 r/min. $I_A = \pi D^3 t/8$ for a thin-walled tube.

13.31 The steel quill shaft with splined ends shown in Fig. 13.40 is to be used as a drive shaft at 10,000 r/min in an aircraft installation. To keep the weight of the shaft to a

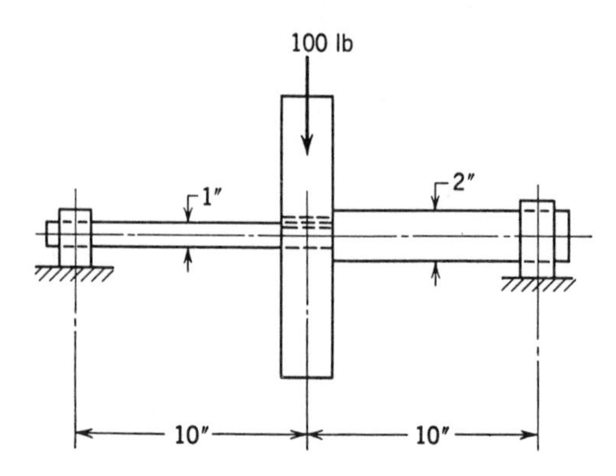

Figure 13.36

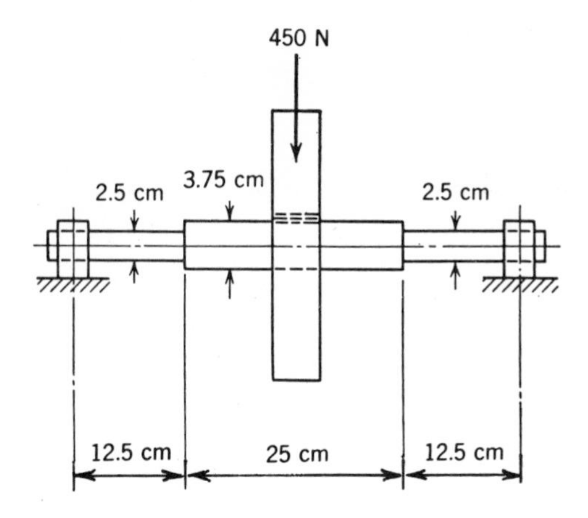

Figure 13.37

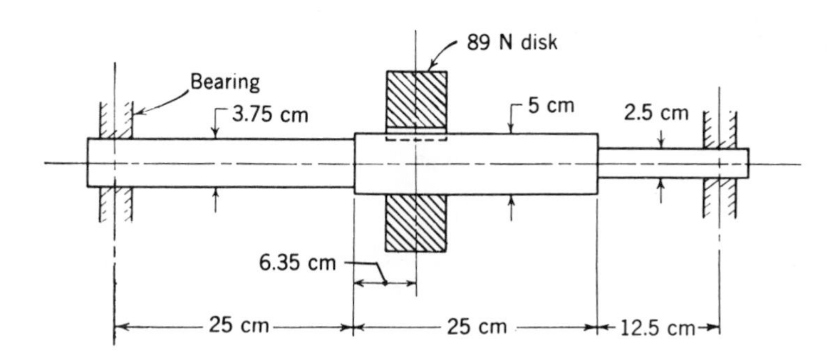

Figure 13.38

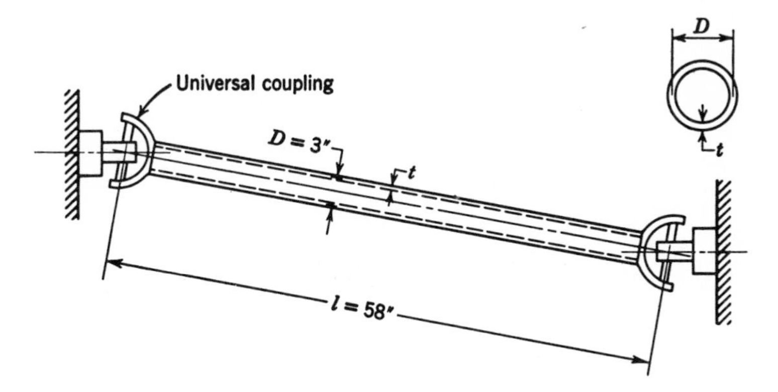

Figure 13.39

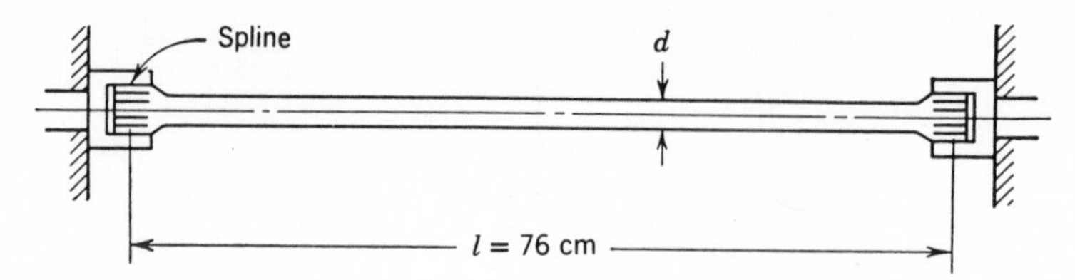

Figure 13.40

minimum, it is decided to operate well above the first critical speed but below the second critical speed at a point 75% of the speed from first to second critical speeds. Determine the required diameter d of the shaft, and calculate the revolutions per minute at which resonance will first occur. $I_A = \pi d^4/64$.

13.32 A solid circular steel disk of the dimensions shown in Fig. 13.41 is suspended as a torsional pendulum by steel wire 50 cm long. Determine the size of wire d to give a torsional pendulum frequency of 1 Hz.

13.33 Determine the natural torsional frequency of the steel shaft with two gears shown in Fig. 13.42. Determine the two critical speeds of the shaft caused by tooth meshing disturbances.

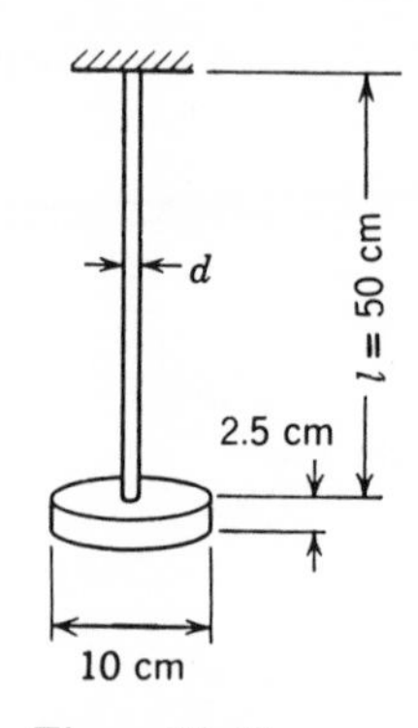

Figure 13.41

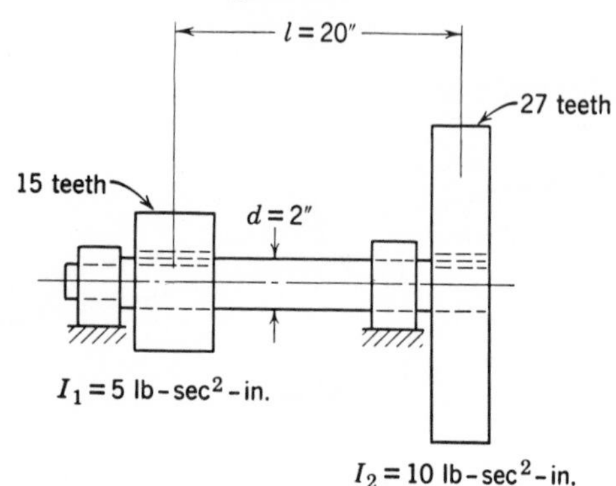

Figure 13.42

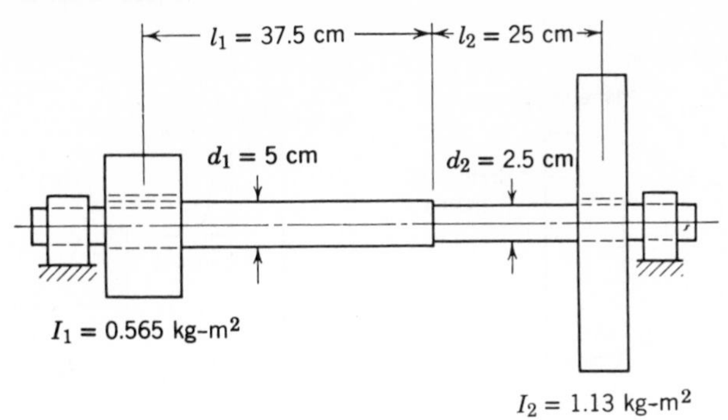

Figure 13.43

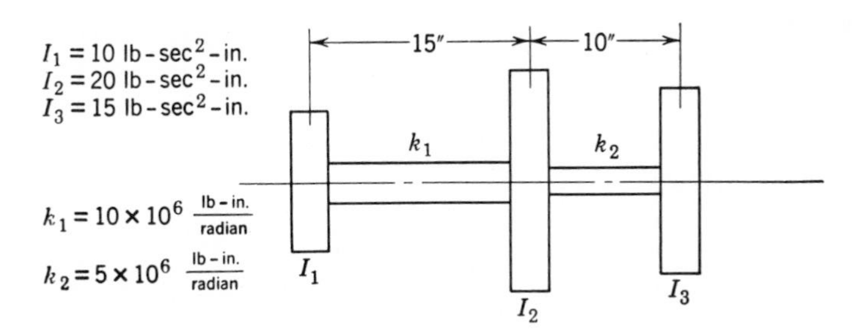

Figure 13.44

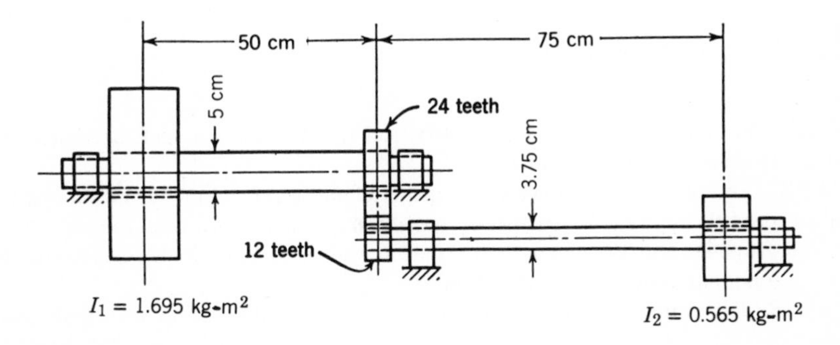

Figure 13.45

13.34 Determine the natural torsional frequency of the stepped steel shaft with two disks shown in Fig. 13.43.

13.35 Determine the two natural torsional frequencies of the system shown in Fig. 13.44. The shaft is of steel.

13.36 Determine the equivalent nongeared torsional system of the geared system shown in Fig. 13.45. Calculate the natural torsional frequency of the system for shafts

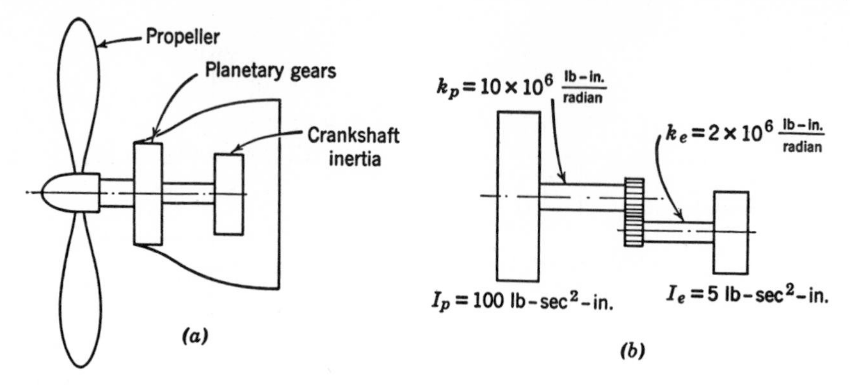

Figure 13.46

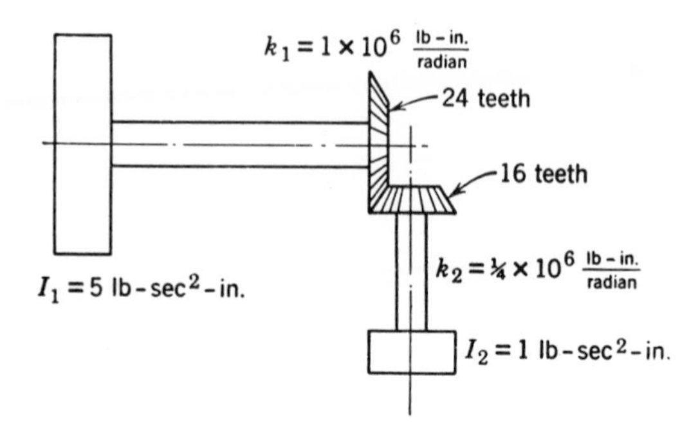

Figure 13.47

of steel. Determine the highest torsional critical speed of the left shaft because of the forced disturbance by the meshing gears.

13.37 In Fig. 13.46*a* is shown the propeller of an aircraft engine turning at one-half engine speed through the 2:1 planetary reduction gears. The principal lumped masses are the propeller and the engine crankshaft as shown in Fig. 13.46*b*. Determine the natural torsional frequency of the system.

13.38 Determine the natural torsional frequency of the system with bevel gears shown in Fig. 13.47. Determine the critical speeds of the high-speed shaft based on gear-meshing disturbances.

APPENDIX 1

TABLE OF INVOLUTE FUNCTIONS

Degrees	0.0	0.1	0.2	0.3	0.4	0.5	0.6	0.7	0.8	0.9
0	0.000000	0.000000	0.000000	0.000000	0.000000	0.000000	0.000000	0.000000	0.000000	0.000001
1	0.000002	0.000002	0.000003	0.000004	0.000005	0.000006	0.000007	0.000009	0.000010	0.000012
2	0.000014	0.000016	0.000019	0.000022	0.000025	0.000028	0.000031	0.000035	0.000039	0.000043
3	0.000048	0.000053	0.000058	0.000064	0.000070	0.000076	0.000083	0.000090	0.000097	0.000105
4	0.000114	0.000122	0.000132	0.000141	0.000151	0.000162	0.000173	0.000184	0.000197	0.000209
5	0.000222	0.000236	0.000250	0.000265	0.000280	0.000296	0.000312	0.000329	0.000347	0.000366
6	0.000384	0.000404	0.000424	0.000445	0.000467	0.000489	0.000512	0.000536	0.000560	0.000586
7	0.000612	0.000638	0.000666	0.000694	0.000723	0.000753	0.000783	0.000815	0.000847	0.000880
8	0.000914	0.000949	0.000985	0.001022	0.001059	0.001098	0.001137	0.001178	0.001219	0.001262
9	0.001305	0.001349	0.001394	0.001440	0.001488	0.001536	0.001586	0.001636	0.001688	0.001740
10	0.001794	0.001849	0.001905	0.001962	0.002020	0.002079	0.002140	0.002202	0.002265	0.002329
11	0.002394	0.002461	0.002528	0.002598	0.002668	0.002739	0.002812	0.002894	0.002962	0.003039
12	0.003117	0.003197	0.003277	0.003360	0.003443	0.003529	0.003615	0.003712	0.003792	0.003883
13	0.003975	0.004069	0.004164	0.004261	0.004359	0.004459	0.004561	0.004664	0.004768	0.004874
14	0.004982	0.005091	0.005202	0.005315	0.005429	0.005545	0.005662	0.005782	0.005903	0.006025
15	0.006150	0.006276	0.006404	0.006534	0.006665	0.006799	0.006934	0.007071	0.007209	0.007350
16	0.007493	0.007637	0.007784	0.007932	0.008082	0.008234	0.008388	0.008544	0.008702	0.008863
17	0.009025	0.009189	0.009355	0.009523	0.009694	0.009866	0.010041	0.010217	0.010396	0.010577
18	0.010760	0.010946	0.011133	0.011323	0.011515	0.011709	0.011906	0.012105	0.012306	0.012509
19	0.012715	0.012923	0.013134	0.013346	0.013562	0.013779	0.013999	0.014222	0.014447	0.014674
20	0.014904	0.015137	0.015372	0.015609	0.015850	0.016092	0.016337	0.016585	0.016836	0.017089
21	0.017345	0.017603	0.017865	0.018129	0.018395	0.018665	0.018937	0.019212	0.019490	0.019770
22	0.020054	0.020340	0.020630	0.020921	0.021216	0.021514	0.021815	0.022119	0.022426	0.022736
23	0.023049	0.023365	0.023684	0.024006	0.024332	0.024660	0.024992	0.025326	0.025664	0.026005
24	0.026350	0.026697	0.027048	0.027402	0.027760	0.028121	0.028485	0.028852	0.029223	0.029598
25	0.029975	0.030357	0.030741	0.031130	0.031521	0.031917	0.032315	0.032718	0.033124	0.033534
26	0.033947	0.034364	0.034785	0.035209	0.035637	0.036069	0.036505	0.036945	0.037388	0.037835
27	0.038287	0.038696	0.039201	0.039664	0.040131	0.040602	0.041076	0.041556	0.042039	0.042526

28	0.043017	0.043513	0.044012	0.044516	0.045024	0.045537	0.046054	0.046575	0.047100	0.047630
29	0.048164	0.048702	0.049245	0.049792	0.050344	0.050901	0.051462	0.052027	0.052597	0.053172
30	0.053751	0.054336	0.054924	0.055519	0.056116	0.056720	0.057267	0.057940	0.058558	0.059181
31	0.059809	0.060441	0.061079	0.061721	0.062369	0.063022	0.063680	0.064343	0.065012	0.065685
32	0.066364	0.067048	0.067738	0.068432	0.069133	0.069838	0.070549	0.071266	0.071988	0.072716
33	0.073449	0.074188	0.074932	0.075683	0.076439	0.077200	0.077968	0.078741	0.079520	0.080305
34	0.081097	0.081974	0.082697	0.083506	0.084321	0.085142	0.085970	0.086804	0.087644	0.088490
35	0.089342	0.090201	0.091066	0.091938	0.092816	0.093701	0.094592	0.095490	0.096395	0.097306
36	0.098224	0.099149	0.100080	0.101019	0.101964	0.102916	0.103875	0.104841	0.105814	0.106795
37	0.107782	0.108777	0.109779	0.110788	0.111805	0.112828	0.113860	0.114899	0.115945	0.116999
38	0.118060	0.119130	0.120207	0.121291	0.122384	0.123484	0.124592	0.125709	0.126833	0.127965
39	0.129106	0.130254	0.131411	0.132576	0.133749	0.134931	0.136122	0.137320	0.138528	0.139743
40	0.140968	0.142201	0.143443	0.144694	0.145954	0.147222	0.148500	0.149787	0.151082	0.152387
41	0.153702	0.155025	0.156358	0.157700	0.159052	0.160414	0.161785	0.163165	0.164556	0.165956
42	0.167366	0.168786	0.170216	0.171656	0.173106	0.174566	0.176037	0.177518	0.179009	0.180511
43	0.182023	0.183546	0.185080	0.186625	0.188180	0.189746	0.191324	0.192912	0.194511	0.196122
44	0.197744	0.199377	0.201022	0.202678	0.204346	0.206026	0.207717	0.209420	0.211135	0.212863
45	0.214602	0.216353	0.218117	0.219893	0.221682	0.223483	0.225296	0.227123	0.228962	0.230814
46	0.232678	0.234557	0.236448	0.238352	0.240270	0.242202	0.244147	0.246105	0.248077	0.250064
47	0.252064	0.254078	0.256106	0.258149	0.260206	0.262277	0.264363	0.266463	0.268578	0.270709
48	0.272855	0.275015	0.277191	0.279381	0.281588	0.283810	0.286047	0.288300	0.290570	0.292855
49	0.295157	0.297474	0.299809	0.302160	0.304527	0.306912	0.309313	0.311731	0.314166	0.316619
50	0.319088	0.321577	0.324082	0.326605	0.329146	0.331706	0.334284	0.336879	0.339493	0.342127
51	0.344779	0.347451	0.350141	0.352850	0.355579	0.358328	0.361096	0.363885	0.366693	0.369522
52	0.372371	0.375241	0.378130	0.381041	0.383974	0.386927	0.389903	0.392899	0.395917	0.398958
53	0.402021	0.405105	0.408213	0.411343	0.414495	0.417671	0.420870	0.424094	0.427340	0.430610
54	0.433905	0.437222	0.440566	0.443933	0.447326	0.450744	0.454187	0.457655	0.461150	0.464670
55	0.468217	0.471790	0.475390	0.479017	0.482670	0.486351	0.490060	0.493797	0.497562	0.501355
56	0.505177	0.509027	0.512908	0.516817	0.520755	0.524724	0.528724	0.532753	0.536814	0.540905
57	0.545027	0.549182	0.553368	0.557586	0.561836	0.566120	0.570436	0.574789	0.579173	0.583591
58	0.588043	0.592530	0.597053	0.601609	0.606203	0.610832	0.615498	0.620200	0.624940	0.629717
59	0.634535	0.639387	0.644279	0.649210	0.654181	0.659190	0.664240	0.669331	0.674462	0.679635
60	0.684853	0.690109	0.695409	0.700751	0.706137	0.711567	0.717041	0.722561	0.728126	0.733736

APPENDIX 2

Approximate Method for Drawing Involute Gear Tooth

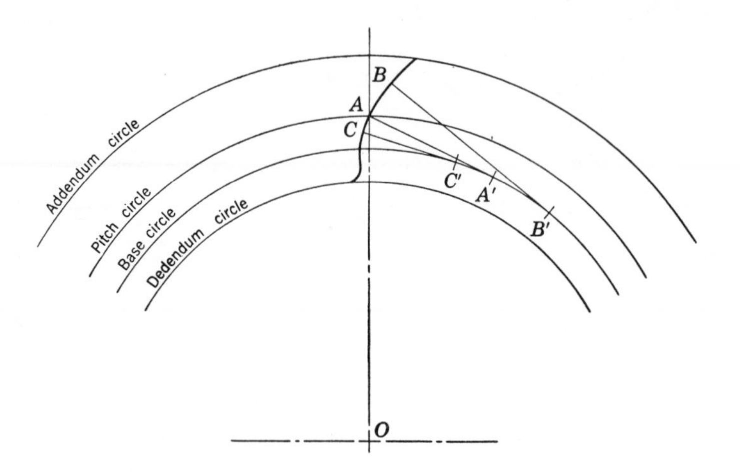

1. Starting with a point A on the pitch circle, pass a line through A tangent to the base circle obtaining the point of tangency A'.
2. Using A' as a center and a radius AA', swing an arc approximately halfway into the space between the pitch and addendum circles, and halfway into the space between the pitch and base circles. This gives points B and C.
3. Using point B, find the point of tangency B'. Swing an arc of radius BB' with center B' to the addendum circle.
4. Repeat item 3 for point C, swinging the arc to the base circle.
5. The profile between the base circle and addendum circle is a radial straight line (except the fillet).

APPENDIX 3

Conversion Factors for the American and International Systems of Units

Force

1 lb_f = 4.448 N
1 N = 0.2248 lb_f

Mass

1 kg = 6.852×10^{-2} slugs
1 slug = 14.59 kg
1 lb_m = 3.108×10^{-2} slugs

Length

1 m = 3.281 ft
1 ft = 0.3048 m
1 in. = 2.54 cm

Moment of inertia (mass)

1 kg-m^2 = 0.7376 slug-ft^2
1 slug-ft^2 = 1.356 kg-m^2

Frequency

1 cycle/sec = 1 Hz

Other Useful Conversions

1 lb-in. = 11.298 N-cm
1 lb/in. = 1.751 N/cm
1 lb/in.^{2} = 0.6894 N/cm^2
1 lb/in.^{3} = 0.2714 N/cm^3
1 mi/h = 1.61 km/h

Index

$\omega = \frac{2\pi n}{60}$ ω in rad/sec n in rev/min